Environmental Science

Environmental Science

Working with the Earth

TENTH EDITION

G. TYLER MILLER, JR.

President, Earth Education and Research

Adjunct Professor of Human Ecology
St. Andrews Presbyterian College

THOMSON

BROOKS/COLE

Australia • Canada • Mexico • Singapore • Spain • United Kingdom • United States

THOMSON

BROOKS/COLE

Publisher: *Jack Carey*
Developmental Editor: *Mary Arbogast*
Production Project Manager: *Andy Marinkovich*
Technology Project Manager: *Keli Amann*
Assistant Editor: *Suzannah Alexander*
Editorial Assistant: *Jana Davis*
Permissions Editor/Photo Researcher: *Linda Rill*
Marketing Manager: *Ann Caven*
Marketing Assistant: *Sandra Perin*
Advertising Project Manager: *Linda Yip*
Print/Media Buyer: *Karen Hunt*

Production Management, copyediting,
and composition: *Thompson Steele, Inc.*
Interior Illustration: *Precision Graphics; Sarah Woodward; Darwin and
Vally Hennings; Tasa Graphic Arts, Inc.; Alexander Teshin Associates;
John and Judith Walker; Raychel Ciemma; Victor Royer, Electronic Pub-
lishing Services, Inc.; J/B Woolsey Associates; and Kerry Wong*
Cover Image: *Endangered green sea turtle. © David B. Fleetham/ Tom Stack
& Associates*
Text and Cover Printer: *Transcontinental Printing/Interglobe*
Title Page Photograph: *Crater Lake, Oregon (Jack Carey)*

For more information about our products, contact us at:
Thomson Learning Academic Resource Center
1-800-423-0563
For permission to use material from this text, contact us by:
Phone: 1-800-730-2214
Fax: 1-800-730-2215
Web: http://www.thomsonrights.com

Library of Congress Control Number: 2003100672

Student Edition: ISBN 0-534-42408-2

International Student Edition: ISBN 0-534-42407-4
(Not for sale in the United States)

Annotated Instuctor's Edition: ISBN: 0-534-42411-2

Brooks/Cole-Thomson Learning
511 Forest Lodge Road
Pacific Grove, CA 93950
USA

Asia
Thomson Learning
5 Shenton Way #01-01
UIC Building
Singapore 068808

Australia
Nelson Thomson Learning
102 Dodds Street
South Melbourne, Victoria 3205
Australia

Canada
Nelson Thomson Learning
1120 Birchmount Road
Toronto, Ontario M1K 5G4
Canada

Europe/Middle East/Africa
Thomson Learning
High Holborn House
50/51 Bedford Row
London WC1R 4LR
United Kingdom

Latin America
Thomson Learning
Seneca, 53
Colonia Polanco
11560 Mexico D.F.
Mexico

Spain
Paraninfo Thomson Learning
Calle/Magallanes, 25
28015 Madrid, Spain

For Instructors

What Are the Major Trends in Environmental Science Education? This is a *science-based* book designed for introductory courses on environmental science. It is an *interdisciplinary* study of how nature works and how things in nature are interconnected. About 40% of the book is devoted to providing a scientific base needed to understand environmental problems and evaluate possible solutions to these problems, 30% presents environmental problems, and another 30% presents and evaluates solutions to these problems.

This tenth edition continues my efforts to emphasize the following major shifts in environmental science education that have taken place over the past 28 years and are expected to accelerate in this century:

- *Emphasis on science-based approaches to understanding and solving environmental problems.* Since its first edition, this book has led the way in using scientific laws, principles, models, and concepts and critical thinking to help us understand environmental and resource problems and their possible solutions and see how these concepts, problems, and solutions are connected. The first edition had four chapters on basic scientific concepts when other books had a single chapter. In this tenth edition, eight chapters are devoted to the treatment of scientific principles and concepts—far more than in any other introductory environmental science text of this size. This emphasis on basic science will become increasingly important throughout this century. I have introduced only the concepts and principles necessary for understanding the material in the book, and I have tried to present them simply but accurately.

- *Emphasis on solutions.* The emphasis in this century is on finding and implementing scientific, technological, economic, and political solutions to environmental problems. This text has stressed solutions as a major theme for many years. About 30% of this book is devoted to presenting and evaluating solutions to environmental problems—far more than in any other introductory environmental science textbook of this size.

- *Emphasis on prevention solutions.* Since its first edition, this book has categorized proposed solutions to environmental problems as either *input* (prevention) solutions, such as pollution prevention and waste reduction, or *output* (cleanup) solutions, such as pollution control and waste management. Both approaches are needed, but there is a growing awareness of the need to put more emphasis on input or prevention approaches.

- *The shift to decentralized micropower.* I highlight the gradual shift from large centralized sources of electricity (mostly coal and nuclear plants) to a dispersed array of smaller micropower plants, including gas turbines, solar-cell arrays, wind turbines, and fuel cells. This shift is under way and will accelerate during this century for economic and security reasons.

- *Integration of economics and environment.* I emphasize the increased use of emissions trading, environmental accounting, full-cost pricing, phasing in environmentally friendly government subsidies, shifting taxes from wealth and income to pollution and resource waste, and evolving eras of environmental management in businesses. This trend toward using environmental economics, discussed in Chapter 2 (pp. 20–30), is under way and is expected to increase rapidly during this century.

To help ensure that the material is accurate and up to date, I have consulted more than 10,000 research sources in the professional literature and about the same number of Internet sites. I have also benefited from the more than 250 experts and teachers (see list on pp. ix–xi) who have provided detailed reviews of this and my other three books in this field.

How Is the Presentation of Views Balanced? There are always *trade-offs* involved in making and implementing environmental decisions. The challenge for an author is to give a fair and balanced presentation of opposing viewpoints, advantages and disadvantages of various technologies and proposed solutions to environmental problems, and good and bad news about environmental problems without injecting personal bias. This allows readers to make up their own mind about important environmental issues. Studying a subject as important as environmental science and ending up with no conclusions, opinions, and beliefs means that both the teacher and student have failed. However, such conclusions should be based on using critical thinking to evaluate opposing ideas and understand the trade-offs involved.

A few examples of my efforts to provide a balanced presentation are **(1)** the advantages and disadvantages of reducing birthrates (p. 232), **(2)** Case Study: Should Oil and Natural Gas Development Be Allowed in the Arctic National Wildlife Refuge? (p. 486 and Figure 19-15, p. 487) **(3)** the discussion of global warming, (pp. 285-292) **(4)** advantages and disadvantages of pesticides (Section 16-8, pp. 399–402), and **(5)** diagrams (such as Figure 11-29, p. 247, on buses; Figure 14-14, p. 315, on withdrawing groundwater; 15-12, p. 364 on incineration of solid and hazardous waste; Figure 16-13, p. 389, on genetically modified crops and foods; Figure 17-16, p. 423, on clear-cutting forests; Figure 19-19, p. 490, on conventional natural gas; Figure 19-26, p. 496, on conventional nuclear power; Figure 20-18, p. 521, on using solar energy to generate high-temperature heat and electricity; and Figure 20-31, p. 532, on using hydrogen as a fuel).

What Are Some Key Features of This Book?
This book is *science based, solution oriented,* and *flexible.* The book is divided into five major parts (see Brief Contents, p. xvii). A highly flexible format allows instructors to use almost any course outline. I suggest that instructors use Chapter 1 to provide an overview of environmental problems and solutions and Chapters 3 through 8 to provide a base of scientific principles and concepts. The remaining chapters—2, 9, and 10 through 20—can be used in virtually any order or omitted as desired. In addition, sections within chapters can be omitted or rearranged to meet instructor needs.

Each chapter begins with a brief *case study* designed to capture interest and set the stage for the material that follows. In addition to these 20 case studies, 48 other case studies are found throughout the book; they provide a more in-depth look at specific environmental problems and their possible solutions. Nine *Guest Essays* present an individual researcher's or activist's point of view, which readers can evaluate using the Critical Thinking questions.

This book is an integrated study of environmental problems, connections, and solutions. The eight integrative themes are *biodiversity and natural resources (ecosystem services), sustainability, connections in nature, pollution prevention and waste reduction, population and exponential growth, energy and energy efficiency, solutions to environmental problems,* and *the importance of individuals working together to bring about environmental change.*

To get an overview of this book I urge you to look at the Brief Contents (p. xvii) and the two-page concept map showing how the major topics of this book are connected (found inside the back cover).

This book has 504 illustrations, 90 of them new to this edition. In addition, 30 existing figures have been updated or improved for this new edition. These illustrations are designed to present complex ideas in understandable ways and to relate learning to the real world.

I have not cited specific sources of information within the text. This is rarely done for an introductory-level text in any field, and it would interrupt the flow of the material. Instead, on the website material for each chapter you will find readings, Internet site references, and references to complete articles that can be accessed online on the *InfoTrac* supplement available free to qualified users of this book. These sources back up most of the content of this book and serve as springboards to further information and ideas. Placing these references on the website also allows me to update them regularly.

Instructors wanting a book covering this material with a different emphasis, organization, and length can use one of my three other books written for various types of environmental science courses: *Living in the Environment,* 13th edition (757 pages, Brooks/Cole, 2004), *Sustaining the Earth: An Integrated Approach,* 6th edition (349 pages, Brooks/Cole, 2004), and *Essentials of Ecology,* 2nd edition (276 pages, Brooks/Cole, 2004).

What Are the Major Changes in the Tenth Edition? Major changes include the following:

- Updated and revised material throughout the book.

- 90 new figures and 30 improved or updated illustrations.

- A new student preface providing tips on how to study and on critical thinking (pp. xiii–xvi).

- Reduced use of bulleted and numbered lists and boxed material to improve the flow of the book.

- Many new topics including *transfer of energy by convection, conduction, and radiation* (Figure 3-9, p. 55); *life cycle of frogs* (Figure 7-6, p. 145); human ecological footprints on the earth's surface (Figure 8-12, p. 169) and on North America (Figure 17-8, p. 415); four principles of sustainability (Figure 8-13, p. 170 and Figure 8-14, p. 171); possible beneficial effects of global warming in some areas (Figure 13-13, p. 291); sequestration of carbon dioxide to slow global warming (Figure 13-16, p. 294); terrorism and the release of toxic chemicals from industrial plants (pp. 349-50); industrial ecosystem in Denmark (Figure 15-6, p. 356); use of phytoremediation (Figure 15-9, p. 362) and a plasma torch (Figure 15-10, p. 363) to treat hazardous wastes; mercury contamination in the environment (Figure 15-19, p. 369 and Figure 15-20, 370); ten most endangered U.S. songbirds (Figure 18-7, p. 457);

examples of nonnative species found in the United States (Figure 18-8, p. 458); brief history of the age of oil (pp. 487-488); terrorist treats to nuclear wastes stored at nuclear power plants (p. 498); and terrorist threats from dirty radioactive bombs (pp. 501-502).

■ New CD-ROM: Interactive Concepts in Environmental Science integrates concept summaries and almost 100 engaging animations and interactions based on figures from the text with flashcards and quizzing on the web.

What In-Text Study Aids Are Provided? Each chapter begins with a few general questions to reveal how it is organized and what students will be learning. When a new term is introduced and defined, it is printed in boldface type. A glossary of all key terms is located at the end of the book.

Questions are used as titles for all subsections so readers know the focus of the material that follows. In effect, this is a built-in set of learning objectives.

Each chapter ends with a set of Review Questions covering *all* of the material in the chapter as a study guide for students and a set of questions to encourage students to think critically and apply what they have learned to their lives. Many instructors give students a list of the particular review questions they expect their students to answer. This can provide students with a comprehensive study guide and a review of text material tailored to your particular course. The Critical Thinking questions are followed by several projects that individuals or groups can carry out.

What Internet and Online Study Aids Are Available? Qualified users of this textbook have free access to the *Brooks/Cole Biology and Environmental Science Resource Center*. Access the online resource material for this book by logging on at

http://biology.brookscole.com/miller10

At this website you will find the following material for each chapter:

■ Flash Cards, which allow you to test your mastery of the Terms and Concepts to Remember for each chapter

■ Tutorial Quizzes, which provide a multiple-choice practice quiz

■ InfoTrac Periodical Research, which includes critical thinking exercises that use InfoTrac College Edition as a research tool, as well as suggested readings for every chapter.

■ References, which lists the major books and articles consulted in writing this chapter

■ A brief What You Can Do list addressing key environmental problems

■ Hypercontents, which takes you to an extensive list of websites with news, research, and images related to individual sections of the chapter

Qualified adopters of this textbook also have access to *WebTutor Toolbox on WebCT or Blackboard* at

http://e.thomsonlearning.com

It provides access to a full array of study tools, including flashcards (with audio), practice quizzes, online tutorials, and web links.

Teachers and students using *new* copies of this textbook also have free and unlimited access to *Info-Trac College Edition*. This fully searchable online library gives users access to complete environmental articles from several hundred periodicals dating back over the past four years. Each chapter ends with two practice exercises to help students learn how to navigate this valuable source of information.

Other student learning tools include:

■ With access to the Opposing Viewpoints Resource Center (OVRC), students enter a world of intelligent, thoughtfully worded discussion centering on the key issues of our time, such as genetic engineering, environmental policy, biological weapons, global warming, endangered species, and AIDS. This online resource center draws on the acclaimed social issues series published by Greenhaven Press, as well as core reference content from other Gale and Macmillan Reference USA sources. The result is a timely, well-stocked online library that allows students to investigate current issues and topics from a number of perspectives. To take a quick tour of the OVRC, visit

http://www.wadsworth.com/pubco/serv_opposing.html

■ A new interactive CD-ROM: *Interactive Concepts in Environmental Science* that integrates concept summaries and almost 100 engaging animations and interactions based on figures from the text with flashcards and quizzing on the web.

■ *Essential Study Skills for Science Students by Daniel D. Chiras.* This books includes chapters on developing good study habits, sharpening memory, getting the most out of lectures, labs, and reading assignments, improving test-taking abilities, and becoming a critical thinker. Your instructor can have this book bundled FREE with your textbook.

■ *Laboratory Manual by C. Lee Rockett and Kenneth J. Van Dellen.* This manual includes a variety of laboratory exercises, workbook exercises, and projects that require a minimum of sophisticated equipment.

What Supplementary Materials Are Available for Instructors? The following supplementary materials are available to instructors adopting this book:

- *Multimedia Manager.* This CD-ROM, free to adopters, allows you to create custom lectures using over 778 pieces of high-resolution artwork from the book and Flash animations from the student CD, assemble database files, and create Microsoft® Power-Point® lectures using text slides and figures from the textbook. This program's editing tools allow use of slides from other lectures, modification or removal of figure labels and leaders, insertion of your own slides, saving slides as JPEGs, and preparation of lectures for use on the Web.

- *Transparency Acetates and Masters.* Includes 100 color acetates of line art and nearly 450 black and white master sheets of key diagrams for making overhead transparencies. Free to adopters.

- *CNN^{TM} Today Videos.* These videos, updated annually, contain short clips of news stories about environmental news. Adopters can receive one video free each year for 3 years. These videos are now available to professors electronically for classroom presentation as well as on videotape.

- Two videos, *In the Shadow of the Shuttle: Protecting Endangered Species,* and *Costa Rica: Science in the Rainforest,* are available to adopters.

- *Instructor's Manual with Test Bank.* Free to adopters.

- *ExamView.* Allows you to easily create and customize tests, see them on the screen exactly as they will print, and print them out.

Help Me Improve This Book   Let me know how you think this book can be improved; if you find any errors, bias, or confusing explanations, please e-mail them to me at

mtg89@hotmail.com

Most errors can be corrected in subsequent printings of this edition rather than waiting for a new edition.

Acknowledgments   I wish to thank the many students and teachers who have responded so favorably to the 9 previous editions of *Environmental Science,* the 13 editions of *Living in the Environment,* and the 6 editions of *Sustaining the Earth* and who have corrected errors and offered many helpful suggestions for improvement. I am also deeply indebted to the more than 250 reviewers, who pointed out errors and suggested many important improvements in the various editions of these three books. Any errors and deficiencies left are mine.

The members of the talented production team, listed on the copyright page, have made vital contributions as well. My thanks also go to production editors Andy Marinkovich and Andrea Fincke, copy editor Anita Wagner, Thompson Steele's page layout artist Bonnie Van Slyke, Brooks/Cole's hard-working sales staff, and Keli Amann, Chris Evers, Steve Bolinger, Margaret Parks, Joy Westberg, and the other members of the talented team who developed the multimedia, website, and advertising materials associated with this book.

I also thank C. Lee Rockett and Kenneth J. Van Dellen for developing the *Laboratory Manual* to accompany this book; Jane Heinze-Fry for her work on concept mapping and *Environmental Articles, Critical Thinking and the Environment: A Beginner's Guide;* Irene Kokkala for her excellent work on the *Instructor's Manual,* and the people who have translated this book into six different languages for use throughout much of the world.

My deepest thanks go to Jack Carey, biology publisher at Brooks/Cole, for his encouragement, help, 37 years of friendship, and superb reviewing system. It helps immensely to work with the best and most experienced editor in college textbook publishing.

I dedicate this book to the earth and to Kathleen Paul Miller, my wife and research assistant.

G. Tyler Miller, Jr.

Guest Essayists and Reviewers

Guest Essayists The following are authors of Guest Essays: **Robert D. Bullard,** professor of sociology and director of the Environmental Justice Resource Center at Clark Atlanta University; **Lois Marie Gibbs,** director, Center for Health, Environment, and Justice; **Garrett Hardin,** professor emeritus of human ecology, University of California, Santa Barbara; **Paul G. Hawken,** environmental author and business leader; **Amory B. Lovins,** energy policy consultant and director of research, Rocky Mountain Institute; **Peter Montague,** director, Environmental Research Foundation; **Norman Myers,** tropical ecologist and consultant in environment and development; **David W. Orr,** professor of environmental studies, Oberlin College; and **Nancy Wicks,** ecopioneer and director of Round Mountain Organics.

Cumulative Reviewers Barbara J. Abraham, Hampton College; Donald D. Adams, State University of New York at Plattsburgh; Larry G. Allen, California State University, Northridge; Susan Allen-Gil, Ithaca College; James R. Anderson, U.S. Geological Survey; Mark W. Anderson, University of Maine; Kenneth B. Armitage, University of Kansas; Samuel Arthur, Bowling Green State University; Gary J. Atchison, Iowa State University; Marvin W. Baker Jr., University of Oklahoma; Virgil R. Baker, Arizona State University; Ian G. Barbour, Carleton College; Albert J. Beck, California State University, Chico; W. Behan, Northern Arizona University; Keith L. Bildstein, Winthrop College; Jeff Bland, University of Puget Sound; Roger G. Bland, Central Michigan University; Grady Blount II, Texas A&M University, Corpus Christi; Georg Borgstrom, Michigan State University; Arthur C. Borror, University of New Hampshire; John H. Bounds, Sam Houston State University; Leon F. Bouvier, Population Reference Bureau; Daniel J. Bovin, Université; Michael F. Brewer, Resources for the Future, Inc.; Mark M. Brinson, East Carolina University; Dale Brown, University of Hartford; Patrick E. Brunelle, Contra Costa College; Terrence J. Burgess, Saddleback College North; David Byman, Pennsylvania State University, Worthington–Scranton; Lynton K. Caldwell, Indiana University; Faith Thompson Campbell, Natural Resources Defense Council, Inc.; Ray Canterbery, Florida State University; Ted J. Case, University of San Diego; Ann Causey, Auburn University; Richard A. Cellarius, Evergreen State University; William U. Chandler, Worldwatch Institute; F. Christman, University of North Carolina, Chapel Hill; Preston Cloud, University of California, Santa Barbara; Bernard C. Cohen, University of Pittsburgh; Richard A. Cooley, University of California, Santa Cruz; Dennis J. Corrigan; George Cox, San Diego State University; John D. Cunningham, Keene State College; Herman E. Daly, University of Maryland; Raymond F. Dasmann, University of California, Santa Cruz; Kingsley Davis, Hoover Institution; Edward E. DeMartini, University of California, Santa Barbara; Charles E. DePoe, Northeast Louisiana University; Thomas R. Detwyler, University of Wisconsin; Peter H. Diage, University of California, Riverside; Lon D. Drake, University of Iowa; David DuBose, Shasta College; Dietrich Earnhart, University of Kansas; T. Edmonson, University of Washington; Thomas Eisner, Cornell University; Michael Esler, Southern Illinois University; David E. Fairbrothers, Rutgers University; Paul P. Feeny, Cornell University; Richard S. Feldman, Marist College; Nancy Field, Bellevue Community College; Allan Fitzsimmons, University of Kentucky; Andrew J. Friedland, Dartmouth College; Kenneth O. Fulgham, Humboldt State University; Lowell L. Getz, University of Illinois at Urbana–Champaign; Frederick F. Gilbert, Washington State University; Jay Glassman, Los Angeles Valley College; Harold Goetz, North Dakota State University; Jeffery J. Gordon, Bowling Green State University; Eville Gorham, University of Minnesota; Michael Gough, Resources for the Future; Ernest M. Gould Jr., Harvard University; Peter Green, Golden West College; Katharine B. Gregg, West Virginia Wesleyan College; Paul K. Grogger, University of Colorado at Colorado Springs; L. Guernsey, Indiana State University; Ralph Guzman, University of California, Santa Cruz; Raymond Hames, University of Nebraska, Lincoln; Raymond E. Hampton, Central Michigan University; Ted L. Hanes, California State University, Fullerton; William S. Hardenbergh, Southern Illinois University at Carbondale; John P. Harley, Eastern Kentucky University; Neil A. Harriman, University of Wisconsin, Oshkosh; Grant A. Harris, Washington State University; Harry S. Hass, San Jose City College; Arthur N. Haupt, Population Reference Bureau; Denis A. Hayes, environmental consultant; Stephen Heard, University of Iowa; Gene Heinze-Fry, Department of Utilities, Commonwealth of Massachusetts; Jane Heinze-Fry, environmental educator; John G. Hewston, Humboldt State University; David L. Hicks, Whitworth College; Kenneth M. Hinkel, University of Cincinnati; Eric Hirst, Oak Ridge National Laboratory; Doug Hix, University of Hartford; S. Holling, University of British Columbia;

Milwaukee; Frank E. Studnicka, University of Wisconsin, Platteville; Chris Tarp, Contra Costa College; Roger E. Thibault, Bowling Green State University; William L. Thomas, California State University, Hayward; Shari Turney, copy editor; John D. Usis, Youngstown State University; Tinco E. A. van Hylckama, Texas Tech University; Robert R. Van Kirk, Humboldt State University; Donald E. Van Meter, Ball State University; Gary Varner, Texas A&M University; John D. Vitek, Oklahoma State University; Harry A. Wagner, Victoria College; Lee B. Waian, Saddleback College; Warren C. Walker, Stephen F. Austin State University; Thomas D. Warner, South Dakota State University; Kenneth E. F. Watt, University of California, Davis; Alvin M. Weinberg, Institute of Energy Analysis, Oak Ridge Associated Universities; Brian Weiss; Margery Weitkamp, James Monroe High School (Granada Hills, California); Anthony Weston, State University of New York at Stony Brook; Raymond White, San Francisco City College; Douglas Wickum, University of Wisconsin, Stout; Charles G. Wilber, Colorado State University; Nancy Lee Wilkinson, San Francisco State University; John C. Williams, College of San Mateo; Ray Williams, Rio Hondo College; Roberta Williams, University of Nevada, Las Vegas; Samuel J. Williamson, New York University; Ted L. Willrich, Oregon State University; James Winsor, Pennsylvania State University; Fred Witzig, University of Minnesota at Duluth; George M. Woodwell, Woods Hole Research Center; Robert Yoerg, Belmont Hills Hospital; Hideo Yonenaka, San Francisco State University; Malcolm J. Zwolinski, University of Arizona.

For Students

Why Is It Important to Study Environmental Science? Welcome to *environmental science*—an *interdisciplinary* study of how nature works and how things in nature are interconnected. This book is an integrated and science-based study of environmental problems, connections, and solutions.

Environmental issues affect every part of your life and are an important part of the news stories presented on television and in newspapers and magazines. Thus, the concepts, information, and issues discussed in this book and the course you are taking should be useful to you now and in the future.

Understandably, I am biased. But I strongly believe that environmental science is the single most important course in your education. What could be more important than learning how the earth works, how we are affecting its life support system, and how we can reduce our environmental impact?

We live in an incredibly exciting and challenging time. There is growing concern that we may have no more than 50–100 years to make a new cultural transition in which we learn how to live more sustainably by not degrading our life support system.

If this view is correct, we live during a brief hinge of history as we struggle to enter into a new relationship with the earth by bringing about an *environmental* or *sustainability revolution.* It took more than 10,000 years to bring about the *agricultural revolution* and about 275 years to bring about the *industrial revolution.*

Because of our rapidly growing environmental impact, many analysts believe that we have only 50–100 years to bring about an *environmental revolution.* During this short period what each of us does to work with the earth will have an enormous and long-lasting impact—perhaps greater than in any other single human generation.

Thus, each of us has a unique opportunity, leverage, and responsibility to make the earth a better place to live for this and future generations. I hope this book and the course you are taking will help you learn about the exciting challenges we face. More important, I hope it will stimulate you to become involved in this change in the way we view and treat the earth that sustains us, other life, and all economies.

How Did I Become Involved with Environmental Problems? In 1966, I heard a scientist give a lecture on the problems of population growth and pollution. Afterward I went to him and said, "If even a fraction of what you have said is true, I will feel ethically obligated to give up my research on the corrosion of metals and devote the rest of my life to research and education on environmental problems and solutions. Frankly, I do not want to believe a word you have said, and I am going into the literature to try to show that your statements are either untrue or grossly distorted."

After 6 months of study I was convinced of the seriousness of these and other environmental problems. Since then, I have been studying, teaching, and writing about them. This book summarizes what I have learned in more than three decades of trying to understand environmental principles, problems, connections, and solutions.

How Can You Improve Your Study and Learning Skills? Maximizing your ability to learn should be one of your most important educational goals. This involves continually trying to *improve your study and learning skills.*

This has a number of payoffs. You can learn more and do this more efficiently. You will also have more time to pursue other interests besides studying without feeling guilty about always being behind. It can also help you get better grades and live a more fruitful and rewarding life.

Here are some *general study and learning skills.*

Get organized. The more efficient you become at studying, the more time you will have for other interests.

Make daily to-do lists in writing. Put items in order of importance, focus on the most important tasks, and assign a time to work on these items. Because life is full of uncertainties, you will be lucky to accomplish half of the items on your daily list. Shift your schedule as needed to accomplish the most important items. Otherwise, you fall behind and become increasingly frustrated.

Set up a study routine in a distraction-free environment. Develop a written daily study schedule and stick to it. Study in a quiet, well-lighted space. Work sitting at a desk or table—not lying down on a couch or bed. Take breaks every hour or so. During each break take several deep breaths and move around to help you stay more alert and focused.

Avoid procrastination—putting work off until another time. Do not fall behind on your reading and other assignments. Accomplish this by setting aside a particular time for studying each day and making it a part of your daily routine.

Do not eat dessert first. Otherwise, you may never get to the main meal (studying). Jump into studying. When you have accomplished your study goals then reward yourself with play (dessert).

Make hills out of mountains. It is psychologically difficult to climb a mountain such as reading an entire book, reading a chapter in a book, writing a paper, or cramming to study for a test. Instead, break such large tasks (mountains) down into a series of small tasks (hills). Each day read a few pages of a book or chapter, write a few paragraphs of a paper, and review what you have studied and learned.

Look at the big picture first. Get an overview of an assigned reading by looking at the main headings or chapter outline. In this textbook, I provide a list of the main questions that are the focus of each chapter.

Ask and answer questions as you read. For example, what is the main point of this section or paragraph? To help you do this I start each subsection with a question that the material is designed to answer. After reading a subsection, write down a summary answer to that question and use it to review the material and prepare for classes and tests.

Focus on key terms. Use the glossary in your textbook or a dictionary to look up the meaning of terms or words you do not understand. Make flash cards for learning key terms and concepts and review them frequently. This book shows all key terms in **boldfaced** type and lesser but still important terms in *italicized* type. Flash cards for testing your mastery of key terms for each chapter are available on the website for this book.

Interact with what you read. I do this by marking key sentences and paragraphs with a highlighter or pen. I put an asterisk in the margin next to an idea I think is important and double asterisks next to an idea I think is especially important. I write comments in the margins, such as *Beautiful, Confusing, Misleading,* or *Wrong.* I fold down the top corner of pages with highlighted passages and the top and bottom corners of especially important pages. This way, I can flip through a chapter or book and quickly review the key ideas.

Review to reinforce learning. Before each class review the material you learned in the previous class and read the assigned material. Review, fill in, and organize your notes as soon as possible after each class.

Become a better note taker. Do not try to take down everything your instructor says. Instead take down main points and key facts using your own shorthand system. Fill in and organize your notes after class.

Write out answers to questions to focus and reinforce learning. Answer questions at the end of each chapter or those assigned to you. Do this in writing as if you were turning them in for a grade. Save your answers for review and preparation for tests.

Use the buddy system. Study with a friend or become a member of a study group to compare notes, review material, and prepare for tests. Explaining something to someone else is a great way to focus your thoughts and reinforce your learning. If available, attend review sessions offered by instructors or teaching assistants.

Learn your instructor's test style. Does your instructor emphasize multiple choice, fill-in-the-blank, true-or-false, factual, thought, or essay questions? How much of the test will come from the textbook and how much from lecture material? Adapt your learning and studying methods to this style. You may disagree with this style and feel that it does not adequately reflect what you know. But the reality is that your instructor is in charge and your grade (but not always your learning) usually depends heavily on going along with the instructor's system.

Become a better test taker. Avoid cramming. Eat well and get plenty of sleep before a test. Arrive on time or early. Calm yourself and increase your oxygen intake by taking several deep breaths. Do this about every 10–15 minutes. Look over the test and answer the questions you know well first. Then work on the harder ones. Use the process of elimination to narrow down the choices for multiple choice questions. Getting it down to two choices gives you a 50% chance of guessing the right answer. For essay questions, organize your thoughts before you start writing. If you have no idea what a question means, make an educated guess. You might get some partial credit and avoid a zero. Another strategy for getting some credit is to show your knowledge and reasoning by writing: "If this question means so and so, then my answer is _____."

Develop an optimistic outlook. Try to be a "glass is half-full" rather than a "glass is half-empty" person. Pessimism, fear, anxiety, and excessive worrying (especially over things you have no control over) are destructive, feed on themselves, and lead to inaction. Try to keep your energizing feelings of optimism slightly ahead of your immobilizing feelings of pessimism. Then you will always be moving forward.

Take time to enjoy life. Every day take time to laugh and enjoy nature, beauty, and friendship. Becoming an effective and efficient learner is the best way to do this without getting behind and living under a cloud of guilt and anxiety.

How Can You Improve Your Critical Thinking Skills? Every day we are exposed to a sea of information, ideas, and opinions. How do we know what to believe and why? Do the claims seem reasonable or exaggerated?

Critical thinking involves developing skills to help you analyze and evaluate the validity of information and ideas you are exposed to and to make decisions. Critical thinking skills help you decide rationally what to believe or what to do. This involves examining information and conclusions or beliefs in terms of the evidence and chain of logical reasoning that supports them. Critical thinking helps you distinguish between facts and opinions, evaluate evidence and arguments, take and defend an informed position on issues, integrate information and see relationships, and apply your knowledge to dealing with new and different problems. Here are some basic skills for learning how to think more critically.

Question everything and everybody. Be skeptical, as any good scientist is. Do not believe everything you hear or read, including the content of this textbook. Think about and evaluate all information you receive. Seek other sources and opinions.

Do not believe everything you read on the Internet. The Internet is a wonderful and easily accessible source of information. It is also a useful way to find alternative information and opinions on almost any subject or issue—much of it not available in the mainstream media and scholarly articles. However, because the Internet is so open, anyone can write anything they want with no editorial control or peer evaluation—the method in which scientific or other experts in an area review and comment on an article before it is accepted for publication in a scholarly journal. As a result, evaluating information on the Internet is one of the best ways to put into practice the principles of critical thinking discussed in this preface. Use and enjoy the Internet, but think critically and proceed with caution.

Identify and evaluate your personal biases and beliefs. Each of us has biases and beliefs taught to us by sources such as parents, teachers, friends, role models, and experience. What are your basic beliefs and biases? Where did they come from? What basic assumptions are they based on? How sure are you that your beliefs and assumptions are right and why?

Be open-minded, flexible, and humble. Be open to considering different points of view, suspend judgment until you gather more evidence, and be capable of changing your mind. Recognize that there may be a number of useful and acceptable solutions to a problem and that very few issues are black or white. There are usually valid points on both (or many) sides of an issue. One way to evaluate divergent views is to get into another person's head or walk in their shoes. How do they see or view the world? What are their basic assumptions and beliefs? Is their position logically consistent with their assumptions and beliefs?

Evaluate how the information related to an issue was obtained. Are the statements made based on firsthand knowledge or research or on hearsay? Are unnamed sources used? Is the information based on reproducible and widely accepted scientific studies (*consensus science*, p. 49) or preliminary scientific results that may be valid but need further testing (*frontier science*, p. 49)? Is the information based on a few isolated stories or experiences (*anecdotal information*) instead of carefully controlled studies? Is it based on unsubstantiated and widely doubted scientific information or beliefs (*junk science* or *pseudoscience*)?

Question the evidence and conclusions presented. What are the conclusions or claims? What evidence is presented to support them? Does the evidence support them? Is there a need to gather further evidence to test the conclusions? Are there other, more reasonable conclusions?

Try to identify and assess the assumptions and beliefs of those presenting evidence and drawing conclusions. What is their expertise in this area? Do they have any unstated assumptions, beliefs, biases, or values? Do they have a personal agenda? Can they benefit financially or politically from acceptance of their evidence and conclusions? Would investigators with different basic assumptions or beliefs take the same data and come to different conclusions?

Do the arguments used involve common logical fallacies or debating tricks? Here are seven of many examples. *First,* attack the presenter of an argument rather than the argument itself. *Second,* appeal to emotion rather than facts and logic. *Third,* claim that if one piece of evidence or one conclusion is false, then all other pieces of evidence and conclusions are false. *Fourth,* say that a conclusion is false because it has not been scientifically proven (scientists can never prove anything absolutely but they can establish degrees of reliability, as discussed on p. 49). *Fifth,* inject irrelevant or misleading information to divert attention from important points. *Sixth,* appeal to authority by saying that something is true because a certain authority or expert says so. *Seventh,* present only either/or alternatives when there may be a number of alternatives.

Become a seeker of wisdom, not a vessel of information. Develop a written list of principles, concepts, and rules to serve as guidelines in evaluating evidence and claims and making decisions. Continually evaluate and modify this list on the basis of experience. Many people believe that the main goal of education is to learn as much as you can by concentrating on gathering more and more information—much of it useless or misleading. I believe that the primary goal is to know as little as possible. This is done by learning how to sift through mountains of facts and ideas to find the few *nuggets of wisdom* that are the most useful in understanding the world and in making decisions. This takes a firm commitment to learning how to think logically and critically and continually flushing less valuable and thought-clogging information from our minds.

This book is full of facts and numbers, but they are useful only to the extent that they lead to an understanding of key and useful ideas, scientific laws, concepts, principles, and connections. A major goal of the study of environmental science is to find out how nature works and sustains itself (*environmental wisdom*) and use *principles of environmental wisdom* to help make our human societies and economies more sustainable and thus more beneficial and enjoyable.

How Have I Attempted to Achieve Balance?
There are always *trade-offs* involved in making and implementing environmental decisions. My challenge is to give a fair and balanced presentation of opposing viewpoints, advantages and disadvantages of various technologies and proposed solutions to environmental problems, and good and bad news about environmental problems without injecting personal bias. This allows you to make up your own mind about important environmental issues.

Studying a subject as important as environmental science and ending up with no conclusions, opinions, and beliefs means that both the teacher and student have failed. However, such conclusions should be based on using critical thinking to evaluate opposing ideas and understand the trade-offs involved.

I Welcome Your Help Researching and writing a book that covers and connects ideas in such a wide variety of disciplines is a challenging and exciting task. Almost every day I learn about some new connection in nature. I hope you have as much fun learning about such connections and their implications as I have over the past 37 years.

In a book this complex, there are bound to be some errors—some typographical mistakes that slip through and some statements that you might question based on your knowledge and research. My goal is to provide you with an interesting, accurate, balanced, and challenging book that furthers your understanding of this vital subject. I have also attempted to provide balance on presenting various sides of key environmental issues.

I invite you to contact me and point out any remaining bias, correct any errors you find, and suggest ways to improve this book. Over decades of teaching some of my best teachers have been students taking my classes and reading my textbooks. Please E-mail your suggestions to me at

mtg89@hotmail.com

G. Tyler Miller, Jr.

Brief Contents

Detailed Contents

Temperate deciduous forest, Fall, Rhode Island

Temperate deciduous forest, Winter, Rhode Island

Hawaiian monk seal's mouth caught in plastic

PART II
SCIENTIFIC PRINCIPLES
AND CONCEPTS

3 Science, Systems, Matter, and Energy 46

4 Ecosystems: Components, Energy Flow, and Matter Cycling 64

5 Evolution and Biodiversity: Origins, Niches, and Adaptation 92

Cotton-top tamarin

National Park Service

Glacier National Park

National Archives/EPA Documerica

Monoculture cropland, California

Fast growing Kenaf for making paper, Texas

U.S. Department of Agriculture

U.S. Department of Agriculture

Boll weevil

U.S. Fish and Wildlife Service

Wildlife refuge

Ocean Arcs International

John Todd at solar sewage plant, Rhode Island

Coral reef sanctuary, Tortugas Marine Ecological Reserve, Florida Keys

PART IV
SUSTAINING BIODIVERSITY

Wolf spider

U.S. Wind Power

Wind farm, California

PART V
ENERGY RESOURCES

19   Nonrenewable Energy Resources   478

20   Energy Efficiency and Renewable Energy   506

Appendices

Glossary   G1

Index   I1

1 ENVIRONMENTAL PROBLEMS, THEIR CAUSES, AND SUSTAINABILITY

Living in an Exponential Age

Two ancient kings enjoyed playing chess, with the winner claiming a prize from the loser. After one match, the winning king asked the losing king to pay him by placing one grain of wheat on the first square of the chessboard, two on the second, four on the third, and so on. The number of grains was to double each time until all 64 squares were filled.

The losing king, thinking he was getting off easy, agreed with delight. It was the biggest mistake he ever made. He bankrupted his kingdom and still could not produce the incredibly large number of grains of wheat he had promised. In fact, it is probably more than all the wheat that has ever been harvested!

This fictional story illustrates the concept of **exponential growth,** in which a quantity increases by a fixed percentage of the whole in a given time. Anything growing by a fixed percentage undergoes exponential growth. As the losing king learned, exponential growth is deceptive. It starts off slowly, but after only a few doublings it grows to enormous numbers because each doubling is more than the total of all earlier growth.

Here is another example. Fold a piece of paper in half to double its thickness. If you could continue doubling the thickness of the paper 42 times, the stack would reach from the earth to the moon, 386,400 kilometers (240,000 miles) away. If you could double it 50 times, the folded paper would almost reach the sun, 149 million kilometers (93 million miles) away!

Six important environmental issues are *population growth, increasing and wasteful resource use, global climate change, premature extinction of plants and animals and destruction and degradation of wildlife habitats (biodiversity crisis), pollution,* and *poverty.* All these issues are interconnected and growing exponentially. For example, between 1950 and 2003, world population increased from 2.5 billion to 6.3 billion. Unless death rates rise sharply, it may reach 8 billion by 2028, 9 billion by 2050, and 10–14 billion by 2100 (Figure 1-1) Global economic output, some of it environmentally damaging, is a rough measure of resource use. It has increased sevenfold since 1950.

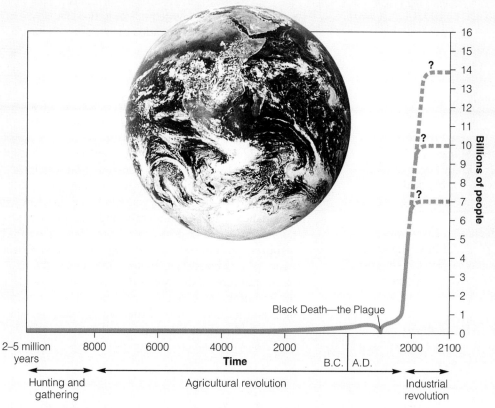

Black Death—the Plague

| 2–5 million years | 8000 | 6000 | 4000 | 2000 | B.C. | A.D. | 2000 | 2100 |

Time

Hunting and gathering — Agricultural revolution — Industrial revolution

Billions of people

Figure 1-1 The J-shaped curve of past exponential world population growth, with projections to 2100. Notice that exponential growth starts off slowly, but as time passes the curve becomes increasingly steep. The current world population of 6.3 billion people is projected to reach 7–14 billion people sometime this century. (This figure is not to scale.) (Data from World Bank and United Nations; photo courtesy of NASA)

Alone in space, alone in its life-supporting systems, powered by inconceivable energies, mediating them to us through the most delicate adjustments, wayward, unlikely, unpredictable, but nourishing, enlivening, and enriching in the largest degree—is this not a precious home for all of us? Is it not worth our love?

BARBARA WARD AND RENÉ DUBOS

This chapter presents an overview of environmental problems, their causes, controversy over their seriousness, and ways we can live more sustainably. It discusses these questions:

- What are natural resources, and why are they important? What is an environmentally sustainable society?

- How fast is the human population increasing?

- What is the difference between economic growth and economic development? What is poverty and what are some harmful environmental effects of poverty?

- What are the earth's main types of resources? How can they be depleted or degraded?

- What are the principal types of pollution? How can pollution be reduced and prevented?

- What are the basic causes of today's environmental problems? How are these causes connected?

- What major beneficial and harmful effects have hunter–gatherer societies, agricultural societies, and industrialized societies had on the environment?

- Is our current course sustainable? What is environmentally sustainable development? How can we live more sustainably?

1-1 LIVING MORE SUSTAINABLY

What Is the Difference between Environment, Ecology, and Environmental Science? **Environment** is everything that affects a living organism (any unique form of life). **Ecology** is a biological science that studies the relationships between living organisms and their environment.

This textbook is an introduction to **environmental science,** an interdisciplinary study of how the earth works, how we are affecting the earth's life-support systems (environment), and how to deal with the environmental problems we face.

What Roles Do Different Groups Play? Many different groups of people are concerned about environmental issues. The cast of major characters you will encounter in this book includes the following:

- **Ecologists,** who are biological scientists studying relationships between living organisms and their environment.

- **Environmental scientists,** who use scientific information and concepts to understand how the earth works, learn how humans interact with the earth, and develop solutions to environmental problems.

- **Conservation biologists,** who in the 1970s created a multidisciplinary science to investigate human impacts on the diversity of life found on the earth (biodiversity) and develop practical plans for preserving biodiversity.

- **Environmentalists,** who are concerned about the impact of people on environmental quality and believe some human actions are degrading parts of the earth's life-support systems for humans and many other forms of life. Some of their beliefs and proposals for dealing with environmental problems are based on scientific information and concepts and some are based on their social and ethical environmental beliefs (environmental worldviews).

- **Preservationists,** concerned primarily with setting aside or protecting undisturbed natural areas from harmful human activities.

- **Conservationists,** concerned with using natural areas and wildlife in ways that sustain them for current and future generations of humans and other forms of life.

- **Restorationists,** devoted to restoration of natural areas that have been degraded by human activities.

Many people consider themselves members of several of these groups.

What Keeps Us Alive? Our existence, lifestyles, and economies depend completely on the sun and the earth, a blue and white island in the black void of space (Figure 1-1). To economists, *capital* is wealth used to sustain a business and to generate more wealth. By analogy, we can think of energy from the sun as **solar capital** and the planet's air, water, soil, wildlife, forests, rangelands, fisheries, minerals, and natural purification, recycling, and pest control processes as **natural resources** or **natural capital** (Figure 1-2 and Guest Essay, p. 4). It consists of *resources* (orange) and *ecological services* (green) that support and sustain the earth's life and economies. **Solar energy** includes direct sunlight and indirect forms of solar energy such as wind power, hydropower (energy from flowing water), and biomass (direct solar energy converted to chemical energy stored in biological sources of energy such as wood).

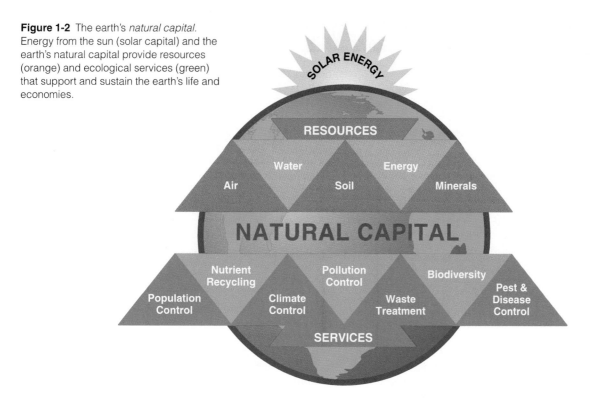

Figure 1-2 The earth's *natural capital.* Energy from the sun (solar capital) and the earth's natural capital provide resources (orange) and ecological services (green) that support and sustain the earth's life and economies.

What Is an Environmentally Sustainable Society? An **environmentally sustainable society** tries to achieve two goals. *First,* it satisfies the basic needs of its people for food, clean water, clean air, and shelter into the indefinite future. *Second,* it does this without depleting or degrading the earth's natural resources and thereby preventing current and future generations of humans and other species from meeting their basic needs. *Living sustainably* means living off natural income replenished by soils, plants, air, and water and not depleting the earth's natural capital that supplies this income (Figure 1-2, top half of back cover, and Guest Essay, p. 4).

For example, imagine you inherit $1 million. Invest this capital at 10% interest per year, and you will have a sustainable annual income of $100,000 without depleting your capital. If you spend $200,000 a year, your $1 million will be gone early in the 7th year; even if you spend only $110,000 a year, you will be bankrupt early in the 18th year. The lesson here is an old one: *Protect your capital and live off the income it provides.* Deplete your capital, and you move from a sustainable to an unsustainable lifestyle.

The same lesson applies to the earth's natural capital. According to many environmentalists and leading scientists, we are living unsustainably by depleting and degrading the earth's natural capital at an accelerating rate as our population (Figure 1-1) and demands on the earth's resources and life-sustaining processes increase exponentially.

Other analysts do not believe we are living unsustainably. They contend that environmentalists have exaggerated the seriousness of population, resource, and environmental problems, and we can overcome these problems by human ingenuity and technological advances.

1-2 POPULATION GROWTH

How Rapidly Is the Human Population Growing? Currently the world's population is growing exponentially at a rate of about 1.26% per year. The ticking of this population clock means that in 2003 the world's population of 6.3 billion grew by 79 million people (6.3 billion × 0.0126 = 79 million), an average increase of 216,000 people a day, or 9,000 an hour.

At this annual rate of exponential growth, it takes about

- 3 days to add the 651,000 Americans killed in battle in all U.S. wars.

- 1.6 years to add the 129 million people killed in all wars fought in the past 200 years.

- 3.7 years to add 291 million people (the population of the United States in 2003).

How much is 79 million? Spending 1 second saying hello to each of 79 million new people added to the

Natural Capital

Paul G. Hawken

Paul G. Hawken understands both business and ecology. In addition to founding Smith & Hawken, a retail company known for its environmental initiatives, he has written seven widely acclaimed books, including Growing a Business *(1987),* The Ecology of Commerce *(1993),* Factor 10, The Next Industrial Revolution *(1998, with Amory and Hunter Lovins), and* Natural Capitalism *(1999, with Amory and Hunter Lovins). He produced and hosted* Growing a Business, *a series for public television shown on 210 U.S. stations and in 115 countries.* The Ecology of Commerce *was hailed as the best business book of 1993 and one of the most important books of the 20th century. In 1987,* Inc. *magazine named Hawken one of the 12 best entrepreneurs of the 1980s, and in 1995 he was named by the* Utne Reader *as one of the 100 visionaries who could change our lives.*

Great ideas, in hindsight, seem obvious. The concept of natural capital is such an idea. *Natural capital* is the myriad necessary and valuable resources and ecological processes that we rely on to produce our food, products, and services [Figure 1-2 and top half of back cover].

The concept of natural capital is not a new one. Economists have long noted that natural capital is a factor in industrial production, but a marginal factor.

A new view is emerging: Our economic systems cannot long endure without taking the flow of renewable and nonrenewable resources through economies into account. This revision of neoclassical economics, yet to be accepted by most mainstream academicians, provides business and public policy with a powerful new tool for the continued prosperity of business and the preservation and restoration of the earth's living natural systems.

Most Americans are filled with cornucopian fantasies of technological prowess, where human ingenuity bypasses natural limits and creates unimagined abundance. Optimism easily intertwines with the belief that nanotechnology, biotechnology, computers, and technologies yet to be developed will eliminate hunger, disease, and want.

Dreams of alleviating human suffering are worthy. However, they usually overlook the absolute necessity of fertile soil, ocean fisheries, a stable climate, biological diversity, and pure water, all of which we are degrading and none of which can be created by any human-made technology known or imagined.

In our pursuit of dominance over the natural world, we have not taken into account the basic principle that industrialism, for all its sophistication, is enormously inefficient with respect to resource use, energy use, and waste production. The hypotheses and theories of neoclassical economists originated in a time of resource abundance. Today it is difficult for many of these economists to understand that the success of linear industrial systems based on increasing economic growth by increasing the rate of flow of materials and energy through economic systems has laid the groundwork for the next stage in economic evolution.

This shift is profoundly biological. It involves incorporating the cycling of material resources that supports natural systems into our ways of making things and our ways of dealing with the waste matter produced by our current linear industrial systems. This shift is going to happen because cyclical industrial systems work better than linear ones. They close the loop and reincorporate

---

earth during 2003 for 24 hours a day would take you about 2.5 years. By then there would be about 198 million more people to shake hands with.

Some *good news* is that the exponential rate of annual population growth slowed from 2.1% in 1963 to 1.26% in 2003. The *bad news* is that the world's population is still growing exponentially at a rapid rate (Figures 1-1 and 1-3).

1-3 ECONOMIC GROWTH, ECONOMIC DEVELOPMENT, POVERTY, AND GLOBALIZATION

What Is Economic Growth? Almost all countries seek **economic growth**: an increase in their capacity to provide people with goods and services. This increase is accomplished by population growth (more con-

Figure 1-3 World population milestones. (Data from United Nations Population Division, *World Population Prospects, 1998* and Population Reference Bureau)

World Population Reached
1 billion in 1804
2 billion in 1927 (123 years later)
3 billion in 1960 (33 years later)
4 billion in 1974 (14 years later)
5 billion in 1987 (13 years later)
6 billion in 1999 (12 years later)

World Population May Reach
7 billion in 2013 (14 years later)
8 billion in 2028 (15 years later)
9 billion in 2050 (22 years later)

wastes as part of the production cycle. There are no landfills in a cyclical society.

If there is so much inefficiency in our current system, why is it not more apparent? The inefficiencies are masked by a financial system in which money, prices, and markets give us inaccurate information. Markets are not giving us correct information about how much our suburbs, cars, and plastic drinking water bottles truly cost based on the environmental harm they cause.

Instead, we are getting warning signals from the beleaguered airsheds and watersheds, the overworked and eroded soils, the life-degrading inner cities and rural counties, and the conflicts based on shortages of water and some other resources in parts of the world. These feedbacks from nature are providing the information that our prices should give us but do not.

Prices do not give us good information for a simple reason: *improper accounting.* Natural capital has never been placed on the balance sheets of companies or the countries of the world. To paraphrase G. K. Chesterton, it could fairly be said that capitalism might be a good idea, but we have not tried it yet. Capitalism cannot be fully attained or practiced until we have an accurate balance sheet, as any accounting student will tell us.

As it stands, our economic system is based on accounting principles that would bankrupt a company. Instead, we need to place capital on the balance sheet, not as a free resource of infinite supply but as an integral and valuable part of the production process. Then we will use resources much more efficiently and reduce the flow of matter and energy resources through economies and the resulting pollution and loss of natural capital. Moving from linear industrial systems to cyclical ones that mimic nature and include the value of natural capi-tal in producing the goods and services we use accomplishes this.

Many people sincerely believe an economic system based on the integrity of natural systems is unworkable. To answer that concern, we may want to reverse the question and ask, "How have we created an economic system that tells us it is cheaper to destroy natural capital than to maintain it?" We know this is not the way to take care of our cars, houses, and bridges, but somehow we have managed to overlook a pricing system that discounts the future and sells off the past. Or to put it another way, "How did we create an economic system that confuses capital with income?"

Can we devise and implement a more rational economic system? I think so. It is right before us. It requires no new theories, only common sense. It is based on the simple but powerful proposition that *all capital must be valued.*

There may be no *right* way to value a forest or a river, but there is a *wrong* way, which is to give it little or no value. If we have doubts about how to value a 500-year-old tree, we need only ask how much it would cost to make a new one from scratch. Or a new river. Or a new atmosphere.

Our goal should be to estimate and integrate the worth of living systems into every aspect of our culture and commerce so that human systems mimic natural systems. Only if we do this can our cultures reflect growth and harmony rather than damage and discord.

Critical Thinking

If you were in charge of the world's economy, what are the three most important things you would do? Compare your answers with those of other members of your class.

---

sumers and producers), more consumption per person, or both.

Economists use several indicators to measure changes in economic growth:

- **Gross national income (GNI)**—formerly called **gross national product (GNP):** the market value in current dollars of all goods and services produced *within* and *outside* a country during a year plus net income earned abroad by a country's citizens.

- **Gross national income in purchasing power parity (GNI PPP):** the market value of a country's GNI in terms of the goods and services it would buy in the United States. This is a better way to compare the standards of living among countries.

- **Gross domestic product (GDP):** the market value in current dollars of all goods and services produced *within* a country during a year.

- **Gross world product (GWP):** the market value in current dollars of all goods and services produced in the world during a year.

- **Per capita GNI**—formerly called **per capita GNP:** the GNI divided by the total population at midyear. It gives the average slice of the economic pie per person.

- **Per capita GNI in purchasing power parity (per capita GNI PPP):** the GNI PPP divided by the total population at midyear. This is a better way to make comparisons of people's economic welfare among countries.

What Is Economic Development? Economic de-velopment is the improvement of living standards by economic growth. The United Nations (UN) classifies the world's countries as economically developed or de-veloping based primarily on their degree of industrial-ization and their per capita GNI (GNP).

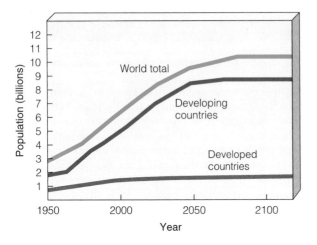

Figure 1-4 Past and projected population size for developed countries, developing countries, and the world, 1950–2100. Developing countries are expected to account for 97% of the 3.6 billion people projected to be added to the world's population between 1990 and 2030. (Data from United Nations)

The **developed countries** (with 1.2 billion people) include the United States, Canada, Japan, Australia, New Zealand, and the countries of Europe. Most are highly industrialized and have high average per capita GNI PPPs (above $10,750 per year, except for industrialized countries in eastern Europe and some in northern and southern Europe). These countries with 19% of the world's population have about 85% of the world's wealth and income, use about 88% of its natural resources, and generate about 75% of its wealth and waste.

All other nations (with 5.1 billion people) are classified as **developing countries,** most of them in Africa, Asia, and Latin America. These countries with 81% of the world's population have about 15% of the world's wealth and income, use about 12% of its natural resources, and generate about 25% of its wealth and waste. Some are *middle-income, moderately developed countries* with average per capita GNI PPPs of $2,701 to $10,750 per year and others are *low-income countries* with per capita GNI PPPs less than $2,701 per year.

About 97% of the projected increase in the world's population is expected to take place in developing countries (Figure 1-4), where about 1 million people are added every 4 days and many key resources are already under serious strain. The primary reason for such rapid population growth in developing countries (1.6% compared to 0.1% in developed countries) is the *large percentage of people who are under age 15* (33% compared to 18% in developed countries in 2003).

Connections: What Are Some Harmful Environmental Effects of Poverty? Daily life is a harsh struggle for the nearly half of the world's people who try to survive on an income of $1–3 (U.S.) per day.

Such poverty is related to environmental quality and people's quality of life. For example, many poor people deplete and degrade local forests, soil, grasslands, wildlife, and water supplies for short-term survival. They do not have the luxury of worrying about long-term supplies of natural resources when their daily life is focused on getting enough food and water to survive.

Another problem is that the poor must often live in areas with high levels of air and water pollution and with the greatest risk of natural disasters such as floods, earthquakes, hurricanes, and volcanic eruptions. Also, they must take jobs (if they can find them) with unhealthy and unsafe working conditions at very low pay.

Poor people often have many children as a form of economic security. Their children help them grow food, gather fuel (mostly wood and dung), haul drinking water, tend livestock, work, and beg in the streets, and help them survive in their old age (typically their 50s or 60s).

Finally, many of the world's poor die prematurely from preventable health problems. According to the World Health Organization (WHO) and the UN Food and Agriculture Organization, each year at least 9 million of the world's desperately poor die prematurely from malnutrition (lack of protein and other nutrients needed for good health), infectious diseases caused by drinking contaminated water, and increased susceptibility to infectious diseases because of their weakened condition from malnutrition. *This premature death of at least 25,000 human beings per day is equivalent to 63 jumbo jet planes, each carrying 400 passengers, accidentally crashing every day with no survivors.* Two-thirds of those dying are children under age 5 (Figure 1-5).

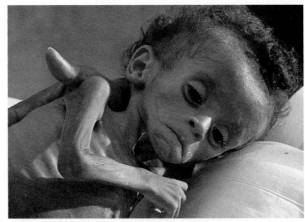

Figure 1-5 One in every three children under age 5, such as this Brazilian child, suffers from malnutrition. According to the World Health Organization, each day at least 13,700 children under age 5 die prematurely from malnutrition and infectious diseases, most from drinking contaminated water and a weakened condition from malnutrition—an average of 10 preventable deaths each minute.

1-4 RESOURCES

What Is a Resource? From a human standpoint, a **resource** is anything obtained from the environment to meet human needs and wants. Examples include food, water, shelter, manufactured goods, transportation, communication, and recreation. On our short human time scale, we classify the material resources we get from the environment as *perpetual, renewable,* or *nonrenewable* (Figure 1-6).

Some resources, such as solar energy, fresh air, wind, fresh surface water, fertile soil, and wild edible plants, are directly available for use. Other resources, such as petroleum (oil), iron, groundwater (water found underground), and modern crops, are not directly available. They become useful to us only with some effort and technological ingenuity. For example, petroleum was a mysterious fluid until we learned how to find, extract, and convert (refine) it into gasoline, heating oil, and other products that we could sell at affordable prices.

What Are Perpetual and Renewable Resources? Solar energy is called a **perpetual resource** because on a human time scale it is renewed continuously. It is expected to last at least 6 billion years as the sun completes its life cycle.

On a human time scale, a **renewable resource** can be replenished fairly rapidly (hours to several decades) through natural processes as long as it is not used up faster than it is replaced. Examples are forests, grasslands, wild animals, fresh water, fresh air, and fertile soil.

However, renewable resources can be depleted or degraded. The highest rate at which a renewable resource can be used *indefinitely* without reducing its available supply is called its **sustainable yield.**

If we exceed a resource's natural replacement rate, the available supply begins to shrink, a process known as **environmental degradation.** Examples of such degradation include urbanization of productive land, waterlogging and salt buildup in soil, excessive topsoil erosion, deforestation, groundwater depletion, overgrazing of grasslands by livestock, and reduction in the earth's forms of wildlife (biodiversity) by elimination of habitats and species, and pollution.

Connections: What Is the Tragedy of the Commons? One cause of environmental degradation is the overuse of **common-property** or **free-access resources.** Such resources are owned by no one (or jointly by everyone in a country or area) but are available to all users at little or no charge.

Examples include clean air, the open ocean and its fish, migratory birds, wildlife species, publicly owned lands (such as national forests, national parks,

and wildlife refuges), gases of the lower atmosphere, and space.

In 1968, biologist Garrett Hardin called the degradation of renewable free-access resources the **tragedy of the commons.** It happens because each user reasons, "If I do not use this resource, someone else will. The little bit I use or pollute is not enough to matter, and such resources are renewable."

With only a few users, this logic works. However, the cumulative effect of many people trying to exploit a free-access resource eventually exhausts or ruins it. Then no one can benefit from it, and therein lies the tragedy.

One solution to this problem is to use free-access resources at rates well below their estimated sustainable yields or overload limits by *reducing population, regulating access, or both.* Some communities have established a set of rules and traditions to regulate and share their access to a common-property resource such as ocean fisheries, grazing lands, and forests. However, it is difficult and expensive to determine the sustainable yield of a forest, grassland, or animal population because such yields vary with weather, climate, and unpredictable biological factors.

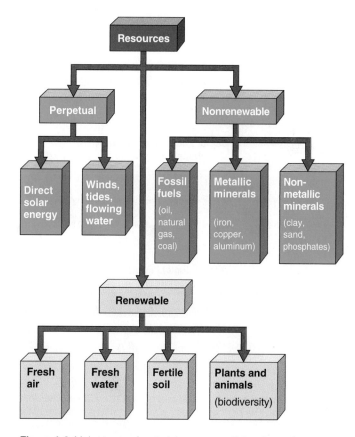

Figure 1-6 Major types of material resources. This scheme is not fixed; renewable resources can become nonrenewable if used for a prolonged period at a faster rate than they are renewed by natural processes.

Another solution is to *convert free-access resources to private ownership*. The reasoning is that owners of land or some other resource have a strong incentive to protect their investment. But this approach is not practical for global common resources (such as the atmosphere, the open ocean, most wildlife species, and migratory birds) that cannot be divided up and converted to private property.

What Is Our Ecological Footprint? The **per capita ecological footprint** is the amount of biologically productive area per person required to produce the renewable resources (such as food and wood), supply space for infrastructure, and absorb the greenhouse gas carbon dioxide emitted from burning fossil fuels. The ecological footprint of each person in developed countries is large compared to that in developing countries (Figure 1-7).

In 1999, the average global ecological footprint per person was 2.3 hectares (5.6 acres). Between 1961 and 1999, the average global ecological footprint per person grew from 70% of the earth's estimated biological capacity of 1.9 hectares (4.7 acres) per person to 120% of this capacity. In other words, humanity's ecological footprint per person exceeds the earth's biological capacity by about 20%.

The United States and many other developed countries are running even larger ecological deficits. For example, the 9.7 hectare (24 acre) per person ecological footprint of the United States is nearly double the country's biological capacity of 5.3 global hectares (13 acres) per person. According to William Rees and

Mathis Wackernal, developers of the ecological footprint concept, it would take the land area of about three planet earths if the world's 6.3 billion people used the same amount of the earth's biologically productive area per person as in the United States.

If these estimates are correct, we are depleting natural capital (Figure 1-2) instead of living off the biological income provided by this capital. Other analysts such as Jesse Ausbel contend that our ecological footprint is likely to decrease in this century because of technological progress in using renewable resources more efficiently and a shift to more environmentally benign forms of farming, forestry, and energy.

What Are Nonrenewable Resources? **Nonrenewable resources** exist in a fixed quantity or stock in the earth's crust. On a time scale of millions to billions of years, geological processes can renew such resources. However, on the much shorter human time scale of hundreds to thousands of years, these resources can be depleted much faster than they are formed.

These exhaustible resources include *energy resources* (such as coal, oil, and natural gas, which cannot be recycled), *metallic mineral resources* (such as iron, copper, and aluminum, which can be recycled), and *nonmetallic mineral resources* (such as salt, clay, sand, and phosphates, which usually are difficult or too costly to recycle).

Figure 1-8 shows the production and depletion cycle of a nonrenewable energy or mineral resource. We never completely exhaust a nonrenewable mineral resource. But such a resource becomes *economically depleted* when the costs of extracting and using what is left exceed its economic value. At that point, we have six choices: try to find more, recycle or reuse existing supplies (except for nonrenewable energy resources, which cannot be recycled or reused), waste less, use less, try to develop a substitute, or wait millions of years for more to be produced.

Some nonrenewable material resources, such as copper and aluminum, can be recycled or reused to extend supplies. **Recycling** involves collecting and reprocessing a resource into new products. For example, discarded aluminum cans can be crushed and melted to make new aluminum cans or other aluminum items. **Reuse** involves using a resource over and over in the same form. For example, glass bottles can be collected, washed, and refilled many times.

Recycling nonrenewable metallic resources takes much less energy, water, and other resources and produces much less pollution and environmental degradation than exploiting vir-

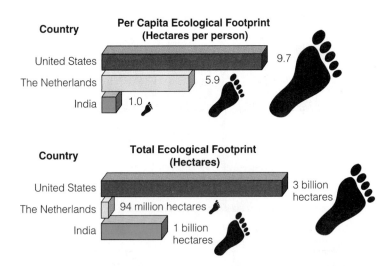

Figure 1-7 Relative *ecological footprints* of the United States, the Netherlands, and India. The *per capita ecological footprint* is a measure of the biologically productive areas of the earth required to produce the renewable resources required per person. Currently, humanity's average ecological footprint per person is 20% higher than the earth's biological capacity per person. (Data from William Rees and Mathis Wackernagel, Redefining Progress)

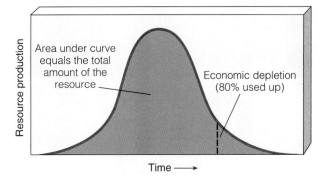

Figure 1-8 Full production and exhaustion cycle of a nonrenewable resource such as copper, iron, oil, or coal. Usually, a nonrenewable resource is considered *economically depleted* when 80% of its total supply has been extracted and used. Normally, it costs too much to extract and process the remaining 20%.

gin metallic resources. Reusing such resources takes even less energy and other resources and produces less pollution and environmental degradation than recycling.

Nonrenewable energy resources, such as coal, oil, and natural gas, cannot be recycled or reused. Once burned, the useful energy in these fossil fuels is gone, leaving behind waste heat and polluting exhaust gases.

1-5 POLLUTION

What Is Pollution, and Where Do Pollutants Come From?   **Pollution** is any addition to air, water, soil, or food that threatens the health, survival, or activities of humans or other living organisms. Pollutants can enter the environment naturally (for example, from volcanic eruptions) or through human (anthropogenic) activities (for example, from burning coal). Most pollution from human activities occurs in or near urban and industrial areas, where pollutants are concentrated. Industrialized agriculture also is a major source of pollution. Some pollutants contaminate the areas where they are produced; others are carried by wind or flowing water to other areas.

Pollutants come from two types of sources, point and nonpoint. **Point sources** of pollutants are single, identifiable sources. Examples are the smokestack of a coal-burning power plant, the drainpipe of a factory, or the exhaust pipe of an automobile. **Nonpoint sources** of pollutants are dispersed and often difficult to identify. Examples are runoff of fertilizers and pesticides (from farmlands, golf courses, and suburban lawns and gardens) into streams and lakes, and pesticides sprayed into the air or blown by the wind into the atmosphere.

It is much easier and cheaper to identify and control pollution from point sources than from widely dispersed nonpoint sources.

What Types of Harm Do Pollutants Cause?   Unwanted effects of pollutants include **(1)** disruption of life-support systems for humans and other species, **(2)** damage to wildlife, human health, and property, and **(3)** nuisances such as noise and unpleasant smells, tastes, and sights.

Solutions: What Can We Do about Pollution? We use two basic approaches to deal with pollution. One is **pollution prevention,** or **input pollution control,** which reduces or eliminates the production of pollutants. The other is **pollution cleanup,** or **output pollution control,** which involves cleaning up or diluting pollutants after they have been produced.

Environmentalists have identified three problems with relying primarily on pollution cleanup. First, *it is only a temporary bandage as long as population and consumption levels grow without corresponding improvements in pollution control technology.* For example, adding catalytic converters to car exhaust systems has reduced air pollution. But increases in the number of cars and in the total distance each travels have reduced the effectiveness of this cleanup approach.

Second, *it often removes a pollutant from one part of the environment only to cause pollution in another.* For example, we can collect garbage, but the garbage is then *burned* (perhaps causing air pollution and leaving a toxic ash that must be put somewhere), *dumped* into streams, lakes, and oceans (perhaps causing water pollution), or *buried* (perhaps causing soil and groundwater pollution).

Finally, *once pollutants have entered and become dispersed into the environment at harmful levels, it usually costs too much to reduce them to acceptable levels.*

Both pollution prevention (front-of-the-pipe) and pollution cleanup (end-of-the-pipe) solutions are needed. However, environmentalists and some economists urge us to put more emphasis on prevention because it works better and is cheaper than cleanup. As Benjamin Franklin observed long ago, "An ounce of prevention is worth a pound of cure." An increasing number of businesses have found that *pollution prevention pays.*

1-6 ENVIRONMENTAL AND RESOURCE PROBLEMS: CAUSES AND CONNECTIONS

Connections: What Are Key Environmental Problems and Their Basic Causes? We face a number of interconnected environmental and resource

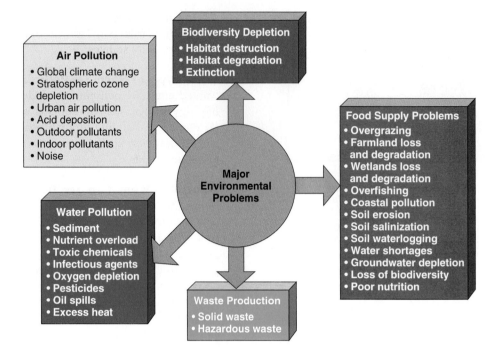

Figure 1-9 Major environmental and resource problems.

problems (Figure 1-9). The first step in dealing with these problems is to identify their underlying causes (Figure 1-10).

Once we have identified environmental problems and their root causes, the next step is to understand how they are connected to one another. The three-factor model in Figure 1-11 is a starting point.

According to this simple model, the environmental impact (I) of a population on a given area depends on three key factors: the number of people (P), the average resource use per person (affluence, A), and the beneficial and harmful environmental effects of the technologies (T) used to provide and consume each

• Rapid population growth

• Unsustainable resource use

• Poverty

• Not including the environmental costs of economic goods and services in their market prices

• Trying to manage and simplify nature with too little knowledge about how it works

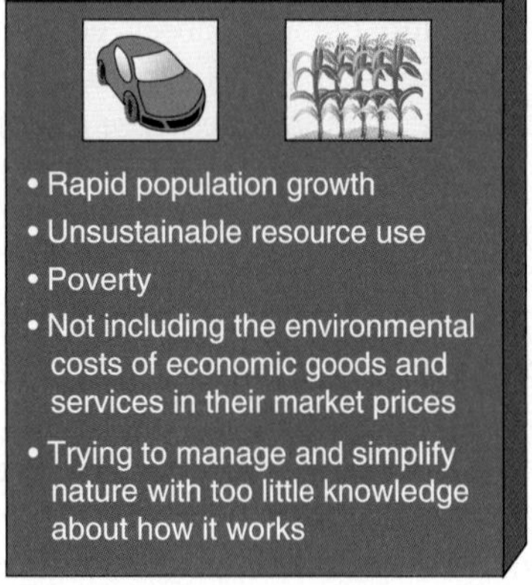

Figure 1-10 Environmentalists have identified five basic causes of the environmental problems we face.

unit of resource and control or prevent the resulting pollution and environmental degradation.

In developing countries, population size and the resulting degradation of renewable resources (as the poor struggle to stay alive) tend to be the key factors in total environmental impact (Figure 1-11, top). In such countries per capita resource use is low.

In developed countries, high rates of per capita resource use and the resulting high levels of pollution and environmental degradation per person usually are the key factors determining overall environmental impact (Figure 1-11, bottom) and a country's ecological footprint per person (Figure 1-7). For example, the average U.S. citizen consumes about 35 times as much as the average citizen of India and 100 times as much as the average person in the world's poorest countries. *Thus poor parents in a developing country would need 70–200 children to have the same lifetime resource consumption as 2 children in a typical U.S. family.*

Some forms of technology, such as polluting factories and motor vehicles and energy-wasting devices, increase environmental impact by raising the T factor in the equation. Other technologies, such as pollution control and prevention, solar cells, and energy-saving devices, lower environmental impact by decreasing the T factor in the equation. In other words, some forms of technology are *environmentally harmful* and some are *environmentally beneficial*.

Developing Countries

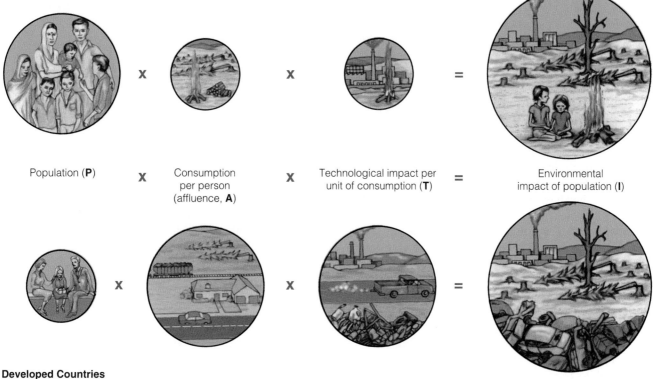

Population (**P**) $\times$ Consumption per person (affluence, **A**) $\times$ Technological impact per unit of consumption (**T**) = Environmental impact of population (**I**)

Developed Countries

Figure 1-11 Simplified model of how three factors—number of people, affluence, and technology—affect the environmental impact of the population in developing countries (top) and developed countries (bottom).

The three-factor model in Figure 1-11 can help us understand how key environmental problems and some of their causes are connected and guide us in seeking solutions to these problems. However, these problems involve a number of poorly understood interactions between many more factors than those in this simplified model, as outlined in Figure 1-12.

1-7 CULTURAL CHANGES AND SUSTAINABILITY

What Major Human Cultural Changes Have Taken Place? Evidence from fossils and studies of ancient cultures suggests that the current form of our species, *Homo sapiens sapiens*, has walked the earth for

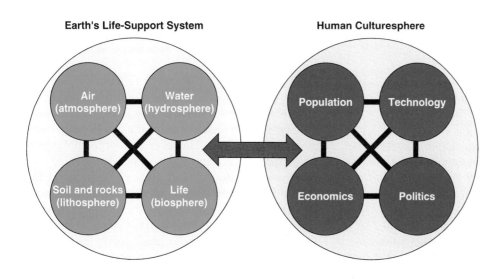

Figure 1-12 Major components and interactions within and between the earth's life-support system and the human sociocultural system (culturesphere). The goal of environmental science is to learn as much as possible about these complex interactions.

only about 60,000 years (some recent evidence suggests 90,000–176,000 years)—an instant in the planet's estimated 4.6-billion-year existence.

Until about 12,000 years ago, we were mostly hunter–gatherers who typically moved as needed to find enough food for survival. Since then, three major cultural changes have occurred: the *agricultural revolution* (which began 10,000–12,000 years ago), the *industrial revolution* (which began about 275 years ago), and the *information and globalization revolution* (which began about 50 years ago).

These major cultural changes have greatly increased human impact on the environment in three ways. *First,* they have given us much more energy and new technologies with which to alter and control more of the planet to meet our basic needs and increasing wants. *Second,* they have allowed expansion of the human population, mostly because of increased food supplies and longer life spans. *Third,* they have greatly increased our resource use, pollution, and environmental degradation.

How Did Ancient Hunting-and-Gathering Societies Affect the Environment? During most of our 60,000-year existence, we were **hunter–gatherers** who survived by collecting edible wild plant parts, hunting, fishing, and scavenging meat from animals killed by other predators. Our hunter–gatherer ancestors typically lived in small bands (of fewer than 50 people) who worked together to get enough food to survive. Many groups were nomadic, picking up their few possessions and moving seasonally from place to place to find enough food.

The earliest hunter–gatherers (and those still living this way today) survived through expert knowledge and understanding of their natural surroundings. They discovered which plants and animals could be eaten and used as medicines, where to find water, how plant availability changed throughout the year, and how some game animals migrated to get enough food. Because of high infant mortality and an estimated average life span of 30–40 years, hunter–gatherer populations grew very slowly.

Advanced hunter–gatherers had a greater impact on their environment than did early hunter–gatherers. They used more advanced tools and fire to convert forests into grasslands, contributed to the extinction of some large animals (including the mastodon, saber-toothed tiger, giant sloth, cave bear, mammoth, and giant bison), and altered the distribution of plants (and animals feeding on such plants) as they carried seeds and plants to new areas.

Early and advanced hunter–gatherers exploited their environment to survive. But their environmental impact usually was limited and local because of their small population sizes, low resource use per person, migration (which allowed natural processes to repair most of the damage they caused), and lack of technology that could have expanded their impact.

What Is the Agricultural Revolution? Some 10,000–12,000 years ago, a cultural shift known as the **agricultural revolution** began in several regions of the world. It involved a gradual move from usually nomadic hunting-and-gathering groups to settled agricultural communities in which people domesticated wild animals and cultivated wild plants.

Plant cultivation probably developed in many areas, especially in the tropical forests of Southeast Asia, northeast Africa, and Mexico. People discovered how to grow various wild food plants from roots or tubers (fleshy underground stems). To prepare the land for planting, they cleared small patches of tropical forests by cutting down trees and other vegetation and then burning the underbrush (Figure 1-13). The ashes fertilized the often nutrient-poor soils in this **slash-and-burn cultivation.**

Early growers also used various forms of **shifting cultivation** (Figure 1-13), primarily in tropical regions. After a plot had been used for several years, the soil became depleted of nutrients or reinvaded by the forest. Then the growers cleared a new plot. They learned that each abandoned patch normally had to be left fallow (unplanted) for 10–30 years before the soil became fertile enough to grow crops again. While patches were regenerating, growers used them for tree crops, medicines, fuelwood, and other purposes. In this manner, early growers practiced *sustainable cultivation.*

These early farmers had fairly little impact on the environment for three reasons. *First,* their dependence mostly on human muscle power and crude stone or stick tools meant they could cultivate only small plots. *Second,* their population size and density were low. *Third,* normally enough land was available so they could move to other areas and leave abandoned plots unplanted for the several decades needed to restore soil fertility.

As more advanced forms of agriculture grew and spread they led to various beneficial (*good news*) and harmful (*bad news*) effects (Figure 1-14, p. 14). There have been many benefits from the development and spread of agriculture. However, environmental degradation from soil erosion and livestock overgrazing was a factor in the downfall of great civilizations in the Middle East, North Africa, and the Mediterranean region. In addition, armies and their leaders rose to power and conquered large areas of valuable land and water resources needed to grow food. These rulers forced powerless people (slaves and landless peas-

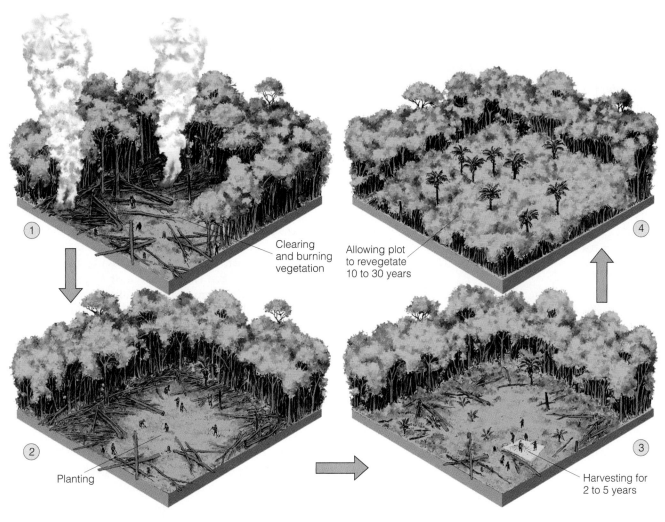

Figure 1-13 The first crop-growing technique may have been a combination of *slash-and-burn* and *shifting cultivation* in tropical forests. This method is sustainable only if small plots of the forest are cleared, cultivated for no more than 5 years, and then allowed to regenerate for 10–30 years to renew soil fertility. Indigenous cultures have developed many variations of this technique and have found ways to use some former plots nondestructively while they regenerate.

(Labels in figure: ① Clearing and burning vegetation; ② Planting; ③ Harvesting for 2 to 5 years; ④ Allowing plot to revegetate 10 to 30 years)

ants) to do the hard, disagreeable work of producing food and constructing irrigation systems, temples, and walled fortresses.

What Is the Industrial Revolution? The next cultural shift, the **industrial revolution,** began in England in the mid-1700s and spread to the United States in the 1800s. It led to a rapid expansion in the production, trade, and distribution of material goods.

The industrial revolution represented a shift from dependence on *renewable* wood (with supplies dwindling in some areas because of unsustainable cutting) and flowing water to dependence on machines running on *nonrenewable* fossil fuels (first coal and later oil and natural gas). This led to a switch from small-scale, localized production of handmade goods to large-scale production of machine-made goods in centralized factories in rapidly growing industrial cities.

Factory towns grew into cities as rural people came to the factories for work. There they worked long hours under noisy, dirty, and hazardous conditions. Other workers toiled in dangerous coal mines.

In these early industrial cities, coal smoke belching out of chimneys was so heavy that many people died of lung ailments. Ash and soot covered everything, and on some days the smoke was so thick that it blotted out the sun.

Fossil fuel–powered farm machinery, commercial fertilizers, and new plant-breeding techniques increased per acre crop yields. This helped protect biodiversity by reducing the need to expand the area of cropland to grow food. Because fewer farmers were needed, more people migrated to cities. With a larger and more reliable food supply and longer life spans, the size of the human population began the sharp increase that continues today (Figure 1-1).

Agricultural Revolution

Good News

More food

Supported a larger population

Longer life expectancy

Higher standard of living for many people

Formation of villages, towns, and cities.

Towns and cities served as centers for trade, government, and religion

Bad News

Destruction of wildlife habitats from clearing forests and grasslands

Killing of wild animals feeding on grass or crops

Fertile land turned into desert by livestock overgrazing

Soil eroded into streams and lakes

Towns and cities concentrated wastes and pollution and increased spread of diseases

Increase in armed conflict and slavery over ownership of land and water resources

Figure 1-14 Trade-offs of the agricultural revolution: good and bad news about the shift from hunting and gathering to agriculture.

After World War I (1914–18), more efficient machines and mass production techniques were developed. These technologies became the basis of today's advanced industrial societies in places such as the United States, Canada, Japan, Australia, and western Europe. Figure 1-15 lists some of the beneficial (*good news*) and harmful (*bad news*) effects of the advanced industrial revolution.

How Might the Information and Globalization Revolution Affect the Environment? One of the major trends since 1950, and especially since 1970, is **globalization,** the process of global social, economic, and environmental change that leads to an increasingly integrated world. This cultural shift is being facilitated by improvements in transmitting information. In this **information and globalization revolution,** new technologies such as the telephone, radio, television, computers, the Internet, automated databases, and remote sensing satellites mean we have increasingly rapid access to much more information on a global scale. Scientific information now doubles about

every 12 years, and general information doubles about every 2.5 years. The World Wide Web contains at least 3 billion pages of electronic pages and grows by roughly a million electronic pages per day. Figure 1-16 lists some of the possible beneficial (*good news*) and harmful (*bad news*) effects of the information and globalization revolution.

What Are the Major Eras of Environmental History in the United States? The environmental history of the United States can be divided into four eras. The first was the *tribal era,* when North America was occupied by 5–10 million tribal people (now called Native Americans) for at least 10,000 years before European settlers began arriving in the early 1600s. With some exceptions, most Native American cultures had a deep respect for the land and its animals and did not believe in land ownership.

The was followed by the *frontier era* (1607–1890), when European colonists began settling North America. Faced with a continent containing seemingly inexhaustible forest and wildlife resources and rich soils, the early colonists developed a **frontier environ-**

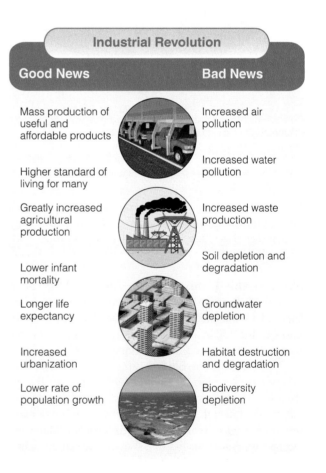

Industrial Revolution

Good News

Mass production of useful and affordable products

Higher standard of living for many

Greatly increased agricultural production

Lower infant mortality

Longer life expectancy

Increased urbanization

Lower rate of population growth

Bad News

Increased air pollution

Increased water pollution

Increased waste production

Soil depletion and degradation

Groundwater depletion

Habitat destruction and degradation

Biodiversity depletion

Figure 1-15 Trade-offs of the industrial revolution: good and bad news about the effects of the advanced industrial revolution.

Information/Globalization Revolution

Good News	Bad News
Computer-generated models and maps of the earth's environmental systems	Information overload can cause confusion and sense of hopelessness
Remote sensing satellite surveys of world's environmental systems	Globalized economy can increase environmental degradation by homogenizing the earth's surface
Ability to respond to environmental problems more effectively and rapidly	Globalized economy can decrease cultural diversity

Figure 1-16 Trade-offs of the information and globalization revolution: good and bad news about the effects of this latest cultural revolution.

mental worldview. They viewed most of the continent as having vast resources and as a wilderness to be conquered by clearing and planting.

Next came the *early conservation era* (1832–1870), during which some people became alarmed at the scope of resource depletion and degradation in the United States. They urged that part of the unspoiled wilderness on public lands owned jointly by all people (but managed by the government) be protected as a legacy to future generations.

This was followed by an era—lasting from 1870 to the present—with an increased role *of the federal government and private citizens in resource conservation, public health, and environmental protection.* Appendix 2 (p. A3) summarizes some of the major events during each of these periods.

1-8 IS OUR PRESENT COURSE SUSTAINABLE?

Are Things Getting Better or Worse? Experts disagree about how serious our population and environmental problems are and what we should do about them. Some analysts believe human ingenuity and technological advances will allow us to clean up pollution to acceptable levels, find substitutes for any re-

sources that become scarce, and keep expanding the earth's ability to support more humans, as we have done in the past. They accuse most scientists and environmentalists of exaggerating the seriousness of the problems we face and failing to appreciate the progress we have made in improving quality of life and protecting the environment.

Environmentalists and many leading scientists, however, contend we are disrupting the earth's life-support system for us and other forms of life at an accelerating rate. They are greatly encouraged by the progress we have made in increasing average life expectancy, reducing infant mortality, increasing food supplies, and reducing many forms of pollution—especially in developed countries (Figure 1-15). But they point out that we need to use the earth in a way that is more sustainable for present and future human generations and other species that support us and other forms of life.

On November 18, 1992, some 1,680 of the world's senior scientists from 70 countries (including 102 of the 196 living scientists who are Nobel laureates) signed and sent an urgent warning to government leaders of all nations. According to this warning,

Our massive tampering with the world's interdependent web of life—coupled with the environmental damage inflicted by deforestation, species loss, and climate change—could trigger widespread adverse effects, including unpredictable collapses of critical biological systems whose interactions and dynamics we only imperfectly understand. . . . No more than one or a few decades remain before the chance to avert the threats we now confront will be lost and the prospects for humanity immeasurably diminished.

Also in 1992, the prestigious U.S. National Academy of Sciences and the Royal Society of London issued a joint report, their first ever, which began,

If current predictions of population growth prove accurate and patterns of human activity on the planet remain unchanged, science and technology may not be able to prevent either irreversible degradation of the environment or continued poverty for much of the world. . . . Sustainable development can be achieved, but only if irreversible degradation of the environment can be halted in time.

These warnings are not the views of a small number of scientists but the consensus of the mainstream scientific community, consisting of most of the world's key researchers on environmental problems.

The most useful answer to the question of whether things are getting better or worse is *both.* Some things are getting better and some are getting worse. Our challenge is not to get trapped into confusion and inaction by listening primarily to two groups of people. One group consists of *technological optimists.* They tend to overstate the situation by telling us to be happy and not to worry, because technological innovations and conventional economic growth and development will lead to a wonderworld for everyone.

The second group consists of *environmental pessimists* who overstate the problems to the point where our environmental situation seems hopeless. According to the noted conservationist Aldo Leopold, "I have no hope for a conservation based on fear."

Solutions: What Is Environmentally Sustainable Economic Development?

A growing number of analysts have called for a shift during this century from traditional economic development fueled by economic growth of essentially any type to much greater emphasis on **environmentally sustainable economic development.** A reward and penalty system can foster this type of development. *Economic rewards* (government subsidies, tax breaks, and emissions trading) are used to *encourage* environmentally beneficial and sustainable forms of economic development. *Economic penalties* (government taxes and regulations) *discourage* environmentally harmful and unsustainable forms of economic growth.

According to the environmental leader Lester R. Brown,

> *The goal is to develop a new type of economy to replace our eventually unsustainable fossil fuel–based, automobile-centered, throwaway economy. This new eco-economy is a solar-powered, bicycle- and rail-centered, reuse and recycle economy that uses energy, water, land, and materials much more efficiently and wisely than we do today. In addition to helping sustain the earth's life-support systems, such an economy can lead to greater economic security, healthier lifestyles, and a worldwide improvement in the human condition.*

Figure 1-17 lists some of the shifts involved in implementing an *environmental or sustainability revolution* during this century.

This chapter has presented an overview of the problems most environmentalists and many of the world's most prominent scientists believe we face and their root causes. It has also summarized the controversy over how serious environmental problems are and the cultural changes that have influenced how we treat the earth.

Try not to be overwhelmed or immobilized by the bad environmental news, because there are also three pieces of great environmental news. *First,* we have made immense progress in improving the human con-

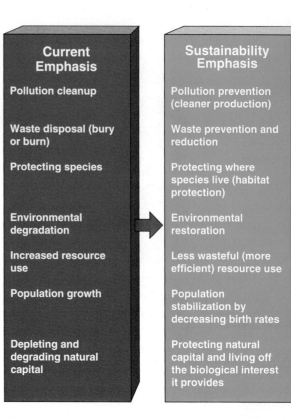

Figure 1-17 Solutions: some shifts involved in the *environmental or sustainability revolution.*

dition and dealing with many environmental problems. *Second,* we are learning a great deal about how nature works and sustains itself. *Third,* we have numerous scientific, technological, and economic solutions available to deal with the environmental problems we face, as you will learn in this book.

The challenge is to make creative use of our economic and political systems to implement such solutions. One key is to recognize that most economic and political change comes about as a result of individual actions and individuals acting together to bring about change. Social scientists suggest it takes only about 5–10% of the population of a country or of the world to bring about major social change. Anthropologist Margaret Mead summarized our potential for change: "Never doubt that a small group of thoughtful, committed citizens can change the world. Indeed, it is the only thing that ever has."

We live in exciting times during what might be called a *hinge of cultural history.* Indeed, if I had to pick a time to live, it would be the next 50 years as we face the challenge of developing more environmentally sustainable societies.

What's the use of a house if you don't have a decent planet to put it on?
HENRY DAVID THOREAU

REVIEW QUESTIONS

1. Define the boldfaced terms in this chapter.

2. What is *exponential growth*? Give two examples of exponential growth.

3. Distinguish among *environment, ecology,* and *environmental science*. Distinguish among *ecologists, environmental scientists, conservation biologists, environmentalists, preservationists, conservationists,* and *restorationists*.

4. Distinguish between *solar capital* and *natural capital* (*natural resources*).

5. What is an *environmentally sustainable society*? Distinguish between living on the earth's natural capital and living on the renewable biological interest provided by this capital. How is this related to the sustainability of **(a)** the earth's life-support system and **(b)** your lifestyle?

6. How rapidly is the world's population growing? How many people does this growth add each year?

7. Define *economic growth, gross national income, gross national purchasing power parity, gross domestic product, per capita GNI, per capita GNI PPP,* and *economic development.* Distinguish between *developed countries* and *developing countries.*

8. List five ways in which poverty is related to environmental quality, people's quality of life, and premature deaths of poor people. Why does it make sense for a poor family to have a large number of children?

9. What are *perpetual resources* and *renewable resources*? Give an example of each.

10. What are *sustainable yield* and *environmental degradation*? Give eight examples of environmental degradation.

11. Define and give three examples of *common-property resources.* What is the *tragedy of the commons*? Give three examples of this tragedy on a global scale. List two ways to deal with the tragedy of the commons.

12. What is the *ecological footprint per person*? What useful information does it give us about the use of renewable resources?

13. What is a *nonrenewable resource*? Distinguish between a *physically depleted resource* and an *economically depleted resource*.

14. Distinguish between *reuse* and *recycling.* Draw a depletion curve for a nonrenewable resource, and explain how recycling and reuse affect depletion time.

15. What is *pollution*? Distinguish between *point sources* and *nonpoint sources* of pollution. List three types of harm caused by pollution.

16. Distinguish between *pollution prevention* (*input pollution control*) and *pollution cleanup* (*output pollution control*). What are three problems with relying primarily on pollution cleanup?

17. According to environmentalists, what are five basic causes of the environmental problems we face?

18. Describe a simple model of relationships between population, resource use, environmental degradation, and overall environmental impact. How do these factors differ in developed and developing countries?

19. Distinguish among *hunter–gatherer societies, agricultural societies, industrial societies,* and *information and globalization societies.* Define *slash-and-burn cultivation* and *shifting cultivation.*

20. Summarize the major beneficial (good news) and harmful (bad news) effects of *agricultural societies, industrial societies,* and *information and globalization societies.*

21. What are four major eras of environmental history in the United States?

22. List three reasons why some analysts believe that the environmental problems we face are not serious.

23. Summarize the 1992 position statement of about 1,700 leading scientists about the state of the environment.

24. What is *environmentally sustainable economic development*? How does it differ from traditional economic growth and economic development?

25. List seven major shifts that environmentalists believe should take place over the next 50 years as part of an *environmental* or *sustainability revolution.*

CRITICAL THINKING

1. Do you believe the society you live in is on an unsustainable path? Explain. Do you believe it is possible for the society you live in to become a more sustainable society within the next 50 years? Explain.

2. Do you favor instituting policies designed to reduce population growth and stabilize **(a)** the size of the world's population as soon as possible and **(b)** the size of the U.S. population (or the population of the country where you live) as soon as possible? Explain. If you agree that population stabilization is desirable, what three major policies would you implement to accomplish this goal?

3. Explain why you agree or disagree with the following propositions:
   a. The economic growth from high levels of resource use in developed countries provides money for more financial aid to developing countries for reducing pollution, environmental degradation, and poverty.
   b. Stabilizing population is not desirable because without more consumers, economic growth would stop.
   c. The world will never run out of *renewable* resources because we can use technology to find substitutes.
   d. We will not run out of currently used *nonrenewable* resources because technological innovations will produce substitutes, reduce resource waste, or allow use of lower grades of scarce nonrenewable resources.

4. List **(a)** three forms of economic growth that you believe are environmentally sustainable and **(b)** three forms that you believe are environmentally unsustainable.

5. Give three examples of how you cause environmental degradation as a result of the tragedy of the commons.

6. When you read that at least 25,000 human beings die prematurely each day (17 per minute) from preventable malnutrition and infectious disease, do you **(a)** doubt whether it is true, **(b)** not want to think about it, **(c)** feel hopeless, **(d)** feel sad, **(e)** feel guilty, or **(f)** want to do something about this problem?

7. How do you feel when you read that **(1)** the average American consumes about 35 times more resources than the average Indian citizen, **(2)** human activities lead to the premature extinction of at least 10 species per day, and **(3)** human activities are projected to make the earth's climate warmer: **(a)** skeptical about their accuracy, **(b)** indifferent, **(c)** sad, **(d)** helpless, **(e)** guilty, **(f)** concerned, or **(g)** outraged? Which of these feelings help perpetuate such problems, and which can help alleviate these problems?

8. Do you agree or disagree with the five basic causes of environmental problems listed in Figure 1-10, p. 10? Explain. List any other root causes you believe should be added.

PROJECTS

1. What are the major resource and environmental problems in the city, town, or rural area where you live? Which of these problems affect you directly? Have these problems gotten better or worse during the last 10 years?

2. Write two-page scenarios describing what your life and that of any children you choose to have might be like 50 years from now if **(a)** we continue on our present path; **(b)** we shift to more sustainable societies throughout most of the world.

3. Make a list of the resources you truly need. Then make another list of the resources you use each day only because you want them. Finally, make a third list of resources you want and hope to use in the future. Compare your lists with those compiled by other members of your class, and relate the overall result to the tragedy of the commons (p. 7).

4. Make a concept map of this chapter's major ideas using the section heads and subheads and the key terms (in boldface type). Look on the website for this book for information about making concept maps.

INTERNET STUDY RESOURCES AND RESOURCES FOR FURTHER READING AND RESEARCH

The website for this book contains helpful study aids and many ideas for further reading and research. Log on to

http://biology.brookscole.com/miller10

and click on the Chapter-by-Chapter area. Choose Chapter 1 and select a resource:

■ Flash Cards allows you to test your mastery of the Terms and Concepts to Remember for this chapter.

■ Tutorial Quizzes provides a multiple-choice practice quiz.

■ Student Guide to InfoTrac will lead you to Critical Thinking Projects that use InfoTrac College Edition as a research tool.

■ References lists the major books and articles consulted in writing this chapter.

■ Hypercontents takes you to an extensive list of sites with news, research, and images related to individual sections of the chapter.

INFOTRAC COLLEGE EDITION

Improve your skills with InfoTrac College Edition, a searchable online database of articles from more than 700 periodicals. Log on to

http://www.infotrac-college.com

or access InfoTrac through the website for this book. Try to find the following articles:

1. Clark, W. C. 2001. A transition toward sustainability. *Ecology Law Quarterly* 27: 1021. *Keywords:* "Clark" and "sustainability." This comprehensive article details how human society and its interaction with the environment must adjust to ensure future sustainability.

2. Jones, K., A. Scholtz, and A. G. Lehmer. 2002. Rio + 10: Let the people be heard. *Earth Island Journal* 17: 12. *Keywords:* "Rio" and "sustainability." Ten years after the Rio Earth Summit, there appears to have been little action on the part of participating countries to realize the recommendation of the summit. In fact, we may actually be going in the other direction.

2 ENVIRONMENTAL ECONOMICS, POLITICS, AND WORLDVIEWS

Biosphere 2: A Lesson in Humility

In 1991, eight scientists (four men and four women) were sealed into Biosphere 2, a $200 million facility designed to be a self-sustaining life-support system (Figure 2-1).

The project, financed with private capital, was designed to provide information and experience in designing self-sustaining stations in space or on the moon or other planets and increase our understanding of the earth's life-support system: Biosphere 1.

The 1.3-hectare (3.2-acre) closed and sealed system was built in the desert near Tucson, Arizona. It had a variety of natural living systems, each built from scratch. They included a tropical rain forest, lakes, a desert, streams, freshwater and saltwater wetlands, and a mini-ocean with a coral reef.

The system was designed to mimic the earth's natural chemical recycling systems. Water that evaporated from its ocean and other aquatic systems condensed to provide rainfall over the tropical rain forest. This water then trickled through soil filters into the marshes and the ocean to provide fresh water for the crew and other living organisms before evaporating again.

The facility was stocked with more than 4,000 species of organisms selected to maintain life-support functions. Sunlight and external natural gas–powered generators provided energy.

The Biospherians were supposed to be isolated for 2 years and to raise their own food, breathe air recirculated by plants, and drink water cleansed by natural recycling processes. From the beginning they encountered numerous unexpected problems.

The life-support system began unraveling. Large amounts of oxygen disappeared mysteriously. Additional oxygen had to be pumped in from the outside to keep the Biospherians from suffocating.

The nitrogen and carbon recycling systems also failed to function properly. Levels of nitrous oxide rose high enough to threaten the occupants with brain damage and had to be controlled by outside intervention. Carbon dioxide skyrocketed to levels that threatened to poison the humans and spurred the growth of weedy vines that choked out food crops. Plant nutrients leached from the soil and polluted the water systems.

Disruption of the facility's chemical recycling systems and leaks in the seals from the outside disrupted populations of the system's life forms. Tropical birds disappeared after the first freeze. An Arizona ant species got into the enclosure, proliferated, and killed off most of the system's introduced insect species. After the majority of the introduced insect species became extinct, the facility was overrun with cockroaches and katydids. All together, 19 of the Biosphere's 25 small animal species became extinct. Before the 2-year period was up, all plant-pollinating insects became extinct, thereby dooming to extinction most of the plant species.

Despite many problems, the facility's waste and wastewater were recycled, and the Biospherians were able to produce 80% of their food supply.

Scientists Joel Cohen and David Tilman, who evaluated the project, concluded, "No one yet knows how to engineer systems that provide humans with life-supporting services that natural ecosystems provide for free." In other words, an expenditure of $200 million failed to maintain a life-support system for eight people. The earth—Biosphere 1—does this every day for 6.3 billion people and millions of other species at no cost. If we had to pay for these services at the same annual cost of $12.5 million per person in Biosphere 2, the total bill for the earth's 6.3 billion people would be 1,900 times the annual world national product.

Today Columbia University uses Biosphere 2, the world's largest ecological laboratory—renamed the Lamont-Doherty Earth Observatory—to carry out climate and ecological research.

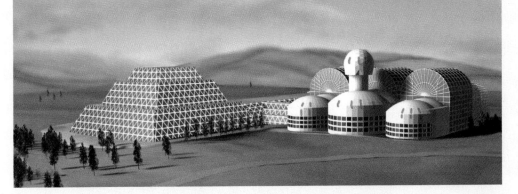

Figure 2-1 Biosphere 2, constructed near Tucson, Arizona, was designed to be a self-sustaining life-support system for eight people sealed into the facility in 1991. The experiment failed because of a breakdown in its nutrient cycling systems.

The main ingredients of an environmental ethic are caring about the planet and all of its inhabitants, allowing unselfishness to control the immediate self-interest that harms others, and living each day so as to leave the lightest possible footprints on the planet.

ROBERT CAHN

This chapter addresses the following questions:

- What are the different types of economic systems, and what types of resources support them?

- How can we monitor economic and environmental progress?

- How can we use economics to help control pollution and manage resources?

- How does poverty reduce environmental quality? How can we reduce poverty?

- How can we shift to more environmentally sustainable economies over the next few decades?

- How is environmental policy made in the United States?

- What are some guidelines for making environmental policy? How can people affect such decisions?

- What human-centered environmental worldviews guide most industrial societies?

- What are some life-centered and earth-centered environmental worldviews?

- How can we live more sustainably?

2-1 ECONOMIC SYSTEMS AND ENVIRONMENTAL PROBLEMS

What Supports and Drives Economies? An **economy** is a system of production, distribution, and consumption of goods and services that satisfies people's wants or needs. In an economy, individuals, businesses, and governments make **economic decisions** about what goods and services to produce, how to produce them, how much to produce, and how to distribute them.

The kinds of resources (or capital) that produce goods and services in an economy are called **economic resources.** They fall into four groups. One group is **natural resources,** or **natural capital,** the goods and services produced by the earth's natural processes, which support all economies and all life (Figure 1-2, p. 3, and Guest Essay, p. 4). There are no substitutes for many of these natural resources, such as the planet's air, water, and land; chemicals (nutrients) in the soil and water used to support all life; chemicals (minerals) found in the earth's crust that can be extracted and used to manufacture useful items; wild and domesticated plants and animals (biodiversity);

and nature's dilution, waste disposal, pest control, and recycling services.

A second resource type is **human resources,** people's physical and mental talents that provide labor, innovation, culture, and organization. A third type is **financial resources,** the cash, investments, and monetary institutions used to support the use of natural resources and human resources to provide goods and services. Finally, there are **manufactured resources,** items made from natural resources with the help of human and financial resources. This type of capital includes tools, machinery, equipment, factories, and transportation and distribution facilities used to provide goods and services.

What Are the Major Types of Economic Systems? The two major types of economic systems are command and market. In a **pure command economic system,** the government makes all *economic decisions* about what, how, and how much goods and services are produced, and for whom they are produced.

Market economic systems can be divided into two types: pure free market and capitalist market.

There are four major components of a **pure free-market economic system,** which so far exists only in theory. *First,* all economic decisions would be made in *markets,* in which buyers (demanders) and sellers (suppliers) of economic goods freely interact without any government or interference. *Second,* all buying and selling would be based on *pure competition,* in which no seller or buyer can control or manipulate the market.

Third, all sellers and buyers would have full access to the market and enough information about the beneficial and harmful aspects of economic goods to make informed decisions. *Fourth,* prices would reflect all harmful costs to society and the environment (*full-cost pricing*).

The **capitalist market economic systems** found in the real world are designed to subvert many of the theoretical conditions of a truly free market. Here are six rules for maximizing success for a company operating in the world's capitalist market economies: *First,* drive out all competition and gain monopolistic control of market prices on a global scale. *Second,* lobby for unrestricted *global free trade* that allows anything to be manufactured anywhere in the world and sold anywhere else. *Third,* lobby for government subsidies, tax breaks, or regulations that give the company's products a market advantage over their competitors and for governments to bail the company out if it makes bad investments.

Fourth, try to withhold information from consumers about dangers posed by products. This makes it difficult for consumers to make informed choices about what to buy. *Fifth,* maximize profits by passing harmful costs resulting from production and sale of goods and services on to the public, the environment, and in some cases future generations. *Finally*, recognize that a company's primary obligation is to produce the highest profit for the owners or stockholders whose financial capital the company is using to do business.

Why Have Governments Intervened in Market Economic Systems? Governments intervene in economies for a number of reasons. They might do it to help level the economic playing field by preventing a single seller or buyer (monopoly) or a single group of sellers or buyers (oligopoly) from dominating the market and thus controlling supply or demand and price. Or they might aim to ensure economic stability by trying to control boom-and-bust cycles that occur in market systems.

Some other reasons for government interventions are to

- Provide basic services such as national security, education, and health care
- Provide an economic safety net for people who because of health, age, and other factors cannot work and meet their basic needs.

- Protect people from fraud, trespass, theft, and bodily harm.
- Protect the health and safety of workers and consumers.
- Help compensate owners for large-scale destruction of their assets by natural disasters.
- Protect common-property resources such as the atmosphere and open oceans.
- Prevent or reduce pollution and depletion of natural resources.
- Manage public land resources such as national forests, parks, and wildlife reserves.

How Do Conventional and Ecological Economists Differ in Their View of Market-Based Economic Systems? *Conventional economists* often depict a market-based economic system as a circular flow of economic goods and money between households and businesses operating essentially independently of the earth's life-support systems or natural resources (Figure 2-2). They assume the environment is a subsystem of the economic system. They consider the earth's natural resources important but not vital because of our ability to find substitutes for scarce resources and ecosystem services. They also believe that human ingenuity will keep depletion and degradation of natural resources from limiting future economic growth.

Capitalist Market Economic System

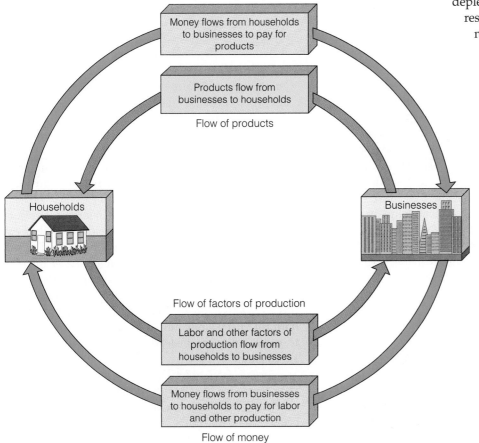

Flow of money

Money flows from households to businesses to pay for products

Products flow from businesses to households

Flow of products

Households

Businesses

Flow of factors of production

Labor and other factors of production flow from households to businesses

Money flows from businesses to households to pay for labor and other production

Flow of money

Figure 2-2 *Conventional* view of economic activity. In a market economic system, economic goods and money flow between households and businesses in a closed loop. People in households spend money to buy goods that firms produce, and firms spend money to buy factors of production (natural, human, financial, and manufactured resources). In many economics textbooks, such market economic systems are shown, as here, as if they were self-contained and thus independent of the natural resources that support all economies and all life. This model reinforces the idea that unlimited economic growth of any kind is sustainable.

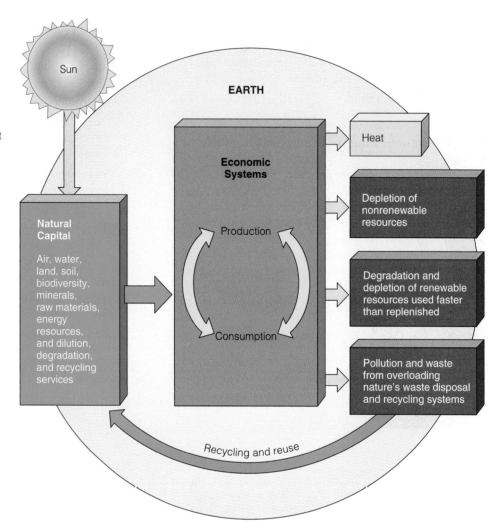

Figure 2-3 *Ecological* view of economic activity. Ecological economists see all economies as human subsystems that depend on resources and services provided by the sun and the earth's natural resources. A consumer society devoted to economic growth to satisfy ever-expanding wants assumes that our technological cleverness will allow us to find substitutes to overcome any limits on resources and ways to keep pollution and environmental degradation at acceptable levels. To ecological economists, such a society eventually is unsustainable because of its depletion and degradation of natural resources, many of which have no substitutes.

Ecological economists disagree with this view for two reasons. *First,* economic systems are subsystems of the environment (Figure 2-3). *Second,* conventional economic growth eventually is unsustainable because it depletes or degrades the natural resources or natural capital (Figure 1-2, p. 3, and top half of back cover) on which economic systems depend (Guest Essay, p. 4).

Ecological economists distinguish between unsustainable economic growth and environmentally sustainable economic development (Figure 2-4). They call for making a shift from our current economy based on unlimited economic growth to a more *environmentally sustainable economy,* or *eco-economy.*

2-2 MONITORING ECONOMIC AND ENVIRONMENTAL PROGRESS

Are GNI and GDP Useful Measures of Economic and Environmental Health and Human Well-Being? *Gross national income (GNI), gross do-*

mestic product (GDP), and *per capita GNI and GDP* indicators (p. 5) provide a standardized method for comparing the economic outputs of nations. Economists who developed these indicators many decades ago never intended them to be used as measures of economic health, human well-being, or environmental quality. However, most governments and business leaders use them for such purposes.

GNI and GDP indicators are poor measures of economic health and human well-being as well as environmental quality for several reasons. *First, GNI and GDP hide the harmful environmental and social effects of producing goods and services.* Funds spent to deal with pollution, crime, sickness, and death are counted as positive gains in the GNI or GDP. In other words, these harmful effects are counted as *benefits* that are added to these indicators instead of as *costs* that should be subtracted.

Second, GNI and GDP do not include the depletion and degradation of natural resources or assets on which all economies depend. A country can be headed toward ecological bankruptcy, exhausting its mineral re-

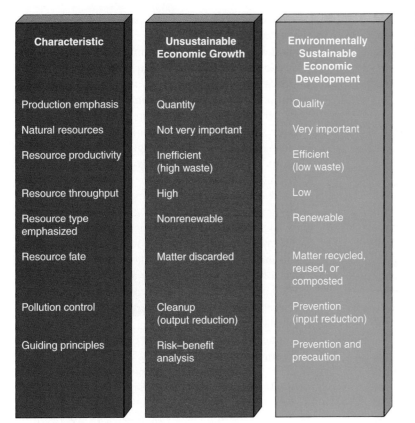

Characteristic	Unsustainable Economic Growth	Environmentally Sustainable Economic Development
Production emphasis	Quantity	Quality
Natural resources	Not very important	Very important
Resource productivity	Inefficient (high waste)	Efficient (low waste)
Resource throughput	High	Low
Resource type emphasized	Nonrenewable	Renewable
Resource fate	Matter discarded	Matter recycled, reused, or composted
Pollution control	Cleanup (output reduction)	Prevention (input reduction)
Guiding principles	Risk–benefit analysis	Prevention and precaution

Figure 2-4 Comparison of unsustainable economic growth and environmentally sustainable economic development.

that give a more realistic picture of environmental quality, human welfare, and economic health.

Basically, such indicators would do two things. *First, subtract* from the GNI and GDP things that lead to a lower quality of life and depletion of natural resources. *Second, add* to the GNI and GDP things that enhance environmental quality and human well-being but are currently being left out.

One such indicator is the *genuine progress indicator (GPI)*. This indicator, developed by Redefining Progress, evaluates economic output by *subtracting* expenses that do not improve environmental quality and human well-being from the GDP and *adding* services that improve environmental quality and human well-being not currently included in the GDP. Figure 2-5 compares the GDP and GPI for the United States between 1950 and 2000.

This and other environmental indicators are far from perfect and include many crude estimates. However, without such indicators, we do not know much about what is happening to people, the environment, and the planet's natural resource base or have an effective way to measure what policies work. In effect, according to ecological

sources, eroding its soils, cutting down its forests, destroying its wetlands and estuaries, and depleting its wildlife and fisheries. At the same time, it can have a rapidly rising GNI and GDP (at least for a while) until its ecological debts come due when the country has lost much of its environmental assets and the income from these assets.

Third, these indicators do not include many beneficial transactions that meet basic needs in which no money changes hands. Examples are the labor we put into volunteer work, health care and childcare we give loved ones, food we grow for ourselves, and cooking, cleaning, and repairs we do for ourselves.

Finally, GNI and GDP tell us nothing about income distribution and economic justice. They do not reveal how resources, income, or the harmful effects of economic growth (pollution, waste dumps, and land degradation) are distributed among the people in a country. The United Nations Children's Fund (UNICEF) suggests that countries should be ranked not by average per capita GNI or GDP but by average or median income of the poorest 40% of their people.

Solutions: Can Environmental Accounting Help? Ecological economists call for supplementing GNI and GDP indicators with *environmental indicators*

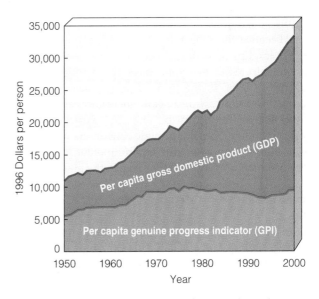

Figure 2-5 Comparison of the per capita gross domestic product (GDP) and per capita genuine progress indicator (GPI) in the United States between 1950 and 2000. (Data from Redefining Progress, 2002)

economists we are trying to guide national and global economies through treacherous economic and environmental waters at ever-increasing speeds using faulty radar.

2-3 SOLUTIONS: USING ECONOMICS TO IMPROVE ENVIRONMENTAL QUALITY

What Are Internal and External Costs? All economic goods and services have both internal and external costs. For example, the price a consumer pays for a car reflects the costs of the factory, raw materials, labor, marketing, and shipping, as well as a markup to allow the car company and its dealers some profits. After a car is purchased, the buyer must pay for gasoline, maintenance, and repair. All these direct and indirect costs, which are paid for by the seller and the buyer of an economic good, are called **internal costs.**

Making, distributing, and using any economic good or service also involve external costs or benefits not included in the market price. For example, if a car dealer builds an aesthetically pleasing showroom and grounds, it is an **external benefit** to people who enjoy the sight at no cost.

However, extracting and processing raw materials to make and propel cars depletes nonrenewable energy and mineral resources, produces solid and hazardous wastes, disturbs land, pollutes the air and water, contributes to global climate change, and reduces biodiversity. These harmful effects are **external costs** passed on to the public, the environment, and in some cases future generations.

Because these harmful costs are not included in the market price, people do not connect them with car ownership. Still, everyone pays these hidden costs sooner or later, in the form of poorer health, higher costs for health care and health insurance, and higher taxes for pollution control. Thus the world's current pricing system does not provide consumers with accurate information about the environmental impacts of the products and services they buy in the marketplace.

What Is Full-Cost Pricing? For most economists, the solution to the harmful costs of goods and services is to include or internalize such costs in the market prices of goods and services. However, *internalizing external costs* will not occur unless it is required by government action. As long as businesses receive subsidies and tax breaks for extracting and using virgin resources and are not taxed for the pollutants they produce, few will volunteer to reduce short-term profits by becoming more environmentally responsible.

Assume you own a company and believe it is wrong to subject your workers to hazardous conditions and to pollute the environment beyond what natural processes can handle. Suppose you voluntarily improve safety conditions for your workers and install pollution controls, but your competitors do not. Then your product will cost more to produce than similar products produced by your competitors, and you will be at a competitive disadvantage. Your profits will decline and you may eventually go bankrupt and have to lay off your employees.

Governments can deal with the problem of harmful external costs in several ways. They can levy taxes, pass laws and develop regulations, provide subsidies, or use other strategies that encourage or force producers to include all or most of these costs in the market prices of their economic goods and services. Then the market price would be the **full cost** of these goods and services: internal costs plus short- and long-term external costs.

Internalizing the external costs of pollution and environmental degradation has three important benefits. *First,* it would make preventing pollution more profitable than cleaning it up. *Second,* waste reduction, recycling, and reuse would be more profitable than burying or burning most of the waste we produce. *Third,* it provides consumers with information needed to make informed economic decisions about the effects of their purchases and lifestyles on the planet's life-support systems and on human health.

The *bad news* is that when external costs are internalized, the market prices for most goods and services would rise. However, the *good news* is that the total price people pay would be about the same because the hidden external costs related to each product would be included in its market price.

However, as external costs are internalized, governments must reduce taxes on income, payroll, and profits and withdraw subsidies formerly used to hide and pay for these external costs. Otherwise, consumers will face higher market prices without tax relief—a politically unacceptable policy guaranteed to fail.

Some more *good news* is that some goods and services would cost less. The reason is that internalizing external costs encourages producers to cut costs by inventing more resource-efficient and less-polluting methods of production and to offer more environmentally beneficial (or *green*) products. Jobs would be lost in *environmentally harmful* businesses. But at least as many and probably more jobs would be created in *environmentally beneficial* businesses. If a shift to full-cost pricing took place over several decades, most current environmentally harmful businesses would have time to transform themselves into profitable environmentally beneficial businesses.

What Is Holding Back the Shift to Full-Cost Pricing? Full-cost pricing seems to make a lot of sense. Why is it not used more widely? There are several reasons. Many producers of harmful and wasteful goods would have to charge more. Some would go out of business. For other, more desirable products, the higher price would raise consumer objections.

Another problem with full-cost pricing is the difficulty of putting a price tag on many environmental and health costs. It is also hard to make such a change because many consumers are unaware they are paying these costs in ways not connected to the market prices of goods and services.

Finally, a major obstacle to full-cost pricing is that huge government subsidies, in the form of funding and tax breaks, distort the marketplace and hide many of the harmful environmental and social costs of producing and using some goods and services. Studies by Norman Myers and other analysts estimate that governments around the globe spend about $2 trillion per year to help subsidize several categories of environmentally harmful activities.

How Much Are We Willing to Pay to Control Pollution? The cleanup cost for removing a specific pollutant in gases or wastewater discharged into the environment rises with each additional unit of pollutant that is removed. As Figure 2-6 shows, a large percentage of the pollutants emitted by a smokestack or wastewater discharge can be removed fairly cheaply.

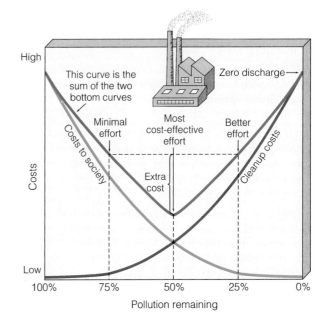

Figure 2-7 Finding the optimum level of pollution. This graph shows the optimum level at 50%, but the actual level varies depending on the pollutant.

However, as more and more pollutants are removed, the cost of removing each additional unit of pollution rises sharply. Beyond a certain point, the cleanup costs exceed the harmful costs of pollution. The *breakeven point* is the level of pollution control at which the harmful costs to society and the costs of cleanup are equal. Pollution control beyond this point costs more than it is worth, and not controlling pollution to this level causes environmental harm we can afford to avoid.

To find the breakeven point, economists plot two curves. One is a curve of the estimated economic costs of cleaning up pollution. The second is a curve of the estimated harmful external costs of pollution to society. They then add the two curves together to get a third curve showing the total costs. The lowest point on this third curve is the breakeven point, or the *optimum level of pollution* (Figure 2-7).

On a graph, determining the optimum value of pollution looks neat and simple. However, there are three problems with this approach. *First*, ecological economists, health scientists, and business leaders often disagree in their estimates of the harmful costs of pollution.

Second, some critics raise environmental justice questions about who benefits and who suffers from allowing optimum levels of pollution. The levels may be optimum for an entire country or large area. But this is not the case for the people who live near or downwind or downriver from a polluting power plant, incinerator, or factory and who are exposed to higher levels of pollution.

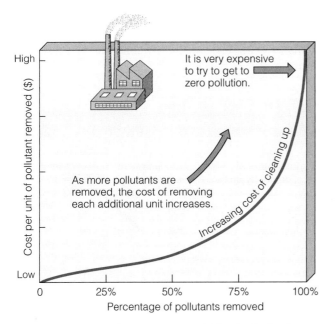

Figure 2-6 The cost of removing each additional unit of pollution rises. Cleaning up a certain amount of pollution is affordable, but at some point the cost of pollution control is greater than the harmful costs of the pollution to society.

Finally, assigning monetary values to things such as lost lives, forests, wetlands, and ecological services is difficult and controversial. But assigning little or no value to such things means they will not be counted in determining optimum pollution levels.

Some analysts point out that we should not be overly influenced by curves showing the sharply increasing costs of improving pollution control (Figure 2-6). The reason is that stricter regulation and higher pollution control costs can stimulate business to find cheaper ways to control pollution or redesign manufacturing processes to sharply reduce or eliminate pollution.

For example, in the 1970s the U.S. chemical industry predicted that controlling benzene emissions would cost $350,000 per plant. Shortly after these predictions were made, however, the companies developed a process that substituted other chemicals for benzene and almost eliminated control costs.

Should We Rely Mostly on Regulations or Market Forces? Most economists agree that government intervention in the marketplace is needed to control or prevent pollution and reduce resource waste. Such government action can take the form of regulation, the use of market forces, or some combination of these approaches. Each of the various regulatory and market approaches to reducing pollution and resource waste has advantages and disadvantages (Table 2-1).

Solutions: How Can Economic Incentives Be Used to Improve Environmental Quality and Reduce Resource Waste? *Market forces* can help improve environmental quality and reduce resource waste. This is achieved mostly by prompting the internalization of external costs through *economic incentives*

(rewards) or *economic disincentives* (punishments). This is based on a basic principle of the marketplace in today's capitalist market economic systems: *What we reward* (mostly by government subsidies and tax breaks) *we tend to get more of, and what we discourage* (mostly by regulations and taxes) *we tend to get less of.*

One way to put this principle into practice is to *phase in* government subsidies and tax breaks that encourage environmentally beneficial behavior and *phase out* government subsidies and tax breaks that encourage environmentally harmful behavior. However, this involves political decisions that often are opposed successfully by powerful economic interests.

Solutions: How Can We Use Economic Disincentives to Improve Environmental Quality and Reduce Resource Waste? Here are two examples of economic disincentives. One is to *use green taxes or effluent fees* to help internalize many of the harmful environmental costs of production and consumption. Taxes can be levied on each unit of pollution discharged into the air or water, hazardous or nuclear waste produced, virgin resources used, and fossil fuel used. Economists point out two requirements for successful implementation of *green taxes. First,* they should reduce or replace income, payroll, or other taxes. *Second,* the poor and middle class need to be given a safety net to reduce the regressive nature of consumption taxes on essentials such as food, fuel, and housing.

A second disincentive is to *charge user fees* that cover all or most costs for activities. Such fees can be charged for extracting lumber and minerals from public lands, using water provided by government-financed projects, and using public lands for livestock grazing.

Table 2-1 Economic Solutions to Pollution and Resource Waste					
Solution	Internalizes External Costs	Innovation	International Competitiveness	Administrative Costs	Increases Government Revenue
Regulation	Partially	Can encourage	Decreased*	High	No
Subsidies	No	Can encourage	Increased	Low	No
Withdrawing harmful subsidies	Yes	Can encourage	Decreased*	Low	Yes
Tradable rights	Yes	Encourages	Decreased*	Low	Yes
Green taxes	Yes	Encourages	Decreased*	Low	Yes
User fees	Yes	Can encourage	Decreased*	Low	Yes
Pollution-prevention bonds	Yes	Encourages	Decreased*	Low	No

*Unless more cost-effective and productive technologies are developed.

Solutions: Should We Shift the Tax Burden from Wages and Profits to Pollution and Waste? According to a number of economists, important goals of a more environmentally sustainable economic system are to *decrease* taxes on labor, income, and wealth and *increase* taxes on the pollution and waste produced by an economy.

To many analysts the tax system in most countries is backwards. It *discourages* what we want more of (jobs, income, and profit-driven innovation) and *encourages* what we want less of (pollution and resource depletion and degradation). It also leads to tremendous waste of natural and human capital. Such tax systems persist primarily because of political resistance to change and the political power of businesses that profit from them.

Such a tax shift would have to be phased in over 15 to 20 years to allow businesses to plan for the future and depreciate existing capital investments over their useful lives. Because such consumption taxes place a larger burden on the poor than do income taxes, governments would need to provide an economic safety net for poor and lower-middle-class people.

Trial versions of such tax shifts are taking place in Germany, Sweden, Great Britain, Norway, Denmark, Italy, France, Spain, and the Netherlands. If these countries continue moving toward this type of tax shift over the next two decades, their labor costs will be lowered and their resource productivity will increase. To remain competitive in the global marketplace, the United States and other countries may be forced to follow.

Solutions: Should We Rely More on Tradable Pollution and Resource-Use Rights? Another market approach is for the government to *grant tradable pollution and resource-use rights*. With this approach, the government sets a total limit or cap on emissions of a pollutant or use of a resource such as a fishery. Then the government issues permits to allocate or auction the total among manufacturers or users. Permit holders not using their entire allocation could use it as a credit against future expansion, use it in another part of their operation, or sell it to other companies.

Some environmentalists and economists support tradable pollution rights as an improvement over the current regulatory approach. Other environmentalists oppose allowing companies to buy and trade rights to pollute for several reasons. *First,* it allows the wealthiest companies to continue polluting and thereby excludes smaller companies from the market. *Second,* it tends to concentrate pollutants at the dirtiest plants and thus jeopardizes the health of people downwind or downstream from the plant.

Third, it creates an incentive for fraud because most pollution control regulations are based on self-reporting of pollution outputs, and government monitoring and enforcement of such outputs are inadequate.

Finally, this approach often has no built-in incentives to reduce overall pollution and resource use unless the system requires a gradual decrease in the total pollution or resource use allowed (which is rarely the case).

Solutions: How Has Environmental Management Changed? Figure 2-8 shows the evolution of several phases of environmental management, with the long-term goal of achieving more environmentally sustainable economies and societies.

The period between 1970 and 1985 can be viewed as the *resistance-to-change management era*. Two things happened in the 1970s. *First,* many companies and government environmental regulators developed an adversarial approach in which companies resented and actively resisted environmental regulations. *Second,* many government regulators thought they had to prescribe ways for reluctant companies to clean up their pollution emissions. At this stage, most companies

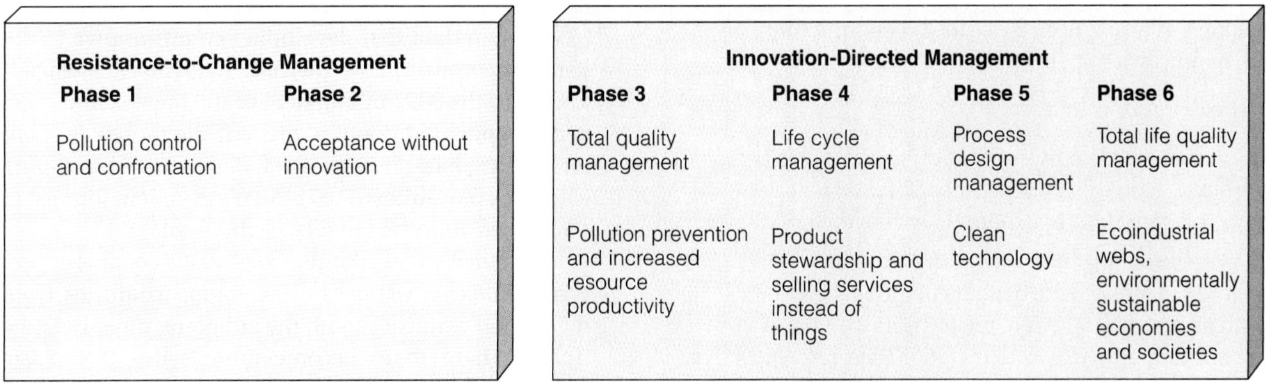

Resistance-to-Change Management		Innovation-Directed Management			
Phase 1	**Phase 2**	**Phase 3**	**Phase 4**	**Phase 5**	**Phase 6**
Pollution control and confrontation	Acceptance without innovation	Total quality management	Life cycle management	Process design management	Total life quality management
		Pollution prevention and increased resource productivity	Product stewardship and selling services instead of things	Clean technology	Ecoindustrial webs, environmentally sustainable economies and societies

Figure 2-8 Solutions: evolution in environmental management.

dealt with environmental regulations by hiring outside environmental consultants (who usually favored end-of-pipe pollution control solutions), using lawyers to oppose or find legal loopholes in the regulations, and lobbying elected officials to have environmental laws and regulations overthrown or weakened.

By 1985, most company managers accepted environmental regulations and continued to rely mostly on pollution control. However, they placed little emphasis on trying to find innovative solutions to pollution and resource waste problems because they believed them to be too costly.

In the 1990s, a growing number of company managers began realizing that environmental improvement is an economic and competitive opportunity instead of a cost to be resisted. This was the beginning of the *innovative management era*, which environmental and business visionaries project we will go through in several phases over the next 40–50 years (Figure 2-8, right).

2-4 REDUCING POVERTY TO IMPROVE ENVIRONMENTAL QUALITY AND HUMAN WELL-BEING

Is Economic Growth the Solution to Reducing Poverty? Most economists believe a growing economy is the best way to help the poor. They argue that economic growth can create more jobs, enable more of the increased wealth to reach workers, and provide greater tax revenues that can be used to help the poor help themselves.

Some *good news* is that

■ The percentage of people in developing countries suffering from extreme poverty (living on less than $1 per day) fell from 29% in 1990 to 23% in 2000. Also, the World Bank and other international monetary agencies hope to reduce the rate of extreme poverty to 14.5% by 2015.

■ Rich countries and individuals could essentially eliminate poverty within a decade by sharing more of their wealth.

The *bad news* is that

■ About one of every five people on the planet is trying to survive on the equivalent of $1 (U.S.) a day and one of every two is struggling to live on less than $3 (U.S) a day.

■ Since 1960, most of the benefits of global economic growth as measured by income have flowed up to the rich rather than down to the poor (Figure 2-9). According to Ismail Serageldin, the planet's richest three persons have more wealth than the combined GDP of the world's 47 poorest countries.

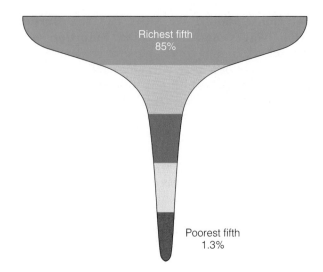

Figure 2-9 Data on the *global distribution of income* show that most of the world's income has flowed up; the richest 20% of the world's population receive more of the world's income than all of the remaining 80%. Each horizontal band in this diagram represents one-fifth of the world's population. This upward flow of global income has accelerated since 1960 and especially since 1980. This trend can increase environmental degradation by increasing average per capita consumption by the richest 20% of the population and causing the poorest 20% of the world's people to use renewable resources faster than they are replenished to survive. (Data from UN Development Programme and Ismail Serageldin, 2002, "World Poverty and Hunger: A Challenge for Science," *Science*, 296, 54–58)

Solutions: How Can We Reduce Poverty? Analysts point out that reducing poverty requires the governments of most developing countries to make policy changes. One way is to shift more of the national budget to help the rural and urban poor work their way out of poverty. Another is to give villages, villagers, and the urban poor title to common lands and to crops and trees they plant on them.

Analysts also urge developed countries and the wealthy in developing countries to help reduce poverty. Analysts propose forgiving at least 60% of the $2.6 trillion debt that developing countries owe to developed countries and international lending agencies, and all of the $422 billion debt of the poorest and most heavily indebted poor countries if the money is spent on meeting basic human needs. Currently, developing countries pay almost $300 billion per year in interest to developed countries to service this debt.

Another suggestion is to *increase nonmilitary government and private aid* to developing countries from developed countries, with the aid going directly to the poor to help them become more self-reliant. Also, banks and other organizations should be encouraged to *make small loans* to poor people wanting to increase their income (Solutions, p. 29).

Microloans to the Poor

Most of the world's poor desperately want to earn more, become more self-reliant, and have a better life. However, they have no credit record. Also, they have few if any assets to use for collateral to secure a loan to buy seeds and fertilizer for farming or tools and materials for a small business.

During the last 27 years an innovative tool called *microlending* or *microfinance* has increasingly helped deal with this problem. For example, since economist Muhammad Yunus started it in 1976, the Grameen (Village) Bank in Bangladesh has provided more than $2 billion in microloans (varying from $50 to $500) to 2.3 million mostly poor, rural, and landless women in 40,000 villages. About 94% of the loans are to women (making up 70% of the world's poor) who start their own small businesses as sewers, weavers, bookbinders, peanut fryers, or vendors.

To stimulate repayment and provide support, the Grameen Bank organizes microborrowers into five-member "solidarity" groups. If one member of the group misses a weekly payment or defaults on the loan, the other members of the group must make the payments.

The Grameen Bank's experience has shown that microlending is both successful and profitable. For example, less than 3% of microloan repayments to the Grameen Bank are late, and the repayment rate on its loans is an astounding 95%, much higher than the repayment rate for conventional loans by commercial banks throughout most of the world.

About half of Grameen's borrowers move above the poverty line within 5 years, and domestic violence, divorce, and birth rates are lower among borrowers. Microloans to the poor by the Grameen Bank are being used to develop day-care centers, health clinics, reforestation projects, drinking water supply projects, literacy programs, and group insurance programs. They are also being used to bring small-scale solar and wind power systems to rural villages.

Grameen's model has inspired the development of microcredit projects in more than 58 countries that have reached 36 million people (including dependents). In 1997, some 2,500 representatives of microlending organizations from 113 countries met at a Microcredit Summit in Washington, D.C., and adopted a goal of reaching 100 million of the world's poorest people by 2005.

Critical Thinking

Why do you think there has been little use of microloans by international development and lending agencies such as the World Bank and the International Monetary Fund? How might this situation be changed?

Finally, poverty can be reduced if governments of both developed and developing countries establish policies to slow population growth and *stabilize their populations.*

According to the United Nations Development Program (UNDP), it will cost about $40 billion a year to provide universal access to basic services such as education, health, nutrition, family planning, reproductive health, safe water, and sanitation. The UNDP notes that this is less than 0.1% of the world's annual income.

2-5 MAKING THE TRANSITION TO MORE ENVIRONMENTALLY SUSTAINABLE ECONOMIES

Solutions: How Can We Make Working with the Earth Profitable? Figure 2-10 (p. 30) lists principles that Paul Hawken (Guest Essay, p. 4) and several other business leaders and economists have suggested for making the transition from environmentally unsustainable to more environmentally sustainable economies during this century.

Hawken's simple golden rule for such an economy is this: *"Leave the world better than you found it, take*

no more than you need, try not to harm life or the environment, and make amends if you do."

Can We Make the Transition to a More Environmentally Sustainable Economy? Even if people believe a more environmentally sustainable economy is desirable, is it possible to make such a drastic change in the way we think and act? Some environmentalists, economists, and business leaders say it is not only possible but imperative, and it can be done over the next 40–50 years.

According to Paul Hawken, this new approach to economic thinking and actions recognizes that most business leaders are not evil, earth-degrading villains. Instead, they are trapped in a system that by design rewards them (with the highest profits and salaries and best chances for promotion) for maximizing short-term profits for owners and investors, regardless of the harmful short- and long-term environmental and social impacts.

Hawken argues that shifting to more environmentally sustainable economies throughout the world would free business leaders, workers, and investors from this ethical dilemma. In such economies, they would be financially compensated and respected for

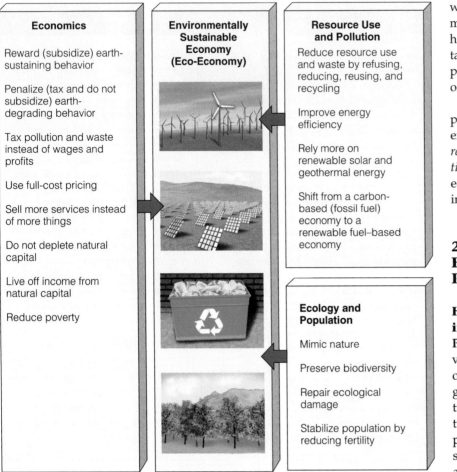

Economics	Environmentally Sustainable Economy (Eco-Economy)	Resource Use and Pollution
Reward (subsidize) earth-sustaining behavior		Reduce resource use and waste by refusing, reducing, reusing, and recycling
Penalize (tax and do not subsidize) earth-degrading behavior		Improve energy efficiency
Tax pollution and waste instead of wages and profits		Rely more on renewable solar and geothermal energy
Use full-cost pricing		Shift from a carbon-based (fossil fuel) economy to a renewable fuel–based economy
Sell more services instead of more things		
Do not deplete natural capital		
Live off income from natural capital		**Ecology and Population**
Reduce poverty		Mimic nature
		Preserve biodiversity
		Repair ecological damage
		Stabilize population by reducing fertility

Figure 2-10 Solutions: principles for shifting to more environmentally sustainable economies or eco-economies during this century.

doing socially and ecologically responsible work, improving environmental quality, and still making hefty profits for owners and stockholders.

The problem in making this shift is not economics but politics. It involves the difficult task of convincing more business leaders, elected officials, and voters to begin changing current government systems of economic rewards and penalties.

In moving toward an environmentally sustainable economy, there are three pieces of great news. *First,* making such a shift could be an extremely profitable enterprise that will create many jobs, greatly improve environmental quality, and sharply reduce poverty. According to environmental leaders Lester R. Brown and Christopher Flavin, "Converting the economy of the 21st century into one that is environmentally sustainable represents the greatest investment opportunity in history."

Second, we have most of the technologies needed to implement this economic shift. *Third,* governments would not have to spend more money. Such a shift

would be revenue neutral if governments replaced environmentally harmful subsidies with environmentally beneficial subsidies and taxed pollution and resource waste instead of wages and income.

Forward-looking investors, corporate executives, and political leaders recognize that *the environmental revolution is also an economic revolution.* This involves recognizing that economic and ecological systems are interdependent (Figure 2-3).

2-6 POLITICS AND ENVIRONMENTAL POLICY

How Does Social Change Occur in Representative Democracies? **Politics** is the process by which individuals and groups try to influence or control the policies and actions of governments at the local, state, national, or international levels. Politics is concerned with who has power over the distribution of resources and who gets what, when, and how.

Democracy is government by the people through elected officials and representatives. In a *constitutional democracy,* a constitution provides the basis of the government's authority, limits government power by mandating free elections, and guarantees freely expressed public opinion.

Political institutions in constitutional democracies are designed to allow gradual change to ensure economic and political stability. In the United States, for example, rapid and destabilizing change is curbed by the system of checks and balances that distributes power among the three branches of government—*legislative, executive,* and *judicial*—and among federal, state, and local governments.

In passing laws, developing budgets, and formulating regulations, elected and appointed government officials must deal with pressure from many competing *special-interest groups.* Each group advocates passing laws, providing subsidies, or establishing regulations favorable to its cause and weakening or repealing laws, subsidies, and regulations unfavorable to its position.

Some special-interest groups (such as corporations) are *profit-making organizations,* and others are *nonprofit nongovernment organizations (NGOs).* Examples of NGOs are educational institutions, labor unions, and mainstream and grassroots environmental organizations.

What Factors Hinder the Ability of Democracies to Deal with Environmental Problems?
The deliberately stable design of democracies is highly desirable. However, it has several related disadvantages for dealing with environmental problems. *First,* emphasis is on reacting to short-term environmental problems in isolation from one another instead of acting to prevent them from occurring in the future. However, many important environmental problems such as climate change, biodiversity loss, and long-lived hazardous waste have long-range effects, are related to one another, and require integrated long-term solutions emphasizing prevention.

Second, elections are held every few years. This means that most politicians wanting to get reelected focus on short-term individual problems rather than on the more complex, time-consuming, and often politically unrewarding job of finding integrated solutions to long-term problems.

Third, children, unborn generations, and wild species do not vote. *Fourth,* most politicians will no longer be in office when any harmful long-term effects from many environmental problems appear. Thus no powerful political constituency exists for the future or for long-term environmental sustainability.

Finally, whether they like it or not, most elected officials must spend much of their time raising money to get reelected.

What Principles Can Be Used as Guidelines in Making Environmental Policy Decisions?
Analysts have suggested that legislators and individuals evaluating existing or proposed environmental policy should be guided by the following principles:

- *The humility principle:* Recognize that our understanding of nature and of the consequences of our actions is quite limited.

- *The reversibility principle:* Try not to do something that cannot be reversed later if the decision turns out to be wrong. For example, most biologists believe the current large-scale destruction and degradation of forests, wetlands, wild species, and other components of the earth's biodiversity is unwise because much of it could be irreversible on a human time scale.

- *The precautionary principle:* When much evidence indicates that an activity raises threats of harm to human health or the environment, take precautionary measures to prevent or reduce such harm even if some of the cause-and-effect relationships are not fully es-

tablished scientifically. In such cases, it is better to be safe than sorry.

- *The prevention principle:* Whenever possible, make decisions that help prevent a problem from occurring or becoming worse.

- *The integrative principle:* Make decisions that involve integrated solutions to environmental and other problems.

- *The environmental justice principle:* Establish environmental policy so no group of people bears an unfair share of the harmful environmental risks from industrial, municipal, and commercial operations or the execution of laws, regulations, and policies. The EPA defines **environmental justice** as "the fair treatment and meaningful involvement of all people" in establishing and enforcing policies.

How Can Individuals Affect Environmental Policy?
A major theme of this book is that *individuals matter.* History shows that significant change usually comes from the *bottom up* when individuals join with others to bring about change. Without grassroots political action by millions of individual citizens and organized groups, the air you breathe and the water you drink today would be much more polluted, and much more of the earth's biodiversity would have disappeared. Figure 2-11 lists ways individuals can influence and change government policies in constitutional democracies.

According to most political analysts, *campaign financing* is the biggest problem that keeps elected officials from being more responsive to the environmental and other needs and problems of ordinary citizens.

Making a Difference

- Become informed on issues

- Run for office (especially at local level)

- Make your views known at public hearings

- Make your views known to elected representatives

- Contribute money and time to candidates for office

- Vote

- Form or join nongovernment organizations (NGOs) seeking change

- Support reform of election campaign financing

Figure 2-11 *Individuals matter:* ways you can influence environmental policy.

One suggestion for reducing undue influence by powerful special interests is to let the people (taxpayers) *alone* finance all federal, state, and local election campaigns, with low spending limits.

With such a reform, elected officials could spend their time governing instead of raising money and catering to powerful special interests. Office seekers would not need to be wealthy. Special-interest groups would be heard because of the validity of their ideas, not the size of their pocketbooks.

What Is Environmental Leadership? Individuals can provide leadership on environmental (or other) issues in three ways. One way is to *lead by example.* This involves using one's lifestyle to show others that change is possible and beneficial.

A second approach is to *work within existing economic and political systems to bring about environmental improvement.* People can influence political decisions by campaigning and voting for candidates and by communicating with elected officials. People can also work within the system by choosing environmental careers (Individuals Matter, at right). Many employers are actively seeking environmentally educated graduates. They are especially interested in people with scientific and engineering backgrounds and double majors (business and ecology, for example) or double minors.

A third form of environmental leadership involves *proposing and working for better solutions to environmental problems.* Leadership is more than being against something. It also involves coming up with better ways to accomplish various goals and getting people to work together to achieve such goals.

2-7 CASE STUDY: ENVIRONMENTAL POLICY IN THE UNITED STATES

How Is Environmental Policy Made in the United States? The major function of the federal government in the United States is to develop and implement *policy* for dealing with various issues. This policy typically is composed of various *laws* passed by the legislative branch, *regulations* instituted by the executive branch to put laws into effect, and *funding* to implement and enforce the laws and regulations. Figure 2-12 is a greatly simplified overview of how individuals and lobbyists for and against a particular environmental law interact with the three branches of government in the United States.

Several steps are involved in establishing federal environmental policy (or any other policy) in the United States. The *first step* is to persuade lawmakers that an environmental problem exists and the government has a responsibility to address it.

INDIVIDUALS MATTER

Environmental Careers

In the United States (and in other developed countries), economists claim the *green job market* is one of the fastest growing segments of the economy.

Many employers are actively seeking environmentally educated graduates. They are especially interested in people with scientific and engineering backgrounds and double majors (business and ecology, for example) or double minors.

Environmental career opportunities exist in a number of fields. They include environmental engineering, sustainable forestry and range management, parks and recreation, air and water quality control, solid waste and hazardous waste management, recycling, urban and rural land-use planning, computer modeling, ecological restoration, and soil, water, fishery, and wildlife conservation and management.

Environmental careers can also be found in education, environmental planning, environmental management, environmental health, toxicology, geology, ecology, conservation biology, chemistry, climatology, population dynamics and regulation (demography), law, risk analysis, risk management, accounting, environmental journalism, design and architecture, energy conservation and analysis, renewable-energy technologies, hydrology, consulting, public relations, activism and lobbying, economics, diplomacy, development and marketing, publishing (environmental magazines and books), and teaching and law enforcement (pollution detection and enforcement teams).

Critical Thinking

Have you considered an environmental career? Why or why not?

The *second step* is to influence how laws are written and to pass laws to deal with the problem. Most environmental bills are evaluated by as many as ten committees in the House of Representatives and the Senate. Effective proposals often are weakened by this fragmentation and by lobbying from groups opposing the law. Nonetheless, since the 1970s, a number of environmental laws (see list on the website for this chapter) have been passed in the United States.

The *third step* is to get enough funds appropriated to implement and enforce each law. Indeed, developing and adopting a budget is the most important and controversial activity of the executive and legislative branches. Developing a budget involves

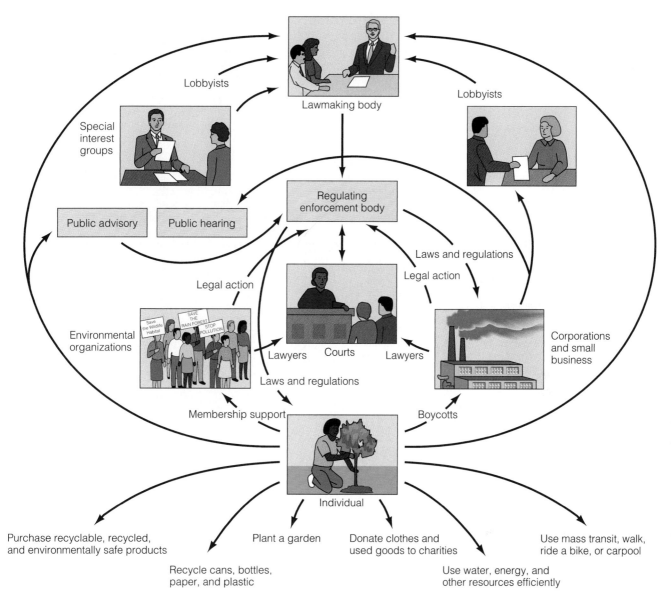

Lobbyists

Lawmaking body

Special interest groups

Lobbyists

Public advisory

Public hearing

Regulating enforcement body

Laws and regulations

Legal action

Legal action

Environmental organizations

SAVE the Wildlife Habitat

SAVE THE RAIN FOREST

STOP POLLUTION

Lawyers   Courts   Lawyers

Laws and regulations

Corporations and small business

Membership support

Boycotts

Individual

Purchase recyclable, recycled, and environmentally safe products

Plant a garden

Donate clothes and used goods to charities

Use mass transit, walk, ride a bike, or carpool

Recycle cans, bottles, paper, and plastic

Use water, energy, and other resources efficiently

Figure 2-12 Greatly simplified overview of how individuals and lobbyists for and against a particular environmental law interact with the legislative, executive, and judicial branches of government in the United States. The bottom of this diagram also shows some ways in which individuals can bring about environmental change through their own lifestyles.

answering two key questions: what programs will be funded and how much money will be used to address each problem?

Fourth, the appropriate government department or agency must draw up regulations for implementing each law.

The *final step* is to implement and enforce the approved regulations. Proponents or affected groups may take the agency to court for failing to implement and enforce the regulations or for enforcing them too rigidly.

According to social scientists, the development of public policy in democracies often goes through a *policy*

life cycle consisting of four stages: *recognition, formulation, implementation,* and *control.* Figure 2-13 (p. 34) shows the general position of several major environmental problems in the policy life cycle in the United States and most other developed countries.

What Are the Roles of Mainstream Environmental Groups? The spearhead of the global conservation and environmental movement consists of more than 30,000 NGOs working at the international, national, state, and local levels.

In the United States, more than 8 million U.S. citizens belong to at least 10,000 NGOs dealing with

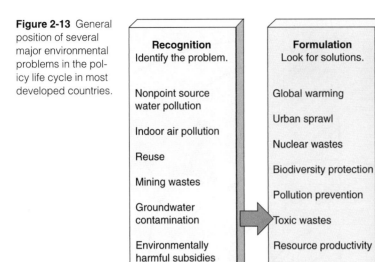
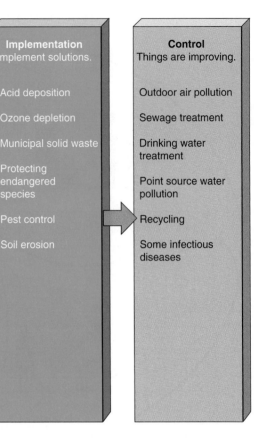

Figure 2-13 General position of several major environmental problems in the policy life cycle in most developed countries.

Recognition
Identify the problem.

Nonpoint source water pollution

Indoor air pollution

Reuse

Mining wastes

Groundwater contamination

Environmentally harmful subsidies

Market prices do not include environmentally harmful costs

Integrated environmental management

Formulation
Look for solutions.

Global warming

Urban sprawl

Nuclear wastes

Biodiversity protection

Pollution prevention

Toxic wastes

Resource productivity

Aquifer depletion

Environmental justice

Sustainable economic development

Implementation
Implement solutions.

Acid deposition

Ozone depletion

Municipal solid waste

Protecting endangered species

Pest control

Soil erosion

Control
Things are improving.

Outdoor air pollution

Sewage treatment

Drinking water treatment

Point source water pollution

Recycling

Some infectious diseases

environmental issues. Some of these environmental organizations are multimillion-dollar *mainstream* groups, led by chief executive officers and staffed by expert lawyers, scientists, and economists. In the United States, mainstream environmental groups are active primarily at the national level and to a lesser extent at the state level. Sometimes they also form coalitions to work together on issues.

Mainstream groups work within the political system. Many of these NGOs have been major forces in persuading the U.S. Congress to pass and strengthen environmental laws (see the list on the website for this chapter) and fighting off attempts to weaken or repeal such laws. However, mainstream environmental groups must guard against being subverted by the political system they work to improve. They must also avoid losing touch with ordinary people and nature in the insulated atmosphere of national and state capitals.

Instead of acting as adversaries, some industries and environmental groups have worked together to find solutions to environmental problems. For example, the Environmental Defense Fund has worked with McDonald's to redesign its packaging system to eliminate polyethylene foam clamshell hamburger containers. It has also worked with General Motors to help remove high-pollution cars from the road and with nine multinational corporations (including the Canadian aluminum company Alcan and the Mexican oil company Pemex) to set targets for reducing their carbon dioxide emissions (which can lead to global warming).

Some environmental groups have also shifted some of their resources away from demonstrating and litigating to publicizing research on innovative solutions to environmental problems. For example, to promote the use of chlorine-free paper, Greenpeace Germany printed a magazine using such paper and encouraged readers to demand that magazine publishers switch to chlorine-free paper. Shortly thereafter, several major magazines made such a shift.

What Are the Roles of Grassroots Environmental Groups? The base of the environmental movement in the United States and throughout the world consists of thousands of grassroots citizens' groups organized to improve environmental quality, often at the local level. According to political analyst Konrad von Moltke, "There isn't a government in the world that would have done anything for the environment if it weren't for the citizen groups."

These groups carry out a number of environmental roles. One is to work with individuals and communities to oppose harmful projects such as landfills, waste incinerators, nuclear waste dumps, clear-cutting of forests, and various development projects. They have also spurred government officials to take action when group members have been victims of environmental harm or of environmental injustice because of the unequal distribution of environmental risks.

Grassroots groups have also formed land trusts and other local organizations to save wetlands, forests, farmland, and ranchland from development, restore degraded rivers and wetlands, and convert abandoned urban lots into community gardens and parks.

Other groups are coalitions of workers and environmentalists who aim to improve worker safety and health. Local activists have also set up Internet service providers and networks to improve environmental education and health, provide individuals and NGOs with information on toxic releases and other harmful activities by local industries, and publicize successful environmental projects.

Case Study: Environmental Action on Campuses Since 1988, there has been a boom in environmental awareness on a number of college campuses and public schools across the United States.* Most student environmental groups work with members of the faculty and administration to bring about environmental improvements on their own campuses and in their communities.

Many of these groups make environmental audits of their campuses or schools. Then they use the data gathered to propose changes that will make their campuses or schools more environmentally sustainable, usually saving them money in the process.†

Such audits have resulted in numerous improvements. For example, Morris A. Pierce, a graduate student at the University of Rochester in New York, developed an energy management plan adopted by that school's board of trustees. Under this plan, a capital investment of $33 million is projected to save the university $60 million over 20 years. Students have also helped convince almost 80% of universities and colleges in the United States to develop recycling programs.

At Bowdoin College in Maine, chemistry professor Dana Mayo and student Caroline Foote developed the concept of *microscale experiments*, in which smaller amounts of chemicals are used. This has reduced toxic

wastes and saved the chemistry department more than $34,000. Today more than 50% of all undergraduates in chemistry in the United States use such microscale techniques, as do universities in a growing number of other countries.

Students at Oberlin College in Ohio helped design a more sustainable environmental studies building. At Northland College in Wisconsin, students helped design a "green" dorm that features a large wind generator, panels of solar cells, recycled furniture, and waterless (composting) toilets.

A 1997 report by the National Wildlife Foundation's Campus Ecology Program found that 23 student-researched and student-motivated projects had saved the participating universities and colleges $16.3 million. According to this study, implementing similar programs in the nation's 3,700 universities and colleges could help improve environmental quality and environmental education and lead to a savings of more than $14.6 billion.

Such student-spurred environmental activities and research studies are spreading to universities in at least 42 other countries.

How Successful Have Environmental Groups Been? During the past 32 years, a variety of environmental groups have achieved three important accomplishments. *First,* they raised understanding of environmental issues by the general public and some business and government leaders. *Second,* they have gained public support for an array of environmental and resource-use laws in the United States (see the list on the website for this chapter) and other developed countries. *Third,* they have helped individuals deal with a number of local environmental problems.

Polls show that three-fourths of U.S. citizens are strong supporters of environmental laws and regulations and do not want them weakened. However, polls also show that less than 10% of the U.S. public views the environment as one of the nation's most pressing problems. As a result, environmental concerns often do not get transferred to the ballot box. As one political scientist put it, "Environmental concerns are like the Florida Everglades, a mile wide but only a few inches deep."

What Are the Goals of the Anti-Environmental Movement in the United States? Despite general public approval, many environmental proposals are strongly opposed by the following groups:

■ Leaders of some corporations and people in positions of economic and political power who see environmental laws and regulations as threats to their wealth and power.

---

*These efforts are described in the book *Ecodemia: Campus Environmental Stewardship at the Turn of the 21st Century* (Washington, D.C.: National Wildlife Federation, 1995) and in the *Campus Environmental Yearbook,* published annually by the National Wildlife Federation.

†Details for conducting such audits are found in April Smith and the Student Environmental Action Coalition, *Campus Ecology: A Guide to Assessing Environmental Quality and Creating Strategies for Change* (Los Angeles, Calif.: Living Planet Press, 1993), and Jane Heinze-Fry, *Green Lives, Green Campuses,* available free on the website for this textbook.

■ Citizens who see environmental laws and regulations as threats to their private property rights and jobs.

■ Some state and local government officials who are tired of having to implement federal environmental laws and regulations without federal funding (unfunded mandates) or disagree with certain federal environmental regulations.

Since 1980, businesses, individuals, and grassroots groups in the United States have mounted a strong campaign to weaken or repeal existing environmental laws, allow more extraction of resources from public lands with less regulation, and destroy the reputation and effectiveness of the environmental movement. Some of the tactics used by this anti-environmental movement are listed on the website for this chapter. Because of the anti-environmental movement, since 1980 mainstream environmental groups have spent most of their time and money trying to prevent existing environmental laws and regulations from being weakened or repealed.

2-8 GLOBAL ENVIRONMENTAL POLICY

Solutions: Should We Expand the Concept of National and Global Security? Countries are legitimately concerned with *military security* and *economic security*. However, ecologists point out that all economies are supported by the earth's natural resources or natural capital (Figure 1-2, p. 3, top half of back cover, and Figure 2-3) and many environmental problems do not recognize political boundaries. Thus military and economic security also depend on national and global environmental security.

Proponents call for all countries to make environmental security a major focus of diplomacy and government policy at all levels. This should be implemented by having a council of advisers made up of highly qualified experts in environmental, economic, and military security who integrate all three security concerns in making major decisions.

Figure 2-14 Trade-offs: good and bad news about international efforts to deal with global environmental problems.

What Progress Has Been Made in Developing International Environmental Cooperation and Policy? Since the 1972 UN Conference on the Human Environment in Stockholm, Sweden, some progress has been made in addressing environmental issues at the global level. Figure 2-14 lists some of the good and bad news about international efforts to deal with global environmental problems such as poverty, climate change, biodiversity loss, and ocean pollution.

2-9 HUMAN-CENTERED ENVIRONMENTAL WORLDVIEWS

What Is an Environmental Worldview? There are conflicting views about how serious our environmental problems are and what we should do about them. These conflicts arise mostly out of differing **environmental worldviews:** how people think the world works, what they think their role in the world should be, and what they believe is right and wrong environmental behavior (**environmental ethics**).

People who have widely differing environmental worldviews can take the same data, be logically consistent, and arrive at quite different conclusions because they start with different assumptions and values.

The many different types of environmental worldviews are summarized in Figure 2-15. Most can be divided into two groups according to whether they are *individual centered* (atomistic) or *earth centered* (holistic).

Global Efforts on Environmental Problems

Good News

Environmental protection agencies in 115 nations

Over 500 international environmental treaties and agreements

UN Environment Programme (UNEP) created in 1972 to negotiate and monitor international environmental treaties

1992 Rio Earth Summit adopted key principles for dealing with global environmental problems

2002 Johannesburg Earth Summit attempted to implement policies and goals of 1992 Rio summit and find ways to reduce poverty

Bad News

Most international environmental treaties lack criteria for monitoring and evaluating their effectiveness

1992 Rio Earth Summit led to nonbinding agreements without enough funding to implement

By 2003 there was little improvement in the major environmental problems discussed at the 1992 Rio Earth Summit

2002 Johannesburg Earth Summit failed to provide adequate goals, deadlines, and funding for dealing with global environmental problems such as climate change, biodiversity loss, and poverty

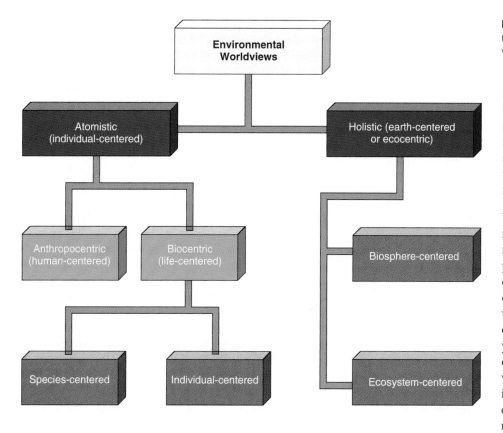

Figure 2-15 General types of environmental worldviews. (Diagram developed by Jane Heinze-Fry)

ronmental worldviews focus on sustaining the earth's natural systems (ecosystems), life forms (biodiversity), and life-support systems (biosphere) for the benefit of humans and other forms of life.

What Are the Major Human-Centered Environmental Worldviews? Most people in today's industrial consumer societies have a **planetary management worldview,** which has become increasingly common during the past 50 years. According to this human-centered environmental worldview, we are the planet's most important and dominant species and we can and should manage the planet mostly for our own benefit. Other species and parts of nature are seen as having only *instrumental value* based on how useful they are to us.

Atomistic environmental worldviews tend to be *human centered* (anthropocentric) or *life centered* (biocentric, with the primary focus on individual species or individual organisms). Holistic or ecocentric envi-

Figure 2-16 (left) summarizes the four major beliefs or assumptions of one version of this worldview.

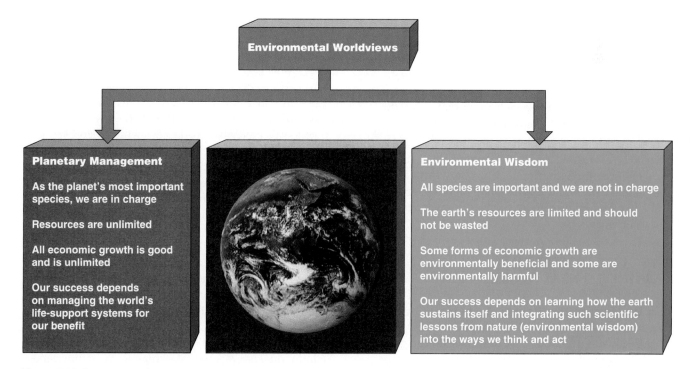

Figure 2-16 Comparison of two opposing environmental worldviews.

All or most aspects of this worldview are widely supported because it is said to be the primary driving force behind the major improvements in the human condition since the beginning of the industrial revolution.

There are several variations of this environmental worldview:

- The *no-problem school.* We can solve any environmental, population, or resource problems with more economic growth and development, better management, and better technology.

- The *free-market school.* The best way to manage the planet for human benefit is through a free-market global economy with minimal government interference and regulations. Free-market advocates would convert all public property resources to private property resources and let the global marketplace, governed by pure free-market competition (p. 20), decide essentially everything.

- The *responsible planetary management school.* We have serious environmental problems, but we can sustain our species with a mixture of market-based competition, better technology, and some government intervention that promotes environmentally sustainable forms of economic development (Figure 2-4), protects environmental quality and private property rights, and protects and manages public and common property resources. People holding this view follow the pragmatic principle of *enlightened self-interest:* Better earth care is better self-care.

- The *spaceship-earth school.* The earth is seen as a spaceship: a complex machine that we can understand, dominate, change, and manage to prevent environmental overload and provide a good life for everyone. This view developed as a result of photographs taken from space showing the earth as a finite planet or island "floating" in space (Figure 1-1, p. 1). This powerful image led many people to see that the earth is our only home and we had better treat it right.

- The *stewardship school.* We have an ethical responsibility to be caring and responsible managers, or stewards, of the earth. According to this view, we can and should make the world a better place for our species and other species through love, caring, knowledge, and technology.

2-10 LIFE-CENTERED ENVIRONMENTAL WORLDVIEWS

Can We Manage the Planet?   Some people believe any human-centered worldview will eventually fail because it wrongly assumes we now have (or can gain) enough knowledge to become effective managers or stewards of the earth.

According to these analysts, the unregulated global free-market approach will not work for two reasons. *First,* it is based on increased losses or degradation of natural capital (Guest Essay, p. 4) that support all life and economies. *Second,* it focuses on short-term economic benefits regardless of the harmful long-term environmental and social consequences.

The image of the earth as an island or spaceship in space has played an important role in raising global environmental awareness. However, critics argue that thinking of the earth as a spaceship we should and can manage is an oversimplified and misleading way to view an incredibly complex and ever-changing planet. This view was supported by the failure of Biosphere 2 (p. 19). As biologist David Ehrenfeld puts it, "In no important instance have we been able to demonstrate comprehensive successful management of the world, nor do we understand it well enough to manage it even in theory."

Even if we had enough knowledge and wisdom to manage spaceship earth, some critics see this approach as requiring us to give up individual freedom to survive. Life on spaceship earth under a comprehensive system of planetary management or world government might be much like the regimented life of astronauts in their capsule. The astronauts have almost no individual freedom; essentially all their actions are dictated by a central command (ground control).

What Are Some Major Life-Centered Worldviews?   Critics of human-centered environmental worldviews believe such worldviews should be expanded to recognize the *inherent* or *intrinsic value* of all forms of life regardless of their potential or actual use to us.

Most people who have a life-centered (biocentric) worldview believe we have an ethical responsibility not to cause the premature extinction of a species because of our activities. Others believe we must go beyond this biocentric worldview, which focuses mostly on species. They believe we have an ethical responsibility not to degrade the earth's natural systems (ecosystems), life-forms (biodiversity), and life-support systems for this and future generations of humans and other species. In other words, they have an *earth-centered,* or *ecocentric,* environmental worldview, devoted to preserving the earth's biodiversity and the functioning of its life-support systems for all forms of life.

According to the ecocentric worldview, we are part of, not apart from, the community of life and the ecological processes that sustain all life. Aldo Leopold (Individuals Matter, p. 39) summed up this idea in 1948: "All ethics rest upon a single premise: that the individual is a member of a community of interdependent parts."

Aldo Leopold and His Land Ethic

INDIVIDUALS MATTER

Aldo Leopold (see figure) is best known as a strong proponent of *land ethics*, a philosophy in which humans as part of nature have an ethical responsibility to preserve wild nature.

After earning a master's degree in forestry from Yale University, he joined the U.S. Forest Service. Overgrazing and land deterioration on public lands where he worked alarmed him and convinced him that the United States was losing too much of its mostly untouched wilderness lands.

In 1933, Leopold became a professor at the University of Wisconsin and founded the profession of game management. In 1935, he was one of the founders of the Wilderness Society.

As years passed, he developed a deep understanding and appreciation for wildlife and urged us to include nature in our ethical concerns. Through his writings and teachings he became one of the founders of the *conservation* and *en-vironmental movements* of the 20th century.

Leopold died in 1948 while fighting a brush fire at a neighbor's farm. His weekends of planting, hiking, and observing nature at his own nearby farm provided material he used to write his most famous book, *A Sand County Almanac*, published after his death in 1949. Since then more than 2 million copies of this important book have been sold.

The following quotes from his writings reflect Leopold's land ethic and form the basis of many of the beliefs of the modern *environmental wisdom worldview* (Figure 2-16, right):

That land is a community is the basic concept of ecology, but that land is to be loved and respected is an extension of ethics.

The land ethic changes the role of Homo sapiens *from conqueror of the land-community to plain member and citizen of it.*

We abuse land because we regard it as a commodity belonging to us. When we see land as a community to which we belong, we may begin to use it with love and respect.

A thing is right when it tends to preserve the integrity, stability, and beauty of the biotic community. It is wrong when it tends otherwise.

Aldo Leopold (1887–1948) was a forester, writer, and conservationist. His book, *A Sand County Almanac* (published after his death), is considered an environmental classic that inspired the modern environmental movement. His *land ethic* expanded the role of humans as protectors of nature.

There are many life-centered and earth-centered environmental worldviews, and several of them overlap in some of their beliefs. One ecocentric environmental worldview is the **environmental wisdom worldview** (Figure 2-16, right), which is the opposite of the planetary management worldview (Figure 2-16, left).

Others say we do not need to be biocentrists or ecocentrists to value life or the earth. They point out that human-centered stewardship and planetary management environmental worldviews also call for us to value individuals, species, and the earth's life-support systems as part of our responsibility as the earth's caretakers.

Should We Care about Future Generations?
Some people believe our only ethical obligation is to the present human generation. They ask, "What has the future done for me?" or believe we cannot know enough about the condition of the earth for future generations to be concerned about it.

According to biologist David W. Ehrenfeld, caring about future generations enough not to degrade the earth's life-support systems is important because it gives future generations more options for dealing with the problems they will face. He points out that if our ancestors had left for us the ecological degradation we appear to be leaving our descendants, our options for enjoyment—perhaps even for survival—would be quite limited.

In response to the question, "What can future generations do for us?" Ehrenfeld gives the following answer: "They give us a reason for treating our ecological home respectfully, so that our lives as well as theirs will be enriched."

According to this view, as we use the earth's natural resources we are borrowing from the earth and from future generations and have an ethical responsibility to leave the earth in as good a condition or better than it is now.

In thinking about our responsibility toward future generations, some analysts believe we should consider

the wisdom given to us in the 18th century by the Iroquois Confederation of Native Americans: *In our every deliberation, we must consider the impact of our decisions on the next seven generations.*

2-11 SOLUTIONS: LIVING MORE SUSTAINABLY

What Are the Main Components of an Environmental Literacy? Most environmentalists believe learning how to live sustainably takes a foundation of environmental education. According to its proponents, the most important goals of such an education are these:

- *Develop respect or reverence for all life.*

- *Understand as much as we can about how the earth works and sustains itself and use such knowledge to guide our lives, communities, and societies.*

- *Become seekers of environmental wisdom instead of vessels of environmental information.*

- *Understand and evaluate our environmental worldview and see this as a lifelong process.* Evaluating and perhaps changing our environmental worldview can set off a cultural *mindquake* because it involves examining many of our most basic beliefs. However, once we change our worldview, it no longer makes sense for us to do things in the old ways. If enough people do this and put their beliefs into action (probably no more than 10% of the population), then tremendous cultural change, once considered impossible, can take place rapidly.

- *Learn how to evaluate the beneficial and harmful consequences of our choices of lifestyle and profession on the earth, today and in the future.*

- *Foster a desire to make the world a better place and act on this desire.* As environmental educator David Orr puts it, education should help students "make the leap from 'I know' to 'I care' to 'I'll do something.'"

According to environmental educator Mitchell Thomashow, four basic questions should be at the heart of environmental literacy.

First, where do the things I consume come from? *Second*, what do I know about the place where I live? *Third*, how am I connected to the earth and other living things? *Fourth*, what is my purpose and responsibility as a human being?

How we answer these questions determines our *ecological identity.*

How Can We Learn from the Earth? Seeking Environmental Wisdom Formal environmental education is important, but is it enough? Many analysts say

no and urge us to take the time to escape the cultural and technological body armor we use to insulate ourselves from nature and to experience nature directly.

They suggest we kindle a sense of awe, wonder, mystery, and humility by standing under the stars, sitting in a forest, taking in the majesty and power of an ocean, or experiencing a stream, lake, coral reef, or other part of nature.

We might pick up a handful of soil and try to sense the teeming microscopic life in it that keeps us alive. We might look at a tree, mountain, rock, or bee and try to sense how they are a part of us and we a part of them as interdependent participants in the earth's life-sustaining recycling processes.

Philosopher and naturalist Michael J. Cohen suggests we recognize who we really are by saying,

> I am a desire for water, air, food, love, warmth, beauty, freedom, sensations, life, community, place, and spirit in the natural world. . . . I have two mothers: my human mother and my planet mother, Earth. The planet is my womb of life.

Many psychologists believe that consciously or unconsciously we spend much of our lives in a search for roots: something to anchor us in a bewildering and frightening sea of change. As philosopher Simone Weil observed, "To be rooted is perhaps the most important and least recognized need of the human soul."

Earth-focused philosophers say that to be rooted, each of us needs to find a *sense of place*: a stream, a mountain, a yard, a neighborhood lot, or any piece of the earth we feel at one with as a place we know, experience emotionally, and love. It can be a place where we live or a place we occasionally visit and experience in our inner being. When we become part of a place, it becomes a part of us. Then we are driven to defend it from harm and to help heal its wounds.

How Can We Live More Simply? Many analysts urge us to *learn how to live more simply*. Looking for happiness through the pursuit of material things is considered folly by almost every major religion and philosophy. Yet, it is preached incessantly by modern advertising that encourages us to buy more and more things. As humorist Will Rogers said, "Too many people spend money they haven't earned to buy things they don't want, to impress people they don't like." Figure 2-17 summarizes some ethical guidelines proposed by various ethicists and philosophers for living more sustainably or simply on the earth.

Some affluent people in developed countries are adopting a lifestyle of *voluntary simplicity*, doing and enjoying more with less by learning to live more simply. Voluntary simplicity is based on Mahatma Gandhi's *principle of enoughness:* "The earth provides enough to

Biosphere and Ecosystems	Species and Cultures	Individual Responsibility
Help sustain the earth's natural capital and biodiversity	Avoid premature extinction of any species mostly by protecting and restoring its habitat	Do not inflict unnecessary suffering or pain on any animal
Do the least possible environmental harm when altering nature	Avoid premature extinction of any human culture	Use no more of the earth's resources than you need

satisfy every person's need but not every person's greed. . . . When we take more than we need, we are simply taking from each other, borrowing from the future, or destroying the environment and other species."

Implementing this principle means asking oneself, "How much is enough?" This is not easy because people in affluent societies are conditioned to want more and more, and often to think of such wants as vital needs.

Voluntary simplicity begins by asking a series of questions before buying anything: Do I really need this? Can I buy it secondhand (reuse)? Can I borrow, rent, lease, or share it? Can I build it myself?

The decision to buy something triggers another set of questions: Is the product produced in an environmentally sustainable manner? Did the workers producing it get fair wages for their work, and did they have safe and healthful working conditions? Is it designed to last as long as possible? Is it easy to repair, upgrade, reuse, and recycle?

We should not confuse voluntary simplicity by those who have more than they need with the *forced simplicity* of the poor, who do not have enough to meet their basic needs for food, clothing, shelter, clean water and air, and good health.

After a lifetime of studying the growth and decline of the world's human civilizations, historian Arnold Toynbee summarized the true measure of a civilization's growth in what he called the *law of progressive simplification:* "True growth occurs as civilizations transfer an increasing proportion of energy and attention from the material side of life to the nonmaterial side and thereby develop their culture, capacity for compassion, sense of community, and strength of democracy."

How Can We Move beyond Blame, Guilt, and Denial to Responsibility? When we first encounter an environmental problem, our initial response often is to find someone or something to blame: greedy industrialists, uncaring politicians, environmentalists, and people with misguided worldviews or beliefs. It is the fault of such villains, and we are the victims. This response can lead to despair, denial, and inaction because we feel powerless to stop or influence these forces.

Upon closer examination we may realize we all make some direct or indirect contributions to the environmental problems we face. As the comic strip character Pogo said, "We have met the enemy and it is us." We do not want to feel guilty or bad about the environmental harm our lifestyle may be doing. Thus we try not to think about this too much—another path to denial and inaction.

How do we move beyond blame, fear, denial, and guilt to more responsible environmental actions in our daily lives? Analysts have suggested several ways to do this.

One is to recognize and avoid common mental traps that lead to denial, indifference, and inaction. These traps include *gloom-and-doom pessimism* (it is hopeless), *blind technological optimism* (science and technofixes will save us), *fatalism* (we have no control over our actions and the future), *extrapolation to infinity* (if I cannot change the entire world quickly, I will not try to change any of it), *paralysis by analysis* (searching for the perfect worldview, philosophy, solutions, and scientific information before doing anything), and *faith in simple, easy answers.*

We also need to keep our empowering feelings of hope slightly ahead of our immobilizing feelings of despair. In working with the earth we should be guided by historian Arnold Toynbee's observation, "If you make the world ever so little better, you will have done splendidly, and your life will have been worthwhile," and by George Bernard Shaw's reminder that "indifference is the essence of inhumanity."

It is important not to use guilt and fear to motivate other people to work with the earth and other people. We need to nurture, reassure, understand, and care for one another.

Recognizing that there is no single correct or best solution to the environmental problems we face is also important. Indeed, one of nature's most important lessons is that preserving diversity or a rainbow of flexible and adaptable solutions to our problems is the best way to adapt to earth's largely unpredictable, ever-changing conditions.

Finally, we should have fun and take time to enjoy life. Laugh every day and enjoy nature, beauty, friendship, and love.

What Are the Major Components of the Environmental Revolution? The *environmental revolution* that many environmentalists call for us to bring about during this century would have several components:

▪ An *efficiency revolution* that involves not wasting matter and energy resources.

▪ A *solar–hydrogen revolution* based on decreasing our dependence on carbon-based nonrenewable fossil fuels and increasing our dependence on forms of renewable solar energy that can be used to produce hydrogen fuel from water.

▪ A *pollution prevention revolution* that reduces pollution and environmental degradation from harmful chemicals. This means avoiding their release into the environment by recycling or reusing them within industrial processes, trying to find less polluting substitutes, or not producing them at all.

▪ A *biodiversity protection revolution* devoted to protecting and sustaining the genes, species, natural systems, and chemical and biological processes that make up the earth's biodiversity.

▪ A *sufficiency revolution.* This involves trying to meet the basic needs of all people on the planet and asking how many material things we really need to have a decent and meaningful life.

▪ A *demographic revolution* based on reducing fertility to bring the size and growth rate of the human population into balance with the earth's ability to support humans and other species without serious environmental degradation.

▪ An *economic and political revolution* in which we use economic systems to reward environmentally beneficial behavior and discourage environmentally harmful behavior.

Opponents of such a cultural change like to paint environmentalists as messengers of gloom, doom, and hopelessness. However, *the real message of environmentalism is not gloom and doom, fear, and catastrophe but hope and a positive vision of the future* (Figure 2-18).

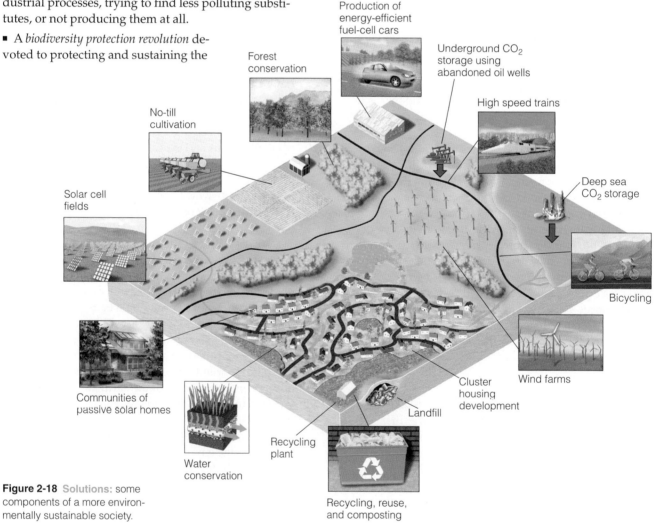

Figure 2-18 Solutions: some components of a more environmentally sustainable society.

We should rejoice in our environmental accomplishments, but the real question is *where do we go from here?* How do we transfer the environmental advances of the past 35 years to developing countries? How during this century do we make a new cultural transition to more environmentally sustainable societies?

It is an incredibly exciting time to be alive as we struggle to enter into a new relationship with the earth that keeps us all alive and supports our economies. Jump in and get involved in guiding this new wave of cultural change.

Envision the earth's life-sustaining processes as a beautiful and diverse web of interrelationships—a kaleidoscope of patterns, rhythms, and connections whose very complexity and multitude of possibilities remind us that cooperation, sharing, honesty, humility, and love should be the guidelines for our behavior toward one another and the earth.

When there is no dream, the people perish.
PROVERBS 29:18

REVIEW QUESTIONS

1. Define the boldfaced terms in this chapter.

2. Describe the Biosphere 2 project and the lessons it taught us.

3. Distinguish among *natural, human, financial,* and *manufactured resources* used in an economic system.

4. Distinguish among *pure command, pure free-market,* and *capitalist market* systems. List ten reasons why governments intervene in economic systems.

5. Explain how conventional economists and ecological economists differ in their view of market-based economic systems. Distinguish between *economic growth* and *environmentally sustainable economic development.* What are three characteristics of environmentally sustainable economic development?

6. List four reasons why GNI and GDP are not useful measures of economic health, environmental health, or human well-being. Describe an environmental indicator that could be used to provide such information.

7. Distinguish between *internal costs* and *external costs,* and give an example of each. What is *full-cost pricing,* and what are the advantages and disadvantages of using this approach to internalize external environmental costs? List four reasons why full-cost pricing has not been widely used.

8. How do economists determine the *optimum level of pollution* for a particular chemical? What are the advantages and disadvantages of using this approach?

9. Describe two types of *economic incentives* (rewards) that can be used to improve environmental quality and reduce resource waste. Describe two types of *economic*

disincentives (punishments) that can be used to improve environmental quality and reduce resource waste.

10. What are the advantages and disadvantages of using tradable pollution and resource-use rights to reduce pollution and resource waste?

11. List six phases in the evolution of environmental management.

12. List two ways in which the governments of developing countries can reduce poverty. List four ways in which governments of developed countries can help reduce poverty. What are *microloans,* and how are they being used to reduce poverty?

13. List 16 principles for shifting to more environmentally sustainable economies over the next several decades. List three great pieces of news about making the shift to more environmentally sustainable economies over the next several decades. Explain the idea that the environmental revolution is also an economic revolution.

14. What is *politics?* What is a *democracy?*

15. List four factors that hinder the ability of democracies to deal with environmental problems.

16. List six principles that can be used as guidelines in making environmental policy decisions. What is *environmental justice?*

17. List eight ways in which individuals can influence the environmental policies of local, state, and federal governments.

18. Describe three types of environmental leadership.

19. Describe the five steps used to develop federal environmental policy in the United States. Describe the four phases of a *policy life cycle.*

20. Describe the key roles of mainstream and grassroots environmental groups. Describe some of the environmentally beneficial activities that have been carried out by high school and college students.

21. List three reasons why some people oppose environmental reform.

22. Explain the importance of making environmental security a key priority of governments.

23. Summarize five pieces of good and four pieces of bad news about international efforts to deal with major environmental problems.

24. What is an *environmental worldview?* What are the four major beliefs of the *planetary management environmental worldview?* Describe the **(a)** *no problem,* **(b)** *free-market,* **(c)** *responsible planetary management,* **(d)** *spaceship earth,* and **(e)** *stewardship* variations of the planetary management worldview.

25. Explain why some analysts believe planetary management worldviews will not work.

26. What are the four major beliefs of the *environmental wisdom worldview?*

27. Give two reasons for caring about future generations.

28. List six components of environmental literacy and four questions that should be answered to determine one's *ecological identity.* Explain why some people believe it is important to learn more about the earth from direct experience.

29. List six ethical guidelines for living more sustainably.

30. What is *voluntary simplicity?*

31. List five ways to move beyond blame, guilt, and denial and become more environmentally responsible. Describe six mental traps that can lead to denial, indifference, and inaction about environmental (and other) problems.

32. What are seven major components of an environmental revolution that most environmentalists believe we should bring about during this century?

CRITICAL THINKING

1. Some analysts argue that the problems with Biosphere 2 resulted mostly from inadequate design and that a better team of scientists and engineers could make it work. Explain why you agree or disagree with this view.

2. The primary goal of current economic systems is to maximize economic growth by producing and consuming more and more economic goods. Do you agree with that goal? Explain. What are the alternatives?

3. Suppose that over the next 20 years the current harmful environmental and health costs of goods and services are internalized so their market prices reflect their total costs. What harmful and beneficial effects might this have on **(a)** your lifestyle and **(b)** any child you might have?

4. Do you believe we should establish zero-discharge levels for toxic chemicals we release into the environment? Explain.

5. Do you favor shifting the tax burden from wages and profits to pollution and waste? Explain.

6. Do you agree or disagree with the proposals that various analysts have made for sharply reducing poverty as discussed on pp. 28–29? Explain.

7. Explain why you agree or disagree with each of the major principles for shifting to a more environmentally sustainable economy listed in Figure 2-10, p. 30.

8. What are the greatest strengths and weaknesses of the system of government in your country with respect to **(a)** protecting the environment and **(b)** ensuring environmental justice for all? What three major changes, if any, would you make in this system?

9. Explain why you agree or disagree with each of the six principles for making environmental policy listed on p. 31.

10. This chapter has summarized a number of different environmental worldviews. Analyze these worldviews and find the beliefs you agree with to describe your own environmental worldview. Which of your beliefs were added or modified as a result of taking this course?

11. Explain why you agree or disagree with the following ideas: **(a)** Everyone has the right to have as many children as they want, **(b)** each member of the human species has a right to use as many resources as he or she wants, and **(c)** individuals should have the right to do anything they want with land they own. Relate your answers to the beliefs of your environmental worldview that you described in question 10.

12. Some analysts believe learning environmental wisdom by experiencing the earth and forming an emotional bond with its life-forms and processes is unscientific, mystical nonsense based on a romanticized view of nature. They believe better scientific understanding of how the earth works and improved technology are the only ways to achieve sustainability. Do you agree or disagree? Explain.

PROJECTS

1. List all the goods you use, and then identify those that meet your basic needs and those that satisfy your wants. Identify any economic wants **(a)** you would be willing to give up, **(b)** you believe you should give up but are unwilling to give up, and **(c)** you hope to give up in the future. Relate the results of this analysis to your personal impact on the environment. Compare your results with those of your classmates.

2. What student environmental groups (if any) are active at your school? How many people actively participate in these groups? What environmentally beneficial things have they done? What actions (if any) taken by such groups do you disagree with? Why?

3. Does your school's curriculum provide *all* graduates with the basic elements of environmental literacy? To what extent are the funds in its financial endowments invested in enterprises that are working to develop or encourage environmental sustainability? Over the past 20 years, what important roles have its graduates played in making the world a better and more sustainable place to live? Using such information, rate your school on a 1–10 scale in terms of its contributions to environmental awareness and sustainability. Develop a detailed plan illustrating how your school could become better at achieving such goals and present this information to school officials, alumni, parents, and financial backers.

4. If you knew you were going to die and had an opportunity to address everyone in the world for 5 minutes, what would you say? Write out your 5-minute speech and compare it with those of other members of your class.

5. Make a concept map of this chapter's major ideas, using the section heads and subheads and the key terms (in boldface type). Look on the website for this book for information about making concept maps.

INTERNET STUDY RESOURCES AND RESOURCES FOR FURTHER READING AND RESEARCH

The website for this book contains helpful study aids and many ideas for further reading and research. Log on to

http://biology.brookscole.com/miller10

and click on the Chapter-by-Chapter area. Choose Chapter 2 and select a resource:

- Flash Cards allows you to test your mastery of the Terms and Concepts to Remember for this chapter.

- Tutorial Quizzes provides a multiple-choice practice quiz.

- Student Guide to InfoTrac will lead you to Critical Thinking Projects that use InfoTrac College Edition as a research tool.

- References lists the major books and articles consulted in writing this chapter.

- Hypercontents takes you to an extensive list of sites with news, research, and images related to individual sections of the chapter.

INFOTRAC COLLEGE EDITION

Improve your skills with InfoTrac College Edition, a searchable online database of articles from more than 700 periodicals. Log on to

http://www.infotrac-college.com

or access InfoTrac through the website for this book. Try to find the following articles:

1. Holly, C. 2001. Study shows green taxes can benefit economy. *Energy Daily* 29: 4. *Keywords:* "green taxes." These days it seems that when people start talking about taxes and regulations to benefit the environment, they are met with dire predictions of economic catastrophe. But is this really the case?

2. Costanza, R. 2001. Visions, values, valuation, and the need for an ecological economics. *BioScience* 51: 459. *Keywords:* "ecological economics." This article looks deeply at how we view the world and how we view ourselves in it.

3 SCIENCE, SYSTEMS, MATTER, AND ENERGY

Two Islands: Can We Treat This One Better?

Easter Island (Rapa Nui) is a small, isolated island in the great expanse of the South Pacific. It was first colonized by Polynesians about 2,500 years ago.

The civilization they developed was based on the island's towering palm trees, which were used for shelter, tools, fishing boats, fuel, food, rope, and clothing. Using these resources, they developed an impressive civilization and a technology capable of making and moving large stone structures, including their famous statues (Figure 3-1).

The people flourished, with the population peaking at about 10,000 (with estimates ranging from 7,000 to 20,000) by 1400. However, they used up the island's precious trees faster than they were regenerated—an example of the tragedy of the commons (p. 7). Each person who cut a tree reaped immediate personal benefits while helping doom the civilization in the long run.

Once the trees were gone, the islanders could not build canoes for hunting porpoises and catching fish. Without the forest to absorb and slowly release water, springs and streams dried up, exposed soils eroded, crop yields plummeted, and famine struck.

The starving people turned to warfare and possibly cannibalism. Both the population and the civilization collapsed. When Dutch explorers first reached the island on Easter Day, 1722, they found only about 2,000 inhabitants, struggling under primitive conditions on a mostly barren island.

Like Easter Island at its peak, the earth is an isolated island (in the vastness of space) with no other suitable planet to migrate to. As on Easter Island, our population and resource consumption are growing.

Will the humans on Earth Island recreate the tragedy of Easter Island on a grander scale, or will we learn how to live more sustainably on this planet that is our only home? Some analysts believe that we will not run out of resources and are not living unsustainably. Other analysts disagree and warn that we are already depleting or degrading some of the earth's natural capital (Figure 1-2, p. 3, and top half of back cover) in parts of the world and need to learn how to live more sustainably over the next few decades.

Scientific knowledge is a key in evaluating these conflicting claims and learning how to live more sustainably. Thus we need to know what science is, understand the behavior of complex systems studied by scientists, and have a basic knowledge of the nature of the matter and energy that make up the earth's living and nonliving resources.

Figure 3-1 These massive stone figures on Easter Island are the remains of the technology created by an ancient civilization of Polynesians. Their civilization collapsed because the people used up the trees (especially large palm trees) that were the basis of their livelihood. More than 200 of these stone statues once stood on huge stone platforms lining the coast. At least 700 additional statues were abandoned in rock quarries or on ancient roads between the quarries and the coast. No one knows how the early islanders (with no wheels, no draft animals, and no sources of energy except their own muscles) transported these gigantic structures for miles before erecting them. We presume they accomplished it by felling large trees and using them to roll and erect the statues.

This chapter addresses the following questions:

- What is science, and what do scientists do?
- What are major components and behaviors of complex systems?
- What are the basic forms of matter? What is matter made of? What makes matter useful to us as a resource?
- What are the major forms of energy? What makes energy useful to us as a resource?
- What are physical and chemical changes? What scientific law governs changes of matter from one physical or chemical form to another?
- What three main types of nuclear changes can matter undergo?
- What are two scientific laws governing changes of energy from one form to another?
- How are the scientific laws governing changes of matter and energy from one form to another related to resource use and environmental disruption?

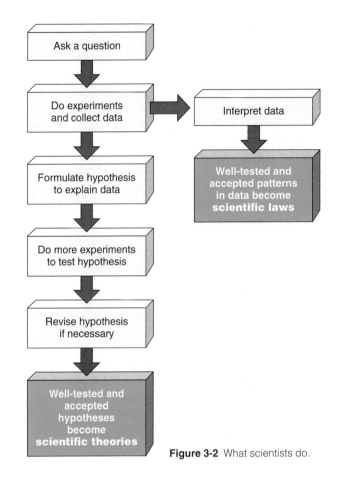

Figure 3-2 What scientists do.

3-1 THE NATURE OF SCIENCE

What Is Science and What Do Scientists Do?
Science is an attempt to discover order in nature and use that knowledge to make predictions about what is likely to happen in nature. Figure 3-2 summarizes the systematic version of the critical thinking process that scientists use.

The first thing scientists do is ask a question or identify a problem to be investigated. Then scientists working on this problem collect **scientific data,** or facts, by making observations and measurements. Repeated observations and measurements, ideally by several different investigators, must confirm the resulting scientific data or facts.

The primary goal of science is not facts themselves. Instead, it is to develop a new idea, principle, or model that connects and explains certain scientific data and leads to useful predictions about what is likely to happen in nature. Scientists working on a particular problem try to come up with a variety of possible or tentative explanations, or **scientific hypotheses,** of what they (or other scientists) observe in nature.

To be accepted, a scientific hypothesis must explain scientific data and phenomena and make predictions that can be tested by further experiments. One

method scientists use to test a hypothesis is to develop a **model,** an approximate representation or simulation of a system being studied.

If repeated experiments or tests using models support a particular hypothesis or a group of related hypotheses, it becomes a **scientific theory.** In other words, a *scientific theory* is a verified, highly reliable, and widely accepted scientific hypothesis or a related group of scientific hypotheses.

To scientists, scientific theories are not to be taken lightly. They are not guesses, speculations, or suggestions. Instead, scientific theories are useful explanations of processes or natural phenomena that have a high degree of certainty because they are supported by extensive evidence.

A scientific theory is the closest thing to the "truth" or "absolute" proof that science can provide. New evidence or a better explanation may modify, or in rare cases overturn, a particular scientific theory. But unless or until this happens a scientific theory is the best and most reliable knowledge we have about how nature works.

Nonscientists often use the word *theory* incorrectly when they mean to refer to a *scientific hypothesis,* a tentative explanation that needs further evaluation. The statement, "Oh, that's just a theory," made in everyday conversation, implies a lack of knowledge and careful testing—the opposite of the scientific meaning of the word.

Another important result of science is a **scientific, or natural, law:** a description of what we find happening in nature over and over in the same way. For example, after making thousands of observations and measurements over many decades, scientists discovered the *second law of thermodynamics*. Simply stated, this law says that heat always flows spontaneously from hot to cold—something you learned the first time you touched a hot object. *Scientific laws* describe what we find happening in nature in the same way, whereas *scientific theories* are widely accepted explanations of data and laws.

A scientific law is no better than the accuracy of the observations or measurements upon which it is based. New or more accurate data may result in a scientific law being modified, or in rare cases overturned. However, scientific laws are highly reliable and well-tested descriptions of what we find occurring in nature.

How Do Scientists Learn about Nature? We often hear about *the* scientific method. In reality, many **scientific methods** exist: they are ways scientists gather data and formulate and test scientific hypotheses, models, theories, and laws (Figure 3-2).

Here is an example of applying the scientific method to an everyday situation.

Observation: You walk into your bedroom at night and flick on the light switch. The light does not come on.

Question: Why did the light not come on?

Hypothesis: Maybe the power for the house is out.

Test the hypothesis: If the power is out, the lights in other rooms should also be out.

Experiment: To check this prediction, go to other rooms and click light switches.

Results: Lights in other rooms come on when their switches are clicked.

Conclusion: Power to whole house is not out.

New hypothesis: Maybe the light bulb is burned out.

Experiment: Replace bulb with a new bulb.

Results: Light comes on when switch is flicked.

Conclusion: Second hypothesis is verified.

Situations in nature are usually much more complicated than this. Many *variables* or *factors* influence most processes or parts of nature that scientists seek to understand. Ideally, scientists conduct a *controlled experiment* to isolate and study the effect of a single variable. To do *single-variable analysis* scientists set up two groups: an *experimental group,* in which the chosen

CONNECTIONS

What Is Harming the Robins?

Suppose a scientist observes an abnormality in the growth of robin embryos in a certain area. She knows the area has been sprayed with a pesticide and suspects the chemical may be causing the abnormalities she has observed.

To test this hypothesis, the scientist carries out a *controlled experiment.* She maintains two groups of robin embryos of the same age in the laboratory. Each group is exposed to exactly the same conditions of light, temperature, food supply, and so on, except the embryos in the experimental group are exposed to a known amount of the pesticide in question.

The embryos in both groups are then examined over an identical period of time for the abnormality. If she finds a significantly larger number of the abnormalities in the experimental group than in the control group, the results support the idea that the pesticide is the culprit.

To be sure no errors occur during the procedure, the original researcher should repeat the experiment several times. Ideally one or more other scientists should repeat the experiment on an independent basis.

Critical Thinking

Can you find flaws in this experiment that might lead you to question the scientist's conclusions? (*Hint:* What other factors in nature—not the laboratory—and in the embryos themselves could possibly explain the results?)

variable is changed in a known way, and a *control group,* in which the chosen variable is not changed. If the experiment is designed properly, any difference between the two groups should result from a variable that was changed in the experimental group (Connections, above).

A basic problem is that many of the problems environmental scientists investigate involve a huge number of interacting variables. This limitation is overcome in some cases by using *multivariable analysis.* This involves using mathematical models run on high-speed computers to analyze the interactions of many variables without having to carry out traditional controlled experiments.

Scientists use critical thinking skills (p. xv) to develop and evaluate scientific ideas. According to physicist Albert Einstein, however, "There is no completely logical way to a new scientific idea." Intuition, imagination, and creativity are as important in science as they are in poetry, art, music, and other great adventures of the human spirit.

How Valid Are the Results of Science? Scientists can do two major things. *First,* they can disprove things. *Second,* they can establish that a particular model, theory, or law has a very high probability or degree of certainty of being true. However, like scholars in any field, scientists cannot prove that their theories, models, and laws are *absolutely* true.

When people say something has or has not been "scientifically proven," they can mislead us by falsely implying that science yields absolute proof or certainty. Although it may be extremely low, some degree of uncertainty is always involved in any scientific theory, model, or law.

How Does Frontier Science Differ from Consensus Science? News reports often focus on new so-called scientific breakthroughs and disputes between scientists over the validity of preliminary (untested) data, hypotheses, and models. These preliminary results, called **frontier science,** are controversial because they have not been widely tested and accepted. At the preliminary frontier stage, it is normal and healthy for reputable scientists in a field to disagree about the meaning and accuracy of scientific data and the validity of various hypotheses.

By contrast, **consensus science** consists of data, theories, and laws that scientists who are considered experts in the field involved widely accept. This aspect of science is very reliable but is rarely considered newsworthy. One way to find out what scientists generally agree on is to seek out reports by scientific bodies such as the U.S. National Academy of Sciences and the British Royal Society that attempt to summarize consensus among experts in key areas of science.

The history of science shows that occasionally the scientific consensus about an idea can be modified or overturned by new information or better ideas. However, until such an event occurs, current scientific consensus is our most useful guideline.

3-2 MODELS AND BEHAVIOR OF SYSTEMS

What Is a System, and What Are Its Major Components? A **system** is a set of components that function and interact in some regular and theoretically predictable manner and can be isolated for the purposes of observation and study. The environment consists of a vast number of interacting systems involving living and nonliving things.

Most *systems* have the following key components: **inputs** (from the environment), **flows** or **throughputs**

within the system at certain rates, and **outputs** (to the environment).

Why Are Models of Complex Systems Useful? Over time, people have learned the value of using models as approximate representations or simulations of real systems to find out how systems work and evaluate which ideas or hypotheses work.

Some of the most powerful and useful technologies invented by humans are mathematical models, which are used to supplement our mental models. *Mathematical models* consist of one or more equations used to describe the behavior of a system and make predictions about the behavior of a system.

Making a mathematical model usually requires going many times through three steps. *First,* make a guess and write down some equations. *Second,* compute the predictions implied by the equations. *Third,* compare the predictions with observations, the predictions of mental models, existing experimental data, and scientific hypotheses, laws, and theories.

Mathematical models are important because they can give us improved perceptions and predictions, especially in situations where our mental models are weak and unreliable. Research shows that this occurs when there are many interacting variables, consequences follow actions only after long delays, consequences of actions lead to other consequences, responses vary from one time to the next, and controlled experiments (Connections, p. 48) are impossible, too slow, or too expensive to conduct.

After building and testing a mathematical model, scientists use it to predict what is *likely* to happen under a variety of conditions. In effect, they use mathematical models to answer *if–then* questions: "*If* we do such and such, *then* what is likely to happen now and in the future?"

Despite its usefulness, a mathematical model is nothing more than a set of hypotheses or assumptions about how we think a certain system works. Such models (like all other models) are no better than the assumptions built into them and the data fed into them to make projections about the behavior of complex systems.

How Do Feedback Loops Affect Systems? Systems undergo change as a result of feedback loops. A **feedback loop** occurs when an output of matter, energy, or information is fed back into the system as an input that changes the system.

A **positive feedback loop** causes a system to change further in the same direction. One example involves depositing money in a bank at compound interest and leaving it there. The interest increases the balance, which through a positive feedback loop leads to more interest and an even higher balance.

A **negative feedback loop** causes a system to change in the opposite direction. An example is recycling aluminum cans. This involves melting aluminum and feeding it back into an economic system to make new aluminum products. This negative feedback loop of matter reduces the need to find, extract, and process virgin aluminum ore and the flow of waste matter (discarded aluminum cans) into the environment.

Most systems contain one or a series of *coupled positive and negative feedback loops.* For example, the temperature-regulating system of your body involves coupled negative and positive feedback loops (Figure 3-3). Normally a negative feedback loop regulates your body temperature. However, if your body temperature exceeds 42°C (108°F), your built-in negative feedback temperature control system breaks down as your body produces more heat than your sweat-dampened skin can get rid of. Then a positive feedback loop caused by overloading the system (Figure 3-3) overwhelms the negative or corrective feedback loop. These conditions produce a net gain in body heat, which produces even more body heat, and so on, until you die from heatstroke.

The tragedy on Easter Island discussed at the beginning of the chapter also involved the coupling of positive and negative feedback loops. As the abundance of trees turned to a shortage of trees, the positive feedback loop (more births than deaths) became weaker as death rates rose, and the negative feedback loop (more deaths than births) eventually dominated and caused a dieback of the human population.

Connections: How Do Time Delays Affect Complex Systems? Complex systems often show **time delays** between the input of a stimulus and the response to it. A long time delay can mean that corrective action comes too late. For example, a smoker exposed to cancer-causing chemicals in cigarette smoke may not get lung cancer for 20 years or more.

Time delays can allow a problem to build up slowly until it reaches a *threshold level* and causes a fundamental shift in the behavior of a system. Examples in which prolonged delays dampen the negative feedback mechanisms that might slow, prevent, or halt environmental problems are population growth, leaks from toxic waste dumps, and degradation of forests from prolonged exposure to air pollutants.

Connections: What Is Synergy, and How Can It Affect Complex Systems? In arithmetic, 1 plus 1 always equals 2. However, in some of the complex systems found in nature, 1 plus 1 may add up to more than 2 because of synergistic interactions. A **synergistic interaction** occurs when two or more processes interact so the combined effect is greater than the sum of their separate effects.

Synergy can result when two people work together to accomplish a task. For example, suppose you and I need to move a 140-kilogram (300-pound) tree that has fallen across the road. By ourselves, each of us can lift only, say, 45 kilograms (100 pounds). If we cooperate and use our muscles properly, however, together we can move the tree out of the way. That is using syn-

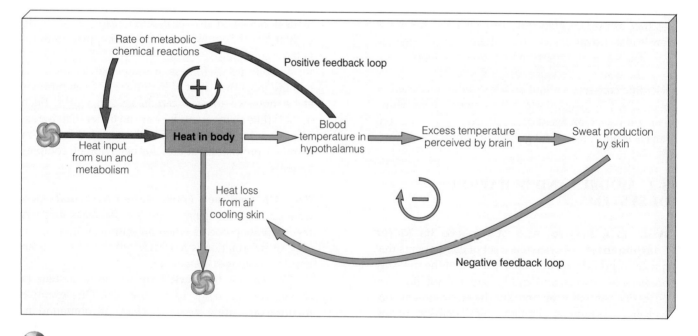

Figure 3-3 Coupled negative and positive feedback loops involved in temperature control of the human body.

ergy to solve a problem. Research in the social sciences suggests that most political changes or changes in cultural beliefs are brought about by only about 5% (and rarely more than 10%) of a population working together (synergizing) and expanding their efforts to influence other people.

3-3 MATTER: FORMS, STRUCTURE, AND QUALITY

What Are Nature's Building Blocks? **Matter** is anything that has mass (the amount of material in an object) and takes up space. Matter is found in two *chemical* forms. One is **elements**: the distinctive building blocks of matter that make up every material substance. The other consists of **compounds**: two or more different elements held together in fixed proportions by attractive forces called *chemical bonds.*

All matter is built from the 115 known chemical elements (92 of them occur naturally and the other 23 have been synthesized in laboratories). To simplify things, chemists represent each element by a one- or two-letter symbol. Examples used in this book are hydrogen (H), carbon (C), oxygen (O), nitrogen (N), phosphorus (P), sulfur (S), chlorine (Cl), fluorine (F), bromine (Br), sodium (Na), calcium (Ca), lead (Pb), mercury (Hg), arsenic (As), and uranium (U).

If you had a supermicroscope capable of looking at individual elements and compounds, you could see they are made up of three types of building blocks. The first is **atoms**: the smallest units of matter that are unique to a particular element.

The second is **ions**: electrically charged atoms or combinations of atoms. Two types of ions are discussed in this book. One is *positive* ions such as hydrogen ions (H^+), sodium ions (Na^+), calcium ions (Ca^{2+}), and ammonium ions (NH_4^+). A second type consists of *negative* ions such as chloride ions (Cl^-), nitrate ions (NO_3^-), sulfate ions (SO_4^{2-}), and phosphate ions (PO_4^{3-}).

A third building block consists of the structural units of *compounds*. Chemists use a shorthand **chemical formula** to show the number of atoms (or ions) of each type in a compound. The formula contains the symbols for each of the elements present and uses subscripts to represent the number of atoms or ions of each element in the compound's basic structural unit. Examples of compounds and their formulas encountered in this book are water (H_2O, read as "H-two-O"), oxygen (O_2), ozone (O_3), nitrogen (N_2), nitrous oxide (N_2O), nitric oxide (NO), hydrogen sulfide (H_2S), carbon monoxide (CO), carbon dioxide (CO_2), nitrogen dioxide (NO_2), sulfur dioxide (SO_2), sulfur trioxide (SO_3), ammonia (NH_3), sulfuric acid (H_2SO_4), nitric acid (HNO_3), methane (CH_4), glucose ($C_6H_{12}O_6$), sodium chloride (NaCl), and hydrogen chloride (HCl).

Table sugar, vitamins, plastics, aspirin, penicillin, and many other important materials have one thing in common: they are **organic compounds**, containing carbon atoms combined with each other and with atoms of one or more other elements such as hydrogen, oxygen, nitrogen, sulfur, phosphorus, chlorine, and fluorine. All other compounds are called **inorganic compounds.**

The millions of known organic (carbon-based) compounds include the following:

- *Hydrocarbons:* compounds of carbon and hydrogen atoms. An example is methane (CH_4), the main component of natural gas.

- *Chlorinated hydrocarbons:* compounds of carbon, hydrogen, and chlorine atoms. An example is the insecticide DDT ($C_{14}H_9Cl_5$).

- *Simple carbohydrates* (simple sugars): certain types of compounds of carbon, hydrogen, and oxygen atoms. An example is glucose ($C_6H_{12}O_6$), which most plants and animals break down in their cells to obtain energy.

Larger and more complex organic compounds, called *polymers,* consist of a number of basic structural or molecular units (*monomers*) linked by chemical bonds, somewhat like cars linked in a freight train. The three major types of organic polymers are *complex carbohydrates* consisting of two or more monomers of simple sugars (such as glucose) linked together, *proteins* formed by linking together monomers of amino acids, and *nucleic acids,* such as DNA and RNA, made by linked sequences of monomers called nucleotides.

Genes consist of specific sequences of nucleotides in a DNA molecule. Each gene carries codes (each consisting of three nucleotides) needed to make various proteins. These coded units of genetic information about specific traits are passed on from parents to offspring during reproduction.

Chromosomes are combinations of genes that make up a single DNA molecule, together with a number of proteins. Genetic information coded in your chromosomal DNA is what makes you different from an oak leaf, an alligator, or a flea and from your parents. The relationships of genetic material to cells are depicted in Figure 3-4 (p. 52).

What Are Atoms Made Of? If you increased the magnification of your supermicroscope, you would find that each different type of atom contains a certain number of *subatomic particles.* The main building blocks of an atom are positively charged **protons** (p), uncharged **neutrons** (n), and negatively charged **electrons** (e).

Each atom consists of an extremely small center, or **nucleus,** containing protons and neutrons, and one or more electrons in rapid motion somewhere outside the

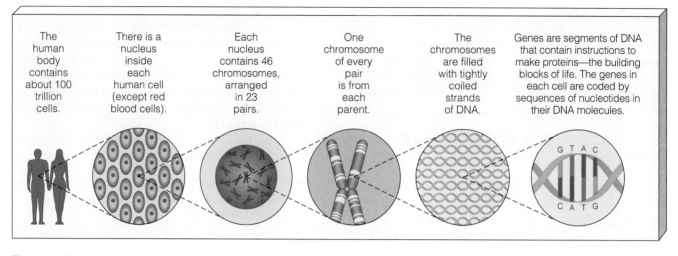

| The human body contains about 100 trillion cells. | There is a nucleus inside each human cell (except red blood cells). | Each nucleus contains 46 chromosomes, arranged in 23 pairs. | One chromosome of every pair is from each parent. | The chromosomes are filled with tightly coiled strands of DNA. | Genes are segments of DNA that contain instructions to make proteins—the building blocks of life. The genes in each cell are coded by sequences of nucleotides in their DNA molecules. |

Figure 3-4 Relationships among cells, nuclei, chromosomes, DNA, and genes.

nucleus. Atoms are incredibly small. For example, more than 3 million hydrogen atoms could sit side by side on the period at the end of this sentence.

Each atom has an equal number of positively charged protons (inside its nucleus) and negatively charged electrons (outside its nucleus). Because these electrical charges cancel one another, *the atom as a whole has no net electrical charge.*

Each element has its own specific **atomic number,** equal to the number of protons in the nucleus of each of its atoms. The simplest element, hydrogen (H), has only 1 proton in its nucleus, so its atomic number is 1. Carbon (C), with 6 protons, has an atomic number of 6, whereas uranium (U), a much larger atom, has 92 protons and an atomic number of 92.

Because atoms are electrically neutral, the atomic number of an atom tells us the number of positively charged protons in its nucleus and the equal number of negatively charged electrons outside its nucleus. For example, an atom of uranium with an atomic number of 92 has 92 protons in its nucleus and 92 electrons outside, and thus no net electrical charge.

Because electrons have so little mass compared with the mass of a proton or a neutron, *most of an atom's mass is concentrated in its nucleus.* The mass of an atom is described in terms of its **mass number:** the total number of neutrons and protons in its nucleus. For example, a hydrogen atom with 1 proton and no neutrons in its nucleus has a mass number of 1, and an atom of uranium with 92 protons and 143 neutrons in its nucleus has a mass number of 235 (92 + 143 = 235).

All atoms of an element have the same number of protons in their nuclei.

However, they may have different numbers of uncharged neutrons in their nuclei, and thus may have different mass numbers. Various forms of an element having the same atomic number but a different mass number are called **isotopes** of that element. Scientists identify isotopes by attaching their mass numbers to the name or symbol of the element. For example, hydrogen has three isotopes: hydrogen-1 (H-1), hydrogen-2 (H-2, common name *deuterium*), and hydrogen-3 (H-3, common name *tritium*). A natural sample of an element contains a mixture of its isotopes in a fixed proportion or percentage abundance by weight (Figure 3-5).

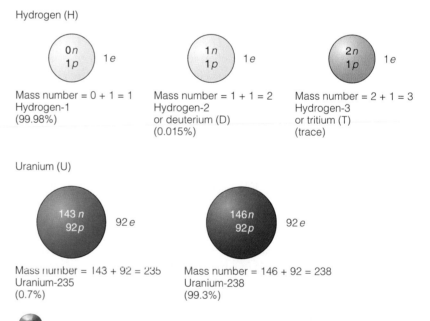

Hydrogen (H)

Mass number = 0 + 1 = 1
Hydrogen-1
(99.98%)

Mass number = 1 + 1 = 2
Hydrogen-2
or deuterium (D)
(0.015%)

Mass number = 2 + 1 = 3
Hydrogen-3
or tritium (T)
(trace)

Uranium (U)

Mass number = 143 + 92 = 235
Uranium-235
(0.7%)

Mass number = 146 + 92 = 238
Uranium-238
(99.3%)

Figure 3-5 Isotopes of hydrogen and uranium. All isotopes of hydrogen have an atomic number of 1 because each has one proton in its nucleus; similarly, all uranium isotopes have an atomic number of 92. However, each isotope of these elements has a different mass number because its nucleus contains a different number of neutrons. Figures in parentheses indicate the percentage abundance by weight of each isotope in a natural sample of the element.

What Are Matter Quality and Material Efficiency? **Matter quality** is a measure of how useful a form of matter is to us as a resource, based on its availability and concentration. **High-quality matter** is concentrated, usually is found near the earth's surface, and has great potential for use as a matter resource. **Low-quality matter** is dilute, often is deep underground or dispersed in the ocean or the atmosphere, and usually has little potential for use as a matter resource (Figure 3-6).

An aluminum can is a more concentrated, higher-quality form of aluminum than aluminum ore containing the same amount of aluminum. That is why it takes less energy, water, and money to recycle an aluminum can than to make a new can from aluminum ore.

Material efficiency, or **resource productivity,** is the total amount of material needed to produce each unit of goods or services. Although resource productivity has been improving, only about 2–6% of the matter resources flowing through the economies of developed countries ends up providing useful goods and services.

Here is some *good news*. Because of such waste, resource productivity in developed countries could be improved by 75–90% within two decades using existing technologies, say business expert Paul Hawken (Guest Essay, p. 4) and physicist Amory Lovins.

3-4 ENERGY: FORMS AND QUALITY

What Is Energy and How Is It Transmitted? **Energy** is the capacity to do work and transfer heat. Work is performed when an object—be it a grain of sand, this book, or a giant boulder—is moved over some distance. Work, or matter movement, also is needed to boil water or burn natural gas to heat a house or cook food. Energy is also the heat that flows automatically from a hot object to a cold object when they come in contact. **Radiation** is the transmission of energy through space as particles or waves.

What Is the Difference between Kinetic Energy and Potential Energy? There are two major types of energy. One is **kinetic energy,** which matter has because of its mass and its speed or velocity. Examples of this energy in motion are wind (a moving mass of air), flowing streams, heat flowing from a body at a high temperature to one at a lower temperature, and electricity (flowing electrons).

The second type is **potential energy,** which is stored and potentially available for use. Examples of this stored energy are a rock held in your hand, an unlit stick of dynamite, still water behind a dam, the chemical energy stored in gasoline molecules, and the nuclear energy stored in the nuclei of atoms.

Potential energy can be changed to kinetic energy. When you drop a rock, its potential energy changes into kinetic energy. When you burn gasoline in a car engine, the potential energy stored in the chemical bonds of its molecules changes into heat, light, and mechanical (kinetic) energy that propels the car.

What Is Electromagnetic Radiation? Another type of energy is **electromagnetic radiation:** energy radiated in the form of a *wave* as a result of the changing electric and magnetic fields. There are many different forms of electromagnetic radiation, each with a different *wavelength* (distance between successive peaks or troughs in the wave) and *energy content* (Figure 3-7, p. 54). Such radiation travels through space at the speed of light, which is about 300,000 kilometers per second (186,000 miles per second).

Cosmic rays, gamma rays, X rays, and ultraviolet radiation (Figure 3-7, left side) are called **ionizing radiation** because they have enough energy to knock

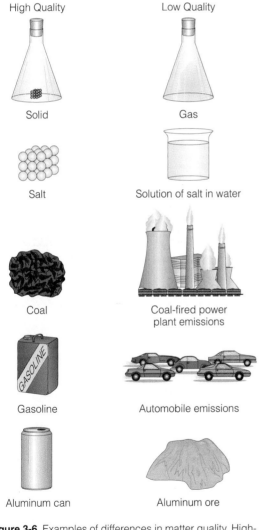

Figure 3-6 Examples of differences in matter quality. High-quality matter (left-hand column) is fairly easy to extract and concentrated; low-quality matter (right-hand column) is more difficult to extract and more dispersed than high-quality matter.

High Quality — Solid — Salt — Coal — Gasoline — Aluminum can

Low Quality — Gas — Solution of salt in water — Coal-fired power plant emissions — Automobile emissions — Aluminum ore

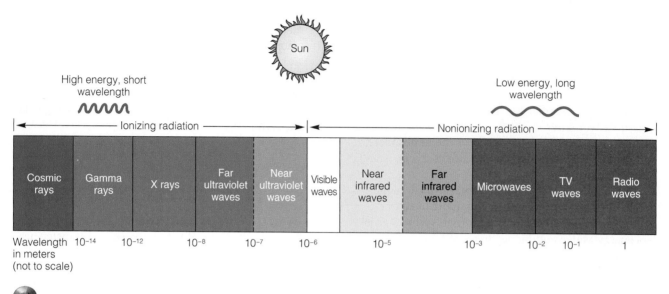

Figure 3-7 The *electromagnetic spectrum:* the range of electromagnetic waves, which differ in wavelength (distance between successive peaks or troughs) and energy content.

electrons from atoms and change them to positively charged ions. The resulting highly reactive electrons and ions can disrupt living cells, interfere with body processes, and cause many types of sickness, including various cancers. The other forms of electromagnetic radiation (Figure 3-7, right side) do not contain enough energy to form ions and are called **nonionizing radiation.**

The visible light that we can detect with our eyes is a form of nonionizing radiation that occupies only a small portion of the full range or spectrum of different types of electromagnetic radiation (Figure 3-7). Figure 3-8 shows that visible light makes up most of the spectrum of electromagnetic radiation emitted by the sun.

What Is Heat and How Is It Transferred? Heat is the total kinetic energy of all the moving atoms, ions, or molecules within a given substance, excluding the overall motion of the whole object. **Temperature** is the average speed of motion of the atoms, ions, or molecules in a sample of matter at a given moment.

Heat can be transferred from one place to another by three different methods (Figure 3-9).

What Is Energy Quality? **Energy quality** is a measure of an energy source's ability to do useful work (Figure 3-10). **High-quality energy** is concentrated and can perform much useful work. Examples are electricity, the chemical energy stored in coal and gasoline, concentrated sunlight, and nuclei of uranium-235 used as fuel in nuclear power plants.

By contrast, **low-quality energy** is dispersed and has little ability to do useful work. An example is heat dispersed in the moving molecules of a large amount

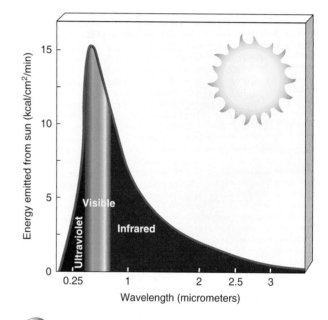

Figure 3-8 The spectrum of electromagnetic radiation released by the sun consists mostly of visible light.

of matter (such as the atmosphere or a large body of water) so that its temperature is low. For example, the total amount of heat stored in the Atlantic Ocean is greater than the amount of high-quality chemical energy stored in all the oil deposits of Saudi Arabia. Yet the ocean's heat is so widely dispersed, it cannot be used to move things or to heat things to high temperatures. It makes sense to match the quality of an energy source with the quality of energy needed to perform a particular task (Figure 3-10) because doing so saves energy and usually money (unless government subsidies or taxes have distorted the energy marketplace).

Convection

Heating water in the bottom of a pan causes some of the water to vaporize into bubbles. Because they are lighter than the surrounding water, they rise. Water then sinks from the top to replace the rising bubbles.This up and down movement (convection) eventually heats all of the water.

Conduction

Heat from a stove burner causes atoms or molecules in the pan's bottom to vibrate faster. The vibrating atoms or molecules then collide with nearby atoms or molecules, causing them to vibrate faster. Eventually, molecules or atoms in the pan's handle are vibrating so fast it becomes too hot to touch.

Radiation

As the water boils, heat from the hot stove burner and pan radiate into the surrounding air, even though air conducts very little heat.

Figure 3-9 Three ways in which heat can be transferred from one place to another.

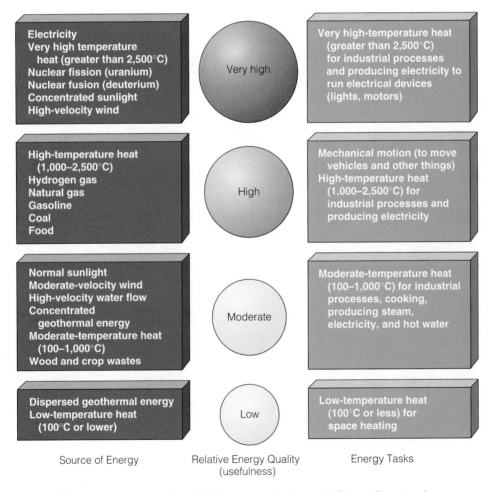

Source of Energy | Relative Energy Quality (usefulness) | Energy Tasks

Figure 3-10 Categories of the quality of different sources of energy. *High-quality energy* is concentrated and has great ability to perform useful work. *Low-quality energy* is dispersed and has little ability to do useful work. To avoid unnecessary energy waste, it is best to match the quality of an energy source with the quality of energy needed to perform a task.

3-5 PHYSICAL AND CHEMICAL CHANGES AND THE LAW OF CONSERVATION OF MATTER

What Is the Difference between a Physical and a Chemical Change? A **physical change** involves no change in chemical composition. Cutting a piece of aluminum foil into small pieces is one example. Changing a substance from one physical state to another is a second example. When solid water (ice) is melted or liquid water is boiled, none of the H_2O molecules involved are altered; instead, the molecules are organized in different spatial (physical) patterns.

In a **chemical change,** or **chemical reaction,** the chemical compositions of the elements or compounds are altered. Chemists use shorthand chemical equations to represent what happens in a chemical reaction. For example, when coal burns completely, the solid carbon (C) it contains combines with oxygen gas (O_2) from the atmosphere to form the gaseous compound carbon dioxide (CO_2):

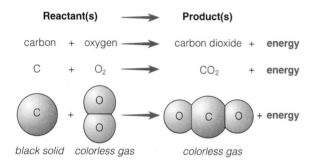

Energy is given off in this reaction, making coal a useful fuel. The reaction also shows how the complete burning of coal (or any of the carbon-containing compounds in wood, natural gas, oil, and gasoline) gives off carbon dioxide gas, which is a key gas that can lead to warming of the lower atmosphere (troposphere).

The Law of Conservation of Matter: Why Is There No "Away"? *We may change various elements and compounds from one physical or chemical form to another, but in no physical and chemical change can we create or destroy any of the atoms involved.* All we can do is rearrange them into different spatial patterns (physical changes) or different combinations (chemical changes). The italicized statement, based on many thousands of measurements, is known as the **law of conservation of matter.** In describing chemical reactions, chemists use a shorthand system to account for all the atoms, since none are created or destroyed.

The law of conservation of matter means there is no "away" in "to throw away." *Everything we think we have thrown away is still here with us in some form.* We can

collect dust and soot from the smokestacks of industrial plants, but these solid wastes must then be put somewhere. We can remove substances from polluted water at a sewage treatment plant. But then we must *burn them* (producing some air pollution), *bury them* (possibly contaminating underground water supplies), or *clean them up* and apply the gooey sludge to the land as fertilizer (dangerous if the sludge contains nondegradable toxic metals such as lead and mercury). Banning use of the pesticide DDT in the United States but still selling it abroad means it can return to the United States as DDT residues in imported coffee, fruit, and other foods, or fallout from air masses moved long distances by winds.

How Do Chemists Keep Track of Atoms? In keeping with the law of conservation of matter, each side of a chemical equation must have the same number of atoms of each element involved. Doing this leads to what chemists call a *balanced chemical equation.* The equation for the burning of carbon (C + O_2 $\longrightarrow$ CO_2) is balanced because one atom of carbon and two atoms of oxygen are on both sides of the equation.

Consider the following chemical reaction: When electricity is passed through water (H_2O), the latter can be broken down into hydrogen (H_2) and oxygen (O_2), as represented by the following equation:

$$H_2O \longrightarrow H_2 + O_2$$
2 H atoms   2 H atoms   2 O atoms
1 O atom

This equation is unbalanced because one atom of oxygen is on the left but two atoms are on the right.

We cannot change the subscripts of any of the formulas to balance this equation because that would change the arrangements of the atoms so that different substances are involved. Instead, we could use different numbers of the *molecules* involved to balance the equation. For example, we could use two water molecules:

$$2\,H_2O \longrightarrow H_2 + O_2$$
4 H atoms   2 H atoms   2 O atoms
2 O atoms

This equation is still unbalanced because although the numbers of oxygen atoms on both sides are now equal, the numbers of hydrogen atoms are not.

We can correct this by having the reaction produce two hydrogen molecules:

$$2\,H_2O \longrightarrow 2\,H_2 + O_2$$
4 H atoms   4 H atoms   2 O atoms
2 O atoms

Now the equation is balanced, and the law of conservation of matter has been observed. We see that for every two molecules of water through which we pass electricity, two hydrogen molecules and one oxygen molecule are produced.

Try to balance the chemical equation for the reaction of nitrogen gas (N_2) with hydrogen gas (H_2) to form ammonia gas (NH_3).

How Harmful Are Pollutants? We can make the environment cleaner and convert some potentially harmful chemicals into less harmful physical or chemical forms. However, *the law of conservation of matter means we will always face the problem of what to do with some quantity of wastes and pollutants.*

Thus, regardless of what we do, we will always have certain pollutants that can cause harm to humans or other forms of life. Three factors determining the severity of a pollutant's harmful effects are its chemical nature, its concentration, and its persistence.

The second of these factors, *concentration,* is sometimes expressed in *parts per million (ppm);* 1 ppm corresponds to 1 part pollutant per 1 million parts of the gas, liquid, or solid mixture in which the pollutant is found. Smaller concentration units are parts per billion (ppb) and parts per trillion (ppt). We can reduce the concentration of a pollutant by dumping it into the air or a large volume of water, but there are limits to the effectiveness of this dilution approach.

The third factor, **persistence,** is how long the pollutant stays in the air, water, soil, or body. Pollutants can be classified into three categories based on their persistence. **Degradable,** or **nonpersistent, pollutants,** are broken down completely or reduced to acceptable levels by natural physical, chemical, and biological processes. Complex chemical pollutants that living organisms (usually specialized bacteria) break down (metabolize) into simpler chemicals are called **biodegradable pollutants.** Human sewage in a river, for example, is biodegraded fairly quickly by bacteria if the sewage is not added faster than it can be broken down.

Slowly degradable, or **persistent, pollutants** take decades or longer to degrade. Examples include the insecticide DDT and most plastics.

Nondegradable pollutants cannot be broken down by natural processes. Examples include the toxic elements lead, mercury, and arsenic.

3-6 NUCLEAR CHANGES

What Is Natural Radioactivity? In addition to physical and chemical changes, matter can undergo a third type of change known as a **nuclear change.** This occurs when nuclei of certain isotopes spontaneously change or are made to change into one or more different isotopes. Three types of nuclear change are natural radioactive decay, nuclear fission, and nuclear fusion.

Natural radioactive decay is a nuclear change in which unstable isotopes spontaneously emit fast-moving chunks of matter (called particles), high-energy radiation, or both at a fixed rate. The unstable isotopes are called **radioactive isotopes** or **radioisotopes.** Radioactive decay into various isotopes continues until the original isotope is changed into a stable isotope that is not radioactive.

Radiation emitted by radioisotopes is damaging ionizing radiation. The most common form of ionizing energy released from radioisotopes is **gamma rays,** a form of high-energy electromagnetic radiation (Figure 3-7). High-speed ionizing particles emitted from the nuclei of radioactive isotopes are most commonly of two types: **alpha particles** (fast-moving, positively charged chunks of matter that consist of two protons and two neutrons) and **beta particles** (high-speed electrons).

Each type of radioisotope spontaneously decays at a characteristic rate into a different isotope. This rate of decay can be expressed in terms of **half-life:** the time needed for *one-half* of the nuclei in a radioisotope to decay and emit their radiation to form a different isotope. The decay continues, often producing a series of different radioisotopes, until a nonradioactive isotope is formed. Each radioisotope has a characteristic half-life, which may range from a few millionths of a second to several billion years (Table 3-1).

An isotope's half-life cannot be changed by temperature, pressure, chemical reactions, or any other known factor. Half-life can be used to estimate how long a sample of a radioisotope must be stored in a

Table 3-1 Half-Lives of Selected Radioisotopes		
Isotope	**Radiation Half-Life**	**Emitted**
Potassium-42	12.4 hours	Alpha, beta
Iodine-131	8 days	Beta, gamma
Cobalt-60	5.27 years	Beta, gamma
Hydrogen-3 (tritium)	12.5 years	Beta
Strontium-90	28 years	Beta
Carbon-14	5,370 years	Beta
Plutonium-239	24,000 years	Alpha, gamma
Uranium-235	710 million years	Alpha, gamma
Uranium-238	4.5 billion years	Alpha, gamma

safe container before it decays to what is considered a safe level. A general rule is that such decay takes about 10 half-lives. Thus people must be protected from radioactive waste containing iodine-131 (which concentrates in the thyroid gland and has a half-life of 8 days) for 80 days (10 × 8 days). Plutonium-239, which has a half-life of 24,000 years and is produced in nuclear reactors and used as the explosive in some nuclear weapons, can cause lung cancer when its particles are inhaled in minute amounts. Thus it must be stored safely for 240,000 years (10 × 24,000 years)—about four times longer than the latest version of our species has existed.

What Is Nuclear Fission? Splitting Nuclei **Nuclear fission** is a nuclear change in which nuclei of certain isotopes with large mass numbers (such as uranium-235) are split apart into lighter nuclei when struck by neutrons; each fission releases two or three more neutrons and energy (Figure 3-11). Each of these neutrons, in turn, can cause an additional fission. For these multiple fissions to take place, enough fissionable nuclei must be present to provide the **critical mass** needed for efficient capture of these neutrons.

Multiple fissions within a critical mass form a **chain reaction,** which releases an enormous amount of energy (Figure 3-12). Living cells can be damaged by the ionizing radiation released by the radioactive lighter nuclei and by high-speed neutrons produced by nuclear fission.

In an atomic bomb, an enormous amount of energy is released in a fraction of a second in an uncontrolled nuclear fission chain reaction. This reaction is initiated by an explosive charge, which suddenly pushes two masses of fissionable fuel together and causes the fuel to reach the critical mass needed for a chain reaction.

In the reactor of a nuclear power plant, the rate at which the nuclear fission chain reaction takes place is controlled so that under normal operation only one of every two or three neutrons released is used to split

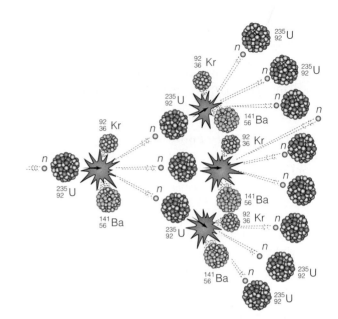

Figure 3-12 A *nuclear chain reaction* initiated by one neutron triggering fission in a single uranium-235 nucleus. This figure illustrates only a few of the trillions of fissions caused when a single uranium-235 nucleus is split within a critical mass of uranium-235 nuclei. The elements krypton (Kr) and barium (Ba), shown here as fission fragments, are only two of many possibilities.

another nucleus. In conventional nuclear fission reactors, the splitting of uranium-235 nuclei releases heat, which produces high-pressure steam to spin turbines and thus generate electricity.

What Is Nuclear Fusion? Forcing Nuclei to Combine **Nuclear fusion** is a nuclear change in which two isotopes of light elements, such as hydrogen, are forced together at extremely high temperatures until they fuse to form a heavier nucleus, releasing energy in the process. Temperatures of at least 100 million °C are needed to force the positively charged nuclei (which strongly repel one another) to fuse.

Nuclear fusion is much more difficult to initiate than nuclear fission, but once started it releases far more energy per unit of fuel than does fission. Fusion of hydrogen nuclei to form helium nuclei is the source of energy in the sun and other stars.

After World War II, the principle of *uncontrolled nuclear fusion* was used to develop extremely powerful hydrogen, or thermonuclear, weapons. These weapons use the D–T fusion reaction, in which a hydrogen-2, or deuterium (D), nucleus and a hydrogen-3 (tritium, T) nucleus are fused to form a larger, helium-4 nucleus, a neutron, and energy (Figure 3-13).

Scientists have also tried to develop *controlled nuclear fusion,* in which the D–T reaction is used to produce heat that can be converted into electricity. Despite more than 50 years of research, this process is still

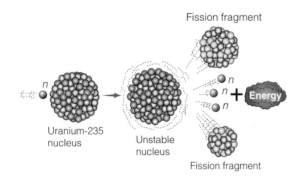

Figure 3-11 Fission of a uranium-235 nucleus by a neutron (*n*).

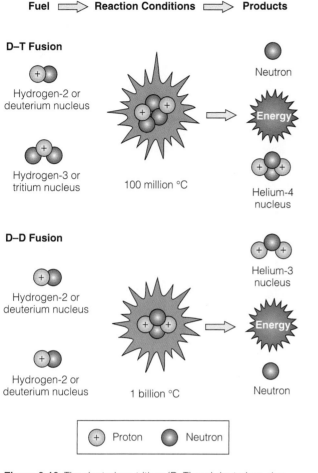

Fuel ⟹ Reaction Conditions ⟹ Products

D–T Fusion

Hydrogen-2 or
deuterium nucleus

Hydrogen-3 or
tritium nucleus

100 million °C

Neutron

Energy

Helium-4
nucleus

D–D Fusion

Hydrogen-2 or
deuterium nucleus

Hydrogen-2 or
deuterium nucleus

1 billion °C

Helium-3
nucleus

Energy

Neutron

| + Proton | ● Neutron |

Figure 3-13 The deuterium–tritium (D–T) and deuterium–deuterium (D–D) nuclear fusion reactions, which take place at extremely high temperatures.

in the laboratory stage. Even if it becomes technologically and economically feasible, many energy experts do not expect it to be a practical source of energy until 2030, if then.

3-7 TWO LAWS GOVERNING ENERGY CHANGES

What Is the First Law of Thermodynamics? You Cannot Get Something for Nothing Scientists have observed energy being changed from one form to another in millions of physical and chemical changes, but they have never been able to detect the creation or destruction of any energy (except in nuclear changes). The results of their experiments have been summarized in the **law of conservation of energy,** also known as the **first law of thermodynamics:** *In all physical and chemical changes, energy is neither created nor destroyed, but it may be converted from one form to another.*

This scientific law tells us that when one form of energy is converted to another form in any physical or chemical change, *energy input always equals energy output.* No matter how hard we try or how clever we are, we cannot get more energy out of a system than we put in; in other words, *we cannot get something for nothing in terms of energy quantity.*

What Is the Second Law of Thermodynamics? You Cannot Even Break Even Because the first law of thermodynamics states that energy can be neither created nor destroyed, we may be tempted to think we will always have enough energy. Yet if we fill a car's tank with gasoline and drive around or use a flashlight battery until it is dead, something has been lost. If it is not energy, what is it? The answer is *energy quality* (Figure 3-10), the amount of energy available that can perform useful work.

Countless experiments have shown that when energy is changed from one form to another, a decrease in energy quality always occurs. The results of these experiments have been summarized in what is called the **second law of thermodynamics:** *When energy is changed from one form to another, some of the useful energy is always degraded to lower quality, more dispersed, less useful energy.* This degraded energy usually takes the form of heat given off at a low temperature to the surroundings (environment). There it is dispersed by the random motion of air or water molecules and becomes even more disorderly and less useful.

Basically, this law says that in any energy conversion, we always end up with *less* usable energy than we started with. So not only can we not get something for nothing in terms of energy quantity, *we cannot even break even in terms of energy quality because energy always goes from a more useful to a less useful form when energy is changed from one form to another.* No one has ever found a violation of this fundamental scientific law (see quote at the end of this chapter).

Here are three examples of the second law of thermodynamics in action. *First,* when a car is driven, only about 10% of the high-quality chemical energy available in its gasoline fuel is converted into mechanical energy (to propel the vehicle) and electrical energy (to run its electrical systems). The remaining 90% is degraded to low-quality heat that is released into the environment and eventually lost into space.

Second, when electrical energy flows through filament wires in an incandescent light bulb, it is changed into about 5% useful light and 95% low-quality heat that flows into the environment. In other words, this so-called *light bulb* is really a *heat bulb.*

Third, in living systems, solar energy is converted into chemical energy (food molecules) and then into mechanical energy (moving, thinking, and living). During each of these conversions, high-quality energy is degraded and flows into the environment as low-quality heat (Figure 3-14, p. 60).

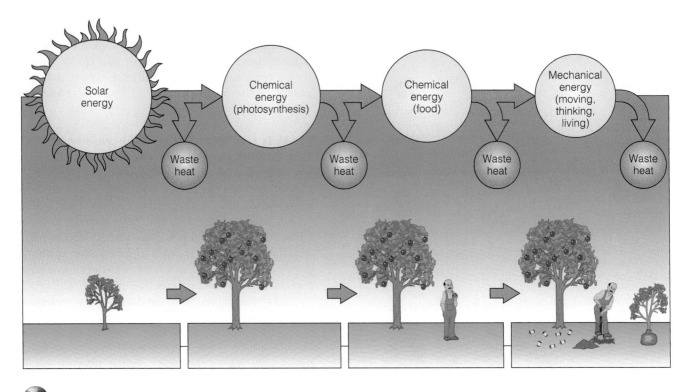

Figure 3-14 The second law of thermodynamics in action in living systems. Each time energy is changed from one form to another, some of the initial input of high-quality energy is degraded, usually to low-quality heat that disperses into the environment.

The second law of thermodynamics also means that *we can never recycle or reuse high-quality energy to perform useful work.* Once the concentrated energy in a serving of food, a liter of gasoline, a lump of coal, or a chunk of uranium is released, it is degraded to low-quality heat that is dispersed into the environment. We can heat air or water at a low temperature and upgrade it to high-quality energy, but the second law of thermodynamics tells us that it will take more high-quality energy to do this than we get in return.

Energy efficiency, or **energy productivity,** is a measure of how much useful work is accomplished by a particular input of energy into a system. Some *good news* is that there is plenty of room for improving energy efficiency. For example, scientists estimate that only about 16% of the energy used in the United States ends up performing useful work. The remaining 84% is either unavoidably wasted because of the second law of thermodynamics (41%) or unnecessarily wasted (43%).

Connections: How Does the Second Law of Thermodynamics Affect Life? To form and maintain the highly ordered arrangement of molecules and the organized biochemical processes in your body, you must continually get and use high-quality matter and energy resources from your surroundings. As you use these resources, you add low-quality heat and low-quality waste matter to your surroundings. Your body continuously gives off heat roughly equal to that of a 100-watt incandescent light bulb, which is why a closed room full of people gets warm. You also continuously break down solid large molecules (such as glucose) into smaller molecules of carbon dioxide gas and water vapor, which are dispersed in the atmosphere.

Planting, growing, processing, and cooking food all use high-quality energy and matter resources that add low-quality heat and waste materials to the environment. In addition, enormous amounts of low-quality heat and waste matter are added to the environment when concentrated deposits of minerals and fuels are extracted from the earth's crust, processed, and used.

3-8 CONNECTIONS: MATTER AND ENERGY CHANGE LAWS AND ENVIRONMENTAL PROBLEMS

What Is a High-Throughput Economy? As a result of the law of conservation of matter and the second law of thermodynamics, individual resource use automatically adds some waste heat and waste matter to the environment. Most of today's advanced industrialized countries have **high-throughput (high-waste) economies** that attempt to sustain ever-increasing economic growth by increasing the one-way flow of matter and energy resources through their economic systems (Figure 3-15). These resources flow through their economies into planetary *sinks* (air, water, soil, organisms), where pollutants and wastes end up and can accumulate to harmful levels.

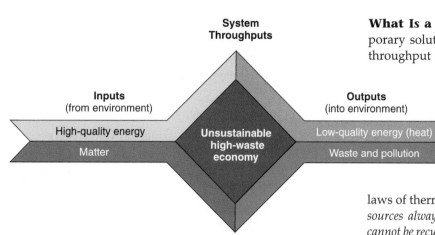

Figure 3-15 The *high-throughput economies* of most developed countries are based on maximizing the rates of energy and matter flow. This process rapidly converts high-quality matter and energy resources into waste, pollution, and low-quality heat.

What happens if more and more people continue to use and waste more and more energy and matter resources at an increasing rate? The law of conservation of matter and the two laws of thermodynamics discussed in this chapter tell us that eventually this will exceed the capacity of the environment to dilute and degrade waste matter and absorb waste heat. However, they do not tell us how close we are to reaching such limits.

What Is a Matter-Recycling Economy? A temporary solution to this problem is to convert a high-throughput economy to a **matter-recycling economy.** The goal of such a conversion is to allow economic growth to continue without depleting matter resources or producing excessive pollution and environmental degradation.

Even though recycling matter saves energy, the two laws of thermodynamics tell us that *recycling matter resources always requires using high-quality energy (which cannot be recycled) and adds waste heat to the environment.*

Changing to a matter-recycling economy is an important way to buy some time. However, it does not allow more and more people to use more and more resources indefinitely, even if all of them were somehow perfectly recycled.

What Is a Low-Throughput Economy? Learning from Nature The three scientific laws governing matter and energy changes suggest that the best long-term solution to our environmental and resource problems is to shift from an economy based on maximizing matter and energy flow (throughput) to a more sustainable **low-throughput (low-waste) economy,** as summarized in Figure 3-16.

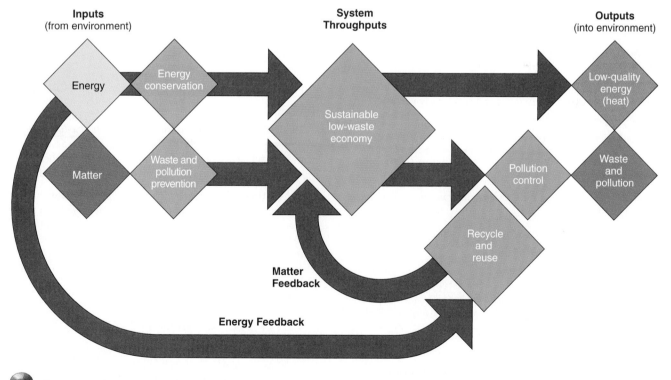

Figure 3-16 Solutions: lessons from nature. A *low-throughput economy,* based on energy flow and matter recycling, works with nature to reduce the throughput of matter and energy resources (items shown in green). This is done by reusing and recycling most nonrenewable matter resources, using renewable resources no faster than they are replenished, using matter and energy resources efficiently, reducing unnecessary consumption, emphasizing pollution prevention and waste reduction, and controlling population growth.

The next four chapters apply the three basic scientific laws of matter and thermodynamics to living systems and look at some *biological principles* that can teach us how to live more sustainably by learning from and working with nature.

The second law of thermodynamics holds, I think, the supreme position among laws of nature. . . . If your theory is found to be against the second law of thermodynamics, I can give you no hope.

ARTHUR S. EDDINGTON

REVIEW QUESTIONS

1. Define the boldfaced terms in this chapter.

2. Describe what happened to the people on Easter Island and how it may relate to the current situation on the earth.

3. Define *science* and explain how it works. Distinguish among *scientific data, scientific hypothesis, scientific model, scientific theory,* and *scientific law.* Explain why we should take a scientific theory seriously.

4. What is a *controlled experiment?* What is *multivariable analysis?*

5. What does "scientifically proven" mean? If scientists cannot establish absolute proof, what do they establish?

6. Distinguish between *frontier science* and *consensus science.*

7. What is a *mathematical model,* and how is such a model made? Why are mathematical models important?

8. What is a *system?* Distinguish among the *inputs, flows* or *throughputs,* and *outputs* of a system.

9. What is a *feedback loop?* Distinguish between a *positive feedback loop* and a *negative feedback loop,* and give an example of each.

10. Define and give an example of a *time delay* in a system.

11. Define *synergy,* and give an example of how it can change a system.

12. Distinguish among *matter, elements, compounds.* What is a *chemical formula?*

13. Distinguish between *atoms* and *ions,* and give an example of each.

14. Distinguish between *organic compounds* and *inorganic compounds,* and give an example of each type. Distinguish among *hydrocarbons, chlorinated hydrocarbons, simple carbohydrates, polymers, complex carbohydrates, proteins, nucleic acids,* and *nucleotides.*

15. Distinguish between *genes* and *chromosomes.*

16. What three major types of subatomic particles are found in atoms? Which two of these particles are found in the nucleus, and which is found outside the nucleus?

17. Distinguish between *atomic number* and *mass number.* What is an *isotope* of an atom?

18. Distinguish between *high-quality matter* and *low-quality matter,* and give an example of each. What is *material efficiency?*

19. What is *energy?* What is *radiation?*

20. Distinguish between *kinetic energy* and *potential energy,* and give an example of each.

21. What is *electromagnetic radiation?* List three types of electromagnetic radiation.

22. Distinguish between *heat* and *temperature.* Explain how *convection, conduction,* and *radiation* can transmit heat.

23. Distinguish between *high-quality energy* and *low-quality energy,* and give an example of each. What is *energy efficiency?*

24. Distinguish between a *physical change* and a *chemical change,* and give an example of each.

25. What is the *law of conservation of matter?* Explain why there is no "away" as a repository for pollution. What is a *balanced chemical equation,* and how is it related to the law of conservation of matter?

26. What three factors determine the harm that a pollutant causes? What is the *persistence* of a pollutant? Distinguish among *degradable (nonpersistent), biodegradable, slowly degradable (persistent),* and *nondegradable pollutants,* and give an example of each type.

27. What is a *nuclear change?* Distinguish among *natural radioactive decay, radioisotopes, gamma rays, alpha particles,* and *beta particles.* What is the *half-life* of a radioactive isotope? For how many half-lives should radioactive material be stored safely before it decays to an acceptable level of radioactivity?

28. Distinguish between *nuclear fission* and *nuclear fusion.* Distinguish between *critical mass* and a *nuclear chain reaction.*

29. Distinguish between the *first law of thermodynamics* and the *second law of thermodynamics,* and give an example of each law in action. Use the second law of thermodynamics to explain why energy cannot be recycled.

30. Distinguish among a *high-throughput (high-waste) economy,* a *low-throughput (low-waste) economy,* and a *matter-recycling economy.* Use the law of conservation of matter and the first and second laws of thermodynamics to explain the need to shift from a high-throughput economy to a matter-recycling economy and eventually to a low-throughput economy.

CRITICAL THINKING

1. Respond to the following statements:
   a. Scientists have not absolutely proven that anyone has ever died from smoking cigarettes.
   b. The greenhouse theory—that certain gases (such as water vapor and carbon dioxide) warm the atmosphere—is not a reliable idea because it is only a scientific theory.

2. See whether you can find an advertisement or an article describing some aspect of science in which **(a)** the concept of scientific proof is misused, **(b)** the term *theory* is used when it should have been *hypothesis*, and **(c)** a consensus scientific finding is dismissed or downplayed because it is "only a theory."

3. How does a scientific law (such as the law of conservation of matter) differ from a societal law (such as one imposing maximum speed limits for vehicles)? Can each be broken?

4. Explain why we do not really consume anything and why we can never really throw matter away.

5. A tree grows and increases its mass. Explain why this is not a violation of the law of conservation of matter.

6. Balancing equations is based on the law of conservation of matter. Do you believe this is an ironclad law of nature or one that through new scientific discoveries could be overthrown? Explain.

7. Imagine you have the power to revoke the law of conservation of matter. List the three major ways this would affect your life.

8. If there is no "away," why isn't the world filled with waste matter?

9. Methane (CH_4) gas is the major component of natural gas. Write and balance the chemical equation for the burning of methane when it combines with oxygen gas in the atmosphere to form carbon dioxide and water. Use this equation to explain why burning natural gas can contribute to projected warming of the atmosphere (global warming).

10. Someone wants you to invest money in an automobile engine that will produce more energy than the energy in the fuel (such as gasoline or electricity) you use to run the motor. What is your response? Explain.

11. Use the second law of thermodynamics to explain why a barrel of oil can be used only once as a fuel.

12. **(a)** Imagine you have the power to violate the law of conservation of energy (the first law of thermodynamics) for 1 day. What are the three most important things you would do with this power? **(b)** Repeat this process, imagining you have the power to violate the second law of thermodynamics for 1 day.

PROJECTS

1. If you have the use of a sensitive balance, try to demonstrate the law of conservation of mass in a physical change. Weigh a container with a lid (a glass jar will do), add an ice cube and weigh it again, and then allow the ice to melt and weigh it again.

2. Use the library or Internet to find examples of various perpetual motion machines and inventions that allegedly violate the two laws of thermodynamics by producing more high-quality energy than the high-quality energy needed to make them run. What has happened to these schemes and machines (many of them developed by scam artists to attract money from investors)?

3. Make a concept map of this chapter's major ideas using the section heads and subheads and the key terms (in boldface). Look on the website for this book for information about making concept maps.

INTERNET STUDY RESOURCES AND RESOURCES FOR FURTHER READING AND RESEARCH

The website for this book contains helpful study aids and many ideas for further reading and research. Log on to

http://biology.brookscole.com/miller10

and click on the Chapter-by-Chapter area. Choose Chapter 3 and select a resource:

- Flash Cards allows you to test your mastery of the Terms and Concepts to Remember for this chapter.

- Tutorial Quizzes provides a multiple-choice practice quiz.

- Student Guide to InfoTrac will lead you to Critical Thinking Projects that use InfoTrac College Edition as a research tool.

- References lists the major books and articles consulted in writing this chapter.

- Hypercontents takes you to an extensive list of sites with news, research, and images related to individual sections of the chapter.

INFOTRAC COLLEGE EDITION

Improve your skills with InfoTrac College Edition, a searchable online database of articles from more than 700 periodicals. Log on to

http://www.infotrac-college.com

or access InfoTrac through the website for this book. Try to find the following articles:

1. Scully, M. 2001. Quantum mechanics boosts engine efficiency. *Inside R&D* (February 1). *Keywords:* "quantum mechanics" and "efficiency." This article explores issues in frontier science by examining the theory of coupling the energy of atomic structure with the energy of atomic motion in the internal combustion engine. This type of system could increase the efficiency of current energy use.

2. Beals, G. 2001. Pining for a breakthrough: Despite years of ostracism, a small and dwindling army of cold-fusion faithful are ever hopeful. *Newsweek International* (October 15): 57. *Keywords:* "cold fusion" and "breakthrough." Twelve years after the first announcement of "cold fusion," and the subsequent lack of repeatability, people are still searching for this elusive "unlimited energy source." However, life on the frontier of cold fusion is starting to get lonelier.

4 ECOSYSTEMS: COMPONENTS, ENERGY FLOW, AND MATTER CYCLING

Have You Thanked the Insects Today?

Insects have a bad reputation. We classify many insect species as *pests* because they compete with us for food, spread human diseases (such as malaria), and invade our lawns, gardens, and houses. Some people have "bugitis," fear all insects, and think the only good bug is a dead bug. However, this view fails to recognize the vital roles insects play in helping sustain life on earth.

A large proportion of the earth's plant species (including many trees) depend on insects to pollinate their flowers (Figure 4-1, right). In turn, we and other land-dwelling animals depend on plants for food, either by eating them or by consuming animals that eat them. Without pollinating insects, very few fruits and vegetables would be available for us and plant-eating animals to eat.

Insects such as the praying mantis (Figure 4-1, left), which eat other insects, help control the populations of at least half the species of insects we call pests. This free pest control service is an important part of the earth's natural capital that helps sustain us.

Suppose all insects disappeared today. Within a year most of the earth's animals would become extinct because of the disappearance of so much plant life. The earth would be covered with rotting vegetation and animal carcasses being decomposed by unimaginably huge hordes of bacteria and fungi.

Fortunately, this is not a realistic scenario because insects, which have been around for at least 400 million years, are phenomenally successful forms of life. They were the first animals to invade the land and, later, the air. Today they are by far the planet's most diverse, abundant, and successful animals.

Insects can rapidly evolve new genetic traits, such as resistance to pesticides. They also have an exceptional ability to evolve into new species when faced with new environmental conditions and are extremely resistant to extinction.

For example, we can apply chemical pesticides to protect crops from pests and help grow more food. However, the pesticides can also contaminate the air and water far from where they are applied, harm beneficial insects that help protect crops from other insects, and threaten the health of wildlife and people. Widespread use of chemical pesticides can also accelerate the natural ability of rapidly reproducing insect pests to develop genetic resistance (immunity) to such chemicals. Thus in the long run, chemical pesticides can backfire and become less effective in reducing crop losses from insect pests.

The environmental lesson is that although insects can thrive without newcomers such as us, we and most other land organisms would perish quickly without them. Learning about the roles insects play in nature requires us to understand how insects and other organisms living in a biological *community* (such as a forest or pond) interact with one another and with the nonliving environment. *Ecology* is the science that studies such relationships and interactions in nature, as I discuss in this chapter and the four chapters that follow.

Figure 4-1 Insects play important roles in helping sustain life on earth. The bright green caterpillar moth feeding on pollen in a crocus (right) and other insects pollinate flowering plants that serve as food for many plant eaters. The praying mantis eating a monarch butterfly (left) and many other insect species help control the populations of at least half of the insect species we classify as pests.

The earth's thin film of living matter is sustained by grand-scale cycles of energy and chemical elements.

G. Evelyn Hutchinson

This chapter addresses the following questions:

- What is ecology?

- What basic processes keep us and other organisms alive?

- What are the major components of an ecosystem?

- What happens to energy in an ecosystem?

- What happens to matter in an ecosystem?

- How do scientists study ecosystems?

- What are ecosystem services, and how do they affect the sustainability of the earth's life-support systems? What are two principles of sustainability?

4-1 THE NATURE OF ECOLOGY

What Is Ecology? Ecology (from the Greek words *oikos*, "house" or "place to live," and *logos*, "study of") is the study of how organisms interact with one an-

other and with their nonliving environment. In effect, it is a study of *connections in nature*.

What Are Organisms? Ecologists focus on trying to understand the interactions among organisms, populations, communities, ecosystems, and the biosphere (Figure 4-2, p. 66).

An **organism** is any form of life. The **cell** is the basic unit of life in organisms. Organisms may consist of a single cell (bacteria, for instance) or many cells.

On the basis of their cell structure, organisms can be classified as either *eukaryotic* or *prokaryotic*. Each cell of a **eukaryotic** organism is surrounded by a membrane, has a distinct *nucleus* (a membrane-bounded structure containing genetic material in the form of DNA), and has several other internal parts called *organelles* (Figure 4-3a, p. 67). All organisms except bacteria are eukaryotic.

A membrane surrounds the cell of a **prokaryotic** organism. But inside the cell there is no distinct nucleus, and a membrane does not enclose other internal parts (Figure 4-3b, p. 67).

All bacteria are single-celled prokaryotic organisms. Although most familiar organisms are eukaryotic, they could not exist without hordes of prokaryotic organisms (called microbes; Connections, below). Most

---

Microbes: The Invisible Rulers of the Earth

CONNECTIONS

They are everywhere and there are trillions of them. Billions are found inside your body, on your body, in a handful of soil, and in a cup of river water.

These mostly invisible rulers of the earth are *microbes*, a catchall term for many thousands of species of bacteria, protozoa, fungi, and yeasts, most of which are too small for us to see with the naked eye.

Most microbes do not get the respect they deserve. Most of us think of them as threats to our health in the form of infectious bacteria or "germs," fungi that cause athlete's foot and other skin diseases, and protozoa that cause killer diseases such as malaria. However, these potentially harmful microbes are in the minority.

Most of the earth's hordes of microbes not only are harmless but

also make the rest of life possible. Some of them play a vital role in producing foods such as bread, cheese, yogurt, vinegar, tofu, soy sauce, beer, and wine. Others provide us with food by converting nitrogen gas in the atmosphere into forms that plants can take up from the soil as nutrients.

Bacteria and fungi in the soil decompose organic wastes into nutrients that can be taken up by plants. Bacteria in your intestinal tract break down the food you eat. Some microbes in your nose prevent harmful bacteria from reaching your lungs. Other microbes have been the source of disease-fighting antibiotics, including penicillin, erythromycin, and streptomycin.

Another vital ecological service that some microbes provide is controlling some plant diseases and populations of insect species that attack food crops. Enlisting some of these microbes for pest control can

reduce the use of potentially harmful chemical pesticides.

In addition, genetic engineers are developing microbes that can extract metals from ores, break down various pollutants, and help clean up toxic waste sites.

Harvard biologist Edward O. Wilson, who has developed many important ecological theories and is one of the world's experts on ants, says that if he were starting over he would study microbes.

Critical Thinking

1. A bumper sticker reads, "Have You Thanked Microbes Today?" Give reasons for doing so, and explain why microbes are the real rulers of the earth.

2. What are some potentially harmful effects of **(a)** using genetic engineering to design microbes to break down oil and toxic chemicals and **(b)** overusing antibacterial soaps, sprays, and antibiotics?

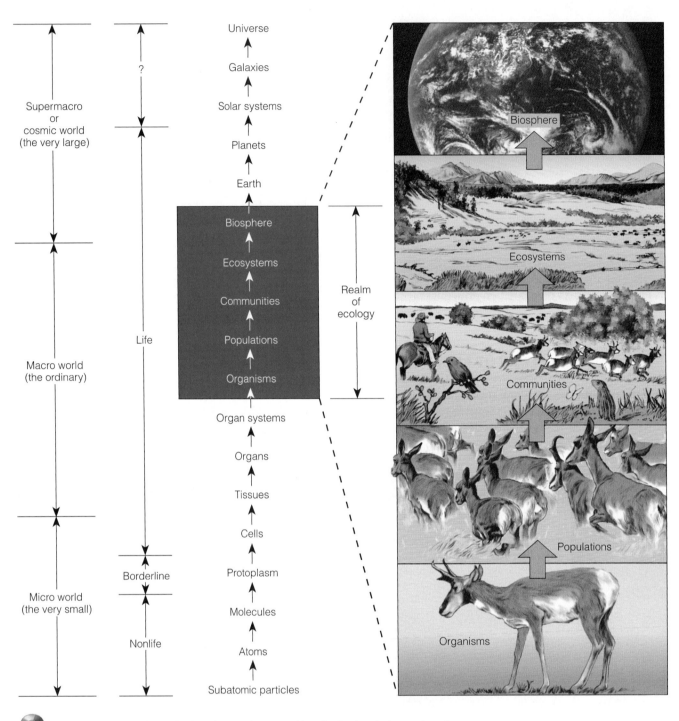

Figure 4-2 Levels of organization of matter in nature. Note the five levels that ecology focuses on.

microbes are *microorganisms,* so small they can be seen only with the aid of a microscope.

What Are Species? Organisms can be classified into **species,** or groups of organisms that resemble one another in appearance, behavior, chemistry, and genetic makeup. Species differ in how they produce offspring. **Asexual reproduction** is common in species such as bacteria with only one cell, which divides to produce two identical cells that are clones or replicas of the original cell.

Sexual reproduction occurs in organisms that produce offspring by combining sex cells or gametes (such as ovum and sperm) from both parents. This produces offspring that have combinations of genetic traits from each parent. Sexual reproduction usually gives the species a greater chance of survival under changing environmental conditions than the genetic clones produced by asexual reproduction. Organ-

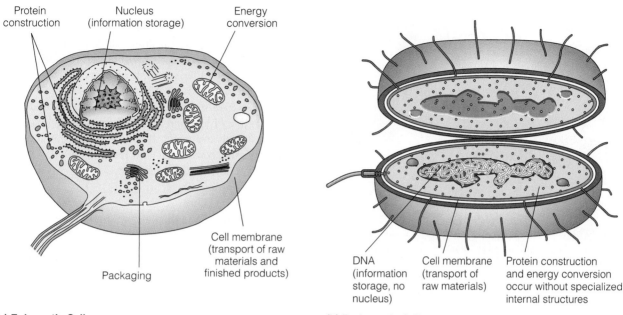

Protein construction   Nucleus (information storage)   Energy conversion

Cell membrane (transport of raw materials and finished products)

Packaging

(a) Eukaryotic Cell

DNA (information storage, no nucleus)   Cell membrane (transport of raw materials)   Protein construction and energy conversion occur without specialized internal structures

(b) Prokaryotic Cell

Figure 4-3 **(a)** Generalized structure of a eukaryotic cell. The parts and internal structure of cells in various types of organisms such as plants and animals differ somewhat from this generalized model. **(b)** Generalized structure of a prokaryotic cell. Note that prokaryotic cells lack a distinct nucleus. (Adapted from Cecie Starr and Ralph Taggart, *Biology: The Unity and Diversity of Life,* 8th ed., Belmont, Calif.: Wadsworth, 1998)

isms that reproduce sexually are classified as members of the same species if, under natural conditions, they can actually or potentially breed with one another and produce live, fertile offspring.

We do not know how many species exist on the earth. Estimates range from 3.6 million to 100 million. Most are insects and microorganisms too small to be seen with the naked eye (Connections, p. 65). Excluding hordes of bacterial species, a best guess of the number of species is about 10–14 million.

So far biologists have identified and named about 1.5–1.8 million species, not including bacteria. Biologists know a fair amount about roughly one-third of the known species but understand the detailed roles and interactions of only a few.

What Is a Population?
A **population** consists of a group of interacting individuals of the same species

that occupy a specific area at the same time (Figure 4-4). Examples are all sunfish in a pond, white oak trees in a forest, and people in a country. In most natural populations, individuals vary slightly in their genetic makeup, which is why they do not all look or behave exactly alike—a phenomenon called **genetic diversity** (Figure 4-5, p. 68). In response to changes in environmental conditions, populations change in *size, age distribution* (number of individuals in each age group), *density* (number of individuals per unit of space), and *genetic composition.*

The place where a population (or an individual organism) normally lives is its **habitat.** It may be as large as an ocean or prairie or as small as the underside of a rotting log or the intestine of a termite.

Figure 4-4 A population of monarch butterflies. The geographic distribution of this butterfly coincides with that of the milkweed plant, on which monarch larvae and caterpillars feed.

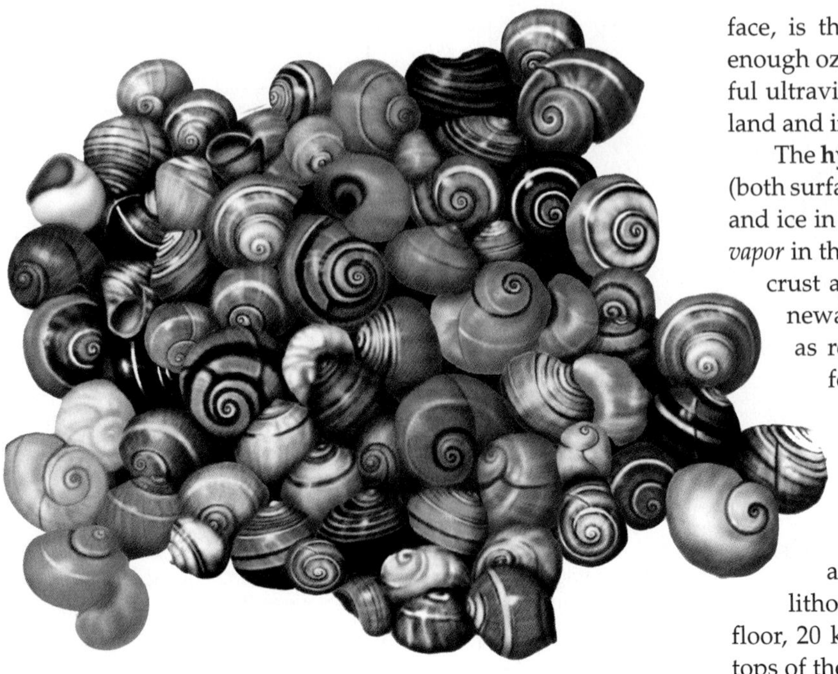

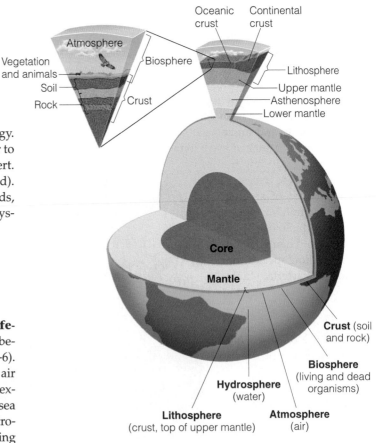

Figure 4-5 The *genetic diversity* among individuals of one species of Caribbean snail is reflected in the variations in shell color and banding patterns.

face, is the **stratosphere.** Its lower portion contains enough ozone (O_3) to filter out most of the sun's harmful ultraviolet radiation, thus allowing life to exist on land and in the surface layers of bodies of water.

The **hydrosphere** consists of the earth's *liquid water* (both surface and underground), *ice* (polar ice, icebergs, and ice in frozen soil layers, or permafrost), and *water vapor* in the atmosphere. The **lithosphere** is the earth's crust and upper mantle; the crust contains nonrenewable fossil fuels and minerals we use as well as renewable soil chemicals (nutrients) needed for plant life.

The **biosphere** is the portion of the earth in which living (biotic) organisms exist and interact with one another and with their nonliving (abiotic) environment. The biosphere includes most of the hydrosphere and parts of the lower atmosphere and upper lithosphere. It reaches from the deepest ocean floor, 20 kilometers (12 miles) below sea level, to the tops of the highest mountains. If the earth were an apple, the biosphere would be no thicker than the apple's skin. *The goal of ecology is to understand the interactions in this thin, life-supporting global skin or membrane of air, water, soil, and organisms.*

What Are Communities, Ecosystems, and the Biosphere? Populations of the different species occupying a particular place make up a **community,** or **biological community.** It is a complex interacting network of plants, animals, and microorganisms.

An **ecosystem** is a community of different species interacting with one another and with their nonliving environment of matter and energy. Ecosystems can range in size from a puddle of water to a stream, a patch of woods, an entire forest, or a desert. Ecosystems can be natural or artificial (human created). Examples of human-created ecosystems are cropfields, farm ponds, and reservoirs. All of the earth's ecosystems together make up what we call the *biosphere.*

4-2 CONNECTIONS: THE EARTH'S LIFE-SUPPORT SYSTEMS

What Are the Major Parts of the Earth's Life-Support Systems? We can think of the earth as being made up of several spherical layers (Figure 4-6). The **atmosphere** is a thin envelope or membrane of air around the planet. Its inner layer, the **troposphere,** extends only about 17 kilometers (11 miles) above sea level but contains most of the planet's air, mostly nitrogen (78%) and oxygen (21%). The next layer, stretching 17–48 kilometers (11–30 miles) above the earth's sur-

Figure 4-6 The general structure of the earth.

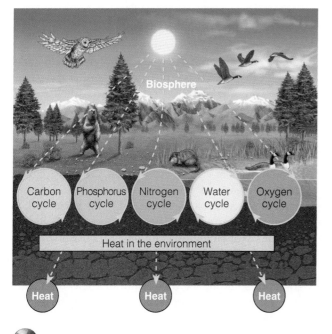

Carbon cycle Phosphorus cycle Nitrogen cycle Water cycle Oxygen cycle

Heat in the environment

Heat   Heat   Heat

Figure 4-7 Life on the earth depends on the *one-way flow of energy* (dashed lines) from the sun through the biosphere, the *cycling of crucial elements* (solid lines around circles), and *gravity*, which keeps atmospheric gases from escaping into space and draws chemicals downward in the matter cycles. This simplified model depicts only a few of the many cycling elements.

What Sustains Life on Earth? Life on the earth depends on three interconnected factors (Figure 4-7):

■ The *one-way flow of high-quality energy* (Figure 3-10, p. 55) from the sun, through materials and living things in their feeding interactions, into the environment as low-quality energy (mostly heat dispersed into air or water molecules at a low temperature), and eventually back into space as heat.

■ The *cycling of matter* (the atoms, ions, or molecules needed for survival by living organisms) through parts of the biosphere. The earth is closed to significant inputs of matter from space. Thus essentially all the nutrients used by organisms are already present on earth and must be recycled again and again for life to continue.

■ *Gravity*, which allows the planet to hold onto its atmosphere and causes the downward movement of chemicals in the matter cycles.

How Does the Sun Help Sustain Life on Earth? The sun is a middle-aged star whose energy supports life on the earth. It does this by lighting and warming the planet, supporting *photosynthesis* (the process used by green plants and some

bacteria to make compounds such as carbohydrates that keep them alive and feed most other organisms), powering the cycling of matter, and driving the climate and weather systems that distribute heat and fresh water over the earth's surface.

What Happens to Solar Energy Reaching the Earth? Because the earth is a tiny sphere in the vastness of space, it receives only about one-billionth of the sun's output of energy. Much of this energy is either reflected away or absorbed by chemicals in its atmosphere (Figure 4-8).

Most of what reaches the atmosphere is *visible light* (Figure 3-8, p. 54), *infrared radiation,* and the small amount of *ultraviolet radiation* that is not absorbed by ozone in the stratosphere (Figure 4-8). This incoming energy warms the troposphere and land, evaporates water and cycles it through the biosphere, and generates winds. Green plants, algae, and bacteria capture a tiny fraction of this incoming solar energy and use it to fuel photosynthesis and make the organic compounds that most forms of life need to survive.

Most unreflected solar radiation is degraded into infrared radiation (which we experience as heat) as it interacts with the earth. Greenhouse gases (such as water vapor, carbon dioxide, methane, nitrous oxide, and ozone) in the atmosphere reduce this flow of heat back into space. This helps warm the earth's lower

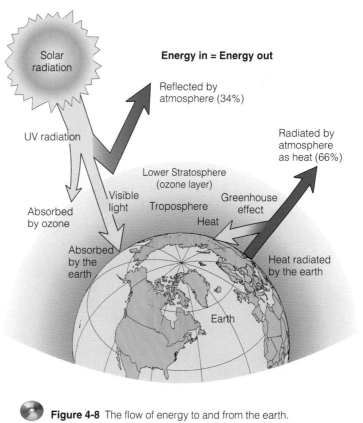

Solar radiation

Energy in = Energy out

Reflected by atmosphere (34%)

UV radiation

Radiated by atmosphere as heat (66%)

Lower Stratosphere (ozone layer)

Visible light   Troposphere   Greenhouse effect

Absorbed by ozone

Heat

Absorbed by the earth

Heat radiated by the earth

Earth

Figure 4-8 The flow of energy to and from the earth.

atmosphere and surface. Without this **natural green-house effect,** the earth would be too cold for life as we know it to exist.

4-3 ECOSYSTEM CONCEPTS AND COMPONENTS

What Are Biomes and Aquatic Life Systems? Viewed from outer space, the earth resembles an enormous jigsaw puzzle consisting of large masses of land and vast expanses of ocean. Biologists have classified the terrestrial (land) portion of the biosphere into **biomes** ("BY-ohms"). They are large regions such as forests, deserts, and grasslands characterized by a distinct climate and specific life-forms (especially vegetation) adapted to it (Figure 4-9).

Climate—long-term patterns of weather—is the main factor determining what type of life, especially what plants, will thrive in a given land area. Each biome consists of a patchwork of many different ecosystems whose communities have adapted to differences in climate, soil, and other factors throughout the biome.

Marine and freshwater portions of the biosphere can be divided into **aquatic life zones,** each containing numerous ecosystems. Aquatic life zones are the aquatic equivalent of biomes. Examples include *fresh-water life zones* (such as lakes and streams) and *ocean* or *marine life zones* (such as estuaries, coastlines, coral reefs, and the deep ocean).

What Are the Two Major Components of Ecosystems? Two types of components are found in the biosphere and its ecosystems. One type, called **abiotic,** consists of nonliving chemical and physical components such as water, air, nutrients in the soil or water, and solar energy. The other type, called **biotic,** is made up of biological components consisting of living and dead plants, animals, and microorganisms.

Figures 4-10 and 4-11 are greatly simplified diagrams of some of the biotic and abiotic components in a freshwater aquatic ecosystem and a terrestrial ecosystem.

What Are the Major Physical and Chemical Components of Ecosystems? The nonliving, or

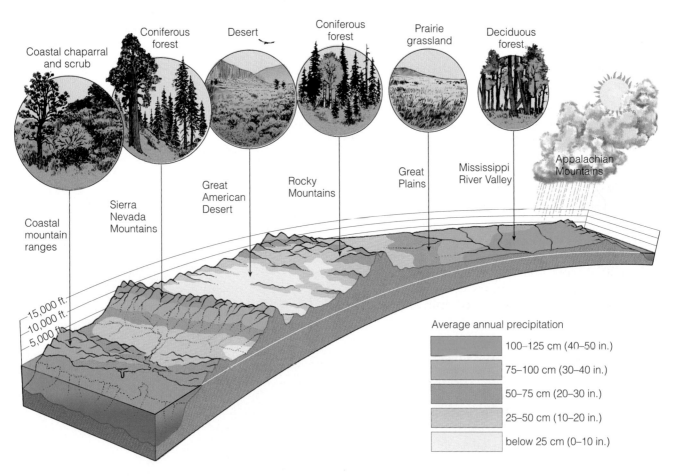

Figure 4-9 Major biomes found along the 39th parallel across the United States. The differences reflect changes in climate, mainly differences in average annual precipitation and temperature (not shown).

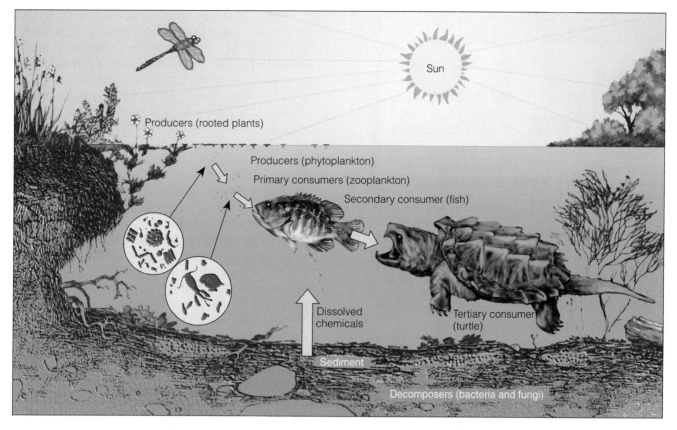

Figure 4-10 Major components of a freshwater ecosystem.

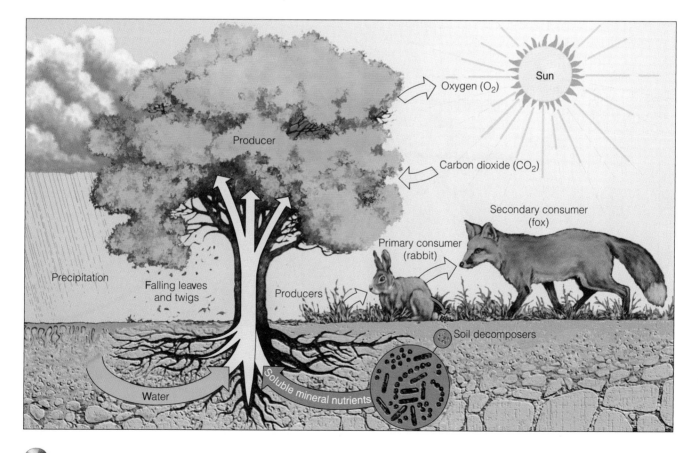

Figure 4-11 Major components of an ecosystem in a field.

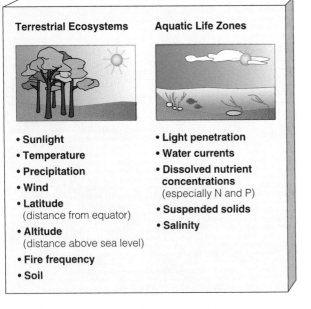

Terrestrial Ecosystems

Aquatic Life Zones

- Sunlight
- Temperature
- Precipitation
- Wind
- Latitude
  (distance from equator)
- Altitude
  (distance above sea level)
- Fire frequency
- Soil

- Light penetration
- Water currents
- Dissolved nutrient
  concentrations
  (especially N and P)
- Suspended solids
- Salinity

Figure 4-12 Key physical and chemical or abiotic factors affecting terrestrial ecosystems (left) and aquatic life zones (right).

abiotic, components of an ecosystem are the physical and chemical factors that influence living organisms in land (terrestrial) ecosystems and aquatic life zones (Figure 4-12).

Different species thrive under different physical conditions. Some need bright sunlight, and others thrive better in shade. Some need a hot environment and others a cool or cold one. Some do best under wet conditions and others under dry conditions.

Each population in an ecosystem has a **range of tolerance** to variations in its physical and chemical environment (Figure 4-13). Individuals within a population may also have slightly different tolerance ranges for temperature or other factors because of small differences in genetic makeup, health, and age. Thus although a trout population may do best within a narrow band of temperatures (*optimum level* or *range*), a few individuals can survive above and below that band. As Figure 4-13 shows, tolerance has its limits, beyond which none of the trout can survive.

These observations are summarized in the **law of tolerance:** *The existence, abundance, and distribution of a species in an ecosystem are determined by whether the levels of one or more physical or chemical factors fall within the range tolerated by that species. A species may have a wide range of tolerance to some factors and a narrow range of tolerance to others. Most organisms are least tolerant during juvenile or reproductive stages of their life cycles. Highly tolerant species can live in a variety of habitats with widely different conditions.*

A variety of factors can affect the number of organisms in a population. However, sometimes one fac-

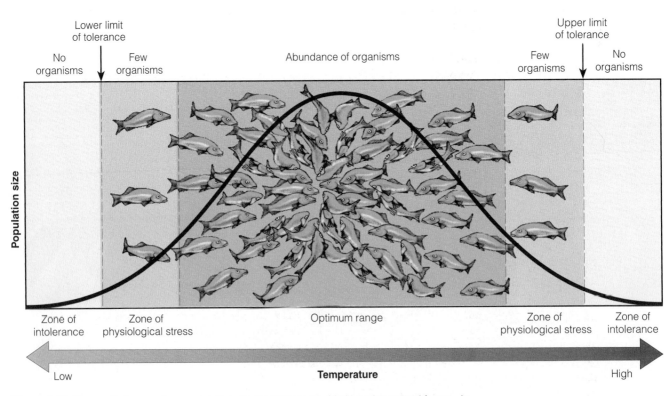

Figure 4-13 Range of tolerance for a population of organisms to an abiotic environmental factor—in this case, temperature.

tor, known as a **limiting factor,** is more important in regulating population growth than other factors. This ecological principle, related to the law of tolerance, is called the **limiting factor principle:** *Too much or too little of any abiotic factor can limit or prevent growth of a population, even if all other factors are at or near the optimum range of tolerance.*

On land, precipitation often is the limiting factor. Lack of water in a desert limits plant growth. Soil nutrients also can act as a limiting factor on land. Suppose a farmer plants corn in phosphorus-poor soil. Even if water, nitrogen, potassium, and other nutrients are at optimum levels, the corn will stop growing when it uses up the available phosphorus.

Too much of an abiotic factor can also be limiting. For example, too much water or too much fertilizer can kill plants, a common mistake of many beginning gardeners.

Important limiting factors for aquatic ecosystems include temperature, sunlight, **dissolved oxygen (DO) content** (the amount of oxygen gas dissolved in a given volume of water at a particular temperature and pressure), and nutrient availability. Another limiting factor in aquatic ecosystems is **salinity** (the amounts of various inorganic minerals or salts dissolved in a given volume of water).

What Are the Major Biological Components of Ecosystems?

Living organisms in ecosystems usually are classified as *producers* or *consumers* based on how they get food. **Producers,** sometimes called **autotrophs** (self-feeders), make their own food from compounds that are obtained from their environment. All other organisms are *consumers,* which depend directly or indirectly on food provided by producers. Thus *producers are the source of all food in an ecosystem.*

On land, most producers are green plants. In freshwater and marine ecosystems, algae and plants are the major producers near shorelines. In open water, the dominant producers are *phytoplankton* (most of them microscopic) that float or drift in the water.

Most producers capture sunlight to make carbohydrates (such as glucose, $C_6H_{12}O_6$) by **photosynthesis.** Although hundreds of chemical changes take place during photosynthesis, the overall reaction can be summarized as follows:

$$\text{carbon dioxide} + \text{water} + \textbf{solar energy} \longrightarrow \text{glucose} + \text{oxygen}$$
$$6\,CO_2 + 6\,H_2O + \textbf{solar energy} \longrightarrow C_6H_{12}O_6 + 6\,O_2$$

A few producers, mostly specialized bacteria, can convert simple compounds from their environment into more complex nutrient compounds without sunlight, a process called **chemosynthesis.**

All other organisms in an ecosystem are **consumers,** or **heterotrophs** ("other feeders"), which get their energy and nutrients by feeding on other organisms or their remains. **Decomposers** (mostly certain types of bacteria and fungi) are specialized consumers that recycle organic matter in ecosystems. They do this by breaking down (*biodegrading*) dead organic material to get nutrients and releasing the resulting simpler inorganic compounds into the soil and water, where they can be taken up as nutrients by producers.

Figure 4-14 (p. 74) shows the feeding relationships among producers, consumers feeding on live organisms, consumers feeding on dead organisms or **detritus** ("di-TRI-tus") consisting of parts of dead organisms and cast-off fragments and wastes of living organisms, and decomposers. Detritus feeders and decomposers are also called *detritivores* (Figure 4-15, p. 75). Some organisms are **omnivores** that can feed on both plants and animals. Examples are pigs, rats, foxes, bears, cockroaches, and humans.

Producers, consumers, and decomposers use the chemical energy stored in glucose and other organic compounds to fuel their life processes. In most cells, this energy is released by **aerobic respiration,** which uses oxygen to convert organic nutrients back into carbon dioxide and water. The net effect of the hundreds of steps in this complex process is represented by the following reaction:

$$\text{glucose} + \text{oxygen} \longrightarrow \text{carbon dioxide} + \text{water} + \textbf{energy}$$
$$C_6H_{12}O_6 + 6\,O_2 \longrightarrow 6\,CO_2 + 6\,H_2O + \textbf{energy}$$

Although the detailed steps differ, the net chemical change for aerobic respiration is the opposite of that for photosynthesis.

Some decomposers get the energy they need by breaking down glucose (or other organic compounds) in the absence of oxygen. This form of cellular respiration is called **anaerobic respiration,** or **fermentation.** Instead of carbon dioxide and water, the end products of this process are compounds such as methane gas (CH_4, the main component of natural gas), ethyl alcohol (C_2H_6O), acetic acid ($C_2H_4O_2$, the key component of vinegar), and hydrogen sulfide (H_2S, when sulfur compounds are broken down).

The survival of any individual organism depends on the *flow of matter and energy* through its body. However, an ecosystem as a whole survives primarily through a combination of *matter recycling* (rather than one-way flow) and *one-way energy flow* (Figure 4-16, p. 75).

Decomposers complete the cycle of matter by breaking down detritus into inorganic nutrients that are used by producers. Without decomposers, the entire world would be knee-deep in plant litter, dead animal bodies, animal wastes, and garbage, and most life as we know it would no longer exist.

What Is Biodiversity and Why Is It Important? One important renewable resource is **biological diversity,** or **biodiversity:** the different forms of life and life-sustaining processes that can best survive the variety of conditions currently found on the earth. Kinds of biodiversity include the following:

- **Genetic diversity** (variety in the genetic makeup among individuals within a species; Figure 4-5)
- **Species diversity** (variety among the species or distinct types of living organisms found in different habitats of the planet
- **Ecological diversity** (variety of forests, deserts, grasslands, streams, lakes, oceans, coral reefs, wetlands, and other biological communities; Figure 4-9)
- **Functional diversity** (biological and chemical processes or functions such as energy flow and matter cycling needed for the survival of species and biological communities; Figure 4-16)

This rich variety of genes, species, biological communities, and life-sustaining biological and chemical processes gives us food, wood, fibers, energy, raw materials, industrial chemicals, and medicines—all of which pour hundreds of billions of dollars into the world economy each year. The earth's biodiversity also provides us with free recycling, purification, and natural pest control services.

Loss of biodiversity reduces the availability of ecosystem services (Figure 1-2, p. 3, and top half of back cover) and decreases the ability of species, communities, and ecosystems to adapt to changing environmental conditions. Thus biodiversity is nature's insurance policy against disasters.

Some people also include *human cultural diversity* as part of the earth's biodiversity. The variety of human cultures represents numerous social and technological solutions to changing environmental conditions.

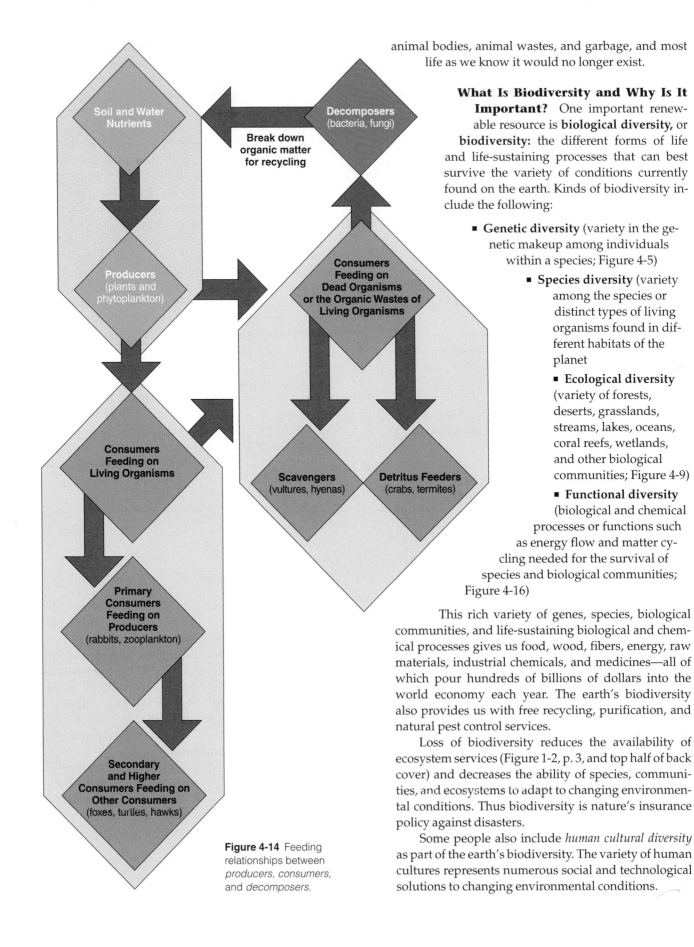

Figure 4-14 Feeding relationships between *producers, consumers,* and *decomposers.*

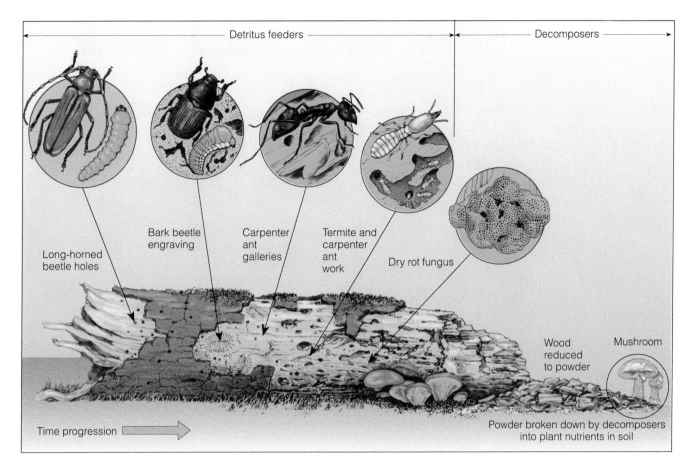

Long-horned
beetle holes

Bark beetle
engraving

Carpenter
ant
galleries

Termite and
carpenter
ant
work

Dry rot fungus

Wood
reduced
to powder

Mushroom

Powder broken down by decomposers
into plant nutrients in soil

Time progression

Figure 4-15 Some detritivores, called *detritus feeders*, directly consume tiny fragments of this log. Other detritivores, called *decomposers* (mostly fungi and bacteria), digest complex organic chemicals in fragments of the log into simpler inorganic nutrients. Producers can use these nutrients again if they are not washed away or otherwise removed from the system.

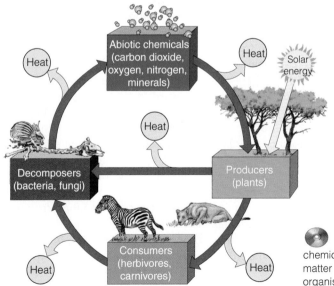

Abiotic chemicals
(carbon dioxide,
oxygen, nitrogen,
minerals)

Heat

Heat

Solar
energy

Heat

Decomposers
(bacteria, fungi)

Producers
(plants)

Heat

Consumers
(herbivores,
carnivores)

Heat

4-4 CONNECTIONS: FOOD WEBS AND ENERGY FLOW IN ECOSYSTEMS

What Are Food Chains and Food Webs?
All organisms, whether dead or alive, are potential sources of food for other organisms. A caterpillar eats a leaf, a robin eats the caterpillar, and a hawk eats the robin. Decomposers consume the leaf, caterpillar, robin, and hawk after they die. As a result, *there is little matter waste in natural ecosystems.*

Figure 4-16 The main structural components (energy, chemicals, and organisms) of an ecosystem are linked by matter recycling and the flow of energy from the sun, through organisms, and then into the environment as low-quality heat.

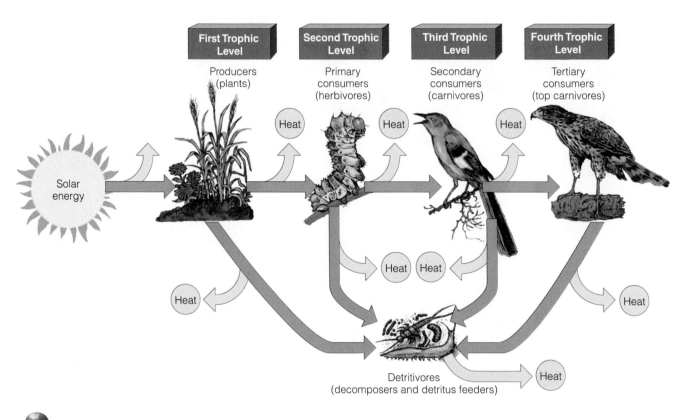

Figure 4-17 Model of a *food chain*. The arrows show how chemical energy in food flows through various *trophic levels*, or energy transfers; most of the energy is degraded to heat, in accordance with the second law of thermodynamics. Food chains rarely have more than four trophic levels.

The sequence of organisms, each of which is a source of food for the next, is called a **food chain.** It determines how energy and nutrients move from one organism to another through an ecosystem (Figure 4-17).

Ecologists assign each organism in an ecosystem to a *feeding level,* or **trophic level** (from the Greek word *trophos,* "nourishment"), depending on whether it is a producer or a consumer and on what it eats or decomposes. Producers belong to the first trophic level, primary consumers to the second trophic level, secondary consumers to the third, and so on. Detritivores (detritus feeders and decomposers) process detritus from all trophic levels.

Real ecosystems are more complex than this. Most consumers feed on more than one type of organism, and most organisms are eaten by more than one type of consumer. Because most species participate in several different food chains, the organisms in most ecosystems form a complex network of interconnected food chains called a **food web** (Figure 4-18). Trophic levels can be assigned in food webs just as in food chains.

How Can We Represent the Energy Flow in an Ecosystem? Pyramids of Energy Flow

Each trophic level in a food chain or web contains a certain amount of **biomass,** the dry weight of all organic matter contained in its organisms. In a food

chain or web, chemical energy stored in biomass is transferred from one trophic level to another.

With each transfer some usable energy is degraded and lost to the environment as low-quality heat. Thus only a small portion of what is eaten and digested is actually converted into an organism's bodily material or biomass, and the amount of usable energy available to each successive trophic level declines.

The percentage of usable energy transferred as biomass from one trophic level to the next is called **ecological efficiency.** It ranges from 5% to 20% (that is, a loss of 80–95%) depending on the types of species and the ecosystem involved, but 10% is typical.

Assuming 10% ecological efficiency (90% loss) at each trophic transfer, if green plants in an area manage to capture 10,000 units of energy from the sun, then only about 1,000 units of energy will be available to support herbivores and only about 100 units to support carnivores.

The more trophic levels or steps in a food chain or web, the greater the cumulative loss of usable energy as energy flows through the various trophic levels. The **pyramid of energy flow*** in Figure 4-19 (p. 78) illustrates this energy loss for a simple food chain,

---

*Because such pyramids represent energy flows, not energy storage, they should not be called pyramids of energy (a common error in some biology and environmental science textbooks).

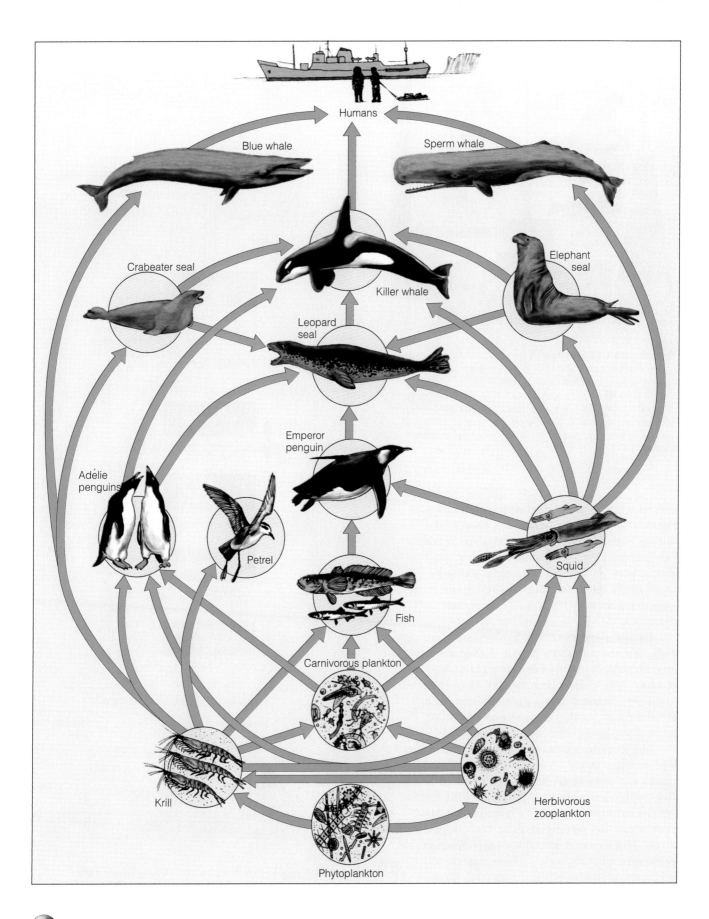

Figure 4-18 Greatly simplified *food web* in the Antarctic. Many more participants in the web, including an array of decomposer organisms, are not depicted here.

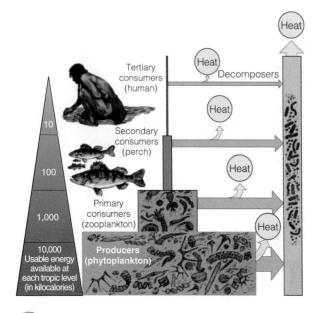

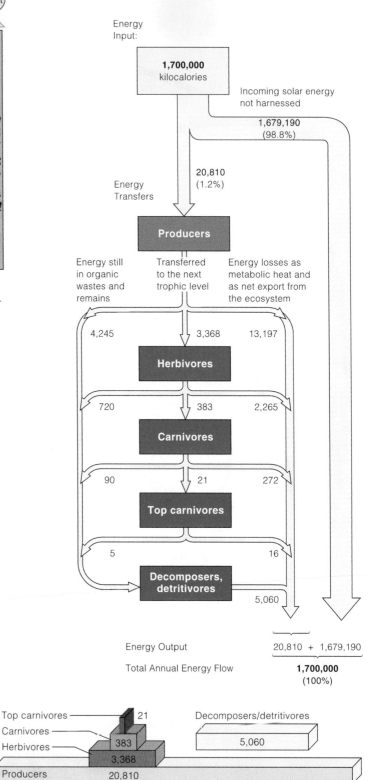

Figure 4-19 Generalized *pyramid of energy flow* showing the decrease in usable energy available at each succeeding trophic level in a food chain or web. In nature, ecological efficiency varies from 5% to 20%, with 10% efficiency being common. This model assumes a 10% ecological efficiency (90% loss in usable energy to the environment, in the form of low-quality heat) with each transfer from one trophic level to another.

assuming a 90% energy loss with each transfer. Figure 4-20 shows the pyramid of energy flow during 1 year for an aquatic ecosystem in Silver Springs, Florida. Pyramids of energy flow *always* have an upright pyramidal shape because of the automatic degradation of energy quality required by the second law of thermodynamics.

Energy flow pyramids explain why the earth can support more people if they eat at lower trophic levels by consuming grains, vegetables, and fruits directly (for example, grain → human) rather than passing such crops through another trophic level and eating grain eaters (grain → steer → human).

The large loss in energy between successive trophic levels also explains why food chains and webs rarely have more than four or five trophic levels. In most cases, too little energy is left after four or five transfers to support organisms feeding at these high trophic levels. This explains why there are so few top carnivores such as eagles, hawks, tigers, and white sharks. It also explains why such species usually are the first to suffer when the ecosystems that support them are disrupted and why they are so vulnerable to extinction.

Figure 4-20 Annual energy flow (in kilocalories per square meter per year) for an aquatic ecosystem in Silver Springs, Florida. The pyramid in the bottom drawing contains layers proportioned according to the energy flow through each trophic level. (From Cecie Starr, *Biology: Concepts and Applications*, 4th ed., Pacific Grove, Calif.: Brooks/Cole, 2000)

4-5 PRIMARY PRODUCTIVITY OF ECOSYSTEMS

How Rapidly Do Producers in Different Ecosystems Produce Biomass? The *rate* at which an ecosystem's producers convert solar energy into chemical energy as biomass is the ecosystem's **gross primary productivity (GPP).** In effect, it is the rate at which plants or other producers use photosynthesis to make more plant material (biomass).

Figure 4-21 shows how this productivity varies across the earth. This figure shows areas with especially high GPP values. They include *shallow waters near continents, coral reefs* (where abundant light, heat, and nutrients stimulate the growth of algae), and *upwelling currents* that bring nitrogen and phosphorus from the ocean bottom to the surface. The lowest GPP is in deserts and other arid regions because of their low precipitation and high temperatures and in the open ocean because of a lack of nutrients and sunlight except near the surface.

To stay alive, grow, and reproduce, an ecosystem's producers must use some of the total biomass they produce for their own respiration. Only what is left, called **net primary productivity (NPP),** is available for use as food by other organisms (consumers) in an ecosystem:

$$
\begin{array}{c}
\text{Net primary} \\
\text{productivity}
\end{array}
=
\begin{array}{c}
\text{Rate at which} \\
\text{producers store} \\
\text{chemical energy} \\
\text{as biomass} \\
\text{(produced by} \\
\text{photosynthesis)}
\end{array}
-
\begin{array}{c}
\text{Rate at which} \\
\text{producers use} \\
\text{chemical energy} \\
\text{stored as biomass} \\
\text{(through aerobic} \\
\text{respiration)}
\end{array}
$$

NPP is the *rate* at which energy for use by consumers is stored in new biomass (cells, leaves, roots, and stems). It is measured in units of the energy or biomass available to consumers in a specified area over a given time. Various ecosystems and life zones differ in their NPP (Figure 4-22, p. 80). The most productive are estuaries, swamps and marshes, and tropical rain forests. The least productive are open ocean, tundra (arctic and alpine grasslands), and desert. Despite its low net primary productivity, there is so much open ocean that it produces more of the earth's NPP per year than any of the other ecosystems and life zones shown in Figure 4-22.

Agricultural land is a highly modified and managed ecosystem. The goal is to increase the NPP and biomass of selected crop plants by adding water

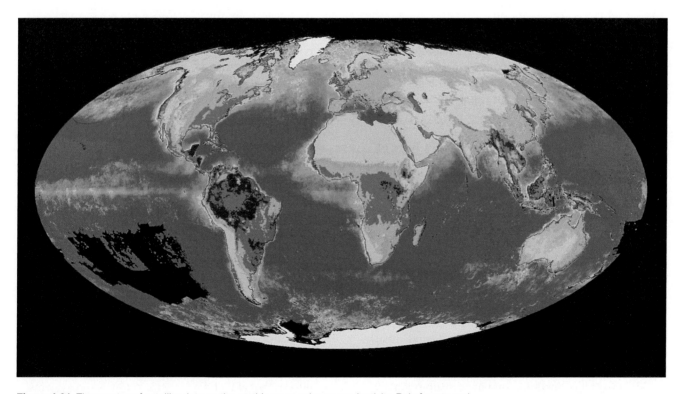

Figure 4-21 Three years of satellite data on the earth's gross primary productivity. Rain forests and other highly productive areas appear as dark green and deserts as yellow. The concentration of phytoplankton, a primary indicator of ocean productivity, ranges from red (highest) to orange, yellow, green, and blue (lowest). (Gene Carl Feldman, Compton J. Tucker—NASA/Goddard Space Flight Center)

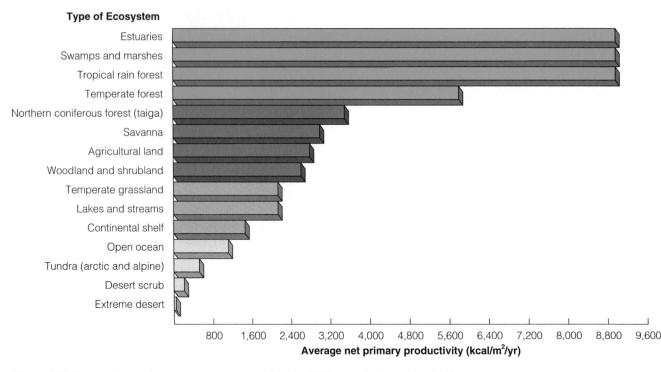

Figure 4-22 Estimated annual average *net primary productivity* (NPP) per unit of area in major life zones and ecosystems, expressed as kilocalories of energy produced per square meter per year (kcal/m²/yr). (Data from R. H. Whittaker, *Communities and Ecosystems*, 2nd ed., New York: Macmillan, 1975)

(irrigation) and nutrients (fertilizers). Nitrogen as nitrate (NO_3^-) and phosphorus as phosphate (PO_4^{3-}) are the most common nutrients in fertilizers because they are most often the nutrients limiting crop growth. Despite such inputs, the NPP of agricultural land is not particularly high compared with that of many other ecosystems (Figure 4-22).

Connections: How Does the World's Net Rate of Biomass Production Limit the Populations of Consumer Species? As we have seen, producers are the source of all food in an ecosystem. Thus the planet's NPP ultimately limits the number of consumers (including humans) that can survive on the earth.

It is tempting to conclude from Figure 4-22 that a good way to feed the world's hungry millions would be to harvest plants in estuaries, swamps, and marshes. But ecologists point out that people cannot eat most plants in these areas. In addition, these plants are vital food sources (and spawning areas) for fish, shrimp, and other aquatic species that provide us and other consumers with protein.

We might also conclude from Figure 4-22 that we could grow more food for human consumption by clearing tropical rain forests and planting food crops. According to most ecologists, this is also a bad idea. In tropical rain forests most of the nutrients needed to grow food crops are stored in the vegetation rather

than in the soil. When the trees are removed, frequent rains and growing crops rapidly deplete the nutrient-poor soils. Thus crops can be grown only for a short time without massive and expensive applications of commercial fertilizers.

Because the earth's vast open oceans provide the largest percentage of the earth's net primary productivity, why not harvest its primary producers (floating and drifting phytoplankton) to help feed the rapidly growing human population? The problem is that harvesting the widely dispersed, tiny floating producers in the open ocean would take much more fossil fuel and other types of energy than the food energy we would get. In addition, this would disrupt the food webs of the open ocean (Figure 4-18) that provide us and other consumer organisms with important sources of energy and protein from fish and shellfish.

How Much of the World's Net Rate of Biomass Production Do We Use? Peter Vitousek and other ecologists estimate that humans now use, waste, or destroy about 27% of the earth's total potential NPP and 40% of the NPP of the planet's terrestrial ecosystems.

They contend this is the main reason we are crowding out or eliminating the habitats and food supplies of a growing number of other species. What might happen to us and to other consumer species if the human population doubles over the next 40–50 years and per capita consumption of resources such as food, timber, and grassland rises sharply?

4-6 CONNECTIONS: MATTER CYCLING IN ECOSYSTEMS

What Are Biogeochemical Cycles? The nutrient atoms, ions, and molecules that organisms need to live, grow, and reproduce are continuously cycled from the nonliving environment (air, water, soil, and rock) to living organisms (biota) and then back again. This takes place in what are called **nutrient cycles,** or **biogeochemical cycles** (literally, life–earth–chemical cycles). These cycles, driven directly or indirectly by incoming solar energy and gravity, include the carbon, oxygen, nitrogen, phosphorus, and hydrologic (water) cycles (Figure 4-7).

The earth's chemical cycles also connect past, present, and future forms of life. Some of the carbon atoms in your skin may once have been part of a leaf, a dinosaur's skin, or a layer of limestone rock. Your grandmother, Plato, or a hunter–gatherer who lived 25,000 years ago may have inhaled some of the oxygen molecules you just inhaled.

How Is Water Cycled in the Biosphere? The **hydrologic cycle,** or **water cycle,** which collects, purifies, and distributes the earth's fixed supply of water, is shown in simplified form in Figure 4-23.

The main processes in this water recycling and purifying cycle are *evaporation* (conversion of water into water vapor), *transpiration* (evaporation from plant leaves after water is extracted from soil by roots and transported throughout the plant), *condensation* (conversion of water vapor into droplets of liquid water), *precipitation* (rain, sleet, hail, and snow), *infiltration* (movement of water into soil), *percolation* (downward flow of water through soil and permeable rock formations to groundwater storage areas called aquifers), and *runoff* (downslope surface movement back to the sea to resume the cycle).

The water cycle is powered by energy from the sun and by gravity. Incoming solar energy evaporates water from oceans, streams, lakes, soil, and vegetation. About 84% of water vapor in the atmosphere comes from the oceans, and the rest comes from land.

Some of the fresh water returning to the earth's surface as precipitation becomes locked in glaciers. Most of the precipitation falling on terrestrial ecosystems becomes *surface runoff* flowing into streams and lakes, which eventually carry water back to the oceans, where it can be evaporated to cycle again.

Besides replenishing streams, lakes, and wetlands, surface runoff also causes soil erosion, which moves soil and weathered rock fragments from one place to another. Water is thus the primary sculptor of the

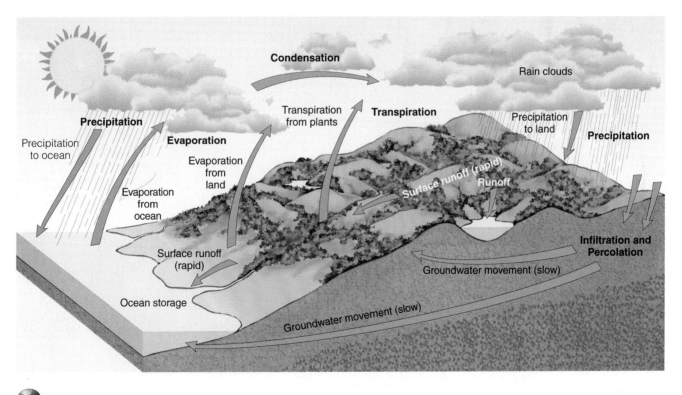

Figure 4-23 Simplified model of the *hydrologic cycle*.

earth's landscape. Because water dissolves many nutrient compounds, it is a major medium for transporting nutrients within and between ecosystems.

Throughout the hydrologic cycle, many natural processes act to purify water. Evaporation and subsequent precipitation act as a natural distillation process that removes impurities dissolved in water. Water flowing above ground through streams and lakes and below ground in aquifers is naturally filtered and purified by chemical and biological processes. Thus the hydrologic cycle can also be viewed as a cycle of natural renewal of water quality.

How Are Human Activities Affecting the Water Cycle? During the past 100 years, we have been intervening in the earth's current water cycle in three major ways. *First,* we withdraw large quantities of fresh water from streams, lakes, and underground sources. In some heavily populated or heavily irrigated areas, withdrawals have led to groundwater depletion or intrusion of ocean salt water into underground water supplies.

Second, we clear vegetation from land for agriculture, mining, road and building construction, and other activities and sometimes cover the land with buildings, concrete, or asphalt. This increases runoff, reduces infiltration that recharges groundwater supplies, increases the risk of flooding, and accelerates soil erosion and landslides.

Third, we modify water quality by adding nutrients (such as phosphates and nitrates found in fertilizers) and other pollutants. This overload of plant nutrients can change or impair natural ecological processes that purify water.

How Is Carbon Cycled in the Biosphere?
Carbon is essential to life as we know it. It is the basic building block of the carbohydrates, fats, proteins, DNA, and other organic compounds that are necessary for life.

The **carbon cycle** (Figure 4-24) is based on carbon dioxide gas, which makes up 0.037% of the volume of the troposphere and is also dissolved in water. Carbon dioxide is a key component of nature's thermostat. If the carbon cycle removes too much CO_2 from the atmosphere, the atmosphere will cool; if the cycle generates too much, the atmosphere will get warmer. Thus even slight changes in the carbon cycle can affect cli-

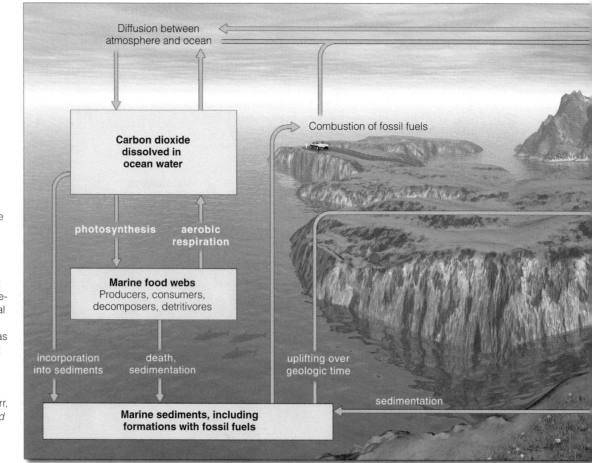

Figure 4-24
Simplified model of the global *carbon cycle.* The left portion shows the movement of carbon through marine systems, and the right portion shows its movement through terrestrial ecosystems. Carbon reservoirs are shown as boxes; processes that change one form of carbon to another are shown in unboxed print. (From Cecie Starr, *Biology: Concepts and Applications,* 4th ed., Pacific Grove, Calif.: Brooks/Cole, 2000)

Diffusion between atmosphere and ocean

Combustion of fossil fuels

Carbon dioxide dissolved in ocean water

photosynthesis   **aerobic respiration**

Marine food webs
Producers, consumers, decomposers, detritivores

incorporation into sediments

death, sedimentation

uplifting over geologic time

sedimentation

Marine sediments, including formations with fossil fuels

mate and ultimately the types of life that can exist on various parts of the planet.

Terrestrial producers remove CO_2 from the atmosphere, and aquatic producers remove it from the water. They then use photosynthesis to convert CO_2 into simple carbohydrates such as glucose ($C_6H_{12}O_6$).

The cells in oxygen-using producers, consumers, and decomposers then carry out aerobic respiration. This breaks down glucose and other complex organic compounds and converts the carbon back to CO_2 in the atmosphere or water for reuse by producers. This linkage between *photosynthesis* in producers and *aerobic respiration* in producers, consumers, and decomposers (Figure 4-16) circulates carbon in the biosphere and is a major part of the global carbon cycle. Oxygen and hydrogen, the other elements in carbohydrates, cycle almost in step with carbon.

Over millions of years, buried deposits of dead plant matter and bacteria are compressed between layers of sediment, where they form carbon-containing *fossil fuels* such as coal and oil (Figure 4-24). This carbon is not released to the atmosphere as CO_2 for recycling until these fuels are extracted and burned or long-term geological processes expose these deposits

to air. In only a few hundred years, we have extracted and burned fossil fuels that took millions of years to form. This is why fossil fuels are nonrenewable resources on a human time scale.

The oceans are a major carbon-storage reservoir in the carbon cycle. Oceans also play a major role in regulating the level of carbon dioxide in the atmosphere. Some carbon dioxide gas, which is readily soluble in water, stays dissolved in the sea and some is removed by photosynthesizing producers. In addition, some reacts with seawater to form carbonate ions (CO_3^{2-}) and bicarbonate ions (HCO_3^-). As water warms, more dissolved CO_2 returns to the atmosphere, just as more carbon dioxide fizzes out of a carbonated beverage when it warms.

In marine ecosystems, some organisms take up dissolved CO_2 molecules, carbonate ions, or bicarbonate ions from ocean water. These ions can then react with calcium ions (Ca^{2+}) in seawater to form slightly soluble carbonate compounds such as calcium carbonate ($CaCO_3$) to build the shells and skeletons of marine organisms. When these organisms die, tiny particles of their shells and bone drift slowly to the ocean depths. There they are buried for eons (as long as 400 million

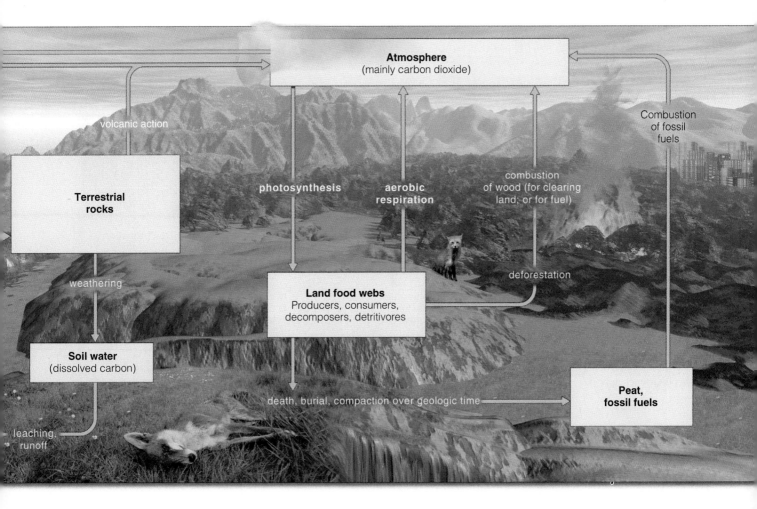

years) in deep bottom sediments (Figure 4-24, left), where under immense pressure they are converted into limestone rock.

How Are Human Activities Affecting the Carbon Cycle?
Since 1800 and especially since 1950, we have been intervening in the earth's carbon cycle in two ways that add carbon dioxide to the atmosphere. *First,* we clear trees and other plants that absorb CO_2 through photosynthesis. *Second,* we add large amounts of CO_2 by burning fossil fuels and wood.

Computer models of the earth's climate systems suggest that increased concentrations of atmospheric CO_2 and other gases we are adding to the atmosphere could enhance the planet's *natural greenhouse effect* that helps warm the lower atmosphere (troposphere) and the earth's surface (Figure 4-8). The resulting *global warming* could disrupt global food production and wildlife habitats, alter temperatures and precipitation patterns, and raise the average sea level in various parts of the world.

How Is Nitrogen Cycled in the Biosphere? Bacteria in Action
Nitrogen is the atmosphere's most abundant element, with chemically unreactive nitrogen gas (N_2) making up 78% of the volume of the troposphere. However, N_2 cannot be absorbed and used (metabolized) directly as a nutrient by multicellular plants or animals.

Fortunately, two natural processes convert N_2 gas in the atmosphere into compounds that can enter food webs as part of the **nitrogen cycle** (Figure 4-25). One process consists of atmospheric electrical discharges in the form of lightning. This causes nitrogen and oxygen in the atmosphere to react and produce nitrogen oxide ($N_2 + O_2 \longrightarrow 2$ NO). The other is the presence of certain types of bacteria in the soil and aquatic systems that can convert N_2 into compounds useful as nutrients for plants and animals.

The nitrogen cycle consists of several major steps (Figure 4-25). In *nitrogen fixation,* specialized bacteria convert gaseous nitrogen (N_2) to ammonia (NH_3) that can be used by plants; the reaction is $N_2 + 3 H_2 \longrightarrow$

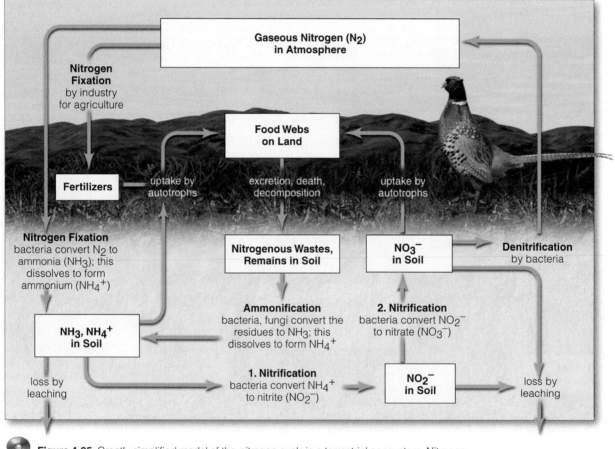

Figure 4-25 Greatly simplified model of the *nitrogen cycle* in a terrestrial ecosystem. Nitrogen reservoirs are shown as boxes; processes changing one form of nitrogen to another are shown in unboxed print. (Adapted from Cecie Starr and Ralph Taggart, *Biology: The Unity and Diversity of Life,* 9th ed., Pacific Grove, Calif.: Brooks/Cole, 2001)

2 NH_3. This is done mostly by cyanobacteria in soil and water and *Rhizobium* bacteria living in small nodules (swellings) on the root systems of a wide variety of plant species, including soybeans and alfalfa.

Ammonia not taken up by plants may undergo *nitrification*. It is a two-step process in which most of the ammonia in soil is converted by specialized aerobic bacteria to *nitrite ions* (NO_2^-), which are toxic to plants, and *nitrate ions* (NO_3^-), which are easily taken up by plants as a nutrient.

Nitrogen fixation and nitrification add inorganic ammonia, ammonium ions, and nitrate ions to soil water. Then plants roots can absorb these dissolved substances in a step called *assimilation*. Plants use these ions to make nitrogen-containing organic molecules such as DNA, amino acids, and proteins. Animals in turn get their nitrogen by eating plants or plant-eating animals.

Plants and animals return nitrogen-rich organic compounds to the environment as wastes, cast-off particles, and dead bodies. In the *ammonification* step, vast armies of specialized decomposer bacteria convert these into simpler nitrogen-containing inorganic compounds such as ammonia (NH_3) and water-soluble salts containing ammonium ions (NH_4^+).

Nitrogen leaves the soil in the *denitrification* step. Other specialized bacteria (mostly anaerobic bacteria in waterlogged soil or in the bottom sediments of lakes, oceans, swamps, and bogs) convert NH_3 and NH_4^+ back into nitrite (NO_2^-) and nitrate (NO_3^-) ions and then into nitrogen gas (N_2) and nitrous oxide gas (N_2O). These are then released to the atmosphere to begin the cycle again.

How Are Human Activities Affecting the Nitrogen Cycle? In the past 100 years, human activities have had several effects on the earth's current nitrogen cycle.

One such intervention has been the addition of large amounts of nitric oxide (NO) to the atmosphere when we burn any fuel ($N_2 + O_2 \longrightarrow$ 2 NO). In the atmosphere, this nitric oxide combines with oxygen to form nitrogen dioxide gas (NO_2), which can react with water vapor to form nitric acid (HNO_3). Droplets of HNO_3 dissolved in rain or snow are components of *acid deposition,* commonly called *acid rain*. Nitric acid, along with other air pollutants, can damage and weaken trees, upset aquatic ecosystems, corrode metals, and damage marble, stone, and other building materials.

This accelerated deposition of acidic nitrogen compounds from the atmosphere onto terrestrial ecosystems also stimulates the growth of weedy plant species, which can outgrow and perhaps eliminate other plant species that cannot take up nitrogen as efficiently.

A *second* intervention is the addition of nitrous oxide (N_2O) to the atmosphere. This occurs through the action of anaerobic bacteria on livestock wastes and commercial inorganic fertilizers applied to the soil. When N_2O reaches the stratosphere, it can help warm the atmosphere by enhancing the natural greenhouse effect and contribute to depletion of the earth's ozone shield (which filters out harmful ultraviolet radiation from the sun).

Third, nitrogen is removed from topsoil when we harvest nitrogen-rich crops, irrigate crops, and burn or clear grasslands and forests before planting crops.

Fourth, agricultural runoff and the discharge of municipal sewage add nitrogen compounds to aquatic ecosystems. This excess of plant nutrients stimulates rapid growth of photosynthesizing algae and other aquatic plants. These dense algal growths prevent sunlight from entering. The subsequent breakdown of dead algae by aerobic decomposers can deplete the water of dissolved oxygen and disrupt aquatic ecosystems by killing some types of fish and other oxygen-using (aerobic) organisms.

Fifth, the destruction of wetlands and forests releases large quantities of nitrogen that had been sequestered (stored) in soils and plants. Overall, human activities are estimated to have more than doubled the annual release of fixed nitrogen from the terrestrial portion of the earth into the environment.

How Is Phosphorus Cycled in the Biosphere? Phosphorus circulates through water, the earth's crust, and living organisms in the **phosphorus cycle** (Figure 4-26, p. 86). Bacteria are less important here than in the nitrogen cycle. Very little phosphorus circulates in the atmosphere because at the earth's normal temperatures and pressures, phosphorus and its compounds are not gases. Phosphorus is found in the atmosphere only as small particles of dust. In contrast to the carbon cycle, the phosphorus cycle is slow, and on a short human time scale much phosphorus flows one way from the land to the oceans.

Phosphorus is typically found as phosphate salts containing phosphate ions (PO_4^{3-}) in terrestrial rock formations and ocean bottom sediments. Because most soils contain little phosphate, it is often the *limiting factor* for plant growth on land unless phosphorus (as phosphate salts mined from the earth) is applied to the soil as a fertilizer.

Phosphorus also limits the growth of producer populations in many freshwater streams and lakes because phosphate salts are only slightly soluble in water. This explains why adding phosphate compounds to lakes greatly increases their biological productivity.

How Are Human Activities Affecting the Phosphorus Cycle? We have been intervening in the

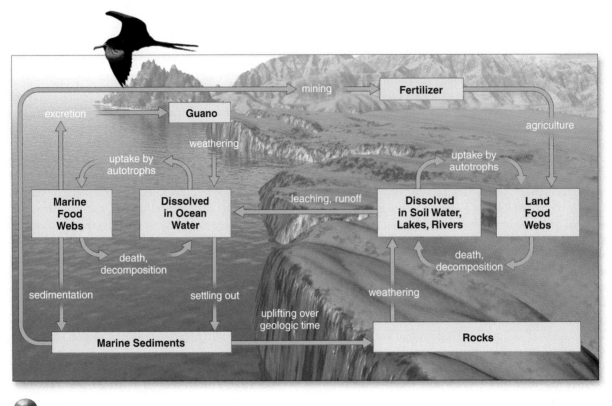

Figure 4-26 Simplified model of the *phosphorus cycle*. Phosphorus reservoirs are shown as boxes; processes that change one form of phosphorus to another are shown in unboxed print. (From Cecie Starr, *Biology: Concepts and Applications*, 4th ed., Pacific Grove, Calif.: Brooks/Cole, 2000)

earth's phosphorus cycle in three ways. *First,* we mine large quantities of phosphate rock for use in commercial inorganic fertilizers and detergents. *Second,* we reduce the available phosphate in tropical forests by removing trees. When such forests are cut and burned, most remaining phosphorus and other soil nutrients are washed away by heavy rains, and the land becomes unproductive.

Third, we add excess phosphate to aquatic ecosystems in runoff of animal wastes from livestock feedlots, runoff of commercial phosphate fertilizers from cropland, and discharge of municipal sewage. Too much of this nutrient causes explosive growth of cyanobacteria, algae, and aquatic plants. When these plants die and are decomposed, they use up dissolved oxygen and disrupt aquatic ecosystems.

Scientists estimate that human activities have increased the natural rate of phosphorus release into the environment about 3.7-fold.

How Is Sulfur Cycled in the Biosphere?
Sulfur circulates through the biosphere in the **sulfur cycle** (Figure 4-27). Much of the earth's sulfur is stored underground in rocks and minerals, including sulfate (SO_4^{2-}) salts buried deep under ocean sediments.

Sulfur also enters the atmosphere from several natural sources. Hydrogen sulfide (H_2S) is a colorless, highly poisonous gas with a rotten-egg smell. It is released from active volcanoes and from organic matter in swamps, bogs, and tidal flats broken down by decomposers that do not use oxygen (anaerobic decomposers). Sulfur dioxide (SO_2), a colorless, suffocating gas, also comes from volcanoes. Particles of sulfate (SO_4^{2-}) salts, such as ammonium sulfate, enter the atmosphere from sea spray.

Certain marine algae produce large amounts of volatile dimethyl sulfide, or DMS (CH_3SCH_3). Tiny droplets of DMS serve as nuclei for the condensation of water into droplets found in clouds. Thus changes in DMS emissions can affect cloud cover and climate.

In the atmosphere, sulfur dioxide reacts with oxygen to produce sulfur trioxide gas (SO_3). Some of the sulfur trioxide then reacts with water droplets in the atmosphere to produce tiny droplets of sulfuric acid (H_2SO_4). Sulfur dioxide also reacts with other chemicals in the atmosphere such as ammonia to produce tiny particles of sulfate salts. These droplets and particles fall to the earth as components of *acid deposition,* which along with other air pollutants can harm trees and aquatic life.

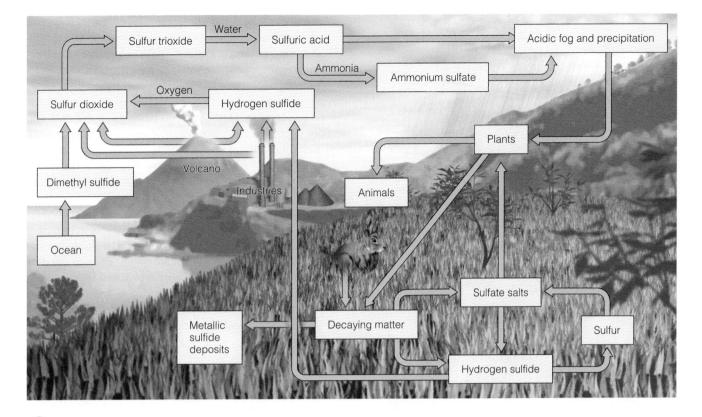

Figure 4-27 Simplified model of the *sulfur cycle*. Green shows the movement of sulfur compounds in living organisms, blue in aquatic systems, and orange in the atmosphere.

How Are Human Activities Affecting the Sulfur Cycle? We intervene in the atmospheric phase of the earth's sulfur cycle three ways. *First,* we burn sulfur-containing coal and oil to produce electric power. This produces about two-thirds of the human inputs of sulfur dioxide into the atmosphere. *Second,* we refine sulfur-containing petroleum to make gasoline, heating oil, and other useful products. *Third,* we use smelting to convert sulfur compounds of metallic minerals into free metals such as copper, lead, and zinc. This releases large amounts of sulfur dioxide into the environment.

4-7 HOW DO ECOLOGISTS LEARN ABOUT ECOSYSTEMS?

What Is Field Research? *Field research,* sometimes called muddy-boots biology, involves going into nature and observing and measuring the structure of ecosystems and what happens in them. Most of what we know about the structure and functioning of ecosystems described in this chapter has come from such research.

Increasingly, ecologists are using new technologies to collect field data. These include *remote sensing* from aircraft and satellites and *geographic information systems (GISs),* in which information gathered from broad geographic regions is stored in spatial databases (Figure 4-28, p. 88). Then computers and GIS software can analyze and manipulate the data and combine them with ground and other data to produce computerized maps of forest cover, water resources, air pollution emissions, coastal changes, relationships between cancer and other health effects and sources of pollution, and changes in global sea temperatures.

How Are Ecosystems Studied in the Laboratory? In the past 50 years, ecologists have increasingly supplemented field research by using *laboratory research* to set up, observe, and make measurements of model ecosystems and populations under laboratory conditions. Such simplified systems have been set up in containers such as culture tubes, bottles, aquarium tanks, greenhouses, and indoor and outdoor chambers where temperature, light, CO_2, humidity, and other variables can be controlled carefully.

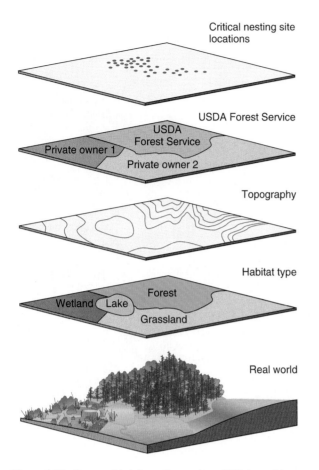

Figure 4-28 *Geographic information systems* (GISs) provide the computer technology for organizing, storing, and analyzing complex data collected over broad geographic areas. GISs enable scientists to overlay many layers of data (such as soils, topography, distribution of endangered populations, and land protection status).

In such systems, it is easier for scientists to carry out controlled experiments. In addition, such laboratory experiments often are quicker and cheaper than similar experiments in the field.

However, we must consider whether what scientists observe and measure in a simplified, controlled system under laboratory conditions takes place in the same way in the more complex and dynamic conditions found in nature. Thus the results of laboratory research must be coupled with and supported by field research.

What Is Systems Analysis? Since the late 1960s, ecologists have made increasing use of *systems analysis* to develop mathematical and other models that simulate ecosystems. Computer simulation of such models can help us understand large and very complex systems (such as rivers, oceans, forests, grasslands, cities, and climate) that cannot be adequately studied and modeled in field and laboratory research. Figure 4-29 outlines the major stages of systems analysis.

Researchers can change values of the variables in their computer models to project possible changes in environmental conditions, help anticipate environmental surprises, and analyze the effects of various alternative solutions to environmental problems.

However, simulations and predictions made using ecosystem models are no better than the data and assumptions used to develop the models. Thus careful field and laboratory ecological research must be used to provide the baseline data and determine the causal relationships between key variables needed to develop and test ecosystem models.

According to a 2002 ecological study published by the Heinz Foundation, scientists have less than half of the basic ecological data they need to evaluate the status of ecosystems in the United States. Even fewer data are available for most other parts of the world. Without such data, it is difficult to determine the seriousness of environmental problems and develop effective strategies for dealing with them.

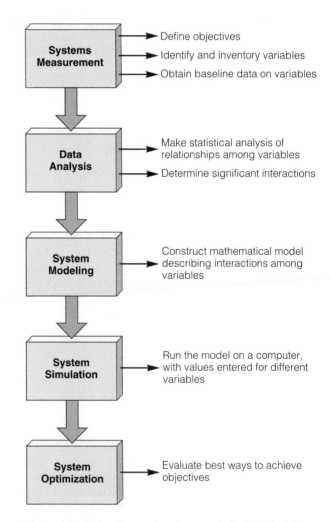

Figure 4-29 Major stages of *systems analysis.* (Modified data from Charles Southwick)

4-8 CONNECTIONS: ECOSYSTEM SERVICES AND SUSTAINABILITY

What Are Ecosystem Services? We depend on nature for food, air, water, and almost everything else we use. Ecosystems provide us and other species with a number of **ecosystem services** that are a vital part of the earth's natural capital (Figure 1-2, p. 3, and top half of back cover). Without these services performed by diverse communities of species, we would be starving, gasping for breath, and drowning in our own wastes.

It is very expensive to build sewage treatment plants to replace free water purification services provided by wetlands, to rely mostly on pesticides rather than natural biological controls to control crop and forest pests, to try to save species whose premature extinction could have been prevented, and to try to restore ecosystems that we have degraded. In addition, these costly replacements are rarely as effective as the free ecological services that nature provides.

What Are the Two Basic Principles of Ecosystem Sustainability? In this chapter we have seen that almost all natural ecosystems and the biosphere itself achieve sustainability in two ways. *First,* they use *renewable solar energy* as their energy source. *Second,* they *recycle the chemical nutrients* their organisms need for survival, growth, and reproduction.

These two principles for sustainability arise from the structure and function of natural ecosystems (Figures 4-7 and 4-16), law of conservation of matter (p. 56), and the two laws of thermodynamics (p. 59). Thus the results of basic research in both the physical and biological sciences provide us with the same guidelines or lessons from nature on how we can live more sustainably on the earth, as summarized in Figure 3-16 (p. 61).

This chapter on ecosystems covered how the earth's life-support systems work, what the living and nonliving components of ecosystems are, how energy flows through ecosystems, how rapidly biomass is produced in different ecosystems, how matter cycles in ecosystems, how ecologists learn about ecosystems, and how ecosystem services sustain life. The next chapter examines how organisms fit into an ecosystem.

All things come from earth, and to earth they all return.
MENANDER (342–290 B.C.)

REVIEW QUESTIONS

1. Define the boldfaced terms in this chapter.

2. Why are insects important for many forms of life and for you and your lifestyle?

3. What is *ecology?* What five levels of the organization of matter are the main focus of ecology?

4. Distinguish among *organism, eukaryotic organism, prokaryotic organism, species, population, genetic diversity, habitat, community, ecosystem,* and *biosphere.*

5. Explain why microbes (microorganisms) are so important.

6. Distinguish among the *atmosphere, troposphere, stratosphere, hydrosphere, lithosphere,* and *biosphere.*

7. What three processes sustain life on earth?

8. How does the sun help sustain life on the earth? How is this related to the earth's natural greenhouse effect?

9. What are *biomes,* and how are they related to climate? What are *aquatic life zones?*

10. Distinguish between the *abiotic* and *biotic* components of ecosystems, and give three examples of each.

11. Distinguish among *range of tolerance* for a population in an ecosystem, the *law of tolerance,* and *tolerance limits.* How does each of these factors affect the composition (structure) of ecosystems? What is a *limiting factor,* and how do such factors affect the composition of ecosystems? What are two important limiting factors for terrestrial ecosystems? For aquatic ecosystems?

12. Distinguish between *producers* and *consumers* in ecosystems, and give three examples of each type. What is *photosynthesis,* and why is it important to both producers and consumers? What is *chemosynthesis?*

13. Distinguish among *primary consumers (herbivores), secondary consumers (carnivores), tertiary consumers, omnivores, scavengers, detritivores, detritus feeders,* and *decomposers.* Why are decomposers important, and what would happen without them?

14. Distinguish between *aerobic respiration* and *anaerobic respiration.*

15. What are the four components of biodiversity? Why is biodiversity important to **(a)** the earth's life-support systems and **(b)** the economy?

16. Distinguish between a *food chain* and a *food web.*

17. What is *biomass?* What is the *pyramid of energy flow* for an ecosystem? What is the effect of the second law of thermodynamics on **(a)** the flow of energy through an ecosystem and **(b)** the amount of food energy available to the top carnivores and humans?

18. Distinguish between *gross primary productivity* and *net primary productivity.* Explain how net primary productivity affects the number of consumers in an ecosystem and on the earth. List two of the most productive ecosystems or aquatic life zones and two of the least productive ecosystems or aquatic life zones. Use the concept of net primary productivity to explain why harvesting plants from estuaries, clearing tropical rain forests to grow crops, and harvesting the primary

producers in oceans to feed the human population are not good ideas.

19. About what percentages of total potential net primary productivity of **(a)** the entire earth and **(b)** the earth's terrestrial ecosystems are used, wasted, or destroyed by humans?

20. What is a *biogeochemical cycle*? How do such cycles connect past, present, and future forms of life?

21. Describe the *water cycle*. What is *groundwater*? What is an *aquifer*?

22. List three human activities that alter the water cycle.

23. Describe the *carbon cycle*, and list two human activities that alter this cycle.

24. Describe the *nitrogen cycle*. Distinguish among *nitrogen fixation, nitrification, assimilation, ammonification,* and *denitrification*. Explain why the level of nitrogen in soil often limits plant growth. List five ways in which humans alter this cycle.

25. Describe the *phosphorus cycle*. Explain why the level of phosphorus in soil often limits plant growth on land and why phosphorus also limits the growth of producers in many freshwater streams and lakes. List three ways in which humans alter this cycle.

26. Describe the *sulfur cycle*, and list three ways in which humans alter this cycle.

27. Distinguish among *field research, laboratory research,* and *systems analysis* as methods for learning about ecosystems. What are *geographic information systems*, and how are they used to learn about ecosystems?

28. Define *ecosystem services*, and give three examples of such services.

29. What are two basic principles of ecosystem sustainability?

CRITICAL THINKING

1. (a) A bumper sticker asks, "Have you thanked a green plant today?" Give two reasons for appreciating a green plant. **(b)** Trace the sources of the materials that make up the bumper sticker, and decide whether the sticker itself is a sound application of the slogan. **(c)** Explain how decomposers help keep you alive.

2. (a) How would you set up a self-sustaining aquarium for tropical fish? **(b)** Suppose you have a balanced aquarium sealed with a clear glass top. Can life continue in the aquarium indefinitely as long as the sun shines regularly on it? **(c)** A friend cleans out your aquarium and removes all the soil and plants, leaving only the fish and water. What will happen? Explain.

3. Using the second law of thermodynamics, explain why there is such a sharp decrease in usable energy as energy flows through a food chain or web. Doesn't an energy loss at each step violate the first law of thermodynamics? Explain.

4. Using the second law of thermodynamics, explain why many poor people in developing countries live on a mostly vegetarian diet.

5. Why do farmers not need to apply carbon to grow their crops but often need to add fertilizer containing nitrogen and phosphorus?

6. Carbon dioxide (CO_2) in the atmosphere fluctuates significantly on a daily and seasonal basis. Why are CO_2 levels higher during the day than at night?

7. Which causes a larger loss of energy from an ecosystem: a herbivore eating a plant or a carnivore eating an animal? Explain.

8. Why are there more mice than lions in an African ecosystem supporting both types of animals?

9. What would happen to an ecosystem if all its decomposers and detritus feeders were eliminated or all its producers were eliminated?

PROJECTS

1. Visit several types of nearby aquatic life zones and terrestrial ecosystems. For each site, try to determine the major producers, consumers, detritivores, and decomposers.

2. Make a concept map of this chapter's major ideas using the section heads and subheads and the key terms (in boldface). Look on the website for this book for information about making concept maps.

INTERNET STUDY RESOURCES AND RESOURCES FOR FURTHER READING AND RESEARCH

The website for this book contains helpful study aids and many ideas for further reading and research. Log on to

http://biology.brookscole.com/miller10

and click on the Chapter-by-Chapter area. Choose Chapter 4 and select a resource:

■ Flash Cards allows you to test your mastery of the Terms and Concepts to Remember for this chapter.

■ Tutorial Quizzes provides a multiple-choice practice quiz.

■ Student Guide to InfoTrac will lead you to Critical Thinking Projects that use InfoTrac College Edition as a research tool.

■ References lists the major books and articles consulted in writing this chapter.

• Hypercontents takes you to an extensive list of sites with news, research, and images related to individual sections of the chapter.

INFOTRAC COLLEGE EDITION

Improve your skills with InfoTrac College Edition, a searchable online database of articles from more than 700 periodicals. Log on to

http://www.infotrac-college.com

or access InfoTrac through the website for this book. Try to find the following articles:

1. Pauly, D., V. Christensen, R. Froese, and M. L. Palomares. 2000. Fishing down aquatic food webs. *American Scientist* 88: 46. *Keywords:* "fishing" and "aquatic food webs." As pressure on fishing becomes more intense, humans are catching and keeping ever-smaller fish. This has implications for the entire aquatic food web, including the recovery of higher-trophic-level species. This work presents a new way of looking at fisheries science using basic ecological principles.

2. Lockaby, B. G., and W. H. Conner. 1999. N:P balance in wetland forests: Productivity across a biogeochemical continuum. *The Botanical Review* 65: 171. *Keywords:* "wetland forests" and "productivity." This paper integrates the concepts of biogeochemical cycling, limiting factors, and productivity by looking at differences among wetland forests.

5 EVOLUTION AND BIODIVERSITY: ORIGINS, NICHES, AND ADAPTATION

Earth: The Just-Right, Resilient Planet

Life on the earth as we know it (Figure 5-1) needs a certain temperature range: Venus is much too hot and Mars is much too cold, but the earth is *just right*. (Otherwise, you would not be reading these words.)

Life as we know it depends on the liquid water that dominates the earth's surface. Again, temperature is crucial; life on the earth needs average temperatures between the freezing and boiling points of water, between 0°C and 100°C (32°F and 212°F) at the earth's atmospheric pressure at sea level.

The earth's orbit is the right distance from the sun to provide these conditions. If the earth were much closer, it would be too hot—like Venus—for water vapor to condense to form rain. If it were much farther away, its surface would be so cold—like Mars—that its water would exist only as ice. The earth also spins; if it did not, the side facing the sun would be too hot and the other side too cold for water-based life to exist.

The earth is also the right size; that is, it has enough gravitational mass to keep its iron and nickel core molten and to keep the gaseous molecules in its atmosphere from flying off into space. (A much smaller earth would be unable to hold onto an atmosphere consisting of such light molecules as N_2, O_2, CO_2, and H_2O.)

The slow transfer of its internal heat (geothermal energy) to the surface also helps keep the planet at the right temperature for life. And thanks to the development of photosynthesizing bacteria more than 2 billion years ago, an ozone sunscreen protects us and many other forms of life from an overdose of ultraviolet radiation.

On a time scale of millions of years, the earth is enormously resilient and adaptive. During the 3.7 billion years since life arose, the average surface temperature of the earth has remained within the narrow range of 10–20°C (50–68°F), even with a 30–40% increase in the sun's energy output. In short, the earth is just right for life as we know it.

We can summarize the 3.7-billion-year biological history of the earth in one sentence: *Organisms convert solar energy to food, chemicals cycle, and a variety of species with different biological roles (niches) have evolved in response to changing environmental conditions.*

Each species here today represents a long chain of evolution, and each plays a unique ecological role in the earth's communities and ecosystems. These species, communities, and ecosystems also are essential for future evolution as the earth continues its ongoing history of environmental change.

This chapter discusses how the earth's species evolved and the nature of their niches or biological roles. This information is important for helping us understand the effects of human actions on wild species and for protecting species—including the human species—from premature extinction.

Figure 5-1 The earth is a blue and white planet in the black void of space. Currently, it has the right physical and chemical conditions to allow the development of life as we know it today.

There is a grandeur to this view of life . . . that, whilst this planet has gone cycling on . . . endless forms most beautiful and most wonderful have been, and are being, evolved.

CHARLES DARWIN

This chapter addresses the following questions:

- How do scientists account for the emergence of life on the earth?
- What is evolution, and how has it led to the current diversity of organisms on the earth?
- How does evolution affect the way organisms fit into their environment?
- What is an ecological niche, and how does it relate to adaptation to changing environmental conditions?
- How do extinction of species and formation of new species affect biodiversity?

5-1 ORIGINS OF LIFE

How Did Life Emerge on the Earth? How did a barren planet become a living jewel in the vastness of space (Figure 5-1)? How did life on the earth evolve to its present incredible diversity of species, living in an interlocking network of matter cycles, energy flows, and species interactions? We do not know the full answer to these questions, but a growing body of evidence suggests what might have happened.

Evidence about the earth's early history comes from chemical analysis and measurements of radioactive elements in primitive rocks and fossils. Chemists have also conducted laboratory experiments showing how simple inorganic compounds in the earth's early atmosphere might have reacted to produce amino acids, simple sugars, and other organic molecules used as building blocks for the protein, complex carbohydrate, RNA, and DNA compounds needed for life.

From this diverse evidence scientists have hypothesized that life on the earth developed in two phases over the past 4.7–4.8 billion years (Figure 5-2): The first phase was *chemical evolution* of the organic molecules, biopolymers, and systems of chemical reactions needed to form the first protocells (taking about 1 billion years).

This was followed by *biological evolution* from single-celled prokaryotic bacteria (Figure 4-3b, p. 67), to single-celled eukaryotic creatures (Figure 4-3a, p. 67), and then to multicellular organisms (taking about 3.7–3.8 billion years) (Figure 5-3, p. 94).

How Do We Know What Organisms Lived in the Past? Most of what we know of the earth's life history comes from **fossils:** mineralized or petrified replicas of skeletons, bones, teeth, shells, leaves, and seeds, or impressions of such items. Such fossils give us physical evidence of organisms that lived long ago and show us what their internal structures looked like.

Despite its importance, the fossil record is uneven and incomplete. Some forms of life left no fossils, some fossils have decomposed, and others are yet to be found. So far we have found fossils representing only about 1% of the species believed to have ever lived.

Other sources of information include chemical and radioactive dating of fossils, nearby ancient rocks, material in cores drilled out of buried ice, and the DNA of organisms alive today.

5-2 EVOLUTION AND ADAPTATION

What Is Evolution? According to scientific evidence, the major driving force of adaptation to changes in environmental conditions is **biological evolution,** or **evolution:** the change in a population's

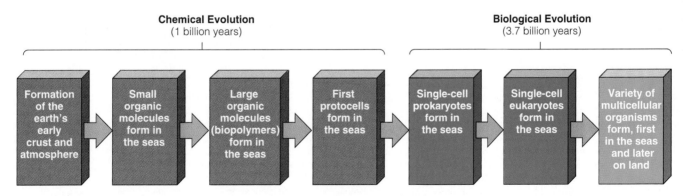

Figure 5-2 Summary of the hypothesized chemical and biological evolution of the earth. This drawing is not to scale. Note that the time span for biological evolution is almost four times longer than that for chemical evolution.

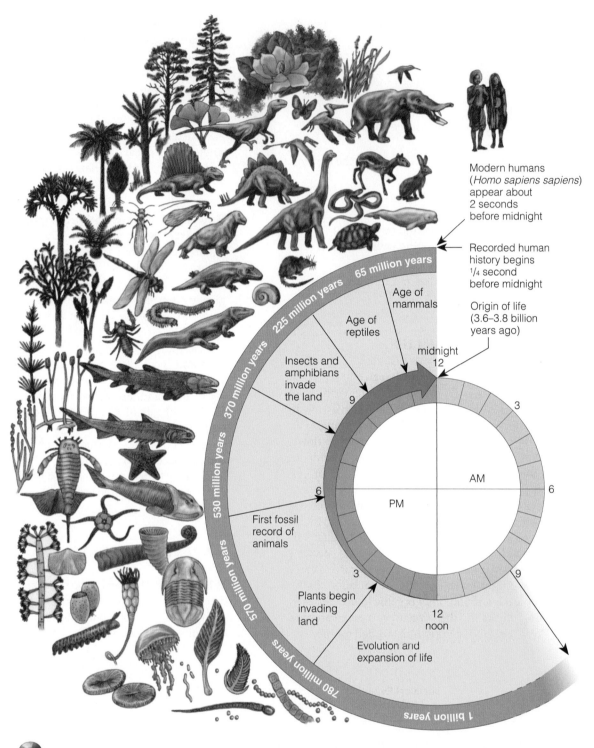

Figure 5-3 Greatly simplified overview of the biological evolution of life on the earth, which was preceded by about 1 billion years of chemical evolution. Evidence indicates that the early span of biological evolution on the earth, between about 3.7 billion and 1 billion years ago, was dominated by microorganisms (mostly bacteria and, later, protists) that lived in water. Plants and animals evolved first in the seas. Fossil and recent DNA evidence suggests that plants began moving onto land about 780 million years ago, and animals began invading the land about 370 million years ago. Humans arrived on the scene only a very short time ago. If we compress the earth's roughly 3.7-billion-year history of biological evolution to a 24-hour time scale, the first human species (*Homo habilis*) appeared about 47–94 seconds before midnight, and our species (*Homo sapiens sapiens*) appeared about 2 seconds before midnight. Agriculture began only 0.25 second before midnight, and the industrial revolution has been around for only 0.007 second.

genetic makeup (gene pool) through successive generations. Note that *populations, not individuals, evolve by becoming genetically different.*

According to the **theory of evolution,** all species descended from earlier, ancestral species. This widely accepted scientific theory explains how life has changed over the past 3.7 billion years (Figure 5-3) and why life is so diverse today.

Biologists use the term **microevolution** to describe the small genetic changes that occur in a population. The term **macroevolution** is used to describe long-term, large-scale evolutionary changes through which new species are formed from ancestral species and other species are lost through extinction.

How Does Microevolution Work? The first step in evolution is the development of *genetic variability* in a population. Recall that genetic information in *chromosomes* is contained in genes, which are sequences of chemical units (called *nucleotides*) in DNA molecules. A gene's sequence encodes certain traits that can be passed on to offspring (Figure 3-4, p. 52).

A population's **gene pool** is the set of all genes in the individuals of the population of a species. *Microevolution* is a change in a population's gene pool over time.

Although members of a population generally have the same number and kinds of genes, a particular gene may have two or more different molecular forms, called **alleles.** Sexual reproduction leads to a random shuffling or recombination of alleles. As a result, each individual in a population (except identical twins) has a different combination of alleles.

Genetic variability in a population originates as a result of **mutations:** random changes in the structure or number of DNA molecules in a cell.

Mutations can occur in two ways. One is by exposure of DNA to external agents such as radioactivity, X rays, and natural and human-made chemicals (called *mutagens*). The other is the result of random mistakes that sometimes occur in coded genetic instructions when DNA molecules are copied each time a cell divides and whenever an organism reproduces. Mutations can occur in any cells, but only those in reproductive cells are passed on to offspring.

Some mutations are harmless, but most are harmful and alter traits so that an individual cannot survive (lethal mutations). Every so often, a mutation is beneficial. The result is new genetic traits that give their bearer and its offspring better chances for survival and reproduction under existing environmental conditions or when such conditions change.

Mutations are random and unpredictable, are the only source of totally new genetic raw material (alleles), and are rare events. Once created by mutation, new alleles can be shuffled together or recombined *randomly* to create new combinations of genes in populations of sexually reproducing species.

What Role Does Natural Selection Play in Microevolution? The process of **natural selection** occurs when some individuals of a population have genetically based traits that increase their chances of survival and their ability to produce offspring. Three conditions are necessary for evolution of a population by natural selection to occur

First, there must be natural *variability* for a trait in a population. *Second,* the trait must be *heritable,* meaning it must have a genetic basis such that it can be passed from one generation to another. *Third,* the trait must somehow lead to **differential reproduction,** meaning it must enable individuals with the trait to leave more offspring than other members of the population.

Natural selection causes any allele or set of alleles that result in a beneficial trait to become more common in succeeding generations and other alleles to become less common. A heritable trait that enables organisms to better survive and reproduce under a given set of environmental conditions is called an **adaptation,** or **adaptive trait.** When faced with a change in environmental conditions, a population of a species can adapt to the new conditions through natural selection, migrate (if possible) to an area with more favorable conditions, or become extinct.

The process of microevolution can be summarized as follows: *Genes mutate, individuals are selected, and populations evolve.*

What Is an Example of Microevolution by Natural Selection? One of the best-documented examples of microevolution by natural selection involves camouflage coloration in the peppered moth, which is found in England (Figure 5-4, p. 96).

Natural selection occurred because there were two color forms (*variability*), color form was genetically based (*heritability*), and there was greater survival and reproduction by one of the color forms (*differential reproduction*). First an environmental change in the form of soot caused a change in the background color of tree trunks. This change allowed bird predators to find and eat the moths with the coloration that no longer blended in with the background (Figure 5-4).

What Are Three Types of Natural Selection? Biologists recognize three types of natural selection (Figure 5-5, p. 97). In *directional natural selection* (Figure 5-5, left) changing environmental conditions cause allele frequencies to shift so individuals with traits at one end of the normal range become more common than midrange forms. One example of this "it pays to be different" type of natural selection is the

Figure 5-4 Two varieties of peppered moths found in England illustrate one kind of adaptation: camouflage. Before the industrial revolution in the mid-1800s, the speckled light-gray form of this moth was prevalent. When these night-flying moths rested on light-gray lichens on tree trunks during the day, their color camouflaged them from their predators (left). A dark-gray form also existed but was quite rare. During the industrial revolution, soot and other pollutants from factory smokestacks began killing lichens and darkening tree trunks. As a result, the dark form of moth became the common one, especially near industrial cities. In this new environment, the dark form of moth blended in with the blackened trees, whereas the light form of moth was highly visible to predators (right). Through natural selection, the dark form began to survive and reproduce at a greater rate than its light-colored kin. (Both varieties appear in each photo. Can you spot them?)

changes in the varieties of peppered moths (Figure 5-4). Another is the evolution of genetic resistance to pesticides among insects and to antibiotics among disease-carrying bacteria. This type of natural selection is most common during periods of environmental change or when members of a population migrate to a new habitat with different environmental conditions.

The second type, called *stabilizing natural selection*, tends to eliminate individuals on both ends of the genetic spectrum and favor individuals with an average genetic makeup (Figure 5-5, center). This "it pays to be average" type of natural selection occurs when there is little change in environmental conditions, and most members of the population are well adapted to that environment.

The third type is *diversifying natural selection*. It occurs when environmental conditions favor individuals at both extremes of the genetic spectrum. This process eliminates or sharply reduces numbers of individuals with normal or intermediate genetic traits (Figure 5-5, right). In this "it does not pay to be normal" type of natural selection, a population is split into two groups.

What Is Artificial Selection? The genetic characteristics of populations of a species also can be changed through **artificial selection.** In this two-step process, humans select one or more desirable genetic traits in the population of a plant or animal. Then they use *selective breeding* to end up with populations of the

species containing large numbers of individuals with the desired traits. This process has been used to develop essentially all of the domesticated breeds of plants and animals from wild populations.

Artificial selection results in many different domesticated breeds or hybrids of the same species, all originally developed from a particular wild species. For example, despite their widely different genetic traits, all of the hundreds of different breeds of dogs are members of the same species because they can potentially interbreed and produce fertile offspring.

What Is Coevolution? Some biologists have proposed that *interactions between species* can also result in microevolution in each of their populations. According to this hypothesis, when populations of two different species interact over a long time, changes in the gene pool of one species can lead to changes in the gene pool of the other species. This process is called **coevolution.**

Suppose that certain individuals in a population of carnivores (such as owls) become better at hunting prey (such as mice). Because of genetic variation, certain individuals of the prey have traits that allow them to escape or hide from their predators, and they pass these adaptive traits on to some of their offspring. However, a few individuals in the predator population may have traits (such as better eyesight or quicker reflexes) that allow them to hunt the better-adapted prey

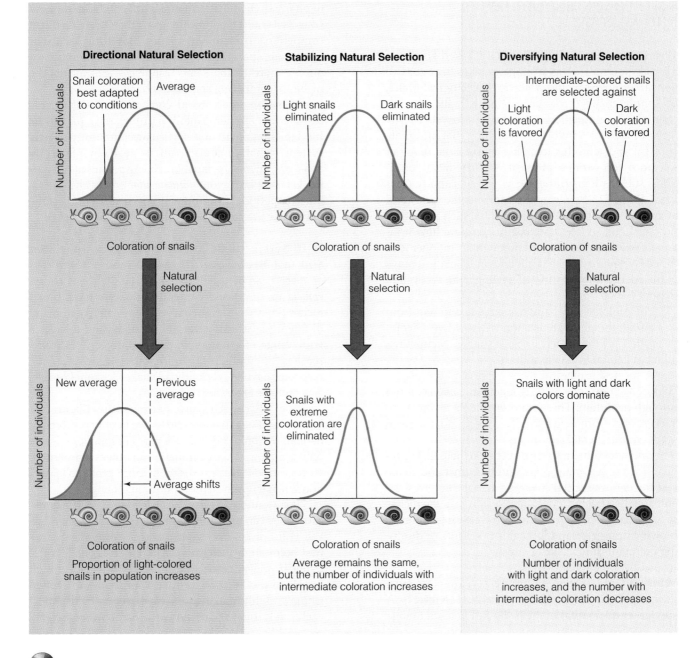

Figure 5-5 Three ways in which natural selection can occur, using the trait of coloration in a population of snails. In *directional natural selection*, changing environmental conditions select organisms with alleles that deviate from the norm so their offspring (lighter-colored snails) make up a larger proportion of the population. In *stabilizing selection*, environmental factors eliminate fringe individuals (light- and dark-colored snails) and increase the number of individuals with average genetic makeup (intermediate-colored snails). In *diversifying natural selection*, environmental factors favor individuals with uncommon traits (light- and dark-colored snails) and greatly reduce those with average traits (intermediate-colored snails).

successfully. They would then pass these traits on to some of their offspring.

Similarly, individual plants in a population may evolve defenses, such as camouflage, thorns, or poisons, against efficient herbivores. In turn, some herbivores in the population may have genetic characteris-

tics that enable them to overcome these defenses and produce more offspring than those without such traits.

In coevolution, adaptation follows adaptation in something like an ongoing, long-term arms race between individuals in interacting populations of different species.

5-3 ECOLOGICAL NICHES AND ADAPTATION

What Is an Ecological Niche? If asked what role a certain species such as an alligator plays in an ecosystem, an ecologist would describe its **ecological niche,** or simply **niche** (pronounced "nitch"), the species' way of life or functional role in an ecosystem. The niche of a species involves everything that affects its survival and reproduction. One aspect is its range of tolerance for various physical and chemical conditions, such as temperature or water availability (Figure 4-13, p. 72). Its niche also includes the types and amounts of resources the species uses, such as food or nutrients and space. Other components are how it interacts with other living and nonliving components of the ecosystems in which it is found and the role it plays in the energy flow and matter cycling in an ecosystem (Figure 4-16, p. 75).

The ecological niche of a species is different from its **habitat,** the physical location where it lives. Ecologists often say that a niche is like a species' occupation, whereas habitat is like its address.

A species' ecological niche represents the *adaptations* or *adaptive traits* that its members have acquired through evolution. These traits enable its members to survive and reproduce more effectively under a given set of environmental conditions.

Understanding a species' niche can help us to prevent it from becoming prematurely extinct and to assess the environmental changes we make in terrestrial and aquatic systems. For example, how will the niches of various species be changed by clearing a forest, plowing up a grassland, filling in a wetland, or dumping pollutants into a lake or stream?

What Is the Difference between a Species' Fundamental Niche and Its Realized Niche? A species' **fundamental niche** is the full potential range of physical, chemical, and biological conditions and resources it could theoretically use if there were no direct competition from other species. But in a particular ecosystem, species often compete with one another for one or more of the same resources. This means the niches of competing species overlap.

To survive and avoid competition for the same resources, a species usually occupies only part of its fundamental niche in a particular community or ecosystem—what ecologists call its **realized niche.** By analogy, you may be capable of being president of a particular company (your *fundamental professional niche*), but competition from others may mean you become only a vice president (your *realized professional niche*).

Is It Better to Be a Generalist or a Specialist Species? Broad and Narrow Niches The niches of species can be used to broadly classify them as *generalists* or *specialists*. **Generalist species** have broad niches (Figure 5-6, right curve). They can live in many different places, eat a variety of foods, and tolerate a wide range of environmental conditions. Flies, cockroaches (Spotlight, p. 99), mice, rats, white-tailed deer, raccoons, coyotes, copperhead snakes, channel catfish, and humans are generalist species.

Specialist species have narrow niches (Figure 5-6, left curve). They may be able to live in only one type of habitat, use only one or a few types of food, or tolerate only a narrow range of climatic and other environmental conditions. This makes them more prone to extinction when environmental conditions change.

For example, *tiger salamanders* are specialists because they can breed only in fishless ponds so their larvae will not be eaten. *Red-cockaded woodpeckers* carve nest holes almost exclusively in old (at least 75 years) longleaf pines, and *spotted owls* need old-growth forests in the Pacific Northwest for food and shelter. China's highly endangered *giant pandas* feed almost exclusively on various types of bamboo. Figure 5-7

Figure 5-6 Overlap of the niches of two different species: a *specialist* and a *generalist*. In the overlap area the two species compete for one or more of the same resources. As a result, each species can occupy only a part of its *fundamental niche*; the part it occupies is its *realized niche*. Generalist species have a broad niche (right), and specialist species have a narrow niche (left).

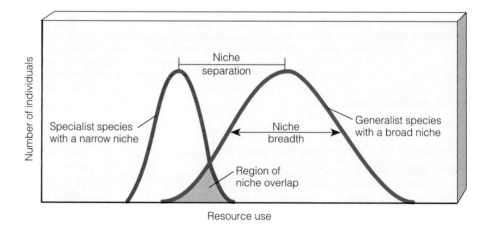

Cockroaches: Nature's Ultimate Survivors

Cockroaches, the bugs many people love to hate, have been around for about 350 million years and are one of the great success stories of evolution. They are so successful because they are *generalists*.

The earth's 4,000 cockroach species can eat almost anything (including algae, dead insects, fingernail clippings, salts in tennis shoes, electrical cords, glue, paper, and soap) and live and breed almost anywhere except in polar regions.

Some species can go for months without food, survive for a month on a drop of water from a dishrag, and withstand massive doses of radiation. One species can survive being frozen for 48 hours.

They can usually evade their predators and a human foot in hot pursuit. Most cockroach species can do this because they have antennae that can detect minute movements of air, vibration sensors in their knee joints, and rapid response times (faster than you can blink). Some even have wings.

They also have high reproductive rates. In only a year, a single female Asian cockroach (especially prevalent in Florida) and its young can add about 10 million new cockroaches to the world. Their high reproductive rate also helps them quickly develop genetic resistance to almost any poison we throw at them.

Most cockroaches also sample food before it enters their mouth and learn to shun foul-tasting poi-

sons. They also clean up after themselves by eating their own dead and, if food is scarce enough, their living.

Only about 25 species of cockroach live in homes. However, such species can carry viruses and bacteria that cause diseases such as hepatitis, polio, typhoid fever, plague, and salmonella and cause people to have allergic reactions ranging from watery eyes to severe wheezing. Indeed, about 60% of the 12 million Americans suffering from asthma are allergic to dead or live cockroaches.

Critical Thinking

If you could, would you exterminate all cockroach species? What might be some ecological consequences of doing this?

(p. 100) shows shorebirds that feed in specialized niches on crustaceans, insects, and other organisms on sandy beaches and their adjoining coastal wetlands.

Is it better to be a generalist than a specialist? It depends. When environmental conditions are fairly constant, as in a tropical rain forest, specialists have an advantage because they have fewer competitors. But under rapidly changing environmental conditions, a generalist species usually is better off than a specialist species.

What Limits Adaptation? Shouldn't evolution lead to perfectly adapted organisms? Shouldn't adaptations to new environmental conditions allow the human skin to become more resistant to the harmful effects of ultraviolet radiation, the lungs to cope with air pollutants, and the liver to better detoxify pollutants?

The answer to these questions is *no* because of three limits to adaptations in nature. First, *a change in environmental conditions can lead to adaptation only for traits already present in the gene pool of a population.*

Second, *even if a beneficial heritable trait is present in a population, that population's ability to adapt can be limited by its reproductive capacity.* Populations of genetically diverse species that reproduce quickly—such as weeds, mosquitoes, rats, bacteria, or cockroaches—can often adapt to a change in environmental conditions in

a short time. In contrast, populations of species that cannot produce large numbers of offspring rapidly, such as elephants, tigers, sharks, and humans, take a long time (typically thousands or even millions of years) to adapt through natural selection.

Third, *even if a favorable genetic trait is present in a population, most of the population would have to die or become sterile so individuals with the trait could predominate and pass the trait on.* This is hardly a desirable solution to the environmental problems the human species faces.

What Are Two Common Misconceptions about Evolution? There are two common misconceptions about evolution. One is that "survival of the fittest" means "survival of the strongest." To biologists, *fitness* is a measure of reproductive success, not strength. Thus the fittest individuals are those that leave the most descendants.

The other misconception is that evolution involves some grand plan of nature in which species become progressively more perfect. From a scientific standpoint, no plan or goal of perfection exists in the evolutionary process. However, some people (creationists) believe there is a conflict between the scientific theory of evolution and their religious beliefs about how life was created on the earth.

Black skimmer seizes small fish at water surface

Scaup and other diving ducks feed on mollusks, crustaceans, and aquatic vegetation

Brown pelican dives for fish, which it locates from the air

Avocet sweeps bill through mud and surface water in search of small crustaceans, insects, and seeds

Flamingo feeds on minute organisms in mud

Louisiana heron wades into water to seize small fish

Oystercatcher feeds on clams, mussels, and other shellfish into which it pries its narrow beak

Figure 5-7 Specialized feeding niches of various bird species in a coastal wetland. Such resource partitioning reduces competition and allows sharing of limited resources.

5-4 SPECIATION, EXTINCTION, AND BIODIVERSITY

How Do New Species Evolve? Under certain circumstances, natural selection can lead to an entirely new species. In this process, called **speciation,** two species arise from one.

The most common mechanism of speciation (especially among animals) takes place in two phases: geographic isolation and reproductive isolation. **Geographic isolation** occurs when groups of the same population of a species become physically separated for long periods. For example, part of a population may migrate in search of food and then begin living in another area with different environmental conditions (Figure 5-8). Populations also may become separated

by a physical barrier (such as a mountain range, stream, lake, or road), by a change such as a volcanic eruption or earthquake, or when a few individuals are carried to a new area by wind or water.

The second phase of speciation is **reproductive isolation.** It occurs when mutation and natural selection operate independently in two geographically isolated populations and change the allele frequencies in different ways. If this process, called *divergence,* continues long enough, members of the geographically and reproductively isolated populations may become so different in genetic makeup that they cannot interbreed—or if they do, they cannot produce live, fertile offspring. Then one species has become two, and *speciation* has occurred through *divergent evolution.*

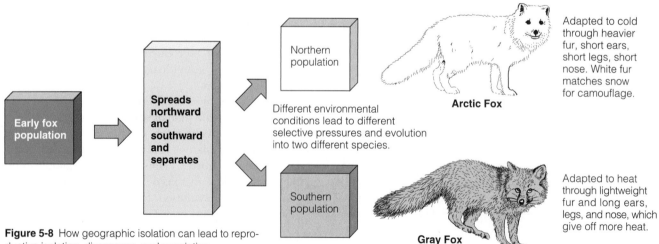

Early fox population

Spreads northward and southward and separates

Northern population

Different environmental conditions lead to different selective pressures and evolution into two different species.

Arctic Fox

Adapted to cold through heavier fur, short ears, short legs, short nose. White fur matches snow for camouflage.

Southern population

Gray Fox

Adapted to heat through lightweight fur and long ears, legs, and nose, which give off more heat.

Figure 5-8 How geographic isolation can lead to reproductive isolation, divergence, and speciation.

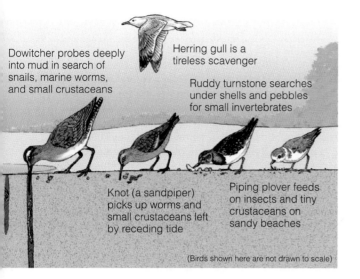

Dowitcher probes deeply into mud in search of snails, marine worms, and small crustaceans

Herring gull is a tireless scavenger

Ruddy turnstone searches under shells and pebbles for small invertebrates

Knot (a sandpiper) picks up worms and small crustaceans left by receding tide

Piping plover feeds on insects and tiny crustaceans on sandy beaches

(Birds shown here are not drawn to scale)

For some rapidly reproducing organisms, this type of speciation may occur within hundreds of years. However, for most species such speciation takes from tens of thousands to millions of years. Given this time scale, it is difficult to observe and document the appearance of a new species. Thus there are many controversial hypotheses about the details of speciation.

How Do Species Become Extinct? After speciation, the second process affecting the number and types of species on the earth is **extinction.** When environmental conditions change, a species must evolve (become better adapted), move to a more favorable area (if possible), or cease to exist (become extinct).

The earth's long-term patterns of speciation and extinction have been affected by three major factors. *One* is large-scale movement of the continents (continental drift) over millions of years (Figure 5-9). The *second* is gradual climate changes caused by continental drift and slight shifts in the earth's orbit around the sun.

The *third* is rapid climate change caused by catastrophic events. Examples include large volcanic eruptions, huge meteorites and asteroids crashing into the earth, and release of large amounts of methane trapped beneath the ocean floor. Some of these events create dust clouds that shut down or sharply reduce photosynthesis long enough to eliminate huge numbers of

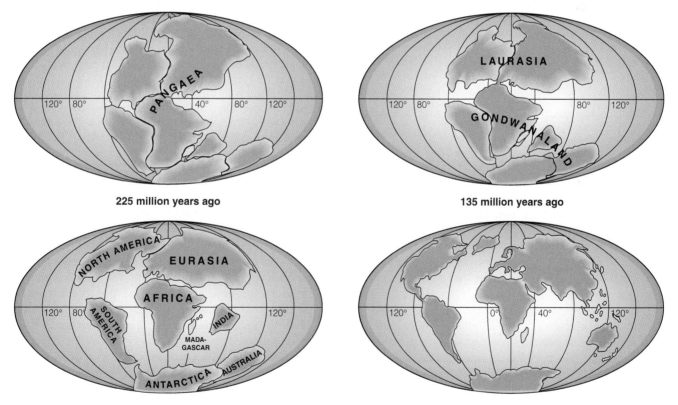

Figure 5-9 *Continental drift*, the extremely slow movement of continents over millions of years on several gigantic plates. This process plays a role in the extinction of species and the rise of new species. Populations are geographically and eventually reproductively isolated as landmasses float apart and new coastal regions are created. Rock and fossil evidence indicates that about 200–250 million years ago all of the earth's present-day continents were locked together in a supercontinent called Pangaea (top left). About 180 million years ago, Pangaea began splitting apart as the earth's huge plates separated and eventually resulted in today's locations of the continents (bottom right).

The Future of Evolution

Norman Myers

GUEST ESSAY

Norman Myers is a tropical ecologist and international consultant in environment and development, with emphasis on conservation of wildlife species and tropical forests. His research and consulting have taken him to 80 countries. A leading environmental expert, he has consulted for many development agencies and research organizations, including the U.S. National Academy of Sciences, the World Bank, the Organization for Economic Cooperation and Development, UN agencies, and the World Resources Institute. Among his many publications (see Further Readings on the website for this book) are The Primary Source: Tropical Forests and Our Future *(1992),* Ultimate Security: The Environmental Basis of Political Security *(1996), and* Perverse Subsidies *(2001).*

Human activities have brought the earth to a biotic crisis. Many biologists have commented that this crisis will result in the loss of large numbers of species, possibly 25–50%, within the lifetime of students reading this book. However, surprisingly few biologists have recognized that in the longer term these extinctions will impoverish evolution's course for several million years.

So the future of evolution should be regarded as one of the most challenging issues humankind has ever encountered because currently we are the world's greatest evolutionary force. After all, we are effectively conducting a planet-scale experiment, with little clue as to how it might turn out, except that it will prove irreversible and

For scientific details, see Norman Myers, "The Biodiversity Crisis and the Future of Evolution," *The Environmentalist* 16 (1996): 37–47.

will severely reduce human well-being. We could get by without half of all mammals and other vertebrates, but if we lost half of all insects with their pollinating functions [p. 64], let alone their many other services, we would be in trouble in the first crop-growing season.

In addition, the mass extinction or depletion under way is the biggest of our environmental problems in terms of the duration of its impact and the numbers of people to be affected. All our other problems are potentially reversible. If we wanted to clean up acid deposition, we could do it within a few decades. We could push back the deserts, restore topsoil, and allow the ozone layer to be repaired within a century or so. We could probably restore climate stability in the wake of global warming within a thousand years. But once a species is gone, it is gone for good.

Of course, in the long run evolution will generate replacement species with numbers and variety to match today's. But that is likely to take millions of years. We are witnessing the gross reduction if not the elimination of entire sectors of biomes, notably tropical forests, coral reefs, and wetlands, all of which may have served as powerhouses of evolution—centers of new speciation—in the prehistoric past.

Suppose, as has happened after the mass extinctions and depletions of the prehistoric past, that the bounce-back period lasts at least 5 million years. This would be 20 times longer than humans have been a species. Suppose that the average number of people on the earth during that period is 2.5 billion people, as opposed to the 6.3 billion today. Then the total number of people affected by what we do (or don't do) to protect the biosphere during the next few decades will be about 500 trillion, in con-

producers and, soon thereafter, the consumers that fed on them.

Extinction is the ultimate fate of all species, just as death is for all individual organisms. Biologists estimate that 99.9% of all the species that have ever existed are now extinct.

As local environmental conditions change, a certain number of species disappear at a low rate, called **background extinction.** In contrast, **mass extinction** is a significant rise in extinction rates above the background level. It is a catastrophic, widespread (often global) event in which large groups of existing species (perhaps 25–70%) are wiped out. Scientists have also identified periods of **mass depletion** in which extinction rates are higher than normal but not high enough to classify as a mass extinction. Recent fossil and geological evidence casts doubt on the hypothesis that there have been five mass extinctions

over the past 500 million years. The new evidence suggests that there have been two mass extinctions and three mass depletions during this period.

A mass extinction or mass depletion crisis for one species is an opportunity for another. The existence of millions of species today means that speciation, on average, has kept ahead of extinction. Evidence shows that the earth's mass extinctions and depletions have been followed by periods of recovery called **adaptive radiations** in which numerous new species evolve to fill new or vacated ecological roles or niches in changed environments. Fossil records suggest that it takes 5 million years or more for adaptive radiations to rebuild biological diversity after a mass extinction or depletion.

Connections: How Do Speciation and Extinction Affect Biodiversity? Speciation minus ex-

trast with the 50 billion people who have ever existed. Even 1 trillion is a big number: Figure out how long a period of time in years is made up of 1 trillion seconds.

In short, we are engaged in by far the biggest decision ever made by one human community on behalf of future human communities. Yet the issue is almost entirely disregarded, whether scientifically, ethically, or otherwise. How often do you hear leading scientists even mention the issue?

Despite our gross ignorance of what lies ahead, we can venture a few hypotheses:

- *A temporary outburst of speciation.* As large numbers of niches are vacated, there could be an outburst of speciation, although not nearly enough to match the extinction spasm.

- *A proliferation of opportunistic species* such as cockroaches [Spotlight, p. 99], rats, flies, and others that prosper when new niches open up. This proliferation will be enhanced by the likely elimination of species that naturally control opportunistic species.

- *An end to large vertebrates.*

- *An end to speciation of large vertebrates.* Even if larger vertebrates were to survive the extinction spasm ahead, our largest protected areas will prove far too small for further speciation of elephants, rhinoceroses, apes, bears, and the bigger cats, among other large vertebrates.

What does all this imply for our conservation efforts? By far the predominant strategy of conservationists is to save as many species as possible. But we now need to safeguard evolutionary processes as well. A prime goal is to look out especially for endemic species (found only in a particular place) or species confined to small habitats. Examples include the California condor, the black-footed ferret, the giant panda, and the gorilla.

However, the fossil record shows that endemic species often turn out to be evolutionary dead ends: Generally they do not throw off new species. So should we shift our conservation priority from endemic species to broader-ranging species in the hope that they have more genetic variability and thus more of a diversified resource stock on which natural selection can work its creative impact?

Similarly, should we devote more attention to protecting the evolutionary powerhouses such as the forests, coral reefs, and wetlands of the tropics? All these are in dire trouble and may be all but eliminated within just a few decades. Do they deserve preferential treatment ahead of, say, temperate-zone woodlands and grasslands and boreal forests with their lack of species, ecological complexity, and evolutionary potential?

Suppose that within your lifetime we allow the current biotic crisis to proceed unchecked—which is what the recent record suggests. Then it is possible that your children will ask you a key question: "When the evolutionary debacle was becoming all too plain at the start of the new millennium, what did you do to help ward off this disaster?" I hope you will engage yourself in dealing with this crucial issue.

Critical Thinking

1. Do you agree or disagree with the thesis of this essay? Explain.

2. If you agree, list three things you could do to help prevent the outcomes described in this essay.

tinction equals *biodiversity,* the planet's genetic raw material for future evolution in response to changing environmental conditions. In this long-term give-and-take between extinction and speciation, mass extinctions and mass depletions temporarily reduce biodiversity. However, they also create evolutionary opportunities for surviving species to undergo adaptive radiations to fill unoccupied and new biological roles or niches.

Although extinction is a natural process, much evidence indicates that humans have become a major force in the premature extinction of species. Biologist Stuart Primm estimates that during the 20th century, extinction rates increased by 100–1,000 times the natural background rate. As human population and resource consumption increase over the next 50–100 years, we are expected to take over more of the earth's surface and net primary productivity (Figure 4-22, p. 80). During this century, this human expansion may cause the premature extinction of up to a quarter of the earth's current species. This could constitute a new *mass depletion* and possibly a new *mass extinction* (Guest Essay, above left).

On our short time scale, such major losses cannot be recouped by formation of new species; it took millions of years after each of the earth's past mass extinctions and depletions for life to recover to the previous level of biodiversity. Genetic engineering cannot stop this loss of biodiversity because genetic engineers do not create new genes. Rather, they transfer existing genes or gene fragments from one organism to another. This means they rely on natural biodiversity for their raw material.

All we have yet discovered is but a trifle in comparison with what lies hid in the great treasury of nature.

ANTOINE VAN LEEUWENHOCK

REVIEW QUESTIONS

1. Define the boldfaced terms in this chapter.

2. Describe the conditions that make life on the earth just right for life as we know it.

3. Distinguish between *chemical evolution* and *biological evolution.*

4. What are *fossils,* and how do they help us formulate ideas about how life developed on the earth?

5. Distinguish among *biological evolution, the theory of evolution, microevolution,* and *macroevolution.*

6. Distinguish among *genes, gene pool, alleles, mutations, natural selection,* and *differential reproduction,* and explain their roles in microevolution.

7. Give an example of microevolution by natural selection. Describe three types of natural selection.

8. What is *artificial selection?* What is *coevolution,* and what is its importance?

9. What is the *ecological niche* of a species, and why is it important to understand the niches of species? What is the difference between a species' *habitat* and its *niche?* What is the difference between a species' *fundamental niche* and its *realized niche?*

10. Distinguish between the niches of specialist and generalist species. Explain why cockroaches have been such a successful species.

11. List three factors that limit adaptation.

12. What are two common misconceptions about evolution?

13. What is *speciation?* Distinguish between *geographic isolation* and *reproductive isolation,* and explain how they can lead to speciation through divergent evolution.

14. What is *extinction?* List three factors that have affected the earth's long-term patterns of speciation and extinction.

15. Distinguish among *background extinction, mass extinction,* and *mass depletion.*

16. What is an *adaptive radiation?* How can such a radiation lead to recovery after a mass extinction or depletion?

17. Explain how speciation and extinction result in the planet's biodiversity.

CRITICAL THINKING

1. **(a)** How would you respond to someone who tells you that he or she does not believe in biological evolution because it is "just a theory"? **(b)** How would you respond to a statement that we should not worry about air pollution because through natural selection the human species will develop lungs that can detoxify pollutants?

2. What would happen if extinction had never occurred during evolutionary history? Do you think we would have more species than we do now? Why?

3. How would you respond to someone who says that because extinction is a natural process, we should not worry about the loss of biodiversity?

4. Why is it important for somebody studying environmental science to understand the basics of evolution?

5. Why is the realized niche of a species narrower, or more specialized, than its fundamental niche?

6. As well as you can, describe the major differences between the ecological niches of humans and cockroaches. Are these two species in competition? If so, how do they manage to coexist?

7. In what ways do humans occupy generalist niches and in what ways do they occupy specialist niches?

8. By analogy, use the concepts of generalist and specialist to evaluate the roles of humans in today's societies. In general, the role of college and graduate education is to create specialists in a particular field or a narrow portion of a field. What are the pros and cons of relying mainly on this approach? Is there a need for more generalists? Or are they people who may know a lot about many things (and about connections between things) but not enough about anything in particular? Can they serve a useful role and make a satisfactory living in today's increasingly specialized societies? Explain your answers.

PROJECTS

1. An important adaptation of humans is a strong opposable thumb, which allows us to grip and manipulate things with our hands. As a demonstration of the importance of this trait, fold each of your thumbs into the palm of its hand and then tape them securely in that position for an entire day. After the demonstration, make a list of the things you could not do without the use of your thumbs.

2. Visit a local forest, pond, or lake, choose a particular organism, and then use the library or the Internet to describe as much as you can about its niche, including what species it depends on and what species help support its survival. Predict what might happen if your selected species disappeared from the local environment.

3. Make a concept map of this chapter's major ideas, using the section heads and subheads and the key terms (in boldface). Look on the website for this book for information about making concept maps.

INTERNET STUDY RESOURCES AND RESOURCES FOR FURTHER READING AND RESEARCH

The website for this book contains helpful study aids and many ideas for further reading and research. Log on to

http://biology.brookscole.com/miller10

and click on the Chapter-by-Chapter area. Choose Chapter 5 and select a resource:

- Flash Cards allows you to test your mastery of the Terms and Concepts to Remember for this chapter.

- Tutorial Quizzes provides a multiple-choice practice quiz.

- Student Guide to InfoTrac will lead you to Critical Thinking Projects that use InfoTrac College Edition as a research tool.

- References lists the major books and articles consulted in writing this chapter.

- Hypercontents takes you to an extensive list of sites with news, research, and images related to individual sections of the chapter.

INFOTRAC COLLEGE EDITION

Improve your skills with InfoTrac College Edition, a searchable online database of articles from more than 700 periodicals. Log on to

http://www.infotrac-college.com

or access InfoTrac through the website for this book. Try to find the following articles:

1. Becker, L. 2002. Repeated blows: Did extraterrestrial collisions capable of causing widespread extinctions pound the earth not once, but twice—or even several times? *Scientific American* 286: 76. *Keywords:* "collisions" and "extinction." Mounting evidence suggests that collisions with space debris have occurred many times, and new techniques are demonstrating close links between such collisions and mass8 extinctions.

2. Peterson, A. T., and D. A. Vieglais. 2001. Predicting species invasions using ecological niche modeling: New approaches from bioinformatics attack a pressing problem. *BioScience* 51: 363. *Keywords:* "invasive species" and "niche." Invasive exotic species are rapidly becoming one of the greatest threats to biodiversity. This paper shows how invasiveness of a species can be predicted based upon studies of its ecological niche.

6 CLIMATE, TERRESTRIAL BIODIVERSITY, AND AQUATIC BIODIVERSITY

Connections: Blowing in the Wind

Wind, a vital part of the planet's circulatory system, connects most life on the earth. Without wind, the tropics would be unbearably hot and most of the rest of the planet would freeze.

Winds also transport nutrients from one place to another. Dust rich in phosphates blows across the Atlantic from the Sahara Desert in Africa (Figure 6-1). This helps replenish rain forest soils in Brazil and build up agricultural soils in the Bahamas. Iron-rich dust blowing from China's Gobi Desert falls into the Pacific Ocean between Hawaii and Alaska. This input of iron stimulates the growth of phytoplankton, the minute producers that support ocean food webs. This is the *good news*.

The *bad news* is that wind also transports harmful viruses, bacteria, fungi, and particles of long-lived pesticides and toxic metals. Particles of reddish-brown soil and pesticides banned in the United States are blown from Africa's deserts and eroding farmlands into the sky over Florida. This makes it difficult for the state to meet federal air pollution standards during summer months. Also, some types of fungi in this dust are a suspected factor in degrading or killing coral reefs in the Florida Keys and the Caribbean.

Particles of iron-rich dust from Africa that enhance the productivity of algae have been linked to outbreaks of toxic algal blooms—referred to as *red tides*—in Florida's coastal waters. People who eat shellfish contaminated by a toxin produced in red

tides can become paralyzed or even die. Europe and the Middle East also receive African dust.

Pollution and dust from rapidly industrializing China and central Asia blow across the Pacific Ocean and degrade air quality over the western United States. In 2001, climate scientists reported that a huge dust storm of soil particles blown from northern China had blanketed areas from Canada to Arizona with a layer of dust. Studies show that Asian pollution contributes as much as 10% to West Coast smog, a threat expected to increase as China industrializes.

We have *mixed news* as well. Particles from volcanic eruptions ride the winds, circle the globe, and change the earth's climate for a while. Emissions from the 1991 eruption of Mount Pinatubo in the Philippines cooled the earth slightly for 3 years, temporarily masking signs of global warming. On the other hand, volcanic ash, like the blowing desert dust, adds valuable trace minerals to the soil where it settles.

The lesson, once again, is that *there is no away* because *everything is connected*. Wind acts as part of the planet's circulatory system for heat, moisture, plant nutrients, and long-lived pollutants we put into the air. Movement of soil particles from one place to another by wind and water is a natural phenomenon, but when we disturb the soil and leave it unprotected we hasten the process.

Wind is also an important factor in climate through its influence on global air circulation patterns. Climate, in turn, is crucial for determining what kinds of plant and animal life are found in the major biomes of the biosphere, as we shall see in this chapter.

Figure 6-1 Some of the dust shown here blowing from Africa's Sahara Desert can end up as soil nutrients in Amazonian rain forests and particles of toxic air pollutants in Florida and the Caribbean.

NASA/Goddard Space Flight Center, The SeaWiFs Project and ORBIMAGE, Scientific Visualization Studio.

To do science is to search for repeated patterns, not simply to accumulate facts, and to do the science of geographical ecology is to search for patterns of plant and animal life that can be put on a map.

ROBERT H. MACARTHUR

This chapter addresses the following broad questions about geographic patterns of ecology:

- What key factors determine the earth's weather and climate?

- How does climate determine where the earth's major biomes are found?

- What are the major types of desert and grassland biomes, and how do human activities affect them?

- What are the major types of forest and mountain biomes, and how do human activities affect them?

- What are the major types of saltwater life zones, and how do human activities affect them?

- What are the major types of freshwater life zones, and how do human activities affect them?

6-1 WEATHER AND CLIMATE: A BRIEF INTRODUCTION

What Is Weather? Every moment at any spot on the earth, the *troposphere* (the inner layer of the atmosphere containing most of the earth's air) has a particular set of physical properties. Examples are temperature, pressure, humidity, precipitation, sunshine, cloud cover, and wind direction and speed. These short-term properties of the troposphere at a particular place and time are **weather.**

Meteorologists use equipment on weather balloons, aircraft, ships, and satellites, as well as radar and stationary sensors, to obtain data on weather variables. These variables include atmospheric pressures, precipitation, temperatures, wind speeds, and locations of air masses and fronts.

The data are fed into computer models to draw weather maps for each of seven levels of the troposphere, ranging from the ground to 19 kilometers (12 miles) up. Computer models use the map data to forecast the weather in each box of the seven-layer grid for the next 12 hours. Other computer models project the weather for the next several days by calculating the probabilities that air masses, winds, and other factors will move and change in certain ways.

What Is Climate? Climate is a region's general pattern of atmospheric or weather conditions over a long period. *Average temperature* and *average precipitation* are the two main factors determining a region's climate and its effects on people (Figure 6-2). Figure 6-3 (p. 108) is a generalized map of the earth's major climate zones.

How Does Global Air Circulation Affect Regional Climates? The temperature and precipitation patterns that lead to different climates (Figure 6-3) are caused primarily by the way air circulates over the earth's surface. Five major factors determine global air circulation patterns *First* is the *uneven heating of the earth's surface.* Air is heated much more at the equator (where the sun's rays strike directly throughout the year) than at the poles (where sunlight strikes at an angle and thus is spread out over a

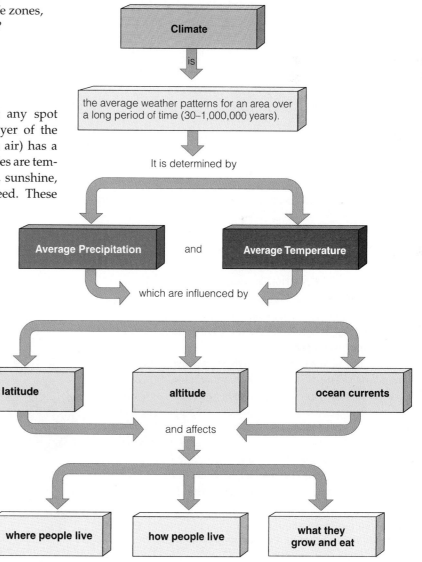

Figure 6-2 Climate and its effects. (Data from National Oceanic and Atmospheric Administration)

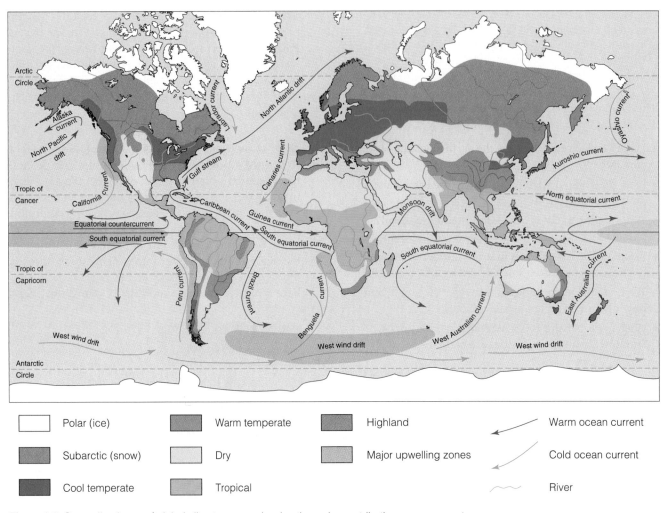

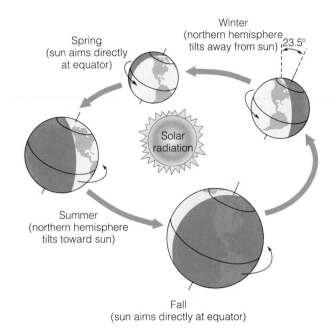

Figure 6-3 Generalized map of global climate zones, showing the major contributing ocean currents and drifts.

Legend:
- Polar (ice)
- Subarctic (snow)
- Cool temperate
- Warm temperate
- Dry
- Tropical
- Highland
- Major upwelling zones
- Warm ocean current
- Cold ocean current
- River

much greater area). These differences in incoming solar energy help explain why tropical regions near the equator are hot, polar regions are cold, and temperate regions in between generally have intermediate average temperatures.

A *second* factor is *seasonal changes in temperature and precipitation*. The earth's axis (an imaginary line connecting the north and south poles) is tilted. As a result, various regions are tipped toward or away from the sun as the earth makes its yearlong revolution around the sun (Figure 6-4). This creates opposite seasons in the northern and southern hemispheres.

A *third* factor is the *rotation of the earth on its axis*. As the earth rotates, its surface turns faster beneath air

Figure 6-4 The effects of the earth's tilted axis on climate. As the planet makes its annual revolution around the sun on an axis tilted about 23.5°, various regions are tipped toward or away from the sun. The resulting variations in the amount of solar energy reaching the earth create the seasons.

masses at the equator and slower beneath those at the poles. This deflects air masses moving north and south to the west or east. This results in the formation of six huge convection cells of swirling air masses (three north and three south of the equator) that transfer heat and water from one area to another (Figure 6-5). The direction of air movement in these cells sets up belts of prevailing winds that distribute air and moisture over the earth's surface.

Fourth, there are *long-term variations in the amount of solar energy striking the earth.* These are caused by occasional changes in solar output and slight planetary shifts in which the earth's axis wobbles (22,000-year cycle) and tilts (44,000-year cycle) as it revolves around the sun.

Finally, properties of air and water affect global air circulation patterns. Heat from the sun evaporates ocean water and transfers heat from the oceans to the atmosphere, especially near the hot equator. This creates cyclical convection cells that transport heat and water from one area to another. The resulting convection cells circulate air, heat, and moisture both vertically and from place to place in the troposphere. This leads to different climates and patterns of vegetation (Figure 6-5).

How Do Ocean Currents Affect Regional Climates? The factors just listed, plus differences in water density, create warm and cold ocean currents (Figure 6-3). These currents, driven by winds and the earth's rotation redistribute heat received from the sun and thus influence climate and vegetation, especially near coastal areas.

For example, without the warm Gulf Stream, which transports 25 times more water than all the world's rivers, the climate of northwestern Europe would be subarctic. If the ocean's currents suddenly stopped flowing, there would be deserts in the tropics and thick ice sheets over northern Europe, Siberia, and Canada. Currents also help mix ocean waters and distribute nutrients and dissolved oxygen needed by aquatic organisms.

What Are Upwellings? The winds blowing along some steep western coasts of continents toward the equator push surface water away from the land. This outgoing surface water is replaced by an **upwelling** of cold, nutrient-rich bottom water (Figure 6-6, p. 110). Upwellings, whether far from shore or near shore (Figure 6-3), bring plant nutrients from the deeper parts of the ocean to the surface. In turn, these nutrients support large populations of phytoplankton, zooplankton, fish, and fish-eating seabirds.

What Is the El Niño–Southern Oscillation? Every few years in the Pacific Ocean, normal

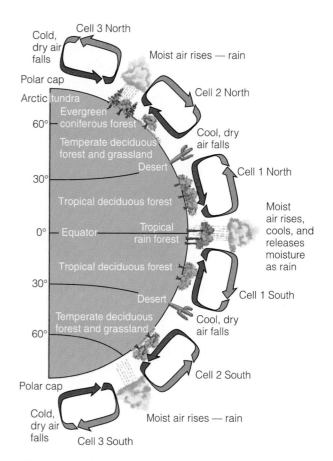

Figure 6-5 Model of global air circulation and biomes. Heat and moisture are distributed over the earth's surface by vertical currents that form six large convection cells (called Hadley cells) at different latitudes. The direction of air flow and the ascent and descent of air masses in these convection cells determine the earth's general climatic zones. The resulting uneven distribution of heat and moisture over the planet's surface leads to the forests, grasslands, and deserts that make up the earth's biomes.

shore upwellings (Figure 6-7, p. 110, left) are affected by changes in climate patterns called the *El Niño–Southern Oscillation,* or *ENSO* (Figure 6-7, right). In an ENSO, often called *El Niño,* the prevailing westerly winds weaken or cease and surface water along the South and North American coasts becomes warmer. This suppresses the normal upwellings of cold, nutrient-rich water. The decrease in nutrients reduces primary productivity and causes a sharp decline in the populations of some fish species.

A strong ENSO can trigger extreme weather changes over at least two-thirds of the globe, especially in lands along the Pacific and Indian Oceans (Figure 6-8, p. 111).

What Is La Niña? The cooling counterpart of El Niño is *La Niña.* Typically La Niña means more Atlantic Ocean hurricanes, colder winters in Canada and the northeastern United States, warmer and drier

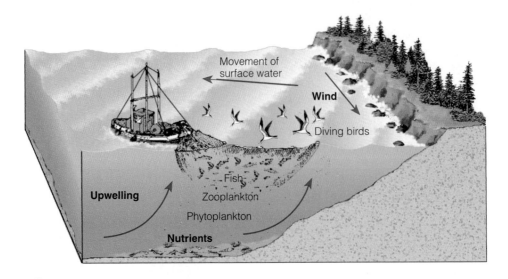

Figure 6-6 A *shore upwelling* (shown here) occurs when deep, cool, nutrient-rich waters are drawn up to replace surface water moved away from a steep coast by wind flowing along the coast toward the equator. Such areas support large populations of phytoplankton, zooplankton, fish, and fish-eating birds. *Equatorial upwellings* occur in the open sea near the equator when northward and southward currents interact to push deep waters and their nutrients to the surface, thus greatly increasing primary productivity in such areas.

winters in the southeastern and southwestern United States, wetter winters in the Pacific Northwest, torrential rains in Southeast Asia, lower wheat yields in Argentina, and more wildfires in Florida.

How Does the Chemical Makeup of the Atmosphere Lead to the Greenhouse Effect?
Small amounts of certain gases play a key role in determining the earth's average temperatures and thus its

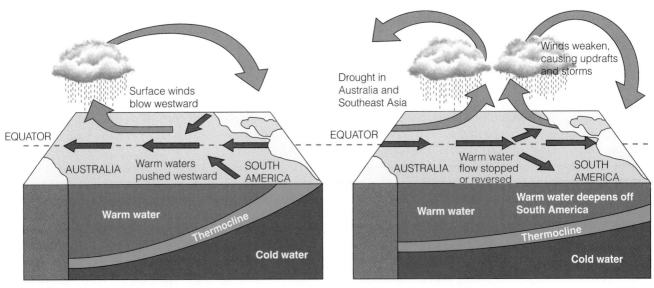

Normal Conditions　　　　　　　　**El Niño Conditions**

Figure 6-7 Normal surface winds blowing westward cause shore upwellings of cold, nutrient-rich bottom water in the tropical Pacific Ocean near the coast of Peru (left). A zone of gradual temperature change called the *thermocline* separates the warm and cold water. Every few years a climate shift known as the *El Niño–Southern Oscillation* (*ENSO*) disrupts this pattern. Westward surface winds weaken, which depresses the coastal upwellings and warms the surface waters off South America (right). When an ENSO lasts 12 months or longer, it severely disrupts populations of plankton, fish, and seabirds in upwelling areas and can trigger extreme weather changes over much of the globe (Figure 6-8).

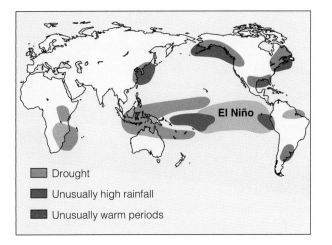

Figure 6-8 Typical global climatic effects of an El Niño–Southern Oscillation. During the 1996–98 ENSO, huge waves battered the California coast, and torrential rains caused widespread flooding and mudslides. In Peru, floods and mudslides killed hundreds of people, left about 250,000 people homeless, and ruined harvests. Drought in Brazil, Indonesia, and Australia led to massive wildfires in tinder-dry forests. India and parts of Africa also experienced severe drought. A catastrophic ice storm hit Canada and the northeastern United States, but the southeastern United States had fewer hurricanes. (Data from United Nations Food and Agriculture Organization)

climates. These gases include water vapor (H_2O), carbon dioxide (CO_2), methane (CH_4), nitrous oxide (N_2O), and synthetic chlorofluorocarbons (CFCs).

Together these gases, known as **greenhouse gases**, allow mostly visible light and some infrared radiation and ultraviolet (UV) radiation from the sun (Figure 3-8, p. 54) to pass through the troposphere. The earth's surface absorbs much of this solar energy. This transforms it to longer wavelength infrared radiation (heat), which then rises into the troposphere (Figure 4-8, p. 69).

Some of this heat escapes into space, and some is absorbed by molecules of greenhouse gases and emitted into the troposphere as even longer wavelength infrared radiation, which warms the air. This natural warming effect of the troposphere is called the **greenhouse effect** (Figure 6-9).

The basic principle behind the natural greenhouse effect is well established. Indeed, without its current greenhouse gases (especially water vapor, which is found in the largest concentration), the earth would be a cold and mostly lifeless planet.

Human activities such as burning fossil fuels, clearing forests, and growing crops release carbon dioxide, methane, and nitrous oxide into the atmosphere. There is concern that large inputs of these greenhouse gases into the troposphere can enhance the earth's natural greenhouse effect and lead to *global warming.* If so, this could alter precipitation patterns, shift areas where we can grow crops, raise average sea levels, and shift areas where some types of plants and animals can live.

How Does the Chemical Makeup of the Atmosphere Create the Ozone Layer?   In a band of the stratosphere 16–26 kilometers (11–16 miles) above the earth's surface, oxygen (O_2) is continuously converted to ozone (O_3) and back to oxygen by a sequence of reactions initiated by UV radiation from the sun

(a) Rays of sunlight penetrate the lower atmosphere and warm the earth's surface.

(b) The earth's surface absorbs much of the incoming solar radiation and degrades it to longer-wavelength infrared radiation (heat), which rises into the lower atmosphere. Some of this heat escapes into space and some is absorbed by molecules of greenhouse gases and emitted as infrared radiation, which warms the lower atmosphere.

(c) As concentrations of greenhouse gases rise, their molecules absorb and emit more infrared radiation, which adds more heat to the lower atmosphere.

Figure 6-9 The *greenhouse effect.* Without the atmospheric warming provided by this natural effect, the earth would be a cold and mostly lifeless planet. According to the widely accepted greenhouse theory, when concentrations of greenhouse gases in the atmosphere rise, the average temperature of the troposphere rises. (Modified by permission from Cecie Starr, *Biology: Concepts and Applications*, 4th ed., Pacific Grove, Calif.: Brooks/Cole, 2000)

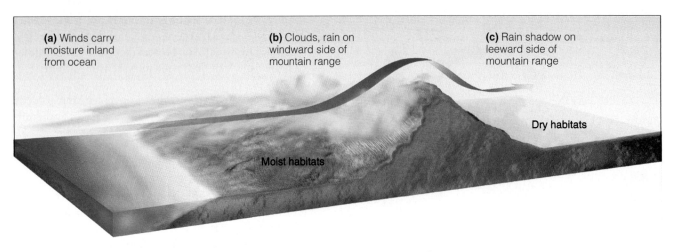

(a) Winds carry moisture inland from ocean

(b) Clouds, rain on windward side of mountain range

(c) Rain shadow on leeward side of mountain range

Dry habitats

Moist habitats

Figure 6-10 The *rain shadow effect* is a reduction of rainfall on the side of high mountains facing away from prevailing surface winds. It occurs when warm, moist air in prevailing onshore winds loses most of its moisture as rain and snow on the windward (wind-facing) slopes of a mountain range. This leads to semiarid and arid conditions on the leeward side of the mountain range and the land beyond. The Mojave Desert, east of the Sierra Nevada in California, is produced by this effect.

$(3 O_2 + UV \rightleftharpoons 2 O_3)$. The result is a thin veil of protective ozone at very low concentrations (up to 12 parts per million).

Normally, the average levels of ozone in this lifesaving layer do not change much because the rate of ozone destruction is equal to its rate of formation. This stratospheric ozone prevents at least 95% of the sun's harmful UV radiation from reaching the earth's surface.

The ozone layer also creates warm layers of air that prevent churning gases in the troposphere from entering the stratosphere. This *thermal cap* is important in determining the average temperature of the troposphere and thus the earth's current climates. Much evidence indicates that chemicals added to the atmosphere by human activities are decreasing levels of protective ozone in the stratosphere.

How Do Topography and Other Features of the Earth's Surface Create Microclimates? Various topographic features of the earth's surface create local climatic conditions, or **microclimates,** that differ from the general climate of a region. For example, mountains interrupt the flow of prevailing surface winds and the movement of storms. When moist air blowing inland from an ocean reaches a mountain range, it cools as it is forced to rise and expand. This causes the air to lose most of its moisture as rain and snow on the windward (wind-facing) slopes. As the drier air mass flows down the leeward (away from the wind) slopes, it draws moisture out of the plants and soil over which it passes. The lower precipitation and the resulting semiarid or arid conditions on the leeward side of high mountains are called the **rain shadow effect** (Figure 6-10).

Cities also create distinct microclimates. Bricks, concrete, asphalt, and other building materials absorb and hold heat, and buildings block wind flow. Motor vehicles and the climate control systems of buildings release large quantities of heat and pollutants. As a result, cities tend to have more haze and smog, higher temperatures, and lower wind speeds than the surrounding countryside.

Land–ocean interactions affect the local climates of coastal areas by creating ocean-to-land breezes (called *sea breezes*) during the day and land-to-ocean breezes (called *land breezes*) at night.

6-2 BIOMES: CLIMATE AND LIFE ON LAND

Why Do Different Organisms Live in Different Places? Why is one area of the earth's land surface a desert, another a grassland, and another a forest? Why do different types of deserts, grasslands, and forests exist?

The general answer to these questions is differences in *climate* (Figure 6-3), caused mostly by differences in average temperature and precipitation caused by global air circulation (Figures 6-5). Different climates promote different communities of organisms.

Figure 6-11 shows global distributions of *biomes:* terrestrial regions with characteristic types of natural, undisturbed ecological communities adapted to the climate of the region. By comparing Figure 6-11 with Figure 6-3, you can see how the world's major biomes vary with climate. Figure 4-9 (p. 70) shows major bi-

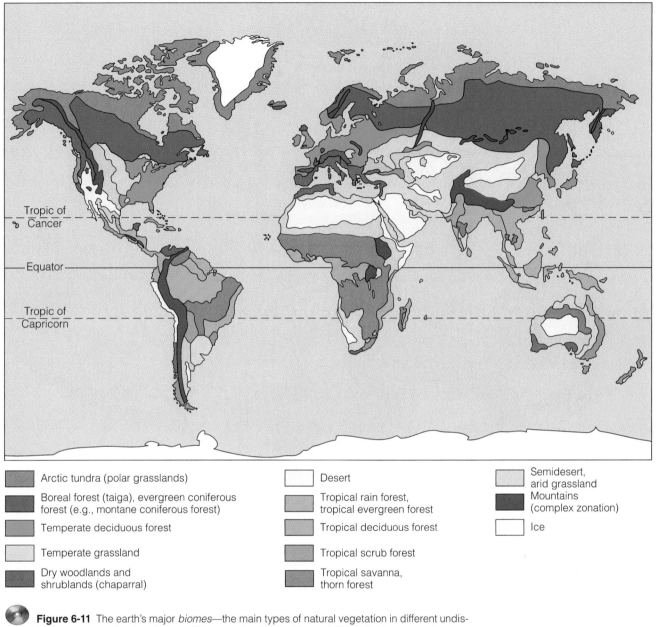

Arctic tundra (polar grasslands)

Boreal forest (taiga), evergreen coniferous forest (e.g., montane coniferous forest)

Temperate deciduous forest

Temperate grassland

Dry woodlands and shrublands (chaparral)

Desert

Tropical rain forest, tropical evergreen forest

Tropical deciduous forest

Tropical scrub forest

Tropical savanna, thorn forest

Semidesert, arid grassland

Mountains (complex zonation)

Ice

Figure 6-11 The earth's major *biomes*—the main types of natural vegetation in different undisturbed land areas—result primarily from differences in climate (Figure 6-3). Each biome contains many ecosystems whose communities have adapted to differences in climate, soil, and other environmental factors. In reality, people have removed or altered much of this natural vegetation in some areas for farming, livestock grazing, lumber and fuelwood, mining, and construction, thereby altering the biomes.

omes in the United States as one moves through different climates along the 39th parallel. Taken together, average annual precipitation and temperature (along with soil type) are the most important factors in producing tropical, temperate, or polar deserts, grasslands, and forests (Figure 6-12, p. 114).

On maps such as the one in Figure 6-11, biomes are presented as having sharp boundaries and being covered with the same general type of vegetation. In reality, *biomes are not uniform.* They consist of a *mosaic of patches,* all with somewhat different biological communities but with similarities unique to the biome.

Figure 6-13 (p. 114) shows how climate and vegetation vary with **latitude** (distance from the equator) and **altitude** (elevation above sea level). If you travel from the equator toward either pole, you will generally encounter colder climates and zones of vegetation adapted to those climates (Figure 6-13, right). Similarly, as elevation above sea level increases, climate becomes colder (Figure 6-13, left). Thus, if you climb a tall mountain from its base to its summit, you can observe changes in plant life similar to those you would encounter in traveling from the equator to the earth's poles.

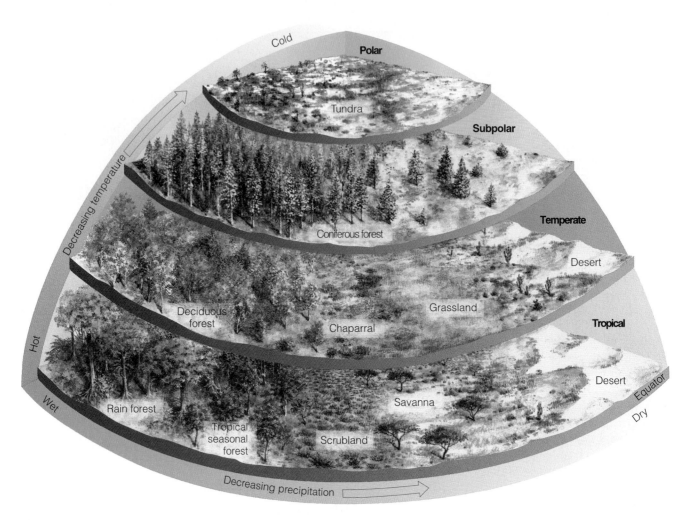

Figure 6-12 Average precipitation and average temperature, acting together as limiting factors over a period of 30 or more years, determine the type of desert, grassland, or forest biome in a particular area. Although the actual situation is much more complex, this simplified diagram explains how climate determines the types and amounts of natural vegetation found in an area left undisturbed by human activities. (Used by permission of Macmillan Publishing Company, from Derek Elsom, *The Earth,* New York: Macmillan, 1992. Copyright © 1992 by Marshall Editions Developments Limited)

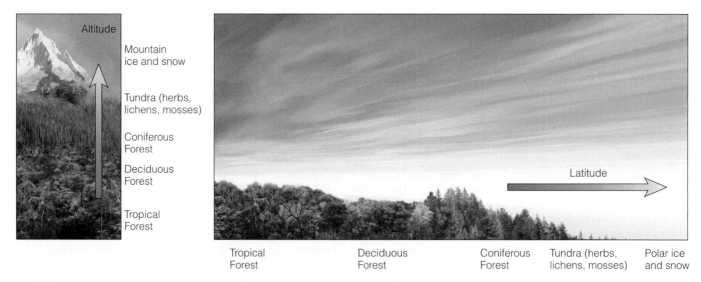

Figure 6-13 Generalized effects of altitude (left) and latitude (right) on climate and biomes. Parallel changes in vegetation type occur when we travel from the equator to the poles or from lowlands to mountaintops.

6-3 DESERT AND GRASSLAND BIOMES

What Are the Major Types of Deserts? A **desert** is an area where evaporation exceeds precipitation. Annual precipitation is low and often scattered unevenly throughout the year. Deserts have sparse, widely spaced, mostly low vegetation.

During the day the baking sun warms the ground in the desert. At night, however, most of the heat stored in the ground radiates quickly into the atmosphere. This occurs because desert soils have little vegetation and moisture to help store the heat and the skies are usually clear. This explains why in a desert you may roast during the day but shiver at night.

A combination of low rainfall and different average temperatures creates tropical, temperate, and cold deserts (Figure 6-12). In *tropical deserts,* such as the southern Sahara in Africa, temperatures usually are high year-round and there is little rain, which typically falls during only 1 or 2 months of the year. These driest places on earth have few plants and a hard, wind-blown surface strewn with rocks and some sand.

In *temperate deserts,* such as the Mojave in southern California (Figure 6-10), daytime temperatures are high in summer and low in winter and there is more precipitation than in tropical deserts. Vegetation is sparse. It consists mostly of widely dispersed, drought-resistant shrubs and cacti or other succulents, and animals are adapted to the lack of water and temperature variations (Figure 6-14). In *cold deserts,* such as the Gobi Desert in China, winters are cold, summers are warm or hot, and precipitation is low.

Figure 6-14 Some components and interactions in a *temperate desert ecosystem.* When these organisms die, decomposers break down their organic matter into minerals that plants use. Colored arrows indicate transfers of matter and energy between producers, primary consumers (herbivores), secondary (or higher-level) consumers (carnivores), and decomposers. Organisms are not drawn to scale.

Producer to primary consumer → Primary to secondary consumer → Secondary to higher-level consumer → All producers and consumers to decomposers

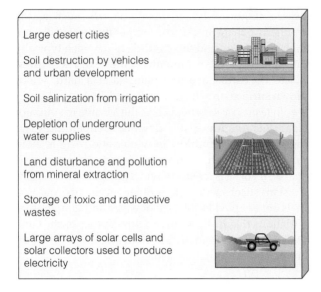

Figure 6-15 Major human impacts on the world's deserts.

- Large desert cities
- Soil destruction by vehicles and urban development
- Soil salinization from irrigation
- Depletion of underground water supplies
- Land disturbance and pollution from mineral extraction
- Storage of toxic and radioactive wastes
- Large arrays of solar cells and solar collectors used to produce electricity

In the semiarid zones between deserts and grasslands we find *semidesert*. This biome is dominated by thorn trees and shrubs adapted to long dry spells followed by brief, sometimes heavy rains.

How Do Desert Plants and Animals Survive?

Adaptations for survival in the desert have two themes. One is *beat the heat* and the other *every drop of water counts*. Desert plants exposed to the sun all day long must be able to conserve enough water for survival and lose enough heat so they do not overheat and die. During long hot and dry spells plants such as mesquite and creosote drop their leaves to survive in a dormant state. During dry spells plants such as mosses and lichens become dormant.

Succulent (fleshy) **plants,** such as the saguaro ("sah-WAH-ro") cactus, survive in dry climates by having no leaves, storing water and synthesizing food in their expandable, fleshy tissue, and reducing water loss by opening their pores (stomata) to take up carbon dioxide (CO_2) only at night. Some desert plants *obtain water* by using deep roots to tap into groundwater. Others such as prickly pear (Figure 6-14) and saguaro cacti use widely spread, shallow roots to collect water after brief showers and store it in their spongy tissue.

Some plants *conserve water* by having wax-coated leaves that minimize transpiration (evergreens such as the creosote bush). Others store much of their biomass in seeds that remain inactive, sometimes for years, until they receive enough water to germinate (annual wildflowers and grasses). Shortly after a rain these seeds germinate, grow, carpet such deserts with a dazzling array of colorful flowers, produce new seed, and die, all in only a few weeks.

Most desert animals are small. Some beat the heat by hiding in cool burrows or rocky crevices by day and coming out at night or in the early morning. Others become dormant during periods of extreme heat or drought.

Some desert animals have physical adaptations for conserving water. For example, insects and reptiles have thick outer coverings to minimize water loss through evaporation, and their wastes are dry feces and a dried concentrate of urine. Some animals found in deserts get their water from dew or from the food they eat (many spiders and insects). Arabian oryxes survive by licking the dew that accumulates at night on rocks and on one another's hair.

Figure 6-15 shows major human impacts on deserts. Deserts take a long time to recover from disturbances because of their slow plant growth, low species diversity, slow nutrient cycling (because of little bacterial activity in their soils), and water shortages. Desert vegetation destroyed by livestock overgrazing and off-road vehicles may take decades to grow back.

What Are the Major Types of Grasslands?

Grasslands are regions with enough average annual precipitation to allow grass (and in some areas, a few trees) to prosper but with precipitation so erratic that drought and fire prevent large stands of trees from growing. Most grasslands are found in the interiors of continents (Figure 6-11).

Grasslands persist because of a combination of seasonal drought, grazing by large herbivores, and occasional fires—all of which keep large numbers of shrubs and trees from invading and becoming established. If not overgrazed by large herbivores, grasses in these biomes are renewable resources because these plants grow out from the bottom. This allows their stems to grow again after being nibbled off by grazing animals.

The three main types of grasslands—tropical, temperate, and polar (tundra)—result from combinations of low average precipitation and various average temperatures (Figure 6-12).

One type of tropical grassland, called *savanna*, usually has warm temperatures year-round, two prolonged dry seasons, and abundant rain the rest of the year. African tropical savannas contain enormous herds of *grazing* (grass- and herb-eating) and *browsing* (twig- and leaf-nibbling) hoofed animals, including wildebeests, gazelles, zebras, giraffes, and antelopes (Figure 6-16). These and other large herbivores have

Beisa oryx

Cape buffalo

Wildebeest

Topi

Warthog

Thompson's
gazelle

Waterbuck

Grant's zebra

Dry Grassland

Moist Grassland

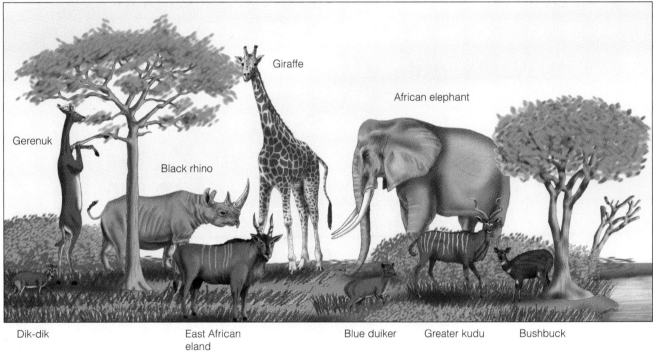

Giraffe

African elephant

Gerenuk

Black rhino

Dik-dik

East African
eland

Blue duiker

Greater kudu

Bushbuck

Dry Thorn Scrub

Riverine Forest

Figure 6-16 Some of the grazing animals found in different parts of the African savanna. These species share vegetation resources by having different feeding niches.

evolved specialized eating habits that minimize competition between species for vegetation. For example, giraffes eat leaves and shoots from the tops of trees, elephants eat leaves and branches further down, Thompson's gazelles and wildebeests prefer short grass, and zebras graze on longer grass and stems.

In *temperate grasslands* winters are bitterly cold, summers are hot and dry, and annual precipitation is fairly sparse and falls unevenly through the year. Drought, occasional fires, and intense grazing inhibit the growth of trees and bushes, except along rivers.

Because the aboveground parts of most of the grasses die and decompose each year, organic matter accumulates to produce a deep, fertile soil. This soil is held in place by a thick network of intertwined roots of drought-tolerant grasses unless the topsoil is plowed

up and allowed to blow away by prolonged exposure to high winds found in these biomes.

Types of temperate grasslands are the *tall-grass prairies* (Figure 6-17) and *short-grass prairies* of the midwestern and western United States and Canada, the South American *pampas*, the African *veldt*, and the *steppes* of central Europe and Asia. Here winds blow almost continuously, and evaporation is rapid, often leading to fires in the summer and fall.

Because of their thick and fertile soils, temperate grasslands are plowed up and widely used to grow crops (Figure 6-18). However, plowing breaks up the soil and leaves it vulnerable to erosion by wind and water.

Polar grasslands, or *arctic tundra,* occur just south of the arctic polar ice cap (Figure 6-11). During most of

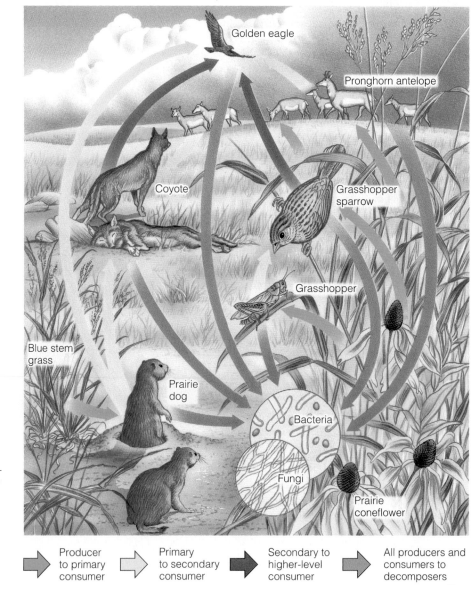

Figure 6-17 Some components and interactions in *a temperate tall-grass prairie ecosystem* in North America. When these organisms die, decomposers break down their organic matter into minerals that plants use. Colored arrows indicate transfers of matter and energy between producers, primary consumers (herbivores), secondary (or higher-level) consumers (carnivores), and decomposers. Organisms are not drawn to scale.

Producer to primary consumer

Primary to secondary consumer

Secondary to higher-level consumer

All producers and consumers to decomposers

Figure 6-18 Replacement of a temperate grassland with a monoculture crop in California. When the tangled root network of natural grasses is removed, the fertile topsoil is subject to severe wind erosion unless it is covered with some type of vegetation.

the year these treeless plains are bitterly cold, swept by frigid winds, and covered with ice and snow. Winters are long and dark, and the scant precipitation falls mostly as snow.

This biome is carpeted with a thick, spongy mat of low-growing plants, primarily grasses, mosses, and dwarf woody shrubs. Trees or tall plants cannot survive in the cold, windy tundra) because they would lose too much of their heat. Most of the annual growth of the tundra's plants occurs during the 6- to 8-week summer, when the sun shines almost around the clock.

One effect of the extreme cold is **permafrost,** a perennially frozen layer of the soil that forms when the water there freezes. In summer, water near the surface thaws, but the permafrost soil layer below stays frozen and prevents liquid water at the surface from seeping into the ground. Thus during the brief summer, the soil above the permafrost layer remains waterlogged, forming a large number of shallow lakes, marshes, bogs, ponds, and other seasonal wetlands. Hordes of mosquitoes, blackflies, and other insects thrive in these shallow surface pools. They feed large colonies of migratory birds (especially waterfowl) that return from the south to nest and breed in the bogs and ponds.

Another type of tundra, called *alpine tundra,* occurs above the limit of tree growth but below the permanent snow line on high mountains (Figure 6-13, left). The vegetation is similar to that found in arctic

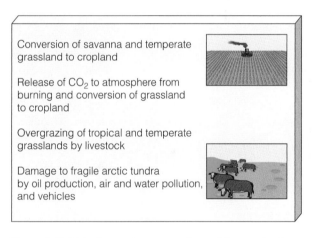

Conversion of savanna and temperate grassland to cropland

Release of CO_2 to atmosphere from burning and conversion of grassland to cropland

Overgrazing of tropical and temperate grasslands by livestock

Damage to fragile arctic tundra by oil production, air and water pollution, and vehicles

Figure 6-19 Major human impacts on the world's grasslands.

tundra, but it gets more sunlight than arctic vegetation and has no permafrost layer.

Figure 6-19 lists the major human impacts on grasslands.

6-4 FOREST AND MOUNTAIN BIOMES

What Are the Major Types of Forests? Undisturbed areas with moderate to high average annual precipitation tend to be covered with **forest,** which contains various species of trees and smaller forms of

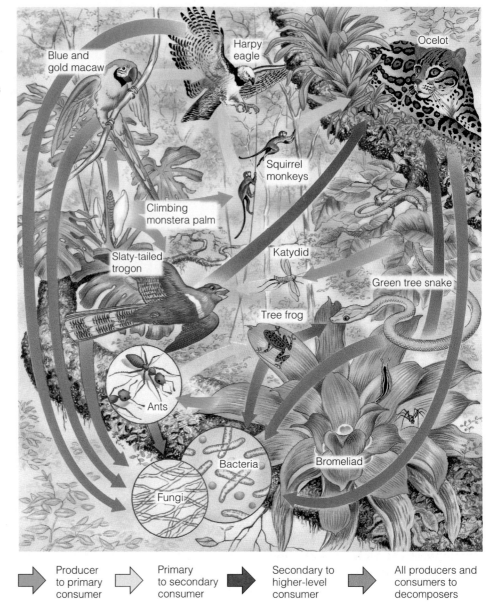

Figure 6-20 Some components and interactions in a *tropical rain forest ecosystem*. When these organisms die, decomposers break down their organic matter into minerals that plants use. Colored arrows indicate transfers of matter and energy between producers, primary consumers (herbivores), secondary (or higher-level) consumers (carnivores), and decomposers. Organisms are not drawn to scale.

Blue and gold macaw

Harpy eagle

Ocelot

Squirrel monkeys

Climbing monstera palm

Katydid

Slaty-tailed trogon

Green tree snake

Tree frog

Ants

Bacteria

Bromeliad

Fungi

Producer to primary consumer

Primary to secondary consumer

Secondary to higher-level consumer

All producers and consumers to decomposers

vegetation. The three main types of forest—*tropical, temperate,* and *boreal* (polar)—result from combinations of this precipitation level and various average temperatures (Figure 6-12).

Tropical rain forests (Figure 6-20) are found near the equator (Figure 6-11), where hot, moisture-laden air rises and dumps its moisture (Figure 6-5). These forests have a warm annual mean temperature (which varies little, daily or seasonally), high humidity, and heavy rainfall almost daily.

Trees of wet tropical rain forests tend to be **broadleaf evergreen plants,** which keep most of their broad leaves year-round. The large surface area of the leaves allows them to collect ample sunlight for photosynthesis and radiate heat during hot weather.

Tropical rain forests have incredible biological diversity. The diverse life-forms occupy a variety of specialized niches in distinct layers, based in the plants' case mostly on their need for sunlight (Figure 6-21). The stratification of specialized plant and animal niches in various layers of a tropical rain forest enables coexistence of a great variety of species (biodiversity). Although tropical rain forests cover only about 2% of the earth's land surface, they are habitats for 50–80% of the earth's terrestrial species.

Dropped leaves, fallen trees, and dead animals decompose quickly because of the warm, moist conditions and hordes of decomposers. This rapid recycling of scarce soil nutrients is why little litter is found on the ground. Instead of being stored in the soil, most miner-

The following labels appear on the figure (with the vertical Height (meters) axis marked from 0 to 45):

- Emergent layer
- Harpy eagle
- Toco toucan
- Canopy
- Wooly opossum
- Understory
- Brazilian tapir
- Shrub layer
- Black-crowned antpitta
- Ground layer

Figure 6-21 Stratification of specialized plant and animal niches in various layers of a *tropical rain forest*. The presence of these specialized niches enables species to avoid or minimize competition for resources and results in the coexistence of a great variety of species (biodiversity).

als released by decomposition are taken up quickly by plants. Thus most of a tropical rain forest's nutrients are stored in the biomass of its living organisms.

Because of the dense vegetation, little wind blows in tropical rain forests, eliminating the possibility of wind pollination. Many of the plants have evolved elaborate flowers that attract particular insects, birds, or bats as pollinators.

When we move a little farther from the equator (Figure 6-11), we find *tropical deciduous forests* (sometimes called *tropical monsoon forests* or *tropical seasonal forests*). These forests are warm year-round and get most of their plentiful rainfall during a wet (monsoon) season that is followed by a long dry season. They contain a mixture of *deciduous trees* (which lose their leaves to survive the dry season) and drought-tolerant *evergreen trees* (which retain most of their leaves year-round). Where the dry season is especially long, we find *tropical scrub forests* (Figure 6-11) containing mostly small deciduous trees and shrubs.

Temperate deciduous forests (Figure 6-22, p. 122) grow in areas with moderate average temperatures that change significantly with the season. These areas have long, warm summers, cold but not too severe winters, and abundant precipitation (often spread fairly evenly throughout the year).

This biome is dominated by a few species of broadleaf deciduous trees such as oak, hickory, maple, poplar, and sycamore. They survive cold winters by dropping their leaves in the fall and becoming dormant.

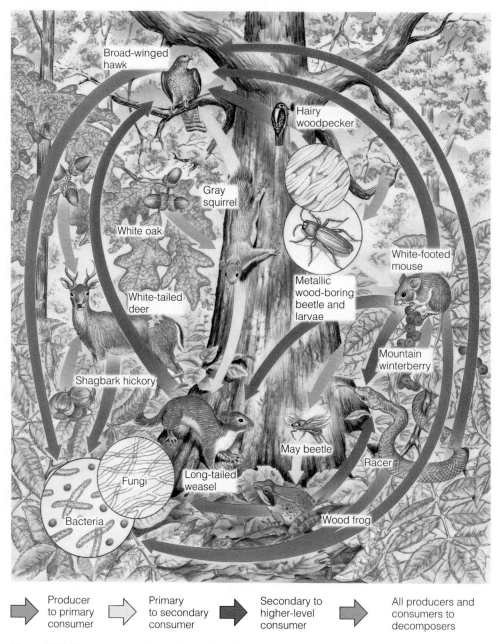

| ⟹ | Producer to primary consumer | ⟹ | Primary to secondary consumer | ⟹ | Secondary to higher-level consumer | ⟹ | All producers and consumers to decomposers |

Figure 6-22 Some components and interactions in a *temperate deciduous forest ecosystem*. When these organisms die, decomposers break down their organic matter into minerals that plants use. Colored arrows indicate transfers of matter and energy between producers, primary consumers (herbivores), secondary (or higher-level) consumers (carnivores), and decomposers. Organisms are not drawn to scale.

Each spring they grow new leaves that change in the fall into an array of reds and golds before dropping (see photos on p. xviii). Because of the fairly low rate of decomposition, these forests accumulate a thick layer of slowly decaying leaf litter that is a storehouse of nutrients.

Evergreen coniferous forests, also called *boreal forests* and *taigas* ("TIE-guhs"), are found just south of the arc-

tic tundra in northern regions across North America, Asia, and Europe (Figure 6-11). In this subarctic climate, winters are long, dry, and extremely cold; in the northernmost taiga, sunlight is available only 6–8 hours a day. Summers are short, with mild to warm temperatures, and the sun typically shines 19 hours a day.

Most boreal forests are dominated by a few species of *coniferous* (cone-bearing) *evergreen trees* such as spruce, fir, cedar, hemlock, and pine that keep some of their narrow-pointed leaves (needles) all year long. The small, needle-shaped, waxy-coated leaves of these trees can withstand the intense cold and drought of winter when snow blankets the ground. The trees are ready to take advantage of the brief summers in these areas without taking time to grow new needles. Plant diversity is low in these forests because few species can survive the winters when soil moisture is frozen.

Beneath the stands of trees is a deep layer of partially decomposed conifer needles and leaf litter. Decomposition is slow because of the low temperatures, waxy coating of conifer needles, and high soil acidity. As the conifer needles decompose, they make the thin, nutrient-poor soil acidic and prevent most other plants (except certain shrubs) from growing on the forest floor.

These biomes contain a variety of wildlife (Figure 6-23). During the brief summer the soil becomes

| Producer to primary consumer | Primary to secondary consumer | Secondary to higher-level consumer | All producers and consumers to decomposers |

Figure 6-23 Some components and interactions in an *evergreen coniferous (boreal* or *taiga) forest ecosystem.* When these organisms die, decomposers break down their organic matter into minerals that plants use. Colored arrows indicate transfers of matter and energy between producers, primary consumers (herbivores), secondary (or higher-level) consumers (carnivores), and decomposers. Organisms are not drawn to scale.

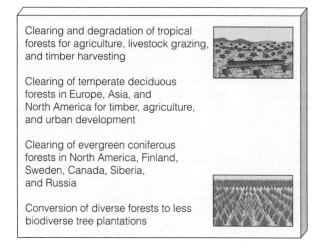

Clearing and degradation of tropical forests for agriculture, livestock grazing, and timber harvesting

Clearing of temperate deciduous forests in Europe, Asia, and North America for timber, agriculture, and urban development

Clearing of evergreen coniferous forests in North America, Finland, Sweden, Canada, Siberia, and Russia

Conversion of diverse forests to less biodiverse tree plantations

Figure 6-24 Major human impacts on the world's forests.

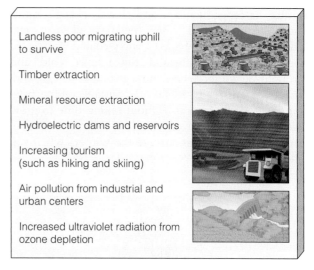

Landless poor migrating uphill to survive

Timber extraction

Mineral resource extraction

Hydroelectric dams and reservoirs

Increasing tourism (such as hiking and skiing)

Air pollution from industrial and urban centers

Increased ultraviolet radiation from ozone depletion

Figure 6-25 Major human impacts on the world's mountains.

waterlogged, forming acidic bogs, or *muskegs*, in low-lying areas of these forests. Warblers and other insect-eating birds feed on hordes of flies, mosquitoes, and caterpillars.

Coastal coniferous forests or *temperate rain forests* are found in scattered coastal temperate areas with ample rainfall or moisture from dense ocean fogs. Dense stands of large conifers such as Sitka spruce, Douglas fir, and redwoods dominate undisturbed areas of biomes along the coast of North America, from Canada to northern California.

Figure 6-24 lists major human impacts on the world's forests.

Why Are Mountains Ecologically Important? Some of the world's most spectacular and important environments are mountains, which make up about 20% of the earth's land surface. Mountains are places where dramatic changes in altitude, climate, soil, and vegetation take place over a very short distance (Figure 6-13, left). Because of the steep slopes, mountain soils are especially prone to erosion when the vegetation holding them in place is removed by natural disturbances (such as landslides and avalanches) or human activities (such as timber cutting and agriculture).

Many freestanding mountains are *islands of biodiversity* surrounded by a sea of lower elevation landscapes transformed by human activities.

Mountains play a number of important ecological roles. They contain the majority of the world's forests, which are habitats for much of the world's terrestrial biodiversity. They often are habitats for endemic species found nowhere else on earth and serve as sanctuaries for animal species driven from lowland areas.

They also help regulate the earth's climate when mountaintops covered with ice and snow reflect solar radiation back into space. Mountains affect sea levels as a result of decreases or increases in glacial ice—most of which is locked up in Antarctica (the most mountainous of all continents).

Finally, mountains play a critical role in the hydrologic cycle by gradually releasing melting ice, snow, and water stored in the soils and vegetation of mountainsides to small streams.

Despite their ecological, economic, and cultural importance, the fate of mountain ecosystems has not been a high priority of governments or many environmental organizations. Mountain ecosystems are coming under increasing pressure from several human activities (Figure 6-25).

6-5 AQUATIC ENVIRONMENTS: TYPES AND CHARACTERISTICS

What Are the Two Major Types of Aquatic Life Zones? The aquatic equivalents of biomes are called *aquatic life zones*. The major types of organisms found in aquatic environments are determined by the water's *salinity* (the amounts of various salts such as sodium chloride [NaCl] dissolved in a given volume of water). As a result, aquatic life zones are divided into two major types: *saltwater* or *marine* (particularly estuaries, coastlines, coral reefs, coastal marshes, mangrove swamps, and oceans) and *freshwater* (particularly lakes and ponds, streams and rivers, and in-

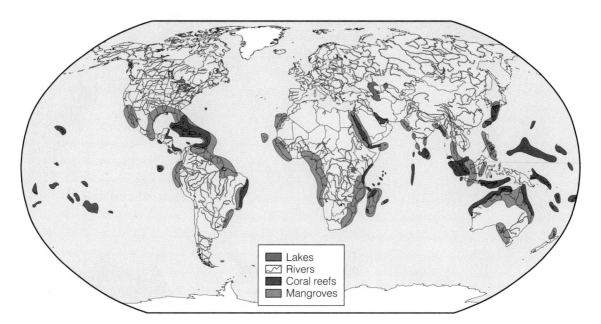

Figure 6-26 The aquatic world. Distribution of the world's major saltwater oceans, coral reefs, mangroves, and freshwater lakes and rivers.

Legend:
- Lakes
- Rivers
- Coral reefs
- Mangroves

land wetlands). Figure 6-26 shows the distribution of the world's major oceans, lakes, rivers, coral reefs, and mangroves.

What Are the Main Kinds of Organisms in Aquatic Life Zones? Saltwater and freshwater life zones contain several major types of organisms. *One* type of aquatic life is weakly swimming, free-floating **plankton** (consisting mostly of *phytoplankton,* or *plant plankton,* and *zooplankton,* or *animal plankton*). A *second* type is strongly swimming consumers (**nekton**) such as fish, turtles, and whales. A *third* type, bottom-dwellers (**benthos**), include barnacles and oysters that anchor themselves to one spot, worms that burrow into the sand or mud, and lobsters and crabs that walk about on the bottom. *Finally,* **decomposers** (mostly bacteria) break down the organic compounds in the dead bodies and wastes of aquatic organisms into simple nutrient compounds for use by producers.

What Factors Limit Life at Different Depths in Aquatic Life Zones? Most aquatic life zones can be divided into three layers: *surface, middle,* and *bottom.* Important environmental factors determining the types and numbers of organisms found in these layers are *temperature, access to sunlight for photosynthesis, dissolved oxygen content,* and *availability of nutrients* such as carbon (as dissolved CO_2 gas), nitrogen (as NO_3^-), and phosphorus (mostly as PO_4^{3-}) for producers.

Photosynthesis is confined mostly to the upper layer, or **euphotic zone,** of deep aquatic systems through which sunlight can penetrate. The depth of the euphotic zone in oceans and deep lakes can be reduced by excessive algal growth (algal blooms) that make water cloudy.

In shallow waters in streams, ponds, and oceans, ample supplies of nutrients for primary producers are usually available. By contrast, in the open ocean, nitrates, phosphates, iron, and other nutrients often are in short supply and limit net primary productivity (NPP) (Figure 4-22, p. 80). However, NPP is much higher in parts of the open ocean where upwellings (Figures 6-3 and 6-6) bring such nutrients from the ocean bottom to the surface for use by producers.

Most creatures living on the bottom of the deep ocean and deep lakes depend on animal and plant plankton that die and fall into deep waters. Because this food is limited, deep-dwelling species tend to reproduce slowly. Thus they are especially vulnerable to depletion from overfishing.

6-6 SALTWATER LIFE ZONES

Why Are the Oceans Important? A more accurate name for Earth would be *Ocean* because saltwater oceans cover about 71% of the planet's surface (Figure 6-27, p. 126). The world's oceans make up 99.5%

Figure 6-27 The ocean planet. The salty oceans cover about 71% of the earth's surface. About 97% of the earth's water is in the interconnected oceans, which cover 90% of the planet's mostly ocean hemisphere (left) and 50% of its land–ocean hemisphere (right).

Ocean hemisphere   Land–ocean hemisphere

Figure 6-28 *Natural capital: marine biodiversity.* Some ocean inhabitants.

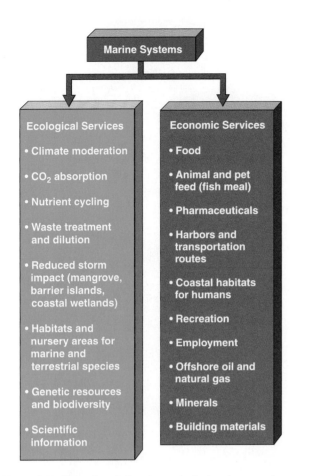

Marine Systems

Ecological Services

- Climate moderation
- CO$_2$ absorption
- Nutrient cycling
- Waste treatment and dilution
- Reduced storm impact (mangrove, barrier islands, coastal wetlands)
- Habitats and nursery areas for marine and terrestrial species
- Genetic resources and biodiversity
- Scientific information

Economic Services

- Food
- Animal and pet feed (fish meal)
- Pharmaceuticals
- Harbors and transportation routes
- Coastal habitats for humans
- Recreation
- Employment
- Offshore oil and natural gas
- Minerals
- Building materials

Figure 6-29 *Natural capital:* major ecological and economic services provided by marine systems.

of the world's habitable volume, contain about 250,000 known species of marine plants and animals (Figure 6-28), and provide many important ecological and economic services (Figure 6-29).

Despite its ecological and economic importance, less than 5% of the earth's global ocean has been explored and mapped with the same level of detail as the surface of the moon and Mars. According to aquatic scientists, the scientific investigation of poorly understood marine and freshwater aquatic systems is a greatly underfunded *research frontier* whose study could result in immense ecological and economic benefits.

What Is the Coastal Zone? Oceans have two major life zones: the *coastal zone* and the *open sea* (Figure 6-30, p. 128). The **coastal zone** is the warm, nutrient-rich, shallow water that extends from the high-tide mark on land to the gently sloping, shallow edge of the *continental shelf* (the submerged part of the continents). This zone has numerous interactions with the land and thus human activities easily affect it.

Although it makes up less than 10% of the world's ocean area, the coastal zone contains 90% of all marine species and is the site of most large commercial marine fisheries. Most ecosystems found in the coastal zone have a very high net primary productivity per unit of area (Figure 4-22, p. 80). This occurs because of the zone's ample supplies of sunlight and plant nutrients (flowing from land and distributed by wind and ocean currents).

What Are Estuaries, Coastal Wetlands, and Mangrove Swamps? One highly productive area in the coastal zone is an **estuary,** a partially enclosed area of coastal water where sea water mixes with fresh water and nutrients from rivers, streams, and runoff from land (Figure 6-31, p. 128). Estuaries and their associated **coastal wetlands** (land areas covered with water all or part of the year) include river mouths, inlets, bays, sounds, mangrove forest swamps in tropical waters (Figure 6-26), and salt marshes in temperate zones (Figure 6-32, p. 129).

The constant water movement stirs up the nutrient-rich silt, making it available to producers. This explains why estuaries and their associated coastal wetlands are some of the earth's most productive ecosystems (Figure 4-22, p. 80) and provide other important ecological and economic services (Figure 6-29).

What Niches Do Rocky and Sandy Shores Provide? The area of shoreline between low and high tides is called the **intertidal zone.** Organisms living in this stressful zone must be able to avoid being swept away or crushed by waves, and avoid or cope with being immersed during high tides and left high and dry (and much hotter) at low tides. They must also cope with changing levels of salinity when heavy rains dilute salt water. To deal with such stresses, most intertidal organisms hold on to something, dig in, or hide in protective shells.

Some coasts have steep *rocky shores* pounded by waves. The numerous pools and other niches in the rocks in the intertidal zone of rocky shores contain a great variety of species (Figure 6-33, top, p. 130).

Other coasts have gently sloping *barrier beaches,* or *sandy shores,* with niches for different marine organisms, including crabs, lugworms, clams, ghost shrimp, sand dollars, and flounder (Figure 6-33, bottom). Most of them are hidden from view and survive by burrowing, digging, and tunneling in the sand. These sandy beaches and their adjoining coastal wetlands are also home to a variety of shorebirds that feed in specialized niches on crustaceans, insects, and other organisms (Figure 5-7, p. 100).

One or more rows of natural sand dunes on undisturbed barrier beaches (with the sand held in place by the roots of grasses) serve as the first line of defense against the ravages of the sea (Figure 6-34, p. 131). But when coastal developers remove the protective dunes

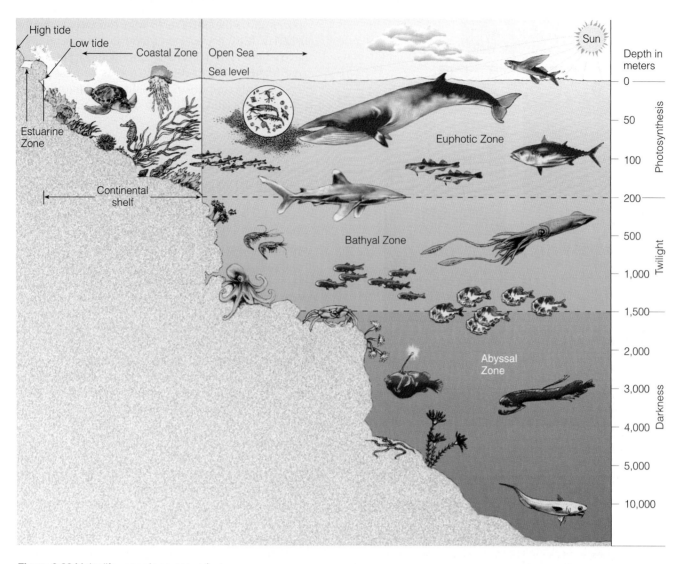

Figure 6-30 Major life zones in an ocean (not drawn to scale). Actual depths of zones may vary.

Figure 6-31 View of an *estuary* taken from space. The photo shows the sediment plume at the mouth of Madagascar's Betsiboka River as it flows through the estuary and into the Mozambique Channel. Because of its topography, heavy rainfall, and the clearing of forests for agriculture, Madagascar is the world's most eroded country.

Herring gulls

Peregrine falcon

Snowy egret

Cordgrass

Short-billed dowitcher

Phytoplankton

Marsh periwinkle

Smelt

Soft-shelled clam

Zooplankton and small crustaceans

Clamworm

Bacteria

Producer to primary consumer

Primary to secondary consumer

Secondary to higher-level consumer

All consumer and producers to decomposers

Figure 6-32 Some components and interactions in a *salt marsh ecosystem* in a temperate area such as the United States. When these organisms die, decomposers break down their organic matter into minerals used by plants. Colored arrows indicate transfers of matter and energy between consumers (herbivores), secondary (or higher-level) consumers (carnivores), and decomposers. Organisms are not drawn to scale.

or build behind the first set of dunes, storms can flood and even sweep away seaside buildings.

What Are Coral Reefs? Coral reefs (photo on p. xxiii) form in clear, warm coastal waters of the tropics and subtropics (Figure 6-26). These beautiful natural wonders are among the world's oldest, most diverse, and productive ecosystems, and they are homes for about one-fourth of all marine species (Figure 6-35, p. 132).

Coral reefs are formed by massive colonies of tiny animals called *polyps* that are close relatives of jellyfish. They slowly build reefs by secreting a protective crust of limestone (calcium carbonate) around their soft bodies. When the corals die, their empty crusts or outer skeletons remain as a platform for more reef growth. The result is an elaborate network of crevices, ledges, and holes that serve as calcium carbonate "condominiums" for a variety of marine animals.

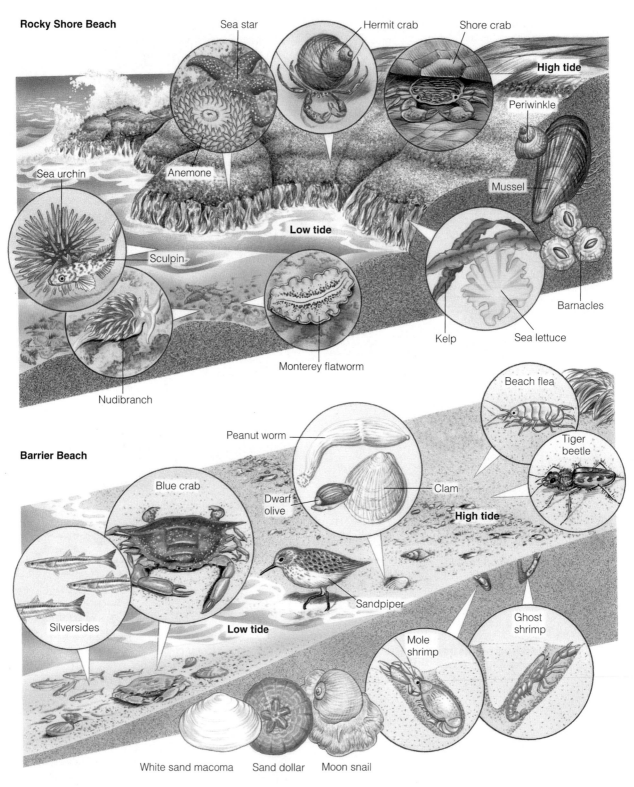

Rocky Shore Beach

Sea star

Hermit crab

Shore crab

High tide

Periwinkle

Anemone

Sea urchin

Mussel

Sculpin

Barnacles

Low tide

Nudibranch

Monterey flatworm

Kelp

Sea lettuce

Barrier Beach

Peanut worm

Beach flea

Tiger beetle

Blue crab

Clam

Dwarf olive

High tide

Silversides

Sandpiper

Low tide

Mole shrimp

Ghost shrimp

White sand macoma   Sand dollar   Moon snail

Figure 6-33 Living between the tides. Some organisms with specialized niches found in various zones on rocky shore beaches (top) and barrier or sandy beaches (bottom). Organisms are not drawn to scale.

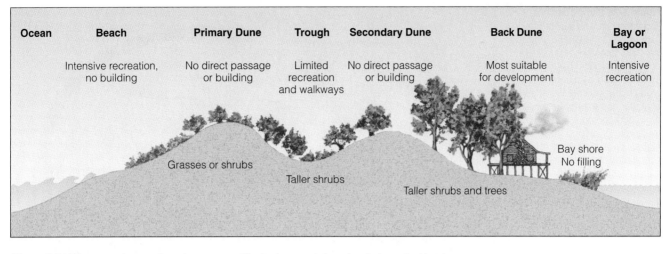

Ocean	Beach	Primary Dune	Trough	Secondary Dune	Back Dune	Bay or Lagoon
Intensive recreation, no building		No direct passage or building	Limited recreation and walkways	No direct passage or building	Most suitable for development	Intensive recreation

Figure 6-34 Primary and secondary dunes on gently sloping sandy beaches help protect land from erosion by the sea. The roots of various grasses that colonize the dunes help hold the sand in place. Ideally, construction is allowed only behind the second strip of dunes, and walkways to the beach are built over the dunes to keep them intact. This helps preserve barrier beaches and protect buildings from damage by wind, high tides, beach erosion, and flooding from storm surges. Such protection is rare because the short-term economic value of ocean-front land is considered much higher than its long-term ecological and economic values.

Coral reefs are a joint and mutually beneficial relationship between the polyps and tiny single-celled algae called *zooxanthellae* ("zoh-ZAN-thel-ee") that live in the tissues of the polyps. The algae provide the polyps with color, food, and oxygen through photosynthesis. The polyps in turn provide a well-protected home for the algae.

Although coral reefs occupy only about 0.1% of the world's ocean area, they provide a number of important ecological and economic services valued conservatively at $375 billion a year. These services include the following:

Ecological

■ Removing some of the carbon dioxide from the atmosphere as part of the carbon cycle (when coral polyps form limestone shells)

■ Acting as natural barriers that help protect 15% of the world's coastlines from erosion by battering waves and storms and allow the ocean to replenish nearby beaches with sand

■ Supporting at least one-fourth of all identified marine species and 65% of marine fish species even though such reefs occupy less than 1% of the ocean floor

Economic

■ Producing roughly 10% of the global fish catch and 25% of the catch in developing countries

■ Providing fish and shellfish, jobs, and building materials for some of the world's poorest countries

■ Supporting fishing and tourism industries worth billions of dollars each year

Coral reefs are vulnerable to damage because they grow slowly and are disrupted easily. They also thrive only in clear, warm, and fairly shallow water of constant high salinity. Corals can live only in water with a temperature of 18–30°C (64–86°F).

The biodiversity of coral reefs can be reduced by natural disturbances such as severe storms, freshwater floods, and invasions of predatory fish. However, throughout their very long geologic history, coral reefs have been able to adapt to such natural environmental changes. Today the biggest threats to the biodiversity of many of the world's coral reefs come from human activities (Figure 6-36, p. 133). Scientists are concerned that these threats are occurring so rapidly (over decades) and over such a wide area that many of the world's coral systems may not have enough time to adapt.

What Biological Zones Are Found in the Open Sea? The sharp increase in water depth at the edge of the continental shelf separates the coastal zone from the vast volume of the ocean called the **open sea.** Primarily on the basis of the penetration of sunlight, it is divided into three vertical zones (Figure 6-30).

The *euphotic zone* is the lighted upper zone of the open sea where photosynthesis occurs, mostly in

Figure 6-35 Some components and interactions in a *coral reef ecosystem.* When these organisms die, decomposers break down their organic matter into minerals used by plants. Colored arrows indicate transfers of matter and energy between producers, primary consumers (herbivores), secondary (or higher-level) consumers (carnivores), and decomposers. Organisms are not drawn to scale.

Gray reef shark

Green sea turtle

Sea nettle

Fairy basslet

Blue tangs

Parrot fish

Sergeant major

Hard corals

Algae

Brittle star

Banded coral shrimp

Symbiotic algae

Phytoplankton

Coney

Zooplankton

Blackcap basslet

Sponges

Moray eel

Bacteria

Producer to primary consumer	Primary to secondary consumer	Secondary to higher-level consumer	All consumer and producers to decomposers

phytoplankton. Nutrient levels are low (except around upwellings), and levels of dissolved oxygen are high. Large, fast-swimming predatory fish such as swordfish, sharks, and bluefin tuna populate this zone.

The *bathyal zone* is the dimly lit middle zone that does not contain photosynthesizing producers because of a lack of sunlight. Various types of zooplankton and smaller fish, many of which migrate to feed on the surface at night, populate this zone.

The *abyssal zone* is the dark lower zone that is very cold, has little dissolved oxygen, and has enough nutrients on the ocean floor to support about 98% of the 250,000 identified species living in the ocean.

Average primary productivity and NPP per unit of area are quite low in the open sea (Figure 4-22, p. 80) except at an occasional equatorial upwelling, where currents bring up nutrients from the ocean bottom. However, because the open sea covers so much of the earth's surface (Figure 6-27), it makes the largest contribution to the earth's overall NPP.

What Impacts Do Human Activities Have on Marine Systems? In their desire to live near the coast, people are destroying or degrading the resources that make coastal areas so enjoyable and economically and ecologically valuable. Currently, about 40% of the

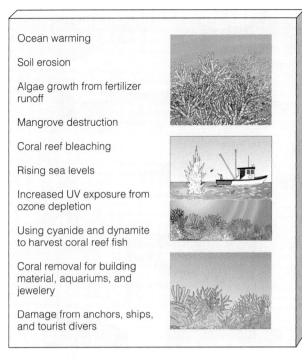

Ocean warming

Soil erosion

Algae growth from fertilizer runoff

Mangrove destruction

Coral reef bleaching

Rising sea levels

Increased UV exposure from ozone depletion

Using cyanide and dynamite to harvest coral reef fish

Coral removal for building material, aquariums, and jewelery

Damage from anchors, ships, and tourist divers

Figure 6-36 Major threats to coral reefs.

world's population live along coasts or within 100 kilometers (62 miles) of a coast. Some 13 of the world's 19 megacities with populations of 10 million or more people are in coastal zones. By 2030, at least 6.3 billion people—equal to the current global population—are expected to live in or near coastal areas. Figure 6-37 lists major human impacts on marine systems.

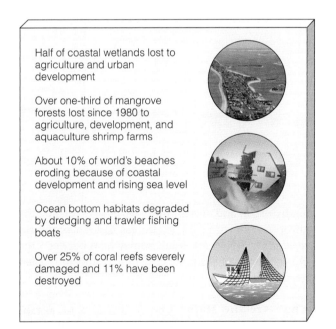

Half of coastal wetlands lost to agriculture and urban development

Over one-third of mangrove forests lost since 1980 to agriculture, development, and aquaculture shrimp farms

About 10% of world's beaches eroding because of coastal development and rising sea level

Ocean bottom habitats degraded by dredging and trawler fishing boats

Over 25% of coral reefs severely damaged and 11% have been destroyed

Figure 6-37 Major human impacts on the world's marine systems.

6-7 FRESHWATER LIFE ZONES

What Are Freshwater Life Zones? **Freshwater life zones** occur where water with a dissolved salt concentration of less than 1% by volume accumulates on or flows through the surfaces of terrestrial biomes. Examples are *standing* (lentic) bodies of fresh water such as lakes, ponds, and inland wetlands and *flowing* (lotic) systems such as streams and rivers. Although freshwater systems cover less than 1% of the earth's surface (Figure 6-26), they contain a variety of species (Figure 6-38, p. 134) and provide a number of important ecological and economic services (Figure 6-39, p. 135).

What Life Zones Are Found in Freshwater Lakes? **Lakes** are large natural bodies of standing fresh water formed when precipitation, runoff, or groundwater seepage fills depressions in the earth's surface. Causes of such depressions include glaciation (the Great Lakes of North America), crustal displacement (Lake Nyasa in East Africa), and volcanic activity (Crater Lake in Oregon; see photo on the title page). Lakes are supplied with water from rainfall, melting snow, and streams that drain the surrounding watershed.

Lakes normally consist of four distinct zones that are defined by their depth and distance from shore (Figure 6-40, p. 135): The top layer is the *littoral zone* ("LIT-tore-el"). It consists of the shallow sunlit waters near the shore to the depth at which rooted plants stop growing and has a high biological diversity.

Next is the *limnetic zone* ("lim-NET-ic"): the open, sunlit water surface layer away from the shore that extends to the depth penetrated by sunlight. As the main photosynthetic body of the lake, it produces the food and oxygen that support most of the lake's consumers.

This is followed by the *profundal zone* ("pro-FUN-dahl"): the deep, open water where it is too dark for photosynthesis. Without sunlight and plants, oxygen levels are low. Fish adapted to its cooler and darker water are found in this zone.

Finally, at the bottom of the lake we find the *benthic zone* ("BEN-thic"). It is inhabited mostly by organisms that tolerate cool temperatures and low oxygen levels.

How Do Plant Nutrients Affect Lakes? Ecologists classify lakes according to their nutrient content and primary productivity. A newly formed lake generally has a small supply of plant nutrients and is called an **oligotrophic** (poorly nourished) **lake** (Figure 6-41, left, p. 136). This type of lake is often deep, with steep banks. Because of its low NPP, such a lake usually has crystal-clear blue or green water and small populations of phytoplankton and fish (such as smallmouth bass and trout).

Figure 6-38 *Natural capital: freshwater biodiversity.* Some inhabitants of freshwater rivers and lakes.

Labels in figure: Bulrush, Bluegill, White bass, Brook trout, White waterlily, Water lettuce, Rainbow trout, Muskellunge, Rainbow darter, Bladderwort, Bowfish, Water hyacinth, Largemouth black bass, Black crappie, White sturgeon, Yellow perch, Velvet cichlid, Walleyed pike, American smelt, Eelgrass, Longnose gar, Duckweed, Common piranha, Carp, Channel catfish, Egyptian white lotus, African lungfish

Over time, sediment washes into an oligotrophic lake, and plants grow and decompose to form bottom sediments. A lake with a large or excessive supply of nutrients (mostly nitrates and phosphates) needed by producers is called a **eutrophic** (well-nourished) **lake** (Figure 6-41, right, p. 146). Such lakes typically are shallow and have murky brown or green water with very poor visibility. Because of their high levels of nutrients, these lakes have a high NPP.

In warm months the bottom layer of a eutrophic lake often is depleted of dissolved oxygen. Human inputs of nutrients from the atmosphere and from nearby urban and agricultural areas can accelerate the eutrophication of lakes—a process called *cultural eutrophication.* Many lakes fall somewhere between the two extremes of nutrient enrichment and are called **mesotrophic lakes.**

What Are the Major Characteristics of Freshwater Streams and Rivers? Precipitation that does not sink into the ground or evaporate is **surface water.** It becomes **runoff** when it flows into streams. The land area that delivers runoff, sediment, and dissolved substances to a stream is called a **watershed,** or

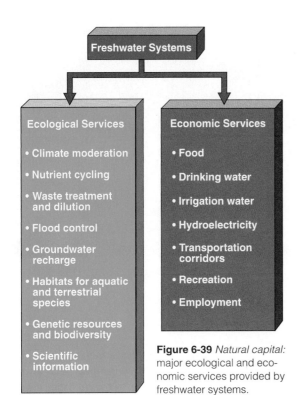

Figure 6-39 *Natural capital:* major ecological and economic services provided by freshwater systems.

Freshwater Systems

Ecological Services

- Climate moderation
- Nutrient cycling
- Waste treatment and dilution
- Flood control
- Groundwater recharge
- Habitats for aquatic and terrestrial species
- Genetic resources and biodiversity
- Scientific information

Economic Services

- Food
- Drinking water
- Irrigation water
- Hydroelectricity
- Transportation corridors
- Recreation
- Employment

drainage basin. Small streams join to form rivers, and rivers flow downhill to the ocean (Figure 6-42, p. 136) as part of the hydrologic cycle.

In many areas, streams begin in mountainous or hilly areas that collect and release water falling to the earth's surface as rain or snow. The downward flow of surface water and groundwater from mountain highlands to the sea takes place in three different aquatic life zones with different environmental conditions: the *source zone,* the *transition zone,* and the *floodplain zone* (Figure 6-42).

As streams flow downhill, they become powerful shapers of land. Over millions of years the friction of moving water levels mountains and cuts deep canyons, and the rock and soil the water removes are deposited as sediment in low-lying areas.

Streams are fairly open ecosystems that receive many of their nutrients from bordering land ecosystems. Such nutrient inputs come from falling leaves, animal feces, insects, and other forms of biomass washed into streams during heavy rainstorms or by melting snow. To protect a stream or river system from excessive inputs of nutrients and pollutants, we must protect its watershed, the land around it.

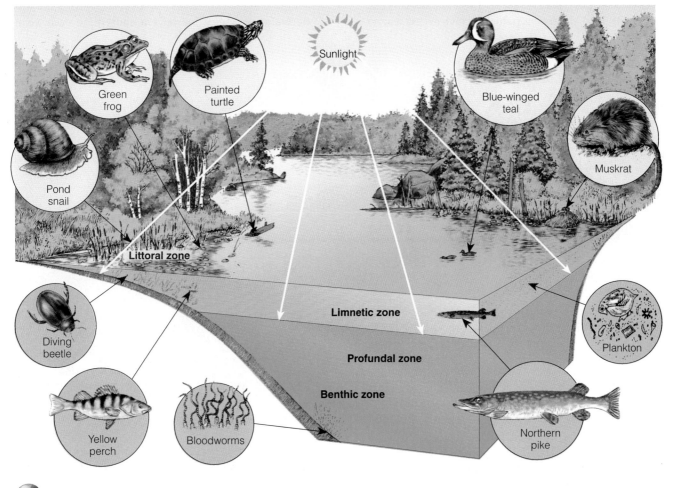

Figure 6-40 The distinct zones of life in a fairly deep temperate zone lake.

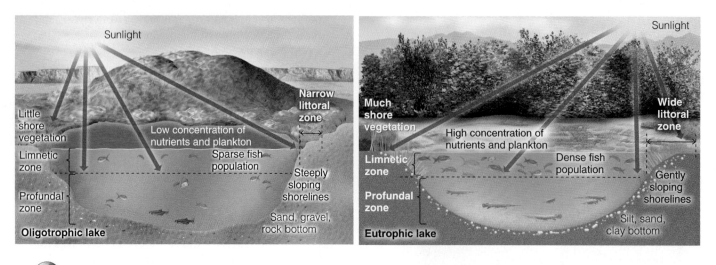

Figure 6-41 An *oligotrophic*, or nutrient-poor, lake (left) and a *eutrophic*, or nutrient-rich, lake (right). Mesotrophic lakes fall between these two extremes of nutrient enrichment. Nutrient inputs from human activities can accelerate eutrophication.

What Are Freshwater Inland Wetlands? Inland wetlands are lands covered with fresh water all or part of the time (excluding lakes, reservoirs, and streams) and located away from coastal areas. They include *marshes* (with few trees), *swamps* (dominated by trees and shrubs), *prairie potholes* (depressions carved out by glaciers), *floodplains* (which receive excess water during heavy rains and floods), *bogs* and *fens* (which have waterlogged soils, tend to accumulate peat, and may or may not have trees), and *wet arctic tundra* in summer. Some wetlands are huge and some are small.

Some wetlands are covered with water year-round. Others, called *seasonal wetlands*, usually are underwater or soggy for only a short time each year.

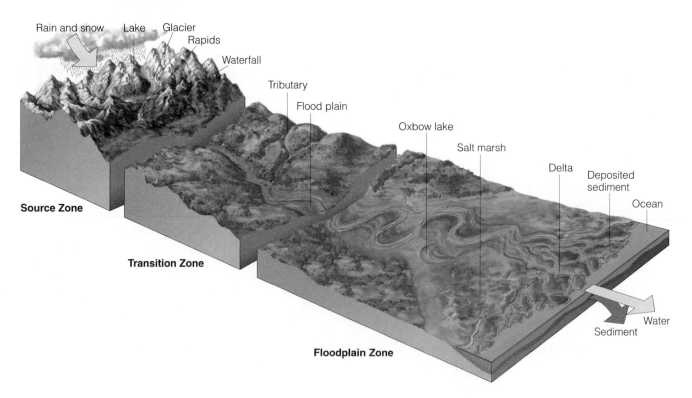

Figure 6-42 The three zones in the downhill flow of water: *source zone* containing mountain (headwater) streams, *transition zone* containing wider, lower-elevation streams, and *floodplain zone* containing rivers, which empty into the ocean.

They include prairie potholes, floodplain wetlands, and bottomland hardwood swamps. Some stay dry for years before being covered with water again. In such cases, scientists must use the composition of the soil or the presence of certain plants (such as cattails, bulrushes, or red maples) to determine that a particular area is really a wetland.

Inland wetlands need to be protected because of the important ecological and economic roles they play. For example, according to conservative estimates by the Audubon Society, inland wetlands in the United States provide water quality protection worth at least $1.6 billion per year and flood control worth $7.7–31 billion per year.

What Are the Impacts of Human Activities on Freshwater Systems?
Here are some major impacts of human activities on freshwater systems:

■ Dams, diversions, or canals fragment almost 60% of the world's 237 large rivers, according to a 2000 study by the World Resources Institute. This alters and destroys wildlife habitats along rivers and in coastal deltas and estuaries by reducing water flow.

■ Flood control levees and dikes built along rivers alter and destroy aquatic habitats, disconnect rivers from their floodplains, and eliminate wetlands and backwaters that are important spawning grounds for fish.

■ Many inland wetlands have been drained or filled to grow crops (about 80% of wetland loss) or have been covered up with concrete, asphalt, and buildings. In the United States, 53% of the inland wetlands estimated to have existed in the lower 48 states during the 1600s no longer exist. This has been an important factor in increased flood and drought damage in the United States. Many other countries have suffered similar losses.

6-8 SUSTAINABILITY OF AQUATIC LIFE ZONES

How Sustainable Are Aquatic Ecosystems?
The *bad news* in terms of human impacts is that each stream, river, and lake reflects the sum of all that occurs in their watersheds. Also, many of the nutrients, wastes, and pollutants produced by human activities end up in the ocean. Recent research also shows that many of the chemicals reaching aquatic systems come from the atmosphere.

The *good news* is that aquatic life zones are constantly renewed because water is purified by natural hydrologic processes, nutrients cycle in and out, and populations of biological organisms can be replenished, given sufficient opportunity and time. How-

ever, these life-sustaining processes work only if they are not overloaded with pollutants and excessive nutrients and overfished.

Scientists have made a good start in understanding important aspects of the ecology of the world's terrestrial and aquatic systems, as discussed in this chapter. One of the main lessons from such research is that in nature *everything is connected*. According to scientists, however, we urgently need more research on how the world's terrestrial and aquatic systems work. With such information we will have a clearer picture of the impacts of our activities on the earth's biodiversity and what we can do to live more sustainably.

When we try to pick out anything by itself, we find it hitched to everything else in the universe.
JOHN MUIR

REVIEW QUESTIONS

1. Define the boldfaced terms in this chapter.

2. How does wind affect climate, global distribution of nutrients, and pollution?

3. What is *weather?*

4. What is *climate?* What two main factors determine a region's climate?

5. What five factors affect global air circulation? How does each factor affect global air circulation? What causes opposite seasons in the northern and southern hemispheres?

6. How do ocean currents affect regional climates? What would happen to the climate of northwestern Europe if the Gulf Stream did not exist? What is an *upwelling*, and why are upwellings important to aquatic life?

7. What is *the El Niño–Southern Oscillation (ENSO)?* How does it affect ocean life and weather in various parts of the world? What is *La Niña*, and how does it affect weather in various parts of the world?

8. What are *greenhouse gases?* List four greenhouse gases. What is the *greenhouse effect*, and how does it affect the earth's climate?

9. What is the *ozone layer?* How does it **(a)** help protect life on the earth and **(b)** affect the earth's climate?

10. What are *microclimates?* What is the *rain shadow effect*, and how can it affect the microclimate on each side of high mountains? Why do urban areas have different microclimates than surrounding areas, and how do these microclimates differ from those of surrounding areas? How do *land breezes* and *sea breezes* affect the microclimates of coastal areas?

11. What is a *biome?* What two factors determine whether an area of the earth's surface is a tropical, temperate, or polar desert, grassland, or forest? How do climate and vegetation vary with latitude and altitude?

12. What is a *desert?* What are the three major types of desert, and how do they differ in climate and biological makeup? How do desert plants and animals survive heat and a lack of water? Why are desert ecosystems vulnerable to disruption? List seven types of human activities that have harmful impacts on deserts.

13. What is *grassland?* What are the three major types of grassland, and how do they differ in climate and biological makeup? Distinguish between *arctic tundra* and *alpine tundra.* Why are grasslands vulnerable to disruption? List four types of human activities that have harmful impacts on grasslands.

14. What is a *forest?* What are the three major types of forest, and how do they differ in climate and biological makeup? Distinguish between *tropical rain forests* and *tropical deciduous forests.* List four human activities that have harmful impacts on forests.

15. Why are *mountains* ecologically important, and what factors make them vulnerable to ecological disruption? List seven types of human activities that have harmful impacts on mountains.

16. What are the two major types of aquatic life zones?

17. List the major ecological and economic services provided by the world's marine systems.

18. What is the *coastal zone?* Distinguish between *estuaries* and *coastal wetlands.* Explain why estuaries and coastal wetlands have such high net primary productivity.

19. What is the *intertidal zone?* Distinguish between *rocky shores* and *sandy beaches,* and describe the major types of aquatic life found in each. Why is it important to preserve the dunes on barrier beaches?

20. What are *coral reefs* and how are they formed? List six ecological and economic services that coral reefs provide. List three reasons why coral reefs are vulnerable to damage. List ten harmful impacts of human activities on coral reefs.

21. What is the *open sea,* and what are its three major vertical zones? Why is the NPP per unit of area so low in the open sea?

22. List five major harmful human impacts on coastal zones.

23. What is a *freshwater life zone,* and what are the two major types of such zones?

24. List the major ecological and economic services provided by freshwater systems.

25. What is a *lake?* Distinguish among the *littoral, limnetic, profundal,* and *benthic zones* of a lake.

26. What are the three types of lakes, based on their nutrient content and primary productivity?

27. Distinguish among *surface water, runoff,* and a *watershed.* Describe the properties and general forms of life found in the three zones of a river as it flows from mountain highlands to the sea.

28. What are *inland wetlands?* List six examples of such wetlands. What are *seasonal inland wetlands?*

29. List three major harmful human impacts on freshwater systems.

30. Summarize *bad* and *good* news about the sustainability of aquatic systems.

31. Explain why a study of terrestrial and aquatic systems reinforces the basic ecological principle that *everything is connected.*

CRITICAL THINKING

1. List a limiting factor for each of the following ecosystems: **(a)** a desert, **(b)** arctic tundra, **(c)** alpine tundra, **(d)** the floor of a tropical rain forest, and **(e)** a temperate deciduous forest.

2. Why do deserts and arctic tundra support a much smaller biomass of animals than do tropical forests?

3. Why do most animals in a tropical rain forest live in its trees?

4. What factors in your lifestyle contribute to the destruction and degradation of some tropical forests?

5. What biomes are best suited for **(a)** raising crops and **(b)** grazing livestock?

6. How would you respond to someone who proposes that we use the deep portions of the world's oceans to deposit our radioactive and other hazardous wastes because the deep oceans are vast and are located far away from human habitats? Give reasons for your response.

7. What factors in your lifestyle contribute to the destruction and degradation of coastal and inland wetlands?

8. Developers want to drain a large area of inland wetlands in your community and build a large housing development. List **(a)** the main arguments the developers would use to support this project and **(b)** the main arguments ecologists would use in opposing this project. If you were an elected city official, would you vote for or against this project? Can you come up with a compromise plan?

9. You are an attorney arguing in court for sparing an undeveloped old-growth tropical rain forest and a coral reef from severe degradation or destruction by development. Write your closing statement for the defense of each of these ecosystems. If the judge decides you can save only one of the ecosystems, which one would you choose, and why?

PROJECTS

1. How has the climate changed in the area where you live during the past 50 years? Investigate the beneficial and harmful effects of these changes. How have these changes benefited or harmed you personally?

2. What type of biome do you live in? What have been the major effects of human activities over the past 50 years on the characteristic vegetation and animal life normally found in the biome you live in? How is your own lifestyle affecting this biome?

3. If possible, visit a nearby lake, pond, or reservoir. Would you classify it as oligotrophic, mesotrophic, or eutrophic? What are the primary factors contributing to its nutrient enrichment? Which of these factors are related to human activities?

4. Examine a topographic map for the area around a stream or lake near where you live to define the watershed for the stream. What human activities occur in the watershed? What influence, if any, do you expect these activities to have on the ecology of the stream or lake?

5. Make a concept map of this chapter's major ideas, using the section heads and subheads and the key terms (in boldface). Look on the website for this book for information about making concept maps.

INTERNET STUDY RESOURCES AND RESOURCES FOR FURTHER READING AND RESEARCH

The website for this book contains helpful study aids and many ideas for further reading and research. Log on to

http://biology.brookscole.com/miller10

and click on the Chapter-by-Chapter area. Choose Chapter 6 and select a resource:

■ Flash Cards allows you to test your mastery of the Terms and Concepts to Remember for this chapter.

■ Tutorial Quizzes provides a multiple-choice practice quiz.

■ Student Guide to InfoTrac will lead you to Critical Thinking Projects that use InfoTrac College Edition as a research tool.

■ References lists the major books and articles consulted in writing this chapter.

■ Hypercontents takes you to an extensive list of sites with news, research, and images related to individual sections of the chapter.

INFOTRAC COLLEGE EDITION

Improve your skills with InfoTrac College Edition, a searchable online database of articles from more than 700 periodicals. Log on to

http://www.infotrac-college.com

or access InfoTrac through the website for this book. Try to find the following articles:

1. Petit, C. W. 2002. Perilous waters: Influx of fresh water in North Atlantic Ocean could disrupt world climate patterns. *U.S. News & World Report* (April 1): 64. *Keywords:* "North Atlantic Ocean" and "climate." Climate change is a major environmental issue at the beginning of the 21st century, but it may not all be due to humans and greenhouse gas emissions. This article looks at the effects of water density on a major ocean current system, and the potential effects on climate of changes in this system.

2. Valiela, I., J. L. Bowen, and J. K. York. 2001. Mangrove forests: One of the world's threatened major tropical environments. *BioScience* 51: 807. *Keywords:* "mangrove" and "deforestation." The loss of mangrove forests has not received the type of media attention that has characterized human activities in other ecosystems. However, the coastal mangrove forests of tropical and subtropical areas support a wide array of important ecological functions.

Flying Foxes: Keystone Species in Tropical Forests

The durian (Figure 7-1, top) is one of the most prized fruits growing in Southeast Asian tropical forests. The odor of this football-sized fruit is so strong, it is illegal to have them on trains and in many hotel rooms in Southeast Asia. However, its custardlike flesh has been described as "exquisite," "sensual," "intoxicating," and "the world's finest fruit."

Durian fruits come from a wild tree that grows in the tropical rain forest. The tree depends on various species of nectar-, pollen-, and fruit-feeding bats called flying foxes (Figure 7-1, left) to pollinate the flowers that hang high in the durian trees (Figure 7-1, bottom right). This pollination by flying foxes is an example of *mutualism:* an interaction between two species in which both species benefit. Hundreds of tropical plant species depend entirely on various flying fox species for pollination and seed dispersal.

Many species of flying foxes are listed as *endangered,* and most populations are much smaller than historic numbers. One reason for these population declines is *deforestation.* Another is *hunting* of these bat species for their meat, which is sold in China and other parts of Asia. The bats also are viewed as pests and are killed to keep them from eating commercially grown fruits (even though these fruits are picked green). Flying foxes are easy to hunt because they tend to congregate in large numbers when they feed or sleep.

According to ecologists, flying foxes are *keystone species* in tropical forest ecosystems. They are important for the plant species they pollinate, the plant seeds they disperse in their droppings (which help maintain forest biodiversity and regenerate deforested areas), and the many other species that depend on them.

Rain forest ecologists are concerned that the decline of flying fox populations could lead to a cascade of linked extinctions. Their decline also has important economic effects. Studies have shown that flying foxes are important for many products including fruits (such as the durian and wild bananas), medicines, timber (such as ebony and mahogany), fibers, dyes, animal fodder, and fuel.

The story of flying foxes and durians illustrates the unique role (niche) of each species in a community or ecosystem and how interactions between species can affect ecosystem structure and function. This chapter discusses these topics and the subjects of *community structure, ecological succession,* and *ecosystem sustainability.*

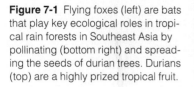

Figure 7-1 Flying foxes (left) are bats that play key ecological roles in tropical rain forests in Southeast Asia by pollinating (bottom right) and spreading the seeds of durian trees. Durians (top) are a highly prized tropical fruit.

This chapter addresses the following questions:

- What determines the number of species in a community?
- How can we classify species according to their roles?
- How do species interact with one another?
- How do communities and ecosystems change as environmental conditions change?
- Does high species diversity increase the stability of ecosystems?

7-1 COMMUNITY STRUCTURE AND SPECIES DIVERSITY

What Is Community Structure? One property of a community or ecosystem is the *structure* or *spatial distribution* of its individuals and populations. Ecologists usually describe the structure of a community or ecosystem in terms of four characteristics:

- *Physical appearance:* relative sizes, stratification, and distribution of its populations and species

- *Species diversity* or *richness:* the number of different species
- *Species abundance:* the number of individuals of each species
- *Niche structure:* the number of ecological niches (Section 5-3, p. 98), how they resemble or differ from each other, and how they interact (species interactions)

How Do Communities Differ in Physical Appearance and Population Distribution? The types, relative sizes, and stratification of plants and animals vary in different terrestrial communities and biomes (Figure 7-2). Marked differences are also apparent in the physical structures of different types of aquatic life zones. Examples are *oceans* (Figure 6-30, p. 128), *rocky shores and sandy beaches* (Figure 6-33, p. 130), *lakes* (Figure 6-40, p. 135), *river systems* (Figure 6-42, p. 136), and *inland wetlands.*

The physical structure within a particular type of community or ecosystem can also vary. A close look at most large terrestrial communities, ecosystems, and biomes reveals that they usually consist of a mosaic of *vegetation patches* of differing size.

Differences in the physical structure and physical properties (such as sunlight, temperature, wind, and humidity) at boundaries and in transition zones between two ecosystems (*ecotones*) are called **edge effects.** For example, the edge area between a forest and an open field may be sunnier, warmer, and drier than the forest interior and have a different combination of species than the forest and field interiors.

Popular wild game animals, such as pheasants and white-tailed deer, often are more plentiful in edges and transition areas between forests and fields. Wild game managers sometimes create patches and edges to increase populations of such species for sport hunters.

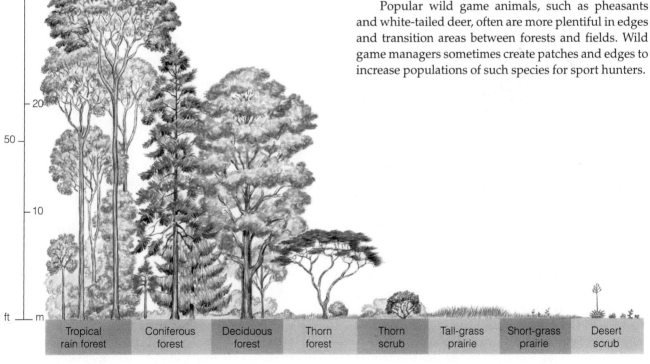

Figure 7-2 Generalized types, relative sizes, and stratification of plant species in various terrestrial communities or ecosystems.

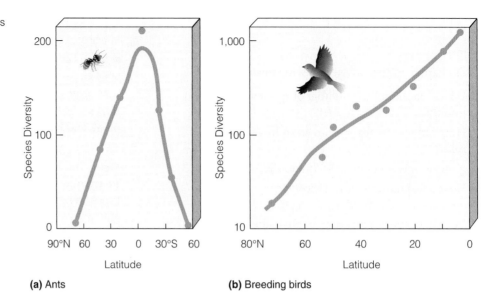

Figure 7-3 Changes in species diversity at different latitudes (distances from the equator) in terrestrial communities for **(a)** ants and **(b)** breeding birds of North and Central America. As a general rule, species diversity steadily declines as we go away from the equator toward either pole. (Modified by permission from Cecie Starr, *Biology: Concepts and Applications,* 4th ed., Pacific Grove, Calif.: Brooks/Cole, 2000)

(a) Ants

(b) Breeding birds

However, increased edge area from habitat fragmentation makes many species more vulnerable to stresses such as predators and fire. It also creates barriers that can prevent some species from colonizing new areas and finding food and mates.

Where Is Most of the World's Biodiversity Found? Studies indicate that the most species-rich environments are tropical rain forests (Figure 6-20, p. 120), coral reefs (Figure 6-35, p. 132), the deep sea, and large tropical lakes.

Communities such as a tropical rain forest or a coral reef with a large number of different species (high species diversity) generally have only a few members of each species (low species abundance).

Field investigations by ecologists have found that three major factors affect species diversity: *latitude* (distance from the equator) in terrestrial communities (Figure 7-3), *depth* in aquatic systems, and *pollution* in aquatic systems (Figure 7-4).

What Determines the Number of Species on Islands? Two factors affecting the species diversity found in an isolated ecosystem such as an island are its *size* and *degree of isolation.* In the 1960s, Robert H. MacArthur and Edward O. Wilson began studying communities on islands to discover why large islands tend to have more species of a certain category (such as insects, birds, or ferns) than do small islands.

To explain these differences in species diversity with island size, MacArthur and Wilson proposed what is called the **species equilibrium model,** or the **theory of island biogeography.** According to this model, a balance between two factors determines the number of species found on an island: the rate at

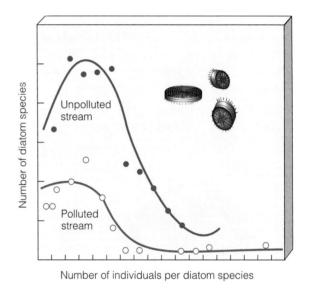

Figure 7-4 Changes in the species diversity and species abundance of diatom species in an unpolluted stream and a polluted stream. Both species diversity and species abundance decrease with pollution.

which new species immigrate to the island and the rate at which species become extinct on the island. The model predicts that at some point the rates of immigration and extinction will reach an equilibrium point (Figure 7-5a) that determines the island's average number of different species (species diversity).

The model also predicts that two features of the island affect immigration and extinction rates (and thus species diversity). One is the island's *size* (Figure 7-5b) and the other is its *distance from the nearest mainland* (Figure 7-5c). According to the model, a small island tends to have lower species diversity than a large one.

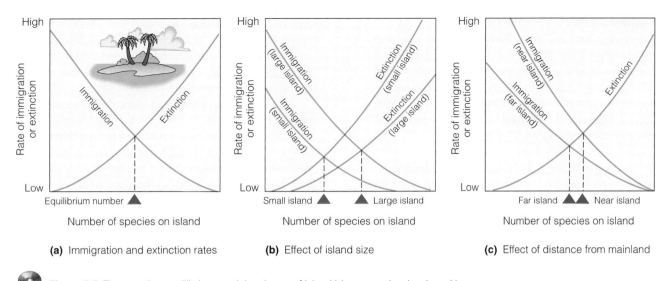

(a) Immigration and extinction rates **(b)** Effect of island size **(c)** Effect of distance from mainland

Figure 7-5 The *species equilibrium model* or *theory of island biogeography*, developed by Robert H. MacArthur and Edward O. Wilson. **(a)** The equilibrium number of species (blue triangle) on an island is determined by a balance between the immigration rate of new species and the extinction rate of species already on the island. **(b)** With time, large islands have a larger equilibrium number of species than smaller islands because of higher immigration rates and lower extinction rates on large islands. **(c)** Assuming equal extinction rates, an island near a mainland will have a larger equilibrium number of species than a more distant island because the immigration rate is greater to a near island than to a more distant one.

One reason is that a small island generally has a lower immigration rate because it is a smaller target for potential colonizers. In addition, a small island should have a higher extinction rate because it generally has fewer resources and less diverse habitats for colonizing species.

The model also predicts that an island's distance from a mainland source of new species is important in determining species diversity. For two islands of about equal size and other factors, the island closer to a mainland source of immigrant species will have the higher immigration rate and thus a higher species diversity (assuming that extinction rates on both islands are about the same).

A series of field experiments have tested and supported MacArthur and Wilson's original model or scientific hypothesis. As a result, biologists have elevated it to the status of an important and useful scientific theory. In recent years, it has also been applied to conservation efforts to protect wildlife in *habitat islands* such as national parks surrounded by a sea of developed and fragmented land.

7-2 GENERAL TYPES OF SPECIES

What Different Roles Do Various Species Play in Ecosystems? When examining ecosystems, ecologists often apply particular labels—such as *native,* *nonnative, indicator,* or *keystone*—to various species to clarify their ecological roles or niches (p. 98). Any given species may function as more than one of these four types in a particular ecosystem.

How Can Nonnative Species Cause Problems? Species that normally live and thrive in a particular ecosystem are known as **native species.** Others that migrate into an ecosystem or are deliberately or accidentally introduced into an ecosystem by humans are called **nonnative species, exotic species,** or **alien species.** Some of these introduced species (such as crops and game species for sport hunting) are beneficial to humans, but some thrive and crowd out native species.

From a human standpoint, introduction of some nonnative species can become a nightmare. In 1957, Brazil imported wild African bees to help increase honey production. Instead, these bees have displaced domestic honeybees and have reduced the honey supply.

Since then, these nonnative bee species, popularly known as "killer bees," have moved northward into Central America. They have become established in Texas, Arizona, New Mexico, Puerto Rico, and California and are heading north at about 240 kilometers (150 miles) per year. They should be stopped eventually by cold winters in the central United States unless they can adapt genetically to cold weather.

Although they are not the killer bees portrayed in some horror movies, these bees are aggressive and unpredictable. They have killed thousands of domesticated animals and an estimated 1,000 people in the western hemisphere. Fortunately, most people not allergic to bee stings can run away. Most of the people killed by these honeybees died because they fell down or became trapped and could not flee.

What Are Indicator Species? Species that serve as early warnings of damage to a community or an ecosystem are called **indicator species.** Birds are excellent biological indicators because they are found almost everywhere and respond quickly to environmental change. Research indicates that a major factor in the current decline of some species of migratory, insect-eating songbirds in North America is habitat loss or fragmentation. The tropical forests of Latin America and the Caribbean that are winter habitats for such birds are disappearing rapidly. Their summer habitats in North America also are disappearing or are being fragmented into patches that make the birds more vulnerable to attack by predators and parasites.

The presence or absence of trout species in water at temperatures within their range of tolerance (Figure 4-13, p. 72) is an indicator of water quality because trout need clean water with high levels of dissolved oxygen.

Case Study: Why Are Amphibians Vanishing?
Some amphibians (frogs, toads, and salamanders), which live part of their lives in water and part on land, are classified as *indicator species.* These cold-blooded creatures range in size from a frog that can sit on your thumb to a Japanese salamander that is about 1.5 meters (5 feet) long.

Since 1980, populations of hundreds of the world's estimated 5,280 amphibian species (including 2,700 frog and toad species) have been vanishing or declining in almost every part of the world, even in protected wildlife reserves and parks.

In some locales, frog deformities (such as extra legs and missing legs) are occurring in unusually high numbers. This information was discovered and published by schoolchildren in Henderson, Minnesota, who were catching frogs in a farm pond.

According to the World Conservation Union, 25% of all known amphibian species are extinct, endangered, or vulnerable. The Nature Conservancy estimates that 38% of the amphibian species in the United States are endangered.

Frogs are especially vulnerable to environmental disruption at various points in their life cycle (Figure 7-6). As tadpoles they live in water and eat plants, and as adults they live mostly on land and eat insects (which can expose them to pesticides). Their eggs have no protective shells to block ultraviolet (UV) radiation or pollution. As adults, they take in water and air through their thin, permeable skins that can readily absorb pollutants from water, air, or soil.

No single cause has been identified to explain the cause of amphibian declines. However, scientists have identified a number of contributing factors that can affect amphibians such as frogs at various points in their life cycle (Figure 7-6). One factor is *habitat loss and fragmentation,* especially because of the draining and filling of inland wetlands, deforestation, and development. Another is *prolonged drought,* which dries up breeding pools so that few tadpoles survive. Dehydration can also weaken amphibians, making them more susceptible to fatal viruses, bacteria, fungi, and parasites.

Pollution can also play a role. Frog eggs, tadpoles, and adults are very sensitive to many pollutants (especially pesticides). Exposure to such pollutants may harm their immune and endocrine systems, make them more vulnerable to bacterial viral, and fungal diseases, and cause an array of sexual abnormalities.

Increases in UV radiation caused by reductions in stratospheric ozone can harm young embryos of amphibians found in shallow ponds.

Increased incidence of parasitism by a flatworm (trematode) may account for many frog deformities but not the worldwide decline of amphibians.

Overhunting can be a factor especially in Asia and France, where frog legs are a delicacy. Populations of some amphibians can also be reduced by *immigration or introduction of nonnative predators and competitors* (such as fish) *and disease organisms.* In most cases, the decline or disappearance of amphibian species is probably caused by a combination of such factors. Scientists are concerned about amphibians' decline for three reasons. *First,* it suggests that the world's environmental health is deteriorating because amphibians are sensitive biological indicators of changes in environmental conditions (such as habitat loss and degradation, pollution, UV exposure, and climate change).

Second, adult amphibians play important roles in the world's ecosystems. For example, amphibians eat more insects (including mosquitoes) than do birds. In some habitats, extinction of certain amphibian species could lead to extinction of other species, such as reptiles, birds, aquatic insects, fish, mammals, and other amphibians that feed on them or their larvae.

Third, from a human perspective, amphibians represent a genetic storehouse of pharmaceutical products waiting to be discovered. Hundreds of secretions from amphibian skin have been isolated, and some of these compounds are used as painkillers and antibiotics and in treating burns and heart disease.

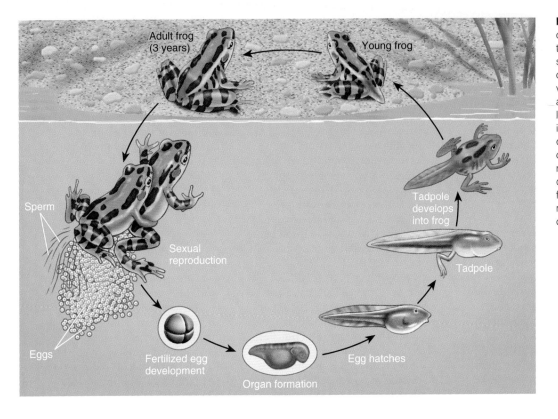

Figure 7-6 Typical life cycle of a frog. Populations of various frog species can decline because of the effects of various harmful factors at different points in their life cycle. Such factors include habitat loss, drought, pollution increased ultraviolet radiation, parasitism, disease, overhunting for food (frog legs), and nonnative predators and competitors.

Adult frog (3 years)

Young frog

Sperm

Sexual reproduction

Tadpole develops into frog

Tadpole

Eggs

Fertilized egg development

Organ formation

Egg hatches

The plight of some amphibian indicator species is a warning signal. They do not need us, but we and other species need them.

What Are Keystone Species? The roles of some species in an ecosystem are much more important than their abundance or biomass suggests. Ecologists call such species **keystone species.** Such species play pivotal roles in the structure and function of an ecosystem because their strong interactions with other species affect the health and survival of these species. They also process material out of proportion to their numbers or biomass.

Keystone species play critical ecological roles. Examples are pollination of flowering plant species by bees, hummingbirds, bats (Figure 7-1), and other species and dispersion of seeds by fruit-eating animals such as bats (Figure 7-1), which scatter undigested seeds with their feces. Other roles include habitat modification and predation by top carnivores that helps control the populations of various species. Some keystone species improve the ability of plant species to obtain soil minerals and water and facilitate the recycling of animal wastes. Some species play more than one of these roles.

Here is an example of beneficial *habitat modifications* by a keystone species. *Elephants* push over, break, break, or uproot trees, creating forest openings in the

savanna grasslands and woodlands of Africa. This promotes the growth of grasses and other forage plants that benefit smaller grazing species such as antelope. It also accelerates nutrient cycling rates.

Top predator keystone species exert a stabilizing effect on their ecosystems by feeding on and helping regulate the populations of certain species. Examples are the wolf, leopard, lion, alligator (Connections, p. 146), and great white shark.

The loss of a keystone species can lead to population crashes and extinctions of other species that depend on it for certain services, a ripple or domino effect that spreads throughout an ecosystem. According to biologist Edward O. Wilson, "The loss of a keystone species is like a drill accidentally striking a power line. It causes lights to go out all over."

7-3 SPECIES INTERACTIONS: COMPETITION AND PREDATION

How Do Species Interact? An Overview When different species in an ecosystem have activities or resource needs in common, they may interact with one another. Members of these species may be harmed by, benefit from, or be unaffected by the interaction. Ecologists identify five basic types of interactions between

Why Should We Care about Alligators?

CONNECTIONS

The American alligator, North America's largest reptile, has no natural predators except humans. This species, which has been around for about 200 million years, has been able to adapt to numerous changes in the earth's environmental conditions.

This changed when hunters began killing large numbers of these animals for their exotic meat and their supple belly skin, used to make shoes, belts, and pocketbooks.

Other people considered alligators to be useless and dangerous and hunted them for sport or out of hatred. Between 1950 and 1960, hunters wiped out 90% of the alligators in Louisiana, and by the 1960s, the alligator population in the Florida Everglades also was near extinction.

People who say "So what?" are overlooking the alligator's important ecological role or *niche* in subtropical wetland ecosystems. Alligators dig deep depressions, or gator holes, that collect fresh water during dry spells, serve as refuges for aquatic life, and supply fresh water and food for many animals.

In addition, large alligator nesting mounds provide nesting and feeding sites for herons and egrets. Alligators also eat large numbers of gar (a predatory fish) and thus help maintain populations of game fish such as bass and bream.

As alligators move from gator holes to nesting mounds, they help keep areas of open water free of invading vegetation. Without these ecosystem services, freshwater ponds and shrubs and trees would fill in coastal wetlands in the alligator's habitat, and dozens of species would disappear.

Some ecologists classify the North American alligator as a *keystone species* because of these important ecological roles in helping maintain the structure and function of its natural ecosystems.

In 1967, the U.S. government placed the American alligator on the endangered species list. Protected from hunters, the alligator population made a strong comeback in many areas by 1975—too strong, according to those who find alligators in their backyards and swimming pools, and to duck hunters, whose retriever dogs sometimes are eaten by alligators.

In 1977, the U.S. Fish and Wildlife Service reclassified the American alligator from an *endangered* species to a *threatened* species in Florida, Louisiana, and Texas, where 90% of the animals live. In 1987, this reclassification was extended to seven other states.

Alligators now number perhaps 3 million, most in Florida and Louisiana. It is generally illegal to kill members of a threatened species, but limited kills by licensed hunters are allowed in some areas of Florida, Louisiana, and South Carolina to control the population. To biologists, the comeback of the American alligator from near premature extinction by overhunting is an important success story in wildlife conservation.

The increased demand for alligator meat and hides has created a booming business in alligator farms, especially in Florida. Such success reduces the need for illegal hunting of wild alligators.

Critical Thinking

Some homeowners in Florida believe they should have the right to kill any alligator found on their property. Others argue this should not be allowed because alligators are a threatened species, and housing developments have invaded the habitats of alligators, not the other way around. What is your opinion on this issue? Explain.

---

species: *interspecific competition, predation, parasitism, mutualism,* and *commensalism.*

How Do Species Compete for Resources? Intraspecific and Interspecific Competition Competition between members of the *same* species for the same resources is called **intraspecific competition.** Intraspecific competition can be intense because members of a particular species compete directly for the same resources.

Competition between members of two or more *different* species for food, space, or any other limited resource is called **interspecific competition.**

As long as commonly used resources are abundant, different species can share them. This allows each species to come closer to occupying the *fundamental niche* it would occupy if there were no competition from other species.

However, most species face competition from other species for one or more limited resources (such as food, sunlight, water, soil nutrients, space, nesting sites, and good places to hide). Because of such *interspecific competition,* parts of the fundamental niches of different species overlap (Figure 5-6, p. 98). The more the niches of two species overlap, the more they compete with one another. With significant niche overlap, one of the competing species must migrate to another area (if possible), shift its feeding habits or behavior through natural selection and evolution, suffer a sharp population decline, or become extinct in that area.

Humans are in competition with other species for space, food, and other resources. As we convert more and more of the earth's land and aquatic resources and net primary productivity (Figure 4-22, p. 80) to our uses, we deprive many other species of resources they need to survive.

How Have Some Species Reduced or Avoided Competition? Over a time scale long enough for evolution to occur, some species that compete for the same resources evolve adaptations that reduce or avoid competition or an overlap of their fundamental niches (Figure 5-6, p. 98). One way this happens is through **resource partitioning,** the dividing up of scarce resources so that species with similar needs use them at different times, in different ways, or in different places (Figure 5-7, p. 100). In effect, they evolve traits that allow them to share available resources.

Each competing species occupies a *realized niche* making up only part of its *fundamental niche*. Thus through evolution the fairly broad niches of two competing species (Figure 7-7, top) become more specialized (Figure 7-7, bottom). Resource partitioning also occurs in the different layers of tropical rain forests (Figure 6-21, p. 121), in coastal wetlands (Figure 5-7, p. 100), grasslands (Figure 6-16, p. 117).

Here are other examples of resource partitioning. When lions and leopards live in the same area, lions take mostly larger animals as prey, and leopards take smaller ones. Hawks and owls feed on similar prey, but hawks hunt during the day and owls hunt at night. Some bird species feed on the ground, whereas others seek food in trees and shrubs.

Ecologist Robert H. MacArthur studied the feeding habits of five species of warblers (small insect-eating birds) that coexist in the forests of the northeastern United States and in the adjacent area of Canada. Although they appear to be competing for the same food resources, MacArthur found that the bird species reduce competition through resource partitioning by spending at least half their time hunting for insects in different parts of trees (Figure 7-8, p. 148).

How Do Predator and Prey Species Interact? In **predation,** members of one species (the *predator*) feed directly on all or part of a living organism of another species (the *prey*). However, they do not live on or in the prey, and the prey may or may not die from the interaction.

In this interaction, the predator benefits and the individual prey is clearly harmed. Together, the two kinds of organisms, such as lions (the predator or hunter) and zebras (the prey or hunted), are said to have a **predator–prey relationship,** as depicted in Figures 4-10 (p. 71), 4-11 (p. 71), 4-17 (p. 76), and 4-18 (p. 77).

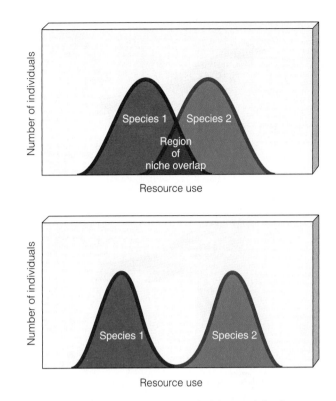

Figure 7-7 *Resource partitioning* and *niche specialization* as a result of competition between two species. The top diagram shows the overlapping niches of two competing species. The bottom diagram shows that through evolution the niches of the two species become separated and more specialized (narrower) so that they avoid competing for the same resources.

At the individual level, members of the prey species are clearly harmed. However, at the population level, predation can benefit the prey species because predators such as tigers and some types of sharks often kill the sick, weak, and aged members (Case Study, p. 149). Reducing the prey population gives remaining prey greater access to the available food supply. This can improve the genetic stock of the prey population, which enhances its chances of reproductive success and long-term survival.

Some people tend to view predators with contempt. When a hawk tries to capture and feed on a rabbit, some root for the rabbit. Yet the hawk (like all predators) is merely trying to get enough food to feed itself and its young; in the process, it is playing an important ecological role in controlling rabbit populations.

How Do Predators Increase Their Chances of Getting a Meal? Predators have a variety of methods that help them capture prey. *Herbivores* can simply walk, swim, or fly up to the plants they feed on.

Carnivores feeding on mobile prey have two main options: *pursuit* and *ambush.* Some, such as the cheetah,

Figure 7-8 *Resource partitioning* of five species of common insect-eating warblers in the spruce forests of Maine. Each species minimizes competition with the others for food by spending at least half its feeding time in a distinct portion (shaded areas) of the spruce trees, and consuming somewhat different insect species. (After R. H. MacArthur, "Population Ecology of Some Warblers in Northeastern Coniferous Forests," *Ecology* 36 (1958): 533–36)

catch prey by running fast; others, such as the American bald eagle, fly and have keen eyesight; still others, such as wolves and African lions, cooperate in capturing their prey by hunting in packs.

Other predators have characteristics or strategies that enable them to hide and ambush their prey. For example, praying mantises (Figure 4-1, left, p. 64) sit in flowers of a similar color and ambush visiting insects. White ermines (a type of weasel) and snowy owls hunt in snow-covered areas. The alligator snapping turtle, camouflaged in its stream-bottom habitat, dangles its worm-shaped tongue to entice fish into its powerful jaws. People camouflage themselves to hunt wild game and use traps to ambush wild game.

How Do Prey Avoid or Defend Themselves against Predators? Species have various characteristics that enable them to avoid predators. They include the ability to run, swim, or fly fast, and a highly developed sense of sight or smell that alerts them to the presence of predators. Other avoidance mechanisms are protective shells (as on armadillos, which roll themselves up into an armor-plated ball,

and turtles), thick bark (giant sequoia), and spines (porcupines) or thorns (cacti and rosebushes). Many lizards have brightly colored tails that break off when they are attacked, often giving them enough time to escape.

Other prey species use *camouflage* by having certain shapes or colors (Figure 7-9a, p. 150) or the ability to change color (chameleons and cuttlefish). A leaf insect may be almost invisible against its background (Figure 7-9b), and an arctic hare in its white winter fur blends into the snow. Some insect species have evolved shapes that look like twigs (Figure 7-9a) or bird droppings on leaves.

Chemical warfare is another common strategy. Some prey species discourage predators with chemicals that are *poisonous* (oleander plants), *irritating* (bombardier beetles, Figure 7-9c), *foul smelling* (skunks, skunk cabbages, and stinkbugs), or *bad tasting* (buttercups and monarch butterflies, Figure 7-9d). Scientists have identified more than 10,000 defensive chemicals made by plants, including cocaine, caffeine, nicotine, cyanide, opium, strychnine, peyote, and rotenone (used as an insecticide).

Why Are Sharks Important Species?

CASE STUDY

The world's 370 shark species vary widely in size. The smallest is the dwarf dog shark, about the size of a large goldfish. The largest is the whale shark, the world's largest fish, which can grow to 15 meters (50 feet) long and weigh as much as two full-grown African elephants.

Various shark species, feeding at the top of food webs, cull injured and sick animals from the ocean and thus play an important ecological role. Without such shark species, the oceans would be overcrowded with dead and dying fish. As top predators such shark species also help control populations of large predators.

Many people—influenced by movies (such as *Jaws*), popular novels, and widespread media coverage of a fairly small number of shark attacks per year—think of sharks as people-eating monsters. However, the three largest species—the whale shark, basking shark, and megamouth shark—are gentle giants that swim through the water with their mouths open, filtering out and swallowing huge quantities of *plankton* (small free-floating sea creatures).

Every year, members of a few species of shark—mostly great white, bull, tiger, gray reef, lemon, hammerhead, shortfin mako, and blue—typically injure about 100 people worldwide and kill between 5 and 15 people. Most attacks are by great white sharks, which feed on sea lions and other marine mammals and sometimes mistake divers and surfers for their usual prey.

Media coverage of shark attacks greatly distort the danger from sharks. You are 30 times more likely to be killed by lightning than by a shark. Each year dogs inflict serious bites on many thousands more people than the 100 or so persons bitten by sharks, and the number of people bitten by other people is more than 10 times the number of people bitten by sharks.

For every shark that injures a person, we kill at least 1 million sharks, a total of 100 million sharks each year. Sharks are killed mostly for their fins, widely used in Asia as a soup ingredient and as a pharmaceutical cure-all and worth as much as $563 per kilogram ($256 per pound). In Hong Kong, a single bowl of shark fin soup can sell for as much as $100.

According to a 2001 study by Wild Aid, shark fins sold in restaurants throughout Asia and in Chinese communities in cities such as New York, San Francisco, and London contain dangerously high levels of toxic mercury. Consumption of high levels of mercury is especially threatening for pregnant women and their babies.

Sharks are also killed for their livers, meat (especially mako and thresher), hides (a source of exotic, high-quality leather), and jaws (especially great whites, whose jaws are worth thousands of dollars to collectors). They are also killed because we fear them. Some sharks (especially blue, mako, and oceanic whitetip) die when they are trapped as bycatch in nets or lines deployed to catch swordfish, tuna, shrimp, and other commercially important species.

However, sharks help save human lives. In addition to providing people with food, they are helping us learn how to fight cancer (which sharks almost never get), bacteria, and viruses. Their highly effective immune system is being studied because it allows wounds to heal without becoming infected.

Sharks have several natural traits that make them prone to population declines from overfishing. They have only a few offspring (between 2 and 10) once every year or two, take 10–24 years to reach sexual maturity and begin reproducing, and have long gestation (pregnancy) periods, up to 24 months for some species.

Sharks are among the most vulnerable and least protected animals on the earth. Eight of the world's shark species, including great whites, Asian camel, sandtigers, and kitefins, are now considered critically endangered, endangered, or vulnerable to extinction. Of the 125 countries that commercially catch more than 100 million sharks per year, only four—Australia, Canada, New Zealand, and the United States—have implemented management plans for shark fisheries, and these plans are hard to enforce.

With more than 400 million years of evolution behind them, sharks have had a long time to get things right. Preserving their evolutionary genetic development begins with the knowledge that sharks do not need us, but we and other species need them.

Critical Thinking

After reading this information, has your attitude toward sharks changed? If so, how has it changed?

Many bad-tasting, bad-smelling, toxic, or stinging prey species have evolved *warning coloration*, brightly colored advertising that enables experienced predators to recognize and avoid them. Examples are *brilliantly colored* poisonous frogs (Figure 7-9e) and red-, yellow-, and black-striped coral snakes and *foul-tasting* monarch butterflies (Figure 7-9d) and grasshoppers.

Other butterfly species, such as the nonpoisonous viceroy (Figure 7-9f), gain some protection by looking and acting like the monarch, a protective device known as *mimicry.*

(a) Span worm

(b) Wandering leaf insect

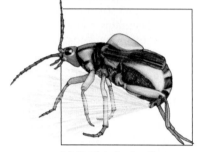

(c) Bombardier beetle

(d) Foul-tasting monarch butterfly

(e) Poison dart frog

(f) Viceroy butterfly mimics monarch butterfly

(g) Hind wings of Io moth resemble eyes of a much larger animal.

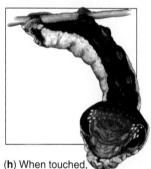

(h) When touched, snake caterpillar changes shape to look like head of snake.

Figure 7-9 Some ways in which prey species avoid their predators by **(a, b)** *camouflage,* **(c–e)** *chemical warfare,* **(d, e)** *warning coloration,* **(f)** *mimicry,* **(g)** *deceptive looks,* and **(h)** *deceptive behavior.*

Some prey species use *behavioral strategies* to avoid predation. Some attempt to scare off predators by puffing up (blowfish), spreading their wings (peacocks), or mimicking a predator (Figure 7-9h). Some moths have wings that look like the eyes of much larger animals (Figure 7-9g). Other prey gain some protection by liv-

ing in large groups (schools of fish, herds of antelope, flocks of birds).

7-4 SPECIES INTERACTIONS: PARASITISM, MUTUALISM, AND COMMENSALISM

What Are Parasites, and Why Are They Important? Parasitism occurs when one species (the *parasite*) feeds on part of another organism (the *host*) by living on or in the host. In this relationship, the parasite benefits and the host is harmed.

Parasitism can be viewed as a special form of predation. But unlike a conventional predator, a parasite usually is smaller than its host (prey). It also rarely kills its host, and remains closely associated with, draws nourishment from, and may gradually weaken its host over time.

Tapeworms, disease-causing microorganisms (or pathogens), and other parasites live *inside* their hosts. Other parasites attach themselves to the *outside* of their hosts. Examples are ticks, fleas, mosquitoes, mistletoe plants, and fungi (that cause diseases such as athlete's foot). Some parasites move from one host to another, as fleas and ticks do; others, such as tapeworms, spend their adult lives with a single host.

From the host's point of view, parasites are harmful, but parasites play important ecological roles. Collectively, the complex matrix of parasitic relationships in an ecosystem acts somewhat like glue that helps hold the species in an ecosystem together. Parasites also promote biodiversity by helping prevent some species from becoming so plentiful that they eliminate other species through competition.

How Do Species Interact So That Both Species Benefit? In mutualism, two species involved in a relationship interact in ways that benefit both. Such benefits include having pollen and seeds dispersed for reproduction, being supplied with food, or receiving protection.

The *pollination* relationship between flowering plants and animals such as insects (Figure 4-1, right, p. 64), birds, and bats (p. 140) is one of the most common forms of mutualism. Some species benefit from *nutritional mutualism. Lichens,* hardy species that can grow on trees or barren rocks, consist of colorful photosynthetic algae and chlorophyll-lacking fungi living together. The fungi provide a home for the algae, and their bodies collect and hold moisture and mineral nutrients used by both species. The algae, through photosynthesis, provide sugars as food for themselves and the fungi.

Another example is the vast armies of bacteria in the digestive systems of animals that break down (digest) their food. The bacteria gain a safe home with a steady food supply and the animal gains more efficient access to a large source of energy.

Here are three examples of mutualistic relationships involving *nutrition* and *protection.* One example is *birds* that ride on the backs of large animals such as African buffalo, elephants, and rhinoceroses (Figure 7-10a). The birds remove and eat parasites from the animal's body and often make noises warning the animal when predators approach.

Another example is *clownfish species,* which live within sea anemones, whose tentacles sting and paralyze most fish that touch them (Figure 7-10b). The clownfish, which are not harmed by the tentacles, gain protection from predators and feed on the detritus left from the meals of the anemones. The sea anemones benefit because the clownfish protect them from some of their predators.

A final example is various *fungi* that combine with plant roots to form mycorrhizae (from the Greek words for fungus and roots). The fungi get nutrition from a plant's roots. In turn the fungi benefit the plant by using their myriad networks of hairlike extensions to improve the plant's ability to extract nutrients and water from the soil (Figure 7-10c).

It is tempting to think of mutualism as an example of cooperation between species, but actually it involves each species benefiting by exploiting the other.

How Do Species Interact So That One Benefits but the Other Is Not Harmed? Commensalism is an interaction that benefits one species but neither harms nor helps the other species much, if at all. One example is a *redwood sorrel,* a small herb. It benefits from growing in the shade of tall redwood trees, with no known negative effects on the redwood trees.

Another example is plants called *epiphytes* (such as some types of orchids and bromeliads), which attach themselves to the trunks or branches of large trees in

(a) Oxpeckers and black rhinoceros

(b) Clownfish and sea anemone

Figure 7-10 Examples of *mutualism.* **(a)** Oxpeckers (or tickbirds) feed on and remove parasitic ticks that infest large thick-skinned animals such as a black rhinoceros. **(b)** A clownfish gains protection and food by living among deadly stinging sea anemones and helps protect the anemones from some of their predators. **(c)** Beneficial effects of mycorrhizal fungi attached to roots of juniper seedlings on plant growth compared to **(d)** growth of such seedlings in sterilized soil without mycorrhizal fungi.

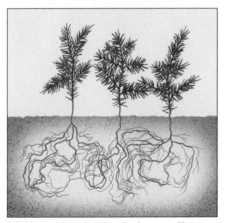

(c) Mycorrhizae fungi on juniper seedlings in normal soil

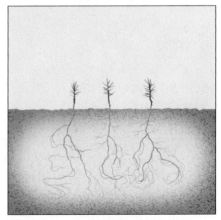

(d) Lack of mycorrhizae fungi on juniper seedlings in sterilized soil

tropical and subtropical forests. These so-called air plants benefit by having a solid base on which to grow and living in an elevated spot that gives them better access to sunlight, water from the humid air and rain, and nutrients falling from the tree's upper leaves and limbs. This apparently does not harm the tree.

7-5 ECOLOGICAL SUCCESSION: COMMUNITIES IN TRANSITION

How Do Ecosystems Respond to Change? One characteristic of all communities and ecosystems is that their structures change constantly in response to changing environmental conditions. The gradual change in species composition of a given area is called **ecological succession.** During succession some species colonize an area and their populations become more numerous, whereas populations of other species decline and may even disappear.

Ecologists recognize two types of ecological succession, depending on the conditions present at the beginning of the process. One is **primary succession,** which involves the gradual establishment of biotic communities on nearly lifeless ground. The other, more common type, called **secondary succession,** involves the establishment of biotic communities in an area where some type of biotic community is already present.

What Is Primary Succession? Establishing Life on Lifeless Ground Primary succession begins with an essentially lifeless area where there is no soil in a terrestrial ecosystem (Figure 7-11) or no bottom sediment in an aquatic ecosystem. Examples include

bare rock exposed by a retreating glacier or severe soil erosion, newly cooled lava, an abandoned highway or parking lot, or a newly created shallow pond or reservoir.

Before a community of plants (producers), consumers, and decomposers can become established on land, there must be *soil:* a complex mixture of rock particles, decaying organic matter, air, water, and living organisms. Depending mostly on the climate, it takes natural processes several hundred to several thousand years to produce fertile soil.

Soil formation begins when hardy **pioneer species** attach themselves to inhospitable patches of bare rock. Examples are wind-dispersed lichens and mosses, which can withstand the lack of moisture and soil nutrients and hot and cold temperature extremes found in such habitats.

As patches of soil build up and spread, eventually the community of lichens and mosses is replaced by a new community. Typically it consists of *small perennial grasses* (plants that live for more than 2 years without having to reseed) and *herbs* (ferns in tropical areas).

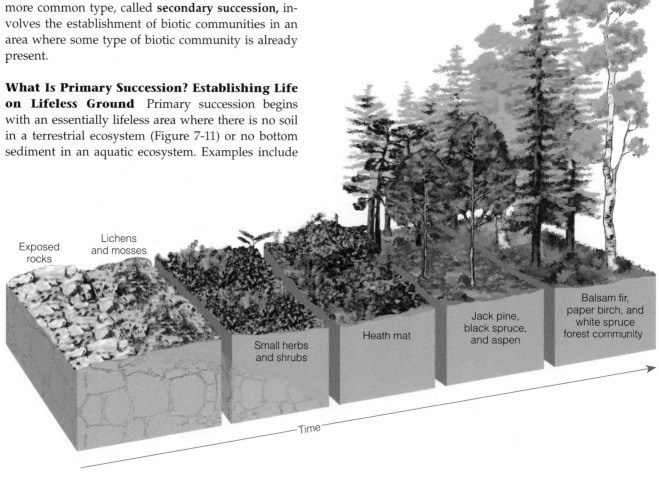

Figure 7-11 Starting from scratch. *Primary ecological succession* over several hundred years of plant communities on bare rock exposed by a retreating glacier on Isle Royal in northern Lake Superior.

The seeds of these plants germinate after arriving on the wind, in the droppings of birds, or on the coats of mammals.

These **early successional plant species** grow close to the ground, can establish large populations quickly under harsh conditions, and have short lives. Some of their roots penetrate the rock and help break it up into more soil particles. The decay of their wastes and dead bodies also adds more nutrients to the soil.

After hundreds of years, the soil may be deep and fertile enough to store enough moisture and nutrients to support the growth of less hardy **midsuccessional plant species** of herbs, grasses, and low shrubs. Trees that need lots of sunlight and are adapted to the area's climate and soil usually replace these species.

As these tree species grow and create shade, they are replaced by **late successional plant species** (mostly trees) that can tolerate shade. Unless fire, flooding, severe erosion, tree cutting, climate change, or other natural or human processes disturb the area, what was once bare rock becomes a complex forest community (Figure 7-11).

What Is Secondary Succession? Secondary succession begins in an area where the natural community of organisms has been disturbed, removed, or destroyed but some soil or bottom sediment remains. Candidates for secondary succession include abandoned croplands, burned or cut forests, heavily polluted streams, and land that has been dammed or flooded. Because some soil or sediment is present, new vegetation can usually begin to germinate within a few weeks. Seeds can be present in soils, or they can be carried from nearby plants by wind or deposited in the droppings of birds and animals.

In the central (Piedmont) region of North Carolina, European settlers cleared the mature native oak and hickory forests and planted the land with crops. Later they abandoned some of this farmland because of erosion and loss of soil nutrients. Figure 7-12 shows one way that such abandoned farmland has undergone secondary succession.

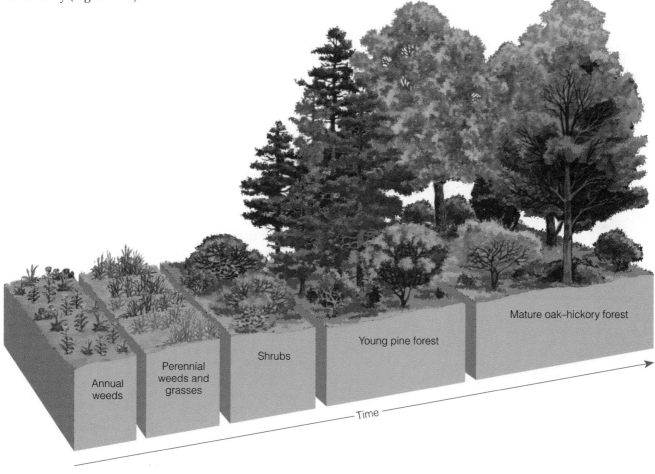

Annual weeds

Perennial weeds and grasses

Shrubs

Young pine forest

Mature oak–hickory forest

Time

Figure 7-12 Natural restoration of disturbed land. *Secondary ecological succession* of plant communities on an abandoned farm field in North Carolina. It took about 150–200 years after the farmland was abandoned for the area to be covered with a mature oak and hickory forest. A new disturbance such as deforestation or fire would create conditions favoring pioneer species. Then, in the absence of new disturbances, secondary succession would recur over time, although not necessarily in the same sequence shown here.

Descriptions of ecological succession usually focus on changes in vegetation. However, these changes in turn affect food and shelter for various types of animals. Thus as succession proceeds, the numbers and types of animals and decomposers also change.

How Do Disturbances Affect Succession and Species Diversity? A **disturbance** is a change in environmental conditions that disrupts an ecosystem or community. Examples are fire, drought, flooding, mining, clearcutting a forest, plowing a grassland, applying pesticides, climate change, and invasion by alien species. Environmental disturbances can range from catastrophic to mild and can be caused by natural changes or human activities. At any time during primary or secondary succession, such disturbances can convert a particular stage of succession to an earlier stage.

Many people think of all environmental disturbances as harmful. Large catastrophic disturbances such as fire, flooding, clearcutting, plowing, overgrazing by livestock, and pesticide application can devastate communities and ecosystems. But many ecologists contend that in the long run some types of disturbances such as fires can be beneficial for the species diversity of some communities and ecosystems. Such disturbances create new conditions that can discourage or eliminate some species but encourage others by releasing nutrients and creating unfilled niches.

According to the *intermediate disturbance hypothesis*, communities that experience fairly frequent but moderate disturbances have the greatest species diversity (Figure 7-13). Researchers hypothesize that in such communities, moderate disturbances are large enough to create openings for colonizing species in disturbed areas. At the same time, they are mild and infrequent enough to allow the survival of some mature species in undisturbed areas. Some field experiments have supported this hypothesis.

How Predictable Is Succession, and Is Nature in Balance? We may be tempted to conclude that ecological succession is an orderly sequence in which each stage leads predictably to the next, more stable stage. According to this classic view, succession proceeds until a generally predictable and stable type of *climax community* occupies an area. Such a community is dominated by a few long-lived plant species and is in balance with its environment. This equilibrium model of succession is what ecologists meant when they talked about the *balance of nature.*

Over the last several decades, many ecologists have changed their views about balance and equilibrium in nature. When these ecologists look at a community or ecosystem, such as a young forest, they see continuous change, instability, and unpredictability instead of equilibrium, stability, and predictability.

Under the old *balance-of-nature* view, a large terrestrial community undergoing succession was expected to eventually have a predictable green blanket of climax vegetation. However, a close look at almost any ecosystem or community reveals that it consists of an ever-changing mosaic of vegetation patches at different stages of succession. These patches result from a variety of mostly unpredictable small and medium-sized disturbances (Figure 7-13).

This irregular quilt of vegetation increases the diversity of plant and animal life and provides sites where early successional species can gain a foothold. Scientists also observe a mosaic of shifting patches of plant and animal species in intertidal communities (Figure 6-33, p. 130), mostly the result of the random actions of waves.

Such research indicates that *we cannot predict the course of a given succession or view it as preordained progress toward an ideally adapted climax community.* Rather, succession reflects the ongoing struggle by different species for enough light, nutrients, food, and space to survive and gain reproductive advantages over other species by occupying as much of their fundamental niches as possible.

This change in the way we view what is happening in nature explains why a growing number of ecologists prefer terms such as *biotic change* instead of *succession* (which implies an ordered and predictable sequence of changes). Many ecologists have also replaced the term *climax community* with terms such as *mature community* or a mosaic of *vegetation patches* at different stages of succession. Ecologists do not consider nature totally chaotic and unpredictable, but they have lowered their expectations of making accurate predictions about the course of succession.

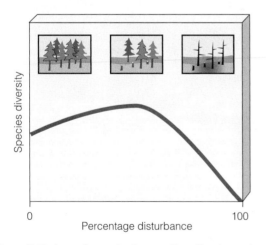

Figure 7-13 According to the *intermediate disturbance hypothesis*, moderate disturbances in communities promote greater species diversity than small or major disturbances.

7-6 ECOLOGICAL STABILITY AND SUSTAINABILITY

What Is Stability?   All living systems, from single-celled organisms to the biosphere, contain complex networks of negative and positive feedback loops (p. 49). These loops interact to provide some degree of stability or sustainability over each system's expected life span.

This stability is maintained only by constant dynamic change in response to changing environmental conditions. For example, in a mature tropical rain forest, some trees die and others take their places. However, unless the forest is cut, burned, or otherwise destroyed, you will still recognize it as a tropical rain forest 50 or 100 years from now.

It is useful to distinguish among three aspects of stability or sustainability in living systems. *One* is **inertia,** or **persistence:** the ability of a living system to resist being disturbed or altered. A *second* is **constancy:** the ability of a living system such as a population to keep its numbers within the limits imposed by available resources. A *third* factor is **resilience:** the ability of a living system to bounce back after an external disturbance that is not too drastic

Does Species Diversity Increase Ecosystem Stability?   In the 1960s, most ecologists believed the greater the species diversity and the accompanying web of feeding and biotic interactions in an ecosystem, the greater its stability. According to this hypothesis, an ecosystem with a diversity of species and feeding paths has more ways to respond to most environmental stresses because it does not have "all its eggs in one basket." However, recent research has found some exceptions to this intuitively appealing idea.

Because no ecosystem can function without some plants and decomposers, there is a minimum threshold of species diversity below which ecosystems cannot function.

Beyond this, it is difficult to know whether simple ecosystems are less stable than complex ones or to identify the threshold of species diversity below which complex ecosystems fail. In part because some species play redundant roles (niches) in ecosystems, we do not know how many or which species can be eliminated before the entire ecosystem begins to lose stability or collapse.

Recent research by ecologist David Tilman and others suggests two hypotheses. *First,* ecosystems with more species tend to have a higher net primary productivity and can be more resilient than simpler ones. *Second,* the populations of individual species can fluctuate more widely in diverse ecosystems than in simpler ones. In 2002, biologists Margaret Palmer and Bradley Cardinale found that when several species of

the caddisfly insect live together in a stream they get more food and are likely to be more productive than when a single caddisfly species lives in the same area.

Such studies support the idea that some level of biodiversity provides insurance against catastrophe. But how much biodiversity is needed in various ecosystems remains uncertain. For example, some recent research suggests that the average annual net primary productivity of an ecosystem reaches a peak with 10–40 producer species. Many ecosystems contain more producer species than this, but it is difficult to distinguish among those that are essential and those that are not.

Part of the problem is that ecologists disagree on how to define *stability* and *diversity.* Does an ecosystem need both high inertia and high resilience to be considered stable? Evidence suggests that some ecosystems have one of these properties but not the other. For example, tropical rain forests have high species diversity and high inertia; that is, they are resistant to significant alteration or destruction.

However, once a large tract of tropical forest is severely degraded, the ecosystem's resilience sometimes is so low that the forest may not be restored. Nutrients (which are stored primarily in the vegetation, not in the soil) and other factors needed for recovery may no longer be present. Such a large-scale loss of forest cover may so change the local or regional climate that forests can no longer be supported.

By contrast, grasslands are much less diverse than most forests and have low inertia because they burn easily. However, because most of their plant matter is stored in underground roots, these ecosystems have high resilience and recover quickly. Grassland can be destroyed only if its roots are plowed up and something else is planted in its place, or it is severely overgrazed by livestock or other herbivores.

Another difficulty is that populations, communities, and ecosystems are rarely, if ever, at equilibrium (balance). Instead, nature is in a continuing state of disturbance, fluctuation, and change.

Why Should We Bother to Protect Natural Systems? The Precautionary Principle Some developers argue that if biodiversity does not necessarily lead to increased ecological stability and if nature is mostly unpredictable, there is no point in trying to preserve and manage old-growth forests and other ecosystems. They conclude that we should cut down diverse old-growth forests, use the timber resources, and replace the forests with tree plantations. Furthermore, they say, we should convert most of the world's grasslands to cropfields, drain and develop inland wetlands, dump our toxic and radioactive wastes into the deep ocean, and not worry about the premature extinction of species.

Figure 7-14 Examples of how some of the earth's natural resources are being depleted and degraded at an accelerating rate as a result of the exponential growth of the human population and resource use by humankind. (Data from the World Conservation Union, World Wildlife Fund, Conservation International, United Nations, Population Reference Bureau, U.S. Fish and Wildlife Service, and Daniel Boivin)

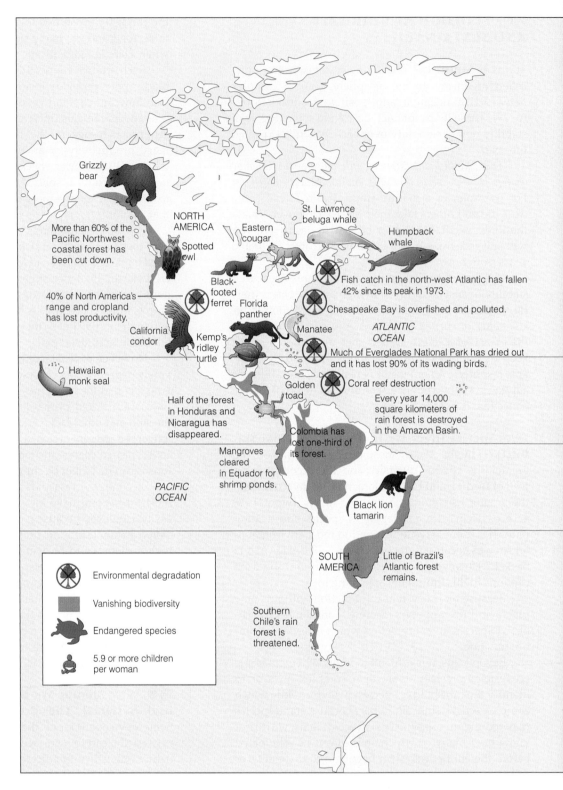

Grizzly bear

St. Lawrence beluga whale

Humpback whale

NORTH AMERICA

Eastern cougar

More than 60% of the Pacific Northwest coastal forest has been cut down.

Spotted owl

Fish catch in the north-west Atlantic has fallen 42% since its peak in 1973.

40% of North America's range and cropland has lost productivity.

Black-footed ferret

Florida panther

Chesapeake Bay is overfished and polluted.

California condor

Manatee

ATLANTIC OCEAN

Kemp's ridley turtle

Much of Everglades National Park has dried out and it has lost 90% of its wading birds.

Hawaiian monk seal

Golden toad

Coral reef destruction

Half of the forest in Honduras and Nicaragua has disappeared.

Every year 14,000 square kilometers of rain forest is destroyed in the Amazon Basin.

Colombia has lost one-third of its forest.

Mangroves cleared in Equador for shrimp ponds.

PACIFIC OCEAN

Black lion tamarin

SOUTH AMERICA

Little of Brazil's Atlantic forest remains.

Southern Chile's rain forest is threatened.

Environmental degradation

Vanishing biodiversity

Endangered species

5.9 or more children per woman

Ecologists and conservation biologists point to the overwhelming evidence that human disturbances (Figure 7-14) are disrupting some of the ecosystem services (Figure 1-2, p. 3, and top half of back cover) that support and sustain all life and all economies. They contend that our ignorance about the effects of our actions means we need to use great caution in making potentially harmful changes to ecosystems.

As an analogy, we know that eating too much of certain types of foods and not getting enough exercise can greatly increase our chances of a heart attack, diabetes, and other disorders. But the exact connections between these health problems, chemicals in various foods, exercise, and genetics are still under study and often debated. We could use this uncertainty and unpredictability as an excuse to continue overeating and

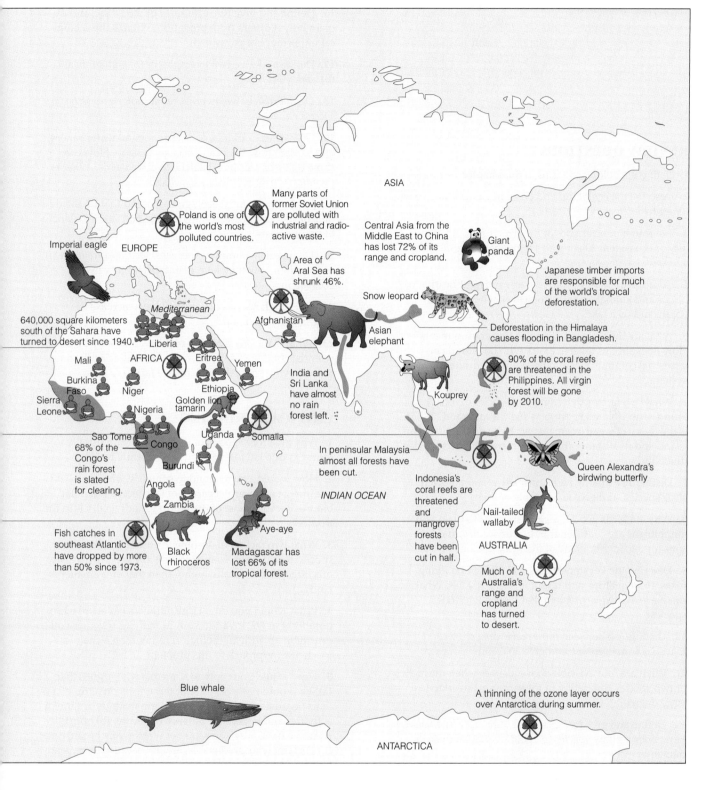

ASIA

Poland is one of the world's most polluted countries.

Many parts of former Soviet Union are polluted with industrial and radio-active waste.

Central Asia from the Middle East to China has lost 72% of its range and cropland.

Giant panda

Imperial eagle

EUROPE

Area of Aral Sea has shrunk 46%.

Japanese timber imports are responsible for much of the world's tropical deforestation.

Snow leopard

Mediterranean

640,000 square kilometers south of the Sahara have turned to desert since 1940.

Afghanistan

Asian elephant

Deforestation in the Himalaya causes flooding in Bangladesh.

Liberia

Mali

AFRICA

Eritrea

Yemen

90% of the coral reefs are threatened in the Philippines. All virgin forest will be gone by 2010.

Burkina Faso

Niger

Ethiopia

India and Sri Lanka have almost no rain forest left.

Kouprey

Sierra Leone

Golden lion tamarin

Nigeria

Sao Tome

Uganda

Somalia

68% of the Congo's rain forest is slated for clearing.

Congo

Burundi

In peninsular Malaysia almost all forests have been cut.

Queen Alexandra's birdwing butterfly

Angola

Zambia

INDIAN OCEAN

Indonesia's coral reefs are threatened and mangrove forests have been cut in half.

Nail-tailed wallaby

Fish catches in southeast Atlantic have dropped by more than 50% since 1973.

Black rhinoceros

Aye-aye

Madagascar has lost 66% of its tropical forest.

AUSTRALIA

Much of Australia's range and cropland has turned to desert.

Blue whale

A thinning of the ozone layer occurs over Antarctica during summer.

ANTARCTICA

not exercising. However, the wise course is to eat better and exercise more to help *prevent* potentially serious health problems.

This approach is based on the **precautionary principle:** When considerable evidence indicates that an activity can harm human health or the environment, we should take precautionary measures to prevent harm even if some of the cause-and-effect relationships have not been fully established scientifically. It is based on the commonsense idea behind many adages such as "Better safe than sorry," "Look before you leap," "First, do no harm," and "Slow down for speed bumps."

What we do know is that the interdependence and connectedness of species, communities, and ecosystems are essential features of life on earth.

The old idea of a static landscape, like a single musical chord sounded forever, must be abandoned, for such a landscape never existed except in our imagination. Nature undisturbed by human influence seems more like a symphony whose harmonies arise from variation and change over every interval of time.

DANIEL B. BOTKIN

REVIEW QUESTIONS

1. Define the boldfaced terms in this chapter.

2. Describe the ecological and economic importance of flying foxes in tropical forests.

3. List four characteristics of the structure of a community or ecosystem. Distinguish between *species diversity* and *species abundance.*

4. Describe the types, relative sizes, and stratification of plants and animals in **(a)** a tropical rain forest, **(b)** an ocean, and **(c)** a lake.

5. What are the three most species-rich environments? How does species diversity vary with **(a)** latitude in terrestrial communities and **(b)** pollution in aquatic systems?

6. What two factors determine the species diversity found in an isolated ecosystem such as an island? What is the *theory of island biogeography?* How do the size of an island and its distance from a mainland affect its species diversity?

7. Distinguish among *native, nonnative, indicator,* and *keystone* species, and give an example of each.

8. Why are birds good indicator species? Explain why amphibians are considered indicator species, and list reasons for declines in many of their populations.

9. Describe the keystone ecological roles of **(a)** flying foxes, **(b)** alligators, and **(c)** some shark species. What can happen in an ecosystem that loses a keystone species?

10. Distinguish between *intraspecific competition* and *interspecific competition,* and give an example of each.

11. What are four possible consequences when the niches of two species competing in the same area overlap to a large degree?

12. Define and give two examples of *resource partitioning.* How does it allow species to avoid overlap of their fundamental niches?

13. What is *predation?* Describe the *predator–prey relationship,* and give two examples of this type of species interaction.

14. Give two examples of how predators increase their chances of finding prey by **(a)** pursuit and **(b)** ambush.

15. List six ways (adaptations) used by prey to avoid their predators, and give an example of each type.

16. Define and give two examples of *parasitism,* and explain how it differs from predation. What is the ecological importance of parasitism?

17. Define and give two examples of **(a)** *mutualism* and **(b)** *commensalism.*

18. Distinguish between *primary succession* and *secondary succession.*

19. Give three examples of environmental disturbances, and explain how they can affect succession. How can some disturbances be beneficial to ecosystems? What is the *intermediate disturbance hypothesis?*

20. Explain why most ecologists contend that **(a)** the details of succession are not predictable and **(b)** no balance of nature exists.

21. Distinguish among *inertia, constancy,* and *resilience,* and explain how they help maintain stability in an ecosystem.

22. Does high species diversity always increase ecosystem stability? Explain.

23. What is the *precautionary principle,* and why do many scientists find it a useful strategy for dealing with some of the environmental problems we face?

CRITICAL THINKING

1. Why are more species of trees per hectare usually found in tropical forests than in temperate forests?

2. How would you respond to someone who claims it is not important to protect areas of temperate and polar biomes because most of the world's biodiversity is in the tropics?

3. What two factors determine the number of different species found on an island? Why is the species diversity of a large island usually higher than that on a smaller island?

4. What would you do if large numbers of cockroaches (Spotlight, p. 99) invaded your home? See whether you can come up with an ecological rather than a chemical (pesticide) approach to this problem.

5. How would you reply to someone who argues that **(a)** we should not worry about our effects on natural systems because succession will heal the wounds of human activities and restore the balance of nature, **(b)** efforts to preserve natural systems are not worthwhile because nature is largely unpredictable, and **(c)** because there is no balance in nature and no stability in species diversity we should cut down diverse old-growth forests and replace them with tree farms?

PROJECTS

1. Use the library or Internet to find and describe two species not discussed in this textbook that are engaged in

(a) a commensalistic interaction, (b) a mutualistic interaction, and (c) a parasite–host relationship.

2. Visit a nearby natural area and identify examples of (a) mutualism and (b) resource partitioning.

3. Visit a nearby land area such as a partially cleared or burned forest or grassland or an abandoned cropfield and record signs of secondary ecological succession. Study the area carefully to see whether you can find patches that are at different stages of succession because of various disturbances.

4. Make a concept map of this chapter's major ideas, using the section heads and subheads and the key terms (in boldface). Look on the website for this book for information about making concept maps.

INTERNET STUDY RESOURCES AND RESOURCES FOR FURTHER READING AND RESEARCH

The website for this book contains helpful study aids and many ideas for further reading and research. Log on to

http://biology.brookscole.com/miller10

and click on the Chapter-by-Chapter area. Choose Chapter 7 and select a resource:

■ Flash Cards allows you to test your mastery of the Terms and Concepts to Remember for this chapter.

■ Tutorial Quizzes provides a multiple-choice practice quiz.

■ Student Guide to InfoTrac will lead you to Critical Thinking Projects that use InfoTrac College Edition as a research tool.

■ References lists the major books and articles consulted in writing this chapter.

■ Hypercontents takes you to an extensive list of sites with news, research, and images related to individual sections of the chapter.

IINFOTRAC COLLEGE EDITION

Improve your skills with InfoTrac College Edition, a searchable online database of articles from more than 700 periodicals. Log on to

http://www.infotrac-college.com

or access InfoTrac through the website for this book. Try to find the following articles:

1. Byers, J. E. 2000. Competition between two estuarine snails: Implications for invasions of exotic species. *Ecology* 81: 1225. *Keywords:* "competition" and "exotic species." The effects of exploitative competition between a native and an exotic species are examined to determine how competition influences the ability of exotic species to drive native species from a habitat.

2. Miller, L. A., and A. Surlykke. 2001. How some insects detect and avoid being eaten by bats: Tactics and countertactics of prey and predator. *BioScience* 51: 570. *Keywords:* "bats," "insects," and "tactics." Echo-locating bats and their prey insects appear to be in an evolutionary relay race. This article describes in detail how these bats find the insects and how insects can avoid the bats.

8 POPULATION DYNAMICS, CARRYING CAPACITY, AND CONSERVATION BIOLOGY

Sea Otters: Are They Back from the Brink of Extinction?

Southern sea otters (Figure 8-1a) live in kelp forests (Figure 8-1c) in shallow waters along the Pacific coast of North America. These tool-using animals use stones to pry shellfish off rocks underwater and to break open the shells while swimming on their backs and using their bellies as a table. Each day a sea otter consumes about 25% of its weight in sea urchins (Figure 8-1b), clams, mussels, crabs, abalone, and about 40 other species of bottom-dwelling organisms.

Before European settlers arrived, about 1 million southern sea otters lived along the Pacific coastline of North America. By the early 1900s, this species was almost extinct mostly because of overhunting for their warm, beautiful fur and removal by abalone fishers.

Biologists have learned that southern sea otters are *keystone species* (p. 145) that help keep sea urchins from depleting kelp forests in offshore waters from Alaska to southern California. Without significant predation by sea otters, sea urchin populations expand and consume much of the kelp. This lowers the diversity of other plants and animals in kelp forest ecosystems.

Preventing the loss of kelp forests has important ecological and economic benefits. For example, they provide essential habitats for a variety of species, in- hibit shore erosion, and reduce the impact of storm waves on coastlines.

Since 1938, the population of southern sea otters has increased from about 300 to around 2,000. Wherever southern sea otters have returned or have been reintroduced, formerly deforested kelp areas recover within a few years and fish populations increase. This pleases biologists. However, it upsets commercial fishers, who argue that sea otters consume too many abalone and compete with commercial fishing.

During the past few years the southern sea otter population has been declining for unknown reasons. One possibility is increased pollution of coastal waters. Shellfish tend to concentrate toxins that are released into coastal waters. High levels of toxins may directly kill sea otters, and exposure to lower levels over long periods may reduce their resistance to disease and parasites.

Another possibility is increased predation of the otters by killer whales (orcas). Evidence indicates that in some areas orcas may consume southern sea otters during declines in their normal prey of Steller sea lions and harbor seals—whose numbers may have dropped because some species they eat have been overfished.

The population dynamics of southern sea otter populations have helped us to better understand the ecological importance of this keystone species. *Population dynamics, conservation biology,* and *human impacts on natural ecosystems* are the subject of this chapter.

(a)

(b)

(c)

Figure 8-1 Three of the species found in a kelp forest ecosystem.

This chapter addresses the following questions:

- How do populations change in size, density, and makeup in response to environmental stress?
- What is the role of predators in controlling population size?
- What different reproductive patterns enhance the survival of different species?
- What is conservation biology?
- What are the major impacts of human activities on populations, communities, and ecosystems?
- What lessons can we learn from ecology about living more sustainably?

8-1 POPULATION DYNAMICS AND CARRYING CAPACITY

What Are the Major Characteristics of a Population? Populations are dynamic and change in response to environmental stress or changes in environmental conditions. They change in their *size* (number of individuals) and *density* (number of individuals in a certain space). They can also change their pattern of *dispersion* in their habitat, depending mostly on resource availability. Three general patterns of population distribution in a habitat are clumping, uniform dispersion, and random dispersion (Figure 8-2). Populations can also experience changes in their *age distribution:* the proportion of individuals of each age in a population. These changes, called **population dynamics,** occur in response to environmental stress and changes in environmental conditions.

What Limits Population Growth? Four variables—*births, deaths, immigration,* and *emigration*—govern changes in population size. A population gains individuals by birth and immigration and loses them by death and emigration:

Population change = (Births + Immigration) − (Deaths + Emigration)

These variables depend on changes in resource availability and other environmental changes (Figure 8-3, p. 162).

Populations vary in their capacity for growth, also known as the **biotic potential** of the population. The **intrinsic rate of increase (r)** is the rate at which a population would grow if it had unlimited resources. Generally, individuals in populations with a high intrinsic rate of increase have certain characteristics. They *reproduce early in life, have short generation times* (the time between successive generations), *can reproduce many times* (have a long reproductive life), and *have many offspring each time they reproduce.*

Some species have an astounding biotic potential. For example, without any controls on its population growth, the descendants of a single female housefly could total about 5.6 trillion flies within about 13 months. If this kept up, within a few years there would be enough flies to cover the earth's entire surface.

Fortunately, this is not a realistic scenario because *no population can grow indefinitely.* In the real world, a rapidly growing population reaches some size limit imposed by a shortage of one or more limiting factors, such as light, water, space, nutrients, or too many competitors or predators. *There are always limits to population growth in nature.*

Environmental resistance consists of all the factors acting jointly to limit the growth of a population. The population size of a species in a given place and time is determined by the interplay between its biotic potential and environmental resistance (Figure 8-3).

Together biotic potential and environmental resistance determine the **carrying capacity (K),** the number of individuals of a given species that can be sustained indefinitely in a given space (area or volume).

The intrinsic rate of increase (r) of many species depends on having a certain minimum population

Clumped
(elephants)

Uniform
(creosote bush)

Random
(dandelions)

Figure 8-2 Generalized *dispersion patterns* for individuals in a population throughout their habitat. The most common pattern is *clumps* of members of a population throughout their habitat (left), mostly because resources are usually found in patches.

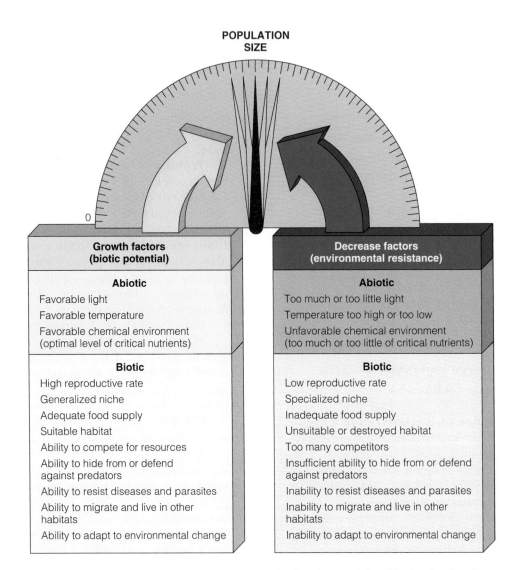

POPULATION SIZE

0

Growth factors (biotic potential)	Decrease factors (environmental resistance)
Abiotic	**Abiotic**
Favorable light	Too much or too little light
Favorable temperature	Temperature too high or too low
Favorable chemical environment (optimal level of critical nutrients)	Unfavorable chemical environment (too much or too little of critical nutrients)
Biotic	**Biotic**
High reproductive rate	Low reproductive rate
Generalized niche	Specialized niche
Adequate food supply	Inadequate food supply
Suitable habitat	Unsuitable or destroyed habitat
Ability to compete for resources	Too many competitors
Ability to hide from or defend against predators	Insufficient ability to hide from or defend against predators
Ability to resist diseases and parasites	Inability to resist diseases and parasites
Ability to migrate and live in other habitats	Inability to migrate and live in other habitats
Ability to adapt to environmental change	Inability to adapt to environmental change

Figure 8-3 Factors that tend to increase or decrease the size of a population. Whether the size of a population grows, remains stable, or decreases depends on interactions between its growth factors (*biotic potential*) and its decrease factors (*environmental resistance*).

size, called the **minimum viable population (MVP).** Suppose a population declines below the MVP needed to support its breeding population. Then certain individuals may not be able to locate mates and genetically related individuals may interbreed and produce weak or malformed offspring. In addition, the genetic diversity may be too low to enable adaptation to new environmental conditions. Then the intrinsic rate of increase falls and extinction is likely.

What Is the Difference between Exponential and Logistic Population Growth? A population that has few if any resource limitations grows exponentially. *Exponential growth* starts out slowly and then proceeds faster and faster as the population increases. If number of individuals is plotted against time, this sequence yields a J-shaped exponential growth curve (Figure 8-4a).

Logistic growth involves exponential population growth followed by a steady decrease in population growth with time as the population encounters environmental resistance and approaches the carrying capacity of its environment and levels off. After leveling off, a population with this type of growth typically fluctuates slightly above and below the carrying capacity. A plot of the number of individuals against time yields a sigmoid, or S-shaped, logistic growth curve (Figure 8-4b). A case of such growth involves the sheep population on the island of Tasmania, south of Australia, in the early 19th century (Figure 8-5).

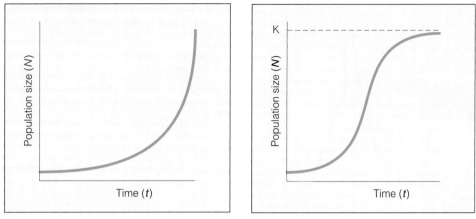

(a) Exponential Growth   **(b) Logistic Growth**

Figure 8-4 Theoretical population growth curves. **(a)** *Exponential growth*, in which the population's growth rate increases with time. Exponential growth occurs when resources are not limiting and a population can grow at its *intrinsic rate of increase* (*r*). Exponential growth of a population cannot continue forever because eventually some factor limits population growth. **(b)** *Logistic growth*, in which the growth rate decreases as the population gets larger. With time, the population size stabilizes at or near the *carrying capacity* (*K*) of its environment and results in a sigmoid (S-shaped) population growth curve.

What Happens If the Population Size Exceeds the Carrying Capacity?

The populations of some species do not make a smooth transition from exponential growth to logistic growth. Instead, such a population uses up its resource base and temporarily *overshoots*, or exceeds, the carrying capacity of its environment. This overshoot occurs because of a *repro-*

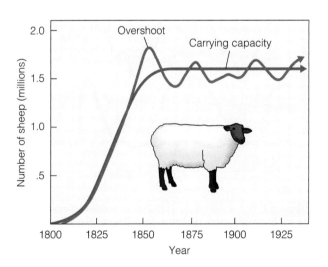

Figure 8-5 *Logistic growth* of a sheep population on the island of Tasmania between 1800 and 1925. After sheep were introduced in 1800, their population grew exponentially because of ample food. By 1855, they overshot the land's carrying capacity. Their numbers then stabilized and fluctuated around a carrying capacity of about 1.6 million sheep.

ductive time lag, the period needed for the birth rate to fall and the death rate to rise in response to resource overconsumption.

In such cases the population suffers a *dieback,* or *crash,* unless the excess individuals switch to new resources or move to an area with more favorable conditions. Such a crash occurred when reindeer were introduced onto a small island off the southwest coast of Alaska (Figure 8-6, p. 164).

Carrying capacity is not a simple, fixed quantity and is affected by many factors. Examples are competition within and between species, immigration and emigration, natural and human-caused catastrophic events, and seasonal fluctuations in the supply of food, water, hiding places, and nesting sites.

Humans are not exempt from population overshoot and dieback, as shown by the tragedy on Easter Island (p. 46). Ireland also experienced a population crash after a fungus destroyed the potato crop in 1845. About 1 million people died, and 3 million people migrated to other countries.

Technological, social, and other cultural changes have extended the earth's carrying capacity for the human species. We have increased food production and used large amounts of energy and matter resources to make normally uninhabitable areas of the earth habitable. However, there is growing concern about how long we will be able to keep doing this on a planet with a finite size and resources and with a human population whose size (Figure 1-1, p. 1) and per capita resource use are growing exponentially.

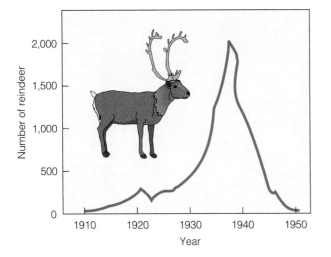

Figure 8-6 Exponential growth, overshoot, and population crash of reindeer introduced to a small island off the southwest coast of Alaska. When 26 reindeer (24 of them female) were introduced in 1910, lichens, mosses, and other food sources were plentiful. By 1935, the herd's population had soared to 2,000, overshooting the island's carrying capacity. This led to a population crash, with the herd plummeting to only 8 reindeer by 1950.

How Does Population Density Affect Population Growth? *Density-independent population controls* affect a population's size regardless of its population density. Examples include floods, hurricanes, unseasonable weather, fire, habitat destruction (such as clearing a forest of its trees or filling in a wetland), and pesticide spraying. For example, a severe freeze in late spring can kill many individuals in a plant population, regardless of its density.

Some limiting factors have a greater effect as a population's density increases. Examples of such *density-dependent population controls* are competition for resources, predation, parasitism, and disease.

Infectious disease is a classic type of density-dependent population control. An example is the *bubonic plague*, which swept through Europe during the 14th century. The bacterium causing this disease normally lives in rodents. It was transferred to humans by fleas that fed on infected rodents and then bit humans. The disease spread like wildfire through crowded cities, where sanitary conditions were poor and rats were abundant. At least 25 million people in European cities died from the disease.

Many health scientists are becoming increasingly alarmed about the possibility of new epidemics of common infectious diseases in crowded urban areas. The primary reason is that many common strains of disease-causing bacteria are becoming genetically resistant to most existing antibiotics through directional natural selection (Figure 5-5, left, p. 97).

What Kinds of Population Change Curves Do We Find in Nature? In nature we find four general types of *population fluctuations*: stable, irruptive, irregular, and cyclic (Figure 8-7). A species whose population size fluctuates slightly above and below its carrying capacity is said to have a fairly *stable* population size (Figures 8-5 and 8-7a). Such stability is characteristic of many species found in undisturbed tropical rain forests, where average temperature and rainfall vary little from year to year.

Some species, such as the raccoon and feral house mouse, normally have a fairly stable population that may occasionally explode, or *irrupt*, to a high peak and then crash to a more stable lower level or in some cases to a very low level (Figures 8-6 and 8-7b). Some factors that temporarily increase carrying capacity for the population, such as favorable weather, more food, or fewer predators, can cause a population explosion.

Some populations appear to have irregular *chaotic behavior* in their changes in population size, with no recurring pattern (Figure 8-7c). Some scientists attribute this behavior to chaos in such systems. Other scientists contend that the behavior may be caused by orderly, nonchaotic behavior whose details and interactions are still poorly understood.

The fourth type consists of *cyclic fluctuations* of population size over a regular time period (Figure 8-7d). Examples are lemmings (whose populations rise and fall every 3–4 years) and lynx and snowshoe hare (whose populations generally rise and fall in a 10-year cycle).

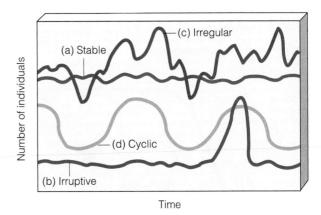

Figure 8-7 General types of simplified population change curves found in nature. **(a)** The population size of a species with a fairly *stable* population fluctuates slightly above and below its carrying capacity. **(b)** The populations of some species may occasionally explode, or *irrupt*, to a high peak and then crash to a more stable lower level. **(c)** The population sizes of some species change irregularly for mostly unknown reasons. **(d)** Other species undergo sharp increases in their numbers, followed by crashes over fairly regular time intervals. Predators sometimes are blamed, but the actual causes of such boom-and-bust cycles are poorly understood.

8-2 CONNECTIONS: THE ROLE OF PREDATION IN CONTROLLING POPULATION SIZE

Do Predators Control Population Size? The Lynx–Hare Cycle Some species that interact as predator and prey undergo cyclic changes in their numbers, with sharp increases in their numbers followed by crashes (Figure 8-8).

For decades, predation has been the explanation for the 10-year population cycles of the snowshoe hare and its predator, the Canadian lynx (Figure 8-8). According to this *top-down control* hypothesis, lynx preying on hares periodically reduce their population. The shortage of hares then reduces the lynx population, which allows the hare population to build up again. At some point the lynx population increases to take advantage of the increased supply of hares, starting the cycle again.

Some research has cast doubt on this appealing hypothesis. Researchers have found that some snowshoe hare populations have similar 10-year boom-and-bust cycles on islands where lynx are absent. These researchers hypothesize that the periodic crashes in the hare population can also be influenced by the food supply of the hares. Large numbers of hares can die when they consume food plants faster than they can be replenished (especially during winter) and have a decrease in the quantity and quality of their food. Once the hare population crashes, the plants recover, and the hare population begins rising again in a hare–plant cycle. If this *bottom-up control* hypothesis is correct, instead of lynx controlling hare populations, the changing hare population size may be causing fluctuations in the lynx population.

These two hypotheses are not mutually exclusive. According to extensive research by Charles J. Krebs, the 10-year population cycle of snowshoe hares in boreal forests is caused by an interaction between predation (by lynx, coyotes, and other predators) and food supplies (especially in winter), and predation is the dominant process.

According to biologists, genuine cases of top-down control systems by predators exist in a number of ecosystems. Two examples already cited are *sharks* (Case Study, p. 146) and *alligators* (Connections, p. 149) controlling some fish populations.

8-3 REPRODUCTIVE PATTERNS AND SURVIVAL

How Do Species Reproduce? Reproductive individuals in populations of all species engage in a struggle for genetic immortality. They do this by trying to have as many members of the next generation as possible carry their genes. Two types of reproduction can pass genes on to offspring. One is **asexual reproduction,** in which all offspring are exact genetic copies (clones) of a single parent. This type of reproduction is common in single-celled species such as bacteria. In such cases, the mother cell divides to produce two identical cells that are genetic clones or replicas of the mother cell.

The second type is **sexual reproduction,** in which organisms produce offspring by combining gametes, or sex cells (such as sperm and ovum), from both parents. This produces offspring that have combinations of traits (chromosomes) from each parent. About 97% of known organisms use sexual reproduction to perpetuate their species.

However, sexual reproduction has several ecological costs and risks. For example, females have to produce twice as many offspring to maintain the same number of young in the next generation as an asexually reproducing organism. This occurs because males do not give birth.

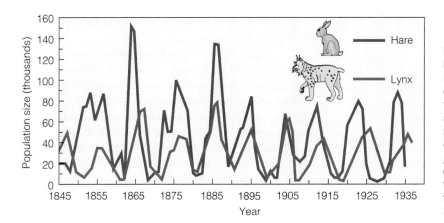

Figure 8-8 Population cycles for the snowshoe hare and Canadian lynx. At one time it was widely believed these curves provided circumstantial evidence that these predator and prey populations regulated one another. More recent research suggests that the periodic swings in the hare population are caused by a combination of predation by lynx and other predators (*top-down population control*) and changes in the availability of the food supply for hares. The rise and fall of the hare population apparently helps determine the lynx population (*bottom-up population control*). (Data from D. A. MacLulich)

A second problem is increased chance of genetic errors and defects during the splitting and recombination of chromosomes. Finally, courtship and mating rituals consume time and energy, can transmit disease, and can inflict injury on males competing for sexual partners.

So if sexual reproduction has a number of costs and risks, why do 97% of the earth's organisms use it? According to biologists, this happens because of two important advantages of sexual reproduction. *First,* it provides a greater genetic diversity in the offspring. A population with many different genetic possibilities has a greater chance of reproducing when environmental conditions change than does a brood of genetically identical clones.

Second, males can gather food for the female and the young and may protect and help train the young in some species. This is especially helpful for species such as mammals and birds that produce only a few young.

What Types of Reproductive Patterns Do Species Have? In 1967, Robert H. MacArthur and Edward O. Wilson suggested that species could be classified into two fundamental reproductive patterns, *r-selected* and *K-selected species.* This classification depends on their position on the sigmoid (S-shaped) or logistic population growth curve (Figure 8-9) and the characteristics of their reproductive patterns (Figure 8-10).

What Are Opportunist or r-Selected Species? Species with a capacity for a high intrinsic rate of increase (*r*) are called **r-selected species** (Figure 8-9 and

Figure 8-10, left). Such species reproduce early and put most of their energy into reproduction. Examples are algae, bacteria, rodents, annual plants (such as dandelions), and most insects (p. 64).

These species have many (usually small) offspring each time they reproduce, reach reproductive age rapidly, and have short generation times. They also give their offspring little or no parental care or protection to help them survive and are short lived (usually with a life span of less than a year). Species with this reproductive pattern overcome the massive loss of their offspring by producing so many unprotected young that a few will survive to reproduce many offspring to begin the cycle again.

Such species tend to be *opportunists.* They reproduce and disperse rapidly when conditions are favorable or when a disturbance opens up a new habitat or niche for invasion, as in the early stages of ecological succession (Figures 7-11, p. 152, and 7-12, p. 153).

Changed environmental conditions from disturbances can allow opportunist species to gain a foothold. However, once established, their populations may crash because of changing or unfavorable environmental conditions or invasion by more competitive species. Therefore, most r-selected or opportunist species go through irregular and unstable boom-and-bust cycles in their population size.

What Are Competitor or K-Selected Species? At the other extreme are *competitor species,* or **K-selected species** (Figure 8-9 and Figure 8-10, right). These species do not put a lot of energy into reproduction, tend to reproduce late in life, and have few offspring with long generation times. They also put most of their energy into nurturing and protecting their young until they reach reproductive age.

Typically the offspring of such species develop inside their mothers (where they are safe), are born fairly large, mature slowly, and are cared for and protected by one or both parents until they reach reproductive age. This reproductive pattern results in a few big and strong individuals that can compete for resources and reproduce a few young to begin the cycle again (Figure 8-10, right).

Such species are called K-selected species because they tend to do well in competitive conditions when their population size is near the carrying capacity (*K*) of their environment. Their populations typically follow a logistic growth curve (Figure 8-4b and Figure 8-9). Examples are most large mammals (such as elephants, whales, and humans), birds of prey, and large and long-lived plants (such as the saguaro cactus, oak trees, redwood trees, and most tropical rain forest trees). Many K-selected species, especially those with long generation times and low reproductive rates (such as ele-

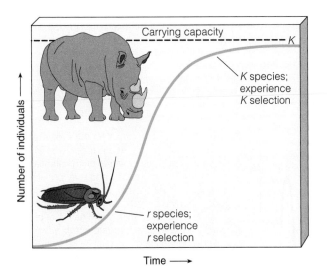

Figure 8-9 Positions of *r-selected* and *K-selected* species on the sigmoid (S-shaped) population growth curve.

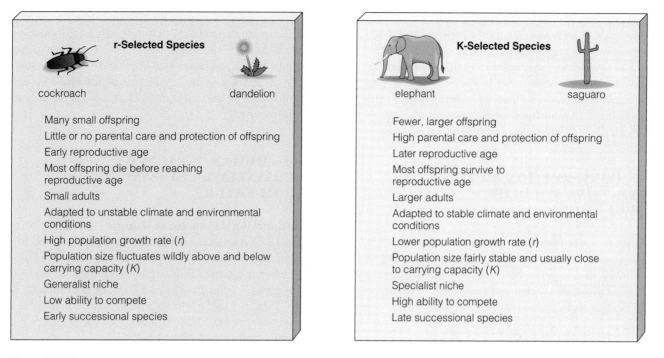

r-Selected Species

cockroach   dandelion

Many small offspring

Little or no parental care and protection of offspring

Early reproductive age

Most offspring die before reaching reproductive age

Small adults

Adapted to unstable climate and environmental conditions

High population growth rate (*r*)

Population size fluctuates wildly above and below carrying capacity (*K*)

Generalist niche

Low ability to compete

Early successional species

K-Selected Species

elephant   saguaro

Fewer, larger offspring

High parental care and protection of offspring

Later reproductive age

Most offspring survive to reproductive age

Larger adults

Adapted to stable climate and environmental conditions

Lower population growth rate (*r*)

Population size fairly stable and usually close to carrying capacity (*K*)

Specialist niche

High ability to compete

Late successional species

Figure 8-10 Generalized characteristics of *r-selected* or opportunist species and *K-selected* or competitor species. Many species have characteristics between these two extremes.

phants, rhinoceroses, and sharks, Case Study, p. 149), are prone to extinction.

Most competitor or K-selected species thrive best in ecosystems with fairly constant environmental conditions. In contrast, opportunists thrive in habitats that have experienced disturbances such as a tree falling, a forest fire, or the clearing of a forest or grassland for raising crops.

Many organisms have reproductive patterns between the extremes of r-selected species and K-selected species, or they change from one extreme to the other under certain environmental conditions. In agriculture we raise both r-selected species (crops) and K-selected species (livestock).

The reproductive pattern of a species may give it a temporary advantage. But *the availability of suitable habitat for individuals of a population in a particular area is what determines its ultimate population size*. Regardless of how fast a species can reproduce, there can be no more dandelions than there is dandelion habitat and no more zebras than there is zebra habitat in a particular area.

What Are Survivorship Curves? Individuals of species with different reproductive strategies tend to have different *life expectancies*. One way to represent the age structure of a population is with a **survivorship curve**, which shows the number of survivors of each age group for a particular species. There are

three generalized types of survivorship curves: *late loss, early loss,* and *constant loss* (Figure 8-11).

A *life table* shows the numbers of individuals at each age from a survivorship curve. It illustrates the

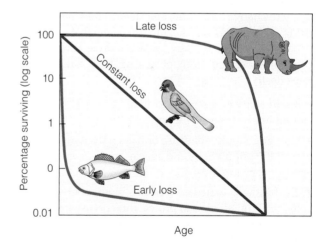

Figure 8-11 Three general *survivorship curves* for populations of different species, obtained by showing the percentages of the members of a population surviving at different ages. A *late loss* population (such as elephants, rhinoceroses, and humans) typically has high survivorship to a certain age, then high mortality. A *constant loss* population (such as many songbirds) shows a fairly constant death rate at all ages. For an *early loss* population (such as annual plants and many bony fish species), survivorship is low early in life. These generalized survivorship curves only approximate the behavior of species.

projected life expectancy and probability of death for individuals at each age. Insurance companies use life tables of human populations in various countries or regions to determine policy costs for their customers. Life tables show that women in the United States survive an average of 6 years longer than men. This explains why a 65-year-old man in the United States normally pays more for life insurance than a 65-year-old woman.

8-4 CONSERVATION BIOLOGY: SUSTAINING WILDLIFE POPULATIONS

What Is Conservation Biology? Conservation **biology** is a multidisciplinary science (originated in the 1970s) that uses the best available science to take action to preserve species and ecosystems. Conservation biology seeks answers to three questions:

- *Which species are in danger of extinction?*
- *How well are various ecosystems functioning, and what ecosystem services* (Figure 1-2, p. 3, and top half of back cover) *of value to humans and other species are we in danger of losing?*
- *What can we do to help sustain ecosystem functions and viable populations of wild species?*

These are challenging questions, and the answers require extensive field research. To understand the status of natural populations, we must measure the current population size, project how population size is likely to change with time, and determine whether existing populations are likely to be sustainable.

Conservation biology rests on three underlying principles derived from observing how nature sustains itself:

- Biodiversity is necessary to all life on earth and should not be reduced by human actions.
- Humans should not cause or hasten the extinction of wildlife populations and species or disrupt vital ecological processes.
- The best way to preserve earth's biodiversity and ecological functions is to protect intact ecosystems that provide sufficient habitat for sustaining natural populations of species.

Conservation biology is based on Aldo Leopold's ethical principle that something is right when it tends to maintain the earth's life-support systems for us and other species and wrong when it does not (Individuals Matter, p. 39).

What Is Bioinformatics? Good conservation biology depends on good information. Increasingly, efforts of conservation biologists are focused on building computer databases that store useful information about biodiversity. Recently, a new discipline called *bioinformatics* has developed. It provides tools for storage of and access to key biological information and builds databases that contain the needed biological information.

8-5 HUMAN IMPACTS ON ECOSYSTEMS: LEARNING FROM NATURE

Connections: How Have Humans Modified Natural Ecosystems? To survive and support growing numbers of people, we have greatly increased the number and area of the earth's natural systems that we have modified, cultivated, built on, or degraded (Figure 7-14, p. 156). Excluding Antarctica, our activities have directly affected to some degree about 83% of the earth's land surface (Figure 8-12).

We have used technology to alter much of the rest of nature in a number of ways. *One is fragmenting and degrading wildlife habitats. A second is simplifying natural ecosystems.* When we plow grasslands and clear forests, we often replace their thousands of interrelated plant and animal species with one crop (Figure 6-18, p. 119) or one kind of tree—called *monocultures*—or with buildings, highways, and parking lots. Then we spend a lot of time, energy, and money trying to protect such monocultures. One threat is an invasion by *opportunist species* of plants (weeds) and *pests* (mostly insects, to which a monoculture crop is like an all-you-can-eat restaurant). Another problem is invasions by *pathogens* (fungi, viruses, or bacteria that harm the plants and animals we want to raise).

A *third* category of alteration is *using, wasting, or destroying an increasing percentage of the earth's net primary productivity that supports all consumer species (including humans)* (Figure 4-22, p. 80). This increasing human consumption of net primary productivity is the main reason we are crowding out or eliminating the habitats and food supplies of a growing number of species.

A *fourth* type of intervention has inadvertently *strengthened some populations of pest species and disease-causing bacteria.* This has occurred through overuse of pesticides and antibiotics that has speeded up directional natural selection (Figure 5-5, left, p. 97) and caused genetic resistance.

A *fifth* effect has been to *eliminate some predators.* Some ranchers want to eradicate bison or prairie dogs that compete with their sheep or cattle for grass. They also want to eliminate wolves, coyotes, eagles, and other predators that occasionally kill sheep. Big game

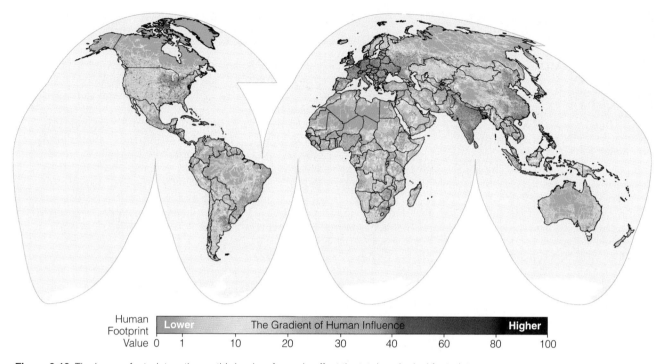

| Human Footprint | **Lower** | The Gradient of Human Influence | | | | | **Higher** | |
| Value | 0 | 1 | 10 | 20 | 30 | 40 | 60 | 80 | 100 |

Figure 8-12 The human footprint on the earth's land surface—in effect the total ecological footprints (Figure 1-7, p. 8) of the human population. Colors represent the percentage of each area influenced by human activities. Excluding Antarctica, human activities have directly affected to some degree about 83% of the earth's land surface and 98% of the area where it is possible to grow rice, wheat, or maize. (Wildlife Conservation Society in collaboration with the Center for International Earth Science Information Network [CIESIN])

hunters push for elimination of predators that prey on game species.

Sixth, new or nonnative species have been deliberately or accidentally introduced into ecosystems. Some of these species are beneficial (such as most food crops) and some harmful to us and other species.

Seventh, some renewable resources have been overharvested. Ranchers and nomadic herders sometimes allow livestock to overgraze grasslands until erosion converts these ecosystems to less productive semideserts or deserts. Farmers sometimes deplete soil nutrients by excessive crop growing. Some fish species are overharvested. Illegal hunting (poaching) endangers wildlife species with economically valuable parts (such as elephant tusks, rhinoceros horns, and tiger skins).

Eighth, some human activities interfere with the normal chemical cycling and energy flows in ecosystems. Soil nutrients can erode from monoculture crop fields, tree plantations, construction sites, and other simplified ecosystems and overload and disrupt other ecosystems such as lakes and coastal ecosystems. Chemicals such as chlorofluorocarbons (CFCs) released into the atmosphere can increase the amount of harmful ultraviolet energy reaching the earth by reducing ozone levels in the stratosphere. Emissions of carbon dioxide and other greenhouse gases—from burning fossil fuels and from clearing and burning forests and grasslands—can trigger global climate change by altering energy flow through the atmosphere.

To survive we must exploit and modify parts of nature. However, we are beginning to understand that any human intrusion into nature has multiple effects, most of them unpredictable (Connections, p. 170).

There are two major challenges. One is to maintain a balance between simplified, human-altered ecosystems and the neighboring, more complex natural ecosystems on which we and other species depend. The other is to slow down the rates at which we are altering nature for our purposes. If we simplify and degrade too much of the planet to meet our needs and wants, what is at risk is not the earth but our own species.

Solutions: What Can We Learn from Ecology about Living More Sustainably? Organisms, populations, and ecosystems are remarkably resilient when exposed to stresses caused by natural or human-induced changes in environmental conditions. However, scientific research indicates that such stresses and changes can affect their health and thier long-term sustainability.

Ecological Surprises

CONNECTIONS

Malaria once infected 9 out of 10 people in Malaysia's North Borneo, now known as Sabah. In 1955, the World Health Organization (WHO) began spraying the island with dieldrin (a DDT relative) to kill malaria-carrying mosquitoes. The program was so successful that the dreaded disease was nearly eliminated.

However, unexpected things began to happen. The dieldrin also killed other insects, including flies and cockroaches living in houses. The islanders applauded this turn of events. But then small lizards

that also lived in the houses died after gorging themselves on dieldrin-contaminated insects.

Next, cats began dying after feeding on the lizards. Then, in the absence of cats, rats flourished and overran the villages. When the people became threatened by sylvatic plague carried by rat fleas, the WHO parachuted healthy cats onto the island to help control the rats.

Then the villagers' roofs began to fall in. The dieldrin had killed wasps and other insects that fed on a type of caterpillar that either avoided or was not affected by the insecticide. With most of its predators eliminated, the caterpillar population exploded, munching its way

through its favorite food: the leaves used in thatched roofs.

Ultimately, this episode ended happily: both malaria and the unexpected effects of the spraying program were brought under control. Nevertheless, the chain of unforeseen events emphasizes the unpredictability of interfering with an ecosystem. It reminds us that when we intervene in nature, we need to ask, "And then what?"

Critical Thinking

Do you believe the beneficial effects of spraying pesticides on Sabah outweighed the resulting unexpected and harmful effects? Explain.

Figure 8-13 summarizes the four major ways that life on earth is sustained based on ecological research and Figure 8-14 (left) gives an expanded description of these principles. Many biologists believe the best way for us to live more sustainably is to learn about the processes and adaptations by which nature sustains itself (Figure 8-14, left) and mimic these lessons from

ecology in designing our societies and economies (Figure 8-14, right).

It is important to understand that the four ways nature sustains itself (Figure 8-14, left) are interconnected. Failure or weakening of any one of these operating principles can disrupt and threaten the long-term sustainability of ecosystems at the local, regional, or global level.

Biologists have used these lessons from ecology to formulate several guidelines for our search for more sustainable societies and lifestyles:

- *Our lives, lifestyles, and economies are totally dependent on the sun and the earth.* We need the earth, but the earth does not need us.

- *Everything is connected to everything else.* The primary goal of ecology is to discover which connections in nature are the strongest, most important, and most vulnerable to disruption for us and other species.

- *We can never do merely one thing.* Any human intrusion into nature has mostly unpredictable side effects (Connections, above). When we alter nature, we need to ask, "And then what?"

- *We cannot indefinitely sustain a civilization that depletes and degrades the earth's natural capital; we can sustain one that lives off the biological income provided by the earth's natural capital* (Figure 1-2, p. 3, and top half of back cover).

Using such guidelines, we can create more ecologically and economically sustainable societies that live within their ecological means. This means taking no

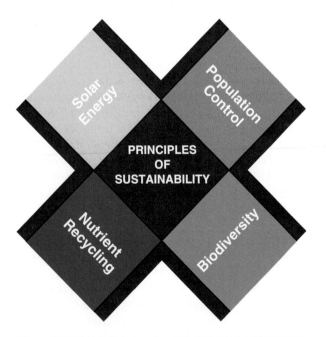

Figure 8-13 Four principles of sustainability derived from learning how nature sustains itself. This diagram also appears on the bottom half of the back cover of this book.

Runs on renewable solar energy.

Rely mostly on renewable solar energy.

Recycles nutrients and wastes. There is little waste in nature.

Prevent and reduce pollution and recycle and reuse resources.

Uses biodiversity to maintain itself and adapt to new environmental conditions.

Preserve biodiversity by protecting ecosystem services and preventing premature extinction of species.

Controls a species' population size and resource use by interactions with its environment and other species.

Reduce births and wasteful resource use to prevent environmental overload and depletion and degradation of resources.

Figure 8-14 Solutions: implications of the four principles of sustainability (left) derived from observing nature for the long-term sustainability of human societies (right). These four operating principles of nature are connected to one another. Failure of any single principle can lead to temporary or long-term unsustainability and disruption of ecosystems and human economies and societies.

more than we need, using renewable resources no faster than nature replaces them, and preserving biodiversity and human cultural diversity.

We cannot command nature except by obeying her.

Sir Francis Bacon

REVIEW QUESTIONS

1. Define the boldfaced terms in this chapter.

2. Explain how the populations of southern sea otters and kelp interact and why the southern sea otter is considered a keystone species.

3. What four factors affect population change?

4. Write an equation showing how population change is related to births, deaths, immigration, and emigration.

5. What is the *biotic potential* of a population? What are four characteristics of a population with a high *intrinsic rate of increase (r)*?

6. What are *environmental resistance* and *carrying capacity*? How do biotic potential and environmental resistance interact to determine carrying capacity? List four factors that can alter an area's carrying capacity.

7. Distinguish between *exponential* and *logistic growth* of a population, and give an example of each type.

8. How can a population overshoot its carrying capacity, and what are the consequences of doing this?

9. Distinguish between *density-dependent* and *density-independent factors* that affect a population's size, and give an example of each.

10. Distinguish among *stable, irruptive, irregular,* and *cyclic* forms of population change.

11. Distinguish between *top-down control* and *bottom-up control* of a population's size. Use these concepts to describe the effects of the predator–prey interactions between the snowshoe hare and the Canadian lynx on the population of each species.

12. Distinguish between *asexual reproduction* and *sexual reproduction.* What are the disadvantages and advantages of sexual reproduction?

13. List the characteristics of **(a)** *r-selected* or *opportunist species* and **(b)** *K-selected* or *competitor* species, and give two examples of each type. Under what environmental conditions are you most likely to find **(a)** r-selected species and **(b)** K-selected species?

14. What is a *survivorship curve,* and how is it used? List three general types of survivorship curves, and give an example of a species with each type.

15. What is *conservation biology?* What three questions does conservation biology try to answer? What are the three underlying principles of conservation biology? What is *bioinformatics,* and what is its importance?

16. List eight potentially harmful ways in which humans modify natural ecosystems.

17. List four *principles of sustainability* observed in natural systems and describe how they can be adapted for developing more sustainable human societies.

18. List four guidelines we could use to help us live more sustainably.

CRITICAL THINKING

1. Why do **(a)** biotic factors that regulate population growth tend to depend on population density and **(b)** abiotic factors that regulate population tend to be independent of population density?

2. What are the advantages and disadvantages of a species undergoing **(a)** exponential growth (Figure 8-4a) and **(b)** logistic growth (Figure 8-4b)?

3. Why are pest species likely to be extreme r-selected species? Why are many endangered species likely to be extreme K-selected species?

4. Predict the type of survivorship curve you would expect given descriptions of the following organisms:

  a. This organism is an annual plant. It lives only 1 year. During that time, it sprouts, reaches maturity, produces many wind-dispersed seeds, and dies.

  b. This organism is a mammal. It reaches maturity after 10 years. It bears one young every 2 years. The parents and the rest of the herd protect the young.

5. How has the human population generally been able to avoid environmental resistance factors that affect other populations? Is this likely to continue? Explain.

6. Explain why you agree or disagree with the four principles of sustainability (Figure 8-14, left) and their lessons for human societies listed in Figure 8-14, right). Identify aspects of your lifestyle that follow or violate each of these four sustainability principles. Would you be willing to change the aspects of your lifestyle that violate these sustainability principles?

PROJECTS

1. Use the four principles of sustainability derived from the scientific study of how nature sustains itself (Figures 8-13 and 8-14) to evaluate the long-term sustainability of the following parts of human systems: **(a)** transportation, **(b)** cities, **(c)** agriculture, **(d)** manufacturing, **(e)** waste disposal, **(f)** economies, and **(g)** your school. Compare your analysis with those made by other members of your class.

2. Use the library and Internet to choose one wild plant species and one animal species and analyze the factors that are likely to limit the population of each species.

3. Make a concept map of this chapter's major ideas, using the section heads and subheads and the key terms (in boldface). Look on the website for this book for information about making concept maps.

INTERNET STUDY RESOURCES AND RESOURCES FOR FURTHER READING AND RESEARCH

The website for this book contains helpful study aids and many ideas for further reading and research. Log on to

http://biology.brookscole.com/miller10

and click on the Chapter-by-Chapter area. Choose Chapter 8 and select a resource:

- Flash Cards allows you to test your mastery of the Terms and Concepts to Remember for this chapter.

- Tutorial Quizzes provides a multiple-choice practice quiz.

- Student Guide to InfoTrac will lead you to Critical Thinking Projects that use InfoTrac College Edition as a research tool.

- References lists the major books and articles consulted in writing this chapter.

- Hypercontents takes you to an extensive list of sites with news, research, and images related to individual sections of the chapter.

INFOTRAC COLLEGE EDITION

Improve your skills with InfoTrac College Edition, a searchable online database of articles from more than 700 periodicals. Log on to

http://www.infotrac-college.com

or access InfoTrac through the website for this book. Try to find the following articles:

1. Wangersky, R. 2000. Too many moose. *Canadian Geographic* 120: 44. *Keywords:* "Newfoundland," "moose," and "problems." Moose were introduced into Newfoundland, where they are now thriving. Although the introduction of moose has some benefits for humans, it is having significant adverse ecological impact.

2. Michigan Technological University. 2002. Longest-running predator-prey study in world links lack of wintry weather to changes in wolf, moose populations on Isle Royale in Lake Superior. *Ascribe Higher Education News Service* (March 12). *Keywords:* "predator-prey" and "Isle Royale." Most population-ecology books refer to the classic case of population regulation of moose and wolves on Isle Royale. This article indicates that other factors may also be affecting the populations of these animals.

9 ENVIRONMENTAL GEOLOGY: PROCESSES, MINERALS, AND SOILS

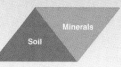

The Dust Bowl: A Lesson from Nature

In the 1930s, Americans learned a harsh environmental lesson when much of the topsoil in several dry and windy midwestern states was lost through a combination of poor cultivation practices and prolonged drought.

Before settlers began grazing livestock and planting crops there in the 1870s, the deep and tangled root systems of native prairie grasses anchored the fertile topsoil firmly in place. Plowing the prairie tore up these roots, and the agricultural crops the settlers planted annually in their place had less extensive root systems.

After each harvest, the land was plowed and left bare for several months, exposing it to high winds. Overgrazing by livestock in some areas also destroyed large expanses of grass, denuding the ground.

The stage was set for severe wind erosion and crop failures; all that was needed was a long drought. Such a drought occurred between 1926 and 1934. In the 1930s, dust clouds created by hot, dry windstorms darkened the sky at midday in some areas; rabbits and birds choked to death on the dust.

During May 1934, a cloud of topsoil blown off the Great Plains traveled some 2,400 kilometers (1,500 miles) and blanketed most of the eastern United States with dust. Journalists gave the Great Plains a new name: the *Dust Bowl* (Figure 9-1).

During the "dirty thirties," large areas of cropland were stripped of topsoil and severely eroded. This triggered one of the largest internal migrations in U.S. history as thousands of displaced farm families from Oklahoma, Texas, Kansas, and Colorado migrated to California or to the industrial cities of the Midwest and East. Most found no jobs because the country was in the midst of the Great Depression.

In May 1934, Hugh Bennett of the U.S. Department of Agriculture (USDA) went before a congressional hearing in Washington to plead for new programs to protect the country's topsoil. Lawmakers took action when Great Plains dust began seeping into the hearing room.

In 1935, the United States passed the Soil Erosion Act, which established the Soil Conservation Service (SCS) as part of the USDA. With Bennett as its first head, the SCS (now called the Natural Resources Conservation Service) began promoting sound conservation practices, first in the Great Plains states and later elsewhere. Soil conservation districts were formed throughout the country, and farmers and ranchers were given technical assistance in setting up soil conservation programs.

We live on a dynamic planet. Energy from the sun and from the earth's interior, coupled with the erosive power of flowing water, have created continents, mountains, valleys, plains, and ocean basins in an ongoing process that continues to change the landscape.

Geology is the science devoted to the study of these dynamic processes. Geologists study and analyze rocks and the features and processes of the earth's interior and surface. Some of these processes lead to geologic hazards such as earthquakes and volcanic eruptions , and others produce the renewable soil and nonrenewable mineral resources and energy resources that support life and economies.

Figure 9-1 The *Dust Bowl* of the Great Plains, where a combination of extreme drought and poor soil conservation practices led to severe wind erosion of topsoil in the 1930s.

Civilization exists by geological consent, subject to change without notice.

WILL DURANT

This chapter addresses the following questions:

- What major geologic processes occur within the earth and on its surface?

- What are rocks, and how are they recycled by the rock cycle?

- How do we find and extract mineral resources from the earth's crust?

- Will there be enough nonrenewable mineral resources for future generations?

- What are the hazards from earthquakes and volcanic eruptions?

- What are soils, and how are they formed?

- What is soil erosion, and how can it be reduced?

9-1 GEOLOGIC PROCESSES

What Is the Earth's Structure? As the primitive earth cooled over eons, its interior separated into three major concentric zones, which geologists identify as the *core*, the *mantle*, and the *crust* (Figure 4-6, p. 68).

What we know about the earth's interior comes mostly from indirect evidence. This includes the results of density measurements, seismic (earthquake) wave studies, measurements of heat flow from the interior, lava analyses, and research on meteorite composition.

The **core** is the earth's innermost zone. It is very hot and has a solid inner part, surrounded by a liquid core of molten material.

A thick, solid zone called the **mantle** surrounds the earth's core. Most of the mantle is solid rock, but under its rigid outermost part is a zone of very hot, partly melted rock that flows like soft plastic. This plastic region of the mantle is called the *asthenosphere.*

The outermost and thinnest zone of the earth is called the **crust.** It consists of the *continental crust,* which underlies the continents (including the continental shelves extending into the oceans, Figure 9-2 and Figure 6-30, p. 128), and the *oceanic crust,* which underlies the ocean basins and covers 71% of the earth's surface (Figure 9-2).

9-2 INTERNAL AND EXTERNAL EARTH PROCESSES

What Geologic Processes Occur within the Earth's Interior? We tend to think of the earth's crust, mantle, and core as fairly static. However, they

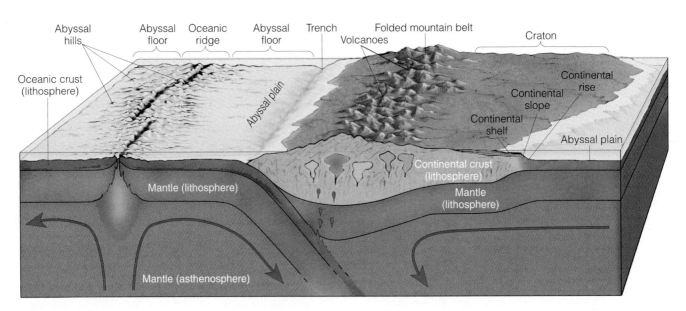

Figure 9-2 Major features of the earth's crust and upper mantle. The *lithosphere,* composed of the crust and outermost mantle, is rigid and brittle. The *asthenosphere,* a zone in the mantle, can be deformed by heat and pressure.

are constantly changed by geologic processes taking place within the earth and on the earth's surface.

Geologic changes originating from the earth's interior are called *internal processes;* generally they build up the planet's surface. Heat from the earth's interior provides the energy for these processes, but gravity also plays a role.

Residual heat from the earth's formation is still being given off as the inner core cools and as the outer core cools and solidifies. Continued decay of radioactive elements in the crust, especially the continental crust, adds to the heat flow from within. This heat from the earth's core causes much of the mantle to deform and flow slowly like heated plastic (in the same way that a red-hot iron horseshoe behaves plastically).

What Is Plate Tectonics? The flows of energy and heated material in the mantle cause movement of rigid plates, called **tectonic plates** (Figure 9-3b, p. 176). These plates are about 100 kilometers (60 miles) thick. They are composed of the continental and oceanic crust and the rigid, outermost part of the mantle (above the asthenosphere)—a combination called the **lithosphere** (Figure 9-2).

These plates move constantly, supported by the slowly flowing asthenosphere like large pieces of ice floating on the surface of a lake. Some plates move faster than others do, but a typical speed is about the rate at which fingernails grow.

The theory explaining the movements of the plates and the processes that occur at their boundaries is called **plate tectonics.** The concept, which became widely accepted by geologists in the 1960s, was developed from an earlier idea called *continental drift.* Throughout the earth's history, continents have split and joined as plates have drifted thousands of kilometers back and forth across the planet's surface (Figure 5-9, p. 101).

Plate motion produces mountains (including volcanoes), the oceanic ridge system, trenches, and other features of the earth's surface (Figure 9-2). Natural hazards such as volcanoes and earthquakes are likely to be found at plate boundaries (Figure 9-3a, p. 176). In addition, plate movements and interactions concentrate many of the minerals we extract and use.

The theory of plate tectonics also helps explain how certain patterns of biological evolution occurred. By reconstructing the course of continental drift over millions of years (Figure 5-9, p. 101), scientists can trace how life-forms migrated from one area to another when continents that are now far apart were still joined together. As the continents separated, populations became geographically and reproductively isolated, and speciation occurred (Figure 5-8, p. 100).

What Types of Boundaries Occur between the Earth's Plates? Lithospheric plates have the following three types of boundaries (Figure 9-4, p. 177):

- **Divergent plate boundaries,** where plates move apart in opposite directions (Figure 9-4, top).

- **Convergent plate boundaries,** where plates are pushed together by internal forces (Figures 9-3b and 9-4, middle). At most convergent plate boundaries, oceanic lithosphere is carried downward (subducted) under the island arc or the continent at a **subduction zone.** A *trench* ordinarily forms at the boundary between the two converging plates (Figure 9-4, middle). Stresses in the plate undergoing subduction cause earthquakes at convergent plate boundaries.

- **Transform faults,** which occur where plates slide past one another along a fracture (fault) in the lithosphere (Figure 9-4, bottom). Most transform faults are on the ocean floor (Figure 9-3b).

What Geologic Processes Occur on the Earth's Surface? Erosion and Weathering Changes in geologic processes based directly or indirectly on energy from the sun and on gravity (rather than on heat in the earth's interior) are called *external processes.* Internal processes generally build up the earth's surface. In contrast, external processes tend to wear it down and produce a variety of landforms and environments formed by the buildup of eroded sediment.

A major external process is **erosion:** the process by which material is dissolved, loosened, or worn away from one part of the earth's surface and deposited in other places. Streams are the most important agent of erosion. They produce valleys and canyons, and may form deltas where streams flow into lakes and oceans (Figure 6-42, p. 136). Wind blowing particles of soil from one area to another (Figure 6-1, p. 106) also causes some erosion. Human activities, particularly those that remove vegetation, accelerate erosion.

Weathering caused by mechanical or chemical processes usually produces loosened material that can be eroded. There are two types of weathering processes. *One* is *mechanical weathering,* in which a large rock mass is broken into smaller fragments of the original material. This is similar to the results you would get by using a hammer to break a rock into small fragments. The most important agent of mechanical weathering is *frost wedging,* in which water collects in pores and cracks of rock, expands upon freezing, and splits off pieces of the rock.

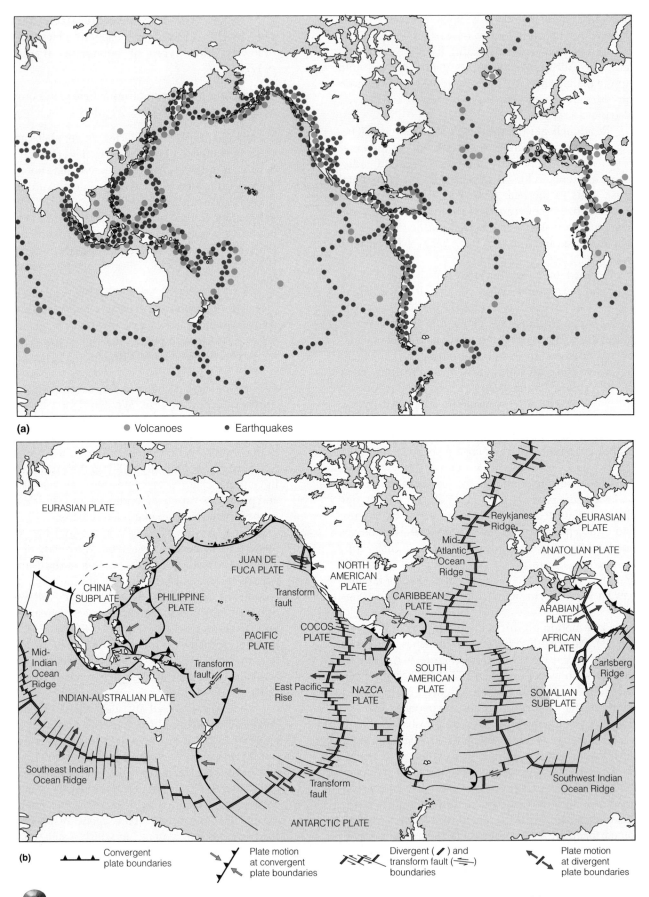

(a) ● Volcanoes ● Earthquakes

(b)

EURASIAN PLATE

JUAN DE FUCA PLATE

NORTH AMERICAN PLATE

CHINA SUBPLATE

PHILIPPINE PLATE

Transform fault

Reykjanes Ridge

Mid-Atlantic Ocean Ridge

EURASIAN PLATE

ANATOLIAN PLATE

CARIBBEAN PLATE

ARABIAN PLATE

PACIFIC PLATE

COCOS PLATE

AFRICAN PLATE

Carlsberg Ridge

Mid-Indian Ocean Ridge

Transform fault

East Pacific Rise

NAZCA PLATE

SOUTH AMERICAN PLATE

SOMALIAN SUBPLATE

INDIAN-AUSTRALIAN PLATE

Southeast Indian Ocean Ridge

Transform fault

Southwest Indian Ocean Ridge

ANTARCTIC PLATE

▲▲▲ Convergent plate boundaries

Plate motion at convergent plate boundaries

Divergent (/) and transform fault (⇥) boundaries

Plate motion at divergent plate boundaries

Figure 9-3 **(a)** Earthquake and volcano sites are distributed mostly in bands along the planet's surface. **(b)** These bands correspond to the patterns for the types of lithospheric plate boundaries shown in Figure 9-4.

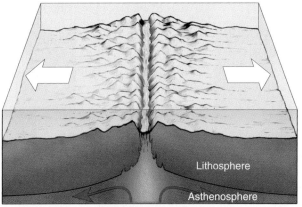

Oceanic ridge at a divergent plate boundary

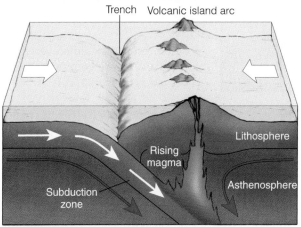

Trench   Volcanic island arc

Lithosphere

Rising
magma

Asthenosphere

Subduction
zone

**Trench and volcanic island arc at a convergent
plate boundary**

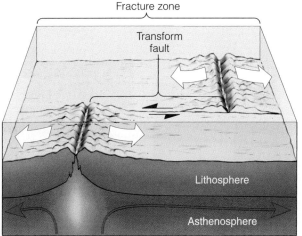

Fracture zone

Transform
fault

Lithosphere

Asthenosphere

Transform fault connecting two divergent plate boundaries

 Figure 9-4 Types of boundaries between the earth's lithospheric plates. All three boundary types occur both in oceans and on continents.

The *second* process is *chemical weathering*, in which one or more chemical reactions decompose a mass of rock. Most chemical weathering involves a reaction of rock material with oxygen, carbon dioxide, and moisture in the atmosphere and the ground.

9-3 MINERALS, ROCKS, AND THE ROCK CYCLE

What Are Minerals and Rocks? The earth's crust, still forming in various places, is composed of minerals and rocks. It is the source of almost all the nonrenewable resources we use: fossil fuels, metallic minerals, and nonmetallic minerals (Figure 1-6, p. 7). It is also the source of soil and of the elements that make up our bodies and those of other living organisms.

A **mineral** is an element or inorganic compound that occurs naturally and is solid. Some minerals consist of a single element, such as gold, silver, diamond (carbon), and sulfur. However, most of the more than 2,000 identified minerals occur as inorganic compounds formed by various combinations of elements. Examples are salt, mica, and quartz.

Rock is any material that makes up a large, natural, continuous part of the earth's crust. Some kinds of rock, such as limestone (calcium carbonate, or $CaCO_3$) and quartzite (silicon dioxide, or SiO_2), contain only one mineral. But most rocks consist of two or more minerals.

What Are the Three Major Rock Types? Based on the way it forms, rock is placed in three broad classes. *One* is **igneous rock,** formed below or on the earth's surface when molten rock material (magma) wells up from the earth's upper mantle or deep crust, cools, and hardens into rock. Examples are granite (formed underground) and lava rock (formed above ground when molten lava cools and hardens). Although often covered by sedimentary rocks or soil, igneous rocks form the bulk of the earth's crust. They also are the main source of many nonfuel mineral resources.

A *second* type is **sedimentary rock,** formed from sediment when preexisting rocks are weathered and eroded into small pieces, transported from their sources, and deposited in a body of surface water. Examples are sandstone and shale (formed from pressure created by deposited layers of sediment), dolomite and limestone (formed from the compacted shells, skeletons, and other remains of dead organisms), and lignite and bituminous coal (derived from plant remains).

The *third* type is **metamorphic rock,** produced when a preexisting rock is subjected to high temperatures (which may cause it to melt partially), high pressures, chemically active fluids, or a combination of these agents. Examples are anthracite (a form of coal), slate, and marble.

Connections: What Is the Rock Cycle? Rocks are constantly exposed to various physical and chemical conditions that can change them over time. The interaction of processes that change rocks from one type to another is called the **rock cycle** (Figure 9-5, p 178).

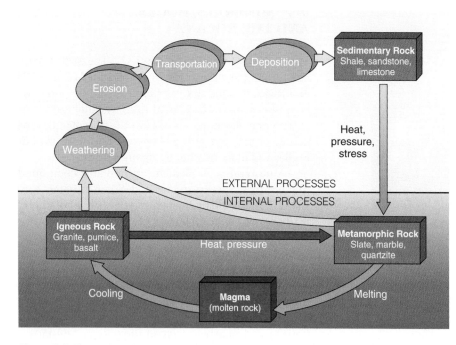

Figure 9-5 The *rock cycle*, the slowest of the earth's cyclic processes. The earth's materials are recycled over millions of years by three processes: *melting, erosion,* and *metamorphism,* which produce *igneous, sedimentary,* and *metamorphic* rocks. Rock of any of the three classes can be converted to rock of either of the other two classes (or can even be recycled within its own class).

The rock cycle recycles material over millions of years and is the slowest of the earth's cyclic processes. It concentrates the planet's nonrenewable mineral resources on which we depend.

What Are Mineral Resources? A **mineral resource** is a concentration of naturally occurring material in or on the earth's crust that can be extracted and processed into useful materials at an affordable cost. Over millions to billions of years the earth's internal and external geologic processes have produced numerous nonfuel mineral resources and energy resources. Because they take so long to produce, they are classified as *nonrenewable resources.*

We know how to find and extract more than 100 nonrenewable minerals from the earth's crust. They include *metallic mineral resources* (iron, copper, aluminum), *nonmetallic mineral resources* (salt, clay, sand, phosphates, and soil), and *energy resources* (coal, oil, natural gas, and uranium).

Ore is rock containing enough of one or more metallic minerals to be mined profitably. We convert about 40 metals extracted from ores into many everyday items that we either use and discard (Figure 3-15, p. 61) or learn to reuse, recycle, or use less wastefully (Figure 3-16, p. 61).

The U.S. Geological Survey divides nonrenewable mineral resources into four major categories (Figure 9-6):

- **Identified resources:** deposits of a nonrenewable mineral resource with a *known* location, quantity, and quality, or whose existence is based on direct geologic evidence and measurements.

- **Undiscovered resources:** potential supplies of a nonrenewable mineral resource assumed to exist on the basis of geologic knowledge and theory but with unknown specific locations, quality, and amounts.

- **Reserves:** identified resources from which a usable nonrenewable mineral can be extracted profitably at current prices.

- **Other resources:** undiscovered resources and identified resources not classified as reserves.

Most published estimates of the supply of a given nonrenewable resource refer to *reserves.* Reserves can increase when new deposits are found or when higher prices or improved mining technology make it profitable to extract deposits that previously were too expensive to extract. Theoretically, all of the *other resources* could eventually be converted to reserves, but this is highly unlikely.

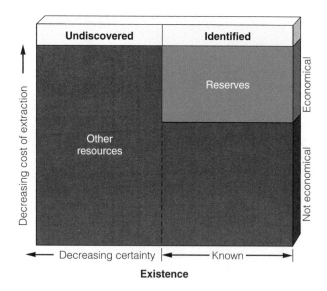

Figure 9-6 General classification of mineral resources. (The area shown for each class does not represent its relative abundance.) In theory, all mineral resources classified as *other resources* could become reserves because of rising mineral prices or improved mineral location and extraction technology. In practice, geologists expect only a fraction of other resources to become reserves.

9-4 FINDING, REMOVING, AND PROCESSING NONRENEWABLE MINERAL RESOURCES

How Are Buried Mineral Deposits Found?
Mining companies use several methods to find promising mineral deposits. One is using aerial photos and satellite images to reveal protruding rock formations (outcrops) associated with certain minerals. Also, planes can be equipped with *radiation-measuring equipment* to detect deposits of radioactive metals such as uranium, and a *magnetometer* to measure changes in the earth's magnetic field caused by magnetic minerals such as iron ore.

Another method uses a *gravimeter* to measure differences in gravity, because the density of an ore deposit usually differs from that of the surrounding rock.

Underground methods include drilling a deep well and extracting core samples. Scientists can also put sensors in existing wells to detect electrical resistance or radioactivity to pinpoint the location of oil and natural gas.

Seismic surveys are conducted on land and at sea by detonating explosive charges and analyzing the resulting shock waves to get information about the makeup of buried rock layers.

Yet another method is to perform *chemical analysis* of water and plants to detect deposits of underground minerals that have leached into nearby bodies of water or have been absorbed by plant tissues.

After suitable mineral deposits are located, several different types of mining techniques are used to remove the deposits, depending on their location and type. Shallow deposits are removed by **surface mining** (Figure 9-7), and deep deposits are removed by **subsurface mining.**

In surface mining, mechanized equipment strips away the **overburden** of soil and rock and usually discards it as waste material called **spoils.** In the United States, surface mining extracts about 90% of

(a) Open Pit Mine

(b) Dredging

(c) Area Strip Mining

(d) Contour Strip Mining

Figure 9-7 Major mining methods used to extract surface deposits of solid mineral and energy resources.

the nonfuel mineral and rock resources and 60% of the coal by weight.

The type of surface mining used depends on the resource being sought and on local topography. There are several types of surface mining. *One* is **open-pit mining** (Figure 9-7a), in which machines dig holes and remove ores (such as iron and copper), sand, gravel, and stone (such as limestone and marble). A *second* is **dredging** (Figure 9-7b), in which chain buckets and draglines scrape up underwater mineral deposits.

A *third* method used where the terrain is fairly flat is **area strip mining** (Figure 9-7c). An earthmover strips away the overburden, and a power shovel digs a cut to remove the mineral deposit. After the mineral is removed, the trench is filled with overburden and a new cut is made parallel to the previous one. The process is repeated over the entire site. If the land is not restored, area strip mining leaves a wavy series of highly erodible hills of rubble called *spoil banks*.

Contour strip mining (Figure 9-7d), a *fourth* method, is used on hilly or mountainous terrain. A power shovel cuts a series of terraces into the side of a hill. An earthmover removes the overburden, a power shovel extracts the coal, and the overburden from each new terrace is dumped onto the one below. Unless the land is restored, a wall of dirt is left in front of a highly erodible bank of soil and rock called a *highwall*.

A *fifth* method is **mountaintop removal.** It uses explosives, massive shovels, and even larger machinery called draglines to remove the top of a mountain and expose seams of coal underneath. The resulting rock and dirt is dumped into the streams and valleys below in many parts of the world. This form of surface mining—increasingly used in West Virginia—causes considerable environmental damage.

Although surface-mined land can be restored (except in arid and semiarid areas), it is expensive and not done in many countries. In the United States, the *Surface Mining Control and Reclamation Act of 1977* requires mining companies to restore most surface-mined land so it can be used for the same purpose as it was before it was mined. The act also levied a tax on mining companies to restore land that was disturbed by surface mining before the law was passed.

Subsurface mining (Figure 9-8) is used to remove coal and various metal ores that are too deep to be extracted by surface mining. Miners dig a deep vertical shaft, blast subsurface tunnels and chambers to get to the deposit, and use machinery to remove the ore or coal and transport it to the surface.

Subsurface mining disturbs less than one-tenth as much land as surface mining and usually produces less waste material. But it leaves much of the resource in the ground and is more dangerous and expensive than surface mining. Hazards include collapse of roofs and walls of underground mines, explosions of dust

and natural gas, and lung diseases caused by prolonged inhalation of mining dust.

What Are the Environmental Impacts of Extracting, Processing, and Using Mineral Resources? The mining, processing, and use of mineral resources takes enormous amounts of energy and often causes land disturbance, soil erosion, and air and water pollution (Figure 9-9, p. 182).

Mining can harm the environment in a number of ways. One is *scarring and disruption of the land surface* (Figure 9-7). The Department of the Interior estimates that about 500,000 mines dot the U.S. landscape, mostly in the West. Cleanup costs are estimated in the tens of billions of dollars.

Another problem is collapse of land above underground mines. This *subsidence* can cause houses to tilt, sewer lines to crack, gas mains to break, and groundwater systems to be disrupted.

Toxin-laced mining wastes can be blown or deposited elsewhere by wind- or water-caused erosion. Another pollution problem is *acid mine drainage*. It occurs when rainwater seeping through a mine or through mine wastes carries sulfuric acid (H_2SO_4, produced when aerobic bacteria act on iron sulfide minerals in spoils) to nearby streams and groundwater (Figure 9-10, p. 182). This contaminates water supplies, and can destroy some forms of aquatic life.

Mining can also result in emission of toxic chemicals into the atmosphere. In the United States, the mining industry produces more toxic emissions than any other industry (typically accounting for almost half of such emissions).

Finally, some forms of wildlife can be exposed to toxic mining wastes stored in holding ponds and leakage of toxic wastes from such ponds.

The typical life cycle of a metal resource is depicted in Figure 9-11 (p. 183). It begins with extracting ore from the earth's crust.

Ore typically has two components. One is the *ore mineral* containing the desired metal and the other is waste material called *gangue* (pronounced "gang"). Removing the gangue from ores produces piles of waste called *tailings*. Particles of toxic metals blown by the wind or leached from tailings by rainfall can contaminate surface water and groundwater.

Most ores consist of one or more compounds of the desired metal. After gangue has been removed, **smelting** is used to separate the metal from the other elements in the ore mineral. Without effective pollution control equipment, smelters emit enormous quantities of air pollutants, which damage vegetation and soils in the surrounding area. Smelters also cause water pollution and produce liquid and solid hazardous wastes that must be disposed of safely. Some companies are using improved technology to reduce pollu-

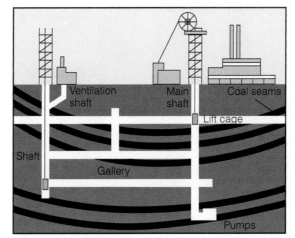

(a) Underground Coal Mine

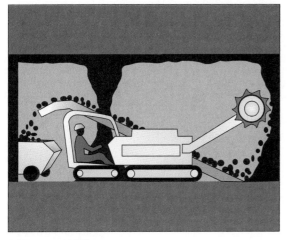

(b) Room-and-Pillar

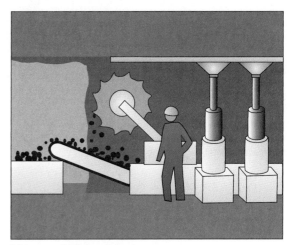

(c) Longwall Mining of Coal

Figure 9-8 Major mining methods used to extract underground deposits of solid mineral and energy resources (primarily coal). **(a)** Mine shafts and tunnels are dug and blasted out. **(b)** In *room-and-pillar* mining, machinery is used to gouge out coal and load it onto a shuttle car in one operation, and pillars of coal are left to support the mine roof. **(c)** In *longwall coal mining*, movable steel props support the roof and cutting machines shear off the coal onto a conveyor belt. As the mining proceeds, roof supports are moved forward and the roof behind is allowed to fall (often causing the land above to sink or subside).

tion from smelting, lower production costs, save costly cleanup bills, and decrease liability for damages.

Once the pure metal has been produced by smelting, it is usually melted and converted to desired products, which are then used and discarded or recycled (Figure 9-11, p. 183).

Are There Environmental Limits to Resource Extraction and Use? Some environmentalists and resource experts believe the greatest danger from the world's continually increasing consumption of nonrenewable mineral resources is not exhaustion of their supplies. Instead, it is more likely to be the environ-

mental damage caused by their extraction, processing, and conversion to products (Figure 9-9, p. 182).

The mineral industry accounts for 5–10% of world energy use. This makes it a major contributor to air and water pollution and to emissions of greenhouse gases such as carbon dioxide (CO_2).

The environmental impacts from mining an ore are affected by its percentage of metal content, or *grade*. The more accessible and higher-grade ores are usually exploited first. As they are depleted, it takes more money, energy, water, and other materials to exploit lower-grade ores. This in turn increases land disruption, mining waste, and pollution.

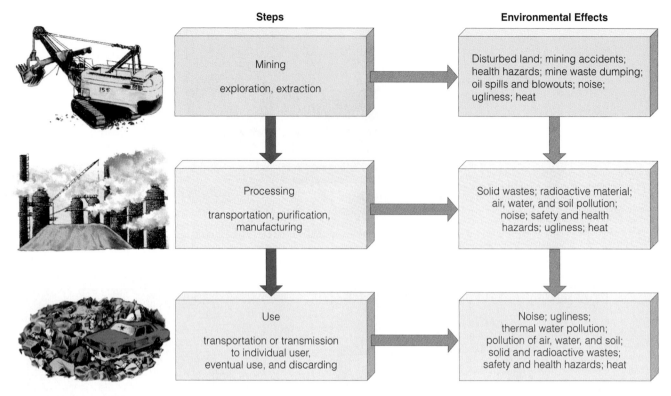

Steps	Environmental Effects
Mining exploration, extraction	Disturbed land; mining accidents; health hazards; mine waste dumping; oil spills and blowouts; noise; ugliness; heat
Processing transportation, purification, manufacturing	Solid wastes; radioactive material; air, water, and soil pollution; noise; safety and health hazards; ugliness; heat
Use transportation or transmission to individual user, eventual use, and discarding	Noise; ugliness; thermal water pollution; pollution of air, water, and soil; solid and radioactive wastes; safety and health hazards; heat

Figure 9-9 Some harmful environmental effects of extracting, processing, and using nonrenewable mineral and energy resources. The energy used to carry out each step causes additional pollution and environmental degradation.

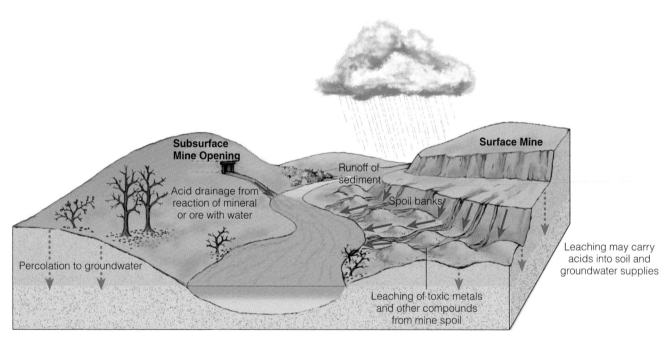

Figure 9-10 Pollution and degradation of a stream and groundwater by runoff of acids—called *acid mine drainage*—and by toxic chemicals from surface and subsurface mining. These substances can kill fish and other aquatic life. Acid mine drainage has damaged more than 19,000 kilometers (12,000 miles) of streams in the United States, mostly in Appalachia and the West.

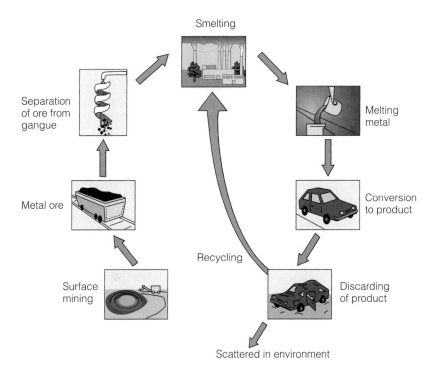

Figure 9-11 Typical *life cycle of a metal resource.* Each step in this process uses energy and produces some pollution and waste heat.

Currently, most of the harmful environmental costs of mining and processing minerals are not included in the prices for processed metals and the resulting consumer products. Instead, these costs are passed on to society and future generations, which gives mining companies and manufacturers little incentive to reduce resource waste and pollution. Environmentalists and some economists call for phasing these external costs into the prices of goods made from minerals through *full-cost pricing* (p. 24).

9-5 SUPPLIES OF MINERAL RESOURCES

Will There Be Enough Mineral Resources? The future supply of nonrenewable minerals depends on two factors. One is the actual or potential supply of the mineral and the other is the rate at which we use that supply.

We never completely run out of any mineral. However, a mineral becomes *economically depleted* when it costs more to find, extract, transport, and process the remaining deposit than it is worth. At that point, there are five choices: *recycle or reuse existing supplies, waste less, use less, find a substitute,* or *do without.*

Depletion time is the time it takes to use up a certain proportion (usually 80%) of the reserves of a min-

eral at a given rate of use (Figure 1-8, p. 9). When experts disagree about depletion times, they are often using different assumptions about supply and rate of use (Figure 9-12, p. 184).

A traditional measure of the projected availability of nonrenewable resources is the **reserve-to-production ratio:** the number of years that proven reserves of a particular nonrenewable mineral will last at current annual production rates. Reserve estimates are continually changing because new deposits are discovered and new mining and processing technology can allow some of the minerals classified as other resources (Figure 9-6) to be converted to reserves. Under these circumstances, the reserve-to-production ratio is the best available projection of the current estimated supply and its estimated depletion time.

The shortest depletion time assumes no recycling or reuse and no increase in reserves (curve A, Figure 9-12). A longer depletion time assumes that recycling will stretch existing reserves and that better mining technology, higher prices, and new discoveries will increase reserves (curve B, Figure 9-12). An even longer depletion time assumes that new discoveries will further expand reserves and that recycling, reuse, and reduced consumption will extend supplies (curve C, Figure 9-12). Finding a substitute for a resource leads to a new set of depletion curves for the new resource.

How Does Economics Affect Mineral Resource Supplies? Geologic processes determine the quantity and location of a mineral resource in the earth's crust. But economics determines what part of the known supply is extracted and used.

According to standard economic theory, in a competitive free market a plentiful mineral resource is cheap when its supply exceeds demand. However, when a resource becomes scarce its price rises. This can encourage exploration for new deposits, stimulate development of better mining technology, and make it profitable to mine lower-grade ores. It can also encourage a search for substitutes and promote resource conservation.

However, according to some economists, this theory may no longer apply to most developed countries. One reason is that industry and government in such countries often control the supply, demand, and prices of minerals to such an extent that a truly competitive free market does not exist.

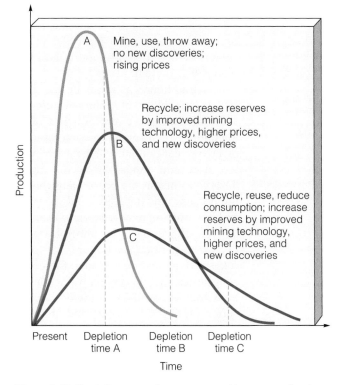

A — Mine, use, throw away; no new discoveries; rising prices

B — Recycle; increase reserves by improved mining technology, higher prices, and new discoveries

C — Recycle, reuse, reduce consumption; increase reserves by improved mining technology, higher prices, and new discoveries

Production

Present | Depletion time A | Depletion time B | Depletion time C

Time

Figure 9-12 *Depletion curves* for a nonrenewable resource (such as aluminum or copper) using three sets of assumptions. Dashed vertical lines represent times when 80% depletion occurs.

Most mineral prices are low because governments subsidize development of their domestic mineral resources to help promote economic growth and national security. In the United States, for instance, mining companies get depletion allowances amounting to 5–22% of their gross income (depending on the mineral). They can also reduce the taxes they pay by deducting much of their costs for finding and developing mineral deposits. In addition, hardrock mining companies operating in the United States can buy public land at 1872 prices and pay no royalties to the government on the minerals they extract (Case Study, right).

Between 1982 and 2001, U.S. mining companies received more than $6 billion in government subsidies. Critics argue that taxing rather than subsidizing the extraction of nonfuel mineral resources would provide governments with revenue, create incentives for more efficient resource use, promote waste reduction and pollution prevention, and encourage recycling and reuse of mineral resources.

Mining company representatives say they need subsidies and low taxes to keep the prices of minerals low for consumers and to encourage the companies not to move their mining operations to other countries with no such taxes and less stringent mining regulations.

Economic problems can also hinder the development of new supplies of mineral resources. because exploring for new mineral resources takes lots of increasingly scarce investment capital and is a risky financial venture. Typically, if geologists identify 10,000 possible deposits of a given resource, only 1,000 sites are worth exploring; only 100 justify drilling, trenching, or tunneling; and only 1 becomes a producing mine or well.

Case Study: Controversy over the General Mining Law of 1872 Some people have gotten rich by using the little known General Mining Law of 1872. This law was designed to encourage mineral exploration and the mining of *hardrock minerals* (such as gold, silver, copper, zinc, nickel, and uranium) on U.S. public lands and to help develop the then sparsely populated West.

Under this 1872 law, a person or corporation can assume legal ownership of parcels of land on essentially all U.S. public land except parks and wilderness areas by *patenting* it. This involves declaring the belief that the land contains valuable hardrock minerals and spending $500 to improve the land for mineral development. Then a claim can be filed. The filer must then pay an annual fee of $100 for each 8 hectares (20 acres) of land to maintain the claim whether or not a mine is in operation. Those with a claim can also pay the federal government $6–12 per hectare ($2.50–5.00 an acre) for the land. Once purchased, the land can be used, leased, or sold for essentially any purpose.

So far, public lands containing an estimated $240–385 billion (adjusted for inflation) of publicly owned mineral resources have been transferred to private interests at 1872 prices. Domestic and foreign mining companies operating under this law remove mineral resources worth at least $2–3 billion per year on once-public land they bought at very low prices.

In addition, the Congressional Budget Office estimates that mining companies remove hardrock minerals worth at least $650 million per year from public land that has not been transferred to private ownership.

Hardrock mining companies pay no royalties on the minerals they extract from such lands. In contrast, oil and natural gas companies pay a gross royalty of 12.5% and coal companies a gross royalty of 8–12.5% on the wholesale price of the resources they remove from public lands.

No provision in the 1872 law requires mining companies to pay for environmental cleanup of any damage they cause to these lands and nearby water resources. Cleanup costs for land and streams damaged by 557,000 abandoned hardrock mines and open pits (mostly in the West) will cost U.S. taxpayers an estimated $33–72 billion.

Mining companies defend the 1872 law. They point out that they must invest large sums of money (often $100 million or more) to locate and develop an ore site before they make any profits from mining hardrock minerals. In addition, their mining operations provide high-paying jobs to miners, supply vital resources for industry, stimulate the national and local economies, reduce trade deficits, and save American consumers money on products produced from such minerals.

The companies also argue that paying royalties on the minerals and requiring them to pay cleanup costs would force them to move their mining operations to other countries.

Environmentalists call for this law to be revised to ban the patenting (sale) of public lands but allow 20-year leases of designated public land for hardrock mining. They would also require mining companies to pay a *gross* royalty of 8–12% on the wholesale value of all minerals removed from public land—similar to what oil, natural gas, and coal companies pay. However, hardrock-mining companies want to continue paying no royalties or pay no more than a 5% *net* royalty based on gross sales minus production costs.

Environmentalists also believe that mining companies should be legally and financially responsible for environmental cleanup and restoration of each site. Some *good news* is that a modification of the 1872 law now requires mining companies to post bonds to cover 100% of the estimated cleanup costs in case they go bankrupt. However, this does not apply to previously mined sites.

Canada, Australia, South Africa, and other countries that are major extractors of hardrock minerals have laws that require royalty payments and responsibility for environmental damage.

Should More Mining Be Allowed on Public Lands in the United States? About one-third of the land in the United States is public land owned jointly by all U.S. citizens. This land, consisting of national forests, parks, resource lands, and wilderness, is managed by various government agencies under laws passed by Congress. About 72% of this public land is in Alaska, and 22% is in western states (where 60% of all land is public land).

For decades, resource developers, environmentalists, and conservationists have argued over how this land should be used. Mineral resource extractors complain that three-fourths of the country's vast public lands, with many areas containing rich deposits of mineral resources, are off limits to mining.

In recent decades, they have stepped up efforts to have Congress open up most of these lands to mineral development, sell off mineral-rich public lands to private interests, or turn their management over to state and local governments (which can often be more readily influenced by mining and development interests). In 2002, the Bush administration began using a mix of these three approaches for management of mineral and timber resources on U.S. public lands.

Conservation biologists and environmentalists strongly oppose such efforts. They argue that this would increase environmental degradation and decrease biodiversity.

Can We Get Enough Minerals by Mining Lower-Grade Ores? Some analysts contend that all we need to do to increase supplies of a mineral is to extract lower grades of ore. They point to the development of new earth-moving equipment, improved techniques for removing impurities from ores, and other technological advances in mineral extraction and processing.

In 1900, for instance, the average copper ore mined in the United States was about 5% copper by weight. Today it is 0.5%, and copper costs less (adjusted for inflation).

However, the mining of lower-grade ores can be limited by several factors. *One* is the increased cost of mining and processing larger volumes of ore. *Second* is the availability of fresh water needed to mine and process some minerals (especially in arid and semiarid areas). A *third* limiting factor is the environmental impact of the increased land disruption, waste material, and pollution produced during mining and processing (Figure 9-9).

Solutions: Can We Use Microbes to Mine Metals? One way to improve mining technology is to use microorganisms for in-place (*in situ*, pronounced "in-SY-too") mining. This biological approach to mining would remove desired metals from ores while leaving the surrounding environment undisturbed. It would also reduce air pollution associated with the smelting of metal ores and reduce water pollution associated with using hazardous chemicals such as cyanides and mercury to extract gold.

Once a commercially viable ore deposit has been identified, wells are drilled into it and the ore is fractured. Then the ore is inoculated with natural or genetically engineered bacteria to extract the desired metal. Next the well is flooded with water, which is pumped to the surface, where the desired metal is removed. Then the water is recycled.

This technique permits economical extraction from low-grade ores, which are increasingly used as high-grade ores are depleted. Since 1958, the copper industry has been using natural strains of the bacterium *Thiobacillus ferroxidans* to remove copper from low-grade copper ore.

Currently, more than 30% of all copper produced worldwide, worth more than $1 billion a year, comes from such *biomining.* If naturally occurring bacteria cannot be found to extract a particular metal, genetic engineering techniques could be used to produce such bacteria.

However, microbiological ore processing is slow. It can take decades to remove the same amount of material that conventional methods can remove within months or years. So far, biological mining methods are economically feasible only with low-grade ore (such as gold and copper) for which conventional techniques are too expensive.

Can We Find Substitutes for Scarce Nonrenewable Mineral Resources? The Materials Revolution

Some analysts believe that even if supplies of key minerals become very expensive or scarce, human ingenuity will find substitutes. They point to the current materials revolution in which silicon and new materials, particularly ceramics and plastics, are being developed and used as replacements for metals.

Ceramics have many advantages over conventional metals. They are harder, stronger, lighter, and longer lasting than many metals, and they withstand intense heat and do not corrode. Within a few decades we may have high-temperature ceramic superconductors in which electricity flows without resistance. Such a development may lead to faster computers, more efficient power transmission, and affordable electromagnets for propelling high-speed magnetic levitation trains.

Plastics also have advantages over many metals. High-strength plastics and composite materials, which are strengthened by lightweight carbon and glass fibers, are likely to transform the automobile and aerospace industries. They cost less to produce than metals because they take less energy, do not need painting, and can be molded into any shape. New plastics and gels are also being developed to provide superinsulation without taking up much space. One new plastic can withstand extremely high temperatures and is not affected by exposure to even the most intense laser beams.

Substitutes undoubtedly can be found for many scarce mineral resources. But finding substitutes for some key materials may be difficult or impossible. Examples are helium, phosphorus for phosphate fertilizers, manganese for making steel, and copper for wiring motors and generators.

In addition, some substitutes are inferior to the minerals they replace. For example, aluminum could replace copper in electrical wiring. But producing aluminum takes much more energy than producing copper, and aluminum wiring is a greater fire hazard than copper wiring.

9-6 NATURAL HAZARDS: EARTHQUAKES AND VOLCANIC ERUPTIONS

What Are Earthquakes? Stress in the earth's crust can cause solid rock to deform until it suddenly fractures and shifts along the fracture, producing a *fault* (Figure 9-4, bottom). The faulting or a later abrupt movement on an existing fault causes an **earthquake.**

An earthquake has certain features and effects (Figure 9-13). When the stressed parts of the earth suddenly fracture or shift, energy is released as shock waves, which move outward from the earthquake's focus like ripples in a pool of water. The *focus* of an earthquake is the point of initial movement, and the *epicenter* is the point on the surface directly above the focus (Figure 9-13).

One way to measure the severity of an earthquake is by its *magnitude* on a modified version of the Richter scale. The magnitude is a measure of the amount of energy released in the earthquake, as indicated by the amplitude (size) of the vibrations when they reach a recording instrument (seismograph). Using this approach, seismologists rate earthquakes as *insignificant* (less than 4.0 on the Richter scale), *minor* (4.0–4.9), *damaging* (5.0–5.9), *destructive* (6.0–6.9), *major* (7.0–7.9), and *great* (over 8.0).

Each unit on the Richter scale represents an amplitude 10 times greater than the next smaller unit. Thus a magnitude 5.0 earthquake is 10 times greater than a magnitude 4.0, and a magnitude 6.0 quake is 100 times greater than a magnitude 4.0 quake.

Earthquakes often have *aftershocks* that gradually decrease in frequency over a period of up to several

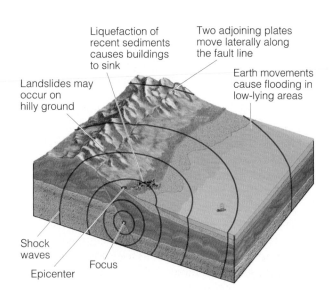

Figure 9-13 Major features and effects of an *earthquake.*

months, and some have *foreshocks* from seconds to weeks before the main shock.

The *primary effects of earthquakes* include shaking and sometimes a permanent vertical or horizontal displacement of the ground. These effects may have serious consequences for people and for buildings, bridges, freeway overpasses, dams, and pipelines.

Secondary effects of earthquakes include rockslides, urban fires, and flooding caused by subsidence (sinking) of land. Coastal areas can be severely damaged by large, earthquake-generated water waves, called *tsunamis* (also called tidal waves although they have nothing to do with tides) that travel as fast as 950 kilometers (590 miles) per hour.

Solutions: How Can We Reduce Earthquake Hazards? One way to reduce loss of life and property from earthquakes is to examine historical records and make geologic measurements to locate active fault zones. We can also map high-risk areas (Figure 9-14), establish building codes that regulate the placement and design of buildings in high-risk areas, and increase research on predicting when and where earthquakes will occur.

Engineers know how to make homes, large buildings, bridges, and freeways more earthquake resistant. However, this can be expensive, especially the reinforcement of existing structures.

What Are Volcanoes? An active **volcano** occurs where magma (molten rock) reaches the earth's surface through a central vent or a long crack (fissure; Figure 9-15). Volcanic activity can release *ejecta* (debris ranging from large chunks of lava rock to ash that may

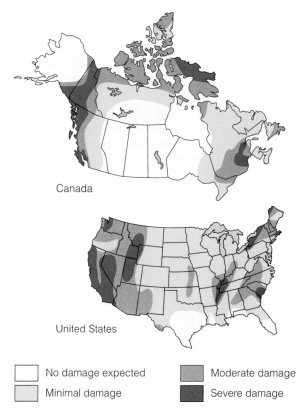

Canada

United States

☐ No damage expected	▨ Moderate damage
▨ Minimal damage	▨ Severe damage

Figure 9-14 Expected damage from earthquakes in Canada and the contiguous United States. This map is based on earthquake records. (Data from U.S. Geological Survey)

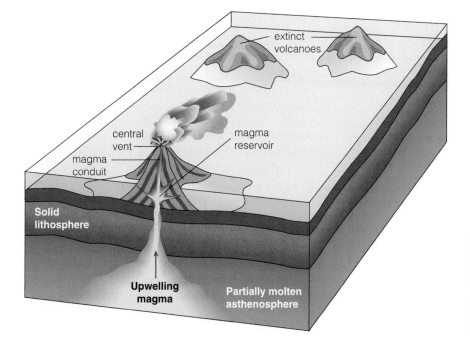

Figure 9-15 A volcano erupts when molten magma in the partially molten asthenosphere rises in a plume through the lithosphere to erupt on the surface as lava that can spill over or be ejected into the atmosphere. Chains of islands can be created by the action of volcanoes that then become inactive.

be glowing hot), *liquid lava,* and *gases* (such as water vapor, carbon dioxide, and sulfur dioxide) into the environment.

Volcanic activity is concentrated for the most part in the same areas as seismic activity (Figure 9-3a). Some volcanoes, such as those at Mount St. Helens in Washington (which erupted in 1980) and Mount Pinatubo in the Philippines (which erupted in 1991), have a steep, flaring cone shape. They usually erupt explosively and eject large quantities of gases and particulate matter (soot and mineral ash) high into the troposphere.

Most of the particles of soot and ash soon fall back to the earth's surface. However, gases such as sulfur dioxide remain in the atmosphere and are converted to tiny droplets of sulfuric acid, many of which stay above the clouds and may not be washed out by rain for up to 3 years. These tiny droplets reflect some of the sun's energy and can cool the atmosphere by as much as 0.5°C (1°F) for 1–4 years.

Other volcanic eruptions at divergent boundaries (for example, in Iceland) and ocean islands (such as the Hawaiian Islands) usually erupt more quietly. They involve primarily lava flows, which can cover roads and villages and ignite brush, trees, and homes.

We tend to think negatively of volcanic activity, but it also provides some benefits. One is outstanding scenery in the form of majestic mountains, some lakes (such as Crater Lake in Oregon; see title page photo), and other landforms. Perhaps the most important benefit of volcanism is the highly fertile soils produced by the weathering of lava.

Solutions: How Can We Reduce Volcano Hazards? We can reduce the loss of human life and sometimes property from volcanic eruptions. Ways to do this include land-use planning, better prediction of volcanic eruptions, and effective evacuation plans. The eruptive history of a volcano or volcanic center can provide some indication of where the risks are.

Scientists are also studying phenomena that precede an eruption. Examples include tilting or swelling of the cone, changes in magnetic and thermal properties of the volcano, changes in gas composition, and increased seismic activity.

9-7 SOIL RESOURCES: FORMATION AND TYPES

What Major Layers Are Found in Mature Soils?
Soil is a complex mixture of eroded rock, mineral nutrients, decaying organic matter, water, air, and billions of living organisms, most of them microscopic decomposers (Figure 9-16). Soil is a renewable resource.

However, the weathering of rock, deposit of sediments, and decomposition of organic matter in dead organisms produces it very slowly..

Mature soils are arranged in a series of zones called **soil horizons,** each with a distinct texture and composition that varies with different types of soils. A cross-sectional view of the horizons in a soil is called a **soil profile.** Most mature soils have at least three of the possible horizons (Figure 9-16).

The top layer is the *surface litter layer,* or *O horizon.* It consists mostly of freshly fallen and partially decomposed leaves, twigs, animal waste, fungi, and other organic materials. Normally, it is brown or black.

The *topsoil layer,* or *A horizon,* is a porous mixture of partially decomposed organic matter, called **humus,** and some inorganic mineral particles. It is usually darker and looser than deeper layers. A fertile soil that produces high crop yields has a thick topsoil layer with lots of humus. This helps topsoil hold water and nutrients taken up by plant roots.

The roots of most plants and most of a soil's organic matter are concentrated in a soil's two upper layers. As long as vegetation anchors these layers, soil stores water and releases it in a nourishing trickle instead of a devastating flood.

The two top layers of most well-developed soils teem with bacteria, fungi, earthworms, and small insects that interact in complex food webs (Figure 9-17, p. 190). Bacteria and other decomposer microorganisms found by the billions in every handful of topsoil recycle the nutrients we and other land organisms need. They break down some complex organic compounds into simpler inorganic compounds soluble in water. Soil moisture carrying these dissolved nutrients is drawn up by the roots of plants and transported through stems and into leaves.

The color of its topsoil tells us a lot about how useful a soil is for growing crops. For example, dark-brown or black topsoil is nitrogen-rich and high in organic matter. Gray, bright yellow, or red topsoils are low in organic matter and need nitrogen enrichment to support most crops.

The *B horizon (subsoil)* and the *C horizon (parent material)* contain most of a soil's inorganic matter, mostly broken-down rock consisting of varying mixtures of sand, silt, clay, and gravel. The C horizon lies on a base of unweathered parent rock called *bedrock.*

The spaces, or pores, between the solid organic and inorganic particles in the upper and lower soil layers contain varying amounts of air (mostly nitrogen and oxygen gas) and water. Plant roots need oxygen for cellular respiration.

Some of the precipitation that reaches the soil percolates through the soil layers and occupies many of the soil's open spaces or pores. This downward

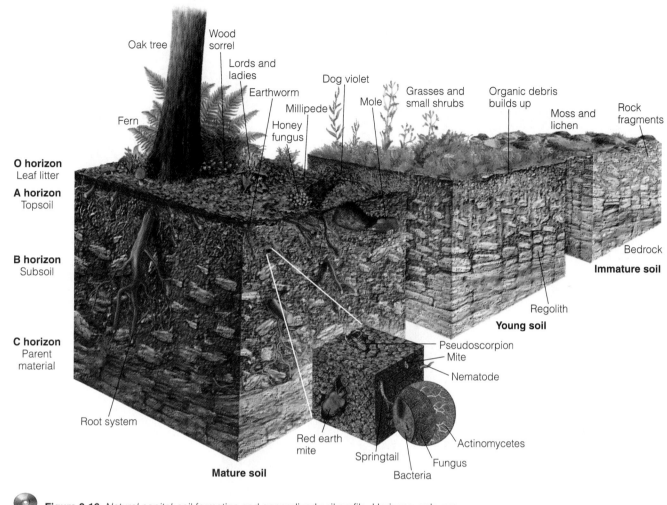

O horizon
Leaf litter

A horizon
Topsoil

B horizon
Subsoil

C horizon
Parent material

Oak tree
Wood sorrel
Lords and ladies
Earthworm
Millipede
Honey fungus
Dog violet
Mole
Grasses and small shrubs
Organic debris builds up
Moss and lichen
Rock fragments
Fern

Bedrock

Immature soil

Regolith

Young soil

Pseudoscorpion
Mite
Nematode
Actinomycetes
Red earth mite
Springtail
Fungus
Bacteria

Root system

Mature soil

Figure 9-16 *Natural capital:* soil formation and generalized soil profile. Horizons, or layers, vary in number, composition, and thickness, depending on the type of soil. (Used by permission of Macmillan Publishing Company from Derek Elsom, *Earth*, New York: Macmillan, 1992. Copyright © 1992 by Marshall Editions Developments Limited)

movement of water through soil is called **infiltration.** As the water seeps down, it dissolves various soil components in upper layers and carries them to lower layers in a process called **leaching.**

Soils develop and mature slowly. Developing an inch (2.5 centimeters) of topsoil (the A horizon) can take several hundred to 1,000 years. Five important soil types, each with a distinct profile, are shown in Figure 9-18 (p. 191). Most of the world's crops are grown on soils exposed when grasslands (Figure 6-18, p. 119) and deciduous forests are cleared.

How Do Soils Differ in Texture, Porosity, and Acidity? Soils vary in their content of *clay* (very fine particles), *silt* (fine particles), *sand* (medium-size particles), and *gravel* (coarse to very coarse particles). The relative amounts of the different sizes and types of mineral particles determine **soil texture.** Soils

with roughly equal mixtures of clay, sand, silt, and humus are called **loams.**

To get an idea of a soil's texture, take a small amount of topsoil, moisten it, and rub it between your fingers and thumb. A gritty feel means that it contains a lot of sand. A sticky feel means a high clay content, and you should be able to roll it into a clump. Silt-laden soil feels smooth, like flour. A loam topsoil, which is best suited for plant growth, has a texture between these extremes—a crumbly, spongy feeling—with many of its particles clumped loosely together.

Soil texture helps determine **soil porosity,** a measure of the volume of pores or spaces per volume of soil and of the average distances between those spaces. Fine particles are needed for water retention and coarse ones for air spaces. A porous soil has many pores and can hold more water and air than a less porous soil. The average size of the spaces or pores in a

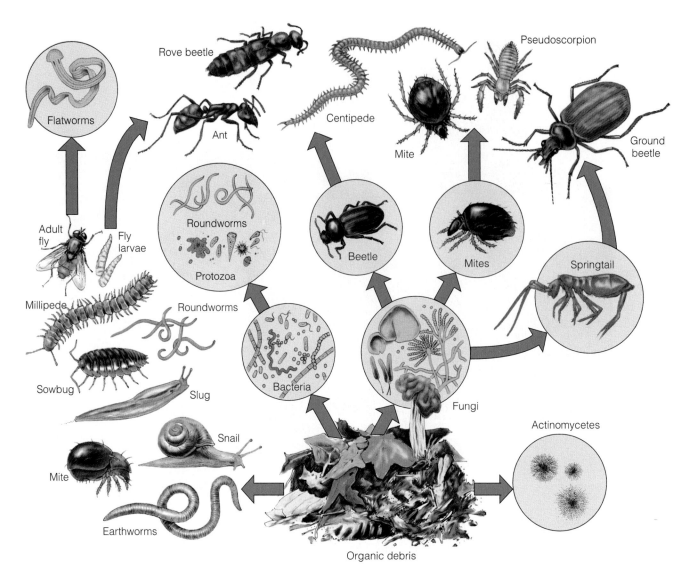

Figure 9-17 *Natural capital:* greatly simplified food web of living organisms found in soil.

soil determines **soil permeability:** the rate at which water and air move from upper to lower soil layers. Soil porosity is also influenced by **soil structure:** the ways in which soil particles are organized and clumped together (Figure 9-19, p. 192).

Loams are the best soils for growing most crops because they hold lots of water but not too tightly for plant roots to absorb. Sandy soils are easy to work, but water flows rapidly through them (Figure 9-19, left). They are useful for growing irrigated crops or those with low water needs, such as peanuts and strawberries.

The particles in clay soils are very small and easily compacted. When these soils get wet, they form large, dense clumps, which is why wet clay can be molded into bricks and pottery. Clay soils are more porous and have a greater water-holding capacity than sandy soils, but the pore spaces are so small that these soils have a low permeability (Figure 9-19,

right). Because little water can infiltrate to lower levels, the upper layers can easily become too waterlogged for most crops.

The acidity or alkalinity of a soil, as measured by its pH (Figure 9-20, p. 192), influences the uptake of soil nutrients by plants. When soils are too acidic, the acids can be partially neutralized by an alkaline substance such as lime. However, lime speeds up the decomposition of organic matter in the soil. Thus manure or another organic fertilizer should be added with the lime to maintain soil fertility.

In dry regions such as much of the western and southwestern United States, rain does not leach away calcium and other alkaline compounds. Therefore, soils in such areas may be too alkaline (pH above 7.5) for some crops. Adding sulfur, which is gradually converted into sulfuric acid by soil bacteria, reduces soil alkalinity.

Figure 9-18 *Natural capital:* soil profiles of the principal soil types typically found in five different biomes.

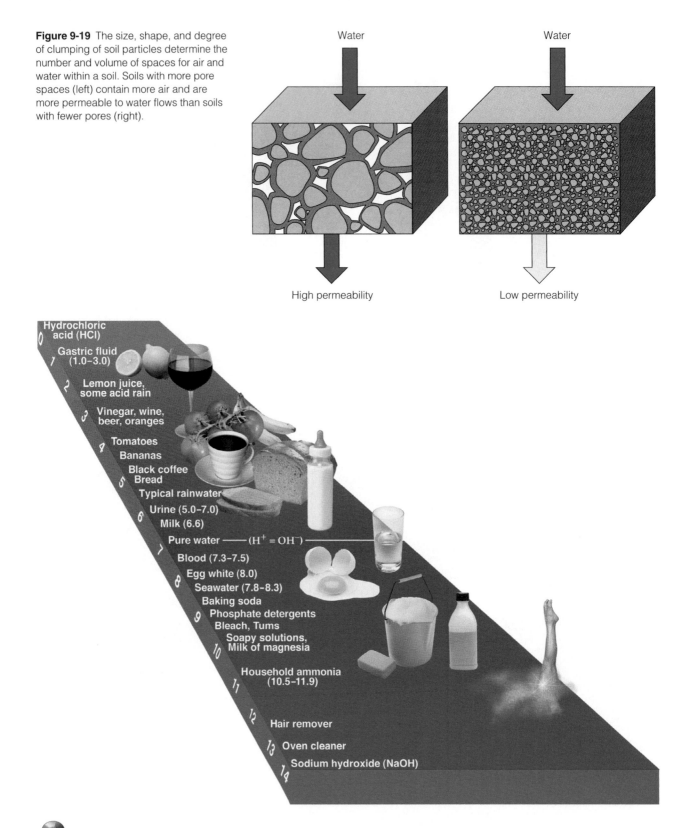

Figure 9-19 The size, shape, and degree of clumping of soil particles determine the number and volume of spaces for air and water within a soil. Soils with more pore spaces (left) contain more air and are more permeable to water flows than soils with fewer pores (right).

Water

Water

High permeability

Low permeability

Hydrochloric acid (HCl)
0

Gastric fluid
(1.0–3.0)
1

Lemon juice,
some acid rain
2

Vinegar, wine,
beer, oranges
3

Tomatoes
4
Bananas

Black coffee
Bread
5

Typical rainwater

Urine (5.0–7.0)
6
Milk (6.6)

Pure water ———— (H⁺ = OH⁻) ————
7

Blood (7.3–7.5)

Egg white (8.0)
8
Seawater (7.8–8.3)

Baking soda
Phosphate detergents
9
Bleach, Tums

Soapy solutions,
Milk of magnesia
10

Household ammonia
(10.5–11.9)
11

Hair remover
12

Oven cleaner
13
Sodium hydroxide (NaOH)
14

Figure 9-20 The pH scale, used to measure acidity and alkalinity of water solutions. Values shown are approximate. A solution with a pH less than 7 is *acidic,* a *neutral solution* has a pH of 7, and one with a pH greater than 7 is *basic.* Each whole-number drop in pH represents a 10-fold increase in acidity. A pH value is a measure of the concentration of hydrogen ions (H⁺) in a water solution. (From Cecie Starr, *Biology: Concepts and Applications,* 4th ed., Pacific Grove, Calif.: Brooks/Cole, 2000)

9-8 SOIL EROSION

What Causes Soil Erosion? **Soil erosion** is the movement of soil components, especially surface litter and topsoil, from one place to another. It results in the buildup of sediments and sedimentary rock on land and in bodies of water. The two main agents of erosion are *flowing water* and *wind,* with moving water causing most soil erosion.

Some soil erosion is natural, and some is caused by human activities. In undisturbed vegetated ecosystems, the roots of plants help anchor the soil, and usually soil is not lost faster than it forms.

Soil becomes more vulnerable to erosion through activities that destroy plant cover, including farming, logging, construction, overgrazing by livestock, off-road vehicle use, and deliberate burning of vegetation. Such human activities can speed up erosion and destroy in a few decades what nature took hundreds to thousands of years to produce.

The two major harmful effects of soil erosion are loss of soil fertility and loss of its ability to hold water

and sediment. Sediment runoff can pollute water, kill fish and shellfish, and clog irrigation ditches, boat channels, reservoirs, and lakes.

Soil, especially topsoil, is classified as a renewable resource because natural processes regenerate it. However, in tropical and temperate areas it takes 200–1,000 years (depending on climate and soil type) for 2.5 centimeters (1 inch) of new topsoil to form. If topsoil erodes faster than it forms on a piece of land, it eventually becomes a nonrenewable resource.

How Serious Is Global Soil Erosion? Here is some *bad news* about the seriousness of soil erosion. A 1992 joint survey by the United Nations (UN) Environment Programme and the World Resources Institute estimated that topsoil is eroding faster than it forms on about 38% of the world's cropland (Figure 9-21). The survey also found that 15% of the world's land (two-thirds of it in Asia and Africa) was degraded to some extent by soil erosion.

Recent studies show that in northwest China, a combination of overplowing and overgrazing is

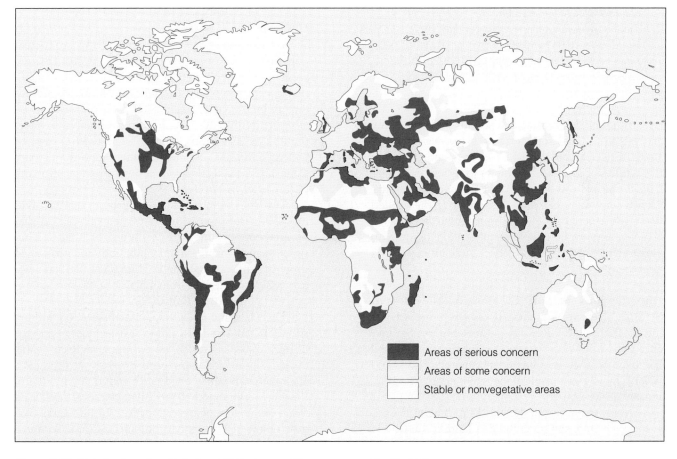

Figure 9-21 Global soil erosion. (Data from UN Environment Programme and the World Resources Institute)

causing massive wind erosion of topsoil. At times the eroded soil rises in huge dust plumes that blot out the sun and reduce visibility in China's northeastern cities. The plumes also reduce visibility and increase air pollution in Japan, the Korean Peninsula, and the northwestern United States.

According to a 2000 study by the Consultative Group on International Agricultural Research, nearly 40% of the world's land (75% in Central America) used for agriculture is seriously degraded by erosion, salt buildup (salinization), and waterlogging. According to this study soil degradation has reduced food production on about 16% of the world's cropland.

The situation is worsening in some developing countries as many poor farmers plow up marginal (easily erodible) lands to survive. Soil expert David Pimentel estimates that worldwide soil erosion causes damages of at least $375 billion per year (an average of $42 million per hour), including direct damage to agricultural lands and indirect damage to waterways, infrastructure, and human health. According to Pimentel,

> One reason that soil erosion is not a high priority for many governments and farmers is that it usually occurs so slowly that its cumulative effects may take decades to become apparent. For example, the loss of 1 millimeter (0.04 inch) of soil is so small that it goes undetected. But over a 25-year period the loss would be 25 millimeters (1 inch), which would take about 500 years to replace by natural processes.

Although soil erosion is a serious problem, some analysts argue that some erosion estimates are overstated. They contend that some surveys underestimate the abilities of some local farmers to restore degraded land. Some analysts and the UN Food and Agriculture Organization also contend that much of the eroded topsoil does not go far and is deposited further down a slope, valley, or plain. In some places, the loss in crop yields in one area could be offset by increased yields elsewhere.

How Serious Is Soil Erosion in the United States? Here are two pieces of *bad news. First,* according to the Natural Resources Conservation Service, about one-third of the nation's original prime topsoil has been washed or blown into streams, lakes, and oceans, mostly as a result of overcultivation, overgrazing, and deforestation.

Second, according to the U.S. Department of Agriculture (USDA), soil on cultivated land in the United States is eroding about 16 times faster than it can form. Erosion rates are even higher in heavily farmed regions. An example is the Great Plains, which has lost

one-third or more of its topsoil in the 150 years since it was first plowed. Some of the country's most productive agricultural lands, such as those in Iowa, have lost about half their topsoil.

There is also *good news.* USDA studies show that from 1982 to 1997, U.S. farmland acreage rated as having the greatest potential for wind erosion decreased by nearly a third and now includes about 15% of U.S. croplands. The area with the greatest potential for water erosion also decreased by nearly a third and includes about 22% of U.S. croplands.

Because of soil conservation efforts, the USDA estimates that soil erosion in the United States decreased by about 40% between 1985 and 1997. Using these data, USDA researchers estimate that soil erosion cost the United States about $30 billion in 1997, an average loss of $3.4 million per hour.

Critics such as Pierre Crosson say these estimates of soil erosion and damages from such erosion are exaggerated and based on inexact models instead of field measurements of soil loss and sedimentation rates in nearby bodies of water. However, other soil scientists point out that current estimates by models and a few on-site measurements do not include all the ecological effects of soil erosion.

What Is Desertification, and How Serious Is This Problem? In **desertification,** the productive potential of arid or semiarid land falls by 10% or more because of a combination of natural climate change that causes prolonged drought and human activities that reduce or degrade topsoil. The process can be *moderate* (with a 10–25% drop in productivity), *severe* (with a 25–50% drop), or *very severe* (with a drop of 50% or more, usually creating huge gullies and sand dunes). Desertification is a serious problem in many parts of the world (Figure 9-22).

Desertification is a complex process that involves multiple natural and human-related causes and that proceeds at varying rates in different climates. It results mainly from a combination of prolonged drought and unsustainable human activities. Figure 9-23 summarizes the major causes and consequences of desertification.

Here is some *bad news.* An estimated 8.1 million square kilometers (3.1 million square miles)—an area the size of Brazil and 12 times the size of Texas—have become desertified in the past 50 years. According to a 1999 UN conference on desertification, about 40% of the world's land and 70% of all drylands are suffering from the effects of desertification. In addition, each year about 150,000 square kilometers (58,000 square miles)—an area larger than Greece—becomes desertified. This threatens the livelihoods of at least 135 million people in 100 countries and causes economic

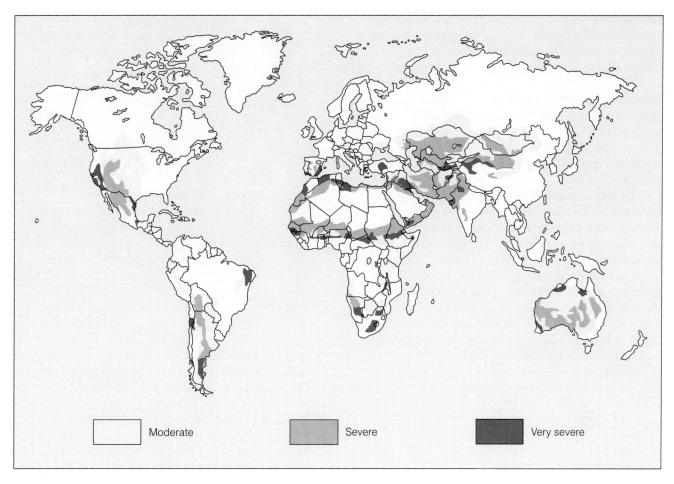

Figure 9-22 *Desertification* of arid and semiarid lands. (Data from UN Environment Programme and Harold E. Drengue)

losses estimated at $42 billion per year. The worst areas are central Asia (where more than 60% of the land is affected by desertification), followed by south Asia (more than 50%), and northeast Asia (about 30%).

Some *good news* is that we know how to slow desertification. Reducing overgrazing, deforestation, and

destructive forms of planting, irrigation, and mining can do this. We can also plant trees and grasses that will anchor the soil, hold water, and help reduce the threat of global warming by increasing uptake of carbon dioxide from the atmosphere. According to the UN Environment Programme, the spread of desertification

Figure 9-23 Causes and consequences of desertification. Natural climate change also plays a role in desertification.

has been stabilized over the past 30 years mostly because of increased plant cover on rangelands in some areas. Now the challenge is to restore much of the land suffering from desertification.

How Do Excess Salts and Water Degrade Soils?
Some *good news* is that the approximately 17% of the world's cropland that is irrigated produces almost 40% of the world's food. Irrigated land can produce crop yields two to three times greater than those from rain watering.

But irrigation also has a downside. Most irrigation water is a dilute solution of various salts, picked up as the water flows over or through soil and rocks. Small quantities of these salts are essential nutrients for plants, but they are toxic in large amounts.

Irrigation water not absorbed into the soil evaporates, leaving behind a thin crust of dissolved salts (such as sodium chloride) in the topsoil. This accumulation of salts is called **salinization** (Figure 9-24), which stunts crop growth, lowers crop yields, and eventually kills plants and ruins the land.

Here is some *bad news* about the extent of salinization. According to a 1995 study, severe salinization has

Figure 9-25 Solutions: methods for preventing and cleaning up soil salinization.

reduced yields on 21% of the world's irrigated cropland, and another 30% has been moderately salinized. The most severe salinization occurs in Asia, especially in China, India, and Pakistan.

In the United States, salinization affects 23% of all irrigated cropland. However, the proportion is much higher in some heavily irrigated western states. For example, salinization affects 66% of the irrigated land in the lower Colorado Basin and 35% of such land in California.

Some *good news* is that we know how to prevent and deal with soil salinization, as summarized in Figure 9-25.

Another problem with irrigation is **waterlogging** (Figure 9-24). Farmers often apply large amounts of irrigation water to leach salts deeper into the soil. Without adequate drainage, however, water accumulates underground and gradually raises the water table. Saline water then envelops the deep roots of plants, lowering their productivity and killing them after prolonged exposure. At least one-tenth of all irrigated land worldwide suffers from waterlogging, and the problem is getting worse.

9-9 SOLUTIONS: SOIL CONSERVATION

How Can Conservation Tillage Reduce Soil Erosion? Soil conservation involves reducing soil erosion and restoring soil fertility. For hundreds of years, farmers have used various methods to reduce soil erosion, mostly by keeping the soil covered with vegetation.

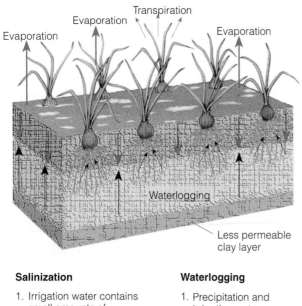

Salinization

1. Irrigation water contains small amounts of dissolved salts.

2. Evaporation and transpiration leave salts behind.

3. Salt builds up in soil.

Waterlogging

1. Precipitation and irrigation water percolate downward.

2. Water table rises.

Figure 9-24 *Salinization* and *waterlogging* of soil on irrigated land without adequate drainage lead to decreased crop yields.

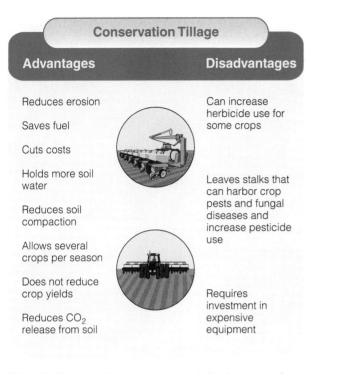

Conservation Tillage	
Advantages	**Disadvantages**
Reduces erosion	Can increase herbicide use for some crops
Saves fuel	
Cuts costs	
Holds more soil water	Leaves stalks that can harbor crop pests and fungal diseases and increase pesticide use
Reduces soil compaction	
Allows several crops per season	
Does not reduce crop yields	Requires investment in expensive equipment
Reduces CO_2 release from soil	

Figure 9-26 Trade-offs: advantages and disadvantages of using *conservation tillage.*

In **conventional-tillage farming,** farmers plow the land and then break up and smooth the soil to make a planting surface. In areas such as the midwestern United States, harsh winters prevent plowing just before the spring growing season. Thus crop fields often are plowed in the fall. This leaves the soil bare during the winter and early spring and makes it vulnerable to erosion.

To reduce erosion, many U.S. farmers use **conservation-tillage farming** (either *minimum-tillage* or *no-till farming*). The idea is to disturb the soil as little as possible while planting crops. With *minimum-tillage farming,* special tillers break up and loosen the subsurface soil without turning over the topsoil, previous crop residues, and any cover vegetation. In *no-till farming,* special planting machines inject seeds, fertilizers, and weed killers (herbicides) into slits made in the unplowed soil. Figure 9-26 lists the advantages and disadvantages of conservation tillage.

Some *good news* is that farmers used conservation tillage on about 45% of U.S. cropland by 2002. In Indiana, the Nature Conservancy is giving farmers money to buy no-till equipment in exchange for a promise to use conservation tillage for at least 3 years. The USDA estimates that using conservation tillage on 80% of U.S. cropland would reduce soil erosion by at least half. Conservation tillage is also used in Brazil, Argentina, Canada, and Paraguay and is beginning to be embraced by more farmers worldwide.

What Other Methods Can Reduce Soil Erosion?

Farmers have developed a number of other ways to reduce soil erosion (Figure 9-27, p. 198). One is **terracing,** which can reduce soil erosion on steep slopes by converting the land into a series of broad, nearly level terraces that run across the land contour (Figure 9-27a). This retains water for crops at each level and reduces soil erosion by controlling runoff. Most farmers of steep land know the risk of not terracing. However, many who are poor have too little time and too few workers to build terraces; they must plant crops without terracing hillsides or starve.

Another method is **contour farming,** which involves plowing and planting crops in rows across the contour of gently sloped land (Figure 9-27b). Each row acts as a small dam to help hold soil and to slow water runoff.

Farmers also use **strip cropping** to reduce soil erosion (Figure 9-27b). It involves planting alternating strips of a row crop (such as corn) and another crop that completely covers the soil (such as a grass or a grass and legume mixture). The strips of cover crop trap soil that erodes from the row crop, catch and reduce water runoff, and help prevent the spread of pests and plant diseases. Planting strips of nitrogen-fixing legumes (such as soybeans or alfalfa) helps restore soil fertility.

Another method for slowing erosion is **alley cropping** or **agroforestry,** in which several crops are planted together in strips or alleys between trees and shrubs that can provide fruit or fuelwood (Figure 9-27c). The trees or shrubs provide shade (which reduces water loss by evaporation) and help retain and slowly release soil moisture. They also can provide fruit, fuelwood, and trimmings that can be used as mulch (green manure) for the crops and as fodder for livestock.

Some farmers establish **windbreaks,** or **shelterbelts,** of trees (Figure 9-27d) to reduce wind erosion, help retain soil moisture, supply some wood for fuel, and provide habitats for birds, pest-eating and pollinating insects, and other animals.

Farmers use **gully reclamation** to restore severely eroded bare land. This involves planting fast-growing shrubs, vines, and trees to stabilize the soil, building small dams at the bottoms of gullies to collect silt and gradually fill in the channels, and building channels to divert water from the gully.

Finally, some governments use *land classification* to identify easily erodible (marginal) land that should be neither planted in crops nor cleared of vegetation. In the United States, the Natural Resources Conservation Service has set up a classification system to identify types of land that are suitable or unsuitable for cultivation.

(a) Terracing

(b) Contour planting and strip cropping

(c) Alley cropping

(d) Windbreaks

Figure 9-27 Solutions: In addition to conservation tillage, soil conservation methods include **(a)** terracing, **(b)** contour planting and strip cropping, **(c)** alley cropping, and **(d)** windbreaks.

Case Study: How Is the United States Slowing Soil Erosion?   Of the world's major food-producing countries, only the United States is sharply reducing some of its soil losses through conservation tillage and government-sponsored soil conservation programs.

The 1985 Farm Act established a strategy for reducing soil erosion in the United States. In the first phase of this program, farmers are given a subsidy for taking highly erodible land out of production and replanting it with soil-saving grass or trees for 10–15 years. By 2002, roughly one-tenth of U.S. cropland was in this Conservation Reserve Program (CRP).

The land in such a *conservation reserve* cannot be farmed, grazed, or cut for hay. Farmers who violate their contracts must pay back all subsidies plus interest.

According to the U.S. Department of Agriculture, since 1985 this program has cut soil losses on cropland in the United States by about 65%—a shining example of *good news*. If lawmakers continue to support this program, it could eventually cut such soil losses as much as 80%.

The second phase of the program required all farmers with highly erodible land to develop government-approved 5-year soil conservation plans for their entire farms by the end of 1990. A third provision of the Farm Act authorizes the government to forgive all or part of farmers' debts to the Farmers Home Administration if they agree not to farm highly erodible cropland or wetlands for 50 years. The farmers must plant trees or grass on this land or restore it to wetland.

The 1985 Farm Act made the United States the first major food-producing country to make soil conservation a national priority. These efforts to slow soil erosion are an important step. However, effective soil conservation is practiced on only about half of all U.S. agricultural land and on less than half of the country's most erodible cropland.

How Can We Maintain and Restore Soil Fertility?   Fertilizers partially restore plant nutrients lost by erosion, crop harvesting, and leaching. Farmers can use **organic fertilizer** from plant and animal materials or **commercial inorganic fertilizer** produced from various minerals.

There are several types of *organic fertilizer. One* is **animal manure:** the dung and urine of cattle, horses, poultry, and other farm animals. It improves soil structure, adds organic nitrogen, and stimulates beneficial soil bacteria and fungi.

However, manure use in the United States has decreased for three reasons. *First,* most operations that mix animal raising and crop farming have been replaced with separate operations for growing crops and raising animals. *Second,* it costs too much to transport animal manure from feedlots near urban areas to distant rural crop-growing areas. *Finally,* tractors and other motorized farm machinery have largely replaced horses and other draft animals that added manure to the soil.

One new source of animal manure fertilizer is being considered. Burning poultry wastes to produce electricity leaves a phosphorus-rich ash. Researchers at the U.S. Department of Agriculture are evaluating its value as an organic fertilizer.

A *second* type of organic fertilizer is **green manure.** It consists of freshly cut or growing green vegetation plowed into the soil to increase the organic matter and humus available to the next crop.

A *third* type of organic fertilizer is **compost:** a sweet-smelling, dark-brown, humuslike material that is rich in organic matter and soil nutrients. It is produced when microorganisms (mostly fungi and aerobic bacteria) in soil break down organic matter such as leaves, food wastes, paper, and wood in the presence of oxygen. Compost is a rich natural fertilizer and soil conditioner that aerates soil, improves its ability to retain water and nutrients, and helps prevent erosion.

Some farmers also use *spores of mushrooms,* including puffballs and truffles, as organic fertilizer. The spores produce rapidly spreading fungal mycelia that attach to plant roots to form mycorrhizae. Such roots can take in more moisture and nutrients from the soil and make plants more disease resistant (Figure 7-10c, p. 151). Unlike typical fertilizers that farmers must apply every few weeks, one application of mushroom fungi lasts all year and costs just pennies per plant.

Crops such as corn, tobacco, and cotton can deplete nutrients (especially nitrogen) in the topsoil if planted on the same land several years in a row. One way to reduce such losses is **crop rotation.** Farmers plant areas or strips with nutrient-depleting crops one year. The next year they plant the same areas with legumes (whose root nodules add nitrogen to the soil). In addition to helping restore soil nutrients, this method reduces erosion by keeping the soil covered with vegetation. It also helps reduce crop losses to insects by presenting them with a changing target.

Can Inorganic Fertilizers Save the Soil?   Today, many farmers (especially in developed countries) rely on *commercial inorganic fertilizers* containing *nitrogen* (as ammonium ions, nitrate ions, or urea), *phosphorus* (as phosphate ions), and *potassium* (as potassium ions). Other plant nutrients may also be present in low or trace amounts.

Worldwide, the use of inorganic commercial fertilizers increased about ninefold between 1950 and 1989 but has leveled off since then. Figure 9-28 (p. 200) lists the advantages and disadvantages of using inorganic fertilizers to enhance or restore soil fertility.

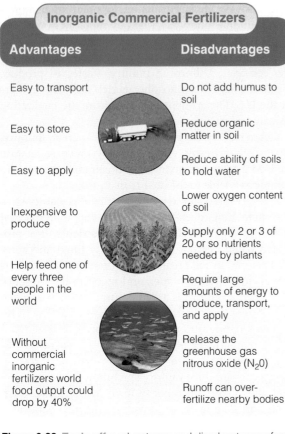

Inorganic Commercial Fertilizers

Advantages	Disadvantages
Easy to transport	Do not add humus to soil
Easy to store	Reduce organic matter in soil
Easy to apply	Reduce ability of soils to hold water
Inexpensive to produce	Lower oxygen content of soil
Help feed one of every three people in the world	Supply only 2 or 3 of 20 or so nutrients needed by plants
	Require large amounts of energy to produce, transport, and apply
Without commercial inorganic fertilizers world food output could drop by 40%	Release the greenhouse gas nitrous oxide (N_2O)
	Runoff can over-fertilize nearby bodies

Figure 9-28 Trade-offs: advantages and disadvantages of using *inorganic commercial fertilizers* to enhance or restore soil fertility.

Who Should Be Involved in Reducing Soil Erosion? According to soil scientists, responsibility for reducing soil erosion should not be limited to farmers. Timber cutting, overgrazing, mining, and urban development that are carried out without proper regard for soil conservation also cause soil erosion.

The challenge is to arrest the excessive loss of topsoil on all land everywhere, reducing it to below the level of new soil formation. The world cannot afford this loss of natural capital. If we cannot preserve the foundation of civilization, we cannot preserve civilization itself.

LESTER R. BROWN

REVIEW QUESTIONS

1. Define all boldfaced terms in this chapter.

2. Describe the *Dust Bowl* event in the United States.

3. What are the main characteristics of the earth's *core, mantle,* and *crust*?

4. What are *tectonic plates?* What is the *lithosphere?* What is the *theory of plate tectonics,* and how is it important to physical and biological processes on the earth?

5. What are the three different types of boundaries between the earth's lithospheric plates?

6. What is *erosion,* and what are its two major causes?

7. Distinguish between a *mineral* and a *rock*. Distinguish among *igneous, sedimentary,* and *metamorphic rock,* and give two examples of each type.

8. Describe the *rock cycle,* and explain its importance.

9. Distinguish between a *mineral resource* and an *ore*. Distinguish among *identified resources, undiscovered resources, reserves,* and *other resources*. List two factors that can increase the reserves of a mineral resource.

10. List seven ways mining companies find mineral deposits. Distinguish among **(a)** *overburden* and *spoils,* **(b)** *surface mining* and *subsurface mining,* and **(c)** *open-pit mining, dredging, area strip mining,* and *contour strip mining.*

11. List five major environmental impacts of mining, processing, and using mineral resources. Describe the life cycle of a metal resource. Distinguish among *ore mineral, gangue,* and *tailings*. What is *smelting,* and what are its major environmental impacts?

12. What is *economic depletion* of a mineral resource? When such depletion occurs, what five choices are available? Distinguish between *depletion time* and the *reserve-to-production ratio* for a mineral resource. How can the estimated depletion time for a resource be extended?

13. Explain how a competitive free market should increase supplies of a scarce mineral resource. List two reasons why this idea may not be effective under today's economic conditions.

14. Describe the basic features of the 1872 Mining Law in the United States and list its advantages and disadvantages.

15. List the advantages and disadvantages of allowing more mineral exploration and mining on public lands in the United States.

16. List the advantages and disadvantages of increasing supplies of mineral resources by **(a)** mining lower-grade ores, and **(b)** finding substitutes for scarce nonrenewable mineral resources.

17. List the advantages and disadvantages of using bacteria to extract metals from ores.

18. What is an *earthquake,* and what are its major harmful effects? List ways to reduce the hazards from earthquakes.

19. What is a *volcanic eruption?* What are some of the hazards and benefits of volcanic eruptions? List ways to reduce the hazards from volcanic eruptions.

20. What is *soil?* Distinguish between a *soil horizon* and a *soil profile.*

21. What is *humus,* and what is its importance? What does the color of topsoil tell you about its usefulness as a soil for growing crops?

22. Distinguish between *soil infiltration* and *leaching.* Distinguish among *soil texture, soil porosity,* and *soil permeability.*

23. What are *loams,* and why are they the best soils for growing most crops?

24. What is *soil erosion,* and what are its major natural and human-related causes?

25. What are the major harmful effects of soil erosion?

26. How serious is soil erosion **(a)** globally and **(b)** in the United States?

27. What is *desertification?* How serious is this problem? What are its major causes and consequences? How can we slow desertification?

28. Distinguish between *salinization* and *waterlogging* of soils. How serious are these problems? List five ways to reduce the threat of soil salinization.

29. What is *soil conservation?* Distinguish between *conventional-tillage farming* and *conservation-tillage farming.* What are the advantages and disadvantages of conservation-tillage farming?

30. Distinguish among *terracing, contour farming, strip cropping, alley cropping, windbreaks, gully reclamation,* and *land classification* as methods for reducing soil erosion.

31. Describe how the U.S. government is reducing soil erosion.

32. Distinguish between *organic fertilizer* and *commercial inorganic fertilizer.* Discuss using *animal manure, green manure, compost,* and *mushroom spores* as methods for fertilizing soil. What is *crop rotation,* and why is it useful in helping maintain soil fertility?

33. List six advantages and eight disadvantages of using commercial inorganic fertilizers to maintain and restore soil fertility.

CRITICAL THINKING

1. List some ways, positive and negative, in which **(a)** the external earth processes of weathering and erosion and **(b)** plate tectonics are important to you.

2. Use the second law of thermodynamics (p. 59) to analyze the scientific and economic feasibility of each of the following processes: **(a)** extracting most minerals dissolved in seawater, **(b)** mining increasingly lower-grade deposits of minerals, **(c)** using inexhaustible solar energy to mine minerals, and **(d)** continuing to mine, use, and recycle minerals at increasing rates.

3. Explain why you support or oppose each of the following proposals concerning extraction of hardrock minerals on public land in the United States: **(a)** not granting title to public lands in the United States for actual or claimed hardrock mining, **(b)** requiring mining companies to pay a royalty of 8–12% on the *gross* revenues they earn from hardrock minerals they extract from public lands, and **(c)** making hardrock mining companies

legally responsible for restoring the land and cleaning up environmental damage caused by their activities.

4. In the area where you live, are you more likely to experience an earthquake or a volcanic eruption? What can you do to escape or reduce the harm if such a disaster strikes? What actions can you take when it occurs?

5. How does your lifestyle directly or indirectly contribute to soil erosion?

6. Why should both inorganic and organic fertilizers be used to restore or increase soil fertility?

7. Congratulations! You are in charge of the world. What are the three most important features of your policy to **(a)** develop mineral resources and **(b)** reduce soil erosion?

PROJECTS

1. Write a brief scenario describing the sequence of consequences to us and to other forms of life if the rock cycle stopped functioning.

2. What mineral resources are extracted in your local area? What mining methods are used, and what have been their environmental impacts? How has mining these resources benefited the local economy?

3. Use the library or the Internet to find out where earthquakes and volcanic eruptions have occurred during the past 30 years, and then stick small flags on a map of the world or place dots on Figure 9-3a (p. 176). Compare their locations with the plate boundaries shown in Figure 9-3b.

4. Conduct a survey of soil erosion and soil conservation in and around your community on cropland, construction sites, mining sites, grazing land, and deforested land. Use these data to develop a plan for reducing soil erosion in your community.

5. Make a concept map of this chapter's major ideas using the section heads and subheads and the key terms (in boldface). Look on the website for this book for information about making concept maps.

INTERNET STUDY RESOURCES AND RESOURCES FOR FURTHER READING AND RESEARCH

The website for this book contains helpful study aids and many ideas for further reading and research. Log on to

http://biology.brookscole.com/miller 10

and click on the Chapter-by-Chapter area. Choose Chapter 9 and select a resource:

■ Flash Cards allows you to test your mastery of the Terms and Concepts to Remember for this chapter.

■ Tutorial Quizzes provides a multiple-choice practice quiz.

- Student Guide to InfoTrac will lead you to Critical Thinking Projects that use InfoTrac College Edition as a research tool.

- References lists the major books and articles consulted in writing this chapter.

- Hypercontents takes you to an extensive list of sites with news, research, and images related to individual sections of the chapter.

INFOTRAC COLLEGE EDITION

Improve your skills with InfoTrac College Edition, a searchable online database of articles from more than 700 periodicals. Log on to

http://www.infotrac-college.com

or access InfoTrac through the website for this book. Try to find the following articles:

1. Pearce, F. 2002. On shaky ground. *Geographical* 74: 32. *Keywords:* "Peru," "earthquake," and "aftermath." Damaging earthquakes are a way of life in Peru. However, the shaking of the earth does not cause all the damage. Now developers are displacing people after areas are considered unsafe for habitation.

2. Sparks, D. L. 2000. Soil as an endangered ecosystem. (Brief Article) *BioScience* 50: 947. *Keywords:* "endangered ecosystem" and "soil." This article explores several issues related to soil erosion and ecosystem damage.

10 RISK, TOXICOLOGY, AND HUMAN HEALTH

The Big Killer

What is roughly the diameter of a 30-caliber bullet, can be bought almost anywhere, is highly addictive, and kills about 13,400 people every day, 560 per hour, or 1 person every 6 seconds? It is a cigarette. *Cigarette smoking is the single most preventable major cause of death and suffering among adults.*

According to the World Health Organization (WHO), tobacco helped kill about 70 million people between 1950 and 2002—more than twice the 30 million people killed in battle in all wars during the 20th century.

WHO estimates that each year tobacco contributes to the premature deaths of at least 4.9 million people from 25 illnesses including *heart disease, lung cancer, other cancers, bronchitis, emphysema,* and *stroke.* The annual death toll from smoking-related diseases is projected to reach 10 million by 2030 (70% of them in developing countries)—an average of about 27,400 preventable deaths per day or 1 death every 3 seconds.

According to a 2002 study by the U.S. Centers for Disease Control and Prevention, smoking kills about 440,000 Americans per year prematurely, an average of 1,205 deaths per day (Figure 10-1). This death toll is roughly equivalent to three fully loaded jumbo (400-passenger) jets crashing accidentally every day with no survivors. According to a 1998 study, secondhand smoke (inhaled by nonsmokers) causes 30,000–60,000 premature deaths per year in the United States. About 22% of U.S. adults and 28% of 9th to 12th graders smoke.

The overwhelming scientific consensus is that the nicotine (and probably the acetaldehyde) inhaled in tobacco smoke is highly addictive. Only 1 in 10 people who try to quit smoking succeed, about the same relapse rate as for recovering alcoholics and those addicted to heroin or crack cocaine. A British government study showed that adolescents who smoke more than one cigarette have an 85% chance of becoming smokers.

According to a 2002 study by the Centers for Disease Control and Prevention, smoking in the United States costs about $158 billion a year for medical bills, increased insurance costs, disability, lost earnings and productivity because of illness, and property damage from smoking-caused fires. This is an average of $7 per pack of cigarettes sold in the United States.

Many health experts urge that a $3–5 federal tax be added to the price of a pack of cigarettes in the United States. Then users of cigarettes, not the rest of society, would pay a much greater share of the health, economic, and social costs associated with their smoking: a *user-pays* approach.

Other suggestions for reducing the death toll and health effects of smoking in the United States include banning all cigarette advertising, prohibiting the sale of cigarettes and other tobacco products to anyone under 21 (with strict penalties for violators), banning cigarette vending machines, classifying nicotine as an addictive and dangerous drug (and placing its use in tobacco or other products under the jurisdiction of the Food and Drug Administration), eliminating all federal subsidies and tax breaks to U.S. tobacco farmers and tobacco companies, and using cigarette tax income to finance an aggressive antismoking advertising and education program.

So far, the U.S. Congress has not enacted such reforms. Between 1996 and 2002, major U.S. tobacco companies donated over $34 million to candidates running for office—about 80% of the money going to Republican and 20% to Democratic candidates and parties.

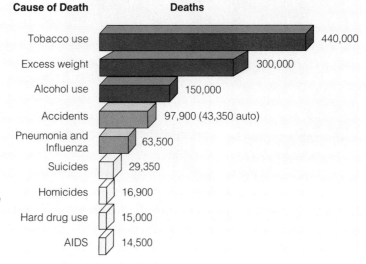

Figure 10-1 Annual deaths in the United States from tobacco use and eight other causes in 2000. Smoking is by far the nation's leading cause of preventable death. Excluding excessive weight, it causes more premature deaths each year than the other categories in this figure combined. (Data from U.S. National Center for Health Statistics and Centers for Disease Control and Prevention and U.S. Surgeon General)

The dose makes the poison.

PARACELSUS, 1540

This chapter addresses the following questions:

- What types of hazards do people face?

- What is toxicology, and how do scientists measure toxicity?

- What chemical hazards do people face, and how can they be measured?

- What types of disease (biological hazards) threaten people in developing countries and developed countries?

- How can risks be estimated, managed, and reduced?

10-1 RISK, PROBABILITY, AND HAZARDS

What Is Risk? Risk is the possibility of suffering harm from a hazard that can cause injury, disease, economic loss, or environmental damage. Risk is expressed in terms of **probability:** a mathematical statement about the likelihood of harm.

Scientists often state probability in terms such as "The lifetime probability of developing cancer from exposure to a certain chemical is 1 in 1 million." This means that 1 of every 1 million people exposed to the chemical at a specified average daily dosage will develop cancer over a typical lifetime (usually considered 70 years). Figure 10-2 summarizes how risks are assessed and managed.

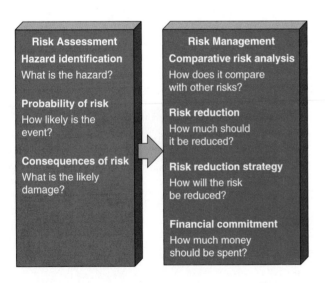

Figure 10-2 *Risk assessment* and *risk management*. These are important, difficult, and controversial processes.

What Are the Major Types of Hazards? Here are some of the major types of hazards we face:

- *Cultural hazards* such as unsafe working conditions, smoking, poor diet, drugs, drinking, driving, criminal assault, unsafe sex, and poverty

- *Chemical hazards* from harmful chemicals in the air, water, soil, and food

- *Physical hazards* such as ionizing radiation, fires, earthquakes, volcanic eruptions, floods, tornadoes, and hurricanes

- *Biological hazards* from pathogens (bacteria, viruses, and parasites), pollen and other allergens, and animals such as bees and poisonous snakes

10-2 TOXICOLOGY

What Determines Whether a Chemical Is Harmful? **Toxicity** measures how harmful a substance is. This depends on several factors: **dose,** the amount a person has ingested, inhaled, or absorbed through the skin; frequency of exposure; who is exposed (adult or child, for example); how well the body's detoxification systems (liver, lungs, and kidneys) work; and genetic makeup that determines an individual's sensitivity to a particular toxin (Figure 10-3).

Several other factors can affect the harm caused by a substance. One is its *solubility. Water-soluble toxins* (which are often inorganic compounds) can move throughout the environment and get into water supplies. *Oil- or fat-soluble toxins* (which are usually organic compounds) can accumulate in body tissues and cells.

A second factor is a substance's *persistence.* Many chemicals, such as chlorofluorocarbons (CFCs), chlorinated hydrocarbons, and plastics, are used widely because of their persistence or resistance to breakdown. However, this persistence also means they can have long-lasting effects on the health of wildlife and people.

Another factor for some substances is **bioaccumulation,** in which some molecules are absorbed and stored in specific organs or tissues at higher than normal levels. A related factor is **biomagnification,** in which levels of some toxins in the environment are magnified as they pass through food chains and webs (Figure 10-4). Examples of chemicals that can be biomagnified include long-lived, fat-soluble organic compounds such as the pesticide DDT, PCBs (oily chemicals used in electrical transformers), and some radioactive isotopes (such as strontium-90; Table 3-1, p. 57). Stored in body fat, such chemicals can be passed along to offspring during gestation or egg laying and as mothers nurse their young.

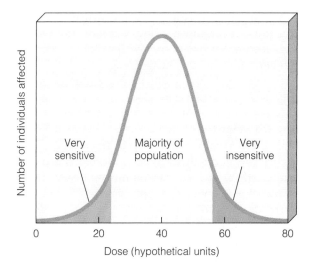

Figure 10-3 Typical variations in sensitivity to a toxic chemical within a population, mostly because of differences in genetic makeup. Some individuals in a population are very sensitive to small doses of a toxin (left), and others are very insensitive (right). Most people fall between these two extremes (middle).

Chemical interactions can decrease or multiply the harmful effects of a toxin. An *antagonistic interaction* can reduce harmful effects. For example, vitamins E and A apparently interact to reduce the body's response to some cancer-causing chemicals. A *synergistic interaction* multiplies harmful effects. For example, workers exposed to tiny fibers of asbestos increase

their chances of getting lung cancer 20-fold. But asbestos workers who also smoke have a 400-fold increase in lung cancer rates.

The type and amount of health damage that result from exposure to a chemical or other agent are called the **response**. An *acute effect* is an immediate or rapid harmful reaction to an exposure—ranging from dizziness or a rash to death. A *chronic effect* is a permanent or long-lasting consequence (kidney or liver damage, for example) of exposure to a harmful substance.

Should We Be Concerned about Trace Levels of Toxic Chemicals in the Environment and in Our Bodies? The answer is that it depends on the chemical and its concentration. The detection of trace amounts of a chemical in air, water, or food does not necessarily mean it is there at a level harmful to most people or to wildlife.

A basic concept of toxicology is that *any synthetic or natural chemical (even water) can be harmful if ingested in a large enough quantity.* Drinking 100 cups of strong coffee one after another would expose most people to a lethal dosage of caffeine. Similarly, downing 100 tablets of aspirin or 1 liter (1.1 quarts) of pure alcohol (ethanol) would kill most people.

The critical question is how much exposure to a particular toxic chemical causes a harmful response. This is the meaning of the quote by German scientist Paracelsus about the dose making the poison (found at the top of p. 204 in this chapter). A basic problem is

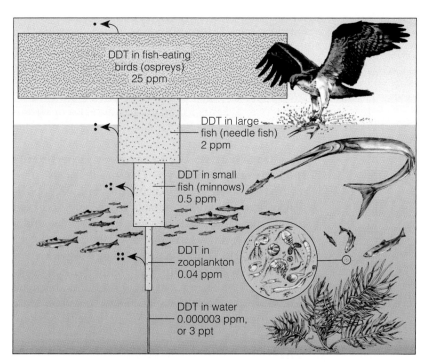

Figure 10-4 *Bioaccumulation* and *biomagnification.* DDT is a fat-soluble chemical that can accumulate in the fatty tissues of animals. In a food chain or web, the accumulated DDT can be biologically magnified in the bodies of animals at each higher trophic level. This diagram shows that the concentration of DDT in the fatty tissues of organisms was biomagnified about 10 million times in this food chain in an estuary near Long Island Sound in New York. If each phytoplankton organism takes up from the water and retains one unit of DDT, a small fish eating thousands of zooplankton (which feed on the phytoplankton) will store thousands of units of DDT in its fatty tissue. Then each large fish that eats ten of the smaller fish will ingest and store tens of thousands of units, and each bird (or human) that eats several large fish will ingest hundreds of thousands of units. Dots represent DDT, and arrows show small losses of DDT through respiration and excretion.

that this question is very difficult to answer. Another problem is that people vary in the dose of a toxin they can tolerate without significant harm (Figure 10-3).

Some people have the mistaken idea that all natural chemicals are safe and all synthetic chemicals are harmful. In fact, many synthetic chemicals are quite safe if used as intended, and many natural chemicals are deadly. For example, the average person is far more likely to be killed by aflatoxin (produced by molds during storage) in peanut butter than by lightning or a shark. However, the chance of dying from eating several spoonfuls of peanut butter a day is quite small.

In addition, the ability of chemists to detect increasingly small amounts of potentially toxic chemicals in air, water, and food can give the false impression that dangers from toxic chemicals are increasing. In 1980, chemists could routinely detect concentrations of substances in parts per million (ppm). By 1990, chemists could detect parts per billion (ppb), and today they can detect concentrations of parts per trillion (ppt) and in some cases parts per quadrillion (ppq).

What Is a Poison? The **median lethal dose** (or **LD$_{50}$**) for a chemical is the amount received in one dose that kills exactly 50% of the animals (usually rats and mice) in a test population within a 14-day period (Figure 10-5). Legally, a **poison** is a chemical that has a median lethal dose of 50 milligrams or less per kilogram of body weight.

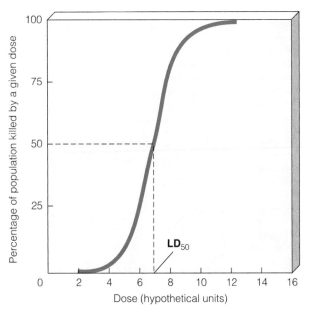

Figure 10-5 Hypothetical *dose-response curve* showing determination of the LD$_{50}$, the dosage of a specific chemical that kills 50% of the animals in a test group.

Chemicals vary widely in their toxicity (Table 10-1). Some poisons can cause serious harm or death after a single acute exposure at very low dosages. Others cause such harm only at such huge dosages that it is nearly impossible to get enough into the body. Most chemicals fall between these two extremes.

How Do Scientists Use Case Reports and Epidemiological Studies to Estimate Toxicity? Scientists use various methods to get information about the harmful effects of chemicals on human health. One is *case reports*, usually made by physicians. They provide information about people suffering some adverse health effect or death after exposure to a chemical. Such information often involves accidental poisonings, drug overdoses, homicides, or suicide attempts.

Most case reports are not a reliable source for determining toxicity because the actual dosage and the exposed person's health status are often not known. However, such reports can provide clues about environmental hazards and suggest the need for laboratory investigations.

Another source of information is *epidemiological studies* in which the health of people exposed to a particular toxic agent (the *experimental group*) is compared with the health of another group of statistically similar people not exposed to the agent (the *control group*). The goal is to determine whether the statistical association between exposure to a toxic chemical and a health problem is strong, moderate, weak, or undetectable.

Such studies are limited because of several factors. In many cases, too few people have been exposed to high enough levels of many toxic agents to detect statistically significant differences. Also, conclusively linking an observed effect with exposure to a particular chemical is very difficult because people are exposed to many different toxic agents throughout their lives. Finally, epidemiological studies cannot be used to evaluate hazards from new technologies or chemicals to which people have not been exposed.

How Do Scientists Use Laboratory Experiments to Estimate Toxicity? The most widely used method for determining toxicity is to expose a population of live laboratory animals (especially mice and rats) to measured doses of a specific substance under controlled conditions. Animal tests take 2–5 years and cost $200,000 to $2 million per substance tested.

Animal welfare groups want to limit or ban use of test animals or ensure that experimental animals are treated in the most humane manner possible. More humane methods for carrying out toxicity tests include using bacteria, cell and tissue cultures, chicken egg membranes, and measurements of changes in the

Table 10-1 Toxicity Ratings and Average Lethal Doses for Humans			
Toxicity Rating	**LD$_{50}$ (milligrams per kilogram of body weight)***	**Average Lethal Dose†**	**Examples**
Supertoxic	Less than 0.01	Less than 1 drop	Nerve gases, botulism toxin, mushroom toxins, dioxin (TCDD)
Extremely toxic	Less than 5	Less than 7 drops	Potassium cyanide, heroin, atropine, parathion, nicotine
Very toxic	5–50	7 drops to 1 teaspoon	Mercury salts, morphine, codeine
Toxic	50–500	1 teaspoon to 1 ounce	Lead salts, DDT, sodium hydroxide, sodium fluoride, sulfuric acid, caffeine, carbon tetrachloride
Moderately toxic	500–5,000	1 ounce to 1 pint	Methyl (wood) alcohol, ether, phenobarbital, amphetamines (speed), kerosene, aspirin
Slightly toxic	5,000–15,000	1 pint to 1 quart	Ethyl alcohol, Lysol, soaps
Essentially nontoxic	15,000 or greater	More than 1 quart	Water, glycerin, table sugar

*Dosage that kills 50% of individuals exposed
†Amounts of substances in liquid form at room temperature that are lethal when given to a 70.4-kilogram (155-pound) human

electrical properties of individual animal cells. Most developed countries are gradually replacing LD$_{50}$ methods with these newer procedures.

These alternatives can greatly decrease the use of animals for testing toxicity. But a number of scientists contend that some animal testing is needed because the alternative methods cannot adequately mimic the complex biochemical interactions of a live animal.

Acute toxicity tests are run to develop a **dose-response curve,** which shows the effects of various dosages of a toxic agent on a group of test organisms (Figure 10-5). Such tests are *controlled experiments* in which the effects of the chemical on a *test group* are compared with the responses of a *control group* of organisms not exposed to the chemical. Care is taken to ensure that organisms in each group are as identical as possible in age, health status, and genetic makeup and exposed to the same environmental conditions.

Fairly high dosages are used to reduce the number of test animals needed, obtain results quickly, and lower costs. Otherwise, tests would have to be run on millions of laboratory animals for many years, and manufacturers could not afford to test most chemicals. For the same reasons, scientists usually use mathematical models to extrapolate the results of high-dose exposures to low-dose levels. Then they extrapolate the low-dose results on the test organisms to humans to estimate LD$_{50}$ values for acute toxicity (Table 10-1).

According to the *nonthreshold dose-response model* (Figure 10-6, left), any dosage of a toxic chemical or ionizing radiation causes harm that increases with the dosage. With the *threshold dose-response model* (Figure 10-6, right), a threshold dosage must be reached before any detectable harmful effects occur, presumably because the body can repair the damage caused by low dosages of some substances. Establishing which of these models applies at low dosages is extremely difficult. To be on the safe side, scientists usually use the nonthreshold dose-response model.

Some scientists challenge the validity of extrapolating data from test animals to humans because human physiology and metabolism often are different

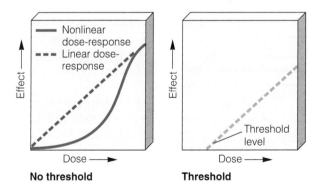

Figure 10-6 Two types of *dose-response curves.* The linear and nonlinear curves in the left graph apply if any dosage of a chemical or ionizing radiation has a harmful effect that increases with the dosage. The curve on the right applies if a harmful effect occurs only when the dosage exceeds a certain *threshold level.* Which of these models applies to various harmful agents is very uncertain because of the difficulty in estimating the response to very low dosages.

from those of the test animals. Other scientists counter that such tests and models work fairly well (especially for revealing cancer risks) when the correct experimental animal is chosen or when a chemical is toxic or harmful to several different test animal species.

How Valid Are Estimates of Toxicity?   As we have seen, all methods for estimating toxicity levels and risks have serious limitations. But they are all we have. To take this uncertainty into account and minimize harm, scientists and regulators typically set standards for allowed exposure to toxic substances and ionizing radiation at levels of 1/100 or even 1/1,000 of the estimated harmful levels.

Despite their many limitations, carefully conducted and evaluated toxicity studies are important sources of information for understanding dose-response effects and estimating and setting exposure standards. However, citizens, lawmakers, and regulatory officials must recognize the huge uncertainties and guesswork involved in all such studies.

10-3   CHEMICAL HAZARDS

What Are Toxic and Hazardous Chemicals? **Toxic chemicals** are substances fatal to more than 50% of test animals (LD_{50}) at given concentrations. **Hazardous chemicals** are flammable or explosive, irritate or damage the skin or lungs, interfere with oxygen uptake, or induce allergic reactions.

What Are Mutagens, Teratogens, and Carcinogens?   There are three major types of potentially toxic agents. *One* consists of **mutagens** that cause random *mutations,* or changes, in the DNA molecules found in cells. *Most mutations are harmless,* probably because all organisms have biochemical repair mechanisms that can correct mistakes or changes in the DNA code. A *second* type consists of **teratogens** that cause birth defects while the human embryo is growing and developing during pregnancy.

The *third* group is **carcinogens** that cause or promote the growth of a malignant (cancerous) tumor, in which certain cells multiply uncontrollably. Many cancerous tumors spread by **metastasis** when malignant cells break off from tumors and travel in body fluids to other parts of the body. There they start new tumors, making treatment much more difficult. Typically, 10–40 years may elapse between the initial exposure to a carcinogen and the appearance of detectable symptoms. Partly because of this time lag, many healthy teenagers and young adults have trouble believing their smoking, drinking, eating, and other lifestyle habits today could lead to some form of cancer before they reach age 50.

How Can Chemicals Harm the Immune, Nervous, and Endocrine Systems?   Since the 1970s, a growing body of research on wildlife and laboratory animals and of epidemiological studies of humans indicates that long-term (often low-level) exposure to various toxic chemicals in the environment can disrupt the body's immune, nervous, and endocrine systems.

The *immune system* consists of specialized cells and tissues that protect the body against disease and harmful substances by forming antibodies that make invading agents harmless. Weakening the human immune system can leave the body vulnerable to attacks by allergens, infectious bacteria, viruses, and protozoans.

Some natural and synthetic chemicals in the environment can harm the human *nervous system* (brain, spinal cord, and peripheral nerves). For example, many poisons are *neurotoxins,* which attack nerve cells (neurons).

The *endocrine system* is a complex network of glands that release very small amounts of *hormones* in the bloodstream of all vertebrate animals. In humans and other vertebrates, low levels of these chemical messengers turn on and off bodily systems that control sexual reproduction, growth, development, learning ability, and behavior.

Each type of hormone has a specific molecular shape that allows it to attach only to certain cell receptors (Figure 10-7, left). Once bonded together, the hormone and its receptor molecule can signal cell mechanisms to execute the chemical message carried by the hormone.

Case Study: Are Hormonally Active Agents a Health Threat?   There is concern that human exposure to low levels of certain synthetic chemicals can mimic and disrupt the effects of natural hormones. Over the last 25 years, experts from a number of disciplines have been piecing together field studies on wildlife, studies on laboratory animals, and epidemiological studies of human populations. This analysis suggests that a variety of human-made chemicals can act as *hormone* or *endocrine disrupters,* known as *hormonally active agents (HAAs).*

Some, called *hormone mimics,* are chemicals similar to estrogens (female sex hormones). They disrupt the endocrine system by attaching to estrogen receptor molecules (Figure 10-7, center). Others, called *hormone blockers,* disrupt the endocrine system by preventing natural hormones such as androgens (male sex hormones) from attaching to their receptors (Figure 10-7, right). There is also growing concern about pollutants that can act as *thyroid disrupters* and cause growth, weight, brain, and behavioral disorders.

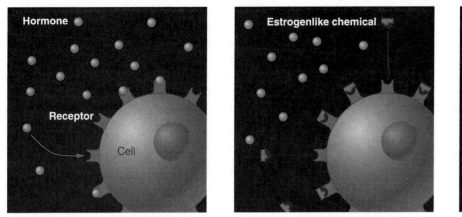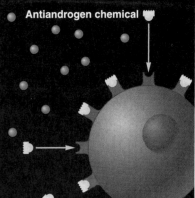

| Normal Hormone Process | Hormone Mimic | Hormone Blocker |

Figure 10-7 Hormones are molecules that act as messengers in the endocrine system to regulate various bodily processes, including reproduction, growth, and development. Each type of hormone has a unique molecular shape that allows it to attach to specially shaped receptors on the surface of or inside cells and transmit its chemical message (left). Molecules of certain pesticides and other synthetic chemicals have shapes similar to those of natural hormones and can affect the endocrine system in people and various other animals. These molecules are called *hormonally active agents* (*HAAs*). Some HAAs, sometimes called *hormone mimics*, disrupt the endocrine system by attaching to estrogen receptor molecules (center) and giving too-strong, too-weak, or mistimed signals. Other HAAs, sometimes called *hormone blockers*, prevent natural hormones such as androgens from attaching to their receptors (right) so that no signal is given. Some pollutants called *thyroid disrupters* may disrupt hormones released by thyroid glands and cause growth and weight disorders and brain and behavioral disorders.

Most natural hormones are broken down or excreted. However, many synthetic hormone impostors are persistent and fat-soluble compounds whose concentrations can be biomagnified as they move through food chains and webs (Figure 10-4). Thus they can pose a special threat to humans and other vertebrate carnivores dining at the top of food webs.

Numerous studies of wildlife and laboratory animals and a small number of epidemiological studies on humans reveal various possible effects of estrogen mimics and hormone blockers. Here are a few of many examples.

Ranch minks fed Lake Michigan fish contaminated with endocrine disrupters (such as DDT and PCBs) failed to reproduce. Exposure to PCBs reduced penis size in some test animals and in 118 boys born to women exposed to PCBs by a 1979 spill in Taiwan.

A 1999 study by Michigan State University zoologists found that female rats exposed to PCBs were reluctant to mate, raising the possibility that such contaminants could lower the sex drive in women.

In 1973, estrogen mimics called PBBs accidentally got into cattle feed in Michigan, and from there into beef. Pregnant women who ate the beef (and whose breast milk had high levels of PBBs) had sons with undersized penises and malformed testicles. A 2000 study found that healthy Dutch preschoolers exposed to low levels of PCBs and dioxins in their mother's breast milk had a diminished number of immune system cells and increased middle ear infections, chest congestion, and coughing.

A 1999 study of this problem by a U.S. National Academy of Sciences panel of scientists came to three major conclusions. *First,* "adverse reproductive and developmental effects have been observed in human populations, wildlife, and laboratory animals as a consequence of exposure to HAAs." *Second,* "there have been only a few studies of the effects of HAAs in humans, but the results of laboratory and wildlife studies suggest that HAAs have the potential to affect human immune functions." *Third,* greatly increased research is needed to come to a more definitive conclusion about whether low levels of most HAAs in the environment pose a threat to human health.

A 2002 study of the literature by the International Program on Chemical Safety reached two major conclusions. *First,* evidence that human health has been adversely affected by exposure to HAAs is generally weak. *Second,* we need much more research to determine whether there is a cause and effect link between HAAs and human health problems.

It will take decades of research to evaluate the effects of HAAs on humans. Meanwhile, some scientists believe we should begin sharply reducing the use of potential hormone disrupters now because they meet the two requirements of the *precautionary principle:* scientific uncertainty and a reasonable suspicion of harm.

Why Do We Know So Little about the Harmful Effects of Chemicals?

According to risk assessment expert Joseph V. Rodricks, "Toxicologists know a great deal about a few chemicals, a little about many, and next to nothing about most." The U.S. National Academy of Sciences estimates that only about 10% of at least 75,000 chemicals in commercial use have been thoroughly screened for toxicity, and only 2% have been adequately tested to determine whether they are carcinogens, teratogens, or mutagens. Hardly any of the chemicals in commercial use have been screened for damage to the nervous, endocrine, and immune systems.

Currently, federal and state governments do not regulate about 99.5% of the commercially used chemicals in the United States. There are three major reasons for this lack of information and regulation. *First,* under existing laws, most chemicals are considered innocent until proven guilty. *Second,* not enough funds, personnel, facilities, and test animals are available to provide such information for more than a small fraction of the many chemicals we encounter in our daily lives. *Finally,* analyzing the combined effects of multiple exposures to various chemicals and the possible interactions of such chemicals is too difficult and expensive. For example, just studying the possible different three-chemical interactions of the 500 most widely used industrial chemicals would take 20.7 million experiments—a physical and financial impossibility.

Because of the difficulty and expense of getting information about the harmful effects of chemicals, an increasing number of scientists and health officials are pushing for much greater emphasis on *pollution prevention.* This strategy greatly reduces the need for statistically uncertain and controversial toxicity studies and exposure standards. It also reduces the risk from exposure to potentially hazardous chemicals and products and their possible but poorly understood multiple interactions.

Under this approach, those proposing to introduce a new chemical or technology would bear the burden of establishing its safety. In other words, new chemicals and technologies would be assumed guilty until proven innocent. Manufacturers and businesses contend that doing this would make it too expensive and almost impossible to introduce any new chemical or technology.

10-4 BIOLOGICAL HAZARDS: DISEASE IN DEVELOPED AND DEVELOPING COUNTRIES

What Are Nontransmissible Diseases? A non-transmissible disease is not caused by living organisms, does not spread from one person to another, and tends to have multiple causes and develop slowly. Examples are cardiovascular (heart and blood vessel) disorders, most cancers, diabetes, asthma, emphysema, and malnutrition.

What Are Transmissible Diseases? A transmissible disease is caused by a living organism (such as a bacterium, virus, protozoa, or parasite) and can spread from one person to another (Figure 10-8). These infectious agents are called *pathogens.* They are spread by air, water, food, body fluids, some insects, and other nonhuman carriers (such as mosquitoes) called *vectors*

Throughout human history, disease transmission by female mosquitoes has probably killed more people than any other single factor. On the other hand, most mosquito species do not transmit infectious diseases and mosquitoes play important ecological roles. Their eggs are a major food source for fish, various insects, and frogs and other amphibians. Adult mosquitoes are an important source of food for bats, spiders, and many insect and bird species.

Typically, a *bacterium* is a one-celled microorganism that can replicate (clone) itself by simple cell division. A *virus* is a microscopic, noncellular infectious

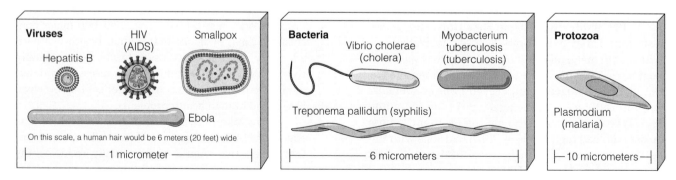

Figure 10-8 Examples of *pathogens* or agents that can cause transmissible diseases. A micrometer is one-millionth of a meter.

agent. Its DNA or RNA contains instructions for making more viruses, but it has no apparatus to do this. To replicate, a virus must invade a host cell and take over the cell's DNA to create a factory for producing more viruses (Figure 10-9).

As a country industrializes, it usually makes an *epidemiological transition* in which deaths from the in-

A typical virus consists of a shell of proteins surrounding genetic material

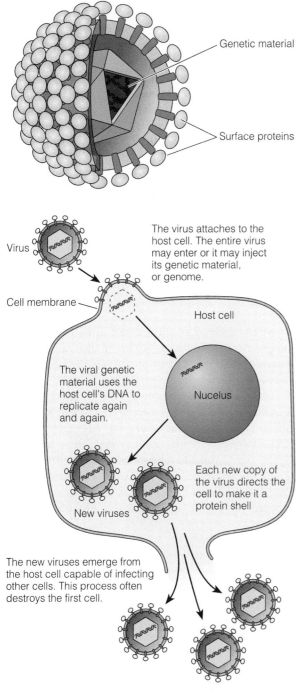

Figure 10-9 How a virus reproduces. (American Medical Association)

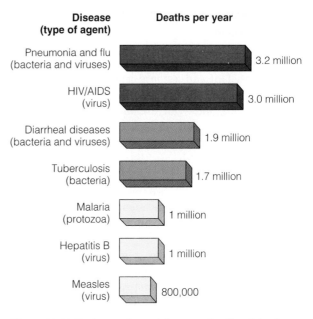

Figure 10-10 Each year the world's seven deadliest infectious diseases kill about 12.6 million people—most of them poor people in developing countries. This amounts to about 42,500 mostly preventable deaths per day. (World Health Organization)

fectious diseases of childhood decrease and those from the chronic diseases of adulthood (heart disease and stroke, cancer, and respiratory conditions) increase.

Some *good news* is that since 1900, and especially since 1950, we have greatly reduced the incidence of infectious diseases and the death rates from such diseases. This has been done mostly by a combination of better health care, using antibiotics to treat infectious disease caused by bacteria, and developing vaccines to prevent the spread of some infectious viral diseases.

Here are three pieces of *bad news. First,* infectious diseases still cause about one of every four deaths each year—mostly in developing countries (Figure 10-10). *Second,* many disease-carrying bacteria have developed genetic immunity to widely used antibiotics. *Third,* many disease-transmitting species of insects (such as mosquitoes) have become immune to widely used pesticides that once helped control their populations.

Are We Losing Ground in Our War against Infectious Bacteria? Growing evidence indicates we may be falling behind in our war against infectious bacterial diseases because bacteria are among the earth's ultimate survivors. The *good news* is that when a colony of bacteria is dosed with an antibiotic such as penicillin, most of the bacteria are killed.

However, the *bad news* is that occasionally a few bacteria in the colony have mutant genes that make them immune to the drug. Through natural selection (Figure 5-5, left, p. 97), a single mutant can pass such traits on to most of its offspring, which can amount to 16,777,216 in only 24 hours.

Even worse, bacteria can become genetically resistant to antibiotics they have never been exposed to. When a resistant and a nonresistant bacterium touch one another (say, on a hospital bedsheet or in a human stomach), they can exchange a small loop of DNA called a plasmid. This enables them to transfer genetic resistance to a disease from one organism to another.

The incredible genetic adaptability of bacteria is one reason the world faces a potentially serious rise in the incidence of some infectious bacterial diseases—such as tuberculosis (Case Study, below)—once controlled by antibiotics.

Other factors also play a key role. One is the spread of harmful bacteria around the globe by human travel and the trade of goods. Another is overuse of antibiotics. According to a 2000 study by Richard Wenzel and Michael Edward, at least half of all antibiotics used to treat humans are prescribed unnecessarily. In many countries antibiotics are available without prescriptions, which also promotes unnecessary use.

Overuse of pesticides is also a problem because it increases populations of pesticide-resistant insects and other carriers of bacterial diseases.

Finally, resistance to some antibiotics can be increased by the widespread use of antibiotics for livestock and dairy animals to control disease and to promote animal growth. Resistant strains of infectious diseases that develop in livestock animals can spread to humans through contact with infected animals or water and through food webs.

The result of these factors acting together is that every major disease-causing bacterium now has strains that resist at least one of the roughly 160 antibiotics we use to treat bacterial infections. In the United States and other countries there has been an increase in patients who contract infectious bacterial diseases while they are in a hospital or other medical facility. According to studies by the U.S. Centers for Disease Control and Prevention (CDC) and a 2002 investigative study by the *Chicago Tribune,* about 70,000 of these patients die prematurely each year from mostly preventable infections in hospitals.

Case Study: The Global Tuberculosis Epidemic
Since 1990, one of the world's most underreported stories has been the rapid spread of tuberculosis (TB). According to the World Health Organization, this highly infectious bacterial disease kills about 1.7 million people and infects about 9 million people per year—most of them in developing countries (Figure 10-11). The WHO projects that between 2000 and 2020, about 35 million people will die of the disease unless current ef-

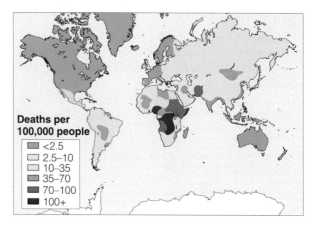

Figure 10-11 The current global tuberculosis epidemic. This easily transmitted disease is spreading rapidly and now kills about 1.7 million people a year—about 84% of them in developing countries. (Data from World Health Organization)

forts and funding to control TB are greatly strengthened and expanded.

The bacterium causing TB infection moves from person to person mainly in airborne droplets produced by coughing, sneezing, singing, or even talking. The TB bacillus now infects about one of every three people in the world.

During their lifetime about 5–10% of these people will become sick or infectious with active TB, especially when their immune system is weakened. Left untreated, each person with active TB will infect 10–15 other people.

Many infected people do not appear to be sick, and about half of them do not know they are infected. As a result, this serious health problem has been called a *silent global epidemic.*

There are several reasons for the recent increase in TB. One is poor TB screening and control programs, especially in developing countries, where about 95% of the new cases occur. Another problem is development of strains of the tuberculosis bacterium that are genetically resistant to almost all effective antibiotics. Also, the incidence of TB tends to increase because population growth and increased urbanization increase contacts between people. Poverty also plays a role. Another factor is the spread of AIDS, which greatly weakens the immune system and allows TB bacteria to multiply.

Slowing the spread of the disease involves early identification and treatment of people with active TB, usually those with a chronic cough. Treatment with a combination of four inexpensive drugs can cure 90% of those with active TB. However, to be successful the drugs must be taken every day for 6–8 months. Because the symptoms disappear after a few weeks, many patients think they are cured and stop taking the drugs. This allows the disease to recur in a hard-to-treat form. It then spreads to other people, and drug-resistant strains of TB bacteria develop.

According to the World Health Organization, a worldwide campaign to help control TB would cost about $360 million to help save at least 20 million lives during the next decade.

How Serious Is the Threat from Viral Diseases? Viral diseases include *influenza* or *flu* (transmitted by the body fluids or airborne emissions of an infected person), *Ebola* (transmitted by the blood or other body fluids of an infected person, Figure 10-8), *HIV/AIDS* (transmitted by unsafe sex, sharing of needles by drug users, infected mothers to offspring before or during birth, and exposure to infected blood, Figure 10-8), *West Nile virus* (transmitted by the bite of a common mosquito that became infected by feeding on birds that carry the virus), and the SARS virus (which emerged in 2003).

Health officials are concerned about the emergence and spread of viral diseases such those caused by Ebola viruses, the West Nile virus, and the SARS virus. However, they point out that in terms of annual deaths the three greatest threats to human health from viruses by far are *AIDS* (caused by the human immunodeficiency virus or HIV), *hepatitis B*, and *flu and viral forms of pneumonia* (Figure 10-10).

Case Study: How Serious Is the Global Threat from HIV and AIDS? The spread of *acquired immune deficiency syndrome (AIDS)*, caused by HIV (Figure 10-8), is one of the world's most serious and rapidly growing health threats. The virus itself is not deadly, but it kills immune cells and leaves the body defenseless against infectious bacteria and other viruses.

According to the WHO, by the beginning of 2003, some 42 million people worldwide (29.4 million of them in sub-Saharan Africa) were infected with HIV—up from 8–10 million in 1990.

Infection rates are increasing rapidly in five countries—Nigeria, Ethiopia, Russia, India, and China—that together have 40% of the world's population. Also, at least 50% of the people infected with HIV develop active TB.

Within 7–10 years, at least half of those with HIV develop AIDS. So far, no cure for AIDS exists, although drugs may help some infected people live longer (if they can afford the treatment, which can cost up to $15,000 per year).

Between 1980 and 2003, about 31 million people (450,000 in the United States) died of AIDS-related diseases—roughly equal to the battlefield deaths in all wars fought during the 20th century. AIDs deaths—mostly among young adults 15–49 years old—has left behind almost 15 million orphans, with 25 million such orphans projected by 2010.

The premature deaths of teachers and health care and other workers in heavily infected African countries leads to diminished education and health care, disintegrating families, and decreased—sometimes backwards—economic development.

Between 2001 and 2020, the WHO estimates 69 million more deaths from AIDS. That 20-year loss is equivalent to a loss of one-fourth the current U.S. population.

How Are Viral Diseases Treated? It is much harder to fight viral infections than infections caused by bacteria and protozoans. Few antiviral drugs exist because most drugs that can kill a virus also harm the cells of its host. Treating viral infections (such as colds, flu, and most mild coughs and sore throats) with antibiotics is useless and increases genetic resistance in disease-causing bacteria.

The best weapons against viruses are *vaccines* that stimulate the body's immune system to produce antibodies to ward off viral infections. Immunization with vaccines has helped reduce the spread of viral diseases such as smallpox, polio, rabies, influenza, measles, and hepatitis B. But developing vaccines takes a long time and is very expensive.

Case Study: Malaria, a Protozoan Parasitic Disease About 40% of the world's population—mostly living in poor tropical countries—is at risk of malaria (Figure 10-12, p. 214). About 90% of the world's malaria cases and deaths occur in Africa, south of the Sahara Desert (Figure 10-12, p. 214). Worldwide, an estimated 300–500 million people are infected with the protozoan parasites that cause malaria, and there are 270–500 million new cases each year.

Malaria's symptoms come and go. They include fever, chills, drenching sweats, anemia, severe abdominal pain and headaches, vomiting, extreme weakness, and greater susceptibility to other diseases. The disease kills about 1 million people each year (about 900,000 of them children under age 5)—an average of 2,700 deaths per day. Malaria kills an African child every 30 seconds. Many children who survive severe malaria have brain damage or impaired learning ability. Pregnant women with malaria usually have severe anemia and are at a higher risk of miscarriages, stillbirths, and having babies with low birth weight.

Malaria is caused by four species of protozoan parasites in the genus *Plasmodium*. Most cases of the disease are transmitted when an uninfected female of any one of about 60 *Anopheles* mosquito species bites an infected person, ingests blood that contains the parasite,

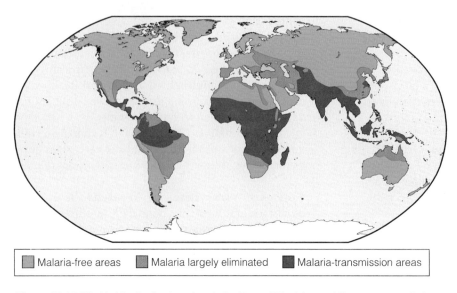

| ■ Malaria-free areas | ■ Malaria largely eliminated | ■ Malaria-transmission areas |

Figure 10-12 Worldwide distribution of *malaria*. About 40% of the world's current population lives in areas in which malaria is present, with the disease killing at least 1 million people a year. In some African countries, 80% of the children are infected with the malaria parasite. (Data from the World Health Organization)

not carry and transmit the parasite to humans, and lead to better drugs, vaccines, insecticides, and insect repellents to counter the disease.

Meanwhile, health experts say prevention is the best approach to slowing the spread of malaria. Methods include increasing water flow in irrigation systems to prevent mosquito larvae from developing (an expensive and wasteful use of water), fixing leaking water pipes, and using mosquito nets dipped in a nontoxic insecticide (permethrin) in windows and doors of homes. Other methods include cultivating fish that feed on mosquito larvae (biological control), clearing vegetation around houses, planting trees that soak up water in low-lying marsh areas where mosquitoes thrive (a method

and later bites an uninfected person (Figure 10-13). When this happens, *Plasmodium* parasites move out of the mosquito and into the human's bloodstream, multiply in the liver, and enter blood cells to continue multiplying. Malaria also can be transmitted by blood transfusions or by sharing needles.

The malaria cycle repeats itself until immunity develops, treatment is given, or the victim dies. *Over the course of human history, malarial protozoa probably have killed more people than all the wars ever fought.*

During the 1950s and 1960s, the spread of malaria was sharply curtailed by draining swamplands and marshes, spraying breeding areas with insecticides, and using drugs to kill the parasites in the bloodstream. Since 1970, malaria has come roaring back. Most species of the malaria-carrying *Anopheles* mosquito have become genetically resistant to widely used insecticides. Worse, the *Plasmodium* parasites have become genetically resistant to common antimalarial drugs.

Researchers are working to develop new antimalarial drugs (such as artemisinins derived from the Chinese herbal remedy qinghaosu), vaccines, and biological controls for *Anopheles* mosquitoes. However, such approaches receive too little funding and have proved more difficult than originally thought.

In 2002, scientists announced they had broken the genetic codes for both the mosquito and the parasite responsible for most malaria cases. Eventually this important information could uncover genetic vulnerabilities in these organisms. This could allow scientists to alter the genetic makeup of mosquitoes so they can-

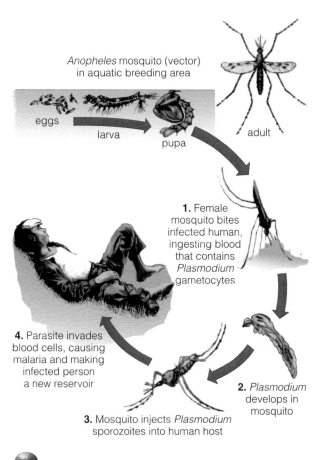

Anopheles mosquito (vector) in aquatic breeding area

eggs
larva
pupa
adult

1. Female mosquito bites infected human, ingesting blood that contains *Plasmodium* gametocytes

2. *Plasmodium* develops in mosquito

3. Mosquito injects *Plasmodium* sporozoites into human host

4. Parasite invades blood cells, causing malaria and making infected person a new reservoir

Figure 10-13 The life cycle of malaria. This life cycle of *Plasmodium* circulates from mosquito to human and back to mosquito.

that can degrade or destroy ecologically important wetlands), and using zinc and vitamin A supplements to boost resistance to malaria in children.

Solutions: How Can We Reduce the Incidence of Infectious Diseases? Here is some *good news* about infectious diseases. According to the WHO, the global death rate from infectious diseases decreased by about 66% between 1970 and 2000 and is projected to continue dropping. Also, between 1971 and 2000, the percentage of children in developing countries immunized against tetanus, measles, diphtheria, typhoid fever, and polio increased from 10% to 84%—saving about 10 million lives a year.

Another breakthrough is the development of simple *oral rehydration therapy* to help prevent death from dehydration for victims of diarrheal diseases, which cause about one-fourth of all deaths of children under age 5. It involves administering a simple solution of boiled water, salt, and sugar or rice, at a cost of only a few cents per person.

In 2001, the WHO began promoting a do-it-yourself technique that uses sunlight to disinfect water. The process is simple: fill a transparent plastic bottle with contaminated water and lay it horizontally on a flat black surface (which absorbs more heat and kills more pathogens). The heat and ultraviolet rays of the sun kill most illness-causing microorganisms in polluted water. This method is especially useful in tropical countries where sunlight is intense.

Finally, in 2002, Bill and Melinda Gates created a fund of more than $24 billion to help fight diseases that affect the world's poorest people.

According to the WHO, there is also some *bad news* about infectious disease. *First,* death rates from infectious diseases in developing countries are unacceptably high (Figure 10-10).

Second, only about 10% of global research and development funds spent on finding cures and treatment for diseases are devoted to infectious diseases in developing countries, even though more people worldwide suffer and die from these diseases than from all other diseases combined. *Finally,* about one-third of the world's people (mostly in developing countries) lack adequate access to clean drinking water and sanitation facilities.

Figure 10-14 lists measures that health scientists and public health officials suggest for preventing or reducing the incidence of infectious diseases that affect humanity (especially in developing countries). According to the WHO, extending primary health care to all the world's people would cost an additional $10 billion per year, about what the world devotes every 4 days and the United States every 10 days to military spending. So far countries have put up only

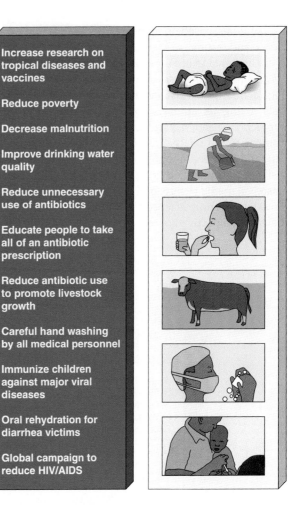

Figure 10-14 Solutions: ways to prevent or reduce the incidence of infectious diseases—especially in developing countries.

$2 billion to help save the lives of at least 10 million people per year—about half of them children under 5 years of age.

10-5 RISK ANALYSIS

How Can We Estimate Risks? Risk analysis involves identifying hazards and evaluating their associated risks (*risk assessment*, Figure 10-2, left), ranking risks (*comparative risk analysis*), determining options and making decisions about reducing or eliminating risks (*risk management*, Figure 10-2, right), and informing decision makers and the public about risks (*risk communication*).

Statistical probabilities based on past experience, animal testing and other tests, and epidemiological studies are used to estimate risks from older technologies and chemicals. To evaluate new technologies and products, risk evaluators use more uncertain statistical probabilities, based on models rather than actual experience and testing.

The left side of Figure 10-15 illustrates results of a *comparative risk analysis,* summarizing the greatest ecological and health risks identified by a panel of scientists acting as advisers to the U.S. Environmental Protection Agency. Note the difference between the comparison of relative risk by scientists (Figure 10-15, left) and the general public (Figure 10-15, right). Some risk experts contend that much of our risk education is based on media reports on the latest risk scare (usually involving frontier science, p. 49) that do not put such risks in perspective.

What Are the Greatest Risks People Face? The greatest risks many people face today are rarely dramatic enough to make the daily news. In terms of the number of premature deaths per year (Figure 10-16) and reduced life span, *the greatest risk by far is poverty* (Figure 10-17, p. 218).

After the health risks associated with poverty and gender, the greatest risks of premature death are mostly the result of voluntary choices people make about their lifestyles (Figures 10-1 and 10-17).

By far the best ways to reduce one's risk of premature death and serious health risks are to not smoke, lose excess weight, reduce consumption of foods con-

Scientists
(Not in rank order
in each category)

Citizens
(In rank order
in each category)

High-Risk Health Problems
- Indoor air pollution
- Outdoor air pollution
- Worker exposure to industrial or farm chemicals
- Pollutants in drinking water
- Pesticide residues on food
- Toxic chemicals in consumer products

High-Risk Ecological Problems
- Global climate change
- Stratospheric ozone depletion
- Wildlife habitat alteration and destruction
- Species extinction and loss of biodiversity

Medium-Risk Ecological Problems
- Acid deposition
- Pesticides
- Airborne toxic chemicals
- Toxic chemicals, nutrients, and sediment in surface waters

Low-Risk Ecological Problems
- Oil spills
- Groundwater pollution
- Radioactive isotopes
- Acid runoff to surface waters
- Thermal pollution

High-Risk Problems
- Hazardous waste sites
- Industrial water pollution
- Occupational exposure to chemicals
- Oil spills
- Stratospheric ozone depletion
- Nuclear power-plant accidents
- Industrial accidents releasing pollutants
- Radioactive wastes
- Air pollution from factories
- Leaking underground tanks

Medium-Risk Problems
- Coastal water contamination
- Solid waste and litter
- Pesticide risks to farm workers
- Water pollution from sewage plants

Low-Risk Problems
- Air pollution from vehicles
- Pesticide residues in foods
- Global climate change
- Drinking water contamination

Figure 10-15 *Comparative risk analysis* of the most serious ecological and health problems according to scientists acting as advisers to the U.S. Environmental Protection Agency (left column). Risks in each of these categories are not listed in rank order. The right side of this figure represents polls showing how U.S. citizens rank the ecological and health risks they perceive as the most serious. Why do you think there is such a great difference between the ranking by risk experts and by the general public? (Data from Science Advisory Board, *Reducing Risks,* Washington, D.C.: Environmental Protection Agency, 1990)

Cause of Death **Annual Deaths**

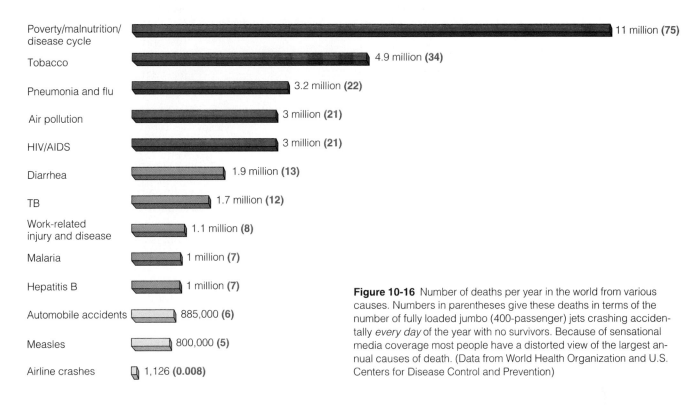

Cause	Deaths
Poverty/malnutrition/disease cycle	11 million **(75)**
Tobacco	4.9 million **(34)**
Pneumonia and flu	3.2 million **(22)**
Air pollution	3 million **(21)**
HIV/AIDS	3 million **(21)**
Diarrhea	1.9 million **(13)**
TB	1.7 million **(12)**
Work-related injury and disease	1.1 million **(8)**
Malaria	1 million **(7)**
Hepatitis B	1 million **(7)**
Automobile accidents	885,000 **(6)**
Measles	800,000 **(5)**
Airline crashes	1,126 **(0.008)**

Figure 10-16 Number of deaths per year in the world from various causes. Numbers in parentheses give these deaths in terms of the number of fully loaded jumbo (400-passenger) jets crashing accidentally *every day* of the year with no survivors. Because of sensational media coverage most people have a distorted view of the largest annual causes of death. (Data from World Health Organization and U.S. Centers for Disease Control and Prevention)

taining cholesterol and saturated fats, eat a variety of fruits and vegetables, exercise regularly, not drink alcohol or drink only in moderation (no more than two drinks in a single day), avoid excess sunlight (which ages skin and causes skin cancer), and for those who can afford a car, drive as safely as possible in a vehicle using the best available safety equipment.

How Can We Estimate Risks for Technological Systems? The more complex a technological system and the more people needed to design and run it, the more difficult it is to estimate the risks. The overall reliability of any technological system (expressed as a percentage) is the product of two factors:

System reliability (%) = Technology reliability × Human reliability

With careful design, quality control, maintenance, and monitoring, a highly complex system such as a nuclear power plant or space shuttle can achieve a high degree of technology reliability. However, human reliability usually is much lower than technology reliability and is almost impossible to predict: To err is human.

Suppose the technology reliability of a nuclear power plant is 95% (0.95) and human reliability is 75% (0.75). Then the overall system reliability is 71% (0.95 × 0.75 = 0.71 = 71%). Even if we could make the technology 100% reliable (1.0), the overall system reliability

would still be only 75% (1.0 × 0.75 = 0.75 = 75%). The crucial dependence of even the most carefully designed systems on unpredictable human reliability helps explain essentially "impossible" tragedies such as the Chernobyl nuclear power plant accident and the *Challenger* and *Columbia* space shuttle accidents.

One way to make a system more foolproof or failsafe is to move more of the potentially fallible elements from the human side to the technical side. However, chance events such as a lightning bolt can knock out an automatic control system, and no machine or computer program can completely replace human judgment. Also, the parts in any automated control system are manufactured, assembled, tested, certified, and maintained by fallible human beings. In addition, computer software programs used to monitor and control complex systems can contain human error or can be deliberately modified by computer viruses to malfunction.

How Can We Evaluate the Results of Risk Analysis? Here are some of the key questions involved in evaluating risk analysis:

■ How reliable are risk assessment data and models?

■ Who profits from allowing certain levels of harmful chemicals into the environment, and who suffers? Who decides this?

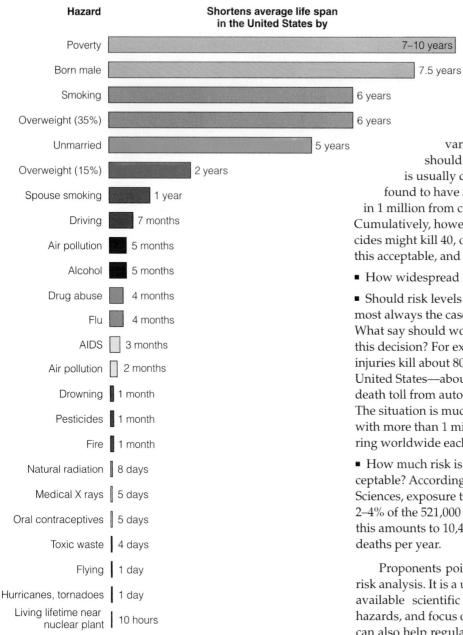

Hazard	Shortens average life span in the United States by
Poverty	7–10 years
Born male	7.5 years
Smoking	6 years
Overweight (35%)	6 years
Unmarried	5 years
Overweight (15%)	2 years
Spouse smoking	1 year
Driving	7 months
Air pollution	5 months
Alcohol	5 months
Drug abuse	4 months
Flu	4 months
AIDS	3 months
Air pollution	2 months
Drowning	1 month
Pesticides	1 month
Fire	1 month
Natural radiation	8 days
Medical X rays	5 days
Oral contraceptives	5 days
Toxic waste	4 days
Flying	1 day
Hurricanes, tornadoes	1 day
Living lifetime near nuclear plant	10 hours

Figure 10-17 Comparison of risks people face, expressed in terms of shorter average life span. After poverty and gender, the greatest risks people face result mostly from voluntary choices they make about their lifestyles. These are only generalized relative estimates. Individual response to some of these risks can vary with factors such as genetic variation (Figure 10-3), family medical history, emotional makeup, stress, and social ties and support. (Data from Bernard L. Cohen)

- Should estimates emphasize short-term risks, or should more weight be put on long-term risks? Who should make this decision?

- Who should do a particular risk analysis, and who should review the results? A government agency? Independent scientists? The public?

- Should the primary goal of risk analysis be to determine how much risk is acceptable (the current approach) or figure out how to do the least damage (a prevention approach)?

- Should cumulative effects of various risks be considered, or should risks be considered separately, as is usually done? Suppose a pesticide is found to have an annual risk of killing 1 person in 1 million from cancer, the current EPA limit. Cumulatively, however, effects from 40 such pesticides might kill 40, or 400, of every 1 million people. Is this acceptable, and to whom is it acceptable?

- How widespread is each risk?

- Should risk levels be higher for workers (as is almost always the case) than for the general public? What say should workers and their families have in this decision? For example, work-related illnesses and injuries kill about 80,000 workers each year in the United States—about twice the country's annual death toll from automobile accidents (Figure 10-1). The situation is much worse in developing countries, with more than 1 million work-related deaths occurring worldwide each year.

- How much risk is acceptable, and to whom is it acceptable? According to the National Academy of Sciences, exposure to toxic chemicals is responsible for 2–4% of the 521,000 cancer deaths in the United States; this amounts to 10,400–20,800 premature cancer deaths per year.

Proponents point to the numerous advantages of risk analysis. It is a useful way to organize and analyze available scientific information, identify significant hazards, and focus on areas that need more research. It can also help regulators decide how money for reducing risks should be allocated and stimulate people to make more informed decisions about health and environmental goals and priorities.

However, critics point out that results of risk analysis are very uncertain. For example, a recent study documented the significant uncertainties involved in even simple risk analysis. Eleven European governments established 11 different teams of their best scientists and engineers (including those from private companies) to assess the hazards and risks from a small plant storing only one hazardous chemical (ammonia). The 11 teams, consisting of world-class experts analyzing this very simple system, disagreed with one another on fundamental points and varied in their assessments of the hazards by a factor of 25,000.

Such inherent uncertainty explains why regulators setting human exposure levels for toxic substances usually divide the best results by 100 to 1,000 to provide the public with a margin of safety.

According to critics, our main decision-making tool should not be to find how much risk is acceptable, which they contend is mostly a political decision. Instead, it should be to find the least damaging reasonable alternatives by asking, "Which alternative will bring sufficient benefits and minimize damage to humans and to the earth?" To these critics, the emphasis should be on *alternative assessment*, not *risk assessment*.

How Should Risks Be Managed? *Risk management* includes the administrative, political, and economic actions taken to decide whether and how to reduce a particular societal risk to a certain level and at what cost.

Risk management involves answering the following questions:

- How reliable is the risk analysis for each risk?

- Which risks should be given the highest priority?

- How much risk is acceptable?

- How much will it cost to reduce each risk to an acceptable level?

- How should limited funds be spent to provide the greatest benefit?

- How will the risk management plan be monitored, enforced, and communicated to the public?

Each step in this process involves making value judgments and weighing trade-offs to find some reasonable compromise among often conflicting political, economic, health, and environmental interests.

How Well Do We Perceive Risks? How much risk is acceptable? Studies indicate that if the chance of death from a chemical or activity is less than 1 in 100,000, most people will not be worried enough to change their ways.

Most of us do poorly in assessing the relative risks from the hazards that surround us (Figures 10-15, 10-16, and 10-17). Also, many people deny or shrug off the high-risk chances of death (or injury) from voluntary activities they enjoy, such as *motorcycling* (1 death in 50 participants), *smoking* (1 in 300 participants by age 65 for a pack-a-day smoker), *hang gliding* (1 in 1,250), and *driving* (1 in 3,300 without a seatbelt and 1 in 6,700 with a seatbelt). Indeed, the most dangerous thing most Americans do each day is drive or ride in a car.

Yet some of these same Americans may be terrified about the possibility of being killed by *a nuclear power plant accident* (1 in 200,000), *West Nile virus* (1 in 1 million), *lightning* (1 in 3 million), *a commercial airplane crash* (1 in 9 million), *a train crash* (1 in 20 million in the United States), *snakebite* (1 in 56 million), *shark attack* (1 in 281 million), or *exposure to trichloroethylene (TCE) in drinking water at the trace levels allowed by the EPA* (1 in 2 billion).

Being bombarded with news about people killed or harmed by various hazards distorts our sense of risk. However, *the most important good news each year is that about 99.1% of the people on the earth did not die.* Despite the greatly increased use of synthetic chemicals in food production and processing, the general health and average life expectancy of people in the United States (and most developed countries) have increased during the past 50 years.

Our perceptions of risk and our responses to perceived risks often have little to do with how risky the experts say something is (Figure 10-15). Under certain conditions, the public tends to see a technology or a product as being riskier than experts do.

One such condition is when the risk is *new or complex rather than familiar.* Examples include genetically modified food or nuclear power, as opposed to traditional plant breeding techniques, large dams, or coal-fired power plants.

Second, the risk is *perceived as mostly involuntary.* Examples include nuclear power plants or food additives, as opposed to driving or smoking.

Third, a risk is *viewed as unnecessary rather than as beneficial or necessary.* Examples might include using chlorofluorocarbons (CFC) propellants in aerosol spray cans or using food additives that increase sales appeal, as opposed to using cars or aspirin.

Another factor involves a risk that can result in a *large, well-publicized death toll from a single catastrophic accident* rather than the same or an even larger death toll spread out over a longer time. Examples might include a severe nuclear power plant accident, an industrial explosion, or an accidental plane crash, as opposed to coal-burning power plants, automobiles, or smoking (Figure 10-16).

Finally, there is concern over the *unfair distribution of risks* from the use of a technology or chemical. Citizens are outraged when government officials decide to put a hazardous waste landfill or incinerator in or near their neighborhood, even when the decision is based on risk analysis. This decision is usually seen as politics, not science. Residents will not be satisfied by estimates that the lifetime risks of cancer death from the facility are not greater than, say, 1 in 100,000. Living near the facility means that they, not the vast majority of people living farther away, have a much higher risk of dying from cancer by having this risk involuntarily imposed on them.

Better education and communication about the nature of risks will help bring the public's perceptions of various risks closer to those of professional risk evaluators. However, such education will not eliminate the emotional, cultural, and ethical factors that regulators and elected officials must take into account in determining the acceptability of a particular risk and evaluating the possible alternatives.

The burden of proof imposed on individuals, companies, and institutions should be to show that pollution prevention options have been thoroughly examined, evaluated, and used before lesser options are chosen.

JOEL HIRSCHORN

REVIEW QUESTIONS

1. Define the boldfaced terms in this chapter.

2. What voluntary human activity kills the largest number of people each year? List seven ways to help reduce the harmful effects of smoking.

3. What are *risk* and *probability?* Distinguish between *risk assessment* and *risk management.*

4. List two **(a)** cultural hazards, **(b)** chemical hazards, **(c)** physical hazards, and **(d)** biological hazards.

5. What is *toxicity?* Distinguish between *dose* and *response* for a potentially harmful substance. List five factors that determine whether a chemical is harmful. Distinguish between *bioaccumulation* and *biomagnification.*

6. Explain what is meant by the phrase, "The dose makes the poison."

7. What is the *median lethal dose?* What is a *poison?*

8. Describe how scientists use *case reports, epidemiological studies,* and *laboratory experiments* to estimate toxicity. List the limitations of each method.

9. What is a *dose-response curve?* Distinguish between a *linear dose-response curve* and a *threshold dose-response curve.*

10. Distinguish between *toxic chemicals* and *hazardous chemicals.* Distinguish among *mutagens, teratogens,* and *carcinogens.*

11. Distinguish among the *immune system, nervous system,* and *endocrine system.* What are *hormone disrupters* and *hormone mimics?* Describe two potentially harmful effects of exposure to *hormonally active agents* (HAAs) on wildlife and two such effects on humans.

12. About what percentage of the 75,000 chemicals in commercial use in the United States have been screened **(a)** to assess toxicity, **(b)** to determine whether they are carcinogens, teratogens, or mutagens, and **(c)** to determine whether they damage the nervous, endocrine, or immune systems?

13. List three reasons for the lack of information about the potentially harmful effects of most chemicals in commercial use. Distinguish between the regulation strategy and the pollution prevention (or precautionary) strategy for protecting the public from potentially harmful chemicals.

14. Distinguish between *nontransmissible* and *transmissible diseases,* and give two examples of each type. Distinguish between a *bacterium* and a *virus,* and explain how a virus replicates itself.

15. Give three pieces of *good news* and three pieces of *bad news* about our efforts to reduce the incidence of infectious diseases.

16. What are the seven deadliest infectious diseases in order of the number of deaths they cause each year?

17. Describe two ways that infectious bacteria become resistant to antibiotics. List six factors that promote such bacterial resistance.

18. What causes tuberculosis, and how is it transmitted? About how many people died of TB during the past year? List five reasons for the increase in TB infections in recent years. How can the spread of this bacterial infectious disease be slowed?

19. Distinguish between *HIV* and *AIDS.* About how many people in the world **(a)** are infected with HIV, **(b)** have died of AIDS, and **(c)** are expected to die of AIDS between 2001 and 2020?

20. What is the best way to treat a viral disease? Explain why antibiotics should not be used to treat such diseases.

21. What causes *malaria?* About how many people die from malaria each year? List six ways to help prevent this protozoal infectious disease.

22. List five pieces of good news and three pieces of bad news about efforts to reduce the incidence of infectious disease in developing countries. List 11 ways to prevent or reduce the incidence of infectious diseases throughout the world.

23. What is *risk analysis?* What is *comparative risk analysis?*

24. List five of the greatest risks people face in terms of **(a)** number of premature deaths per year and **(b)** reduced life span. List eight ways to reduce your risk of premature death and serious health problems.

25. How can we estimate the risks from technological systems?

26. List nine questions to be answered in evaluating the results of risk analysis. What are the major pros and cons of risk analysis?

27. What is *risk management?* What six questions do risk managers try to answer?

28. About what percentage of the people on the earth die each year? List five reasons why people often perceive that certain risks are greater than experts say they are.

CRITICAL THINKING

1. Explain why you agree or disagree with the proposals for reducing the death toll and other harmful effects of smoking listed on p. 203.

2. Do you think chemicals should be regulated based on their effects on the nervous, immune, and endocrine sys-

tems? Explain. What are the financial implications for doing this? What are the alternatives?

3. Evaluate the following statements:
   a. We should not get so worked up about exposure to toxic chemicals because almost any chemical can cause some harm at a large enough dosage.
   b. We should not worry so much about exposure to toxic chemicals because through genetic adaptation we can develop immunity to such chemicals.
   c. We should not worry so much about exposure to toxic chemicals because we can use genetic engineering to reduce or eliminate such problems.

4. Assume you are a government official responsible for regulating chemicals. What information would you require in order to make a decision about the safety of a particular chemical? What level of risk of exposure to the chemical would you consider acceptable for **(a)** society, **(b)** yourself, and **(c)** any child you might have?

5. Should pollution levels be set to protect the most sensitive people in a population (Figure 10-3, left) or the average person (Figure 10-3, middle)? Explain.

6. Do you believe the use of the same antibiotics to treat human illness and to fatten livestock should be banned in the United States (or the country where you live)? Explain. Would you favor using small amounts of such antibiotics to treat disease in livestock? Explain.

7. What are the five major risks you face from **(a)** your lifestyle, **(b)** where you live, and **(c)** what you do for a living? Which of these risks are voluntary and which are involuntary? List the five most important things you can do to reduce these risks. Which of these things do you actually plan to do?

8. How would you answer each of the questions raised about **(a)** risk analysis on pp. 217–218 and **(b)** risk assessment and risk management on p. 219? Explain each of your answers.

PROJECTS

1. Assume that members of your class (or small manageable groups in your class) have been appointed to a technology benefit–risk assessment board. As a group, decide why you would approve or disapprove of widespread use of each of the following: **(a)** drugs to retard aging and **(b)** genetic engineering to produce people with superior intelligence and strength.

2. Use the library or the Internet to find recent articles describing the rise of genetic resistance of disease-causing bacteria to commonly used antibiotics. Evaluate the evidence and claims in these articles.

3. Pick a specific viral disease and use the library or Internet to find out about **(a)** how it spreads, **(b)** its effects,

(c) strategies for controlling its spread, and **(d)** possible treatments.

4. Make a concept map of this chapter's major ideas, using the section heads and subheads and the key terms (in boldface). Look on the website for this book for information about making concept maps.

INTERNET STUDY RESOURCES AND RESOURCES FOR FURTHER READING AND RESEARCH

The website for this book contains helpful study aids and many ideas for further reading and research. Log on to

http://biology.brookscole.com/miller10

and click on the Chapter-by-Chapter area. Choose Chapter 10 and select a resource:

■ Flash Cards allows you to test your mastery of the Terms and Concepts to Remember for this chapter.

■ Tutorial Quizzes provides a multiple-choice practice quiz.

■ Student Guide to InfoTrac will lead you to Critical Thinking Projects that use InfoTrac College Edition as a research tool.

■ References lists the major books and articles consulted in writing this chapter.

■ Hypercontents takes you to an extensive list of sites with news, research, and images related to individual sections of the chapter.

INFOTRAC COLLEGE EDITION

Improve your skills with InfoTrac College Edition, a searchable online database of articles from more than 700 periodicals. Log on to

http://www.infotrac-college.com

or access InfoTrac through the website for this book. Try to find the following articles:

1. Frank, C., A. D. Fix, C. A. Pena, and G. T. Strickland. 2002. Mapping Lyme disease incidence for diagnostic and preventive decisions, Maryland. *Emerging Infectious Diseases* 8: 427. *Keywords:* "Lyme disease" and "mapping." GIS mapping technology was used to illustrate the distribution of Lyme disease in Maryland. Doctors and state health departments can then use this information to assist in diagnosis and distribution of vaccine.

2. Thiele, L. P. 2000. Limiting risks: Environmental ethics as a policy primer. *Policy Studies Journal* 20: 540. *Keywords:* "risk" and "environmental ethics." This article explores the ethical aspects of evaluating involuntary human-caused environmental risks in a technologically driven society.

11 THE HUMAN POPULATION: GROWTH AND DISTRIBUTION

Slowing Population Growth in Thailand

Can a country sharply reduce its population growth in only 15 years? Thailand did.

In 1971, Thailand adopted a policy to reduce its population growth. When the program began, the country's population was growing at a rate of 3.2% per year, and the average Thai family had 6.4 children.

Fifteen years later in 1986, the country's population growth rate had been cut in half to 1.6%. By 2003, the rate had fallen to 0.7%, and the average number of children per family was 1.8. Thailand's population is projected to grow from 63 million in 2003 to 72 million by 2025.

There are a number of reasons for this impressive achievement. They include the creativity of the government-supported family planning program, a high literacy rate among women (90%), an increased economic role for women and advances in women's rights, and better health care for mothers and children. Other factors are the openness of the Thai people to new ideas, support of family planning by the country's religious leaders (95% of Thais are Buddhist), and the willingness of the government to encourage and financially support family planning and to work with the private, nonprofit Population and Community Development Association (PCDA).

Mechai Viravidaiya (Figure 11-1) led the way in reducing the country's population growth rate. This public relations genius and former government economist launched the PCDA in 1974 to help make family planning a national goal. PCDA workers handed out condoms at festivals, movie theaters, and even traffic jams, and they developed ads and witty songs about contraceptive use. Between 1971 and 2003, the percentage of married women using modern birth control rose from 15% to 70%—higher than the 57% usage in developed countries and the 51% usage in developing countries.

Viravidaiya helped establish a German-financed revolving loan plan to enable people participating in family planning programs to install toilets and drinking water systems. Low-rate loans were offered to farmers practicing family planning. The government also offers loans to individuals from a fund that increases as their village's level of contraceptive use rises.

All is not completely rosy. Thailand has done well in slowing population growth and raising per capita income. But it has been less successful in reducing pollution and improving public health. Its capital, Bangkok, is plagued with notoriously high levels of traffic congestion and air pollution (Figure 11-2). The typical motorist in Bangkok spends 44 days per year sitting in traffic. Such traffic delays cost Bangkok about $4 million a day in lost work time.

Figure 11-1 *Individuals matter: Mechai Viravidaiya*, a charismatic leader, played a major role in Thailand's successful efforts to reduce its population growth. In 1974, he established the private, nonprofit Population and Community Development Association (PCDA) to help implement family planning as a national goal.

Figure 11-2 This policeman and schoolchildren in Bangkok, Thailand, are wearing masks to reduce their intake of air polluted mainly by automobiles. Bangkok is one of the world's most car-clogged cities, with car commutes averaging 3 hours per day. Roughly one of every nine of its residents has a respiratory ailment.

The problems to be faced are vast and complex, but come down to this: 6.3 billion people are breeding exponentially. The process of fulfilling their wants and needs is stripping earth of its biotic capacity to produce life; a climactic burst of consumption by a single species is overwhelming the skies, earth, waters, and fauna.

PAUL HAWKEN

This chapter addresses the following questions:

■ How is population size affected by birth, death, fertility, and migration rates?

■ How is population size affected by the percentage of males and females at each age level?

■ How can population growth be slowed?

■ What success have India and China had in slowing population growth?

■ How can global population growth be reduced?

■ How is the world's population distributed between rural and urban areas, and what factors determine how urban areas develop?

■ What are the major resource and environmental problems of urban areas?

■ How do transportation systems shape urban areas and growth, and what are the advantages and disadvantages of various forms of transportation?

■ How can cities be made more sustainable and more desirable places to live?

11-1 FACTORS AFFECTING HUMAN POPULATION SIZE

How Is Population Size Affected by Birth Rates and Death Rates? Populations grow or decline through the interplay of three factors: *births, deaths,* and *migration.* **Population change** is calculated by subtracting the number of people leaving a population (through death and emigration) from the number entering it (through birth and immigration) during a specified period of time (usually a year):

Population change = (Births + Immigration) − (Deaths + Emigration)

When births plus immigration exceed deaths plus emigration, population increases; when the reverse is true, population declines.

Instead of using the total numbers of births and deaths per year, demographers use the **birth rate,** or **crude birth rate** (the number of live births per 1,000 people in a population in a given year), and the **death rate,** or **crude death rate** (the number of deaths per 1,000 people in a population in a given year). Figure 11-3 shows the crude birth and death rates for various groupings of countries in 2003.

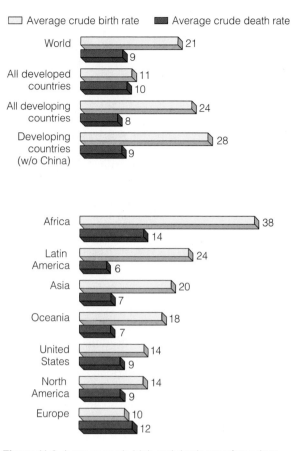

Figure 11-3 Average crude birth and death rates for various groupings of countries in 2003. (Data from Population Reference Bureau)

Birth rates and death rates are coming down worldwide, but death rates have fallen more sharply than birth rates. As a result, more births are occurring than deaths; every time your heart beats, 2.5 more babies are added to the world's population. At this rate, we share the earth and its resources with about 216,000 more people each day (97% of them in developing countries).

The rate of the world's annual population change (excluding migration) usually is expressed as a percentage:

$$\begin{array}{l}\text{Annual rate of}\\\text{natural population}\\\text{change (\%)}\end{array} = \frac{\text{Birth rate} - \text{Death rate}}{1,000\ \text{persons}} \times 100$$

$$= \frac{\text{Birth rate} - \text{Death rate}}{10}$$

Exponential population growth has not disappeared but is occurring at a slower rate. The rate of the world's annual population growth (natural increase) dropped 43% between 1963 and 2003, from 2.2% to 1.26%. This is *good news,* but during the same period the size of the world's population increased by about 97%, from 3.2 billion to 6.3 billion. The *bad news* is that

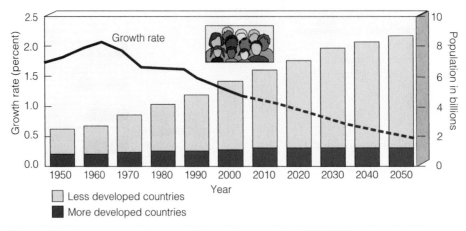

Figure 11-4 Average annual increase in the world's population, 1950–2003, and projected increase 2003–2050 (dotted line). (Data from United Nations)

this drop in the rate of population growth is somewhat like learning that a truck heading straight at you has slowed from 100 kilometers per hour (kph) to 57 kph while its weight has almost doubled.

An exponential growth rate of 1.26% may seem small. However, in 2003 it added about 79 million people to the world's population, compared to 69 million added in 1963, when the world's population growth rate reached its peak. An increase of 79 million people per year is roughly equal to adding another New York City every month, a Germany every year, and a United States every 3.7 years.

Figure 11-4 shows the average annual increase in the world's population and the increase in population size from 1950 to 2003 and the projected increases to 2050. The six nations expected to experience most of this growth are, in order: India, China, Pakistan, Nigeria, Bangladesh, and Indonesia.

In numbers of people, China (1.3 billion people in 2003, about one of every five people in the world) and India (1.1 billion) dwarf all other countries. Together they make up 38% of the world's population. The United States, with 292 million people in 2003, has the world's third largest population but only 4.6% of the world's people.

How Have Global Fertility Rates Changed?
Two types of fertility rates affect a country's population size and growth rate. The first type, **replacement-level fertility,** is the number of children a couple must bear to replace themselves. It is slightly higher than two children per couple (2.1 in developed countries and as high as 2.5 in some developing countries), mostly because some female children die before reaching their reproductive years.

Does reaching replacement-level fertility mean an immediate halt in population growth? No, because so

many future parents are alive. If each of today's couples had an average of 2.1 children and their children also had 2.1 children, the world's population would still grow for 50 years or more (assuming death rates do not rise).

The second type of fertility rate is the **total fertility rate (TFR):** an estimate of the average number of children a woman will have during her childbearing years if between ages 15 and 49 she bears children at the same rate as women did this year. TFRs have dropped sharply since 1950 (Figure 11-5). In 2003, the average global TFR was 2.8 children per woman. It was 1.5 in

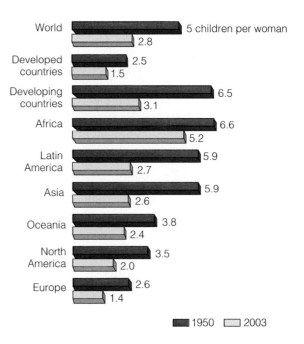

Figure 11-5 *Good news:* decline in total fertility rates for various groupings of countries, 1950–2003. (Data from United Nations)

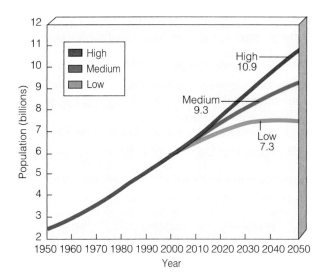

Figure 11-6 UN world population projection, assuming that by 2050 the world's total fertility rate is 2.6 (high), 2.1 (medium), or 1.7 (low) children per woman. (Data from United Nations, *World Population Prospects: The 2000 Revision*, 2001)

developing countries (Figure 1-4, p. 6), where acute poverty (living on less than $1 per day) is a way of life for about 1.4 billion people. Between 2003 and 2050, using medium estimates, the population of developing countries is projected to increase to 8 billion from 5.2 billion.

How Have Fertility Rates Changed in the United States?   The population of the United States has grown from 76 million in 1900 to 292 million in 2003, despite oscillations in the country's TFR (Figure 11-7) and birth rate (Figure 11-8). In 1957, the peak of the baby boom after World War II, the TFR reached 3.7 children per woman. Since then it has generally declined, remaining at or below replacement level since 1972.

The drop in the TFR and birth rates has led to a decline in the rate of population growth in the United States. However, the country's population is still growing faster (1.2% a year, including immigration) than

developed countries (down from 2.5 in 1950) and 3.1 in developing countries (down from 6.5 in 1950).

This is an impressive drop in the average number of children born to women. But this level of fertility is still far above the replacement level of 2.1, especially in developing countries. However, according to UN population experts, if the world's average TFR had not declined from its 1960 level and death rates had stayed the same, more than 8 billion people would be alive today instead of 6.3 billion.

UN population projections to 2050 vary depending on the world's projected average TFR (Figure 11-6). About 97% of this growth is projected to take place in

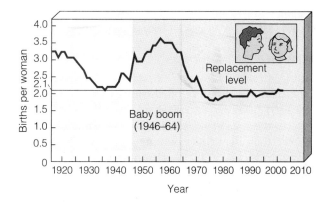

Figure 11-7 Total fertility rates for the United States between 1917 and 2003. (Data from Population Reference Bureau and U.S. Census Bureau)

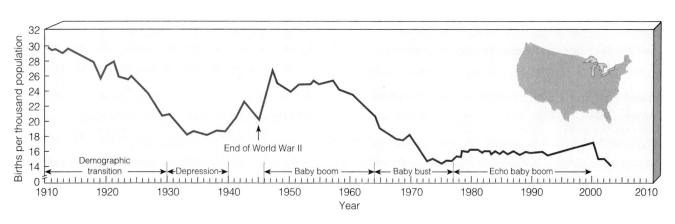

Figure 11-8 Birth rates in the United States, 1910–2003. (Data from U.S. Bureau of Census and U.S. Commerce Department)

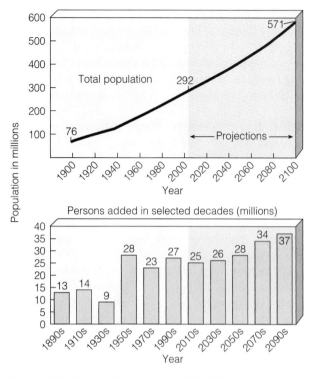

Figure 11-9 U.S. population growth, 1900–2003, and projections to 2100. (Data from U.S. Census Bureau)

that of any other developed country and is not even close to leveling off. In 2003, this growth added about 2.7 million people. This includes 1.5 million more births than deaths (accounting for about 55% of the growth), about 900,000 legal immigrants and refugees, and an estimated 300,000 illegal immigrants (added to the country's 7.5 million illegal immigrants).

According to U.S. Bureau of Census medium projections, the U.S. population will increase from 292 million to 422 million between 2003 and 2050 and reach 571 million by 2100 (Figure 11-9). In contrast, population growth has slowed in other major developed countries since 1950 and overall is projected to decline after 2010. Because of a high per capita rate of resource use, each addition to the U.S. population has an enormous environmental impact (Figure 1-7, p. 8, and Figure 1-11, p. 11).

What Factors Affect Birth Rates and Fertility Rates? Many factors affect a country's average birth rate and TFR. One is the *importance of children as a part of the labor force.* Rates tend to be higher in developing countries (especially in rural areas, where children begin working to help raise crops at an early age).

Another economic factor is the *cost of raising and educating children.* Birth and fertility rates tend to be lower in developed countries, where raising children is much more costly because children do not enter the labor force until their late teens or in their 20s.

The *availability of private and public pension systems* affects how many children couples have. Pensions eliminate parents' need to have many children to help support them in old age.

Urbanization plays a role because people living in urban areas usually have better access to family planning services and tend to have fewer children than those living in rural areas (where children are needed to perform essential tasks).

Another important factor is *the educational and employment opportunities available for women.* TFRs tend to be low when women have access to education and paid employment outside the home. In developing countries, women with no education generally have two more children than women with a secondary school education.

Another factor is the *infant mortality rate.* In areas with low infant mortality rates, people tend to have a smaller number of children because fewer children die at an early age.

Average age at marriage (or, more precisely, the average age at which women have their first child) also plays a role. Women normally have fewer children when their average age at marriage is 25 or older.

Birth rates and TFRs are also affected by the *availability of legal abortions.* Each year about 190 million women become pregnant. The United Nations and the World Bank estimate that about 46 million of these women get an abortion. About 26 million of these abortions are legal and 20 million are illegal (and often unsafe).

The *availability of reliable birth control methods* (Figure 11-10) allows women to control the number and spacing of the children they have. *Religious beliefs, traditions, and cultural norms* also play a role. In some countries, these factors favor large families and strongly oppose abortion and some forms of birth control.

What Factors Affect Death Rates? The rapid growth of the world's population over the past 100 years is not the result of a rise in the crude birth rate. Instead, it has been caused largely by a decline in crude death rates, especially in developing countries (Figure 11-11, p. 228).

More people started living longer (and fewer infants died) because of increased food supplies and distribution, better nutrition, improvements in medical and public health technology (such as immunizations and antibiotics), improved sanitation and personal hygiene, and safer water supplies (which has curtailed the spread of many infectious diseases).

Extremely Effective

Total abstinence

Sterilization

Vaginal ring

100%

99.6%

98–99%

Highly Effective

IUD with slow-release hormones

IUD plus spermicide

Vaginal pouch ("female condom")

IUD

Condom (good brand) plus spermicide

Oral contraceptive

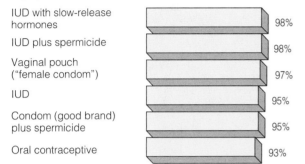

98%

98%

97%

95%

95%

93%

Effective

Cervical cap

Condom (good brand)

Diaphragm plus spermicide

Rhythm method (Billings, Sympto-Thermal)

Vaginal sponge impregnated with spermicide

Spermicide (foam)

89%

86%

84%

84%

83%

82%

Moderately Effective

Spermicide (creams, jellies, suppositories)

Rhythm method (daily temperature readings)

Withdrawal

Condom (cheap brand)

75%

74%

74%

70%

Unreliable

Douche

Chance (no method)

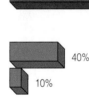

40%

10%

Figure 11-10 Typical effectiveness rates of birth control methods in the United States. Percentages are based on the number of undesired pregnancies per 100 couples using a specific method as their sole form of birth control for a year. For example, a 93% effectiveness rating for oral contraceptives means that for every 100 women using the pill regularly for 1 year, 7 will get pregnant. Effectiveness rates tend to be lower in developing countries, primarily because of lack of education. (Data from Alan Guttmacher Institute)

73 by 2025. Between 1900 and 2003, life expectancy in the United States increased from 47 to 77 years and is projected to reach 81 years by 2025.

Some *bad news* is that in the world's 49 poorest countries (mainly in Africa) life expectancy is 55 years or less. In many African countries life expectancy is expected to fall further because of increased deaths from AIDS.

Because it reflects the general level of nutrition and health care, infant mortality probably is the single most important measure of a society's quality of life. A high infant mortality rate usually indicates insufficient food (undernutrition), poor nutrition (malnutrition), and a high incidence of infectious disease (usually from contaminated drinking water).

Between 1965 and 2003, the world's infant mortality rate dropped from 20 per 1,000 live births to 7 in developed countries and from 118 to 61 in developing countries. This is an impressive achievement, but it still means that at least 8 million infants (most in developing countries) die of preventable causes during their first year of life—an average of 22,000 mostly unnecessary infant deaths per day. This is equivalent to 55 jumbo jets—each loaded with 400 infants under age 1—crashing accidentally each day with no survivors.

Some *good news* is that the U.S. infant mortality rate declined from 165 to 6.5 between 1900 and 2003. This sharp decline in infant mortality rates was a major factor in the marked increase in U.S. average life expectancy during this period.

Despite this improvement, some *bad news* is that 35 countries had lower infant mortality rates than the United States in 2003. Three factors keeping the U.S. infant mortality rate higher than it could be are inadequate health care (for poor women during pregnancy and for their babies after birth), drug addiction among pregnant women, and a high birth rate among teenagers.

Some *good news* is that the U.S. birth rate among girls ages 15–19 in 2003 was lower than at any time since 1940. The *bad news* is that the United States has the highest teenage pregnancy rate of any industrialized country. Each year about 872,000 teenage girls become pregnant in the United States (78% of them

Two useful indicators of overall health of people in a country or region are **life expectancy** (the average number of years a newborn infant can expect to live) and the **infant mortality rate** (the number of babies out of every 1,000 born who die before their first birthday).

Some *good news* is that global life expectancy at birth increased from 48 years to 67 years (76 years in developed countries and 65 years in developing countries) between 1955 and 2003 and is projected to reach

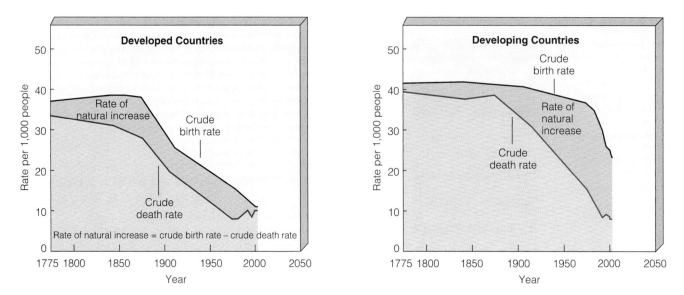

Figure 11-11 Changes in crude birth and death rates for developed and developing countries, 1775–2003. (Data from Population Reference Bureau and United Nations)

unplanned) and about 253,000 of them have abortions. Babies born to teenagers are more likely to have low birth weights, the most important factor in infant deaths.

Case Study: Should the United States Encourage or Discourage Immigration?

Since 1820 the United States has admitted almost twice as many immigrants and refugees as all other countries combined. However, the number of legal immigrants has varied during different periods because of changes in immigration laws and rates of economic growth (Figure 11-12). Currently, legal and illegal immigration account for about 45% of the country's annual population growth.

Between 1820 and 1960, most legal immigrants to the United States came from Europe; since then, most have come from Latin America (53%) and Asia (30%). Between 2002 and 2050, the percentage of Latinos in the U.S. population is projected to almost double from 13% to 24%.

In 1995, the U.S. Commission on Immigration Reform recommended reducing the number of legal immigrants and refugees to about 700,000 per year for a transition period and then to 550,000 a year.

Some demographers and environmentalists go further and call for reducing the number of legal immigrants and refugees allowed into the United States to 300,000–450,000 per year or limiting legal immigration to about 20% of annual population growth. They would accept immigrants only if they can support

themselves, arguing that providing immigrants with public services turns the United States into a magnet for the world's poor.

Most of these analysts also support efforts to sharply reduce illegal immigration. However, some are concerned that a crackdown on illegal immigrants

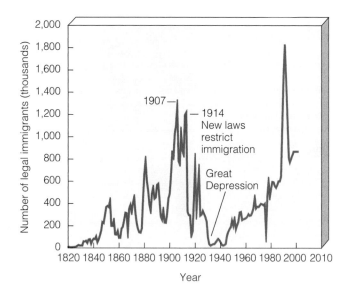

Figure 11-12 Legal immigration to the United States, 1820–2001. The large increase in immigration since 1989 resulted mostly from the Immigration Reform and Control Act of 1986, which granted legal status to illegal immigrants who could show they had been living in the country for several years. (Data from U.S. Immigration and Naturalization Service)

can also lead to discrimination against legal immigrants.

Proponents argue that reducing immigration would allow the United States to stabilize its population sooner and help reduce the country's enormous environmental impact. The public strongly supports reducing U.S. immigration levels. A 1998 *Wall Street Journal*/NBC News poll found that 72% of the people polled were against the country's high immigration rate. A 1993 Hispanic Research Group survey found that 89% of Hispanic Americans supported an immediate moratorium on immigration.

Others oppose reducing current levels of legal immigration. They argue that this would diminish the historical role of the United States as a place of opportunity for the world's poor and oppressed. In addition, immigrants pay taxes, take many menial and low-paying jobs that other Americans shun, open businesses, and create jobs. Moreover, according to the U.S. Census Bureau, after 2020 higher immigration levels will be needed to supply enough workers as baby boomers retire.

11-2 POPULATION AGE STRUCTURE

What Are Age Structure Diagrams? As mentioned earlier, even if the replacement-level fertility rate of 2.1 were magically achieved globally tomorrow, the world's population would keep growing for

at least another 50 years (assuming no large increase in death rates). The reason for this is a population's **age structure:** the proportion of the population (or of each sex) at each age level.

Demographers typically construct a population age structure diagram by plotting the percentages or numbers of males and females in the total population in each of three age categories: *prereproductive* (ages 0–14), *reproductive* (ages 15–44), and *postreproductive* (ages 45 and up). Figure 11-13 presents generalized age structure diagrams for countries with rapid, slow, zero, and negative population growth rates.

How Does Age Structure Affect Population Growth? Any country with many people below age 15 (represented by a wide base in Figure 11-13, left) has a powerful built-in momentum to increase its population size unless death rates rise sharply. The number of births rises even if women have only one or two children because of the large number of girls who will soon be moving into their reproductive years.

In 2003, 30% of the people on the planet were under 15 years old. These 1.9 billion young people are poised to move into their prime reproductive years.

In developing countries the number is even higher: 33%, compared with 18% in developed countries. This powerful force for continued population growth, mostly in developing countries, could be slowed by an effective program to reduce birth rates or a sharp rise in death rates.

Figure 11-13 Generalized population age structure diagrams for countries with rapid (1.5–3%), slow (0.3–1.4%), zero (0–0.2%), and negative population growth rates. (Data from Population Reference Bureau)

Figure 11-14 shows the age structure in developed and developing countries in 2003. We live in a *demographically divided world,* as shown by population data for the United States, Brazil, and Nigeria (Figure 11-15).

How Can Age Structure Diagrams Be Used to Make Population and Economic Projections?

The 79-million-person increase that occurred in the U.S. population between 1946 and 1964 is known as the *baby boom* generation (Figures 11-7 and 11-8). As members of this group grow older they move up through the country's age structure (Figure 11-16).

Baby boomers now make up nearly half of all adult Americans. As a result, they dominate the popu-

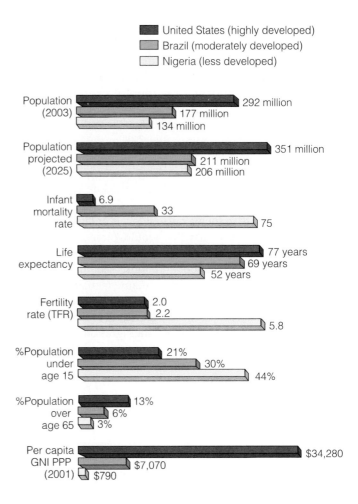

Figure 11-15 Comparison of key demographic indicators in highly developed (United States), moderately developed (Brazil), and less developed (Nigeria) countries in 2003. (Data from Population Reference Bureau)

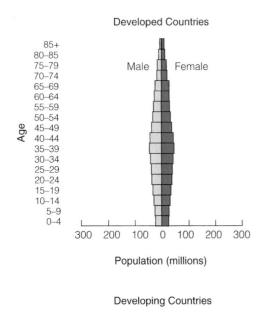

Developed Countries

Population (millions)

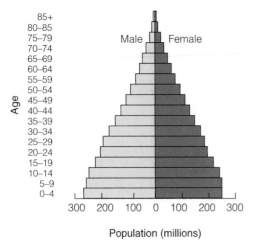

Developing Countries

Population (millions)

Figure 11-14 Population structure by age and sex in developing countries and developed countries, 2003.(Data from United Nations Population Division and Population Reference Bureau)

lation's demand for goods and services and play an increasingly important role in deciding who gets elected and what laws are passed. Baby boomers who created the youth market in their teens and 20s are now creating the 50-something market and will soon move on to create a 60-something market. In 2011, the first baby boomers will turn 65 and their numbers will grow sharply through 2029.

Much of the economic burden of helping support a large number of retired baby boomers will fall on the *baby-bust generation* (also called *generation* X). It consists of people born between 1965 and 1976 (when TFRs fell sharply to below 2.1; Figure 11-7).

Retired baby boomers are likely to use their political clout to force the smaller number of people in the baby-bust generation to pay higher income, healthcare, and Social Security taxes. This could cause resentment and conflicts between the two generations.

In some respects, the baby-bust generation should have an easier time than the baby-boom generation.

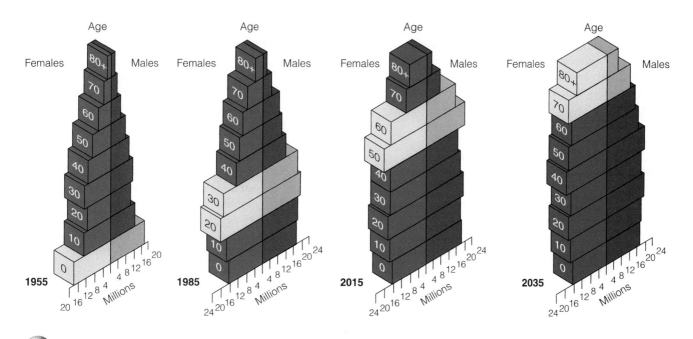

Figure 11-16 Tracking the baby-boom generation in the United States. (Data from Population Reference Bureau and U.S. Census Bureau)

Fewer people will be competing for educational opportunities, jobs, and services. Also, labor shortages may drive up their wages, at least for jobs requiring education or technical training beyond high school.

However, members of the baby-bust group may find it difficult to get job promotions as they reach middle age because members of the much larger baby-boom group will occupy most upper-level positions. Many baby boomers may delay retirement because of improved health, the need to accumulate adequate retirement funds, or extension of the retirement age needed to begin collecting Social Security.

The baby-bust generation is being followed by the *echo-boom generation* (Figure 11-8) consisting of people born from 1977 to 2003.

From these few projections, we can see that any booms or busts in the age structure of a population create social and economic changes that ripple through a society for decades.

What Are Some Effects of Population Decline from Reduced Fertility? The populations of most of the world's countries are projected to grow throughout most of the 21st century. By 2003, however, 38 countries (most of them in Europe) with about 815 million people had roughly stable populations (annual growth rates at or below 0.3%) or declining populations. In other words, about 13% of humanity in industrialized Europe plus Japan have essentially stabilized the size of their populations.

As the age structure of the world's population changes and the percentage of people age 60 or older increases (Figure 11-17), more countries will begin ex-

periencing population declines. If population decline is gradual, its harmful effects usually can be managed.

However, rapid population decline, like rapid population growth, can lead to severe economic and social problems. A country undergoing rapid population

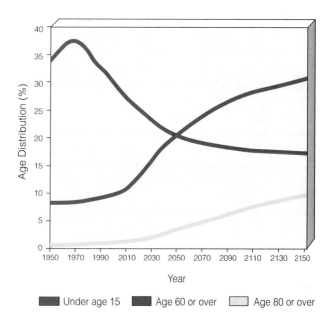

Figure 11-17 *Global aging.* Projected percentage of world population under age 15, age 60 or over, and age 80 or over, 1950–2150, assuming the medium fertility projection shown in Figure 11-6. Between 1998 and 2050 the number of people over age 80 is projected to increase from 66 million to 370 million. The cost of supporting a much larger elderly population will place enormous strains on the world's economy. (Data from the United Nations)

decline has a sharp rise in the proportion of older people, who consume a large share of medical care, Social Security, and other costly public services funded by working taxpayers. Such countries can also face labor shortages unless they rely on greatly increased automation or immigration of foreign workers.

11-3 SOLUTIONS: INFLUENCING POPULATION SIZE

How Is Population Size Affected by Migration? The population of an area is affected by movement of people into (immigration) and out of (emigration) that area. Most countries influence their rates of population growth to some extent by restricting immigration. Only a few countries—chiefly Canada, Australia, and the United States (Figure 11-12)—allow large annual increases in population from immigration.

Migration can help raise the standard of living for the migrants. It can also increase the income of family members left behind when migrant workers send some of their earnings back home. However, countries that migrants leave in search of a better income can suffer from a "brain drain" of some of their most talented individuals. This can hinder the economic development of such countries.

International migration to developed countries absorbs only about 1% of the annual population growth in developing countries. Thus population change for most countries is determined mainly by the difference between their birth rates and death rates.

Migration within countries, especially from rural to urban areas, plays an important role in the population dynamics of cities, towns, and rural areas, as discussed in Section 11-6.

What Are the Advantages and Disadvantages of Reducing Births? The projected increase of the human population from 6.3 to 9.3 billion or more between 2003 and 2050 raises an important question: *Can the world provide an adequate standard of living for 3 billion more people without causing widespread environmental damage?*

There is controversy over this question, whether the earth is already overpopulated, and what measures (if any) should be taken to slow population growth. To some the planet is already overpopulated, but others disagree. Some analysts, mostly economists, argue that we should encourage population growth to help stimulate economic growth.

Others believe that asking how many people the world can support is the wrong question, equivalent to asking how many cigarettes one can smoke before getting lung cancer. Instead, they say, we should be asking what the *optimum sustainable population* of the earth might be, based on the planet's *cultural carrying capacity* (Guest Essay, p. 234).

Such an optimum level would allow most people to live in reasonable comfort and freedom without impairing the ability of the planet to sustain future generations. No one knows what the optimum population size might be. Some consider it a meaningless concept; some put it at 20 billion, others at 8 billion, and others as low as 2 billion.

Those who do not believe the earth is overpopulated point out that the average life span of the world's 6.3 billion people is longer today than at any time in the past. They say that the world can support billions more people. They also point out that people are the world's most valuable resource for solving the problems we face and for stimulating economic growth by becoming consumers.

Some people believe all people should be free to have as many children as they want. Some view any form of population regulation as a violation of their religious beliefs. Others see it as an intrusion into their privacy and personal freedom. Some developing countries and some members of minorities in developed countries regard population control as a form of genocide to keep their numbers and power from rising.

Proponents of slowing and eventually stopping population growth point out that we fail to provide the basic necessities for one out of six people on the earth today. If we cannot (or will not) do this now, they ask, how will we be able to do this for the projected 3 billion more people by 2050?

Proponents of slowing population growth point out that if we do not sharply lower birth rates, we are likely to decide by default to raise death rates for humans (already occurring in parts of Africa) and increase resource use and environmental harm. In 1992, for example, the U.S. National Academy of Sciences and the Royal Society of London issued the following joint statement: "If current predictions of population growth and patterns of human activity on the planet remain unchanged, science and technology may not be able to prevent either irreversible degradation of the environment or continued poverty for much of the world."

Proponents of this view recognize that population growth is not the only cause of environmental and resource problems. However, they argue that adding several hundred million more people in developed countries and several billion more in developing countries can only intensify existing environmental and social problems.

These analysts believe people should have the freedom to produce as many children as they want. However, such freedom would apply only if it did not reduce the quality of other people's lives now and in the future, either by impairing the earth's ability to sustain life or causing social disruption. They point out that limiting the freedom of individuals to do anything they want, in order to protect the freedom of other individuals, is the basis of most laws in modern societies. What is your opinion on this issue?

How Can Economic Development Help Reduce Birth Rates? Demographers have examined the birth and death rates of western European countries that industrialized during the 19th century. From these data they developed a hypothesis of population change known as the **demographic transition:** As countries become industrialized, first their death rates and then their birth rates decline.

According to this hypothesis, the transition takes place in four stages (Figure 11-18). *First* is the *preindustrial stage,* when there is little population growth because harsh living conditions lead to both a high birth rate (to compensate for high infant mortality) and a high death rate.

Next is the *transitional stage,* when industrialization begins, food production rises, and health care improves. Death rates drop and birth rates remain high, so the population grows rapidly (typically 2.5–3% a year).

During the *third* phase, called the *industrial stage,* the birth rate drops and eventually approaches the death rate as industrialization and modernization become widespread. Population growth continues, but at a slower and perhaps fluctuating rate, depending on economic conditions. Most developed countries are now in this third stage (Figure 11-18), and a few developing countries are entering this stage.

The *last* phase is the *postindustrial stage,* when the birth rate declines further, equaling the death rate and thus reaching zero population growth. Then the birth rate falls below the death rate and total population size decreases slowly. Thirty-eight countries (most of them in Europe) containing about 13% of the world's population have entered this stage.

In most developing countries today, death rates have fallen much more than birth rates (Figure 11-11). In other words, these developing countries are still in the transitional stage, halfway up the economic ladder, with high population growth rates. Some economists believe that developing countries will make the demographic transition over the next few decades without increased family planning efforts.

However, some population analysts fear that the still-rapid population growth in many developing countries will outstrip economic growth and degrade local life-support systems. This could cause many of these countries to be caught in a *demographic trap* at stage 2. This is now happening in a number of developing countries, especially in Africa. Indeed, some of the countries in Africa being ravaged by the HIV/AIDS epidemic are falling back to stage 1 as their death rates rise.

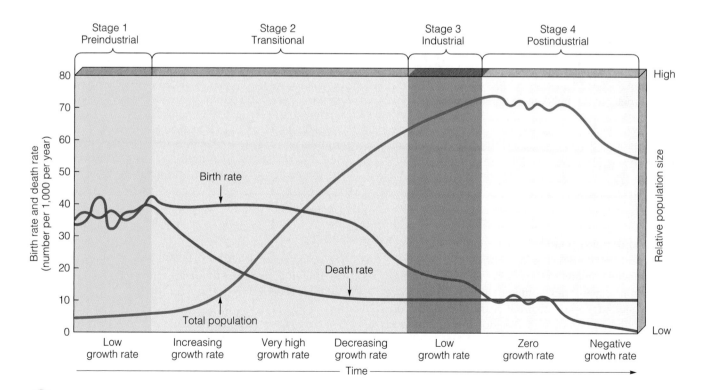

Figure 11-18 Generalized model of the demographic transition.

Moral Implications of Cultural Carrying Capacity

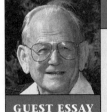

Garrett Hardin

GUEST ESSAY

As a longtime professor of human ecology at the University of California at Santa Barbara, Garrett Hardin made important contributions in relating ethics to biology. He has raised hard ethical questions, sometimes taken unpopular stands, and forced people to think deeply about environmental problems and their possible solutions. He is best known for his 1968 essay "The Tragedy of the Commons," which has had a significant impact on the disciplines of economics and political science and on the management of potentially renewable resources. His 17 books include Filters Against Folly: How to Survive Despite Economists, Ecologists, and the Merely Eloquent; Living Within Limits; *and* The Ostrich Factor: Our Population Myopia.

For many years, Angel Island in San Francisco Bay was plagued with too many deer. A few animals transplanted there in the early 1900s lacked predators and rapidly increased to nearly 300 deer—far beyond the carrying capacity of the island. Scrawny, underfed animals tugged at the heartstrings of Californians, who carried extra food for them from the mainland to the island.

Such well-meaning charity worsened the plight of the deer. Excess animals trampled the soil, stripped the bark from small trees, and destroyed seedlings of all kinds. The net effect was to lower the island's carrying capacity, year by year, as the deer continued to multiply in a deteriorating habitat.

State game managers proposed that the excess deer be shot by skilled hunters. "How cruel!" some people protested. Then the managers proposed that coyotes be introduced onto the island. Though not big enough to kill adult deer, coyotes can kill fawns, thereby reducing the size of the herd. However, the Society for the Prevention of Cruelty to Animals was adamantly opposed to this proposal.

In the end, it was agreed that some deer would be transported to other areas suitable for deer. A total of 203 animals were caught and trucked many miles away. From the fate of a sample of animals fitted with radio col-

lars, it was estimated that 85% of the transported deer died within a year (most of them within 2 months) from various causes: predation by coyotes, bobcats, and domestic dogs, shooting by poachers and legal hunters, and being hit by cars.

The net cost (in 1982 dollars) for relocating each animal surviving for a year was $2,876. The state refused to continue financing the program, and no volunteers stepped forward to pay future bills.

Angel Island is a microcosm of the planet as a whole. Organisms reproduce exponentially, but the environment does not increase at all. The moral is a simple ecological commandment: *Thou shalt not transgress the carrying capacity.*

Now let's examine the situation for humans. A competent physicist has placed global human carrying capacity at 50 billion, about eight times the current world population. Before you give in to the temptation to urge women to have more babies, consider what Robert Malthus said nearly 200 years ago: "There should be no more people in a country than could enjoy daily a glass of wine and piece of beef for dinner."

A diet of grain or bread and water is symbolic of minimum living standards; wine and beef are symbolic of higher living standards that make greater demands on the environment. When land that could produce plants for direct human consumption is used to grow grapes for wine or corn for cattle, more energy is expended to feed the human population. Because carrying capacity is defined as the *maximum* number of animals (humans) an area can support, using part of the area to support such cultural luxuries as wine and beef reduces the carrying capacity. This reduced capacity is called the *cultural carrying capacity,* and it is always smaller than simple carrying capacity.

Energy is the common coin of the realm for all competing demands on the environment. Energy saved by giving up a luxury can be used to produce more food staples and support more people. We could increase the simple carrying capacity of the earth by giving up any (or all) of the following "luxuries": street lighting, vacations, private cars, air conditioning, and artistic performances

Analysts also point out that some of the conditions that allowed developed countries to develop are not available to many of today's developing countries. For example, many developing countries do not have enough skilled workers to produce the high-tech products needed to compete in the global economy and lack the capital and resources needed for rapid economic development.

Another problem is that since 1980 developing countries have experienced a drop in economic assistance from developed countries and a rise in their debt

to such countries. Indeed, since the mid-1980s, developing countries have paid developed countries $40–50 billion a year (mostly in debt interest) more than they have received from these countries.

How Can Family Planning Help Reduce Birth and Abortion Rates and Save Lives? Family **planning** provides educational and clinical services that help couples choose how many children to have and when to have them. Such programs vary from culture to culture, but most provide information on birth

of all sorts. But what we consider luxuries depends on our values as individuals and societies, and values are largely matters of choice. At one extreme, we could maximize the number of human beings living at the lowest possible level of comfort. Or we could try to optimize the quality of life for a much smaller human population.

The carrying capacity of the earth is a scientific question. It may be possible to support 50 billion people at a bread-and-water level. Is that what we choose? The question, "What is the cultural carrying capacity?" requires that we debate questions of value, about which opinions differ.

An even greater difficulty must be faced. So far, we have been treating carrying capacity as a *global* issue, as if there were some global sovereignty capable of enforcing a solution on all people. But there is no global sovereignty ("one world"), nor is there any prospect of one in the foreseeable future. Thus, we must ask how some 200 nations are to coexist in a finite global environment if different sovereignties adopt different standards of living.

Consider a protected redwood forest that produces neither food for humans nor lumber for houses. Because people must travel many kilometers to visit it, the forest is a net loss in the national energy budget. However, for those fortunate enough to wander through the cathedral-like aisles beneath an evergreen vault, a redwood forest does something precious for the human spirit. But then intrudes an appeal from a distant land, where millions are starving because their population has overshot the carrying capacity; we are asked to save lives by sending food. As long as we have surpluses, we may safely indulge in the pleasures of philanthropy. But after we have run out of our surpluses, then what?

A spokesperson for the needy from that land makes a proposal: "If you would only cut down your redwood forests, you could use the lumber to build houses and then grow potatoes on the land, shipping the food to us. Since we are all passengers together on Spaceship Earth, are you not duty bound to do so? Which is more precious, trees or human beings?"

This last question may sound ethically compelling, but let's look at the consequences of assigning a pre-emptive and supreme value to human lives. At least 2 billion people in the world are poorer than the 34 million "legally poor" in America, and their numbers are increasing by about 1 million per year. Unless this increase is halted, sharing food and energy on the basis of need would require the sacrifice of one amenity after another in rich countries. The ultimate result of sharing would be complete poverty everywhere on the earth to maintain the earth's simple carrying capacity. Is that the best humanity can do?

To date, there has been overwhelmingly negative reaction to all proposals to make international philanthropy conditional on the cessation of population growth by overpopulated recipient nations. Foreign aid is governed by two apparently inflexible assumptions:

■ The right to produce children is a universal, irrevocable right of every nation, no matter how hard it presses against the carrying capacity of its territory.

■ When lives are in danger, the moral obligation of rich countries to save human lives is absolute and undeniable.

Considered separately, each of these two well-meaning doctrines might be defensible; taken together, they constitute a fatal recipe. If humanity gives maximum carrying capacity precedence over problems of cultural carrying capacity, the result will be universal poverty and environmental ruin.

Or do you see an escape from this harsh dilemma?

Critical Thinking

1. What population size would allow the world's people to have a good quality of life? What do you believe is the cultural carrying capacity of country where you live? Should your country have a national policy to establish this population size as soon as possible? Explain.

2. Do you support the two principles this essay lists as the basis of foreign aid to needy countries? If not, what changes would you make in the requirements for receiving such aid?

spacing, birth control, and health care for pregnant women and infants.

Here is some *good news* about family planning. It has helped increase the proportion of married women in developing countries who use modern forms of contraception from 10% of married women of reproductive age in the 1960s to 51% of these women in 2003 (40% if China is excluded).

Studies also show that family planning has been responsible for at least 55% of the drop in TFRs in developing countries, from 6 in 1960 to 3.1 in 2003 (3.5 if China is excluded). Family planning has also reduced the number of legal and illegal abortions per year and decreased the risk of death from pregnancy.

Despite such successes, there is also some *bad news. First,* according to John Bongaarts of the Population Council, 42% of all pregnancies in the developing countries are unplanned and 26% end with abortion. *Second,* an estimated 250–350 million family planning women in developing countries want to limit the number and determine the spacing of their children, but they lack access to services. According to the

United Nations, extending family planning services to these women and to those who will soon be entering their reproductive years could prevent an estimated 5.8 million births a year and more than 5 million abortions a year.

How Can Family Planning Be Improved and Expanded? Some analysts call for expanding existing family planning programs to include teenagers and sexually active unmarried women, who are excluded in many existing programs. For teenagers, many advocate much greater emphasis on abstinence.

Another suggestion is to develop programs that educate men about the importance of having fewer children and taking more responsibility for raising them. Proponents also call for greatly increased research on developing new, more effective, and more acceptable birth control methods for men.

Finally, a number of analysts urge pro-choice and pro-life groups to join forces in greatly reducing unplanned births and abortions, especially among teenagers.

According to the United Nations, family planning could be provided in developing countries to all couples who want it for about $17 billion a year, the equivalent of about 8 days' worth of worldwide military expenditures. If developed countries provided one-third of this $17 billion, each person in the developed countries would spend only $4.80 a year to help reduce the world's population.

How Can Empowering Women Help Reduce Birth Rates? Studies show that women tend to have fewer and healthier children and live longer when they have access to education and to paying jobs outside the home and live in societies in which their rights are not suppressed.

Women make up roughly half of the world's population. They do almost all of the world's domestic work and child care and with little or no pay provide more health care than all the world's organized health services combined. They also do 60–80% of the work associated with growing food, gathering fuelwood, and hauling water in rural areas of Africa, Latin America, and Asia. As one Brazilian woman put it, "For poor women the only holiday is when you are asleep." According to some estimates this unpaid labor by women amounts to one-third of the world's annual economic production.

Women work two-thirds of all hours worked but receive only 10% of the world's income and own less than 2% of the world's land. In most developing countries, women do not have the legal right to own land or to borrow money to increase agricultural productivity. Women also make up 70% of the world's poor and two-thirds of the people worldwide who can neither read nor write.

Women's representation in governments has been gradually increasing. However, only 11 of the world's 190 heads of state are women, and women hold just 14% of the seats in the world's parliaments.

Many analysts say that if women everywhere had full legal rights and the opportunity to become educated and earn income outside the home, it would not only slow population growth but also promote human rights and freedom, reduce poverty, and slow environmental degradation. However, empowering women by seeking gender equality will take some major social changes that will be difficult to achieve in male-dominated societies.

How Can Economic Rewards and Penalties Be Used to Help Reduce Birth Rates? Some population experts point out that most couples in developing countries want three or four children. This is well above the replacement-level fertility needed to slow or halt a country's population growth.

These analysts believe we must go beyond family planning and offer economic rewards and penalties to help slow population growth. About 20 countries offer small payments to people who agree to use contraceptives or to be sterilized. However, such payments are most likely to attract people who already have all the children they want.

Some countries, including China, penalize couples who have more than one or two children by raising their taxes, charging other fees, or eliminating income tax deductions for a couple's third child (as is done in Singapore, Hong Kong, and Ghana). Families who have more children than the prescribed limit may also lose health-care benefits, food allotments, and job options.

Research shows that economic rewards and penalties designed to lower birth rates work best if they encourage (rather than coerce) people to have fewer children, reinforce existing customs and trends toward smaller families, do not penalize people who produced large families before the programs were established, and improve a poor family's economic status. Once a country's population growth is out of control, it may be forced to use coercive methods to prevent mass starvation and hardship, as has been the case for China.

11-4 CASE STUDIES: SLOWING POPULATION GROWTH IN INDIA AND CHINA

What Success Has India Had in Controlling Its Population Growth? The world's first national family planning program began in India in 1952, when its population was nearly 400 million. In 2003, after 51 years of population control efforts, India was the world's second most populous country, with a population of 1.1 billion.

In 1952, India added 5 million people to its population. In 2003, it added 17 million. Figure 11-19 compares demographic data for India and China. India is projected to have a larger population than China by 2050, even if China's one-child policy is relaxed.

India faces a number of already serious poverty, malnutrition, and environmental problems that could worsen as its population continues to grow rapidly. By global standards, India's people are poor, with an average per capita GNI PPP of about $2,820 a year; for 30% of the population it is less than $800 a year, or $2.19 a day. Nearly half of India's labor force is unemployed or can find only occasional work.

Some *good news* is that India currently is self-sufficient in food grain production. The *bad news* is that about 40% of its population and 53% of its children suffer from malnutrition, mostly because of poverty.

India faces some serious resource and environmental problems. With 17% of the world's people, it has just 2.3% of the world's land resources and 2% of the world's forests. About half of the country's cropland is degraded as a result of soil erosion, waterlogging, salinization, overgrazing, and deforestation. In addition, about 70% of India's water is seriously polluted and sanitation services often are inadequate.

Without its long-standing family planning program, India's population and environmental problems would be growing even faster. Still, to its supporters the results of the program have been disappointing because of poor planning, bureaucratic inefficiency, the low status of women (despite constitutional guarantees of equality), extreme poverty, and lack of administrative and financial support.

The government has provided information about the advantages of small families for years. Yet Indian women still have an average of 3.1 children. One reason is that most poor couples believe they need many children to do work and care for them in old age. Another is the strong cultural preference for male children, which means some couples keep having children until they produce one or more boys. These factors in part explain why even though 90% of Indian couples know of at least one modern birth control method, only 64% actually use one.

What Success Has China Had in Controlling Its Population Growth? Since 1970, China has made impressive efforts to feed its people and bring its population growth under control. Between 1972 and 2003, China cut its crude birth rate in half and cut its TFR from 5.7 to 1.7 children per woman (Figure 11-19).

To achieve its sharp drop in fertility, China has established the most extensive, intrusive, and strict population control program in the world. Couples are strongly urged to postpone marriage and to have no more than one child. Married couples who pledge to have no more than one child receive extra food, larger pensions, better housing, free medical care, salary

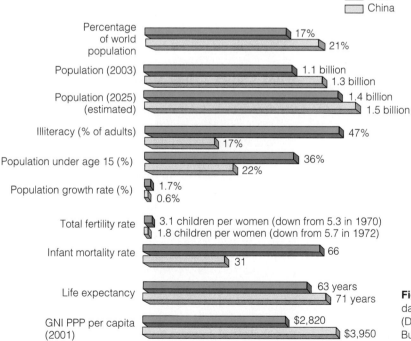

Figure 11-19 Comparison of basic demographic data for India (green) and China (yellow) in 2003. (Data from United Nations and Population Reference Bureau)

bonuses, free school tuition for their one child, and preferential treatment in employment when their child enters the job market. Couples who break their pledge lose all the benefits.

The government also provides married couples with ready access to free sterilization, contraceptives, and abortion. This helps explain why about 83% of married women in China use modern contraception.

Government officials realized in the 1960s that the only alternative to strict population control was mass starvation. China is a dictatorship, and thus, unlike India, it has been able to impose a consistent population policy throughout its society.

China's large and still growing population has an enormous environmental impact that could reduce its ability to produce enough food and threaten the health of many of its people. China has 21% of the world's population but only 7% of its fresh water and cropland, 4% of its forests, and 2% of its oil. Soil erosion in China is serious and apparently getting worse.

In the 1970s, the Chinese government had a system of health clinics that provided moderate health care for its rural farm population. This system collapsed in the 1980s as China embraced capitalist economic reforms. According to government estimates, more than 800 million people—90% of rural Chinese—now have no health insurance or social safety net.

Population experts expect China's population to peak around 2040 and then begin a slow decline. This has led some members of China's parliament to call for amending the country's one-child policy so that some urban couples can have a second child. The goal would be to provide more workers to help support China's aging population.

Perhaps the best lesson for other countries is to act to curb population growth before they must choose between mass starvation and coercive measures that severely restrict human freedom.

11-5 CUTTING GLOBAL POPULATION GROWTH: A NEW VISION

In 1994, the United Nations held its third once-in-a-decade Conference on Population and Development in Cairo, Egypt. One of the conference's goals was to encourage action to stabilize the world's population at 7.8 billion by 2050 instead of the projected 9.3 billion (Figure 11-6). The major goals of the resulting population plan, endorsed by 180 governments, are to do the following by 2015:

- Provide universal access to family planning services and reproductive health care
- Improve the health care of infants, children, and pregnant women

- Encourage development and implementation of national population policies as part of social and economic development policies
- Bring about more equitable relationships between men and women, with emphasis on improving the status of women and expanding education and job opportunities for young women
- Increase access to education, especially for girls and women
- Increase the involvement of men in child-rearing responsibilities and family planning
- Take steps to eradicate poverty
- Reduce and eliminate unsustainable patterns of production and consumption

Some *good news* is that the experience of Japan, Thailand (p. 222), South Korea, Taiwan, Iran, and China indicates that a country can achieve or come close to replacement-level fertility within 15–30 years. Such experience also suggests that the best way to slow population growth is a combination of investing in family planning, reducing poverty, and elevating the status of women.

11-6 POPULATION DISTRIBUTION: URBANIZATION AND URBAN GROWTH

What Is an Urban Area? An **urban,** or **metropolitan, area** often is defined as a town or city plus its adjacent suburban fringes with a population of more than 2,500 people (although some countries set the minimum at 10,000–50,000 residents). A **rural area** usually is defined as an area with a population of less than 2,500 people.

What Causes Urban Growth? A country's **degree of urbanization** is the percentage of its population living in an urban area. **Urban growth** is the rate of increase of urban populations.

Urban areas grow in two ways. One is by *natural increase* (more births than deaths) and the other is by *immigration* (mostly from rural areas).

Each day, about 160,000 people move from rural to urban areas. Such migration to urban areas is influenced by *push factors* that force people out of rural areas and *pull factors* that draw them into the city. People can be *pushed* from rural areas into urban areas by factors such as *poverty, lack of land to grow food, declining agricultural jobs, famine,* and *war.* Agricultural jobs may decline as agriculture is mechanized or government policies set food prices for urban dwellers so low that rural farmers find it uneconomical to grow crops.

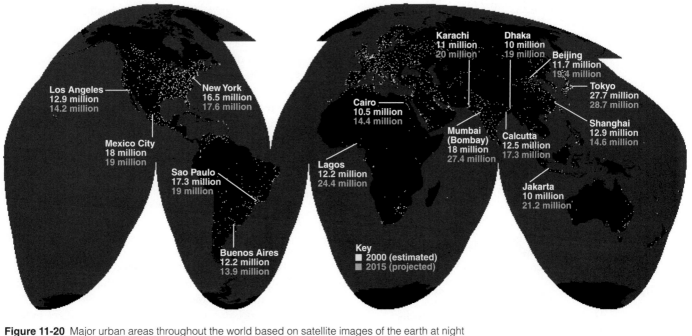

Figure 11-20 Major urban areas throughout the world based on satellite images of the earth at night that show city lights. Currently, the 47% of the world's people living in urban areas occupy about 2% of the earth's land area. Note that most of the world's urban areas are found along the coasts of continents, and most of Africa and much of the interior of South America, Asia, and Australia are dark at night. This figure also shows the populations of 15 of the world's *megacities* with 10 or more million people and their projected populations in 2015. (National Geophysics Data Center, National Oceanic and Atmospheric Administration, and United Nations)

Rural people are *pulled* to urban areas in search of jobs, food, housing, a better life, entertainment, and freedom from religious, racial, and political conflicts.

What Patterns of Urbanization and Urban Growth Are Occurring Throughout the World? Several trends are important in understanding the problems and challenges of urban growth. *First,* the proportion of the global population living in urban areas is increasing. Between 1850 and 2003, the percentage of people living in urban areas increased from 2% to 47% (Figure 11-20). If UN projections are correct, by 2050 about 63% of the world's people will be living in urban areas, with 90% of this urban growth occurring in developing countries.

Second, the number of large cities is mushrooming. In 1900, only 19 cities had a million or more people, and more than 95% of humanity lived in rural communities. In 2003, more than 400 cities had a million or more people (projected to increase to 564 by 2015). Also there were 19 *megacities* (up from 8 in 1985) with 10 million or more people—most of them in developing countries (Figure 11-20).

As they grow and sprawl outward, separate urban areas may merge to form a *megalopolis.* For example, the remaining open space between Boston, Massachusetts, and Washington, D.C., is rapidly urbanizing and coalescing. This 800-kilometer-long (500-mile-long) urban area, sometimes called *Bowash* (Figure 11-21, p. 240), contains almost 60 million people, about twice Canada's entire population.

A *third* trend is that urbanization and the urban population are increasing rapidly in developing countries (Figure 11-20). Currently, about 40% of the people in developing countries live in urban areas. However, 79% of the people in South America live in cities, mostly along the coasts.

Fourth, urban growth is much slower in developed countries (with 75% urbanization) than in developing countries. Still, developed countries are projected to reach 84% urbanization by 2025.

Finally, poverty is becoming increasingly urbanized as more poor people migrate from rural to urban areas. The United Nations estimates that at least 1 billion people live in crowded *slums* (tenements and rooming houses where 3–6 people live in a single room) of central cities and in *squatter settlements* and *shantytowns* (where people build shacks from scavenged building materials) that surround the outskirts of most cities in developing countries.

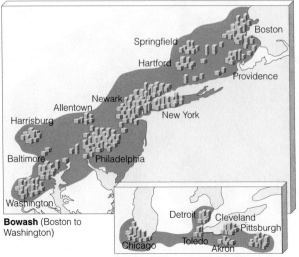

Figure 11-21 Two megalopolises: *Bowash*, consisting of urban sprawl and coalescence between Boston and Washington, D.C., and *Chipitts*, extending from Chicago to Pittsburgh.

Case Study: What Are the Urban Problems of Mexico City? About 18.0 million people—about one of every six Mexicans—live in Mexico City, the world's second most populous city (after Tokyo) (Figure 11-20).

Mexico City suffers from severe air pollution, high unemployment (close to 50%), deafening noise, and a soaring crime rate. More than one-third of its residents live in crowded slums (called *barrios*) or squatter settlements, without running water or electricity.

At least 3 million people have no sewer facilities. This means huge amounts of human waste are deposited in gutters and vacant lots every day, attracting armies of rats and swarms of flies. When the winds pick up dried excrement, a *fecal snow* often falls on parts of the city. This bacteria-laden fallout leads to widespread salmonella and hepatitis infections, especially among children.

Some 3.5 million motor vehicles and 30,000 factories spew pollutants into the atmosphere. Since 1982, the amount of contamination in the city's smog-choked air has more than tripled. Breathing the city's air is said to be roughly equivalent to smoking three packs of cigarettes a day.

The city's air and water pollution cause an estimated 100,000 premature deaths per year. Writer Carlos Fuentes has nicknamed this megacity "Makesicko City."

The Mexican government is industrializing other parts of the country in an attempt to slow migration to Mexico City. Other efforts include banning cars from a 50-block central zone, taking taxis built before 1985 off the streets, having buses and trucks run only on liquefied petroleum gas (LPG), phasing out use of leaded gasoline, and enforcing stricter air pollution emissions standards for industries. The city has also planted more than 25 million trees, bought some land for use as green space, and is considering developing a greatly expanded mass transit system that could help improve air quality.

Some progress has been made. The percentage of days each year in which air pollution standards are violated has fallen from 50% to 20%. However, if the city's population continues to grow its problems are likely to become worse. If you were in charge of Mexico City, what would you do?

How Urbanized Is the United States? Between 1800 and 2003, the percentage of the U.S. population living in urban areas increased from 5% to 79%. This rural-to-urban population shift has taken place in four phases. The first was *migration to large central cities.* Currently, 75% of Americans live in 271 *metropolitan areas* (cities and towns with at least 50,000 people), and nearly half of the country's population lives in consolidated metropolitan areas containing 1 million or more residents (Figure 11-22).

Next, many people *migrated from large central cities to suburbs and smaller cities.* Currently, about 51% of the U.S. population live in suburbs.

Many people also *migrated from the North and East to the South and West.* Since 1980, about 80% of the U.S. population increase has occurred in the South and West, particularly near the coasts. For example, California in the West, with 34.5 million people, is the most populous state—followed by Texas in the Southwest with 21.3 million people. This shift is expected to continue.

In the latest shift, some people *have migrated from urban areas back to rural areas.* This has been going on since the 1970s and especially since 1990.

What Are the Major Urban Problems in the United States? Here is some *good news.* Since 1920, many of the worst urban environmental problems in the United States have been reduced significantly. Most people have better working and housing conditions, and air and water quality have improved.

Also, better sanitation, public water supplies, and medical care have slashed death rates and the prevalence of sickness from malnutrition and transmittable diseases such as measles, diphtheria, typhoid fever, pneumonia, and tuberculosis. Finally, concentrating most of the population in urban areas has helped protect the country's biodiversity by reducing the destruction and degradation of wildlife habitat.

Figure 11-22 Major urban areas in the United States based on satellite images of the earth at night that show city lights (top). About 79% of Americans live in urban areas and occupy about 1.7% of the land area of the lower 48 states. Areas with names in white are the fastest growing metropolitan areas. Nearly half (48%) of Americans live in *consolidated metropolitan areas* with 1 million or more people (bottom). (Data from National Geophysical Data Center/National Oceanic and Atmospheric Administration, U.S. Census Bureau)

Here is some *bad news*. A number of cities in the United States (especially older ones) have *deteriorating services* and *aging infrastructures* (streets, schools, bridges, housing, and sewers). Many also face *budget crunches* (from rising costs as some businesses and people move to the suburbs or rural areas and reduce revenues from property taxes) and *rising poverty in central city areas* (where unemployment typically is 50% or higher).

Another major problem in the United States is **urban sprawl,** the growth of low-density development on the edges of cities and towns that encourages dependence on cars (Figure 11-23, p. 242).

Several factors have promoted urban sprawl in the United States since 1945. Federal government loan guarantees for new single-family homes for World War II veterans stimulated the development of suburbs, and ample land was available for expansion. Low-cost gasoline and federal and state funding of highways have encouraged automobile use and development of once-inaccessible outlying tracts of land. State and local zoning laws requiring large residential lots and separation of residential and commercial use of land in new communities also promote sprawl. Figure 11-23 shows some of the undesirable consequences of urban sprawl.

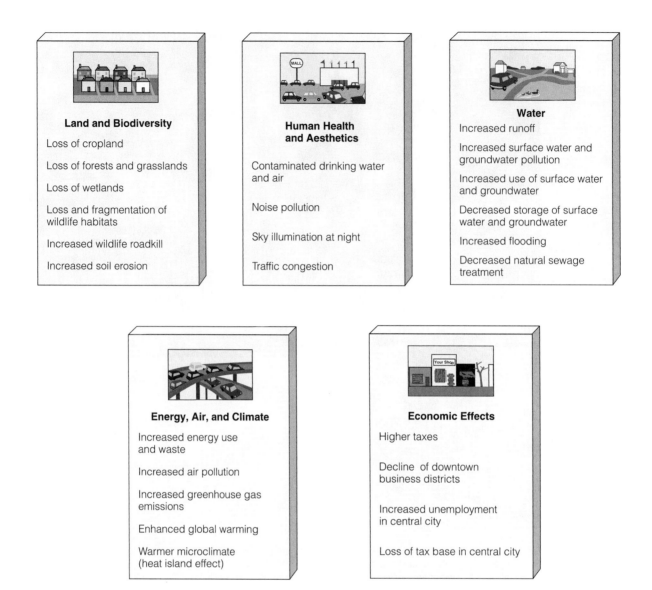

Land and Biodiversity

Loss of cropland

Loss of forests and grasslands

Loss of wetlands

Loss and fragmentation of wildlife habitats

Increased wildlife roadkill

Increased soil erosion

Human Health and Aesthetics

Contaminated drinking water and air

Noise pollution

Sky illumination at night

Traffic congestion

Water

Increased runoff

Increased surface water and groundwater pollution

Increased use of surface water and groundwater

Decreased storage of surface water and groundwater

Increased flooding

Decreased natural sewage treatment

Energy, Air, and Climate

Increased energy use and waste

Increased air pollution

Increased greenhouse gas emissions

Enhanced global warming

Warmer microclimate (heat island effect)

Economic Effects

Higher taxes

Decline of downtown business districts

Increased unemployment in central city

Loss of tax base in central city

Figure 11-23 Some of the undesirable impacts of *urban sprawl* or car-dependent development.

11-7 URBAN RESOURCE AND ENVIRONMENTAL PROBLEMS

What Are the Advantages of Urbanization? Urbanization has many benefits. From an *economic standpoint* cities have spurred economic development, nurtured innovations in science and technology, and served as centers of industry, commerce, transportation, and jobs. Also, the high density of urban populations provides governments and businesses with significant cost advantages in delivering goods and services.

In terms of *health*, urban populations in many parts of the world live longer than do rural populations. They also tend to have lower infant mortality rates and fertility rates. In addition, urban dwellers generally have better access to medical care, family planning, education, and social services than do people in rural areas.

Urban areas also have some *environmental* advantages. For example, recycling is more economically feasible because of the large concentration of recyclable materials and per capita expenditures on environmental protection are higher in urban areas. Also, concentrating people in urban areas helps preserve biodiversity by reducing the stress on wildlife habitats.

What Are the Environmental Disadvantages of Urbanization? Urbanization also has some harmful environmental effects.

In terms of *resource use*, most of the world's cities are not self-sustaining systems because of their high resource input and high waste output (Figure 11-24). Although urban dwellers occupy only about 2% of the earth's land area, they consume about 75% of the earth's resources. In addition, large areas of the earth's land area must be disturbed and degraded (Figure 1-7,

p. 8, and Figure 8-12, p. 169) to provide urban dwellers with food, water, energy, minerals, and other resources. This decreases and degrades the earth's biodiversity. Also as cities expand they destroy rural cropland, fertile soil, forests, wetlands, and wildlife habitats.

Most urban areas lack trees and produce little of the food they use. Most cities have few trees, shrubs, or other plants that absorb air pollutants, give off oxygen, help cool the air as water transpires from their leaves, provide shade, reduce soil erosion, muffle noise, provide wildlife habitats, and give aesthetic pleasure. As one observer remarked, "Most cities are places where they cut down the trees and then name the streets after them."

Many urban areas have water resource problems. For example, many cities have water supply problems. As cities grow and their water demands increase, expensive reservoirs and canals must be built and deeper wells drilled. This can deprive rural and wild areas of surface water and deplete groundwater faster than it is replenished. In urban areas of many developing countries, 50–70% of water is lost or wasted because of leaks and poor management of water distribution systems.

Flooding also tends to be greater in cities. One reason is that many cities are built on floodplain areas or along low-lying coastal areas subject to natural flooding. Another reason is that covering land with buildings, asphalt, and concrete causes precipitation to run off quickly and overload storm drains.

Urban areas experience pollution and health problems. Because of their high population densities and high resource consumption, urban dwellers produce most of the world's air pollution, water pollution, and solid and hazardous wastes. Also, pollution levels normally are higher in urban areas than in rural areas. In addition, high population densities in urban areas can increase the spread of *infectious diseases* (especially if adequate drinking water and sewage systems are not available) and *physical injuries* (mostly from industrial and traffic accidents).

Most urban dwellers are subjected to **noise pollution:** any unwanted, disturbing, or harmful sound that impairs or interferes with hearing, causes stress, hampers concentration and work efficiency, or causes accidents. Noise levels (Figure 11-25, p. 244) above 65 dbA are considered unacceptable, and prolonged exposure to levels above 85 dbA can cause permanent hearing damage.

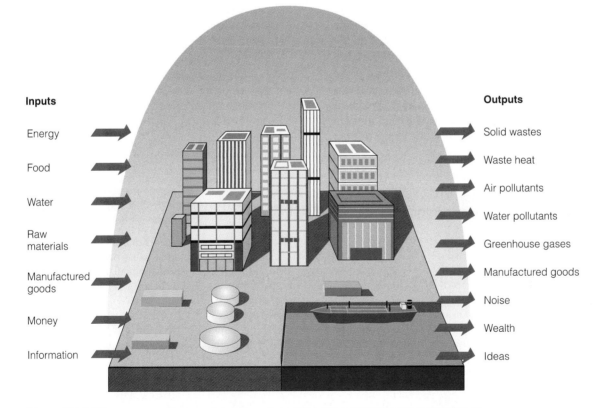

Inputs

Energy

Food

Water

Raw materials

Manufactured goods

Money

Information

Outputs

Solid wastes

Waste heat

Air pollutants

Water pollutants

Greenhouse gases

Manufactured goods

Noise

Wealth

Ideas

Figure 11-24 Urban areas rarely are sustainable systems. The typical city is an open system that depends on other areas for large inputs of matter and energy resources and for large outputs of waste matter and heat. Large areas of nonurban land must be used to supply urban areas with resources. For example, according to an analysis by Mathis Wackernagel and William Rees, 58 times the land area of London is needed to supply its residents with resources. They estimate that meeting the needs of all the world's people at the same rate of resource use as London would take at least three more earths.

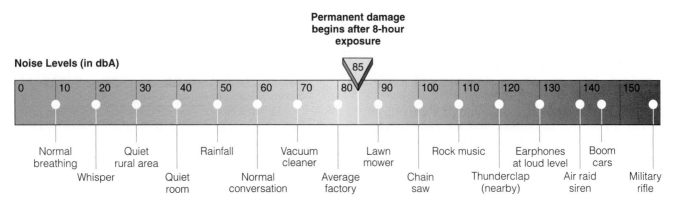

Figure 11-25 *Noise levels* (in decibel-A [dbA] sound pressure units) of some common sounds. Sound pressure becomes damaging at about 75 dbA and painful at around 120 dbA. At 180 dbA it can kill. Because the db and dbA scales are logarithmic, sound pressure is multiplied 10-fold with each 10-decibel rise. Thus a rise from 30 dbA (quiet rural area) to 60 dbA (normal restaurant conversation) represents a 1,000-fold increase in sound pressure on the ear. Ways to control noise include modifying noisy activities and devices to produce less noise, shielding noisy devices or processes, shielding workers from noise, moving noisy operations or machinery away from people, and using anti-noise (a technology that cancels out one noise with another).

You are being exposed to a sound level high enough to cause permanent hearing damage if you need to raise your voice to be heard above the racket, a noise causes your ears to ring, or nearby speech seems muffled.

Cities generally are warmer, rainier, foggier, and cloudier than suburbs and nearby rural areas. The enormous amounts of heat generated by cars, factories, furnaces, lights, air conditioners, and heat-absorbing dark roofs and roads in cities create an *urban heat island* surrounded by cooler suburban and rural areas. As urban areas grow and merge, individual heat islands also merge. This can affect the climate of a large area and keep polluted air from being diluted and cleansed.

Cities can partially counteract the heat island effect in several ways. They can plant trees and use lighter-colored paving, building surfaces, and rooftops to reflect heat away. They can also reduce inputs of waste heat into the atmosphere by establishing high energy-efficiency standards for vehicles, buildings, and appliances. Cooling can also be achieved by establishing gardens on the roofs of large buildings (as is done in parts of the Netherlands, Germany, and Tokyo, Japan).

Finally, urban areas can intensify poverty and social problems. For example, rapid urban growth can increase urban poverty and inequality, which can increase civil unrest and undermine governments. Crime rates also tend be higher in urban areas than in rural areas (Connections, below).

How Can Reducing Crime Help the Environment?

CONNECTIONS

Most people do not realize that reducing crime can also help improve environmental quality. For example, crimes such as robbery, assault, and shootings can have several harmful environmental effects.

It can drive people out of cities, which are our most energy-efficient living arrangements. Every brick in an abandoned urban building represents an energy waste equivalent to burning a 100-watt lightbulb for

12 hours. Each new suburb means replacing farmland or reservoirs of natural biodiversity with dispersed, energy- and resource-wasting roads, houses, and shopping centers.

Crime can also make people less willing to walk, bicycle, and use energy-efficient public transit systems. It also forces many people to use more energy to deter burglars. For example, trees and bushes near a house help save energy by reducing solar heat gain in the summer and providing windbreaks in the winter. However, to help reduce break-ins

many homeowners clear away trees and bushes near their houses. Many homeowners also use more energy by leaving lights, TVs, and radios on to deter burglars.

Finally, the threat of crime causes overpackaging of many items to deter shoplifting or poisoning of food or drug items.

Critical Thinking

Can you think of any environmental benefits of certain types of crimes?

The outward expansion of cities also creates numerous problems for towns in nearby rural areas. Examples are traffic congestion, inability to provide adequate services (such as health care, schools, police, fire, water, and sanitation), and higher taxes to meet the demand for new public services. Some longtime residents are forced out or leave because of rising prices, higher property taxes, decreased environmental quality, and disruption of their way of life.

11-8 TRANSPORTATION AND URBAN DEVELOPMENT

How Do Land Availability and Transportation Systems Affect Urban Development? The two main types of ground transportation are *individual* (such as cars, motor scooters, bicycles, and walking) and *mass* (mostly buses and rail systems). About 90% of all travel in the world is by foot, bicycle, or motor scooter—mostly because only about 10% of the world's people can afford a car.

Land availability is a key factor determining the types of transportation people use. If a city cannot spread outward, it must grow vertically—upward and downward (below ground)—so it occupies a small land area with a high population density. Most people living in such *compact cities* walk, ride bicycles, or use energy-efficient mass transit.

Many European cities and urban areas such as Hong Kong and Tokyo are compact and tend to be more energy efficient than dispersed cities in the United States (Figure 11-21), Canada, and Australia, where ample land often is available for outward expansion.

A combination of cheap gasoline, plentiful land, and a network of highways produces sprawling, automobile-oriented cities with low population density that have a number of undesirable effects (Figure 11-23).

What Is the Role of Motor Vehicles in the United States? With 4.6% of the world's people, the United States has 32% of the world's 700 million cars and trucks (including 555 million cars and 145 million trucks). About two-thirds of the 225 million motor vehicles in the United States are cars and the remaining third are sport utility vehicles (SUVs), pickup trucks, and vans.

In the United States, passenger vehicles are used for 98% of all urban transportation and 91% of travel to work. About 76% of Americans drive to work alone, 4% commute to work on public transit, and 1% bicycle to work. Americans drive 4.3 trillion kilometers (2.7 trillion miles) each year—as far as the rest of the world combined and four times the distance in 1955.

What Are the Advantages and Disadvantages of Motor Vehicles? Motor vehicles have a number of important benefits.

On a personal level, they provide mobility, are convenient for getting from one place to another, and are symbols of power, sex, excitement, social status, and success for many people. They can also serve as islands of privacy in an increasingly hectic world.

From an economic standpoint, much of the world's economy is built on producing motor vehicles and supplying roads, services, and repairs for them. In the United States, for example, $1 of every $4 spent and one of every six nonfarm jobs is connected to the automobile and five of the seven largest U.S. industrial firms produce motor vehicles or their fuel.

Despite their important benefits, motor vehicles have many destructive effects on people and the environment. Motor vehicles have killed almost 18 million people since 1885, when Karl Benz built the first automobile. Throughout the world they kill an estimated 1 million people each year (an average of 2,700 deaths per day) and injure another 15 million people. In the United States, they kill more than 40,000 people a year and injure another 5 million (at least 300,000 of them severely). *More Americans have been killed by cars than have died in all wars in the country's history.* They also kill large numbers of wild animals and pets.

Motor vehicles are the world's largest source of air pollution. They emit six of the seven major air pollutants, which prematurely kill 30,000–60,000 people per year in the United States, according to the Environmental Protection Agency. Motor vehicles are also the fastest growing source of climate-changing emissions of carbon dioxide and now produce almost one-fourth of such emissions. In addition, they account for two-thirds of the oil used in the United States and one-third of the world's total oil consumption.

Motor vehicles have helped create urban sprawl (Figure 11-23). At least a third of urban land worldwide (and half in the United States) is devoted to roads, parking lots, gasoline stations, and other automobile-related uses. This prompted urban expert Lewis Mumford to suggest that the U.S. national flower should be the concrete cloverleaf.

Another problem is congestion. If current trends continue, U.S. motorists will spend an average of 2 years of their lifetime in traffic jams. According to the U.S. General Accounting Office, road congestion in the United States will triple by 2015 even if road capacity is increased by 20% (an unlikely goal). Building more roads may not be the answer. Many analysts agree with the idea, as stated by economist Robert Samuelson, that "cars expand to fill available concrete."

Motor vehicles have harmful economic costs mostly because of deaths and injuries, higher insurance rates, pollution, work time wasted in traffic jams, and decreased property values near roads because of noise and congestion. According to the International Center for Technology Assessment, in the United States such costs amount to $560 billion to $1.7 trillion a year—an average yearly cost of $1,920–5,850 per American.

Is It Feasible to Reduce Automobile Use? Environmentalists and a number of economists suggest that one way to reduce the harmful effects of automobile use is to make drivers pay directly for most of the full costs of automobile use—a *user-pays* approach. One way to phase in such full-cost pricing (p. 24) is to include the estimated harmful costs of driving as a tax on gasoline. Such taxes would amount to about $5–7 per gallon of gasoline in the United States. Another approach is to phase out government subsidies for motor vehicle owners. Such subsidies amount to $1,200–2,400 per vehicle in the United States.

Proponents urge governments to use the gasoline tax revenues and savings from reduced motor vehicle subsidies to lower taxes on income and wages (p. 27) and to help finance mass transit systems, bike paths, and sidewalks. Another way to reduce automobile use and congestion is to raise parking fees and charge tolls on roads, tunnels, and bridges—especially during peak traffic times.

Including the harmful and mostly hidden costs in the market prices of cars, trucks, and gasoline may make economic and environmental sense. Indeed, such an approach has helped reduce car usage in a number of other countries, especially in Europe.

However, most analysts say this approach is not feasible in the United States for several reasons. *First*, it faces strong political opposition from the public (mostly because they are unaware of the huge hidden costs they are already paying) and from powerful transportation-related industries (such as oil and tire companies, road builders, carmakers, and many real estate developers). However, taxpayers might accept sharp increases in gasoline taxes if the extra costs were offset by decreases in taxes on wages and income (p. 27).

A *second* reason is that fast, efficient, reliable, and affordable mass transit options and bike paths are not widely available as alternatives to automobile travel. In addition, the dispersed nature of most U.S. urban areas makes people dependent on the car.

Finally, most people who can afford cars are virtually addicted to them, and most people who cannot afford a car (in the United States and in most other countries) hope to buy one someday.

What Are Alternatives to the Car? Cars are an important form of transportation. However, there are a number of alternatives, each with various advantages and disadvantages. Examples are *bicycles* (Figure 11-26), *motor scooters* (Figure 11-27), *mass transit rail systems in urban areas* (Figure 11-28), *bus systems in urban areas* (Figure 11-29), and *rapid rail systems between urban areas* (Figure 11-30).

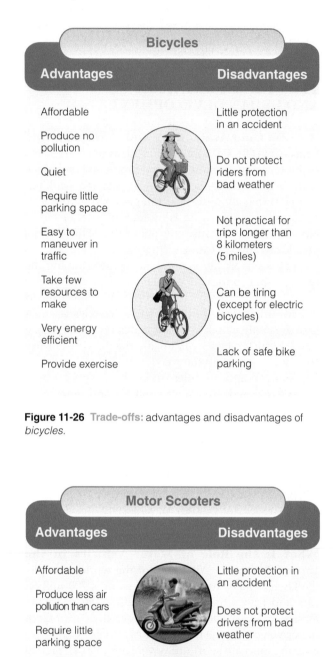

Figure 11-26 Trade-offs: advantages and disadvantages of *bicycles*.

Figure 11-27 Trade-offs: advantages and disadvantages of *motor scooters*.

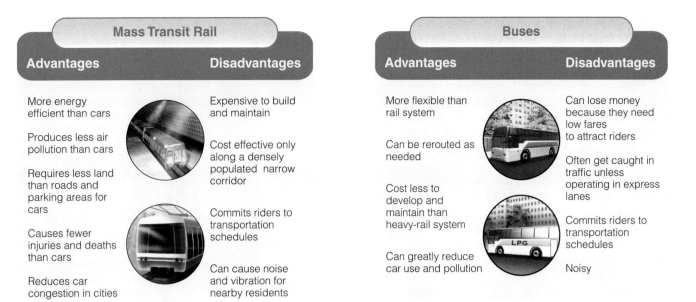

Figure 11-28 Trade-offs: advantages and disadvantages of *mass transit rail systems in urban areas.*

Figure 11-29 Trade-offs: advantages and disadvantages of *bus systems in urban areas.*

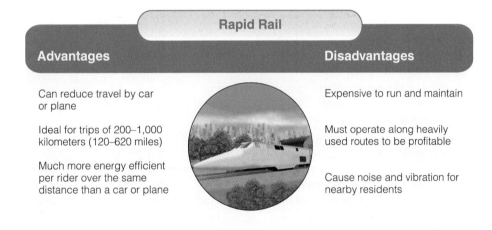

Figure 11-30 Trade-offs: advantages and disadvantages of *rapid rail systems between urban areas.*

11-9 MAKING URBAN AREAS MORE LIVABLE AND SUSTAINABLE

What Is Land-Use Planning? Most urban areas and some rural areas use some form of *land-use planning* to determine the best present and future use of each parcel of land in the area.

Much land-use planning is based on the assumption that considerable future population growth and economic development should be encouraged, regardless of the environmental and other consequences.

Typically this leads to uncontrolled or poorly controlled urban growth and sprawl.

A major reason for this often destructive process in the United States (and some other countries) is that 90% of the revenue that local governments use to provide public services (such as schools, police and fire protection, and water and sewer systems) comes from *property taxes* levied on all buildings and property based on their economic value. Thus local governments often try to raise money by promoting economic growth because they usually cannot raise property taxes enough to meet expanding needs. Typically the long-term result is poorly managed economic growth leading to more environmental degradation.

Environmentalists urge communities to use comprehensive, regional *ecological land-use planning* to anticipate a region's present and future needs and problems. It is a complex process that takes into account geological, ecological, economic, health, and social factors.

Ecological land-use planning sounds good on paper, but is not widely used for several reasons. *One* reason is that local officials seeking reelection every few years usually focus on short-term rather than long-term problems and often can be influenced by economically powerful developers.

Second, officials and the majority of citizens often are unwilling to pay for costly ecological land-use planning and implementation, even though a well-designed plan can prevent or ease many urban problems and save money in the long run.

Third, it is difficult to get municipalities within a region to cooperate in planning efforts. As a result, an ecologically sound development plan in one area may be undermined by unsound development in nearby areas.

Finally, in developing countries, most cities do not have the information or funding to carry out ecological land-use planning. Urban maps often are 20–30 years old and lack descriptions of large areas of cities, especially those growing rapidly because of squatter settlements.

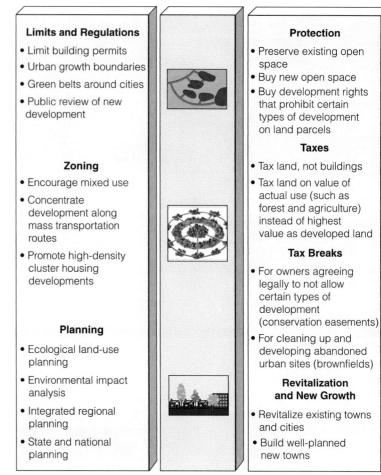

Limits and Regulations
- Limit building permits
- Urban growth boundaries
- Green belts around cities
- Public review of new development

Zoning
- Encourage mixed use
- Concentrate development along mass transportation routes
- Promote high-density cluster housing developments

Planning
- Ecological land-use planning
- Environmental impact analysis
- Integrated regional planning
- State and national planning

Protection
- Preserve existing open space
- Buy new open space
- Buy development rights that prohibit certain types of development on land parcels

Taxes
- Tax land, not buildings
- Tax land on value of actual use (such as forest and agriculture) instead of highest value as developed land

Tax Breaks
- For owners agreeing legally to not allow certain types of development (conservation easements)
- For cleaning up and developing abandoned urban sites (brownfields)

Revitalization and New Growth
- Revitalize existing towns and cities
- Build well-planned new towns

Figure 11-31 Solutions: *smart growth tools* used to prevent and control urban growth and sprawl.

How Is Smart Growth Used to Control Growth and Sprawl? There is growing use of the concept of **smart growth** to encourage development that requires less dependence on cars. It recognizes that urban growth will occur. However, it uses zoning laws and an array of other tools to direct growth to certain areas, discourage urban sprawl, protect ecologically sensitive and important lands and waterways, and develop urban areas that are more environmentally sustainable and more enjoyable places to live. Figure 11-31 lists smart growth tools used to prevent and control urban growth and sprawl.

Some developers, real estate firms, and business interests strongly oppose many of the measures listed in Figure 11-31. They argue that such approaches can hinder economic growth, restrict what private landowners can do with their land, and involve too much regulation by local, state, and federal agencies. However, several studies have shown that most forms of smart growth provide more jobs and spur more economic renewal than conventional economic growth.

Solutions: How Can We Make Urban Areas More Sustainable and More Desirable Places to Live? According to most environmentalists and urban planners, increased urbanization and urban density are better than spreading people out over the countryside, which would destroy more of the planet's biodiversity. To these analysts, the primary problem is not urbanization but our failure to make most cities more sustainable and livable.

Over the next few decades, they call for us to make new and existing urban areas more self-reliant, sustainable, and enjoyable places to live through good ecological design (Guest Essay, p. 250). In a more environmentally sustainable city, called an *ecocity* or *green city*, the demands on the earth's resources are reduced by mimicking the low-throughput circular metabolism of nature (Figure 3-16, p. 61).

In an ecocity, the emphasis is on preventing pollution and reducing waste, using energy and matter resources efficiently, recycling and reusing at least 60% of all municipal solid waste, using solar and other locally available renewable energy resources. Other goals are protecting and encouraging biodiversity, using composting to help create soil and grow food, and using solar-powered living machines and wastewater gardens to treat sewage.

An ecocity is a people-oriented city, not a car-oriented city. Its residents are able to walk or bike to most places (including work) and use low-polluting mass transit.

An ecocity takes advantage of locally available renewable energy sources and requires that all buildings, vehicles, and appliances meet high energy-efficiency standards. Trees and plants adapted to the local climate and soils are planted throughout to provide shade and beauty, supply wildlife habitats, and reduce pollution, noise, and soil erosion. Small organic gardens and a variety of plants adapted to local climate conditions often replace monoculture grass lawns.

Abandoned lots and industrial sites and polluted creeks and rivers are cleaned up and restored. Nearby forests, grasslands, wetlands, and farms are preserved instead of being devoured by urban sprawl. Much of an ecocity's food comes from nearby organic farms, solar greenhouses, community gardens, and small gardens on rooftops, in yards, abandoned lots, and in window boxes. People designing and living in ecocities take seriously the advice Lewis Mumford gave more than

three decades ago: "Forget the damned motor car and build cities for lovers and friends."

The ecocity is not a futuristic dream. Examples of cities working to become more environmentally sustainable and livable include Curitiba, Brazil; Waitakere City, New Zealand; Tapiola, Finland; Portland, Oregon; Davis, California; and Chattanooga, Tennessee.

Solutions: How Has Curitiba, Brazil, Become One of the World's Most Sustainable Major Cities? One of the world's showcase ecocities is Curitiba, Brazil, with a population of 2.3 million. The city is one of Latin America's most livable cities and has a worldwide reputation for its innovative urban planning and environmental protection efforts.

Trees have been planted throughout the city. No tree in the city can be cut down without a permit, and two trees must be planted for each one cut down.

The air is clean because the city is not built around the car and officials have integrated land-use and transportation planning. There are 160 kilometers (100 miles) of bike paths, and more are being built. With the support of shopkeepers, many streets in the downtown shopping district have been converted to pedestrian zones in which no cars are allowed.

Curitiba probably has the world's best bus system. Each weekday a network of clean and efficient buses carries more than 1.9 million passengers—75% of the city's commuters and shoppers—at a low cost (20–40¢ per ride, with unlimited transfers). This privately operated system radiates out from the center city along main routes like spokes in a bicycle wheel along express lanes dedicated to buses (Figure 11-32).

Since 1974, car traffic has declined by 30% even though the city's population has doubled. Two-thirds of all trips in the city are made by bus.

Only high-rise apartment buildings are allowed near major bus routes. Each building must devote the bottom two floors to stores, which reduces the need for residents to travel. Extra-large buses run on popular routes, and tube-shaped bus shelters (Figure 11-32, top right)

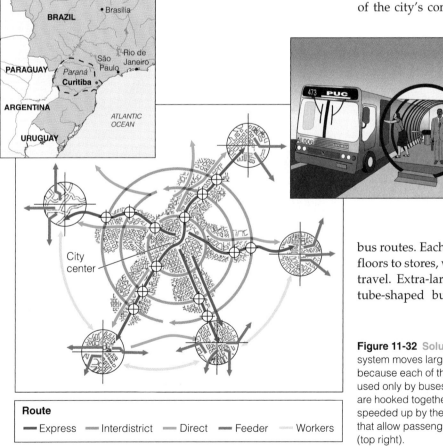

Figure 11-32 Solutions: bus system in Curitiba, Brazil. This system moves large numbers of passengers around rapidly because each of the five major spokes has two express lanes used only by buses. Double- and triple-length bus sections are hooked together and used as needed, and boarding is speeded up by the use of extra-wide doors and raised tubes that allow passengers to pay before getting on the bus (top right).

Route

━ Express ━ Interdistrict ━ Direct ━ Feeder ━ Workers

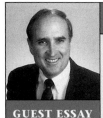

The Ecological Design Arts

David W. Orr

GUEST ESSAY

David W. Orr is chair of environmental studies at Oberlin College in Ohio and one of the nation's most respected environmental educators. He is the author of numerous environmental articles and three books, including Ecological Literacy *and* Earth in Mind. *He is education editor for* Conservation Biology *and a member of the editorial advisory board of* Orion Nature Quarterly. *With help from students, faculty, and townspeople, he used ecological design to build an innovative environmental studies building at Oberlin College.*

If *Homo sapiens sapiens* entered its industrial civilization in an intergalactic design competition, it would be tossed out in the qualifying round. It does not fit. It will not last. The scale is wrong, and even its defenders admit that it's not very pretty. The most glaring design failures of industrial and technologically driven societies are the loss of diversity of all kinds, impending climate change, pollution, and soil erosion.

Of course, industrial civilization was not designed at all; it was mostly imposed by single-minded individuals, armed with one doctrine of human progress or another, each requiring a homogenization of nature and society. These individuals for the most part had no knowledge of ecological design arts.

Good ecological design incorporates understanding about how nature works [Figure 8-14, p. 171] into the ways we design, build, and live in just about anything that directly or indirectly uses energy or materials or governs their use.

When human artifacts and systems are well designed, they are in harmony with the ecological patterns in which they are embedded. When poorly designed, they undermine those larger patterns, creating pollution, higher costs, and social stress.

Good ecological design has certain common characteristics, including correct scale, simplicity, efficient use of resources, a close fit between means and ends, durability, redundancy, and resilience. These characteristics often are place-specific, or, in John Todd's words, "elegant solutions predicated on the uniqueness of place."

Good design also solves more than one problem at a time and promotes human competence, efficient and frugal resource use, and sound regional economies. Where good design becomes part of the social fabric at all levels, unanticipated positive side effects multiply. When people fail to design with ecological competence, unwanted side effects and disasters multiply.

The pollution, violence, social decay, and waste all around us indicate that we have designed things badly, for, I think, three primary reasons. *First,* as long as land and energy were cheap and the world was relatively empty, we did not need to master the discipline of good design. The result was sprawling cities, wasteful economies, waste dumped into the environment, bigger and less efficient automobiles and buildings, and conversion of entire forests into junk mail and Kleenex—all in the name of economic growth and convenience.

Second, design intelligence fails when greed, narrow self-interest, and individualism take over. Good design is a cooperative community process requiring people who share common values and goals that bring them together and hold them together, as has been done in Curitiba, Brazil [p. 249], and Chattanooga, Tennessee. Most American cities, with their extremes of poverty and opulence, are products of people who believe they

where riders pay in advance enhance the system's speed and convenience. As a result, Curitiba has one of Brazil's lowest outdoor air pollution rates.

The city recycles roughly 70% of its paper and 60% of its metal, glass, and plastic, which is sorted by households for collection three times a week. Recovered materials are sold mostly to the city's more than 500 major industries. Litter and graffiti are almost nonexistent because of the civic pride of Curitiba's residents.

The city bought a plot of land downwind of downtown as an industrial park and put in streets, services, housing, and schools. Then it ran a special bus line for workers to the area and enacted stiff air and water pollution control laws. This development has attracted national and foreign corporations and 500 nonpolluting industries that provide one-fifth of the city's jobs. Most of these workers can walk or bike to work from their nearby homes.

To give the poor basic technical training skills needed for jobs, the city set up old buses as roving technical training schools. Each bus gives courses, costing the equivalent of two bus tokens (less than a dollar), for 3 months in a particular area and then moves to another area.

Other retired buses have become classrooms, health clinics, soup kitchens, and some of the city's 200 day-care centers (which are open 11 hours a day and free for low-income parents).

The city's 700,000 poorest residents can swap sorted trash for locally grown vegetables and fruits or bus tokens. They also receive free medical, dental, and child care and 40 feeding centers are available for street children. The city has a *build-it-yourself* system that gives each poor family a plot of land, building materials, two trees, and an hour's consultation with an architect.

In Curitiba, 99.5% of households have electricity and drinking water and 98% have trash collection. About 95% of its citizens can read and write and 83% of the adults have at least a high school education.

have little in common with one another. Greed, suspicion, and fear undermine good community and good design alike.

Third, poor design results from poorly equipped minds. Only people who understand harmony, patterns, and systems can do good design. Industrial cleverness, on the contrary, is evident mostly in the minutiae of things, not in their totality or their overall harmony. Good design requires a breadth of view that causes people to ask how human artifacts and purposes fit within a particular culture and place. It also requires ecological intelligence, by which I mean an intimate familiarity with how nature works [Figure 8-14, p. 171].

An example of good ecological design is found in John Todd's living machines, which are carefully orchestrated ensembles of plants, aquatic animals, technology, solar energy, and high-tech materials to purify wastewater, but without the expense, energy use, and chemical hazards of conventional sewage treatment technology. Todd's living machines resemble greenhouses filled with plants and aquatic animals. Wastewater enters at one end and purified water leaves at the other. In between, an ensemble of organisms driven by sunlight use and remove nutrients, break down toxins, and incorporate heavy metals in plant tissues.

Ecological design standards also apply to the making of public policy. For example, the Clean Air Act of 1970 required car manufacturers to install catalytic converters to remove air pollutants. Three decades later, emissions per vehicle are down substantially, but because more cars are on the road, air quality is about the same—an example of inadequate ecological design. A sounder design approach to transportation would create better access to housing, schools, jobs, stores, and recreation areas, build better public transit systems, restore and improve railroads, and create bike trails and walkways.

An education in the ecological design arts would foster the ability to see things in their ecological context, integrating firsthand experience and practical competence with theoretical knowledge about how nature works. It would equip people to build households, institutions, farms, communities, corporations, and economies that do not emit carbon dioxide or other greenhouse gases, operate on renewable energy, preserve biological diversity, recycle material and organic wastes, and promote sustainable local and regional economies.

The outline of a curriculum in ecological design arts can be found in recent work in ecological restoration, ecological engineering, solar design, landscape architecture, sustainable agriculture, sustainable forestry, energy efficiency, ecological economics, and least-cost, end-use analysis. A program in ecological design would weave these and similar elements together around actual design objectives that aim to make students smarter about systems and about how specific things and processes fit in their ecological context. With such an education we can develop the habits of mind, analytical skills, and practical competence needed to help sustain the earth for us and other species.

Critical Thinking

1. Does your school offer courses or a curriculum in ecological design? If not, suggest some reasons why it does not.

2. Use Orr's ideas about good ecological design to evaluate how well your campus is designed. Suggest ways to improve its design.

Polls show that 99% of the city's inhabitants would not want to live anywhere else.

The entire program is the brainchild of architect and former college teacher Jaime Lerner, an energetic and charismatic leader who has served as the city's mayor three times since the 1970s. Under his leadership, the municipal government has dedicated itself to two principles. *First,* find solutions to problems that are simple, innovative, fast, cheap, and fun. *Second,* establish a government that is honest, accountable, and open to public scrutiny.

The World Bank cites Curitiba as an inspiring example of what can be done to make cities more sustainable and livable through innovative civic leadership and community effort.

The city is not an ecological monstrosity. It is rather the place where both the problems and the opportunities of modern technological civilization are most potent and visible.
PETER SELF

REVIEW QUESTIONS

1. Define the boldfaced terms in this chapter.

2. How did Thailand reduce its birth rate?

3. How is population change calculated? What are the *crude birth rate* and the *crude death rate*? About how many people are added to the world's population each year and each day? How is the *annual rate of population change* calculated? What three countries have the world's largest populations?

4. Distinguish between *replacement-level fertility* and *total fertility rate.* Explain why replacement-level fertility is higher than 2. Explain why reaching replacement-level fertility does not mean an immediate halt in population growth.

5. How have fertility rates and birth rates changed in the United States since 1910? How rapidly is the U.S. population growing?

6. List ten factors that affect birth rates and fertility rates.

7. List five reasons why the world's death rate has declined over the past 100 years.

8. Distinguish between *life expectancy* and *infant mortality rate*. Why is infant mortality the best measure of a society's quality of life? List three factors that keep the U.S. infant mortality higher than it could be.

9. Describe immigration in the United States in terms of numbers. List the major arguments for and against reducing immigration in the United States.

10. What is the *age structure* of a population? Explain why the current age structure of the world's population means the population will keep growing for at least another 50 years even if the replacement-level rate of 2.1 is somehow reached globally tomorrow.

11. Draw the general shape of an age structure diagram for a country undergoing **(a)** rapid population growth, **(b)** moderate population growth, and **(c)** slow or zero population growth.

12. What percentage of population is under age 15 in **(a)** the world, **(b)** developed countries, and **(c)** developing countries? Explain how age structure diagrams can be used to make population and economic projections.

13. What are the benefits and potentially harmful effects of rapid population decline?

14. List the major arguments for and against reducing birth rates globally.

15. What is the *demographic transition,* and what are its four phases? What factors might keep many developing countries from making the demographic transition?

16. What is *family planning,* and what are the advantages of using this approach to reduce the birth rate?

17. Explain how empowering women can help reduce birth rates, poverty, and environmental degradation.

18. What economic rewards and penalties have some countries used to reduce birth rates? What four conditions increase the success of using such economic rewards and penalties?

19. Briefly describe and compare the success China and India have had in reducing their birth rates.

20. List the eight goals of the current UN plan to stabilize the world's population at 7.8 billion by 2050, instead of the projected 9.3 billion.

21. Distinguish between **(a)** an *urban area* and a *rural area* and **(b)** *degree of urbanization* and *urban growth.*

22. List factors that *push* people and *pull* people to migrate from rural areas to urban areas.

23. What are five major trends in urbanization and urban growth? What percentage of the population lives in urban areas in **(a)** the world, **(b)** developed countries, and **(c)** developing countries?

24. Summarize the major urban problems of Mexico City.

25. What four shifts in the U.S. population have taken place since 1800? What percentage of the U.S. population lives in **(a)** urban areas and **(b)** suburban areas?

26. List four pieces of *good news* and four pieces of *bad news* about urban problems in the United States.

27. What is *urban sprawl?* List four factors that have promoted urban sprawl in the United States. List the major harmful effects of urban sprawl.

28. List nine benefits of urbanization.

29. Describe the major resource and environmental problems of urban areas relating to **(a)** sustainability, **(b)** resource use, **(c)** biodiversity preservation, **(d)** land use, **(e)** tree cover, **(f)** food production, **(g)** microclimate (urban heat islands), **(h)** water supply and flooding, **(i)** pollution and human health, **(j)** noise, and **(k)** poverty and social problems.

30. List four ways in which crime can decrease environmental quality.

31. What are the two main types of ground transportation? How does land availability affect urban transportation systems? What are two basic types of cities in terms of population density and transportation systems?

32. Summarize the role that motor vehicles play in the United States.

33. List five personal and economic benefits provided by motor vehicles. List seven harmful effects of motor vehicles in terms of deaths and injuries, air pollution and oil use, land use, congestion, and economic costs.

34. List ways to reduce dependence on the automobile. List three factors promoting dependence on the automobile in the United States.

35. List the major advantages and disadvantages of **(a)** bicycles, **(b)** motor scooters, **(c)** rail systems within urban areas, **(d)** high-speed trains between urban areas, and **(e)** bus systems.

36. What is *land-use planning,* and what are its limitations? What is *smart growth?* List ten tools cities can use to promote smart growth.

37. Describe the major characteristics of a more sustainable *ecocity* or *green city.*

38. Describe how Curitiba, Brazil, has become a global model of a more environmentally sustainable and livable city.

CRITICAL THINKING

1. Why is it rational for a poor couple in a developing country such as India to have four or five children? What changes might induce such a couple to consider their behavior irrational?

2. List what you consider to be a major local, national, or global environmental problem, and describe the role of population growth in this problem.

3. Suppose that all women in the world today began bearing children at replacement-level fertility rates of 2.1 children per woman. Explain why this would not immediately stop global population growth. About how long would it take for population growth to stabilize?

4. Do you believe the population of **(a)** your own country and **(b)** the area where you live is too high? Explain.

5. Evaluate the claims made by those opposing a reduction in births and those promoting a reduction in births, as discussed on p. 232. Which position do you support, and why?

6. Explain why you agree or disagree with each of the following proposals:
   a. The number of legal immigrants and refugees allowed into the United States each year should be reduced sharply.
   b. Illegal immigration into the United States should be decreased sharply. If you agree, how would you go about achieving this?
   c. Families in the United States should be given financial incentives to have more children to prevent eventual population decline.
   d. The United States should adopt an official policy to stabilize its population and reduce unnecessary resource waste and consumption as rapidly as possible.
   e. Everyone should have the right to have as many children as they want.

7. Some people have proposed that the earth could solve its population problem by shipping people off to space colonies, each containing about 10,000 people. Assuming we could build such large-scale, self-sustaining space stations, how many people would have to be shipped off each day to provide living spaces for the 79 million people being added to the earth's population this year? Current space shuttles can handle about 6 to 8 passengers. If this capacity could be increased to 100 passengers per shuttle, how many shuttles would have to be launched per day to offset the 79 million people added this year? According to your calculations, determine whether this proposal is a logical solution to the earth's population problem.

8. Some people believe the most important goal is to sharply reduce the rate of population growth in developing countries, where 97% of the world's population growth is expected to take place. Some people in developing countries agree that population growth in these countries can cause local environmental problems. However, they contend that the most serious environmental problem the world faces is disruption of the global life-support system by high levels of resource consumption per person in developed countries, which use 80% of the world's resources. What is your view on this issue? Explain.

9. Some analysts contend that we have enough food and resources for everyone but these resources are not distributed equitably. To them, poverty, hunger, overpopulation, and environmental degradation are caused mainly by a lack of *social justice*. What is your view on this issue? Explain.

10. If you own a car (or hope to own one), what conditions (if any) would encourage you to rely less on the automobile and travel to school or work by bicycle, on foot, by mass transit, or by a carpool or vanpool?

11. In June 1996, representatives from the world's countries met in Istanbul, Turkey, at the Second UN Conference on Human Settlements (nicknamed the City Summit). One of the areas of controversy was whether housing is a universal *right* (a position supported by most developing countries) or just a *need* (supported by the United States and several other developed countries). What is your position on this issue? Defend your choice.

12. Some analysts suggest phasing out federal, state, and government subsidies that encourage sprawl from building roads, single-family housing, and large malls and superstores. These would be replaced with subsidies that encourage walking and bicycle paths, multifamily housing, high-density residential development, and development with a mix of housing, shops, and offices (mixed-use development). Do you support this approach? Explain.

13. Congratulations! You are in charge of the world. List the three most important features of your **(a)** population policy and **(b)** urban policy.

PROJECTS

1. Assume your entire class (or manageable groups of your class) is charged with coming up with a plan for halving the world's population growth rate within the next 20 years. Develop a detailed plan that would achieve this goal, including any differences between policies in developing countries and developed countries. Justify each part of your plan. Predict what problems you might face in implementing the plan, and devise strategies for dealing with these problems.

2. Prepare an age structure diagram for your community. Use the diagram to project future population growth and economic and social problems.

3. As a class project, **(a)** evaluate land use and land-use planning by your school, **(b)** draw up an improved plan based on ecological principles, and **(c)** submit the plan to school officials.

4. As a class project, use the following criteria to rate your city on a green index from 0 to 10. Are existing trees protected and new ones planted throughout the city? Do you have parks to enjoy? Can you swim in any nearby lakes and rivers? What is the quality of your water and air? Is there an effective noise pollution reduction program? Does your city have a recycling program, a composting program, and a hazardous waste collection program, with the goal of reducing the current solid waste output by at least 60%? Is there an effective mass transit system? Are there bicycle paths? Are all buildings required to meet high energy-efficiency standards? How much of the energy is obtained from locally available renewable resources? Are environmental regulations for existing industry tough enough and enforced well enough to protect citizens? Do local officials look carefully at an industry's environmental record and plans before encouraging it to locate in your city or county? Is ecological planning used to make land-use decisions? Are city officials actively planning to improve the quality of life for all of its citizens? If so, what is their plan?

5. Make a concept map of this chapter's major ideas, using the section heads and subheads and the key terms (in boldface). See material on the website for this book about how to prepare concept maps.

INTERNET STUDY RESOURCES AND RESOURCES FOR FURTHER READING AND RESEARCH

The website for this book contains helpful study aids and many ideas for further reading and research. Log on to

http://biology.brookscole.com/miller10

and click on the Chapter-by-Chapter area. Choose Chapter 11 and select a resource:

■ Flash Cards allows you to test your mastery of the Terms and Concepts to Remember for this chapter.

■ Tutorial Quizzes provides a multiple-choice practice quiz.

■ Student Guide to InfoTrac will lead you to Critical Thinking Projects that use InfoTrac College Edition as a research tool.

■ References lists the major books and articles consulted in writing this chapter.

■ Hypercontents takes you to an extensive list of sites with news, research, and images related to individual sections of the chapter.

INFOTRAC COLLEGE EDITION

Improve your skills with InfoTrac College Edition, a searchable online database of articles from more than 700 periodicals. Log on to

http://www.infotrac-college.com

or access InfoTrac through the website for this book. Try to find the following articles:

1. Brown, R. W. 1999. When we hit the wall: The lesson is likely to be painful. *Free Inquiry* 19: 23. *Keywords:* "world population" and "exponential growth." This article discusses some of the social and ecological implications of the rapidly expanding human population, even though fertility rates are declining.

2. Sheehan, M. O. 2002. What will it take to halt sprawl? *World Watch* 15: 12. *Keyword:* "halt sprawl." Urban sprawl is more than a frustrating commute and rapid development. The implications of sprawl can be far-reaching and most undesirable.

12 AIR AND AIR POLLUTION

When Is a Lichen Like a Canary?

Nineteenth-century coal miners took canaries with them into the mines—not for their songs but for the moment when they stopped singing. Then they knew it was time to get out of the mine because the air contained methane, which could ignite and explode.

Today we use sophisticated equipment to monitor air quality, but living things such as lichens (Figure 12-1) can still warn us of bad air. A lichen consists of a fungus and an alga living together, usually in a mutually beneficial (mutualistic) partnership.

These hardy pioneer species are good air pollution detectors because they are always absorbing air as a source of nourishment. Some lichen species are sensitive to specific air-polluting chemicals. Old man's beard (*Usnea trichodea*) (Figure 12-1, left) and yellow *Evernia* lichens, for example, sicken or die in the presence of too much sulfur dioxide.

Because lichens are widespread, long lived, and anchored in place, they can also help track pollution to its source. The scientist who discovered sulfur dioxide pollution on Isle Royale in Lake Superior, where no car or smokestack had ever intruded, used *Evernia*

lichens to point the finger northward to coal-burning facilities at Thunder Bay, Canada.

In 1986, the Chernobyl nuclear power in Ukraine exploded and spewed radioactive particles into the atmosphere. Some of these particles fell to the ground over northern Scandinavia and were absorbed by lichens that carpet much of Lapland. The area's Saami people depend on reindeer meat for food, and the reindeer feed on lichens. After Chernobyl, more than 70,000 reindeer had to be killed and the meat discarded because it was too radioactive to eat. Scientists helped the Saami identify which of the remaining reindeer to move by analyzing lichens to pinpoint the most contaminated areas.

Last but not least, lichens can replace electronic monitoring stations that cost more than $100,000 each. This is not so much a triumph of nature over technology as a partnership between the two, for technicians use highly sophisticated methods to analyze lichens for pollution and measure their rates of photosynthesis.

We all must breathe air from a global atmospheric commons in which air currents and winds can transport some pollutants long distances (p. 106). Lichens can alert us to the danger. But as with all forms of pollution, the best solution is prevention.

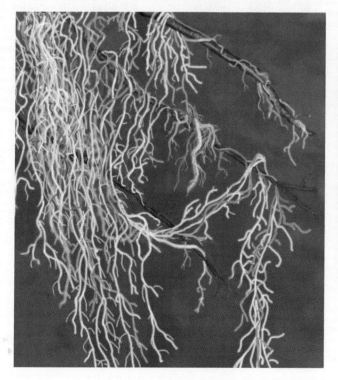

Kenneth W. Fink/Ardea, London

Figure 12-1 Red and yellow crustose lichens growing on slate rock in the foothills of the Sierra Nevada near Merced, California (above), and *Usnea trichodea* lichen growing on a branch of a larch tree in Gifford Pinchot National Park, Washington (left). The vulnerability of various lichen species to specific air pollutants can help researchers detect levels of these pollutants and track down their sources.

*I thought I saw a blue jay this morning. But the smog was so
bad that it turned out to be a cardinal holding its breath.*

MICHAEL J. COHEN

This chapter addresses the following questions:

- What layers are found in the atmosphere?
- What are the major outdoor air pollutants, and
  where do they come from?
- What are two types of smog?
- What is acid deposition, and how can it be reduced?
- What are the harmful effects of air pollutants?
- How can we prevent and control air pollution?

12-1 THE ATMOSPHERE

What Is the Troposphere? Weather Breeder We
live at the bottom of a sea of air called the **atmosphere.**
This sea of life-sustaining gases surrounding the earth
is divided into several spherical layers (Figure 12-2).
Each layer is characterized by abrupt changes in temperature, the result of differences in the absorption of
incoming solar energy.

About 75–80% of the mass of the earth's air is
found in the atmosphere's innermost layer, the **troposphere,** which extends only about 17 kilometers (11
miles) above sea level at the equator and about 8 kilometers (5 miles) over the poles. If the earth were the
size of an apple, this lower layer containing the air we
breathe would be no thicker than the apple's skin. This
thin and turbulent layer of rising and falling air currents and winds is the planet's *weather breeder.*

During several billion years of chemical and biological evolution, the composition of the earth's atmosphere has varied. Today, about 99% of the volume
of the air you inhale consists of two gases: nitrogen
(78%) and oxygen (21%). The remainder consists of
water vapor (varying from 0.01% at the frigid poles to
4% in the humid tropics), slightly less than 1% argon
(Ar), 0.037% carbon dioxide (CO_2), and trace amounts
of several other gases.

**What Is the Stratosphere? Earth's Global Sun-
screen** The atmosphere's second layer is the **stratosphere,** which extends from about 17 to 48 kilometers
(11–30 miles) above the earth's surface (Figure 12-2).
Although the stratosphere contains less matter than
the troposphere, its composition is similar, with two
notable exceptions. *First,* its volume of water vapor is
about 1/1,000 as much. *Second,* its concentration of
ozone (O_3) is much higher (Figure 12-3).

Stratospheric ozone is produced when some of the
oxygen molecules there interact with ultraviolet (UV)
radiation emitted by the sun ($3\ O_2 + UV \rightleftharpoons 2\ O_3$).
This "global sunscreen" of ozone in the stratosphere

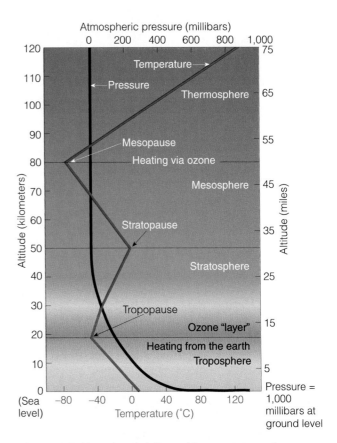

Figure 12-2 *Natural capital:* the earth's current atmosphere
consists of several layers. The average temperature of the atmosphere varies with altitude (red line). The average temperature of the atmosphere at the earth's surface is determined by a
combination of two factors. One is *natural heating* by incoming
sunlight and certain greenhouse gases that release absorbed
energy as heat into the lower troposphere (the natural *greenhouse effect;* Figure 6-9, p. 111). The other is *natural cooling* by
surface evaporation of water and convection processes that
transfer heat to higher altitudes and latitudes (Figure 6-5,
p. 109). Most UV radiation from the sun is absorbed by ozone
(O_3), found primarily in the stratosphere in the *ozone layer*
17–26 kilometers (10–16 miles) above sea level.

keeps about 95% of the sun's harmful UV radiation
from reaching the earth's surface.

This UV filter allows us and other forms of life to
exist on land and helps protect us from sunburn, skin
and eye cancer, cataracts, and damage to our immune
systems. It also prevents much of the oxygen in the
troposphere from being converted to photochemical
ozone, a harmful air pollutant (Figure 12-3).

Evidence indicates that some human activities are
decreasing the amount of beneficial ozone in the stratosphere and *increasing* the amount of harmful ozone in
the troposphere—especially in some urban areas.

12-2 OUTDOOR AIR POLLUTION

**What Are the Major Types and Sources of Air
Pollution?** Air pollution is the presence of one or
more chemicals in the atmosphere in sufficient quanti-

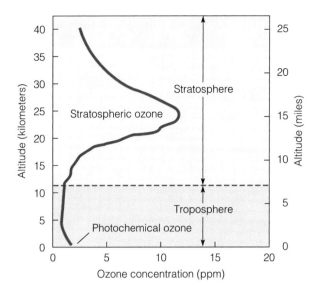

Figure 12-3 Average distribution and concentrations of ozone in the troposphere and stratosphere. *Beneficial ozone* that forms in the stratosphere protects life on earth by filtering out most of the incoming harmful ultraviolet radiation emitted by the sun. *Harmful* or *photochemical ozone* forms in the troposphere when various air pollutants undergo chemical reactions under the influence of sunlight. Ozone in this portion of the atmosphere near the earth's surface damages plants, lung tissues, and some materials such as rubber.

ties and duration to cause harm to us, other forms of life, and materials, or to alter climate. The effects of airborne pollutants range from annoying to lethal. Air pollution is not new (Spotlight, below).

Table 12-1 (p. 258) lists the major classes of pollutants commonly found in outdoor (ambient) air. Such air pollutants come from both natural sources (such as volcanoes and hydrocarbons emitted by some trees) and human (anthropogenic) sources. Most natural sources of outdoor air pollution are spread out. Also, they rarely reach harmful levels, except for those from volcanic eruptions and some forest fires. Most outdoor pollutants in urban areas enter the atmosphere from the burning of fossil fuels in power plants and factories (*stationary sources*) and in motor vehicles (*mobile sources*). Because such pollutants are concentrated in a localized area, they can build up to harmful levels.

Air Pollution in the Past: The Bad Old Days

SPOTLIGHT

Modern civilization did not invent air pollution. It probably began when humans discovered fire and used it to burn wood in poorly ventilated caves.

During the Middle Ages, a haze of wood smoke hung over densely packed urban areas. The industrial revolution brought even worse air pollution from burning coal to power factories and heat homes.

By the 1850s, London had become well known for its "pea soup" fog, a mixture of coal smoke and fog that blanketed the city. In 1880, a prolonged coal fog killed an estimated 2,200 people. Another in 1911 killed more than 1,100 Londoners. The authors of a report on this disaster coined the word *smog* for the deadly mixture of smoke and fog that enveloped the city.

In 1952, an even worse yellow fog lasted for 5 days and killed 4,000 Londoners, prompting Parliament to pass the Clean Air Act of 1956. Additional air pollution disasters in 1956, 1957, and 1962 killed 2,500 more people. Because of strong air

pollution laws, London's air today is much cleaner, and "pea soup" fogs are a thing of the past. Now the major threat is from air pollutants emitted by motor vehicles.

The industrial revolution, powered by coal-burning factories and homes, brought air pollution to the United States. Large industrial cities such as Pittsburgh, Pennsylvania, and St. Louis, Missouri, were known for their smoky air. By the 1940s, the air over some cities was so polluted that people had to use their automobile headlights during the day.

The first documented air pollution disaster in the United States occurred during October 1948, at the town of Donora in Pennsylvania's Monongahela River Valley south of Pittsburgh. Pollutants from the area's industries became trapped in a dense fog that stagnated over the valley for 5 days. About 7,000 of the town's 14,000 inhabitants became sick, and 22 of them died. This killer fog resulted from a combination of mountainous terrain surrounding the valley and weather conditions that trapped and concentrated deadly pollutants emitted by the

community's steel mill, zinc smelter, and sulfuric acid plant.

In 1963, high concentrations of air pollutants accumulated in the air over New York City, killing about 300 people and injuring thousands. Other episodes in New York, Los Angeles, and other large cities in the 1960s led to much stronger air pollution control programs in the 1970s.

Congress passed the original version of the Clean Air Act in 1963, but it did not have much effect until a stronger version of this law was enacted in 1970. Even stricter emission standards were imposed by amendments to the Clean Air Act in 1977 and 1990. Mostly as a result of these laws and actions by states and local areas, the United States has not had any more Donora or New York City incidents.

Critical Thinking

Explain why you agree or disagree with the statement that air pollution in the United States should not be a major concern because of the significant progress in reducing outdoor air pollution since 1970.

Table 12-1 Major Classes of Air Pollutants

Class	Examples
Carbon oxides	Carbon monoxide (CO) and carbon dioxide (CO_2)
Sulfur oxides	Sulfur dioxide (SO_2) and sulfur trioxide (SO_3)
Nitrogen oxides	Nitric oxide (NO), nitrogen dioxide (NO_2), nitrous oxide (N_2O) (NO and NO_2 often are lumped together and labeled NO_x)
Volatile organic compounds (VOCs)	Methane (CH_4), propane (C_3H_8), chlorofluorocarbons (CFCs)
Suspended particulate matter (SPM)	Solid particles (dust, soot, asbestos, lead, nitrate, and sulfate salts), liquid droplets (sulfuric acid, PCBs, dioxins, and pesticides)
Photochemical oxidants	Ozone (O_3), peroxyacyl nitrates (PANs), hydrogen peroxide (H_2O_2), aldehydes
Radioactive substances	Radon-222, iodine-131, strontium-90, plutonium-239 (Table 3-1, p. 57)
Hazardous air pollutants (HAPs), which cause health effects such as cancer, birth defects, and nervous system problems	Carbon tetrachloride (CCl_4), methyl chloride (CH_3Cl), chloroform ($CHCl_3$), benzene (C_6H_6), ethylene dibromide ($C_2H_2Br_2$), formaldehyde (CH_2O_2)

Scientists distinguish between primary and secondary air pollutants in outdoor air. **Primary pollutants** are those emitted directly into the troposphere in a potentially harmful form. While in the troposphere, some of these primary pollutants may react with one another or with the basic components of air to form new pollutants, called **secondary pollutants** (Figure 12-4).

With their large concentrations of cars and factories, cities normally have higher air pollution levels than rural areas. However, prevailing winds can

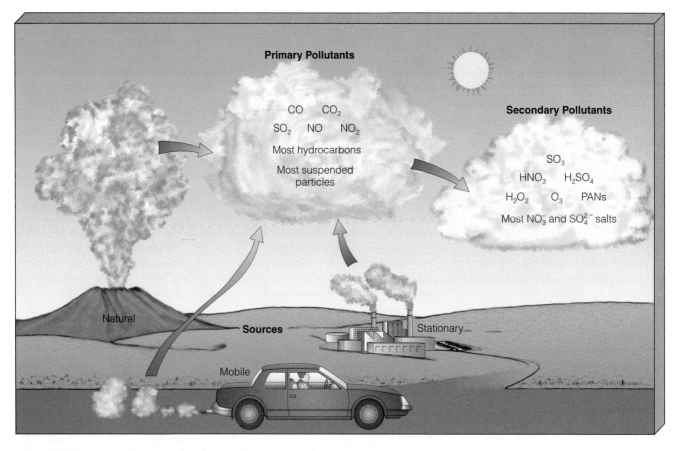

Figure 12-4 Sources and types of air pollutants. Human inputs of air pollutants may come from *mobile sources* (such as cars) and *stationary sources* (such as industrial and power plants). Some *primary air pollutants* may react with one another or with other chemicals in the air to form *secondary air pollutants*.

spread long-lived primary and secondary air pollutants emitted in urban and industrial areas to the countryside and to other downwind urban and rural areas.

Indoor pollutants come from infiltration of polluted outside air and various chemicals used or produced inside buildings, as discussed in more detail in Section 12-5. Experts in risk analysis rate indoor air pollution and outdoor air pollution as high-risk human health problems (Figure 10-15, left, p. 216).

According to the World Health Organization (WHO), more than 1.1 billion people live in urban areas where outdoor air is unhealthy to breathe. Most live in densely populated cities in developing countries where air pollution control laws do not exist or are poorly enforced.

In the United States (and in most other developed countries), government-mandated standards set maximum allowable atmospheric concentrations, or criteria, for six *criteria air pollutants* commonly found in outdoor air (Table 12-2). Some *good news* is that this has helped sharply reduce levels of these pollutants in most of these countries.

Table 12-2 Common Criteria Air Pollutants in the United States*

CARBON MONOXIDE (CO)

Description: Colorless, odorless gas that is poisonous to air-breathing animals; forms during the incomplete combustion of carbon-containing fuels ($2 C + O_2 \longrightarrow 2 CO$).

Major human sources: Cigarette smoking (p. 203), incomplete burning of fossil fuels. About 77% (95% in cities) comes from motor vehicle exhaust.

Health effects: Reacts with hemoglobin in red blood cells and reduces the ability of blood to bring oxygen to body cells and tissues. This impairs perception and thinking; slows reflexes; causes headaches, drowsiness, dizziness, and nausea; can trigger heart attacks and angina; damages the development of fetuses and young children; and aggravates chronic bronchitis, emphysema, and anemia. At high levels it causes collapse, coma, irreversible brain cell damage, and death.

NITROGEN DIOXIDE (NO₂)

Description: Reddish-brown irritating gas that gives photochemical smog its brownish color; in the atmosphere can be converted to nitric acid (HNO_3), a major component of acid deposition.

Major human sources: Fossil fuel burning in motor vehicles (49%) and power and industrial plants (49%).

Health effects: Lung irritation and damage; aggravates asthma and chronic bronchitis; increases susceptibility to respiratory infections such as the flu and common colds (especially in young children and older adults).

Environmental effects: Reduces visibility; acid deposition of HNO_3 can damage trees, soils, and aquatic life in lakes.

Property damage: HNO_3 can corrode metals and eat away stone on buildings, statues, and monuments; NO_2 can damage fabrics.

SULFUR DIOXIDE (SO₂)

Description: Colorless, irritating; forms mostly from the combustion of sulfur-containing fossil fuels such as coal and oil ($S + O_2 \longrightarrow SO_2$); in the atmosphere can be converted to sulfuric acid (H_2SO_4), a major component of acid deposition.

Major human sources: Coal burning in power plants (88%) and industrial processes (10%).

Health effects: Breathing problems for healthy people; restriction of airways in people with asthma; chronic exposure can cause a permanent condition similar to bronchitis. According to the WHO, at least 625 million people are exposed to unsafe levels of sulfur dioxide from fossil fuel burning.

Environmental effects: Reduces visibility; acid deposition of H_2SO_4 can damage trees, soils, and aquatic life in lakes.

Property damage: SO_2 and H_2SO_4 can corrode metals and eat away stone on buildings, statues, and monuments; SO_2 can damage paint, paper, and leather.

SUSPENDED PARTICULATE MATTER (SPM)

Description: Variety of particles and droplets (aerosols) small and light enough to remain suspended in atmosphere for short periods (large particles) to long periods (small particles; Figure 12-8, p. 261); cause smoke, dust, and haze.

Major human sources: Burning coal in power and industrial plants (40%), burning diesel and other fuels in vehicles (17%), agriculture (plowing, burning off fields), unpaved roads, construction.

Health effects: Nose and throat irritation, lung damage, and bronchitis; aggravates bronchitis and asthma; shortens life; toxic particulates (such as lead, cadmium, PCBs, and dioxins) can cause mutations, reproductive problems, cancer.

Environmental effects: Reduces visibility; acid deposition of H_2SO_4 droplets can damage trees, soils, and aquatic life in lakes.

Property damage: Corrodes metal; soils and discolors buildings, clothes, fabrics, and paints.

OZONE (O₃)

Description: Highly reactive, irritating gas with an unpleasant odor that forms in the troposphere as a major component of photochemical smog (Figures 12-3 and 12-6, p. 261).

Major human sources: Chemical reaction with volatile organic compounds (VOCs, emitted mostly by cars and industries) and nitrogen oxides to form photochemical smog (Figure 12-5, p.260).

Health effects: Breathing problems; coughing; eye, nose, and throat irritation; aggravates chronic diseases such as asthma, bronchitis, emphysema, and heart disease; reduces resistance to colds and pneumonia; may speed up lung tissue aging.

Environmental effects: Ozone can damage plants and trees; smog can reduce visibility.

Property damage: Damages rubber, fabrics, and paints.

LEAD

Description: Solid toxic metal and its compounds, emitted into the atmosphere as particulate matter.

Major human sources: Paint (old houses), smelters (metal refineries), lead manufacture, storage batteries, leaded gasoline (being phased out in developed countries).

Health effects: Accumulates in the body; brain and other nervous system damage and mental retardation (especially in children); digestive and other health problems; some lead-containing chemicals cause cancer in test animals.

Environmental effects: Can harm wildlife.

*Data from U.S. Environmental Protection Agency.

12-3 PHOTOCHEMICAL AND INDUSTRIAL SMOG

What Is Photochemical Smog? Brown-Air Smog

A *photochemical reaction* is any chemical reaction activated by light. Air pollution known as **photochemical smog** is a mixture of more than 100 primary and secondary pollutants formed under the influence of sunlight (Figure 12-5).

Its formation begins inside automobile engines and the boilers in coal-burning power and industrial plants. At the high temperatures found there, nitrogen and oxygen in air react to produce colorless nitric oxide (NO). Once in the troposphere, the nitric oxide slowly reacts with oxygen to form nitrogen dioxide (NO_2), a yellowish-brown gas with a choking odor. The NO_2 is responsible for the brownish haze that hangs over many cities during the afternoons of sunny days, explaining why photochemical smog sometimes is called *brown-air smog*.

Hotter days lead to higher levels of ozone and other components of photochemical smog. On a sunny day the photochemical smog (dominated by photochemical ozone, Figure 12-3) builds up to peak levels by early afternoon, irritating people's eyes and respiratory tracts (Figure 12-6).

According to atmospheric chemist Sherwood Rowland, since 1900 the concentration of photochemical ozone near the earth's surface has increased by a factor of 5–8 during summer in the northern hemisphere, 2–4 during the winter in the northern hemisphere, and 2–5 during winter and summer in the southern hemisphere.

All modern cities have photochemical smog, but it is much more common in cities with sunny, warm, dry climates and lots of motor vehicles. Examples of such

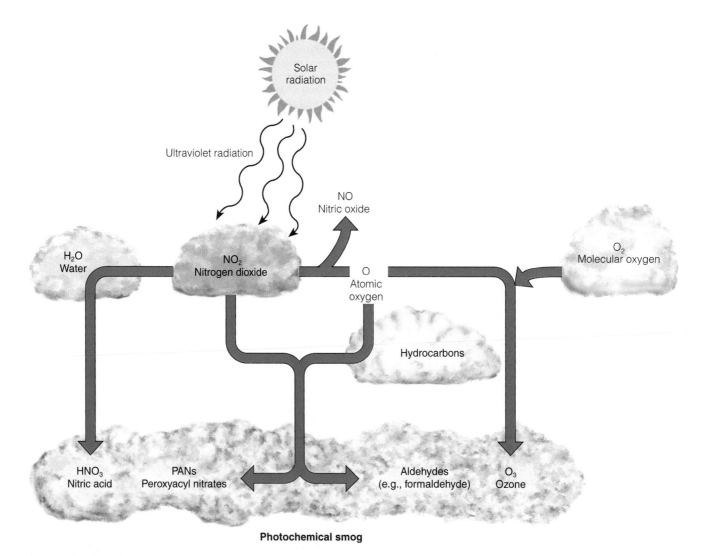

Photochemical smog

Figure 12-5 Simplified scheme of the formation of photochemical smog. The severity of smog generally is associated with atmospheric concentrations of ozone at ground level.

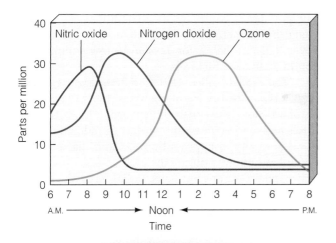

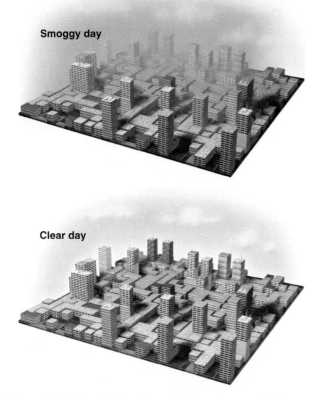

Figure 12-6 Typical daily pattern of changes in concentrations of air pollutants that lead to development of photochemical smog in a city such as Los Angeles, California.

cities are Los Angeles, California; Denver, Colorado; and Salt Lake City, Utah, in the United States, as well as Sydney, Australia; Mexico City, Mexico; São Paulo, Brazil; and Buenos Aires, Argentina.

According to a 1999 article in *Geophysical Research Letters*, if 400 million Chinese drive gasoline-fueled cars by 2050 as projected, the resulting photochemical smog could cover the entire western Pacific in ozone, extending to the United States.

What Is Industrial Smog? Gray-Air Smog Fifty years ago, cities such as London, England, and Chicago and Pittsburgh in the United States burned large amounts of coal and heavy oil (which contain sulfur impurities) in power plants and factories and for space heating. During winter, people in such cities were exposed to **industrial smog** consisting mostly of sulfur dioxide, suspended droplets of sulfuric acid (formed from sulfur dioxide, Figure 12-4), and a variety of suspended solid particles and droplets called aerosols (Figure 12-7).

Some *good news* is that urban industrial smog is rarely a problem today in most developed countries. This has happened because coal and heavy oil are burned only in large boilers with reasonably good pollution control or with tall smokestacks (which

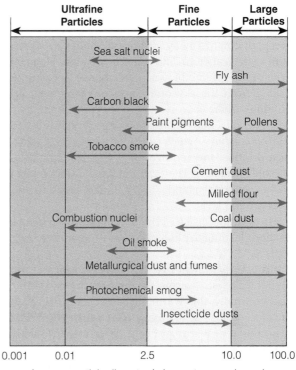

Figure 12-7 Suspended particulate matter consists of particles of solid matter and droplets of liquid that are small and light enough to remain suspended in the atmosphere for short periods (large particles) to long periods (small particles). Suspended particles are found in a wide variety of types and sizes, ranging in diameter from 0.001 micrometer to 100 micrometers (a micrometer, or micron, is one-millionth of a meter, or about 0.00004 inch). Since 1987, the EPA has focused on *fine particles* smaller than 10 microns (known as *PM-10*). In 1997, the agency began focusing on reducing emissions of *ultrafine particles* with diameters less than 2.5 microns (known as *PM-2.5*) because these particles are small enough to reach the lower part of human lungs and contribute to respiratory diseases.

transfer the pollutants to downwind areas). The *bad news* is that industrial smog is a problem in industrialized urban areas of China, India, Ukraine, and some eastern European countries, where large quantities of coal are burned with inadequate pollution controls.

What Factors Influence the Formation of Photochemical and Industrial Smog? The frequency and severity of smog in an area depend on local climate and topography, population density, the amount of industry, and the fuels used in industry, heating, and transportation.

Several natural factors help reduce air pollution. One is *rain and snow,* which help cleanse the air of pollutants. This helps explain why cities with dry climates are more prone to photochemical smog than ones with wet climates.

Another factor is *salty sea spray from the oceans,* which can wash out particulates and other water-soluble pollutants from air that flows from land onto the oceans. A third factor is *winds.* They can help sweep pollutants away, dilute pollutants by mixing them with cleaner air, and bring in fresh air. However, these pollutants are blown somewhere else or are deposited from the sky onto surface waters, soil, and buildings. There is no away.

Several factors can also increase air pollution. One is *urban buildings,* which can slow wind speed and reduce dilution and removal of pollutants. Another is *hills and mountains.* They can reduce the flow of air in valleys below them and allow pollutant levels to build up at ground level.

In addition, *high temperatures* promote the chemical reactions leading to photochemical smog formation.

What Are Temperature Inversions? During daylight, the sun warms the air near the earth's surface. Normally, this warm air and most of the pollutants it contains rise to mix with the cooler air above it. This mixing of warm and cold air creates turbulence, which disperses the pollutants.

Under certain atmospheric conditions, however, a layer of warm air can lie atop a layer of cooler air nearer the ground, a situation known as a **temperature inversion.** Because the cooler air is denser than the warmer air above it, the air near the surface does not rise and mix with the air above it. Pollutants can concentrate in this stagnant layer of cool air near the ground.

There are two types of temperature inversions. *One* is a **subsidence temperature inversion,** which occurs when a large mass of warm air moves into a region at a high altitude and floats over a mass of colder air near the ground. This keeps the air over a city stagnant and prevents vertical mixing and dispersion of air pollutants. Normally such conditions do not last long. But sometimes warmer air masses can remain over cooler air below for days and allow pollutants to build up to harmful levels.

The *second,* called **radiation temperature inversion,** typically occurs at night as the air near the ground cools faster than the air above it. As the sun rises and warms the earth's surface, a radiation inversion normally disappears by noon and disperses the pollutants built up during the night.

Under certain conditions, radiation or subsidence temperature inversions can last for several days and allow pollutants to build up to dangerous concentrations. Areas with two types of topography and weather conditions are especially susceptible to prolonged temperature inversions (Figure 12-8).

One such area is a town or city located in a valley surrounded by mountains that experiences cloudy and cold weather during part of the year (Figure 12-8, top). In such cases, the surrounding mountains along with the clouds block much of the winter sun needed to reverse the nightly radiation temperature inversion, and the mountains block air from moving away laterally. As long as these stagnant conditions persist, concentrations of pollutants in the valley below will build up to harmful and even lethal concentrations. This is what happened during the 1948 air pollution disaster in the valley town of Donora, Pennsylvania (Spotlight, p. 257).

A city with several million people and motor vehicles in an area with a sunny climate, light winds, mountains on three sides, and the ocean on the other has ideal conditions for photochemical smog worsened by frequent subsidence thermal inversions (Figure 12-8, bottom).

This describes California's heavily populated Los Angeles basin, which has prolonged subsidence temperature inversions at least half of the year, mostly during the warm summer and fall. High-pressure air off the coast of California much of the year creates a descending warm air mass that sits atop an air mass below that is cooled by the nearby ocean. When this subsidence thermal inversion persists throughout the day, the surrounding mountains prevent the polluted surface air from being blown away by sea breezes (Figure 12-8, bottom).

12-4 REGIONAL OUTDOOR AIR POLLUTION FROM ACID DEPOSITION

What Is Acid Deposition? Most coal-burning power plants, ore smelters, and other industrial plants in developed countries use tall smokestacks to emit

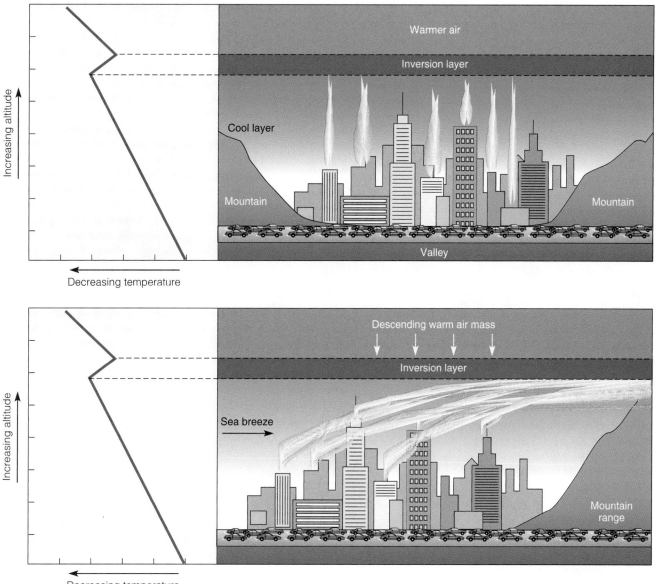

Figure 12-8 Two sets of topography and weather conditions that lead to prolonged *temperature inversions*, in which a warm air layer sits atop a cooler air layer. Air pollutants can build up to harmful levels during an inversion. Top, a *radiation temperature inversion* can occur during cold, cloudy weather in a valley surrounded by mountains. Bottom, frequent and prolonged *subsidence temperature inversions* can occur in an area with a sunny climate, light winds, mountains on three sides, and the ocean on the other. A layer of descending warm air from a high-pressure system prevents ocean-cooled air near the ground from ascending enough to disperse and dilute pollutants.

sulfur dioxide, suspended particles, and nitrogen oxides above the inversion layer (Figure 12-8, top), where mixing, dilution, and removal by wind are more effective. The *good news* is that these tall smokestacks reduce *local* air pollution.

The *bad news* is that this can increase *regional* air pollution downwind. The primary pollutants, sulfur dioxide and nitrogen oxides, emitted into the atmo-

sphere above the inversion layer are transported as much as 1,000 kilometers (600 miles) by prevailing winds. During their trip, they form secondary pollutants such as nitric acid vapor, droplets of sulfuric acid, and particles of acid-forming sulfate and nitrate salts (Figure 12-4).

These acidic substances remain in the atmosphere for 2–14 days, depending mostly on prevailing winds,

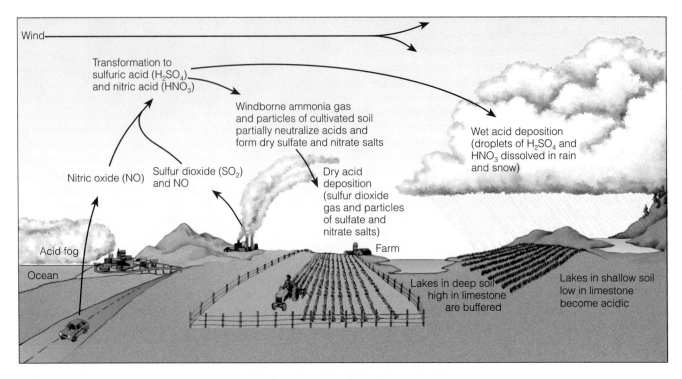

Figure 12-9 *Acid deposition*, which consists of rain, snow, dust, or gas with a pH lower than 5.6, is commonly called acid rain. Soils and lakes vary in their ability to buffer or remove excess acidity.

precipitation, and other weather patterns. During this period they descend to the earth's surface in two forms. One is *wet deposition* as acidic rain, snow, fog, and cloud vapor with a pH less than 5.6 (Figure 9-20, p. 192). The other is *dry deposition* as acidic particles. The resulting mixture is called **acid deposition** (Figure 12-9), sometimes termed *acid rain*. Most dry deposition occurs within about 2–3 days fairly near the emission sources, whereas most wet deposition takes place in 4–14 days in more distant downwind areas.

What Areas Are Most Affected by Acid Deposition? Acid deposition is a regional problem in the eastern United States and in other parts of the world (Figure 12-10). Most of these regions are downwind from coal-burning power plants, smelters, or factories or are major urban areas with large numbers of motor vehicles.

In the United States, coal-burning power and industrial plants in the Ohio Valley emit the largest quantities of sulfur dioxide and other acidic pollutants. Mostly as a result of these emissions, typical precipitation in the eastern United States has a pH of 4.2–4.7. This is 10 times higher than the acidity of natural precipitation, which has a pH of 5.6. Some mountaintop forests in the eastern United States and east of Los Angeles, California, are bathed in fog and dews as acidic as lemon juice, with a pH of 2.3—about 1,000 times the acidity of normal precipitation. According to

the U.S. Environmental Protection Agency (EPA), about 66% of the SO_2 and 25% of the NO_x that are the primary causes of acid deposition come from coal-burning power plants.

In some areas, soils contain basic compounds (such as calcium carbonate or limestone) that can neutralize, or *buffer*, some inputs of acids. Two types of areas are especially sensitive to acid deposition. One consists of areas with thin acidic soils derived mostly from granite rock without such natural buffering (Figure 12-10, green and most red areas). The other is areas where the buffering capacity of soils has been depleted by decades of acid deposition (some red areas in Figure 12-10).

Many acid-producing chemicals generated by power plants, factories, smelters, and cars in one country are exported to other countries by prevailing winds. Here are some examples. Acidic emissions from industrialized areas of western Europe (especially the United Kingdom and Germany) and eastern Europe blow into Norway, Switzerland, Austria, Sweden, the Netherlands, and Finland. Some SO_2 and other emissions from coal-burning power and industrial plants in the Ohio Valley of the United States end up in southeastern Canada (Figure 12-10). Some acidic emissions in China end up in Japan and North and South Korea.

The worst acid deposition is in Asia, especially in China, which gets about 59% of its energy from burn-

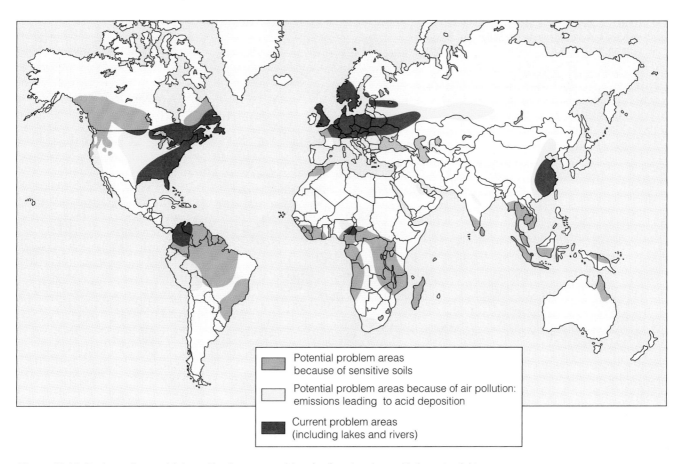

Figure 12-10 Regions where acid deposition is now a problem (red) and regions with the potential to develop this problem (yellow and green). Such regions have large inputs of air pollution (mostly from power plants, industrial plants, and ore smelters), or are sensitive areas with soils and bedrock that cannot neutralize (buffer) inputs of acidic compounds (green areas and most red areas). (Data from World Resources Institute and U.S. Environmental Protection Agency)

Legend:
- Potential problem areas because of sensitive soils
- Potential problem areas because of air pollution: emissions leading to acid deposition
- Current problem areas (including lakes and rivers)

ing coal. Scientists estimate that by 2025 China will emit more sulfur dioxide and carbon dioxide than the United States, Canada, and Japan combined.

In 1999, researchers from California's Scripps Institution of Oceanography found a thick brown haze of air pollution covering much of the Indian Ocean during winter. The haze-covered area is about the size of the continental United States and rises as high as 3,000 meters (10,000 feet). During the late spring and summer, when prevailing winds reverse, some of the haze can be blown back onto the land, where it can combine with monsoon rains and fall as acid deposition.

What Are the Effects of Acid Deposition on Human Health, Materials, and the Economy? Acid deposition has a number of harmful effects. It contributes to human respiratory diseases such as bronchitis and asthma, can leach toxic metals (such as lead and copper) from water pipes into drinking wa-

ter, and damages statues, buildings, metals, and car finishes.

Acid deposition also decreases atmospheric visibility, mostly because of the sulfate particles it contains. It can also lower profits and cause job losses because of lower productivity in fisheries, forests, and farms and loss of tourism in scenic areas.

What Are the Effects of Acid Deposition on Aquatic Ecosystems? Acid deposition has many harmful ecological effects when the pH of most aquatic systems falls below 6 and especially below 5 (Figure 12-11, p. 266). One effect is loss of essentially all fish populations below a pH of 4.5 (Figure 12-11). Another is the release of aluminum ions (Al^{3+}) attached to minerals in nearby soil into lakes, where they can kill many kinds of fish by stimulating excessive mucus formation. This asphyxiates the fish by clogging their gills.

Much of the damage to aquatic life in sensitive areas with little buffering capacity (Figure 12-10) is a

Figure 12-11 Fish and other aquatic organisms vary in their sensitivity to acidity. The figure shows the lowest pH (highest acidity) at which various species can survive. Note that the greatest effects occur when the pH drops below 5.5.

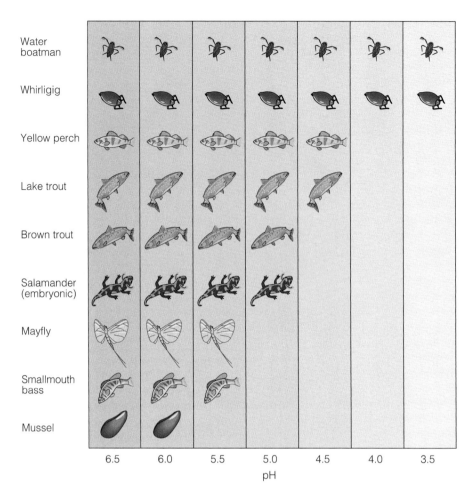

result of *acid shock.* This is caused by the sudden runoff of large amounts of highly acidic water and aluminum ions into lakes and streams when snow melts in the spring or after unusually heavy rains.

Because of excess acidity, at least 16,000 lakes in Norway and Sweden contain no fish, and 52,000 more lakes there have lost most of their acid-neutralizing capacity. In Canada, some 14,000 acidified lakes contain few if any fish, and some fish populations in 150,000 more lakes are declining because of increased acidity. In the United States, about 9,000 lakes (most in the Northeast and upper Midwest) are threatened with excess acidity, one-third of them seriously.

What Are the Effects of Acid Deposition on Plants and Soils? Acid deposition (often along with other air pollutants such as ozone) can harm forests and crops, especially when the soil pH falls below 5.1. The effects of acid deposition on trees and other plants are partly caused by chemical interactions in forest and cropland soils (Figure 12-12).

Direct damage to leaves and needles is one harmful effect. Another results from the leaching of essential plant nutrients such as calcium and magnesium salts from soils. This reduces plant productivity and the ability of the soils to buffer or neutralize acidic inputs.

Acid deposition can also release aluminum ions (Al^{3+}) attached to insoluble soil compounds, which can hinder uptake and use of soil nutrients and water by plants. An input of acids can also dissolve insoluble soil compounds and release ions of metals such as lead, cadmium, and mercury. These ions can be absorbed by plants and are highly toxic to plants and animals.

Another problem is that acid deposition can promote growth of acid-loving mosses that can kill trees. These mosses can hold enough water to drown tree roots and can kill fungi that help tree roots absorb nutrients. Acid deposition also weakens trees and other plants so they become more susceptible to other types of damage such as severe cold, diseases, insect attacks, drought, and harmful mosses.

It is important to note that *most of the world's forests and lakes are not being destroyed or seriously harmed by acid deposition.* Instead, acid deposition is a regional problem that can harm forests and lakes downwind from two types of sources. One consists of power plants, smelters, and industrial plants that burn coal without

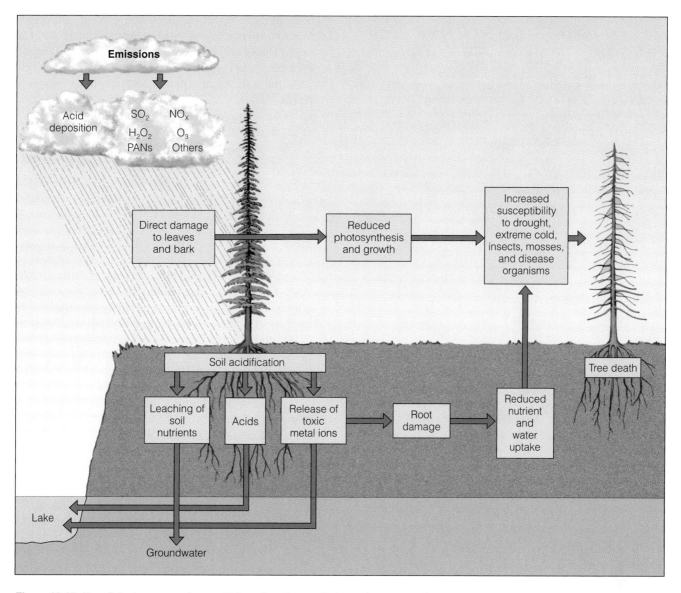

Figure 12-12 Air pollutants are one of several interacting stresses that can damage, weaken, or kill trees and pollute surface water and groundwater.

modern air pollution equipment. The other is large car-dominated cities without adequate controls on emissions of nitrogen oxides (NOx) from motor vehicles.

Mountaintop forests are the terrestrial areas hardest hit by acid deposition. This occurs because these areas tend to have thin soils without much buffering capacity, and trees on mountaintops (especially conifers such as red spruce and balsam fir that keep their leaves year-round) are bathed almost continuously in very acidic fog and clouds.

How Serious Is Acid Deposition in the United States? Since 1980, the federal government has sponsored the National Acid Precipitation Assessment Program (NAPAP). Its goals are to coordinate govern-

ment acid deposition research, and assess the costs, benefits, and effectiveness of the country's acid deposition legislation and control program.

According to the NAPAP and other researchers, the *good news* is that the 1990 amendments to the Clean Air Act have helped reduce some of the harmful impacts of acid deposition in the United States. However, the *bad news* is that there is still a long way to go.

Solutions: What Can Be Done to Reduce Acid Deposition? Figure 12-13 (p. 268) summarizes ways to reduce acid deposition. According to most scientists studying acid deposition, the best solutions are *prevention approaches* that reduce or eliminate emissions of SO_2, NO_x, and particulates.

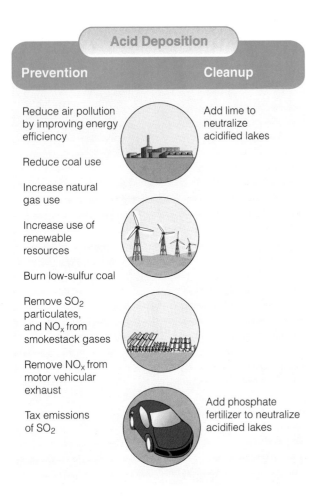

Acid Deposition

Prevention	Cleanup

Reduce air pollution by improving energy efficiency

Reduce coal use

Increase natural gas use

Increase use of renewable resources

Burn low-sulfur coal

Remove SO_2 particulates, and NO_x from smokestack gases

Remove NO_x from motor vehicular exhaust

Tax emissions of SO_2

Add lime to neutralize acidified lakes

Add phosphate fertilizer to neutralize acidified lakes

Figure 12-13 Solutions: methods for reducing acid deposition and its damage.

Controlling acid deposition is a difficult political problem for three reasons. *One* is that the people and ecosystems it affects often are quite distant from those who cause the problem. *Second,* countries with large supplies of coal (such as China, India, Russia, and the United States) have a strong incentive to use it as a major energy resource. *Third,* owners of coal-burning power plants say the costs of adding equipment to reduce air pollution, using low-sulfur coal, or removing sulfur from coal are too high and would increase the cost of electricity for consumers.

Environmentalists respond that much cleaner and affordable ways are available to produce electricity—including wind turbines and burning natural gas in turbines. They also point out that the largely hidden health and environmental costs of burning coal are twice its market cost. They urge inclusion of these costs in the price of producing electricity from coal (full-cost pricing, p. 24). Then consumers would have more realistic knowledge about the effects of using coal to produce electricity.

Large amounts of limestone or lime can be used to neutralize acidified lakes or surrounding soil—the

only cleanup approach now being used. However, there are several problems with liming. One is that it is an expensive and temporary remedy that usually must be repeated annually. Also, it can kill some types of plankton and aquatic plants and can harm wetland plants that need acidic water. Finally, it is difficult to know how much of the lime to put where (in the water or at selected places on the ground).

Researchers in England found that adding a small amount of phosphate fertilizer can neutralize excess acidity in a lake. The effectiveness of this approach is still being evaluated.

12-5 INDOOR AIR POLLUTION

What Are the Types and Sources of Indoor Air Pollution? If you are reading this book indoors, you may be inhaling more air pollutants with each breath than if you were outside (Figure 12-14). According to EPA studies, in the United States levels of 11 common pollutants generally are two to five times higher inside homes and commercial buildings than outdoors and as much as 100 times higher in some cases.

The EPA also found that levels of fine particles (Figure 12-7), which can contain toxins and metals such as lead and cadmium, can be as much as 60% higher indoors than outdoors. Another finding is that concentrations of several pesticides (such as chlordane), approved for outdoor use only, were 10 times greater inside than outside monitored homes (some coming from pesticide dust tracked in on shoes). According to other EPA studies, pollution levels inside cars in traffic-clogged urban areas in the United States can be up to 18 times higher than those outside the vehicles.

The health risks from exposure to such chemicals are magnified because people typically spend 70–98% of their time indoors or inside vehicles. In 1990, the EPA placed indoor air pollution at the top of the list of 18 sources of cancer risk, and it is rated by risk analysis scientists as a high-risk health problem for humans (Figure 10-15, left, p. 216).

The EPA estimates that more than 3,000 cases of cancer per year in the United States may be caused by exposure to indoor air pollutants. At greatest risk are smokers, infants and children under age 5, the old, the sick, pregnant women, people with respiratory or heart problems, and factory workers. According to the EPA and the American Lung Association, air pollution in the United States costs at least $150 billion annually in health care and lost work productivity, with $100 billion of that caused by indoor air pollution.

Danish and U.S. EPA studies have linked pollutants found in buildings to dizziness, headaches, coughing, sneezing, nausea, burning eyes, chronic

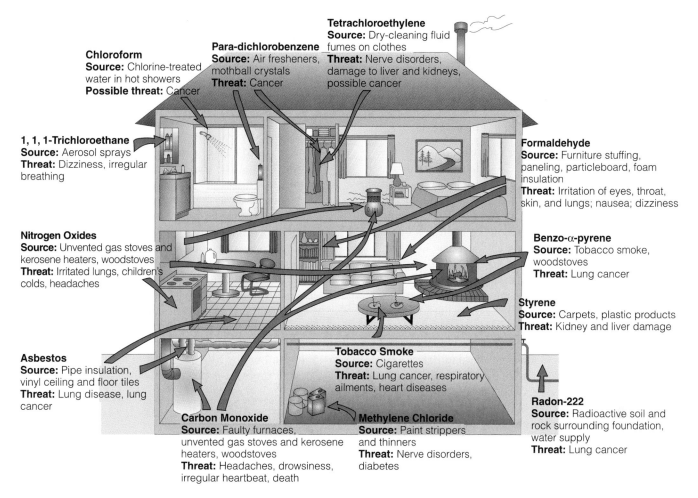

Chloroform
Source: Chlorine-treated water in hot showers
Possible threat: Cancer

Para-dichlorobenzene
Source: Air fresheners, mothball crystals
Threat: Cancer

Tetrachloroethylene
Source: Dry-cleaning fluid fumes on clothes
Threat: Nerve disorders, damage to liver and kidneys, possible cancer

1, 1, 1-Trichloroethane
Source: Aerosol sprays
Threat: Dizziness, irregular breathing

Formaldehyde
Source: Furniture stuffing, paneling, particleboard, foam insulation
Threat: Irritation of eyes, throat, skin, and lungs; nausea; dizziness

Nitrogen Oxides
Source: Unvented gas stoves and kerosene heaters, woodstoves
Threat: Irritated lungs, children's colds, headaches

Benzo-α-pyrene
Source: Tobacco smoke, woodstoves
Threat: Lung cancer

Styrene
Source: Carpets, plastic products
Threat: Kidney and liver damage

Asbestos
Source: Pipe insulation, vinyl ceiling and floor tiles
Threat: Lung disease, lung cancer

Tobacco Smoke
Source: Cigarettes
Threat: Lung cancer, respiratory ailments, heart diseases

Carbon Monoxide
Source: Faulty furnaces, unvented gas stoves and kerosene heaters, woodstoves
Threat: Headaches, drowsiness, irregular heartbeat, death

Methylene Chloride
Source: Paint strippers and thinners
Threat: Nerve disorders, diabetes

Radon-222
Source: Radioactive soil and rock surrounding foundation, water supply
Threat: Lung cancer

Figure 12-14 Some important indoor air pollutants. (Data from U.S. Environmental Protection Agency)

fatigue, and flu-like symptoms, known as the *sick-building syndrome.* New buildings are more commonly "sick" than old ones because of reduced air exchange (to save energy) and chemicals released from new carpeting and furniture. EPA studies indicate that at least 17% of the 4 million commercial buildings in the United States are considered "sick" (including EPA headquarters).

According to the EPA and public health officials, the four most dangerous indoor air pollutants are *cigarette smoke* (p. 203), *formaldehyde, radioactive radon-222 gas,* and *very small fine and ultrafine particles* (Figure 12-7).

Worker exposure to asbestos fibers in mines and in factories making asbestos material also is a serious indoor air pollution problem, especially in developing countries. A number of research studies on laboratory animals have also identified tiny fibers of *fiberglass* as a widespread and potentially potent carcinogen in indoor air.

In developing countries, the burning of wood, dung, crop residues, and coal in open fires or in un-

vented or poorly vented stoves for cooking and heating exposes inhabitants (especially women and young children) to very high levels of particulate air pollution. According to the United Nations, inhaling fumes from indoor cooking fires prematurely kills an estimated 2.5 million women and children a year.

Case Study: Is Your Home Contaminated with Formaldehyde? The chemical that causes most people difficulty is *formaldehyde,* a colorless, extremely irritating gas widely used to manufacture common household materials.

According to the EPA, as many as 20 million Americans suffer from chronic breathing problems, dizziness, rash, headaches, sore throat, sinus and eye irritation, wheezing, and nausea caused by daily exposure to low levels of formaldehyde emitted (outgassed) from common household materials.

There are many sources of formaldehyde. They include building materials (such as plywood, particleboard, paneling, and high-gloss wood used in floors and cabinets), furniture, drapes, upholstery, adhesives in carpeting and wallpaper, urethane-formaldehyde insulation, fingernail hardener, and wrinkle-free coating on permanent-press clothing (Figure 12-14). The EPA estimates that as many as 1 of every 5,000 people who live in manufactured homes for more than 10 years will develop cancer from formaldehyde exposure.

Case Study: Are You Being Exposed to Radioactive Radon Gas? Radon-222—a naturally occurring radioactive gas that you cannot see, taste, or smell—is produced by the radioactive decay of uranium-238. Most soil and rock contain small amounts of uranium-238. But this isotope is much more concentrated in underground deposits of minerals such as uranium, phosphate, granite, and shale.

When radon gas from such deposits seeps upward through the soil and is released outdoors, it disperses quickly in the atmosphere and decays to harmless levels. However, radon gas can enter buildings above such deposits through cracks in foundations and walls, openings around sump pumps and drains, and hollow concrete blocks (Figure 12-15). Once inside, it can build up to high levels, especially in unventilated lower levels of homes and buildings.

Radon-222 gas quickly decays into solid particles of other radioactive elements that, if inhaled, expose lung tissue to a large amount of ionizing radiation from alpha particles. This exposure can damage lung tissue and lead to lung cancer over the course of a 70-year lifetime. Your chances of getting lung cancer from radon depend mostly on how much radon is in your home, how much time you spend in your home, and whether you are a smoker or have ever smoked.

In 1998, the National Academy of Sciences estimated that prolonged exposure for a lifetime of 70 years to low levels of radon or radon acting together with smoking is responsible for 15,000–22,000 (or 12%) of the lung cancer

deaths each year in the United States. This makes radon the second leading cause of lung cancer after smoking. Most of the deaths are among smokers or former smokers, with about 2,100–2,900 deaths among nonsmokers.

Scientists made two assumptions in estimating these risks from radon. *First,* there is no safe threshold dose for radon exposure (Figure 10-6, left, p. 207). *Second,* the incidence of lung cancer in uranium miners exposed to high levels of radon in mines can be extrapolated to estimate lung cancer deaths for people in homes exposed to much lower levels of radon.

Some scientists question these assumptions and say these estimates are too high. They also point to the contradictory results of several epidemiological studies on the risks of lung cancer from radon exposure.

EPA indoor radon surveys suggest that 4–5 million U.S. homes may have annual radon levels above 4 picocuries* per liter of air and that 50,000–100,000 homes may have levels above 20 picocuries per liter.

If the 4 picocuries per liter standard is adopted (as proposed by the EPA), the cost of testing and correcting the problem could run about $50 billion, with a

---

*A *picocurie* is a trillionth of a curie, which is the amount of radioactivity emitted by a gram of radium.

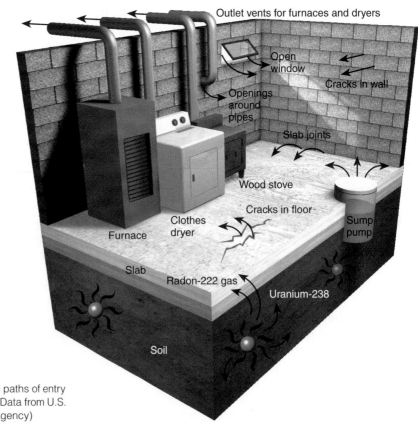

Figure 12-15 Sources and paths of entry for indoor radon-222 gas. (Data from U.S. Environmental Protection Agency)

15–20% reduction in radon-related deaths. Some researchers argue that it makes more sense to spend perhaps only $500 million to find and fix homes and buildings with radon levels above 20 picocuries per liter until more reliable data are available on the threat from exposure to lower levels of radon.

Because radon hot spots can occur almost anywhere, we do not know which buildings have unsafe levels of radon without conducting tests. In 1988, the EPA and the U.S. Surgeon General's Office recommended that everyone living in a detached house, a townhouse, a mobile home, or on the first three floors of an apartment building test for radon.

Ideally, radon levels should be monitored continuously in the main living areas (not basements or crawl spaces) for 2 months to a year. By 2003, only about 6% of U.S. households had conducted radon tests (most lasting only 2–7 days and costing $20–100 per home). In 2003, EPA Administrator Christine Whitman urged Americans to test their homes for indoor radon gas.

For information about radon testing visit the EPA website at *www.epa.gov/iaq/radon* or call the radon hotline at 800-SOS-RADON. According to the EPA, radon control could add $350–500 to the cost of a new home, and correcting a radon problem in an existing house could run $800–2,500.

12-6 EFFECTS OF AIR POLLUTION ON LIVING ORGANISMS AND MATERIALS

How Does Your Respiratory System Help Protect You from Air Pollution? Some *good news* is that your respiratory system (Figure 12-16) has a number of mechanisms that help protect you from much air pollution. Hairs in your nose filter out large particles. Sticky mucus in the lining of your upper respiratory tract captures smaller (but not the smallest) particles and dissolves some gaseous pollutants. Sneezing and coughing expel contaminated air and mucus when pollutants irritate your respiratory system.

In addition, hundreds of thousands of tiny mucus-coated hairlike structures called *cilia* line your upper respiratory tract. They continually wave back and forth and transport mucus and the pollutants they trap to your throat (where they are swallowed or expelled).

The *bad news* is that exposure to air pollutants including tobacco smoke can overload or break down these natural defenses. This can cause or contribute to various respiratory diseases. One that can start early in life is *asthma*, typically an allergic reaction causing sudden episodes of muscle spasms in the bronchial walls, which results in acute shortness of breath.

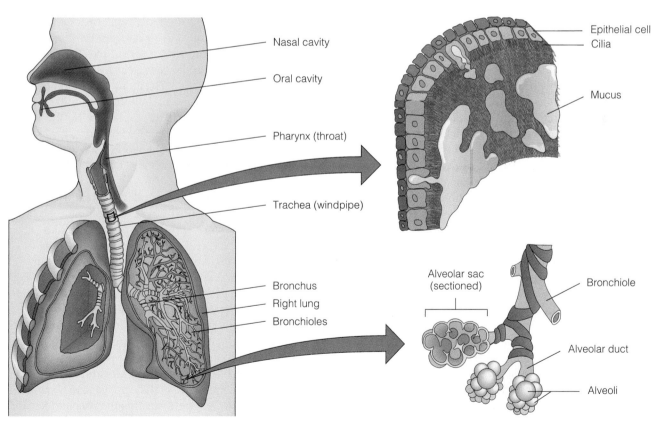

Figure 12-16 Major components of the human respiratory system.

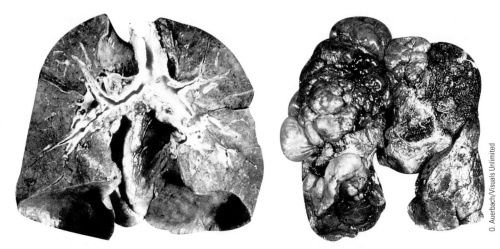

Figure 12-17 Normal human lungs (left) and the lungs of a person who died of emphysema (right). Prolonged smoking and exposure to air pollutants can cause emphysema in anyone, but about 2% of emphysema cases result from a defective gene that reduces the elasticity of the air sacs in the lungs. Anyone with this hereditary condition, for which testing is available, should not smoke and should not live or work in a highly polluted area.

O. Auerbach/Visuals Unlimited

Years of smoking and breathing air pollutants can lead to other respiratory disorders including *lung cancer* and *chronic bronchitis,* which involves persistent inflammation and damage to the cells lining the bronchi and bronchioles. The results are mucus buildup, painful coughing, and shortness of breath. Damage deeper in the lung can cause *emphysema,* which is irreversible damage to air sacs or alveoli leading to abnormal dilation of air spaces, loss of lung elasticity, and acute shortness of breath (Figure 12-17).

People with respiratory diseases are especially vulnerable to air pollution, as are older adults, infants, pregnant women, and people with heart disease.

How Many People Die Prematurely from Air Pollution? It is difficult to estimate by risk analysis how many people die prematurely from respiratory or cardiac problems caused or aggravated by air pollution. The primary reason is that people are exposed to so many different pollutants over a lifetime. As a result, researchers can only provide us with a range of estimated deaths.

In the United States, the EPA estimates that annual deaths related to outdoor air pollution range from 65,000 to 200,000 (most from exposure to very small fine and ultrafine particles). If we include indoor air pollution, estimated annual deaths from air pollution in the United States range from 150,000 to 350,000 people—equivalent to one to two fully loaded 400-passenger jumbo jets crashing accidentally *each day* with no survivors. Millions more become ill and lose work time.

A 2000 study by state and local air pollution officials estimated that each year more than 125,000 Americans (120,000 of them in urban areas) get cancer from breathing diesel fumes from buses, trucks, and other diesel engines. According to a 2002 EPA report, long-term inhalation of particulates from diesel exhaust is likely to pose a lung-cancer risk to humans as

well as other types of lung damage. However, the EPA said there was not enough data to produce a quantitative estimate of the number of people likely to get lung cancer from diesel exhaust.

The EPA says that the study supports its emission standards for diesel-powered vehicles that become effective in 2007, which should reduce diesel emission by as much as 95%. Manufacturers of diesel-powered vehicles dispute the findings of the EPA report and are pressuring the Bush administration to relax or delay the 2007 emission standards.

According to the World Health Organization, worldwide at least 3 million people (most of them in Asia) die prematurely each year from the effects of air pollution. This is more than three times the estimated 885,000 people who die each year in automobile accidents (Figure 10-16, p. 217). About 2.8 million of these deaths are from *indoor* air pollution, and 200,000 are from *outdoor* pollution. This explains why the World Health Organization and the World Bank consider indoor air pollution one of the world's most serious environmental problems.

How Are Plants Damaged by Air Pollutants? The effects of exposure of trees to a combination of air pollutants may not appear for several decades. Then large numbers of trees suddenly begin dying because of soil nutrient depletion and increased susceptibility to pests, diseases, fungi, and drought (Figure 12-12). This phenomenon has degraded areas of forests downwind of coal-burning facilities and some large urban areas.

The greatest harm occurs to downwind coniferous forests at high altitudes whose needles are exposed to air pollution year-round. In the United States, the most seriously affected areas are high-elevation spruce trees that populate the ridges of the Appalachian Mountains from Maine to Georgia, including the Shenandoah and Great Smoky Mountains national parks.

Air pollution, mostly by ozone, also threatens some crops (especially corn, wheat, and soybeans) and reduces U.S. food production by 5–10%. In the United States, estimates of agricultural losses as a result of air pollution (mostly by ozone) range from $2–6 billion per year, with an estimated $1 billion of damages in California alone.

What Are the Harmful Effects of Air Pollutants on Materials? Each year, air pollutants cause billions of dollars in damage to various materials we use. The fallout of soot and grit on buildings, cars, and clothing necessitates costly cleaning. Air pollutants break down exterior paint on cars and houses, and they deteriorate roofing materials. Irreplaceable marble statues, historic buildings, and stained glass windows around the world have been pitted, gouged, and discolored. The EPA estimates damage to buildings in the United States from acid deposition alone at $5 billion per year.

12-7 SOLUTIONS: PREVENTING AND REDUCING AIR POLLUTION

How Have Laws Helped Reduce Air Pollution in the United States? The U.S. Congress passed Clean Air Acts in 1970, 1977, and 1990. These laws require the EPA to establish air pollution regulations for controlling key pollutants that are enforced by each state and by major cities.

Congress directed the EPA to establish *national ambient air quality standards* (*NAAQS*) for six outdoor criteria pollutants (Table 12-2). The EPA regulates these chemicals by using *criteria* developed from risk assessment methods (Section 10-2, p. 204) to set maximum permissible levels in outdoor air. Each standard specifies the maximum allowable level, averaged over a specific period, for a certain pollutant in outdoor (ambient) air. The EPA has also established national emission standards for more than 100 toxic air pollutants known to cause or suspected of causing cancer or other adverse health effects.

Here are several pieces of *great news* based on EPA studies. Between 1970 and 2001, national total emissions of the six criteria pollutants declined 25% while U.S. population increased 32%, gross domestic product increased 161%, energy consumption rose 42%, and vehicle miles traveled rose nearly 150%.

Between 1970 and 2001, mean concentrations of five criteria air pollutants in the troposphere decreased: by 98% for lead, 76% for suspended particulate matter (10 micrometers or less in diameter), 44% for sulfur dioxide, 38% for volatile organic compounds (VOCs), and 19% for carbon monoxide.

Between 1970 and 1990, the Clean Air Act saved $6.4 trillion and prevented an estimated 1.6 million premature deaths and 300 million cases of respiratory disease. Finally, between 1990 and 2010, the 1990 amendments to the Clean Air Act should prevent 23,000 Americans from dying prematurely and avert 1.7 million asthma attacks per year.

According to the EPA there is also some *bad news*. *First*, emissions of nitrogen oxides (NO_x) increased 15% between 1970 and 2001. *Second*, between 1990 and 2001, ground-level concentrations of ozone increased, especially in the southern and northeastern regions of the United States.

Finally, the EPA said that in 2001 more than 133 million people lived in areas where air is unhealthy to breathe during part of the year because of high levels of air pollutants—primarily ozone and fine particles.

How Can U.S. Air Pollution Laws Be Improved? The Clean Air Act of 1990 was an important step in the right direction. But many environmentalists point to several deficiencies in this law. One is *continuing to rely mostly on pollution cleanup rather than prevention.* An example of the power of prevention is that in the United States, the air pollutant with the largest drop (98% between 1970 and 2001) in its atmospheric level was lead, which was largely banned in gasoline.

A second major problem has been *failure to increase fuel-efficiency standards for cars, sport utility vehicles (SUVs), and light trucks.* According to environmental scientists, increased fuel efficiency would reduce air pollution more quickly and effectively than any other method, reduce CO_2 emissions, save energy, and save consumers enormous amounts of money.

Another problem is *inadequate regulation of emissions from inefficient two-cycle gasoline engines.* These engines are used in lawn mowers, leaf blowers, chain saws, and personal marine engines used to power jet skis, outboard motors, snowmobiles, and personal watercraft. According to the California Air Resources Board, a 1-hour ride on a typical jet ski creates more air pollution than the average U.S. car does in a year, and operating a 100-horsepower marine engine for 7 hours emits more air pollutants than a new car driven 160,000 kilometers (100,000 miles).

Some *good news* is that in 2001 the EPA announced plans to reduce emissions from most of these sources by 2004 and establish even stricter emissions by 2007. Some *bad news* is that under pressure from manufacturers the Bush administration extended deadlines for some reductions in emissions from snowmobiles from 2010 to 2012.

A final serious deficiency is *failure of the act to do much about reducing emissions of carbon dioxide and other greenhouse gases.*

Executives of companies affected by implementing such policies strongly oppose such changes in air pollution laws. They claim that implementing such changes would cost too much, harm economic growth, and cost jobs.

Proponents contend that history has shown that most industry estimates of implementing various air pollution control standards in the United States were many times the actual cost of implementation. In addition, implementing such standards has helped increase economic growth and create jobs by stimulating companies to develop new technologies for reducing air pollution emissions. Many of these technologies are sold in the international marketplace.

Case Study: Should We Use the Marketplace to Reduce Pollution? To help reduce SO_2 emissions, the Clean Air Act of 1990 allows an *emissions trading policy.* It enables the 110 most polluting coal-burning power plants in 21 states (primarily in the Midwest and East) to buy and sell SO_2 pollution rights.

Each year a coal-burning power plant is given a certain number of pollution credits or rights that allow it to emit a certain amount of SO_2. A utility that emits less SO_2 than its limit has a surplus of pollution credits. It can use these credits to avoid reductions in SO_2 emissions from some of its other facilities, bank them for future plant expansions, or sell them to other utilities, private citizens, or environmental groups.

The proponents argue that this system allows the marketplace to determine the cheapest, most efficient way to get the job done instead of having the government dictate how to control pollution. Some environmentalists see this *cap-and-trade* market approach as an improvement over the regulatory *command-and-control* approach, as long as it achieves a net reduction in SO_2 pollution. This would be accomplished by limiting the total number of credits and gradually lowering the annual number of credits. However, this reduction in the cap is not required by the 1990 amendments to the Clean Air Act and currently is not being done.

Some environmentalists criticize marketing pollution rights for SO_2 or other pollutants. They contend that it allows utilities with older, dirtier power plants to buy their way out and keep on emitting unacceptable levels of SO_2. In addition, this approach creates incentives to cheat because air quality regulation is based largely on self-reporting of emissions and pollution monitoring is incomplete and imprecise.

Here are two pieces of *good news. First,* between 1990 and 2001, the emissions trading system helped reduce SO_2 emissions in the United States by 33% (with 25% of this drop occurring since 1990). *Second,* the cost of doing this was less than one-tenth the cost projected by industry because this market-based system motivated companies to reduce emissions in more efficient ways.

In 1997, the EPA proposed a voluntary program for trading emission credits for smog-forming nitrogen oxides (NO_x) in 22 eastern states and the District of Columbia. Emissions trading may also be implemented for particulate emissions and volatile organic compounds and the combined emissions of SO_2, NO_x, and mercury from coal-burning power plants.

This combined emissions trading scheme was proposed by the Bush administration in 2001 under its Clear Skies initiative, strongly supported by the electric power industry. The initiative aims for a 70% reduction in emissions of mercury, sulfur dioxide, and nitrogen oxides over a decade.

Critics say the Bush plan does not go far enough to curb harmful emissions and would permit more pollution for longer periods of time than under current law. They charge that the biggest air polluters largely crafted the plan to get themselves off the hook.

Here is some *bad news* about the effectiveness of cap-and-trade emissions programs. In 2002, the EPA reported results from evaluation of the country's oldest and largest emissions trading program, in effect since 1993 in southern California. The EPA study found that this cap and trade model "produced far less emissions reductions than were either projected for the program or could have been expected from" the command-and-control system it replaced. The study also found accounting abuses, including setting emissions caps 60% higher than current emissions.

Environmentalists caution that the success of any emissions trading approach depends on how low the initial cap is set and requiring phased-in reduction of the cap to promote continuing innovation in air pollution prevention and control. Without these elements, emissions trading programs mostly move air pollutants from one area to another without achieving an overall reduction in air quality. None of the existing or proposed cap-and-trade emissions programs meets these two criteria.

How Can We Reduce Outdoor Air Pollution from Coal-Burning Facilities? Figure 12-18 summarizes ways to reduce emissions of sulfur oxides, nitrogen oxides, and particulate matter from stationary sources (such as electric power plants and industrial plants that burn coal).

Until recently, emphasis has been on output approaches that disperse and dilute the pollutants by using tall smokestacks or that add equipment to remove some of the particulate pollutants after they are produced (Figure 12-19). However, under the sulfur reduction requirements of the 1990 amendments to the Clean Air Act, more utilities are switching to low-sulfur coal to reduce SO_2 emissions. Environmentalists call for greater emphasis on such prevention methods.

What Should We Do About Air Pollution from Older Coal-Burning Facilities?

For several decades, environmental scientists, environmentalists, and the attorneys general of a number of states have been fighting to require greatly improved emissions control regulations for about 17,000 older coal-burning power plants and industrial plants and for oil refineries in the United States. Such plants have not been fully regulated by the Clean Air Act because they were already in existence when the law was passed. Most states pushing for improved controls are in the northeastern United States where forests, parks, croplands, and cities are downwind from such plants.

Mostly this has been a losing battle. Owners of such facilities say that meeting stricter regulations would cost too much and raise the price of electricity, gasoline, and many manufactured goods.

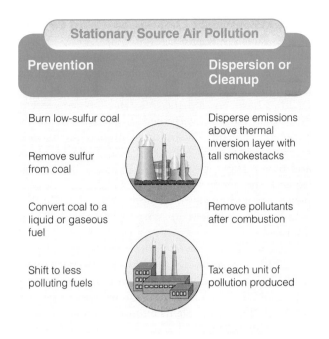

Stationary Source Air Pollution

Prevention	Dispersion or Cleanup
Burn low-sulfur coal	Disperse emissions above thermal inversion layer with tall smokestacks
Remove sulfur from coal	
Convert coal to a liquid or gaseous fuel	Remove pollutants after combustion
Shift to less polluting fuels	Tax each unit of pollution produced

Figure 12-18 (above) Solutions: methods for reducing emissions of sulfur oxides, nitrogen oxides, and particulate matter from stationary sources such as coal-burning electric power plants and industrial plants.

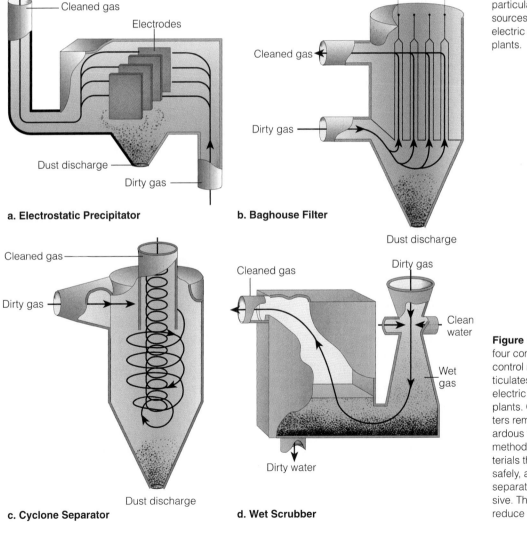

a. **Electrostatic Precipitator**

b. **Baghouse Filter**

c. **Cyclone Separator**

d. **Wet Scrubber**

Figure 12-19 (at left) Solutions: four commonly used output or control methods for removing particulates from the exhaust gases of electric power and industrial plants. Of these, only baghouse filters remove many of the more hazardous fine particles. All these methods produce hazardous materials that must be disposed of safely, and except for cyclone separators, all of them are expensive. The wet scrubber can also reduce sulfur dioxide emissions.

A 1977 rule in the Clean Air Act—called the New Source Review—requires about 17,000 older power plants, refineries, and manufacturers to upgrade pollution control equipment when they expand or modernize their facilities. However, owners of such facilities have resisted this rule and lobbied elected officials to have it overturned. In 2002, their efforts paid off when the Bush administration announced that it was easing the New Source Review restrictions and thus making it easier for these older facilities to expand and modernize without having to add expensive pollution control equipment. Changes in this rule by the president do not require congressional approval.

According to lobbyists for these industries, the revised EPA rule would free them from excessive regulations, which discourage them from modernizing and expanding their plants. They claim that such modernization will make the plants more efficient and reduce pollution emissions.

Opponents say the revised rule would gut the only provision in the Clean Air Act that could be used to force such facilities to reduce their air pollution emissions. They contend that the new rules will increase already serious air pollution for such facilities and result in thousands of premature deaths. In 2003, the attorneys general of nine northeastern states challenged this change in the New Source Review in federal court. Their lawsuit alleges that the EPA is exceeding its authority by enacting rules that weaken the Clean Air Act.

How Can We Reduce Outdoor Air Pollution from Motor Vehicles? Figure 12-20 lists ways to reduce emissions from motor vehicles, the primary culprits in producing photochemical smog. One way to make significant reductions in such air pollution is to get older, high-polluting vehicles off the road. According to EPA estimates, 10% of the vehicles on the road in the United States emit 50–70% of the air pollutants from vehicles.

In 2002, Joseph Norbeck, an engineer at the University of California-Riverside, conducted tests showing that gasoline engines in some new cars can essentially eliminate sulfur dioxide emissions by burning low-sulfur gasoline that is widely available in California. However, it probably will take a regulation by the federal government to make this cleaner-burning gasoline available across the United States.

Some *good news* is that during the past 30 years, outdoor air quality in Tokyo, across Japan, and in most western European cities has improved. The *bad news* is that outdoor air quality has remained about the same or has gotten worse in most rapidly growing urban areas in developing countries.

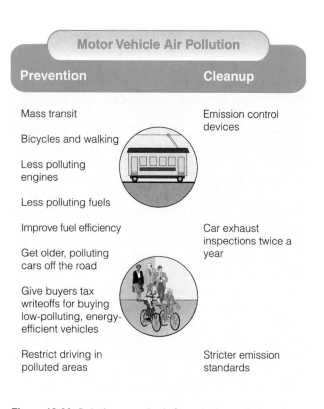

Figure 12-20 Solutions: methods for reducing emissions from motor vehicles.

Case Study: What Should We Do about Fine and Ultrafine Particles? Research indicates that invisible particles—especially *fine particles* with diameters less than 10 microns (PM-10) and *ultrafine particles* with diameters less than 2.5 microns (PM-2.5)—pose a significant health hazard. Such particles are emitted by incinerators, motor vehicles, radial tires, wind erosion, wood-burning fireplaces, and power and industrial plants (Figure 12-7).

Such tiny particles are not effectively captured by most air pollution control equipment (Figure 12-19) and are small enough to penetrate the respiratory system's natural defenses against air pollution (Figure 12-16). They can also bring with them droplets or other particles of toxic or cancer-causing pollutants that become attached to their surfaces.

Once they are lodged deep within the lungs, these fine particles can cause chronic irritation that can trigger asthma attacks, aggravate other lung diseases, and cause lung cancer. These lung problems interfere with the blood's uptake of oxygen and release of CO_2, which strains the heart and increases the risk of death from heart disease.

According to several recent studies of air pollution in U.S. cities, fine and ultrafine particles prematurely kill 65,000–200,000 Americans each year. There is no known threshold level below which the harmful effects of fine particles disappear.

Exposure to particulate air pollution is much worse in most developing countries, where urban air quality has generally deteriorated. The World Bank estimates that reducing particulate levels globally to WHO guidelines would prevent 300,000–700,000 premature deaths per year.

In 1997, the EPA announced stricter emission standards for ultrafine particles with diameters less than 2.5 microns (PM-2.5) in the United States. The EPA estimates the cost of implementing the standards at $7 billion per year, with the resulting health and other benefits estimated at $120 billion per year.

According to industry officials, the new standard is based on flimsy scientific evidence and its implementation will cost $200 billion per year. EPA officials say that their review of the scientific evidence—one of the most exhaustive ones ever undertaken by the agency—supports the need for the new standard for ultrafine particles. Furthermore, a 2000 study by the Health Effects Institute of 90 large American cities supported the link between fine and ultrafine particles and higher rates of death and disease.

How Can We Reduce Indoor Air Pollution? Reducing indoor air pollution does not require setting indoor air quality standards and monitoring the more than 100 million homes and buildings in the United States. Instead, air pollution experts suggest several ways to prevent or reduce indoor air pollution (Figure 12-21). Another possibility for cleaner indoor air in high-rise buildings is rooftop greenhouses through which building air can be circulated.

In developing countries, indoor air pollution from open fires and leaky and inefficient stoves that burn wood, charcoal, or coal (and the resulting high levels of respiratory illnesses) is the major threat. This could be reduced if governments gave people inexpensive clay or metal stoves that burn biofuels more efficiently (which would also reduce deforestation) and that are vented outside, or stoves that use solar energy to cook food (solar cookers).

What Is the Next Step? Individuals Matter It is encouraging that since 1970 most of the world's developed countries have enacted laws and regulations that have significantly reduced outdoor air pollution. However, without strong political pressure by individuals and organized groups of individuals on elected officials in the 1970s and 1980s these laws and regulations would not have been enacted, funded, and implemented. In turn these laws and regulations spurred companies, scientists, and engineers to come up with better ways to control outdoor pollution.

These laws and regulations are a useful *output approach* to controlling pollution. To environmentalists,

Figure 12-21 Solutions: ways to prevent and reduce indoor air pollution.

however, the next step is to shift to *preventing air pollution*. With this approach, the question is not *"What can we do about the air pollutants we produce?"* but *"How can we not produce such pollutants in the first place?"*

In developed countries and a growing number of developing countries, burning fossil fuels is the major cause of most outdoor air pollution. In terms of human health, the most serious problem is indoor air pollution. The WHO cites three reasons for this. *One* is burning wood, charcoal, dung, and crop wastes in open fires or inefficient stoves in developing countries. *Another* is poverty, which is the primary reason people in developing countries cannot afford less polluting ways to cook and heat their dwellings and water. The *third* is indoor smoking in developed and developing countries (p. 203). Figure 12-22 (p. 278) shows ways to prevent outdoor and indoor air pollution over the next 30–40 years.

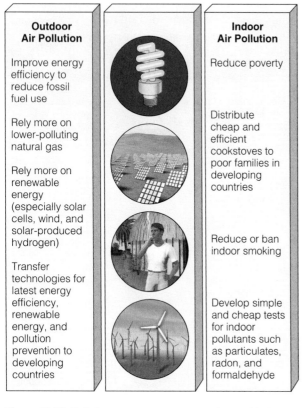

Outdoor Air Pollution	Indoor Air Pollution
Improve energy efficiency to reduce fossil fuel use	Reduce poverty
Rely more on lower-polluting natural gas	Distribute cheap and efficient cookstoves to poor families in developing countries
Rely more on renewable energy (especially solar cells, wind, and solar-produced hydrogen)	
	Reduce or ban indoor smoking
Transfer technologies for latest energy efficiency, renewable energy, and pollution prevention to developing countries	Develop simple and cheap tests for indoor pollutants such as particulates, radon, and formaldehyde

Figure 12-22 Solutions: ways to prevent outdoor and indoor air pollution over the next 30–40 years.

Like the shift to *controlling air pollution* between 1970 and 2000, this new shift to *preventing air pollution* will not take place without political pressure on elected officials by individual citizens and groups of such citizens. See the website material for this chapter for what you can do to reduce your exposure to indoor and outdoor air pollution.

Turning the corner on air pollution requires moving beyond patchwork, end-of-pipe approaches to confront pollution at its sources. This will mean reorienting energy, transportation, and industrial structures toward prevention.

HILARY F. FRENCH

REVIEW QUESTIONS

1. Define the boldfaced terms in this chapter.

2. How can lichens be used to detect air pollutants?

3. Distinguish among *atmosphere, troposphere,* and *stratosphere.* What key role does the stratosphere play in maintaining life on the earth?

4. Distinguish among *air pollution, primary air pollutants,* and *secondary air pollutants.* List the major classes of pollutants found in outdoor air. Distinguish between *stationary* and *mobile* sources of pollution for outdoor air. What are the two major sources of indoor air pollution?

5. Briefly describe the history of air pollution in Europe and the United States.

6. List the six *criteria* air pollutants regulated in the United States (and in most developed countries). For each of these pollutants, summarize its major human sources and health effects.

7. What is *photochemical smog,* and how does it form? What is *industrial smog,* and how does it form?

8. List three factors that can reduce air pollution and three factors that can increase air pollution.

9. What is a *temperature inversion,* and what are its harmful effects? Distinguish between a *subsidence temperature inversion* and a *radiation temperature inversion.* What types of places are most likely to suffer from prolonged inversions of each type?

10. What is *acid deposition,* and what are its major components and causes? Distinguish among *acid deposition, wet deposition,* and *dry deposition.*

11. What areas tend to be most affected by acid deposition? What is a *buffer,* and what types of geologic areas can neutralize or buffer some inputs of acidic chemicals? What two types of areas are most sensitive to acid deposition?

12. What are the major harmful effects of acid deposition on **(a)** human health, **(b)** materials, **(c)** aquatic life, **(d)** trees and other plants, and **(e)** soils?

13. List eight ways to prevent acid deposition and two ways to clean it up. List three reasons why controlling acid deposition is a difficult political problem.

14. How serious is indoor air pollution, and what are some of its sources? What is the *sick-building syndrome?* According to the EPA, what are the four most dangerous indoor air pollutants in the United States? What is the most dangerous indoor air pollutant in most developing countries?

15. Summarize the problems of indoor pollution from **(a)** formaldehyde and **(b)** radioactive radon gas.

16. List four defenses of your body against air pollution. What are the major harmful health effects of **(a)** carbon monoxide, **(b)** suspended particulate matter, **(c)** sulfur dioxide, **(d)** nitrogen oxides, and **(e)** ozone (see Table 12-2, p. 259)? Describe the health dangers from inhaling fine particles.

17. About how many people die prematurely each year from exposure to air pollutants in **(a)** the United States and **(b)** the world? What percentage of these deaths is the result of indoor air pollution?

18. What are the major effects of air pollutants on **(a)** plants (including food crops) and **(b)** materials?

19. What is the Clean Air Act, and how has it helped reduce outdoor air pollution in the United States? What is a *national ambient air quality* standard?

20. List four pieces of *good news* and three pieces of *bad news* about the effectiveness of the Clean Air Act in reducing outdoor air pollution in the United States. According to environmentalists, what are four weaknesses of the current Clean Air Act in the United States?

21. What is an *emissions trading policy,* and what are the advantages and disadvantages of using this approach to help reduce air pollution?

22. List the major prevention and cleanup methods for dealing with air pollution from **(a)** emissions of sulfur oxides, nitrogen oxides, and particulate matter from stationary sources, **(b)** automobile emissions, **(c)** indoor air pollution in developed countries, and **(d)** indoor air pollution in developing countries.

23. Why do environmentalists believe that we must now put much greater emphasis on preventing air pollution? List four ways to prevent outdoor air pollution and four ways to prevent indoor air pollution.

CRITICAL THINKING

1. Evaluate the pros and cons of the following statement: "Because we have not proved absolutely that anyone has died or suffered serious disease from nitrogen oxides, current federal emission standards for this pollutant should be relaxed."

2. Identify climate and topographic factors in your local community that **(a)** intensify outdoor air pollution and **(b)** help reduce outdoor air pollution.

3. Should all tall smokestacks be banned? Explain.

4. Why are most severe air pollution episodes associated with subsidence temperature inversions rather than radiation temperature inversions?

5. Should annual government-held auctions of marketable trading permits be used as a primary way to control and reduce air pollution? Explain. What conditions, if any, would you put on using this approach?

6. Are you for or against establishing a stricter standard for emissions of ultrafine particles? Explain. Use the library or Internet to determine whether stricter standards for ultrafine particles have been implemented in the United States (or the country where you live).

7. Do you agree or disagree with the possible weaknesses of the U.S. Clean Air Act listed on pp. 273–274? Defend each of your choices. Can you identify other weaknesses?

8. Explain why you agree or disagree with the proposals listed in Figure 12-22 (p. 278) for shifting the emphasis to preventing air pollution over the next several decades. List ways to bring about such a shift.

9. Congratulations! You are in charge of reducing air pollution in the country where you live. List the three most important features of your policy for **(a)** outdoor air pollution and **(b)** indoor air pollution.

PROJECTS

1. Evaluate your exposure to some or all of the indoor air pollutants in Figure 12-14 (p. 269) in your school, workplace, and home. Come up with a plan for reducing your exposure to these pollutants.

2. Have buildings at your school been tested for radon? If so, what were the results? What has been done about areas with unacceptable levels? If this testing has not been done, talk with school officials about having it done.

3. Make a concept map of this chapter's major ideas, using the section heads and subheads and the key terms (in boldface). Look on the website for this book for information about making concept maps.

INTERNET STUDY RESOURCES AND RESOURCES FOR FURTHER READING AND RESEARCH

The website for this book contains helpful study aids and many ideas for further reading and research. Log on to

http://biology.brookscole.com/miller10

and click on the Chapter-by-Chapter area. Choose Chapter 12 and select a resource:

■ Flash Cards allows you to test your mastery of the Terms and Concepts to Remember for this chapter.

■ Tutorial Quizzes provides a multiple-choice practice quiz.

■ Student Guide to InfoTrac will lead you to Critical Thinking Projects that use InfoTrac College Edition as a research tool.

■ References lists the major books and articles consulted in writing this chapter.

■ Hypercontents takes you to an extensive list of sites with news, research, and images related to individual sections of the chapter.

INFOTRAC COLLEGE EDITION

Improve your skills with InfoTrac College Edition, a searchable online database of articles from more than 700 periodicals. Log on to

http://www.infotrac-college.com

or access InfoTrac through the website for this book. Try to find the following articles:

1. Becker, R., and V. Henderson. 2000. Effects of air quality regulations on polluting industries. *Journal of Political Economy* 108: 379. *Keywords:* "air quality regulations" and "polluting." Air pollution regulations can have a significant effect on how and where industry can build manufacturing plants.

2. Spake, A. 2002. Don't breathe the air. (Brief article) *U.S. News and World Report* (July 1): 36. *Keyword:* "ozone pollution." Ground-level ozone is quickly becoming the primary air quality concern in many parts of the country. It could potentially have a great impact on the development of respiratory diseases in children.

13 CLIMATE CHANGE AND OZONE LOSS

A.D. 2060: Green Times on Planet Earth

Mary Wilkins sat in the living room of the solar-powered and earth-sheltered house (Figure 13-1) she shared with her daughter Jane and her family. It was July 4, 2060: Independence Day.

She got up and walked into the greenhouse that provided much of her home's heat. There she heard the hum of solar-powered pumps trickling water to rows of organically grown vegetables and glanced at the fish in the aquaculture and waste treatment tanks.

Mary returned to the coolness of her earth-sheltered house and began putting the finishing touches on her grandchildren's costumes for this afternoon's pageant in Rachel Carson Park. It would honor earth heroes who began the Age of Ecology in the 20th century and those who continued this tradition in the 21st century.

She was delighted that her 12-year-old grandson Jeffrey had been chosen to play Aldo Leopold, who in the late 1940s began urging people to work with the earth (Individuals Matter, p. 39). Her pride swelled when her 10-year-old granddaughter Lynn was chosen to play Rachel Carson, who in the 1960s warned us about threats from our increasing exposure to pesticides and other harmful chemicals (Individuals Matter, p. 400). Her neighbor's son Manuel had been chosen to play biologist Edward O. Wilson, who in the last third of the 20th century alerted us to the need to preserve the earth's biodiversity.

The transition to more sustainable societies and economies began around 2010 when people and governments began to mimic the way the earth has sustained itself for billions of years (Figure 8-13, p. 170, and Figure 8-14, p. 171, and top half of back cover) instead of trying to conquer nature. Even in her most idealistic dreams, Jane never guessed she would see the loss of global biodiversity slowed to a trickle. Most air pollution gradually disappeared when energy from the sun, wind, and hydrogen (produced by using solar energy to decompose water) had replaced most use of fossil fuels by 2040. Most food was now grown by more sustainable agriculture.

Preventing pollution and reducing resource waste had become important money-saving priorities for businesses and households based on the four *R*s of resource consumption: *reduce, reuse, recycle,* and *refuse.* Walking and bicycling had increased in cities and towns designed as vibrant communities for people instead of cars (p. 248). Low-polluting and safe eco-cars got up to 128 kilometers per liter (300 miles per gallon), and most places had bike paths and efficient mass transportation.

World population had stabilized at 8 billion in 2028 and then had begun a slow decline. The threat of climate change from atmospheric warming enhanced by human activities and air pollution lessened as the use of fossil fuels declined. International treaties enacted in the 1990s effectively banned the chemicals that had begun depleting ozone in the stratosphere during the last quarter of the 20th century. By 2050, ozone levels in the stratosphere had returned to 1980 levels.

Two hours later, Mary, her daughter Jane, and her son-in-law Gene watched with pride as 40 children honored the leaders of the Age of Ecology. At the end, Lynn stepped forward and said, "Today we have honored many earth heroes, but the real heroes are the ordinary people in this audience and around the world who worked to help sustain the earth's life-support systems for us and other species. Thank you, Grandma, Grandpa, Mom, Dad, and everyone here for giving us such a wonderful gift. We promise to leave the earth even better for our children and grandchildren and all living creatures."

Figure 13-1 An *earth-sheltered house* in the United States. Solar cells on the roof provide most of the house's electricity. About 13,000 families across the United States have built such houses. Mary Wilkins's fictional house in 2060 could be similar to this one.

We are embarked on the most colossal ecological experiment of all time—doubling the concentration of one of the most important gases in the earth's atmosphere—and we really have little idea of what might happen.

PAUL A. COLINVAUX

This chapter addresses the following questions:

- How has the earth's climate changed in the past?

- How might the earth's climate change in the future?

- What factors can affect changes in the earth's average temperature?

- What are some possible effects of climate change from a warmer earth?

- What can we do to slow or adapt to projected climate change caused by natural processes, human activities, or both?

- How have human activities depleted ozone in the stratosphere, and why should we care?

- What can we do to slow and eventually reverse ozone depletion in the stratosphere caused by human activities?

13-1 PAST CLIMATE CHANGE AND THE NATURAL GREENHOUSE EFFECT

How Have the Earth's Temperature and Climate Changed in the Past? Climate change is neither new nor unusual. The earth's average surface temperature and climate have been changing throughout the world's 4.7-billion-year history. Sometimes it has changed gradually (over hundreds to millions of years) and at other times fairly quickly (over a few decades) (Figure 13-2).

Over the past 900,000 years, the average temperature of the atmosphere near the earth's surface has undergone prolonged periods of *global cooling* and *global warming* (Figure 13-2, top left). During each cold period, thick glacial ice covered much of the earth's surface for about 100,000 years. Most of this ice melted during a *warmer interglacial period* lasting 10,000–12,500 years that followed each of these glacial periods.

For the past 10,000 years, we have had the good fortune to live in an interglacial period with a fairly stable climate and average global surface temperature

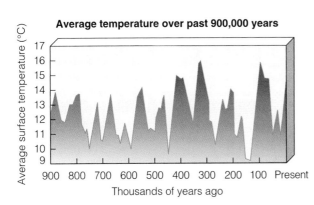

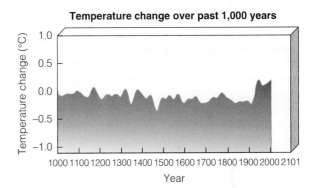

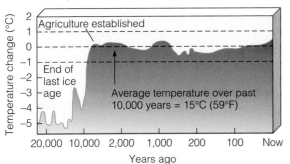

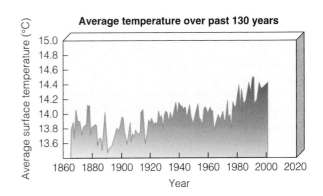

Figure 13-2 Estimated changes in the average global temperature of the atmosphere near the earth's surface over different periods of time. Past temperature changes are estimated by analysis of radioisotopes in rocks and fossils; plankton and radioisotopes in ocean sediments; ice cores from ancient glaciers; temperature measurements at different depths in boreholes drilled deep into the earth's surface; pollen from lake bottoms and bogs; tree rings; historical records; and temperature measurements (since 1860). (Data from Goddard Institute for Space Studies, Intergovernmental Panel on Climate Change, National Academy of Sciences, National Aeronautics and Space Agency, National Center for Atmospheric Research, and National Oceanic and Atmospheric Administration)

(Figure 13-2, bottom left). However, even during this generally stable period, significant changes have occurred in regional climates. For example, about 7,000 years ago, most of the current Sahara desert received almost 20 times more annual rainfall than it does today.

What Is the Greenhouse Effect? For the earth and its atmosphere to remain at a constant temperature, incoming solar energy must be balanced by an equal amount of outgoing energy (Figure 4-8, p. 69). Although the overall average temperature of the atmosphere is constant, its average temperature at various altitudes varies (Figure 12-2, p. 256, red line).

In addition to incoming sunlight, a natural process called the *greenhouse effect* (Figure 6-9, p. 111) warms the earth's lower troposphere and surface. Recall that it occurs because molecules of certain atmospheric gases, called *greenhouse gases,* warm the lower atmosphere by absorbing some of the infrared radiation (heat) radiated by the earth's surface. This causes their molecules to vibrate and transform the absorbed energy into longer wavelength infrared radiation (heat) in the troposphere.

Actually, the atmosphere does not behave like a real greenhouse (or a car with its windows closed). The primary reason air heats up in a greenhouse (or car interior) is that the closed windows keep the air from being carried away by convection (Figure 3-9, left, p. 55) to the outside air. By contrast, convection is the primary way that heat released by molecules of greenhouse gases is spread through the atmosphere. A more scientifically accurate term for this phenomenon would be *tropospheric heating effect.* But the term *greenhouse effect* is widely used and accepted.

Swedish chemist Svante Arrhenius first recognized this natural tropospheric heating effect in 1896. Since then numerous laboratory experiments and measurements of atmospheric temperatures at different altitudes have confirmed this idea. As a result, it is one of the most widely accepted theories in the atmospheric sciences.

The two greenhouse gases with the largest concentrations in the atmosphere are *water vapor* controlled by the hydrologic cycle (Figure 4-23, p. 81) and *carbon dioxide* (CO_2) controlled by the carbon cycle (Figure 4-24, p. 82). Table 13-1 shows the major sources, average time in the troposphere, and the relative warming potential of various greenhouse gases present in the troposphere.

Scientists have analyzed the content of gases in bubbles trapped at various depths in ancient glacial ice. According to these measurements, estimated changes in tropospheric CO_2 levels correlate fairly closely with estimated variations in the atmosphere's average global temperature near the earth's surface during the past 160,000 years (Figure 13-3).

Table 13-1 Major Greenhouse Gases from Human Activities

Greenhouse Gas	Human Sources	Average Time in the Troposphere	Relative Warming Potential (compared to CO_2)
Carbon dioxide (CO_2)	Fossil fuel burning, especially coal (70–75%), deforestation, and plant burning	50–120 years	1
Methane (CH_4)	Rice paddies, guts of cattle and termites, landfills, coal production, coal seams, and natural gas leaks from oil and gas production and pipelines	12–18 years	23
Nitrous oxide (N_2O)	Fossil fuel burning, fertilizers, live-stock wastes, and nylon production	114–120 years	296
Chlorofluorocarbons (CFCs)*	Air conditioners, refrigerators, plastic foams	11–20 years (65–110 years in the stratosphere)	900–8,300
Hydrochloro-fluorocarbons (HCFCs)	Air conditioners, refrigerators, plastic foams	9–390	470–2,000
Hydrofluorocarbons (HFCs)	Air conditioners, refrigerators, plastic foams	15–390	130–12,700
Halons	Fire extinguishers	65	5,500
Carbon tetrachloride	Cleaning solvent	42	1,400

*CFC use is being phased out, but it will take 50–100 years for the ozone layer to recover.

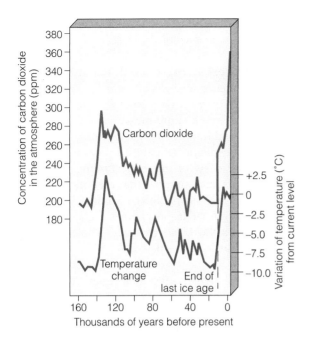

Figure 13-3 Atmospheric carbon dioxide levels and global temperature. Estimated long-term variations in average global temperature of the atmosphere near the earth's surface are graphed along with average tropospheric CO_2 levels over the past 160,000 years. These CO_2 levels were obtained by inserting metal tubes deep into Antarctic glaciers, removing the ice, and analyzing bubbles of ancient air trapped in ice at various depths throughout the past. The rough correlation between CO_2 levels in the troposphere and temperature shown in these estimates based on ice core data suggests a connection between these two variables, although no definitive causal link has been established. In 1999, the world's deepest ice core sample revealed a similar correlation between air temperatures and the greenhouse gases CO_2 and CH_4 going back for 460,000 years. (Data from Intergovernmental Panel on Climate Change and National Center for Atmospheric Research)

13-2 CLIMATE CHANGE AND HUMAN ACTIVITIES

What Is Global Warming? Since the beginning of the industrial revolution around 1750 (and especially since 1950), three human activities have emitted significant amounts of greenhouse gases into the troposphere. One has been the sharp rise in *the use of fossil fuels,* which release large amounts of the greenhouse gases CO_2 and CH_4 into the troposphere. Another is *deforestation and clearing and burning of grasslands to raise crops,* which release CO_2 and N_2O. The third is *cultivation of rice in paddies and use of inorganic fertilizers,* which release N_2O into the troposphere.

Figure 13-4 shows that since 1860, the concentrations of the greenhouse gases CO_2, CH_4, and N_2O in the troposphere have risen sharply, especially since 1950. The two largest contributors to CO_2 emissions are the world's thousands of coal-burning power and industrial plants and more than 700 million gasoline-burning

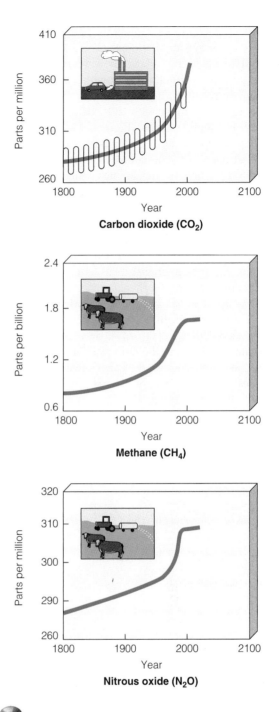

Carbon dioxide (CO₂)

Methane (CH₄)

Nitrous oxide (N₂O)

Figure 13-4 Increases in average concentrations of the greenhouse gases CO_2, methane, and nitrous oxide in the troposphere between 1860 and 2002. (Data from Intergovernmental Panel on Climate Change, National Center for Atmospheric Research, and World Resources Institute)

motor vehicles (555 million of them cars). Emissions of CO_2 from U.S. coal-burning power and industrial plants alone exceed the combined CO_2 emissions of 146 nations where 75% of the world's people live.

According to a 2002 report by the Australia Institute, Australia had the highest greenhouse gas emissions per person in 2001, mostly because of land clearing. The country's emissions per person were double the average for all industrial countries and 35% higher than the United States, which had the second highest per capita emissions of greenhouse gases.

Climate scientists have looked at evidence about past changes in atmospheric CO_2 concentrations and atmospheric temperatures (Figures 13-2 and 13-3). They have also developed sophisticated mathematical models of the earth's climate system to make projections about possible future climate change.

As a result of these studies and models, most climate scientists have come to an important conclusion. They believe that increased inputs of CO_2 and other greenhouse gases into the atmosphere from human activities will enhance the earth's natural greenhouse effect and raise the average global temperature of the atmosphere near the earth's surface. This *enhanced greenhouse effect* is called **global warming.** It should not be confused with the different atmospheric problem of ozone depletion (Table 13-2 and Section 13-8).

Some scientists say the term *global warming* is misleading because although the earth's overall temperature goes up, some parts of the earth will get cooler.

Instead of calling it global warming, they believe a more accurate name would be *global climate instability*.

Are We Experiencing Global Warming? Considerable evidence suggests that some global warming is occurring. This evidence comes from analysis of ice core samples, temperature measurements at different levels in several hundred boreholes in the earth's surface, and atmospheric temperature measurements.

Here are a few of the many findings from such measurements that support the scientific consensus that the earth's atmosphere is warming. The concentration of CO_2 in the troposphere is higher than it has been in the past 420,000 years. The 20th century was the hottest century in the past 1,000 years (Figure 13-2, top right).

Since 1861 (when direct atmospheric temperature measurements began), the average global temperature of the troposphere near the earth's surface has risen $0.6 \pm 0.2°C$ ($1.1 \pm 0.4°F$), with most of this increase taking place since 1946 (Figure 13-2, bottom right). Nine of the ten warmest years since 1861 have occurred since 1990. The hottest year was 1998, followed

Table 13-2 Major Characteristics of Global Warming and Ozone Depletion		
Characteristic	**Global Warming**	**Ozone Depletion**
Region of atmosphere involved	Troposphere.	Stratosphere.
Major substances involved	CO_2, CH_4, N_2O (greenhouse gases).	O_3, O_2, chlorofluorocarbons (CFCs).
Interaction with radiation	Molecules of greenhouse gases absorb infared (IR) radiation from the earth's surface, vibrate, and release longer-wavelength IR radiation (heat) into the lower troposphere. This natural greenhouse effect helps warm the lower troposphere.	About 95% of incoming ultraviolet (UV) radiation from the sun is absorbed by O_3 molecules in the stratosphere and does not reach the earth's surface.
Nature of problem	There is a high probability that increasing concentrations of greenhouse gases in the troposphere from burning fossil fuels, deforestation, and agriculture are enhancing the natural greenhouse effect and raising the earth's average surface temperature (Figure 13-2, bottom right, and Figure 13-8, p. 286).	CFCs and other ozone-depleting chemicals released into the troposphere by human activities have made their way to the stratosphere, where they decrease O_3 concentration. This can allow more harmful UV radiation to reach the earth's surface.
Possible consequences	Changes in climate, agricultural productivity, water supplies, and sea level.	Increased incidence of skin cancer, eye cataracts, and immune system suppression and damage to crops and phytoplankton.
Possible responses	Decrease fossil fuel use and deforestation; prepare for climate change.	Eliminate CFCs and other ozone-depleting chemicals and find acceptable substitutes.

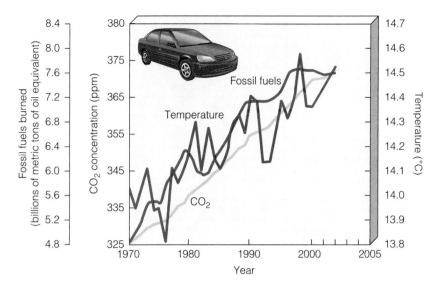

Figure 13-5 Apparent connections between increases in fossil fuel use, atmospheric concentrations of CO_2, and global temperature between 1970 and 2002. (Data from BP Amoco, U.S. Department of Energy, Goddard Institute of Space Studies, Scripps Institution of Oceanography, and Worldwatch Institute)

in order by 2002 and 2001. There is an apparent correlation between increases in fossil fuel use, atmospheric concentrations of CO_2, and global temperature between 1970 and 2002 (Figure 13-5).

Here are some more signs of a warmer troposphere observed during recent decades. At the earth's poles and in Greenland, temperatures have increased and there has been some melting of land-based ice caps and floating ice. Some glaciers on the tops of mountains in the Alps, Andes, Himalayas, northern Cascades of Washington, and Mount Kilimanjaro in Africa have begun shrinking and melting

In addition, some warm-climate fish, tree, and other species have migrated northward. Also, spring has been arriving earlier and autumn frosts have arrived later in many parts of the world.

The global climate system is complex and still poorly understood. As a result, scientists cannot pin down the exact causes of the changes just listed. They could be the result of natural climate fluctuations, changes in climate from human activities, or a combination of both factors—the most likely explanation according to most climate scientists.

Regardless of the cause, significant climate change caused by atmospheric warming or cooling over several decades to 100 years has important implications for human life, wildlife, and the world's economies. Such rapid climate change can affect the availability of water resources by altering rates of evaporation and precipitation. It can also shift areas where crops can be grown, change average sea levels, and alter the structure and location of the world's biomes (Figure 6-11, p. 113).

13-3 PROJECTING FUTURE CHANGES IN THE EARTH'S CLIMATE

How Do Scientists Model Climate Changes? Computer Models as Crystal Balls To project the effects of increases in greenhouse gases on average global temperature and the earth's climate, scientists develop complex *mathematical models* of the global air circulation system (Figure 6-5, p. 109). Then they run the models on supercomputers.

These *general circulation models* (*GCMs*) simulate the earth's atmosphere mathematically. These computer simulations cover the earth's surface with a large number of stacked layers of gigantic cells or boxes, each several hundred kilometers on a side and about 3 kilometers (2 miles) high (Figure 13-6). The flow of various types of matter and energy in and out of the box-filled atmosphere is then simulated with a set of mathematical equations.

Greatly improved climate models and measurements have increased the ability of scientists to account

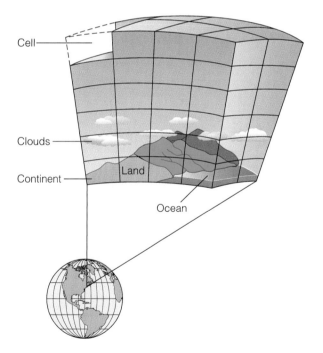

Figure 13-6 Global circulation model (GCM) of climate (and weather) divides the earth's atmosphere into large numbers of gigantic boxes or cells stacked many layers high. The laws of physics and our understanding of global air circulation patterns (Figure 6-5, p. 109) are used to describe numerically what happens to major variables affecting climate in each cell and how they change from one cell to another.

for past changes (Figure 13-7) and use various model scenarios to project future changes in the earth's average surface temperature (Figure 13-8). However, because of the complexity of the global climate system, there are still significant scientific uncertainties, as discussed in Section 13-4.

What Is the Scientific Consensus about Future Climate Change and Its Effects? In 1988, the United Nations and the World Meteorological Organization established the Intergovernmental Panel on Climate Change (IPCC) to study climate change. The IPCC is a network of about 2,000 of the world's leading climate experts from 70 nations. The U.S. National Academy of Sciences and the American Geophysical Union have also evaluated possible future climate changes.

So far, most of the world's climate scientists have come to two major conclusions. *First*, there will be significant increases in the emissions of CO_2, CH_4, and N_2O during the 21st century (Figure 13-9). *Second*, such increases are very likely to enhance the earth's natural greenhouse effect.

In 1990, 1995, and 2001, the IPCC published reports evaluating the available evidence about past changes in global temperatures (Figure 13-2) and climate models projecting future changes in average global temperatures and climate (Figure 13-8). Here are two major findings of the 2001 IPCC report. *First*, there is new and stronger evidence that most of the warning observed

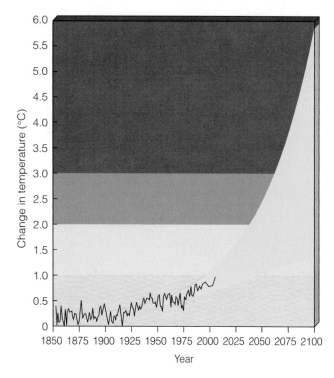

Figure 13-8 Comparison of measured changes in the average temperature of the atmosphere at the earth's surface between 1860 and 2001 and the projected range of temperature increase during this century. Current models indicate that the average global temperature at the earth's surface will rise by 1.4–5.8°C (2.5–10.4°F) sometime during the 21st century. Because of the uncertainty involved in current climate models, the projected warming shown here could be *overestimated* or *underestimated* by a factor of 2. (Data from U.S. National Academy of Sciences, National Center for Atmospheric Research, and Intergovernmental Panel on Climate Change)

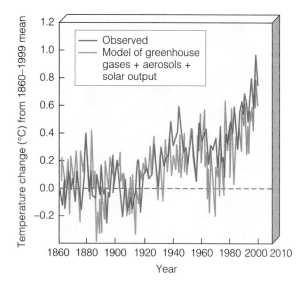

Figure 13-7 Comparison of measured changes in the average atmospheric temperature near the earth's surface (blue line) with those projected by the latest climate models between 1860 and 2000 (orange line). The latest models have been able to reproduce the climate of the past 100 years or so with increasing accuracy. (Data from Tom L. Wigley, Pew Center on Global Climate Change and Intergovernmental Panel on Climate Change)

over the last 50 years is attributable to human activities. *Second*, there is a 90–95% chance that the earth's mean surface temperature will increase 1.4–5.8°C (2.5–10.4°F) between 2000 and 2100 (Figure 13-8).

Here are some additional findings by other major scientific advisory bodies. According to a 2001 report by the National Academy of Sciences, "Greenhouse gases are accumulating in the Earth's atmosphere as a result of human activities, causing surface air temperatures and subsurface ocean temperatures and surface ocean temperatures to rise. . . . The changes observed over the last several decades are likely mostly due to human activities. . . . Global warming could have serious societal and ecological impacts by the end of this century."

In May 2000, 63 national academies of science issued a joint statement that said, "The balance of scientific evidence demands effective steps now to avert damaging changes to the Earth's climate." At the 1998 World Economic Forum, the CEOs of the world's 1,000 largest corporations voted climate change as the most critical problem facing humanity.

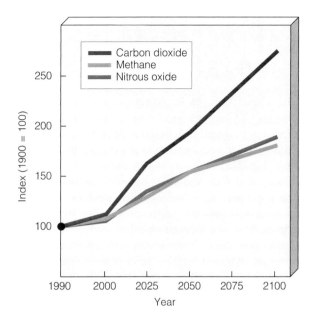

Figure 13-9 Projected emissions of three important greenhouse gases as a result of human activities, 1990–2100. (Data from United Nations Food and Agriculture Organization, 1997)

Finally, a 2002 Bush administration report, prepared by various U.S. government agencies for the United Nations, reached two major conclusions. *First,* recent climate changes "are likely mostly due to human activities." *Second,* by 2100 the average global temperature is likely to rise by about 3°C (5.4°F) and the average sea level is expected to rise by 9–88 centimeters (35 inches).

Why Do a Few Scientists Disagree with the Scientific Consensus about Future Climate Change? A small minority of atmospheric and other scientists inside and outside the IPCC disagree with this general scientific consensus. According to these scientists we still know too little about natural climate variables that could change the assessment (up or down), as discussed in Section 13-4. In addition, computer models used to predict climate changes are improving but still are not very reliable.

They also point out that some signs of global warming may not necessarily be caused by human activities. For example, while many glaciers are shrinking, others are growing. Also, glaciers shrink and grow naturally over long periods of time for reasons that are largely unknown.

Finally, they contend that global warming may be a lot less damaging than many people think and can be beneficial for some regions. For example, some countries may be able to increase crop productivity because of more rainfall and longer growing seasons.

IPCC and other climate scientists holding the consensus view agree that we need to improve climate models to help narrow the range of uncertainty in projected temperature changes (Figure 13-8). To do this, they call for greatly increased research on how the earth's climate system works, how it is affected by natural and human-induced variables, what we can do to slow the projected rate of global warming, and better cost estimates for climate change impacts and options for slowing the rate of climate change and adapting to such changes.

These scientists also point out that the concern is not just a temperature change but how rapidly it occurs, regardless of its cause. They point out that warming or cooling of the earth's surface by even 1°C (1.8°F) per century as a result of natural or human-caused factors would be faster than any temperature change occurring in the last 10,000 years, since the beginning of agriculture (Figure 13-2, bottom left). They warn that such rapid climate change could seriously disrupt the earth's ecosystems and human societies during this century.

13-4 FACTORS AFFECTING CHANGES IN THE EARTH'S AVERAGE TEMPERATURE

Will the Earth Continue to Get Warmer? The average temperature at the earth's surface has increased during the past 20 years (Figures 13-5 and 13-8). Will this trend continue? According to climate scientists, *this is very likely but we cannot know for sure* because science can only establish levels of probability or confidence—not absolute certainty (p. 49).

Scientists have identified a number of natural and human-influenced factors that might *amplify* (positive feedback) or *dampen* (negative feedback) changes in the earth's average surface temperature in various areas.

How Might Changes in Solar Output Affect the Earth's Temperatures? Solar output varies by about 0.1% over the 13-year and 22-year sunspot cycles and over 80-year and other, much longer cycles. These up-and-down changes in solar output can temporarily warm or cool the earth and thus affect the projections of climate models.

Studies by atmospheric scientists indicate that such changes in solar output cannot account for more than 50% of the atmospheric warming between 1900 and 2002 (Figure 13-8).

How Might Changes in the Earth's Reflectivity Affect Atmospheric Temperatures? Different parts of the earth's surface vary in their **albedo,** or ability to reflect light. White or shiny surfaces such as clouds, snow, ice, and sand reflect most of the sunlight that hits them. Much less sunlight is reflected by darker surfaces such as forests, grass, cities, and oceans.

The earth's reflectivity increases when polar ice caps expand during glacial periods and decreases when they melt and expose less reflective land and ocean surfaces. Satellite measurements of how albedo varies throughout the world and projections of how this might change have been incorporated into current climate models used to make the projections in Figure 13-8.

How Might Changes in the Ability of Oceans to Store CO₂ and Heat Affect Climate? Currently, the oceans help moderate the earth's average surface temperature in two ways. *First,* they remove about 29% of the excess CO_2 we pump into the atmosphere as part of the global carbon cycle (Figure 4-24, p. 82). *Second,* they absorb heat from the atmosphere and transfer some of it to the deep ocean, where it is stored temporarily.

Scientists do not know whether over the next few decades the oceans will release some of their stored heat and dissolved CO_2 and amplify global warming.

How Might Global Warming Affect Climate by Changing Ocean Currents? Connected deep and surface ocean currents act like a gigantic conveyor belt to transfer heat from one place to another and store CO_2 and heat in the deep sea (Figure 13-10). Scientists are concerned that an influx of fresh water from thawing ice in the Arctic and the Antarctic might slow or disrupt this conveyor belt and decrease the amount of heat it brings to the North Atlantic region.

If this loop stalls out, evidence from past climate changes indicates that this could trigger atmospheric

temperature changes of more than 5°C (9°F) over periods as short as 40 years. Thus, atmospheric warming at the global level could result in much colder weather in much of Europe and other North Atlantic countries.

How Can Changes in Average Sea Level Affect Climate? Changes in average sea level affect the amount of heat and CO_2 that can be stored in the ocean and the nature and location of the earth's biomes (Figure 6-11, p. 113).

Figure 13-11 shows estimated changes in the earth's average sea level over the past 250,000 years based on data obtained from cores drilled in the ocean floor. Average sea level can rise because of two factors. One is expansion of ocean waters when their average temperature increases. The other is increased input of water from the melting of land-based glaciers or excessive runoff of water from other land surfaces.

According to the 2001 IPCC report, the rate of sea level rise is now faster than at any other time during the past 1,000 years. Even a slight increase in average sea level would flood many coastal wetlands, low-lying islands, and parts of heavily populated coastal cities (Figure 11-20, p. 239, and Figure 11-22, p. 241).

How Might Changes in Cloud Cover Affect Climate? Warmer temperatures increase evaporation of surface water and create more clouds. These additional clouds could have a *warming effect* (positive feedback) by absorbing and releasing heat into the troposphere or a *cooling effect* (negative feedback) by reflecting sunlight back into space.

The net result of these two opposing effects depends on several factors. One is whether it is day or

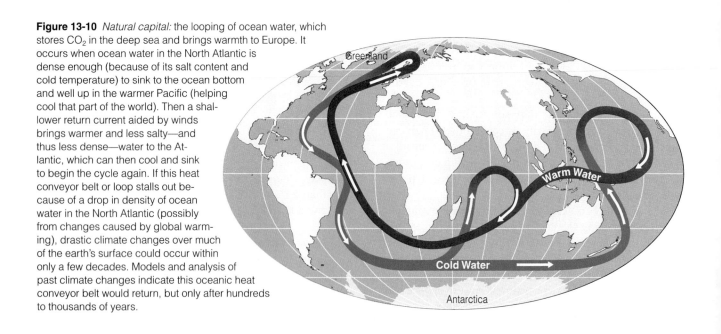

Figure 13-10 *Natural capital:* the looping of ocean water, which stores CO₂ in the deep sea and brings warmth to Europe. It occurs when ocean water in the North Atlantic is dense enough (because of its salt content and cold temperature) to sink to the ocean bottom and well up in the warmer Pacific (helping cool that part of the world). Then a shallower return current aided by winds brings warmer and less salty—and thus less dense—water to the Atlantic, which can then cool and sink to begin the cycle again. If this heat conveyor belt or loop stalls out because of a drop in density of ocean water in the North Atlantic (possibly from changes caused by global warming), drastic climate changes over much of the earth's surface could occur within only a few decades. Models and analysis of past climate changes indicate this oceanic heat conveyor belt would return, but only after hundreds to thousands of years.

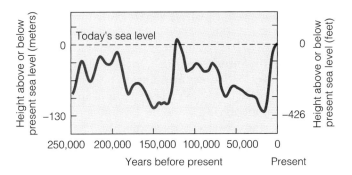

Figure 13-11 Changes in *average sea level* over the past 250,000 years based on data from cores removed from the ocean. The coming and going of glacial periods (ice ages) largely determine the rise and fall of sea level. As glaciers melted and retreated since the peak of the last glacial period about 18,000 years ago, the earth's average sea level has risen about 125 meters (410 feet) and changed the earth's coastal zones significantly. (Adapted from Tom Garrison, *Oceanography: An Invitation to Marine Science*, 3/E, © 1998. Brooks/Cole.)

night. Others include the type (thin or thick), coverage (continuous or discontinuous), altitude of clouds, and the size and number of water droplets formed in clouds.

For example, an increase in thick and continuous clouds at low altitudes can decrease surface warming by reflecting and blocking more sunlight. But an increase in thin and discontinuous cirrus clouds at high altitudes can warm the lower atmosphere and increase surface warming.

Scientists do not know which of these various factors might predominate or how cloud types, cover, heights, and chemical content might vary in different parts of the world.

In 1999, researchers at the University of Colorado reported that the wispy condensation trails (contrails) left behind by jet planes might have a greater impact on the earth's climate than scientists had thought. Using infrared satellite images, they found that jet contrails expand and turn into large cirrus clouds that tend to release heat into the upper atmosphere. If these preliminary results are confirmed, emissions from jet planes could be responsible for as much as half of the atmospheric warming in the northern hemisphere.

How Might Air Pollution Affect Climate? Aerosols (tiny droplets and solid particles, Figure 12-7, p. 261) of various air pollutants are released or formed in the atmosphere by volcanic eruptions and human activities.

Some of the resulting clouds have a high albedo and reflect more incoming sunlight back into space during the day . This could help counteract the heating effects of increased greenhouse gases.

Nights would be warmer because many of the clouds would still be there and prevent some of the heat stored in the earth's surface (land and water) during the day from being radiated into space. These pollutants may explain why most of the recent warming in the northern hemisphere occurs at night.

However, these interactions are complex. Pollutants in the lower troposphere can either warm or cool the air, depending on the reflectivity of the underlying surface and their ability to limit the size of water droplets.

In 2001, Environmental engineer Mark Jacobson of Stanford University reported that tiny particles of *soot* or *black carbon*—produced mainly from incomplete combustion in coal burning, diesel engines, and open fires—may be responsible for 15–30% of the observed global warming over the past 50 years. If this preliminary research is correct, such soot would be the second biggest contributor to global warming, just behind the greenhouse gas CO_2.

How Might Increased CO_2 Levels Affect Photosynthesis? Some studies suggest that more CO_2 in the atmosphere could increase the rate of photosynthesis in areas with adequate amounts of water and other soil nutrients. This would remove more CO_2 from the atmosphere and help slow atmospheric warming.

However, recent studies cast doubt on such a generalization for three reasons. *First,* this effect would slow as the plants reach maturity and take up less CO_2. *Second,* it is a temporary effect. When the plants die and are decomposed or burned, the carbon they stored is returned to the atmosphere as CO_2. *Third,* much of any increased plant growth could be offset by plant-eating insects that would breed more rapidly and year-round through lengthened warm seasons.

Here is another problem. In 2002, a laboratory study of wheat seedlings by Arnold Bloom and his colleagues at the University of California-Davis found that elevated levels of CO_2 inhibited the uptake of nitrate fertilizer. If this preliminary study is verified for wheat and other major crops, the decreased uptake of nitrate fertilizer could offset increased plant growth from higher CO_2 levels.

Also, a 2002 study by scientists at the University of Montana's School of Forestry in Missoula found that between 1950 and 1993, increased rainfall—not increased CO_2—accounted for two-thirds of the additional growth of the forests studied.

Can Soils Absorb More CO_2? When a plant dies and is decomposed, some of its carbon returns to the atmosphere as CO_2 and some is released into the soil. Some scientists suggest we may be able to slow global warming by increasing the amount of carbon stored in soils by planting more trees, improving forest management, and reducing soil erosion.

A 2001 study of the soil around loblolly pines exposed to elevated levels of CO_2 found that the soil did accumulate carbon. But much of the carbon was released back into the air as CO_2 when organic material in the soil decomposed. The researchers, John Lichter and William H. Schlesinger at Duke University, suggest that their preliminary findings "call into question the role of soils as long-term carbon sinks." Other experiments with a young sweetgum forest and a mature forest in Tennessee support these findings.

How Rapidly Could Climate Shift?   If moderate climate change takes place gradually over several hundred years, people in areas with unfavorable climate changes may be able to adapt to the new conditions.

However, suppose global temperature change takes place over several decades. Then we may not have enough time (and money) to switch food-growing regions, relocate millions of people from low-lying coastal areas, and build systems of dikes and levees to help protect the large portion of the world's population living near coastal areas. Such rapid changes in temperature could lead to large numbers of premature deaths from lack of food and social and economic chaos, especially in developing countries.

This concern stems partly from recent analyses of ice cores and deep-sea sediments. These measurements suggest that during the warm interglacial period that began about 125,000 years ago (Figure 13-2, top left), average temperatures varied as much as 10°C (18°F) in only a decade or two and such warming and cooling periods each lasted 1,000 years or more. If these findings are correct and also apply to the current interglacial period, fairly small rises in greenhouse gas concentrations could trigger rapid up-and-down shifts in the earth's average surface temperatures.

13-5   SOME POSSIBLE EFFECTS OF A WARMER WORLD

Why Should We Worry If the Earth's Temperature Rises a Few Degrees?   So what is the big deal? Why should we worry about a possible rise of only a few degrees in the earth's average surface temperature? We often have that much change between May and July, or even between yesterday and today.

The key point is that we are not talking about normal swings in *local weather* but a projected *global* change in *climate*—weather averaged over decades, centuries, and millennia. Once set into motion, such climate change cannot be reversed on a human time scale and would last for hundreds to thousands of years.

What Are Some Possible Effects of Atmospheric Warming? Winners and Losers   A warmer global climate could have a number of harmful and beneficial effects (Figures 13-12 and 13-13) for humans, other species, and ecosystems depending mostly on location and how rapidly the climate changes.

Here are a few of the many projected effects based on current climate models. The largest temperature increases will take place at the earth's poles. This is expected to cause more melting of glaciers and floating ice.

Several factors could accelerate atmospheric warming by releasing large amounts of CO_2 into the lower atmosphere. They include dieback of some forests that cannot spread fast enough to areas with more optimal temperatures (Figure 13-14, p. 292), greatly increased wildfires in forest and grassland areas where the climate becomes drier, and tree deaths from increased disease and pest populations that would thrive in areas with a warmer climate.

For each 1°C (1.8°F) rise in the earth's average temperature, climate belts in middle latitude regions would shift toward the earth's poles by an estimated 100–150 kilometers (60–90 miles) (Figure 6-13, right, p. 114) or upward 150 meters (500 feet) in altitude (Figure 6-13, left, p. 114). Such shifts could change areas where crops could be grown and affect the makeup and location of at least one-third of today's forests.

Changes in the structure and location of wildlife habitats could expand ranges and populations of some plant and animal species adapted to warmer climates. However, such changes would threaten plant and animal species that could not migrate rapidly enough to new areas and species with specialized niches. In addition, shifts in regional climate would threaten many parks, wildlife reserves, wilderness areas, wetlands, and coral reefs—wiping out many current efforts to stem the loss of biodiversity.

According to a 2002 study by the Pew Center on Global Climate Change: "Climate change could likely be the 'sleeper issue' that pushes our already stressed and fragile coastal and marine ecosystems over the edge. Particularly vulnerable are coastal and shallow water areas already stressed by human activity, such as estuaries and coral reefs. The situation is analogous to that faced by a human whose immune system is compromised and who may succumb to a disease that would not threaten a healthy person."

Another problem is a rise in global sea level, mainly because water expands slightly when heated. In their 2001 IPCC report, climate scientists projected a

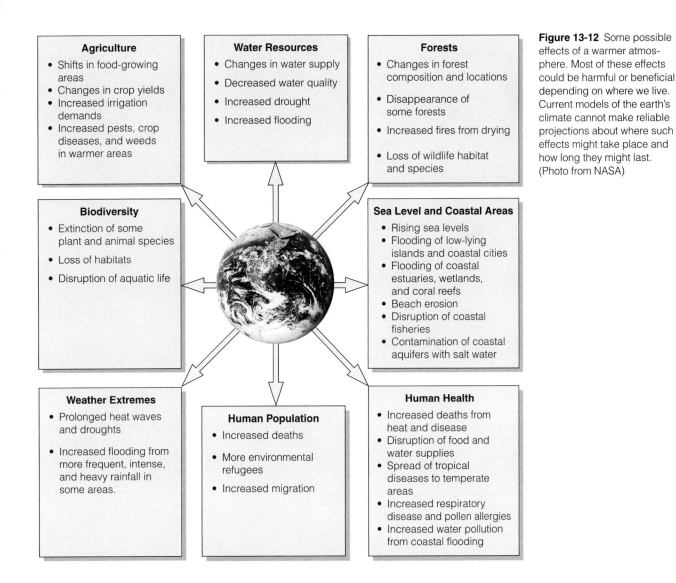

Agriculture
- Shifts in food-growing areas
- Changes in crop yields
- Increased irrigation demands
- Increased pests, crop diseases, and weeds in warmer areas

Water Resources
- Changes in water supply
- Decreased water quality
- Increased drought
- Increased flooding

Forests
- Changes in forest composition and locations
- Disappearance of some forests
- Increased fires from drying
- Loss of wildlife habitat and species

Biodiversity
- Extinction of some plant and animal species
- Loss of habitats
- Disruption of aquatic life

Sea Level and Coastal Areas
- Rising sea levels
- Flooding of low-lying islands and coastal cities
- Flooding of coastal estuaries, wetlands, and coral reefs
- Beach erosion
- Disruption of coastal fisheries
- Contamination of coastal aquifers with salt water

Weather Extremes
- Prolonged heat waves and droughts
- Increased flooding from more frequent, intense, and heavy rainfall in some areas.

Human Population
- Increased deaths
- More environmental refugees
- Increased migration

Human Health
- Increased deaths from heat and disease
- Disruption of food and water supplies
- Spread of tropical diseases to temperate areas
- Increased respiratory disease and pollen allergies
- Increased water pollution from coastal flooding

Figure 13-12 Some possible effects of a warmer atmosphere. Most of these effects could be harmful or beneficial depending on where we live. Current models of the earth's climate cannot make reliable projections about where such effects might take place and how long they might last. (Photo from NASA)

rise of 9–88 centimeters (4–35 inches) during the this century.

The high projected rise in sea level of about 88 centimeters (35 inches) would threaten half of the world's coastal estuaries, wetlands (one-third of those in the United States), and coral reefs and disrupt marine fisheries. It would cause gently sloping coastlines (especially along the U.S. East Coast) to retreat inland by about 1.3 kilometers (0.8 mile) and flood agricultural lowlands and deltas in parts of Bangladesh, India, and China (where much of the world's rice is grown).

It would also flood low-lying coastal regions and put an estimated 200 million people living in 30 of the world's largest coastal cities directly at risk (Figure 11-20, p. 239). Such a rise in sea level would also contaminate freshwater coastal aquifers with salt water and submerge some low-lying islands in the Pacific and Caribbean.

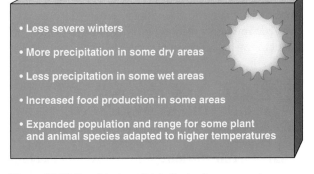

- Less severe winters
- More precipitation in some dry areas
- Less precipitation in some wet areas
- Increased food production in some areas
- Expanded population and range for some plant and animal species adapted to higher temperatures

Figure 13-13 Possible *beneficial effects* of a warmer atmosphere for some countries and people.

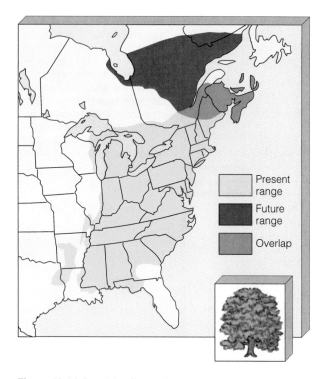

Figure 13-14 Possible effects of global warming on the geographic range of beech trees based on ecological evidence and computer models. According to one projection, if CO_2 emissions doubled between 1990 and 2050, beech trees (now common throughout the eastern United States) would survive only in a greatly reduced range in northern Maine and southeastern Canada. This is only one of a number of tree species whose geographic ranges could be changed drastically by increased atmospheric warming. For example, the ranges of some tree species adapted to a warm climate would spread. (Data from Margaret B. Davis and Catherine Zabinski, University of Minnesota)

One comedian jokes that he plans to buy land in the inland U.S. state of Kansas because it will probably become valuable beachfront property. Another boasts that she is not worried because she lives in a houseboat—the "Noah strategy."

The largest burden of the harmful effects of global warming will fall on people and economies in poorer tropical and subtropical nations, which do not have the economic and technological resources needed to adapt to the harmful impacts of climate change.

13-6 SOLUTIONS: DEALING WITH THE THREAT OF CLIMATE CHANGE

What Are Our Options? There are four schools of thought concerning what we should do about global warming. *One* is to *do nothing*. A dozen or so scientists

contend that climate change from human activities is not a threat, and a few popular media commentators and writers claim that global warming is a hoax.

A *second* approach is to *do more research before acting*. Many scientists and economists call for more research before making far-reaching economic and political decisions like phasing out fossil fuels and sharply reducing deforestation.

A *third* and growing group of scientists, economists, and business leaders believe that we should *act now to reduce the risks from climate change*. They call for us to adopt a *precautionary strategy*. When dealing with risky and far-reaching environmental problems such as climate change, they believe the safest course is to take informed preventive action *before* overwhelming scientific evidence justifies acting. In 1997, more than 2,500 scientists from a variety of disciplines signed a Scientists' Statement on Global Climate Disruption and concluded, "We endorse those [IPCC] reports and observe that the further accumulation of greenhouse gases commits the earth irreversibly to further global climatic change and consequent ecological, economic, and social disruption. The risks associated with such changes justify preventive action through reductions in emissions of greenhouse gases." Also in 1997, 2,700 economists led by eight Nobel laureates declared, "As economists, we believe that global climate change carries with it significant environmental, economic, social, and geopolitical risks and that preventive steps are justified."

A *fourth* strategy is to *act now as part of a no-regrets strategy*. Scientists and economists supporting this approach say we should take the key actions needed to slow projected atmospheric warming even if it may turn out not to be a serious threat because such actions lead to other important environmental, health, and economic benefits (Solutions, p. 293). For example, a reduction in the combustion of fossil fuels, especially coal, will lead to sharp reductions in air pollution that harms and prematurely kills large numbers of people, lowers food and timber productivity, and decreases biodiversity.

How Can We Reduce the Threat of Climate Change from Human Activities? Figure 13-15 presents a variety of prevention and cleanup solutions that analysts have suggested for slowing the rate and intensity of climate change from increased greenhouse gas emissions. Once greenhouse gases enter the atmosphere, most are there for 50–120 years (Table 13-1). So a delay of even a decade or so in reducing these emissions can make it much more difficult and costly to slow the rate and momentum of global warming and avert or lessen the more extreme consequences.

Some economic studies project that slowing global warming will be too costly. However, a number of other economic studies indicate that implementing the strategies listed in Figure 13-15 would boost the global economy, provide much needed jobs (especially in developing countries with large numbers of unemployed and underemployed people), and cost much less than trying to deal with the harmful effects of these problems.

Can We Remove and Store (Sequester) Enough CO_2 to Slow Global Warming? Are Output Approaches the Answer? Figure 13-16 (p. 294) shows several ways that scientists are evaluating for removing CO_2 from the atmosphere or from smoke-

Energy Efficiency to the Rescue

SOLUTIONS

According to energy expert Amory Lovins (Guest Essay, p. 510), *the major remedies for slowing possible global warming are things we should be doing even if there were no threat of global warming.*

According to Lovins, improving energy efficiency is the fastest, cheapest, and surest way to slash emissions of CO_2 and most other air pollutants within two decades, using existing technology. Lovins estimates that increased energy efficiency would also save the world $1 trillion per year in reduced energy costs (as much as the annual global military budget) and $300 billion per year in the United States.

Using energy more efficiently would also reduce pollution, help protect biodiversity, and help prevent arguments between governments about how CO_2 reductions should be divided up and enforced. It would also make the world's supplies of fossil fuel last longer, reduce international tensions over who gets the world's oil supplies, and allow more time to phase in renewable energy sources.

A growing number of companies have gotten the message and are saving money and energy while reducing their CO_2 emissions and selling their emission reductions in the global marketplace.

Critical Thinking

1. Do you agree that improving energy efficiency should be done regardless of its impact on the threat of global warming? Explain.

2. Why do you think there has been little emphasis on improving energy efficiency?

Global Warming

Prevention

- Cut fossil fuel use (especially coal)
- Shift from coal to natural gas
- Transfer energy efficiency and renewable energy technologies to developing countries
- Improve energy efficiency
- Shift to renewable energy resources
- Reduce deforestation
- Use sustainable agriculture
- Limit urban sprawl
- Reduce poverty
- Slow population growth

Cleanup

- Remove CO_2 from smokestack and vehicle emissions
- Store (sequester) CO_2 by planting trees
- Sequester CO_2 underground
- Sequester CO_2 in soil by using no-till cultivation and taking crop land out of production
- Sequester CO_2 in the deep ocean
- Repair leaky natural gas pipelines and facilities
- Use feeds that reduce CH_4 emissions by belching cows

Figure 13-15 Solutions: methods for slowing atmospheric warming during the 21st century.

stacks and storing (sequestering) it in other parts of the environment.

One way is to plant trees to remove CO_2 from the atmosphere. However, studies indicate that a global reforestation program in which each person in the world planted and tended an average of 1,000 trees every year would offset only about 3 years of current CO_2 emissions from burning fossil fuels. Also, this is only a temporary approach because the rate of removal of CO_2 from the atmosphere by photosynthesis decreases as trees mature and grow at a slower pace. In addition, trees release their stored CO_2 back into the atmosphere when they die and decompose or if they catch fire.

A 2002 study led by Duke University environmental scientist Robert Jackson found that previous estimates of the amount of carbon stored when plantations of trees and shrubs grow on former grasslands may have been too high. This team of scientists

Figure 13-16 Solutions: methods for removing carbon dioxide from the atmosphere or from smokestacks and storing (sequestering) it in plants, deep underground reservoirs, and the deep ocean.

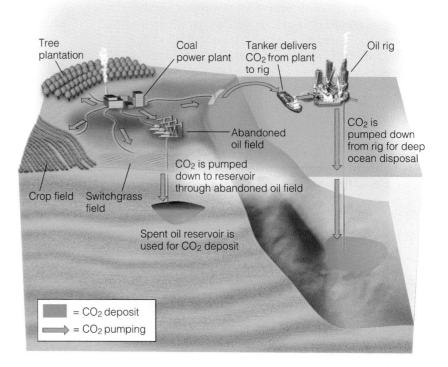

found that in wet locations, the extra carbon saved in the wood of trees and shrubs replacing grasses was more than offset by the carbon lost from the underlying soil.

A *second* approach is to use plants such as switchgrass to remove CO_2 from the air and deposit it in the soil. Farmers could make money by receiving sequestering payments from power companies to grow such plants on land not suitable for ordinary crops. However, warmer temperatures can increase decomposition in soils and return some of the CO_2 they stored to the atmosphere (p. 290).

A *third* strategy is to reduce the release of carbon dioxide and nitrous oxide from soil. Two methods have been suggested. One is to greatly increase use of *no-till cultivation* (Figure 9-26, p. 197), which would also reduce soil erosion. The other is to retire depleted crop fields and leave them untouched for 10 years as conservation reserves (Case Study, p. 199).

A *fourth* possibility is to collect CO_2 from smokestacks (and natural gas wells) and *pump it deep underground* into unminable coal seams and abandoned oil fields or *inject it into the deep ocean* or into the porous rock of *saltwater aquifers* found below the seafloor (Figure 13-16).

Some scientists estimate that saltwater aquifers below the ocean floor might be able to store large amounts of CO_2 for centuries. Tests are under way to evaluate how long deep sea aquifers, the deep ocean, and various underground sites might hold the gas.

These output approaches would take a costly investment in materials and transportation of the CO_2 (presumably by pipeline) to storage sites. According to DOE estimates, the cost of sequestering carbon dioxide in various underground and deep ocean repositories will have to be reduced at least tenfold to make this approach economically feasible. In addition, injecting large quantities of CO_2 into the ocean could upset the global carbon cycle, seawater acidity, and some forms of deep-sea life in unpredictable ways.

In 2003, Japanese scientist Nakamichi Yamasaki reported that his research team had developed a potentially affordable method for turning carbon dioxide into hydrocarbon gases by reacting it with hydrogen gas. Also in 2003 scientist and inventor J. Craig Ventor suggested that a single-cell organism found on the deep ocean floor near hydrothermal vents could convert carbon dioxide to methane by reacting it with hydrogen gas rising through the vents.

Scientists are investigating these and other possibilities. However, a serious problem is that current methods can remove only about 30% of the CO_2 from smokestack emissions, and using them would double to triple the cost of producing electricity by burning coal.

This explains why many scientists believe that *input approaches* (such as reducing fossil fuel use and energy waste and shifting to noncarbon fuels) are quicker, cheaper, and more effective ways to reduce inputs of carbon dioxide into the atmosphere.

13-7 WHAT IS BEING DONE TO REDUCE GREENHOUSE GAS EMISSIONS?

What Is the Kyoto Treaty? In December 1997, more than 2,200 delegates from 161 nations met in Kyoto, Japan, to negotiate a new treaty to help slow global warming. The resulting *Kyoto treaty* would require 38 developed countries to cut greenhouse emissions to an average of about 5.2% below 1990 levels by 2012 because they are responsible for 36% of the world's CO_2 emissions and thus should take the lead in reducing their emissions. It would not require developing countries to make any cuts in their greenhouse gas emissions until a later version of the treaty. Also, it would allow emissions trading among participating countries.

Some analysts praise the Kyoto agreement as a small but important step in dealing with the problem of global warming and hope the conditions of the treaty will be strengthened in future negotiating sessions. However, according to computer models, the 5.2% reduction goal of the Kyoto Protocol would shave only about 0.06°C (0.1°F) off the 0.7–1.7°C (1–3°F) temperature rise projected by 2060 (Figure 13-8).

However, proponents point out that the Kyoto treaty is not viewed as a 50-year treaty. Instead, it is a first step designed to have the countries producing the largest amounts of greenhouse gases agree to make modest reductions by 2012. Further reductions by developed and developing countries would hopefully follow in later versions of the treaty.

Should the United States Participate in the Kyoto Treaty? In 2001, President George W. Bush withdrew U.S. participation in the Kyoto treaty because it did not require emissions reductions by developing countries such as China and would hurt the U.S. economy. This decision set off strong protests by many scientists, citizens, and leaders throughout most of the world who pointed out that strong leadership is needed by the United States because it has the highest total and per capita CO_2 emissions of any country.

Some economic models indicate that the projected costs of the Kyoto treaty for the United States will exceed the projected benefits. But a number of other economic studies show that reducing greenhouse gases by boosting energy efficiency and use of renewable energy would have a net benefit for the U.S. economy by saving money in energy costs (Solutions, p. 293) and creating more jobs than will be lost. A 2000 study by the Department of Energy estimated that by 2012 the United States could meet most of its reductions in greenhouse gases under the Kyoto treaty at *no net cost.*

CEOs of a number of major U.S. companies are concerned that failure of the United States to participate in the treaty will hurt the U.S. economy in two ways. *First,* it would give competitors in other countries (especially the European Union and Japan) a headstart in developing and selling innovative technologies that reduce greenhouse gas emissions. *Second,* it could hinder U.S. companies from entering into the lucrative global market in greenhouse emissions trading, which according to the World Bank could reach $250–500 billion by 2010.

What Progress Is Being Made? Here are several pieces of *good news.* Despite U.S. refusal to support the Kyoto treaty, 178 other countries are going ahead with it and have adopted goals and strategies to reduce their greenhouse gas emissions.

A number of developing countries, including China (the world's third largest emitter of CO_2, after the United States and the European Union), are voluntarily improving energy efficiency and increasing their use of renewable energy resources even though they are exempt from cuts under the first round of the Kyoto treaty. According to a 2001 study by the Natural Resources Defense Council, China reduced its CO_2 emissions by 17% between 1997 and 2000, a period during which CO_2 emissions in the United States rose by 14%. China accomplished the reduction through three kinds of changes. *One* is switching from coal to cleaner energy resources by phasing out all coal subsidies, closing coal mines, and shutting down inefficient coal-fired electric plants. *Second,* it has stepped up its 20-year commitment to promoting energy efficiency. *Third,* China is restructuring its economy to reduce dependence on fossil fuels and increase use of renewable energy resources.

Omitting greenhouse gas emissions by Australia and the United States (neither of which have signed the Kyoto treaty), emissions by most industrial nations have decreased and could meet the Kyoto treaty goals.

Major global companies, such as Alcoa, DuPont, IBM, Toyota, BP, and Shell, have established targets to reduce their greenhouse gas emissions by 10–65% from 1990 levels by 2010. In addition, major automobile companies are investing billions in developing more efficient motor vehicles powered by *hybrid gas–electric* and *fuel cell* engines.

Since 1990, local governments in more than 500 cities around the world (including 110 in the United

States) have established programs to reduce their greenhouse gas emissions. In 2002, California became the first state to require a reduction in CO_2 emissions from motor vehicles beginning in 2009. However, most automobile companies, with support from the Bush administration, hope to have this law overturned in the courts.

Finally, some individuals are taking steps to cut their CO_2 emissions (Figure 13-17). See the website for this chapter for other things you can do to reduce the threat of global warming.

How Can We Prepare for Global Warming? According to the latest global climate models, the world needs to cut current emissions of greenhouse gases (not just CO_2) by at least 50% by 2018 to stabilize concentrations of such gases in the air at their present levels. Such a large reduction in emissions is extemely unlikely for political and economic reasons because it would take rapid, widespread changes in industrial processes, energy sources, transportation options, and individual lifestyles.

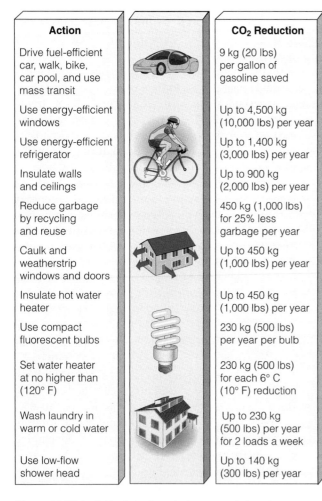

Action		CO_2 Reduction
Drive fuel-efficient car, walk, bike, car pool, and use mass transit		9 kg (20 lbs) per gallon of gasoline saved
Use energy-efficient windows		Up to 4,500 kg (10,000 lbs) per year
Use energy-efficient refrigerator		Up to 1,400 kg (3,000 lbs) per year
Insulate walls and ceilings		Up to 900 kg (2,000 lbs) per year
Reduce garbage by recycling and reuse		450 kg (1,000 lbs) for 25% less garbage per year
Caulk and weatherstrip windows and doors		Up to 450 kg (1,000 lbs) per year
Insulate hot water heater		Up to 450 kg (1,000 lbs) per year
Use compact fluorescent bulbs		230 kg (500 lbs) per year per bulb
Set water heater at no higher than (120° F)		230 kg (500 lbs) for each 6° C (10° F) reduction
Wash laundry in warm or cold water		Up to 230 kg (500 lbs) per year for 2 loads a week
Use low-flow shower head		Up to 140 kg (300 lbs) per year

Figure 13-17 *Individuals matter:* what you can do to reduce your annual emissions of CO_2. (Data from U.S. Department of Energy)

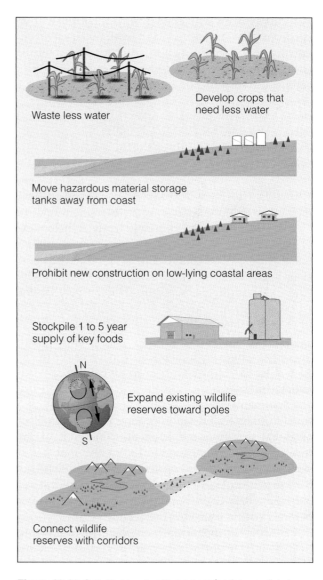

Figure 13-18 Solutions: ways to prepare for the possible long-term effects of climate change caused by increased atmospheric temperatures.

As a result, a growing number of analysts suggest that in addition to slowing global warming (Figure 13-15) we should also begin preparing for the possible effects of long-term atmospheric warming and climate change. Figure 13-18 shows some ways to do this.

13-8 OZONE DEPLETION IN THE STRATOSPHERE

What Is the Threat from Ozone Depletion? A layer of ozone in the lower stratosphere (Figure 12-2, p. 256, and Figure 12-3, p. 257) keeps about 95% of the sun's harmful ultraviolet (UV) radiation from reaching the earth's surface. Measuring instruments on balloons, aircraft, and satellites clearly show seasonal depletion (thinning) of ozone concentrations in the stratosphere above Antarctica and the Arctic. Similar measurements reveal a lower overall thinning of stratospheric ozone everywhere except over the tropics.

Based on these measurements and chemical models, the overwhelming consensus of researchers in this field is that ozone depletion (thinning) in the stratosphere is a serious long-term threat to humans, many other animals, and the sunlight-driven primary producers (mostly plants) that support the earth's food chains and webs (Table 13-2, right).

What Causes Ozone Depletion? From Dream Chemicals to Nightmare Chemicals
This situation started when Thomas Midgley, Jr., a General Motors chemist, discovered the first *chlorofluorocarbon (CFC)* in 1930, and chemists made similar compounds to create a family of highly useful CFCs. The two most widely used are CFC-11 (trichlorofluoromethane, CCl_3F) and CFC-12 (dichlorodifluoromethane, CCl_2F_2), known by their trade name, Freons.

These chemically stable (nonreactive), odorless, nonflammable, nontoxic, and noncorrosive compounds seemed to be dream chemicals. Cheap to make, they became popular as coolants in air conditioners and refrigerators (replacing toxic sulfur dioxide and ammonia), propellants in aerosol spray cans, cleaners for electronic parts such as computer chips, sterilants for hospital instruments, fumigants for granaries and ship cargo holds, and bubbles in plastic foam used for insulation and packaging. Between 1960 and the early 1990s, CFC production rose sharply.

But CFCs were too good to be true. In 1974, calculations by chemists Sherwood Rowland and Mario Molina at the University of California-Irvine indicated that CFCs were lowering the average concentration of ozone in the stratosphere. They shocked both the scientific community and the $28-billion-per-year CFC industry by calling for an immediate ban of CFCs in spray cans (for which substitutes were available).

Rowland and Molina's research led them to four major conclusions. *First,* CFCs remain in the troposphere because they are insoluble in water and are chemically unreactive. *Second,* over 11–20 years they rise into the stratosphere mostly through convection, random drift, and the turbulent mixing of air in the troposphere.

Third, once they reach the stratosphere, the CFC molecules break down under the influence of high-energy UV radiation. This releases highly reactive chlorine atoms, which speed up the breakdown of very reactive ozone (O_3) into O_2 and O in a cyclic chain of chemical reactions (Figure 13-19). This causes ozone in various parts of the stratosphere to be destroyed faster than it is formed.

Finally, each CFC molecule can last in the stratosphere for 65–385 years, depending on its type—the most widely used CFCs last 75–111 years. During that time, each chlorine atom released from these molecules can convert up to 100,000 molecules of O_3 to O_2.

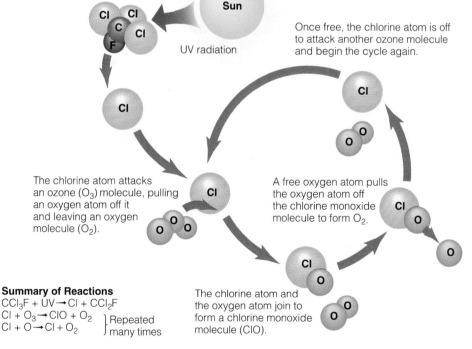

Ultraviolet light hits a chlorofluorocarbon (CFC) molecule, such as $CFCl_3$, breaking off a chlorine atom and leaving $CFCl_2$.

Sun

UV radiation

Once free, the chlorine atom is off to attack another ozone molecule and begin the cycle again.

The chlorine atom attacks an ozone (O_3) molecule, pulling an oxygen atom off it and leaving an oxygen molecule (O_2).

A free oxygen atom pulls the oxygen atom off the chlorine monoxide molecule to form O_2.

The chlorine atom and the oxygen atom join to form a chlorine monoxide molecule (ClO).

Summary of Reactions
$CCl_3F + UV \rightarrow Cl + CCl_2F$
$Cl + O_3 \rightarrow ClO + O_2$
$Cl + O \rightarrow Cl + O_2$ } Repeated many times

Figure 13-19 Simplified summary of how chlorofluorocarbons (CFCs) and other chlorine-containing compounds destroy ozone in the stratosphere. Note that chlorine atoms are continuously regenerated as they react with ozone. Thus they act as *catalysts*, chemicals that speed up chemical reactions without being used up by the reaction. Bromine atoms released from bromine-containing compounds that reach the stratosphere also destroy ozone by a similar mechanism.

According to Rowland and Molina's calculations and later models and atmospheric measurements of CFCs in the stratosphere, these dream molecules turned into a nightmare of global ozone destroyers. In 1995, Rowland and Molina received the Nobel prize in chemistry for their work.

What Other Chemicals Deplete Stratospheric Ozone? CFCs are not the only ozone-depleting compounds (ODCs). Other examples are *halons* and *HBFCs* (used in fire extinguishers), *methyl bromide* (a widely used fumigant), *hydrogen chloride* (emitted into the stratosphere by U.S. space shuttles), and cleaning solvents such as *carbon tetrachloride, methyl chloroform, n-propyl bromide,* and *hexachlorobutadiene.*

The oceans and occasional volcanic eruptions also release chlorine compounds into the troposphere. However, most of these chlorine compounds do not make it to the stratosphere because they easily dissolve in water and are washed out of the troposphere in rain. Bromine compounds may be less likely to be washed out of the troposphere, but further study is needed to confirm this possibility. Measurements and models indicate that 75–85% of the observed ozone losses in the stratosphere since 1976 are the result of ODCs released into the atmosphere by human activities beginning in the 1950s.

What Happens to Ozone Levels over the Earth's Poles During Part of the Year? In 1984, researchers analyzing satellite data discovered that 40–50% of the ozone in the upper stratosphere over Antarctica was being destroyed during the Antarctic spring and early summer (September–December), especially since 1976 (Figure 13-20).

Figure 13-21 shows the seasonal variation of ozone with altitude over Antarctica during 2002. The observed seasonal loss during the summer above Antarctica has been incorrectly called an *ozone hole.*

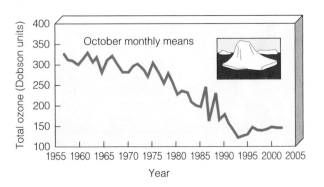

Figure 13-20 Mean total level of ozone for October over the Halley Bay measuring station in Antarctica, 1956–2002. (Data from British Antarctic Survey and World Meteorological Organization)

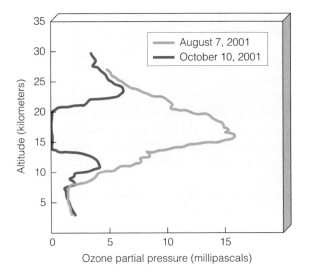

Figure 13-21 Seasonal variation of ozone level with altitude over Antarctica during 2002. Note the severe depletion of ozone during October (during the Antarctic summer, red line) and its return to more normal levels in August (during the Antarctic winter, green line). (Data from National Oceanic and Atmospheric Administration)

A more accurate term is *ozone thinning* because the ozone depletion varies with altitude (Figure 13-21) and location.

The total area of the atmosphere above Antarctica that suffers from ozone thinning during the peak season varies from year to year. In 2000, seasonal ozone thinning above Antarctica was the largest ever and covered an area three times the size of the continental United States. However, in 2001 and 2002 its size decreased somewhat. Measurements and models indicate that CFCs are the primary culprits.

After a dark winter, sunlight returns to Antarctica. This sets into motion reactions that release large numbers of Cl atoms that initiate the catalytic chlorine cycle (Figure 13-19). Within weeks, this typically destroys 40–50% of the ozone above Antarctica (and 100% in some places). Huge masses of ozone-depleted air above Antarctica then flow northward and linger for a few weeks over parts of Australia, New Zealand, South America, and South Africa. This raises biologically damaging UV-B levels in these areas by 3–10%, and in some years by as much as 20%.

In 1988, scientists discovered that similar but usually less severe ozone thinning occurs over the Arctic during the arctic spring and early summer (February–June), with a seasonal ozone loss of 11–38% (compared to a typical 50% loss above Antarctica). When this mass of air above the Arctic breaks up each spring, large masses of ozone-depleted air flow south to linger over parts of Europe, North America, and Asia. Some *good news* is that in 2002 models indicate that the Arctic is unlikely to develop the large-scale ozone thinning found over the Antarctic.

According to a 1998 model developed by scientists at NASA's Goddard Institute for Space Studies, ozone depletion over the Antarctic and Arctic will be at its worst between 2010 and 2019.

Why Should We Be Worried about Ozone Depletion? Life in the Ultraviolet Zone   Why should we care about ozone loss? Figure 13-22 lists some of the expected effects of decreased levels of ozone in the stratosphere. From a human standpoint the answer is that with less ozone in the stratosphere, more biologically damaging UV-A and UV-B radiation will reach the earth's surface. This will give humans worse sunburns, more eye cataracts (a clouding of the eye's lens that reduces vision and can cause blindness if not corrected), and more skin cancers (Figure 13-23, p. 300).

Humans can make cultural adaptations to increased UV-B radiation by staying out of the sun, protecting their skin with clothing, and applying sunscreens. However, UV-sensitive plants and animals that help support us and other forms of life cannot make such changes except through the long process of biological evolution.

Connections: What Cancer Are You Most Likely to Get?   Research indicates that years of exposure to UV-B ionizing radiation in sunlight is the primary cause of *squamous cell* (Figure 13-23, left) and *basal cell* (Figure 13-23, center) *skin cancers.* Together these two types make up 95% of all skin cancers. Typically there is a 15- to 40-year lag between excessive exposure to UV-B and development of these cancers.

Caucasian children and adolescents who get only one severe sunburn double their chances of getting these two types of cancers. Some 90–95% of these types of skin cancer can be cured if detected early enough, although their removal may leave disfiguring scars. These cancers kill 1–2% of their victims, which amounts to about 2,300 deaths in the United States each year.

A third type of skin cancer, *malignant melanoma* (Figure 13-23, right), occurs in pigmented areas such as moles anywhere on the body. Within a few months, this type of cancer can spread to other organs.

It kills about one-fourth of its victims (most under age 40) within 5 years, despite surgery, chemotherapy, and radiation treatments. Each year it kills about 100,000 people (including more than 7,400 Americans), mostly Caucasians. It can be cured if detected early enough, but recent studies show that some melanoma survivors have a recurrence more than 15 years later.

Recent evidence suggests that about 90% of sunlight's melanoma-causing effect may come from exposure to UV-A (which is not blocked by window glass) and 10% from UV-B. Some sunscreens do not protect from UV-A, and tanning booth lights and sunlamps emit mostly UV-A.

Evidence indicates that people (especially Caucasians) who get three or more blistering sunburns before age 20 are five times more likely to develop malignant melanoma than those who have never had severe sunburns. About 10% of those who get malignant melanoma have an inherited gene that makes them especially susceptible to the disease.

To protect yourself, the safest course is to stay out of the sun (especially between 10 A.M. and 3 P.M., when UV levels are highest) and do not use tanning parlors or sunlamps. When you are in the sun, wear tightly woven protective clothing, a wide-brimmed hat, and

Human Health

• Worse sunburn

• More eye cataracts

• More skin cancers

• Immune system suppression

Food and Forests

• Reduced yields for some crops

• Reduced seafood supplies from reduced phytoplankton

• Decreased forest productivity for UV-sensitive tree species

Wildlife

• Increased eye cataracts in some species

• Decreased population of aquatic species sensitive to UV radiation

• Reduced population of surface phytoplankton

• Disrupted aquatic food webs from reduced phytoplankton

Air Pollution and Materials

• Increased acid deposition

• Increased photochemical smog

• Degradation of outdoor paints and plastics

Global Warming

• Accelerated warming because of decreased ocean uptake of CO_2 from atmosphere by phytoplankton and CFCs acting as greenhouse gases

Figure 13-22 Expected effects of decreased levels of ozone in the stratosphere.

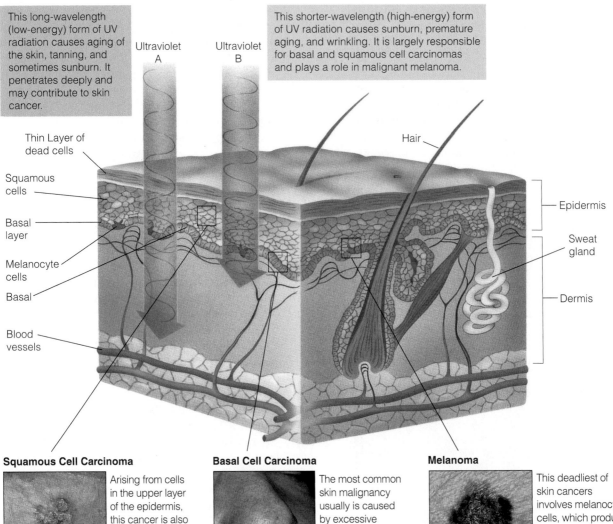

This long-wavelength (low-energy) form of UV radiation causes aging of the skin, tanning, and sometimes sunburn. It penetrates deeply and may contribute to skin cancer.

This shorter-wavelength (high-energy) form of UV radiation causes sunburn, premature aging, and wrinkling. It is largely responsible for basal and squamous cell carcinomas and plays a role in malignant melanoma.

Ultraviolet A

Ultraviolet B

Thin Layer of dead cells

Squamous cells

Basal layer

Melanocyte cells

Basal

Blood vessels

Hair

Epidermis

Sweat gland

Dermis

Squamous Cell Carcinoma

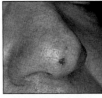

Arising from cells in the upper layer of the epidermis, this cancer is also caused by exposure to sunlight or tanning lamps. It is usually curable if treated early. It grows faster than basal cell carcinoma and can metastasize.

Basal Cell Carcinoma

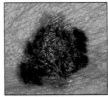

The most common skin malignancy usually is caused by excessive exposure to sunlight or tanning lamps. It develops slowly, rarely metastasizes and is nearly 100% curable if diagnosed early and treated properly.

Melanoma

This deadliest of skin cancers involves melanocyte cells, which produce pigment. It can develop from a mole or on blemished skin, grows quickly, and can spread to other parts of the body (metastasize).

Figure 13-23 Structure of the human skin and the relationships between ultraviolet (UV-A and UV-B) radiation and the three types of skin cancer. The incidence of these types of cancer is rising, mostly because more fair-skinned people have increased their exposure to sunlight by moving to areas with sunnier climates and spending more of their leisure time exposed to sunlight. If ozone-destroying chemicals continue to reduce stratospheric ozone levels, incidence of these types of cancers is expected to rise. (The Skin Cancer Foundation)

sunglasses that protect against UV-A and UV-B radiation (ordinary sunglasses may actually harm your eyes by dilating your pupils so more UV radiation strikes the retina).

Because UV rays can penetrate clouds, overcast skies do not protect you; neither does shade, because UV rays can reflect off sand, snow, water, or patio floors. People who take antibiotics and women who take birth control pills are more susceptible to UV damage.

Use a sunscreen that offers protection against both UV-A and UV-B and has a protection factor of 15 or more (25 if you have light skin). Apply to all exposed skin (including lips) and reapply sunscreens about every 2 hours or immediately after swimming or excessive perspiration. Most people do not realize that the protection factors for sunscreens are based on using one full ounce of the product—about enough to fill a shot glass. Some people also increase their risk of skin cancer by falsely assuming that sunscreens allow them to spend more time in the sun.

Children who use a sunscreen with a protection factor of 15 every time they are in the sun from age 1 to age 18 decrease their chance of getting skin cancer by 80%. Babies under a year old should not be exposed to the sun at all and should not have sunscreens applied until they are at least 6 months old.

Become familiar with your moles and thoroughly examine your skin and scalp at least once a month. Warning signs of skin cancer are a change in the size, shape, or color of a mole or wart (the major sign of malignant melanoma, which must be treated quickly), sudden appearance of dark spots on the skin, or a sore that keeps oozing, bleeding, and crusting over but does not heal. Be alert for precancerous growths (reddish-brown spots with a scaly crust). If you observe any of these signs, consult a doctor immediately.

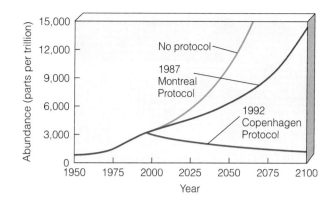

Figure 13-24 Solutions: projected concentrations of ozone-depleting chemicals (ODCs) in the stratosphere under three scenarios: no action, the 1987 Montreal Protocol, and the 1992 Copenhagen Protocol. Stratospheric ozone levels are not expected to return to 1980 levels until about 2050 and to 1950 levels by about 2100, assuming the international agreements are followed and any additional chemicals found to be depleting stratospheric ozone will also be phased out. However, in 2003 the Bush administration and governments of several other developed countries proposed slowing down the phaseout of methyl bromide, an ODC that is widely used as a fumigant in agriculture. (Data from World Meteorological Organization).

13-9 SOLUTIONS: PROTECTING THE OZONE LAYER

How Can We Protect the Ozone Layer? *The scientific consensus of researchers in this field is that we should immediately stop producing all ozone-depleting chemicals (ODCs).* Even with immediate action, the models indicate it will take about 50 years for the ozone layer to return to 1980 levels and about 100 years for recovery to pre-1950 levels.

Some *good news* is that substitutes are available for most uses of CFCs, and others are being developed (Individuals Matter, p. 302).

What Is Being Done to Reduce Ozone Depletion? Some Hopeful Progress In 1987, representatives of 36 nations meeting in Montreal, Canada, developed a treaty, commonly known as the *Montreal Protocol*. Its goal was to cut emissions of CFCs (but not other ozone depleters) into the atmosphere by about 35% between 1989 and 2000. After hearing more bad news about seasonal ozone thinning above Antarctica in 1989, representatives of 93 countries met in London in 1990 and in Copenhagen, Denmark, in 1992 and adopted the *Copenhagen Protocol,* which accelerated the phasing out of key ozone-depleting chemicals.

Here is some *great news.* These landmark international agreements, now signed by 177 countries, are important examples of global cooperation in response to a serious global environmental problem. Without these agreements, ozone depletion would be a much more serious threat (Figure 13-24).

Here is a piece of *disturbing news.* According to a 1998 study by the World Meteorological Organization, ozone depletion in the stratosphere has been cooling the troposphere and has helped offset or disguise as much as 30% of the global warming from our emissions of greenhouse gases. Thus restoring the ozone layer could lead to an increase in global warming.

The ozone treaty set an important precedent for global cooperation and action to avert potential global disaster by using *prevention* to solve a serious environmental problem. Nations and companies agreed to work together to solve this problem for three reasons. *First,* there was convincing and dramatic scientific evidence of a serious problem. *Second,* CFCs were produced by a small number of international companies. *Third,* the certainty that CFC sales would decline over a period of years unleashed the economic and creative resources of the private sector to find even more profitable substitute chemicals.

Why Is It So Difficult to Deal with the Threat of Global Warming? International cooperation in dealing with projected atmospheric warming is much more difficult than dealing with ozone depletion for three reasons. *First,* we lack clear-cut and dramatic evidence that a serious problem exists. *Second,* greenhouse gas emissions result from the actions of hundreds of different large and politically powerful industries (such as coal, oil, chemicals, automobiles, and steel) and billions of consumers. *Third,* reducing greenhouse

Ray Turner and His Refrigerator

INDIVIDUALS MATTER

Ray Turner, an aerospace manager at Hughes Aircraft in California, made an important low-tech ozone-saving discovery by using his head—and his refrigerator. His concern for the environment led him to look for a cheap and simple substitute for the CFCs used as cleaning agents to remove films of oxidation from the electronic circuit boards manufactured at his plant.

He started by looking in his refrigerator. He decided to put drops of various substances on a corroded penny to see whether any of them removed the film of oxidation. Then he used his soldering gun to see whether solder would stick to the surface of the penny, indicating the film had been cleaned off.

First, he tried vinegar. No luck. Then he tried some ground-up lemon peel, also a failure. Next he tried a drop of lemon juice and watched as the solder took hold. The rest, as they say, is history.

Today, Hughes Aircraft uses inexpensive citrus-based solvents that are CFC-free to clean circuit boards. This new cleaning technique has reduced circuit board defects by about 75% at Hughes. And Turner got a hefty bonus. Now other companies, such as AT&T, clean computer boards and chips using acidic chemicals extracted from cantaloupes, peaches, and plums. Maybe you can find a solution to an environmental problem in your refrigerator, grocery, drugstore, or backyard.

gas emissions will take far-reaching and politically controversial changes within most countries and industries and lifestyle changes for billions of consumers.

Thus reducing and adapting to the threat of global warming is a difficult political, economic, and scientific challenge. However, with enough political and business leadership it can be done over the next few decades (Figures 13-15 and 13-18). Moreover, numerous economic studies show that in the long run, meeting this challenge will save huge amounts of money, create many jobs, and be much more profitable for many of the world's major businesses than doing nothing or waiting to act.

The atmosphere is the key symbol of global interdependence. If we can't solve some of our problems in the face of threats to this global commons, then I can't be very optimistic about the future of the world.

MARGARET MEAD

REVIEW QUESTIONS

1. Define the boldfaced terms in this chapter.

2. Summarize briefly how the earth's average temperature has changed over the past 900,000 years and over the past 130 years. Distinguish between *glacial* and *interglacial* periods.

3. What is the earth's *natural greenhouse effect?* How widely is this theory accepted? What are the two major greenhouse gases?

4. What is *global warming?* List three human activities that increase the input of greenhouse gases into the troposphere and could enhance the earth's natural greenhouse effect.

5. List four measurements that suggest increases in atmospheric emissions of greenhouse gas are warming the atmosphere. List four other signs that the troposphere is getting warmer.

6. Describe how scientists develop mathematical models to make projections about future climate change.

7. According to the latest models, between 2000 and 2100 about how much increase is projected for the global average temperature? What is the scientific consensus about how these projected changes are related to human activities such as fossil fuel burning and deforestation? Why does rapid climate change over a few decades to 100 years, whether caused by natural or human-related factors, pose a serious threat to human life, wildlife, and the world's economies? List four reasons why a small minority of climate scientists disagrees with the scientific consensus on climate change.

8. Explain how each of the following factors might enhance or dampen global warming: **(a)** changes in solar output, **(b)** changes in the earth's reflectivity (albedo), **(c)** ability of oceans to store CO_2 and heat, **(d)** changes in ocean currents, **(e)** changes in average sea level, **(f)** changes in cloud cover, **(g)** air pollution, **(h)** effects of increased CO_2 levels on photosynthesis, **(i)** increased methane emissions, and **(j)** ability of soils to absorb CO_2.

9. How rapidly might climates shift? What are the implications for humans if climate shifts rapidly within a few decades?

10. Why should we worry about a possible rise of only one to a few degrees in the average temperature at the earth's surface?

11. List some beneficial effects of atmospheric warming for people and natural ecosystems in some areas. Explain how atmospheric warming might affect each of the following: **(a)** food production, **(b)** water supplies, **(c)** forests, **(d)** biodiversity, **(e)** sea levels and coastal areas, **(f)** weather extremes, **(g)** human population, **(h)** human health, and **(i)** developing countries.

12. What are the four schools of thought about what we should do about global warming?

13. List nine prevention methods and seven cleanup methods for slowing climate change from increased greenhouse gas emissions.

14. Explain how improving energy efficiency could help reduce greenhouse gas emissions.

15. Describe the advantages and disadvantages of removing CO_2 from the atmosphere or smokestacks and storing it in **(a)** immature trees, **(b)** plants that store it in the soil, **(c)** deep underground reservoirs, **(d)** saltwater aquifers underneath the ocean floor, and **(e)** the deep ocean.

16. What is the Kyoto treaty? What are the advantages and disadvantages of this treaty? List seven pieces of good news about progress in reducing greenhouse gas emissions.

17. List the five most important ways you could reduce your emissions of greenhouse gases.

18. List seven ways in which we might prepare for and adjust to the harmful effects of global warming.

19. What is stratospheric *ozone depletion,* and how serious is this problem? What types of chemicals cause ozone depletion? How do these chemicals cause such depletion?

20. What is seasonal ozone thinning over the earth's poles and what are its effects?

21. What are the major harmful effects of ozone depletion on **(a)** human health, **(b)** crop yields, **(c)** forest productivity, **(d)** materials such as plastics and paints, and **(e)** plankton productivity?

22. Distinguish among *squamous cell skin cancer, basal cell skin cancer,* and *malignant melanoma.* List ways in which you can reduce your chances of getting skin cancer.

23. If all ozone-depleting chemicals were banned now, about how long would it take for average concentrations of ozone in the stratosphere to return to levels in **(a)** 1980 and **(b)** 1950?

24. Summarize the progress that has been made in reducing the threat of ozone depletion and explain the importance of such efforts.

25. List three factors that helped countries to agree to an international treaty to phase out ozone-depleting chemicals. List three reasons why getting countries to develop an international treaty and engage in other unified actions to reduce greenhouse gas emissions is much more difficult.

CRITICAL THINKING

1. In preparation for the 1992 UN Conference on the Human Environment in Rio de Janeiro, President George H. W. Bush's top economic adviser gave an address in Williamsburg, Virginia, to representatives of governments from a number of countries. He told his audience not to worry about global warming because the average temperature increases scientists are pre-

dicting were much less than the temperature increase he experienced in coming from Washington, D.C., to Williamsburg, Virginia. What is the fundamental flaw in this reasoning?

2. What changes might occur in the **(a)** global hydrologic cycle (Figure 4-23, p. 81) and **(b)** global carbon cycle (Figure 4-24, p. 82) if the atmosphere experienced significant warming? Explain.

3. What effect should clearing forests and converting them to grasslands and crops have on the earth's **(a)** reflectivity (albedo) and **(b)** average surface temperature? Explain.

4. Explain how global warming could lead to a much cooler climate in Europe and eastern North America.

5. Which of the four schools of thought about what should be done about possible global warming (p. 292) do you favor? Explain.

6. Explain why you agree or disagree with each of the proposals listed in **(a)** Figure 13-15 (p. 293) for slowing down emissions of greenhouse gases into the atmosphere and **(b)** Figure 13-18 (p. 296) for preparing for the effects of global warming. What might be the harmful effects on your life of *not* taking these actions?

7. Do you agree or disagree with the argument by developing countries that developed countries should bear the brunt of reducing CO_2 emissions because they produce much more of these emissions than developing countries? Explain.

8. Should developed countries provide significant aid to help developing countries adapt to the harmful effects of global warming? Explain.

9. What consumption patterns and other features of your lifestyle directly add greenhouse gases to the atmosphere? Which, if any, of these things would you be willing to give up to slow global warming?

10. You have been diagnosed with a treatable basal cell skin cancer (Figure 13-23, middle, p. 300). Explain why you could increase your chances of getting more such cancers within 15–40 years by moving to Australia or Florida.

11. Congratulations! You are in charge of the world. List your three most important actions for dealing with the problems of **(a)** global warming and **(b)** depletion of ozone in the stratosphere.

PROJECTS

1. As a class, conduct a poll of students at your school to determine **(a)** whether they understand the difference between global warming of the troposphere and ozone depletion in the stratosphere (Table 13-2, p. 284) and **(b)** whether they believe global warming from an enhanced greenhouse effect is a very serious problem, a moderately serious problem, or of little concern. Tally the results to see whether there are differences related to year in school, political leaning (liberal, conservative, independent), or sex of poll participants.

2. As a class, conduct a poll of students at your school to determine whether they believe stratospheric ozone depletion is a very serious problem, a moderately serious problem, or of little concern. Tally the results to see whether there are differences related to year in school, political leaning (liberal, conservative, independent), or sex of poll participants.

3. Use the library or the Internet to determine how the current government policy on global warming in the country where you live compares with the policy suggestions made by various analysts and listed in Figures 13-15 (p. 293) and 13-18 (p. 296).

4. Write a 1- to 2-page scenario of what your life could be like by 2060 if nations, companies, and individuals do not take steps to reduce projected global warming caused at least partly by human activities. Contrast your scenario with the positive scenario at the opening of this chapter. Compare and critique scenarios written by different members of your class.

5. Make a concept map of this chapter's major ideas, using the section heads and subheads and the key terms (in boldface). Look on the website for this book for information about making concept maps.

INTERNET STUDY RESOURCES AND RESOURCES FOR FURTHER READING AND RESEARCH

The website for this book contains helpful study aids and many ideas for further reading and research. Log on to

http://biology.brookscole.com/miller10

and click on the Chapter-by-Chapter area. Choose Chapter 13 and select a resource:

■ Flash Cards allows you to test your mastery of the Terms and Concepts to Remember for this chapter.

■ Tutorial Quizzes provides a multiple-choice practice quiz.

■ Student Guide to InfoTrac will lead you to Critical Thinking Projects that use InfoTrac College Edition as a research tool.

■ References lists the major books and articles consulted in writing this chapter.

■ Hypercontents takes you to an extensive list of sites with news, research, and images related to individual sections of the chapter.

INFOTRAC COLLEGE EDITION

Improve your skills with InfoTrac College Edition, a searchable online database of articles from more than 700 periodicals. Log on to

http://www.infotrac-college.com

or access InfoTrac through the website for this book. Try to find the following articles:

1. Global Environmental Change Report. 2001. Amphibian deaths traced to climate changes. *Global Environmental Change Report* 13: 6. *Keywords:* "amphibian" and "climate change." Declines in amphibian populations have been well documented over the last several years. Much of the blame has been placed on water pollution, but there is new evidence that global climate change may significantly contribute to these declines.

2. Lal, R. 2000. A modest proposal for the year 2001: We can control greenhouse gases and feed the world . . . with proper soil management. *Journal of Soil and Water Conservation* 55: 429. *Keywords:* "greenhouse gases" and "soil management." Much of the debate about global warming centers around vehicle and power plant CO_2 emissions, but a substantial amount of CO_2 is generated by the world agricultural industry. At a time when there is greater need for food production, can we have both food and a stable climate?

14 WATER RESOURCES AND WATER POLLUTION

Water Conflicts in the Middle East

Future military wars between countries in the Middle East could be fought over water. Most water in this dry region comes from three shared river basins: the Nile, Jordan, and Tigris-Euphrates (Figure 14-1).

Ten countries share water from the Nile River basin (Figure 14-1), but three countries—Egypt, Sudan, and Ethiopia—use most of the water. Egypt, where it rarely rains, is mostly desert and would not exist without irrigation water from the Nile.

Because of water scarcity, Egypt must import 40% of its grain to feed its current population of 72 million. By 2050, its population is expected to increase to 127 million, greatly increasing its demand for already scarce water.

Most of the precipitation that feeds 85% of the Nile's flow falls in Ethiopia. To meet water needs and help lift many of its 71 million people (173 million projected by 2050) out of poverty, Ethiopia plans to divert more water from the Nile. Sudan also plans to use more of the Nile's water to help feed its 38 million people (projected to more than double to 84 million by 2050).

Such upstream diversions by Ethiopia and Sudan would reduce the amount of water available to water-short Egypt. Its options are to go to war with Sudan and Ethiopia to obtain more water, cut population growth, and improve irrigation efficiency. It can also build the world's longest concrete canal and pump water out of Lake Nasser (the reservoir created from the Nile by the Aswan High Dam). Other options are to irrigate farmland in the middle of the desert, import more grain to reduce the need for irrigation water, work out water-sharing agreements with other countries, or suffer the harsh human and economic consequences of extreme hydrological poverty. Some *promising news* is that the UN Food and Agriculture Organization is working with 10 ten nations bordering the Nile River to help them manage the river's water more equitably as part of the Nile Basin Water Resources project.

Figure 14-1 The Middle East, whose countries have some of the highest population growth rates in the world. Because of the dry climate, food production depends heavily on irrigation. Existing conflicts between countries in this region over access to shared water supplies may soon overshadow both long-standing religious and ethnic clashes and attempts to take over valuable oil supplies.

The Jordan basin is by far the most water-short region, with fierce competition for its water among Jordan, Syria, Palestine (Gaza and the West Bank), and Israel (Figure 14-1). The combined populations of these already water-short countries are projected to more than double from 33 million to 70 million between 2003 and 2050. Some *good news* is that in 1994, Israel and Jordan signed a peace treaty that addressed their disputes over water from the Jordan River basin.

Also, in 2002 Israel and Turkey signed a treaty in which Turkey will supply Israel with water in exchange for arms. However, in 2002 Israel said it would not allow Lebanon to divert water from a river shared by the two countries that feeds into the Jordan River.

Syria plans to build dams and withdraw more water from the Jordan River, decreasing the downstream water supply for Jordan and Israel. Israel warns it may destroy the largest dam that Syria plans to build.

Turkey, located at the headwaters of the Tigris and Euphrates Rivers, controls how much water flows downstream to Syria and Iraq before emptying into the Persian Gulf (Figure 14-1). Turkey is building 24 dams along the upper Tigris and Euphrates Rivers to generate electricity and irrigate a large area of land.

If completed, these dams will reduce the flow of water downstream to Syria and Iraq by up to 35% in normal years and much more in dry years. Syria also plans to build a large dam along the Euphrates River to divert water arriving from Turkey. This will leave little water for Iraq and could lead to war between Syria and Iraq. Indeed, water scientists estimate that separate plans by Turkey, Syria, and Iraq for building dams and irrigating fields would, taken together, withdraw nearly one and a half times more water than the Euphrates usually holds.

Resolving these water distribution problems will require a combination of regional cooperation in allocating water supplies, slowed population growth, improved efficiency in water use, higher water prices to encourage water conservation and improve irrigation efficiency, and increased grain imports to reduce water needs.

Our liquid planet glows like a soft blue sapphire in the hard-edged darkness of space. There is nothing else like it in the solar system. It is because of water.

JOHN TODD

This chapter addresses the following questions:

- What are water's unique physical properties?

- How much fresh water is available to us, and how much of it are we using?

- What causes freshwater shortages, and what can be done about this problem?

- What causes flooding, and what can be done to reduce the risk of flooding and flood damage?

- What pollutes water, where do these pollutants come from, and what effects do they have?

- What are the major water pollution problems of streams, lakes, and groundwater?

- What are the major water pollution problems of oceans?

- How can we prevent and reduce water pollution?

- How can we use the earth's water more sustainably?

14-1 WATER'S IMPORTANCE AND UNIQUE PROPERTIES

Why Is Water So Important? We live on the water planet, with a precious film of water—most of it salt water—covering about 71% of the earth's surface (Figure 6-27, p. 126). All organisms are made up mostly of water; a tree is about 60% water by weight, and most animals are about 50–65% water.

Each of us needs only about a dozen cupfuls of water per day to survive, but we use huge amounts of water to supply us with food, shelter, and our other needs and wants. Water also plays a key role in sculpting the earth's surface, moderating climate, and diluting pollutants.

What Are Some Important Properties of Water? Water is a remarkable substance with a unique combination of properties:

- *There are strong forces of attraction (called hydrogen bonds) between molecules of water.* These attractive forces are the major factor determining water's unique properties.

- *Water exists as a liquid over a wide temperature range because of the strong forces of attraction between water molecules.* Its high boiling point of 100°C (212°F) and low freezing point of 0°C (32°F) mean that water remains a liquid in most climates on the earth.

- *Liquid water changes temperature slowly because it can store a large amount of heat without a large change in temperature.* This high heat capacity helps protect living organisms from temperature fluctuations. It also moderates the earth's climate and makes water an excellent coolant for car engines, power plants, and heat-producing industrial processes.

- *Evaporating liquid water takes large amounts of heat because of the strong forces of attraction between its molecules.* Water absorbs large amounts of heat as it changes into water vapor and releases this heat as the vapor condenses back to liquid water. This helps distribute heat throughout the world (Figure 6-5, p. 109) and determine the climates of various areas. This property also makes water evaporation an effective cooling process—explaining why you feel cooler when perspiration or bathwater evaporates from your skin.

- *Liquid water can dissolve a variety of compounds.* This enables it to carry dissolved nutrients into the tissues of living organisms, flush waste products out of those tissues, serve as an all-purpose cleanser, and help remove and dilute the water-soluble wastes of civilization. Water's superiority as a solvent also means that water-soluble wastes pollute it easily.

- *Water molecules can break down (ionize) into hydrogen ions (H^+) and hydroxide ions (OH^-).* This helps maintain a balance between acids and bases in cells, as measured by the pH of water solutions (Figure 9-20, p. 192).

- *Water filters out wavelengths of ultraviolet radiation* (Figure 3-7, p. 54) *that would harm some aquatic organisms.*

- *The strong attractive forces between the molecules of liquid water cause its surface to contract (high surface tension) and to adhere to and coat a solid (high wetting ability).* These cohesive forces pull water molecules at the surface layer together so strongly that it can support small insects. The combination of high surface tension and wetting ability allow water to rise through a plant from the roots to the leaves (capillary action).

- *Unlike most liquids, water expands when it freezes.* This means that ice has a lower density (mass per unit of volume) than liquid water. Thus ice floats on water. Without this property, lakes and streams in cold climates would freeze solid and lose most of their current forms of aquatic life. Because water expands upon freezing, it can break pipes, crack a car's engine blocks block (which is why we use antifreeze), break up streets, and fracture rocks (thus forming soil).

Water is the lifeblood of the biosphere. It connects us to one another, to other forms of life, and to the entire planet. Despite its importance, water is one of our most poorly managed resources. We waste it and pollute it. We also charge too little for making it available. This encourages still greater waste and pollution of this resource, for which we have no substitute. As Benjamin Franklin said many decades ago: "It is not until the well runs dry that we know the worth of water."

14-2 SUPPLY, RENEWAL, AND USE OF WATER RESOURCES

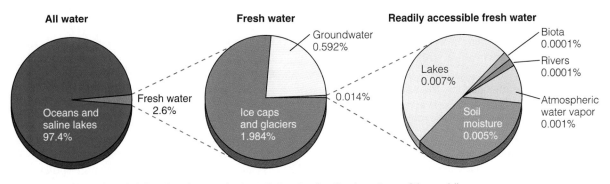 **How Much Fresh Water Is Available?** Only a tiny fraction of the planet's abundant water is available to us as fresh water (Figure 14-2). About 97.4% by volume is found in the oceans and is too salty for drinking, irrigation, or industry (except as a coolant). Most of the remaining 2.6% that is fresh water is locked up in ice caps or glaciers or in groundwater too deep or salty to be used (Figure 14-2).

Thus only about 0.014% of the earth's total volume of water is easily available to us as soil moisture, usable groundwater, water vapor, and lakes and streams (Figure 14-2). If the world's water supply were only 100 liters (26 gallons), our usable supply of fresh water would be only 0.014 liter (2.5 teaspoons).

Some *good news* is that the available fresh water amounts to a generous supply. Moreover, this water is continuously collected, purified, recycled, and distributed in the solar-powered *hydrologic cycle* (Figure 4-23, p. 81). However, this works only as long as we do not overload water systems with slowly degradable and nondegradable wastes or withdraw water from underground supplies faster than it is replenished. The *bad news* is that in parts of the world we are doing both.

Differences in average annual precipitation divide the world's countries and people into water *haves* and *have-nots.* For example, Canada, with only 0.5% of the world's population, has 20% of the world's usable fresh water, whereas China, with 21% of the world's people, has only 7% of the supply.

As population, irrigation, and industrialization increase, water shortages in already water-short regions will intensify and heighten regional tensions between and within countries (p. 305). Global warming can in-

crease global rates of evaporation, shift precipitation patterns, and disrupt water supplies and thus food supplies. Some areas will get more precipitation and some less. Some river flows will change.

What Is Surface Water? **Surface runoff** is precipitation that does not infiltrate the ground or return to the atmosphere by evaporation (including transpiration). This runoff flows into streams, lakes, wetlands, estuaries, and reservoirs.

About two-thirds of the world's annual runoff is lost by seasonal floods and is not available for human use. The remaining one-third is **reliable runoff,** which generally we can generally count on as a stable source of water from year to year.

A **watershed,** also called a **drainage basin,** is a region from which water drains into a stream, lake, reservoir, wetland, estuary, or other body of surface water.

What Is Groundwater? Some precipitation infiltrates the ground and percolates downward through voids (pores, fractures, crevices, and other spaces) in soil and rock (Figure 14-3, p. 308). The water in these voids is called **groundwater.**

Close to the surface, the voids have little moisture in them. However, below some depth, in the **zone of saturation,** the voids are completely filled with water. The **water table** is located at the top of the zone of saturation. It falls in dry weather and rises in wet weather.

Porous, water-saturated layers of sand, gravel, or bedrock through which groundwater flows are called **aquifers** (Figure 14-3). They are like large elongated sponges through which groundwater seeps. Aquifers are replenished naturally by precipitation that percolates downward through soil and rock in what is called **natural recharge,** but some are recharged from the side by *lateral recharge* from nearby streams.

Groundwater normally moves from points of high elevation and pressure to points of lower elevation and pressure. This movement is quite slow, typically

Figure 14-2 *Natural capital:* the planet's water budget. Only a tiny fraction by volume of the world's water supply is fresh water available for human use.

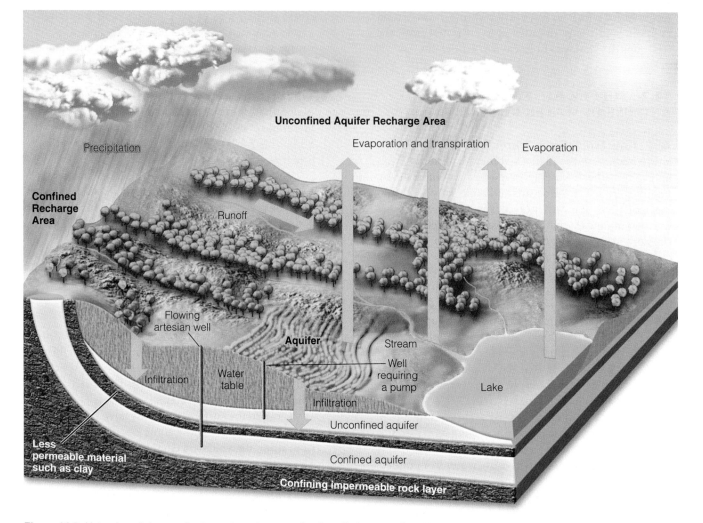

Figure 14-3 *Natural capital:* groundwater system. An *unconfined aquifer* is an aquifer with a water table. A *confined aquifer* is bounded above and below by less permeable beds of rock. Groundwater in this type of aquifer is confined under pressure.

Labels on figure:

Unconfined Aquifer Recharge Area

Precipitation

Evaporation and transpiration

Evaporation

Confined Recharge Area

Runoff

Flowing artesian well

Aquifer

Stream

Well requiring a pump

Infiltration

Water table

Lake

Infiltration

Less permeable material such as clay

Unconfined aquifer

Confined aquifer

Confining impermeable rock layer

only a meter or so (about 3 feet) per year and rarely more than 0.3 meter (1 foot) per day.

Eventually most groundwater flows into rivers, lakes, estuaries, and wetlands. Thus there is a hydrological connection between groundwater and surface water because the two terms merely describe the physical location of some of the earth's water as it circulates through the hydrologic cycle (Figure 4-23, p. 81).

Whether water flows from a stream to a nearby aquifer or vice versa depends of on the level of the water table in the aquifer and on the stream's elevation. If the water table of an aquifer is above a stream's elevation, water will flow laterally from the aquifer into the stream and increase its flow. However, as groundwater overpumping lowers the water table, a point is reached where water flows laterally from the stream into the aquifer. Thus if we disrupt the hydrological cycle by removing groundwater faster than it is replenished, some nearby streams, lakes, and wetlands can dry up. For example, in Arizona groundwater overpumping has dried up or greatly reduced the water in 90% of the state's desert streams and rivers.

Much of the groundwater in U.S. aquifers came from huge freshwater lakes that covered large areas of the country during various ice ages. Water from these lakes percolated into the ground and collected in aquifers.

Some aquifers get very little (if any) recharge and on a human time scale are nonrenewable resources. Typically, they are found fairly deep underground, formed tens of thousands of years ago, and get little or no recharge. Withdrawals from such aquifers amount to *water mining* that, if kept up, will deplete these ancient deposits of water.

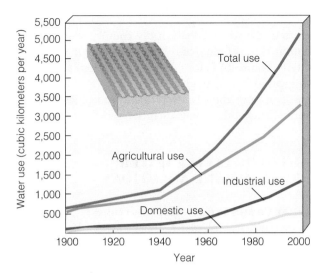

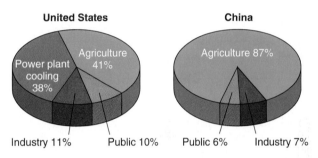

Figure 14-4 Global water withdrawal, 1900–2000. Between 2000 and 2050, the world's population is expected to increase by about 3.3 billion people and greatly increase the demand for water. (Data from World Commission on Water Use in the 21st Century)

How Much of the World's Reliable Water Supply Are We Withdrawing?

During the last century the human population tripled. According to a 2002 report by the United Nations, during this period our global water withdrawal (use) increased sevenfold (Figure 14-4) and per capita withdrawal quadrupled.

As a result, we now withdraw about 35% of the world's reliable runoff. We leave another 20% of this runoff in streams to transport goods by boats, dilute pollution, and sustain fisheries and wildlife. Thus *we directly or indirectly are already using about 55% of the world's reliable runoff of surface water.*

Because of increased population growth and economic development, global withdrawal rates of surface water could reach more than 70% of the reliable surface runoff by 2025 and exceed such runoff in a growing number of areas. In 2001, the National Intelligence Council (an advisory group to the U.S. Central Intelligence Agency) issued the following warning: "As countries press against the limits of available water between now and 2015, the possibility of armed conflict over water supplies will increase."

According to Sandra Postel, director of the Global Water Policy Project, "If we're to have any hope of satisfying the food and water needs of the world's people in the years ahead, we will need a fundamental shift in how we use and manage water."

How Do We Use the World's Fresh Water?

Uses of withdrawn water vary from one region to

another and from one country to another (Figure 14-5). Worldwide, about 69% of all water withdrawn each year from rivers, lakes, and aquifers is used to irrigate 17% of the world's cropland and produce about 40% of the world's food.

Industry uses about 23% of the water withdrawn each year, and cities and residences use the remaining 8%. Agriculture and manufacturing use large amounts of water (Figure 14-6).

Some of the water withdrawn from a source may be returned to that source for reuse. *Consumptive water use* occurs when water withdrawn becomes unavailable for reuse in the basin from which it was removed—mostly because of losses such as evaporation or contamination.

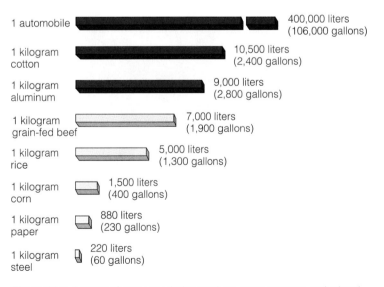

Figure 14-5 Use of water withdrawn in the United States and China. The United States has the world's highest per capita use of water, amounting to an average of 4,800 liters (1,280 gallons) per person per day in 1999. Some *good news* is that between 1980 and 1999, total water use in the United States decreased by 10%, and per capita use dropped by 33%. Most of these decreases in water use occurred because of more efficient irrigation. (Data from Worldwatch Institute and World Resources Institute)

1 automobile	400,000 liters (106,000 gallons)
1 kilogram cotton	10,500 liters (2,400 gallons)
1 kilogram aluminum	9,000 liters (2,800 gallons)
1 kilogram grain-fed beef	7,000 liters (1,900 gallons)
1 kilogram rice	5,000 liters (1,300 gallons)
1 kilogram corn	1,500 liters (400 gallons)
1 kilogram paper	880 liters (230 gallons)
1 kilogram steel	220 liters (60 gallons)

Figure 14-6 Amount of water needed to produce some common agricultural and manufactured products. (Data from U.S. Geological Survey)

Case Study: Freshwater Resources in the United States The United States has plenty of fresh water. But much of it is in the wrong place at the wrong time or contaminated by agricultural and industrial practices. The eastern states usually have ample precipitation, whereas many western states have too little (Figure 14-7, top).

In the East, the largest uses for water are for energy production, cooling, and manufacturing. The largest use by far in the West is for irrigation, which accounts for about 85% of its water use.

In many parts of the eastern United States, the most serious water problems are flooding, occasional urban shortages, and pollution. For example, the 3 million residents of Long Island, New York, get most of their water from an increasingly contaminated aquifer.

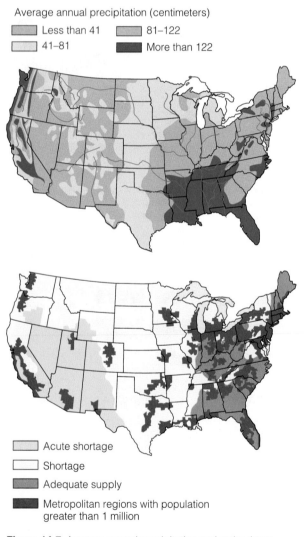

Average annual precipitation (centimeters)

- Less than 41
- 41–81
- 81–122
- More than 122

- Acute shortage
- Shortage
- Adequate supply
- Metropolitan regions with population greater than 1 million

Figure 14-7 Average annual precipitation and major rivers (top) and water-deficit regions in the continental United States and their proximity to metropolitan areas having populations greater than 1 million (bottom). (Data from U.S. Water Resources Council and U.S. Geological Survey)

The major water problem in the arid and semiarid areas of the western half of the country (Figure 14-7, bottom) is a shortage of runoff, caused by low precipitation (Figure 14-7, top), high evaporation, and recurring prolonged drought. Water tables in many areas are dropping rapidly as farmers and cities deplete aquifers faster than they are recharged.

14-3 TOO LITTLE WATER: PROBLEMS AND SOLUTIONS

What Causes Freshwater Shortages? According to Swedish hydrologist Malin Falkenmark, there are four causes of water scarcity: a *dry climate* (Figure 6-3, p. 108), *drought* (a period of 21 days or longer in which precipitation is at least 70% lower and evaporation is higher than normal), *desiccation* (drying of exposed soil because of such activities as deforestation and overgrazing by livestock), and *water stress* (low per capita availability of water caused by increasing numbers of people relying on limited runoff levels).

Figure 14-8 shows the degree of stress on the world's major river systems, based on comparing the amount of reliable runoff of surface water available with the amount people use. A country is said to be *water stressed* when the volume of its reliable runoff per person drops below about 1,700 cubic meters (60,000 cubic feet) per year. A country suffers from *water scarcity* when per capita water availability falls below 1,000 cubic meters (35,000 cubic feet) per year.

According to a 2002 report by the United Nations, about 500 million people live in countries that are water scarce or water stressed. By 2025, there may be 2.4–3.4 billion people in water-scarce or water-stressed river basins, particularly in North Africa and west Asia. According to China's former environmental minister, China's freshwater supplies are capable of sustainably supporting only about half of its current population, In 2002, World Bank officials said that dwindling water supplies in various parts of the world would be a major factor inhibiting economic growth and development.

In 2002 the International Food Policy Research Institute and the International Water Management Institute used computer modeling to project that by 2025, water scarcity will cause annual global losses of 350 million metric tons (385 million tons) of food production—slightly more than the entire current U.S. grain crop. According to Mark Rosegrant, lead author of the report, "Unless we change policies and priorities, in 20 years, there won't be enough water for cities, households, the environment, or growing food. Water is not like oil. There is no substitute. If we continue to take it for granted, much of the earth is going to run short of water or food—or both."

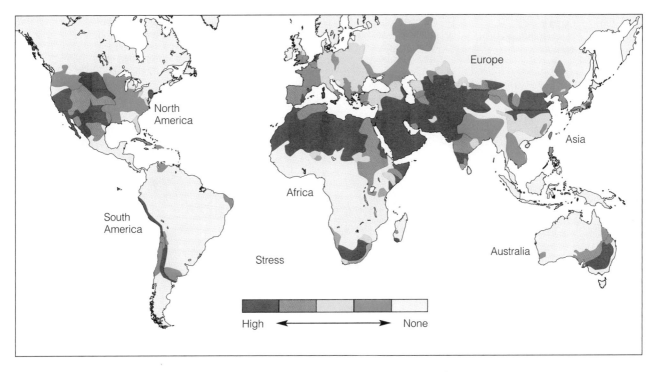

Figure 14-8 Stress on the world's major river basins, based on a comparison of the amount of water available with the amount used by humans. (Data from World Commission on Water Use in the 21st Century)

Even when a plentiful supply of water exists, most of the 1.1 billion poor people living on less than $1 a day cannot afford a safe supply of drinking water and live in *hydrological poverty.* Most are cut off from municipal water supplies and must collect water from unsafe sources or buy water (often coming from polluted rivers) from private vendors at high prices. In water-short rural areas in developing countries, many women and children must walk long distances each day, carrying heavy jars or cans, to get a meager and sometimes contaminated supply of water.

Since the 1970s, water stress and scarcity, intensified by prolonged drought, has killed more than 24,000 people per year and created millions of environmental refugees.

Solutions: How Can We Increase Freshwater Supplies? There are several ways to increase the supply of fresh water in a particular area. One is to *build dams and reservoirs to store runoff for release as needed.* Another is to *bring in surface water from another area.* We can also *withdraw groundwater* and *convert salt water to fresh water (desalination).*

Other strategies are to *reduce water waste* and *import food to reduce water use in growing food.* Each imported metric ton of grain saves roughly 1,000 metric tons of water needed to produce the grain.

In *developed countries,* people tend to live where the climate is favorable and then bring in water from

another watershed. In *developing countries,* most people (especially the rural poor) must settle where the water is and try to capture the precipitation they need. Poverty is the principal cause of lack of access to sufficient water, regardless of how much water is available.

What Are the Advantages and Disadvantages of Large Dams and Reservoirs? Large dams and reservoirs have benefits and drawbacks (Figure 14-9, p. 312). Their main purpose is to capture and store runoff and release it as needed for controlling floods, producing hydroelectric power, and supplying water for irrigation and for towns and cities. Reservoirs also provide recreational activities such as swimming, fishing, and boating.

Some *good news* is that the world's dams have increased the annual reliable runoff available for human use by nearly one-third. Some *bad news* is that a series of dams on a river, especially in arid areas, can reduce downstream flow to a trickle and prevent it from reaching the sea as a part of the hydrologic cycle. According to the World Commission on Water in the 21st Century, half of the world's major rivers are going dry part of the year or are seriously polluted.

In addition to threatening the water supplies for 500 million people, this engineering approach to river management often impairs some of the important ecological and economic services rivers provide (Figure 6-39, p. 135, and Figure 14-10, p. 312).

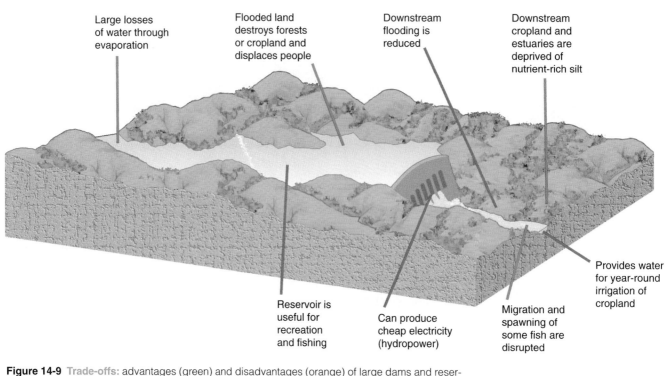

Large losses of water through evaporation

Flooded land destroys forests or cropland and displaces people

Downstream flooding is reduced

Downstream cropland and estuaries are deprived of nutrient-rich silt

Provides water for year-round irrigation of cropland

Reservoir is useful for recreation and fishing

Can produce cheap electricity (hydropower)

Migration and spawning of some fish are disrupted

Figure 14-9 Trade-offs: advantages (green) and disadvantages (orange) of large dams and reservoirs. The world's 45,000 large dams (higher than 15 meters or 50 feet) capture and store about 14% of the world's runoff, provide water for about 45% of irrigated cropland, and supply more than half the electricity used by 65 countries. The United States has more than 70,000 large and small dams, capable of capturing and storing half of the country's entire river flow.

Figure 14-10 *Natural capital:* ecological services provided by rivers. Currently, the services are given little or no monetary value when the costs and benefits of dam and reservoir projects are assessed. According to environmental economists, attaching even crudely estimated monetary values to these ecosystem services would help sustain them.

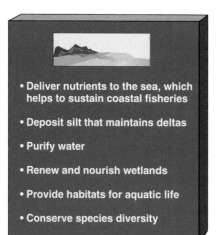

• **Deliver nutrients to the sea, which helps to sustain coastal fisheries**

• **Deposit silt that maintains deltas**

• **Purify water**

• **Renew and nourish wetlands**

• **Provide habitats for aquatic life**

• **Conserve species diversity**

For example, according to water-resource expert Peter H. Gleck, more than 24% of the world's freshwater fish species are threatened or endangered, primarily because dams and water withdrawals have destroyed many free-flowing rivers.

Case Study: China's Three Gorges Dam When completed, China's Three Gorges project on the moun-

tainous upper reaches of the Yangtze River will be the world's largest hydroelectric dam and reservoir. The dam, 2 kilometers (1.2 miles) long, is supposed to be completed by 2013 at a cost of at least $25 billion.

About 1.9 million people will be relocated from the area to be flooded to form the gigantic 600-kilometer-long (370-mile-long) reservoir behind the dam. When completed, the dam will have the electric output of 20 large coal-burning or nuclear power plants. It will also help hold back the Yangtze River's floodwaters, which have killed more than 500,000 people during the past 100 years—including 4,000 people in 1998. Figure 14-11 lists major advantages and disadvantages of this huge dam and reservoir.

What Are the Advantages and Disadvantages of Large-Scale Water Transfers? The Aral Sea Disaster Tunnels, aqueducts, and underground pipes can transfer stream runoff collected by dams and reservoirs from water-rich areas to water-poor areas.

China's Three Gorges Dam

Advantages	Disadvantages
Generates 10% of China's electricity	Floods large areas of cropland and forests
Reduces dependence on coal	Displaces 1.9 million people
Reduces air pollution	Increases water pollution because of reduced water flow
Reduces CO_2 emissions	Reduces deposits of nutrient-rich sediments below dam
Reduces chances of downstream flooding for 15 million people	Increases saltwater introduced into drinking water near mouth of river because of decreased water flow
Reduces river silting below dam by eroded soil	Disrupts spawning and migration of some fish below dam
Increases irrigation water for cropland below dam	High cost

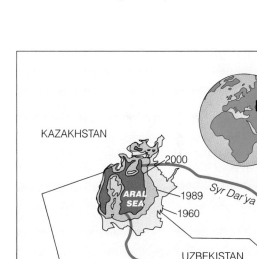

Figure 14-11 Trade-offs: advantages and disadvantages of the Three Gorges dam being built across the Yangtze River in China.

Although such transfers can bring water to water-short areas, they also create environmental problems. Indeed, most of the world's dam projects and large-scale water transfers illustrate the important ecological principle that *you cannot do just one thing.*

An example of this principle in action is the shrinking of the Aral Sea (Figure 14-12). It is a result of a large-scale water transfer project in an area of the former Soviet Union with the driest climate in central Asia. Since 1960, enormous amounts of irrigation water have been diverted from the inland Aral Sea and its two feeder rivers to create one of the world's largest irrigated areas. The irrigation canal, the world's longest, stretches over 1,300 kilometers (800 miles).

This large-scale water diversion project (coupled with droughts) and high evaporation rates in this area's hot, dry climate have caused a regional ecological, economic, and health disaster. Here are some of the problems caused by this massive water-diversion project. The sea's salinity has tripled, its surface area has decreased by 54% (Figure 14-12), and its volume has decreased by 75%. Water withdrawal for agriculture has also reduced the sea's two supply rivers to mere trickles.

This process has converted about 36,000 square kilometers (14,000 square miles) of former lake bottom to a human-made desert covered with glistening white salt. It has also caused the presumed extinction of 20 of the area's 24 native fish species from the increased salt concentration. This has devastated the area's fishing industry, which once provided work for more than 60,000 people. Fishing villages and boats once on the sea's coastline now are in the middle of a salt desert and have been abandoned.

About 85% of the wetlands in the area have been eliminated, which along with increased pollution has greatly reduced waterfowl populations. Roughly half the area's bird and mammal species have disappeared.

Shrinkage of the lake has caused one of the world's worst salinization problems. Winds pick up the salty dust that encrusts the lake's now-exposed bed and blow it onto fields as far as 300 kilometers (190 miles) away. As the salt spreads, it pollutes water and kills wildlife, crops, and other vegetation. Aral Sea dust settling on glaciers in the Himalayas is causing them to melt at a faster than normal rate.

Groundwater and surface water pollution have increased. To raise yields, farmers have increased inputs of herbicides, insecticides, fertilizers, and irrigation water on some crops. Many of these chemicals have

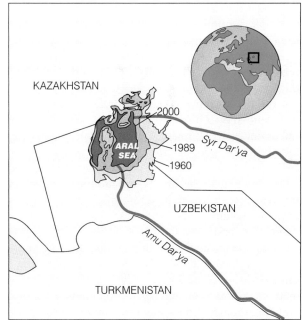

Figure 14-12 Once the world's fourth largest freshwater lake, the Aral Sea has been shrinking and getting saltier since 1960 because most of the water from the rivers that replenish it has been diverted to grow cotton and food crops. As the lake shrinks, it leaves behind a salty desert, economic ruin, increasing health problems, and severe ecological disruption.

percolated downward and accumulated to dangerous levels in the groundwater, from which most of the region's drinking water comes.

Shrinkage of the Aral Sea has altered the area's climate. The once-huge sea acted as a thermal buffer that moderated the heat of summer and the extreme cold of winter. Now there is less rain, summers are hotter and drier, winters are colder, and the growing season is shorter. A combination of such climate change and severe salinization have has reduced crop yields by 20–50% on almost a third of the area's cropland.

Finally, there have been increasing health problems from a combination of toxic dust, salt, and contaminated water for a growing number of the 58 million people living in the Aral Sea's watershed.

Can the Aral Sea be saved, and can the area's serious ecological and human health problems be reduced? Since 1999, the United Nations and the World Bank have spent about $600 million to purify drinking water and upgrade irrigation and drainage systems to improve irrigation efficiency, flush salts from croplands, and boost crop productivity. In addition, wetlands and artificial lakes have been constructed to help restore aquatic vegetation, wildlife, and fisheries. However, this process will take decades and will not prevent the shrinkage of the Aral Sea into a few brine lakes.

Case Study: The California Water Transfer Project One of the world's largest watershed transfer projects is the *California Water Project*. It uses a maze of giant dams, pumps, and aqueducts to transport water from water-rich northern California to heavily populated areas and to arid and semiarid agricultural regions (Figure 14-7), mostly in southern California (Figure 14-13).

For decades, northern and southern Californians have feuded over how the state's water should be allocated under this project. Southern Californians say they need more water from the north to grow more crops and to support Los Angeles, San Diego, and other growing urban areas. Agriculture uses 74% of the water withdrawn in California, much of it for growing water-thirsty crops such as alfalfa under desert-like conditions.

Opponents in the north say that sending more water south would degrade the Sacramento River, threaten fisheries, and reduce the flushing action that helps clean San Francisco Bay of pollutants. They also argue that much of the water sent south is wasted unnecessarily. They point to studies showing that making irrigation just 10% more efficient would provide enough water for domestic and industrial uses in southern California.

However, if water supplies in northern California and in the Colorado River basin drop sharply because

Figure 14-13 Solutions: California Water Project and the Central Arizona Project. These projects involve large-scale water transfers from one watershed to another. Arrows show the general direction of water flow.

of global warming, the amount of water delivered by the huge distribution system will plummet. According to a 2002 joint study by more than two dozen scientists and engineers, projected global warming will have a devastating effect on water availability in California (especially southern California) and other water-short states in the western United States (Figure 14-7, bottom) even under the study's best-case scenario. The study predicts that overall precipitation levels are likely to remain constant, but warmer temperatures will cause what would have fallen as snow to come down as rain. Currently, the annual snowpack acts as a natural reservoir, storing water through the winter so it will melt and be released during the spring and summer when the demand for water is high. If that precipitation falls as winter rain instead of as snow, the rain will fill rivers and streams at a time of year when the demand for water is low. Some analysts project that sometime during this century many of the people living in arid southern California cities and farm areas will have to move somewhere else because of a lack of water.

Pumping out more groundwater is not the answer because it is already being withdrawn faster than it is replenished throughout much of California. To most analysts, quicker and cheaper solutions are improving irrigation efficiency and allowing farmers to sell their legal rights to withdraw certain amounts of water from rivers.

Use of these solutions may increase. For years, California has drawn more water from the Colorado River than its legally allotted amount. This was not challenged because other states did not use the full amount they were entitled to under a 1929 accord allotting withdrawals by states along the river.

This changed in 2003 because of after a combination of rapid population and economic growth in other states along the Colorado and severe drought (which reduced the amount of water flowing in the river). In 2002, the Secretary of the Interior enforced this long-standing water law. Effective January 1, 2003, the Interior Department began cutting the amount of water California can withdraw for from the Colorado River.

What Are the Advantages and Disadvantages of Withdrawing Groundwater? Aquifers (Figure 14-3) provide drinking water for about one-fourth of the world's people. In the United States, water pumped from aquifers supplies about 51% of the drinking water (96% in rural areas and 20% in urban areas) and 43% of irrigation water.

Relying more on groundwater has advantages and disadvantages (Figure 14-14). The *good news* is that aquifers are widely available. They are also renewable sources of water as long as the water is not withdrawn faster than it is replaced and the aquifers do not become contaminated. The *bad news* is that according to Earth Policy Institute and the World Resources Institute, wa-

ter tables are falling in many areas of the world as the rate of water pumping exceeds the rate of recharge from precipitation.

In the United States groundwater is being withdrawn at four times its replacement rate. The most serious overdrafts are occurring in parts of the huge Ogallala Aquifer, extending from southern South Dakota to central Texas (Case Study, p. 316) and in parts of the arid southwestern United States (Figure 14-7, top, p. 310). Serious groundwater depletion is also taking place in California's water-short Central Valley, which supplies about half the country's vegetables and fruits.

These and other withdrawals have led to groundwater overdrafts (Figure 14-15, top, p. 316) and subsidence or sinking of land below aquifers in various parts of the United States (Figure 14-15, bottom). A 2002 study by three environmental groups found that urban sprawl, which paves over large areas of landscape, slows the replenishment of underground aquifers and makes it harder for communities to cope with drought.

Groundwater overdrafts near coastal areas can contaminate groundwater supplies by causing intrusion of salt water into aquifers (Figure 14-16, p. 316). This is an especially serious problem in coastal areas in Florida, California, South Carolina, and Texas.

Overdrafting groundwater also increases costs because more energy is needed to lift water (a heavy substance) from lower levels. In Arizona, for example, the electric energy to run a deep commercial water well can cost up to $2,000 a month. Deeper wells can be drilled at great cost as water tables drop. But water pumped from such deep wells can be polluted because naturally occurring toxic elements such as arsenic, fluorine, and radioactive radon are more prevalent at deeper levels in the earth. Also, more of these substances dissolve in the hotter water found at these deeper levels.

Aquifer depletion is also a problem in Saudi Arabia, central and northern China, northwest and southern India, northern Africa (especially Libya and Tunisia), southern Europe, the Middle East, and parts of Mexico, Thailand, and Pakistan.

According to water resource expert Sandra Postel, about 480 million of the world's 6.3 million billion people are being fed with grain produced with eventually unsustainable water mining from aquifers. This example of the tragedy of the commons (p. 7) is expected to increase as irrigated areas are expanded to help feed the 3.3 billion more people projected to join the ranks of humanity between 2000 and 2050.

In addition to limiting future food production, overpumping aquifers is increasing the gap between the rich and poor in some areas. As water tables drop, farmers must drill deeper wells, buy larger pumps to bring the water to the surface, and use more electricity

Withdrawing Groundwater

Advantages	Disadvantages
Good source of water for drinking and irrigation	Aquifer depletion from overpumping
Available year-round	Sinking of land (subsidence) when water removed
Exists almost everywhere	Polluted aquifers unusable for decades or centuries
Renewable if not overpumped or contaminated	Saltwater intrusion into drinking water supplies near coastal areas
No evaporation losses	Reduced water flows into streams, lakes, estuaries, and wetlands
Cheaper to extract than most surface waters	Increased cost, energy use, and contamination from deeper wells

Figure 14-14 Trade-offs: advantages and disadvantages of withdrawing groundwater.

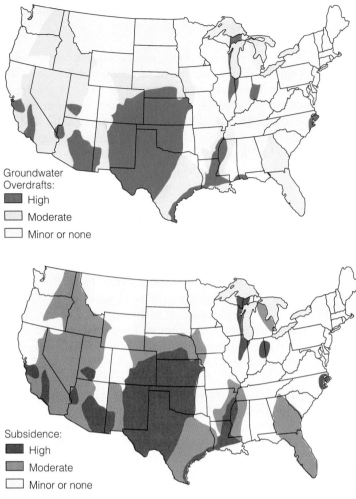

Groundwater
Overdrafts:

- ■ High
- ▨ Moderate
- ☐ Minor or none

Subsidence:

- ■ High
- ▨ Moderate
- ☐ Minor or none

Figure 14-15 Areas of greatest aquifer depletion from ground-water overdraft (top) and ground subsidence (bottom) in the continental United States. Aquifer depletion is also high in Hawaii and Puerto Rico (not shown on map). (Data from U.S. Water Resources Council and U.S. Geological Survey)

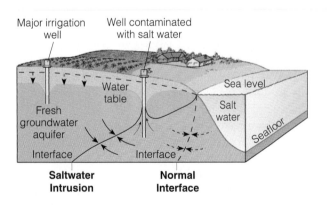

Figure 14-16 *Saltwater intrusion* along a coastal region. When the water table is lowered, the normal interface (dotted line) between fresh and saline groundwater moves inland (solid line), making groundwater drinking supplies unusable.

to run the pumps. Poor farmers cannot afford to do this and many end up losing their land and either working for richer farmers or hoping to survive by migrating to cities.

Figure 14-17 lists ways to prevent or slow the problem of groundwater depletion. Because groundwater moves so slowly, it may take decades of groundwater overpumping before the aquifer is depleted. Then it is too late because replenishing such aquifers can take decades to thousands of years.

Case Study: A Shrinking Ogallala Aquifer

Large amounts of water have been pumped from the Ogallala, the world's largest known aquifer (Figure 14-18). The *good news* is that this has helped transform vast areas of arid high plains prairie land into one of the largest and most productive agricultural regions in the United States. Mostly because of irrigated farming, this region produces 20% of U.S. agricultural output (including 40% of its feedlot beef), valued at $32 billion per year.

The *bad news* is the increasing aquifer depletion in some areas. Although this aquifer is gigantic, it is essentially nonrenewable (stored during the retreat of the last ice age about

Groundwater Depletion

Prevention	Control
Waste less water	Raise price of water to discourage waste
Subsidize water conservation	
Ban new wells in aquifers near surface waters	Tax water pumped from wells near surface waters
Buy and retire groundwater withdrawal rights in critical areas	
Do not grow water-intensive crops in dry areas	Set and enforce minimum stream flow levels
Reduce birth rates	

Figure 14-17 Solutions: ways to prevent or slow groundwater depletion.

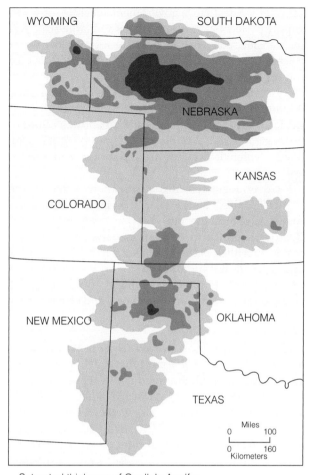

Saturated thickness of Ogallala Aquifer

Less than 61 meters (200 ft.)

61–183 meters (200–600 ft.)

More than 183 meters (600 ft.)
(as much as 370 meters or 1,200 ft. in places)

Figure 14-18 The Ogallala is the world's largest known aquifer. If the water in this aquifer were above ground, it could cover all 50 states with 0.5 meter (1.5 feet) of water. Water withdrawn from this aquifer is used to grow crops, raise cattle, and provide cities and industries with water. As a result, this aquifer, which is renewed very slowly, is being depleted (especially at its thin southern end in parts of Texas, New Mexico, Oklahoma, and Kansas). (Data from U.S. Geological Survey)

15,000–30,000 years ago) with an extremely slow recharge rate. In some areas, water is being pumped out of the aquifer eight to ten times faster than the aquifer's natural recharge rate.

The northernmost states (Wyoming, North Dakota, South Dakota, and parts of Colorado) still have ample supplies. But supplies in parts of the southern states, where the aquifer is thinner (Figure 14-18), are being depleted rapidly, with about two-thirds of the aquifer's depletion taking place in the Texas High Plains.

According to water experts, at the current rate of withdrawal one-fourth of the aquifer's original supply will be depleted by 2020 and much sooner in areas where it is shallow. It will take thousands of years to replenish the aquifer. Drilling more wells in this aquifer, which is like putting more straws in a glass of water, hastens its depletion.

Government subsidies designed to increase crop production also increase depletion of the Ogallala. Such subsidies encourage farmers to grow water-thirsty cotton in the lower basin, give farmers crop-disaster payments, and provide tax breaks in the form of groundwater depletion allowances (with larger breaks for heavier groundwater use). Figure 14-17 (p. 316) lists ways to slow depletion of this essentially nonrenewable water resource.

How Useful Is Desalination? Desalination involves removing dissolved salts from ocean water or from brackish (slightly salty) groundwater. It is another way to increase supplies of fresh water.

One method for desalinating water is *distillation*. It involves heating salt water until it evaporates (and leaves behind salts in solid form) and condenses as fresh water. Another method is *reverse osmosis*. It involves pumping salt water at high pressure through a thin membrane whose pores allow water molecules, but not dissolved salts, to pass through. In effect, high pressure is used to push fresh water out of salt water.

There are about 13,300 desalination plants in 120 countries, especially in the arid, desert nations of the Middle East, North Africa, the Caribbean, and the Mediterranean. These plants meet less than 0.2% of the world's water needs.

Desalination would have to increase 25-fold just to supply 5% of current world water use. This is unlikely because desalination has two major disadvantages. *First, it is expensive because it takes large amounts of energy.* Desalinating water costs 2–3 two to three times as much as the conventional purification of fresh water.

Second, desalination produces large quantities of wastewater (brine) containing high levels of salt and other minerals. Dumping the concentrated brine into the ocean near the plants increases the salinity of nearby ocean water and threatens food resources and aquatic life in estuary waters. Dumping it on land could contaminate groundwater and surface water.

Scientists are working to develop new membranes for reverse osmosis that can separate water from salt more efficiently and under less pressure. If successful, this strategy could bring down the cost of using desalination to produce drinking water. However, desalinated water probably will not be cheap enough to irrigate conventional crops or meet much of the world's demand for fresh water.

Solving this problem would require the development of affordable solar-powered distillation plants and figuring out what to do with the resulting huge quantities of wastewater containing high levels of salt and other minerals.

Can Cloud Seeding and Towing Icebergs Improve Water Supplies? For decades, the United States and several other countries have been experimenting with seeding clouds with tiny particles of chemicals (such as silver iodide). The particles form water condensation nuclei and thus produce more rain over dry regions and more snow over mountains.

Some *bad news* is that cloud seeding is not useful in very dry areas, where it is most needed, because rain clouds rarely are available there. It also introduces large amounts of the cloud-seeding chemicals into soil and water systems, possibly harming people, wildlife, and agricultural productivity.

Another problem is that cloud seeding can result in legal disputes over the ownership of water in clouds. For example, during the 1977 drought in the western United States, the attorney general of Idaho accused officials in neighboring Washington of "cloud rustling" and threatened to file suit in federal court.

Some analysts have proposed towing huge icebergs to arid coastal areas (such as Saudi Arabia and southern California) and pumping fresh water from the melting bergs ashore. But the technology for doing this is not available and the costs may be too high, especially for water-short developing countries.

14-4 REDUCING WATER WASTE

What Are the Benefits of Reducing Waste of Water? Mohamed El-Ashry of the World Resources Institute estimates that *65–70% of the water people use throughout the world is lost through evaporation, leaks, and other losses.* The United States, the world's largest user of water, does slightly better but still loses about 50% of the water it withdraws. El-Ashry believes it is economically and technically feasible to reduce such water losses to 15%, thereby meeting most of the world's water needs for the foreseeable future.

Doing this will require greatly increased use of water-saving technologies and practices. It will also decrease the burden on wastewater plants, reduce the need for expensive dams and water transfer projects that destroy wildlife habitats and displace people, slow depletion of groundwater aquifers, and save energy and money.

Why Do We Waste So Much Water? According to water resource experts, there are two major causes of water waste. One is government subsidies of water supply projects such as large dams and large-scale water transfer schemes that create artificially low water prices. The other is lack of government subsidies for improving water efficiency.

According to water resource expert Sandra Postel, "By heavily subsidizing water, governments give out the false message that it is abundant and can afford to be wasted—even as rivers are drying up, aquifers are being depleted, fisheries are collapsing, and species are going extinct."

Government-subsidized irrigation water in the western United States costs U.S. taxpayers an estimated $2–2.5 billion per year. However, farmers, industries, and others benefiting from government water subsidies argue that they promote settlement and agricultural production in arid and semiarid areas, stimulate local economies, and help lower prices of food, manufactured goods, and electricity for consumers.

Solutions: How Can We Waste Less Water in Irrigation? Some *bad news* is that much of the irrigation water applied throughout the world does not reach the targeted crops. Most irrigation systems obtain water from a groundwater well or a surface water source and allow it to flow by gravity through unlined ditches in crop fields so the water can be absorbed by crops (Figure 14-19, left). This *flood irrigation* method delivers far more water than needed for crop growth and typically allows only 60% of the water to reach crops because of evaporation, seepage, and runoff.

Some *good news* is that more efficient and environmentally sound irrigation technologies exist that could reduce water demands and waste on farms. One of these technologies is a *center-pivot low-pressure sprinkler* (Figure 14-19, right). Typically, it allows 80% of the water input to reach crops and reduces water use over conventional gravity flow systems by 25%.

Another method involves using *low-energy precision application (LEPA) sprinklers.* This form of center-pivot irrigation allows 90–95% of the water input to reach crops by spraying it closer to the ground and in larger droplets than the center-pivot low-pressure system. LEPA sprinklers use 20–30% less energy than low-pressure sprinklers and typically use 37% less water than conventional gravity flow systems.

Farmers also use *surge or time-controlled valves on conventional gravity flow irrigation systems* (Figure 14-19, left). These valves send water down irrigation ditches in pulses instead of in a continuous stream. This can raise irrigation efficiency to 80% and reduce water use by 25%.

Another method is to use *soil moisture detectors to water crops only when they need it.* For example, some

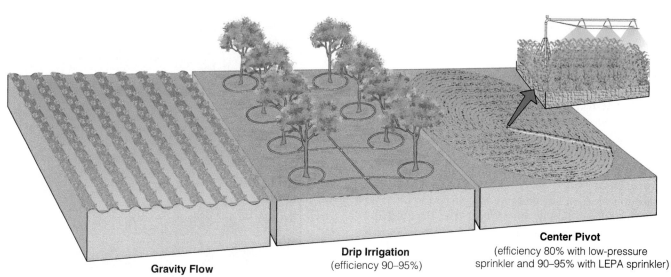

Gravity Flow

(efficiency 60% and 80% with surge valves)

Water usually comes from an
aqueduct system or a nearby river.

Drip Irrigation
(efficiency 90–95%)

Above- or below-ground pipes
or tubes deliver water to
individual plant roots.

Center Pivot
(efficiency 80% with low-pressure
sprinkler and 90–95% with LEPA sprinkler)

Water usually pumped from
underground and sprayed from
mobile boom with sprinklers.

Figure 14-19 Major *irrigation systems*. Because of high initial costs, center-pivot irrigation and drip irrigation are used on only about 11% of the world's irrigated cropland each. However, this may change because of the development of new low-cost drip irrigation systems.

farmers in Texas bury a $1 cube of gypsum, the size of a lump of sugar, at the root zone of crops. Wires embedded in the gypsum are run back to a small portable meter that indicates soil moisture. Farmers using this technique need 33–66% less irrigation water.

Drip irrigation systems are also used (Figure 14-19, center, and Solutions, p. 320) to provide more crop per drop of water.

Figure 14-20 lists other ways to reduce water waste in irrigating crops. Since 1950, water-short Israel has used many of these techniques to slash irrigation water waste by about 84% while irrigating 44% more land. Israel now treats and reuses 30% of its municipal sewage water for crop production and plans to increase this to 80% by 2025. The government also gradually removed most government water subsidies to raise the price of irrigation water to one of the highest in the world. Israelis also import most of their wheat and meat, which are water intensive, and concentrate on growing fruits, vegetables, and flowers that need less water.

Some *bad news* is that more efficient sprinklers are used on only 10% and drip irrigation on just over 1% of the world's irrigated crop fields. This could change with development of cheaper sprinkler and drip irrigation technologies (Solutions, p. 320) and increased government subsidies for farmers using more efficient irrigation methods.

Many of the world's poor farmers cannot afford to use most of the modern technological methods for increasing irrigation and irrigation efficiency. These farmers increase irrigation by using small-scale and low-cost traditional technologies.

- Lining canals bringing water to irrigation ditches

- Leveling fields with lasers

- Irrigating at night to reduce evaporation

- Using soil and satellite sensors and computer systems to monitor soil moisture and add water only when necessary

- Polyculture

- Organic farming

- Growing water efficient crops using drought-resistant and salt-tolerant crop varieties

- Irrigating with treated urban waste water

- Importing water intensive crops and meat

Figure 14-20 Solutions: prevention methods for reducing water waste in irrigation.

Some use pedal-powered treadle pumps to move water through irrigation ditches (widely used in Bangladesh). Other methods include using buckets with holes for drip irrigation and small dams, ponds, and tanks to collect rainwater for irrigation. Farmers also use terracing (Figure 9-27a, p. 198) to reduce water loss on crops grown on steep terrain and grow some crops in seasonally waterlogged wetlands, delta lands, and valley bottoms.

Solutions: The Promise of Drip Irrigation The development of inexpensive, weather-resistant, and flexible plastic tubing after World War II paved the way for a new form of microirrigation called *drip irrigation* (Figure 14-19, middle). It consists of a network of perforated plastic tubing, installed at or below the ground surface. The small holes or emitters in the tubing deliver drops of water at a slow and steady rate close to plant roots.

This technique was developed in Israel in the 1960s and is now used by half of its farmers. It has a number of advantages. One is *adaptability*. The tubing system can easily be fitted to match the patterns of crops in a field and left in place or moved to different locations.

Another is *efficiency*, with 90–95% of the water input reaching crops. This method can also *apply fertilizer solutions in precise amounts*. This reduces fertilizer use and waste, salinization, and water pollution from fertilizer runoff.

Finally, drip irrigation can *increase crop yields by 20–90%*. This occurs because plants get more growth per drop of water and are neither underwatered nor overwatered.

Despite these advantages, drip irrigation is used on just over than 1% of the world's irrigated area. This occurs because the capital cost of conventional drip irrigation systems is too high for most poor farmers and for use on low-value row crops. However, drip irrigation is economically feasible for high-profit fruit, vegetable, and orchard crops and for home gardens.

Some *good news* is that the capital cost of a new type of drip irrigation system is one-tenth as much per hectare as conventional drip systems. Another innovation is DRiWATER®, called "drip irrigation in a box." It consists of 1-liter (1.1-quart) packages of gel-encased water that is released slowly into the soil after being buried near plant roots. It wastes almost no water and lasts about 3 months. Egypt is using it to help grow 17 million trees in a desert community.

These and other low-cost drip irrigation systems could bring about a revolution in more sustainable irrigated agriculture that would increase food yields, reduce water use and waste, and lessen some of the environmental problems associated with agriculture.

- Redesign manufacturing processes

- Landscape yards with plants that require little water

- Use drip irrigation

- Fix water leaks

- Use water meters and charge for all municipal water use

- Raise water prices

- Require water conservation in water-short cities

- Use water-saving toilets, showerheads, and front-loading clothes washers

- Collect and reuse household water to irrigate lawns and nonedible plants

- Purify and reuse water for houses, apartments, and office buildings

Figure 14-21 Solutions: methods of reducing water waste in industries, homes, and businesses.

Solutions: How Can We Waste Less Water in Industry, Homes, and Businesses? Figure 14-21 lists ways to use water more efficiently in industries, homes, and businesses. Many homeowners and businesses in water-short areas are replacing green lawns in arid and semiarid regions with vegetation adapted to a dry climate (Figure 14-22). This form of landscaping, called

Figure 14-22 Solutions: *Xeriscaping*. This technique can reduce water use by as much as 85% by landscaping with rocks and plants that need little water and are adapted to the growing conditions in arid and semiarid areas. The term Xeriscape®—first used in 1978 in Denver, Colorado—means "water conservation through creative landscaping."

Running Short of Water in Las Vegas, Nevada

SPOTLIGHT

Las Vegas, Nevada, located in the Mojave Desert, is an artificial aquatic wonderland of large trees, green lawns and golf courses, waterfalls, and swimming pools. According to water experts, Las Vegas uses more water per person than any other city in the world. It is also one of the fastest growing cities in the United States.

Tucson, Arizona, in the Sonora Desert, is a model of water conservation. It began a strict water conservation program in 1976, including raising water rates 500% for some residents. As a result, low-flush toilets, low-flow showerheads, Xeriscaping, and use of drip irrigation have become the norm.

In contrast, Las Vegas, which gets one-third less rainfall than Tucson, only recently started to encourage water conservation. It raised water rates sharply but they are still less than half those in Tucson. It also began encouraging replacement of lawns with rocks and native plants that survive on little water (Figure 14-22).

Water experts project that even if these recent water conservation efforts are successful, Las Vegas may begin running short of water by 2007.

Critical Thinking

If you were an elected official in charge of Las Vegas, what three actions would you take to improve water conservation? What might be the political implications of doing these things?

Xeriscaping (pronounced "ZEER-i-scaping"), reduces water use by 30–85% and sharply reduces inputs of labor, fertilizer, and fuel. It also reduces polluted runoff, air pollution, and yard wastes.

In Boulder, Colorado, introducing *water meters* reduced water use by more than one-third. About one-fifth of all U.S. public water systems do not have water meters and charge a single low rate for almost unlimited use of high-quality water. Many apartment dwellers have little incentive to conserve water because their water use is included in their rent.

Because of laws requiring water conservation, the desert city of Tucson, Arizona, consumes half as much water per person as Las Vegas, a desert city with much less rainfall and less emphasis on water conservation (Spotlight, above). See the website for this chapter for ways you can reduce your personal water use and waste.

14-5 TOO MUCH WATER: PROBLEMS AND SOLUTIONS

What Are the Causes and Effects of Flooding? Heavy rain or rapid melting of snow is the major cause of natural flooding by streams. This causes water in a stream to overflow its normal channel and flood the adjacent area, called a **floodplain** (Figure 14-23, p. 322).

Floodplains, which include highly productive wetlands, are an important part of the earth's natural capital (Figure 1-2, p. 3, and top half of back cover). They help provide natural flood and erosion control, maintain high water quality, and recharge groundwater.

People settle on floodplains because of their many advantages. They include fertile soil, ample water for irrigation, availability of nearby rivers for transportation and recreation, and flat land suitable for crops, buildings, highways, and railroads. In the United States, 10 million households and businesses with property valued at $1 trillion are located in flood-prone areas.

Floods are a natural phenomenon and have several benefits. They provide the world's most productive farmland because they are regularly covered with nutrient-rich silt left after floodwaters recede. They also recharge groundwater and help refill wetlands.

The *bad news* is that each year floods kill thousands of people and cause tens of billions of dollars in property damage. Between 1985 and 2001, floods prematurely killed about 300,000 people, 96% of them in developing countries. Each year floods also cause billions of dollars in property damage and lost food crops.

Floods, like droughts, are usually considered natural disasters, but since the 1960s human activities have contributed to the sharp rise in flood deaths and damages. Human activities can increase the severity of flood damage in several ways. One is *removal of water-absorbing vegetation,* especially on hillsides (Figure 14-24, p. 322). Another is *draining wetlands* that absorb floodwaters and reduce the severity of flooding.

Living on floodplains also increases the threat of damage from flooding (Case Study, p. 323). Flooding also increases when we *pave or build,* replacing water-absorbing vegetation, soil, and wetlands with highways, parking lots, and buildings that cannot absorb rainwater.

In developed countries, people deliberately settle on floodplains and then expect dams, levees, and other devices to protect them from floodwaters. However, when heavier-than-normal rains occur, these devices can be overwhelmed. In many developing countries, the poor have little choice but to try to survive in flood-prone areas (Case Study, p. 323).

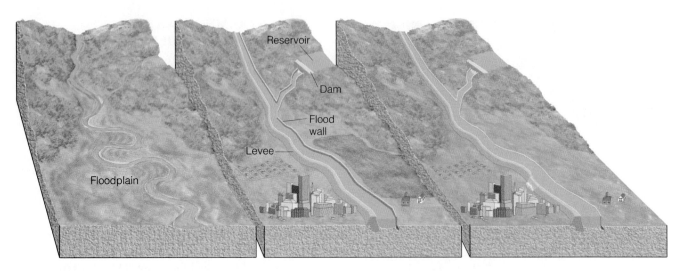

Figure 14-23 Land in a natural floodplain (left) often floods after prolonged rains. When the flood-waters recede, deposits of silt are left behind, creating a nutrient-rich soil. To reduce the threat of flooding (and thus allow people to live in floodplains) to people living in floodplains, rivers have been narrowed and straightened (channelization), equipped with protective levees and walls (middle), and dammed to create reservoirs that store and release water as needed. These alterations can give a false sense of security to floodplain dwellers living in high-risk areas. In the long run, such measures can greatly increase flood damage because they can be overwhelmed by prolonged rains (right), as happened along the Mississippi River in the midwestern United States during the summer of 1993.

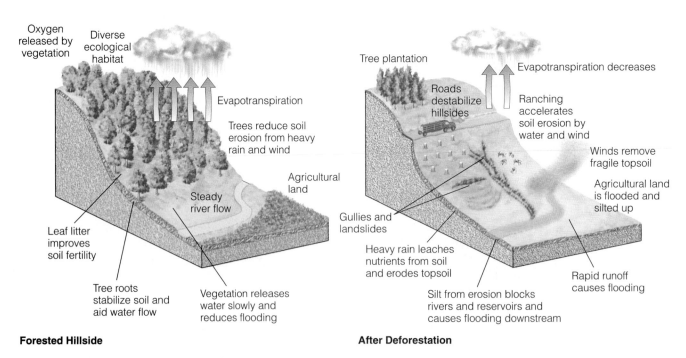

Forested Hillside

After Deforestation

Figure 14-24 A hillside before and after deforestation. Once a hillside has been deforested for tim-ber and fuelwood, livestock grazing, or unsustainable farming, water from precipitation rushes down the denuded slopes, erodes precious topsoil, and floods downstream areas. A 3,000-year-old Chinese proverb says, "To protect your rivers, protect your mountains."

Case Study: Living Dangerously on Floodplains in Bangladesh Bangladesh is one of the world's most densely populated countries, with 147 million people packed into an area that is roughly the size of Wisconsin. It is also one of the world's poorest countries.

The people of Bangladesh depend on moderate annual flooding during the summer monsoon season to grow rice and help maintain soil fertility in the delta basin. The annual floods deposit eroded Himalayan soil on the country's crop fields.

However, excessive flooding can be disastrous. In the past, great floods occurred every 50 years or so, but since the 1970s they have come about every 4 years.

Bangladesh's increased flood problems begin in the Himalayan watershed where several factors have greatly diminished the ability of its mountain soils to absorb water. They include rapid population growth, deforestation, overgrazing, and unsustainable farming on steep and easily erodible mountain slopes. Instead of being absorbed and released slowly, water from the monsoon rains runs off the denuded Himalayan foothills, carrying vital topsoil with it (Figure 14-24).

This increased runoff of soil, combined with heavier-than-normal monsoon rains, has increased the severity of flooding along Himalayan rivers and downstream in Bangladesh. A disastrous flood in 1998 covered two-thirds of Bangladesh's land area for 9 months, leveled 2 million homes, drowned at least 2,000 people, and left 30 million people homeless. It also destroyed more than one-fourth of the country's crops (causing thousands of people to die of starvation) and caused at least $3.4 billion in damages. In 2002, another flood left 5 million people homeless and flooded large areas of rice fields.

Living on Bangladesh's coastal floodplain also carries dangers from storm surges and cyclones. Since 1961, 17 devastating cyclones have slammed into Bangladesh. In 1970, as many as 1 million people drowned in one storm, and another surge killed an estimated 139,000 people in 1991.

In their struggle to survive, the poor in Bangladesh have cleared many of the country's coastal mangrove forests for fuelwood, farming, and aquaculture ponds for raising shrimp. This has led to more severe flooding because these coastal wetlands help shelter Bangladesh's low-lying coastal areas from storm surges and cyclones. Damages and deaths from cyclones in areas of Bangladesh still protected by mangrove forests have been much lower than in areas where the forests have been cleared.

Solutions: How Can We Reduce Flood Risks? We can use several methods to reduce the risk from flooding. One is to *straighten and deepen streams,* a process called *channelization* (Figure 14-23, middle). Channelization can reduce upstream flooding but it removes bank vegetation and increases stream velocity. This increased flow of water can promote upstream bank erosion, increase downstream flooding and sediment deposition, and reduce habitats for aquatic wildlife.

Another approach is to *build levees* (Figure 14-23, middle). Levees contain and speed up stream flow, but this increases the water's capacity for doing damage downstream. They also do not protect against unusually high and powerful floodwaters, as occurred in 1993 when two-thirds of the levees built along the Mississippi River were damaged or destroyed.

Building dams can also reduce the threat of flooding. A flood control dam built across a stream can reduce flooding by storing water in a reservoir and releasing it gradually. Dams have a number of advantages and disadvantages (Figure 14-9). We can also *not destroy existing wetlands* and *restore wetlands* to take advantage of the natural flood control provided by floodplains.

Finally, we can *identify and manage flood-prone areas* (Figure 14-25). Actions include prohibiting certain types of buildings or activities in high-risk flood zones, and elevating or otherwise floodproofing buildings that are allowed on floodplains. In addition, we can construct a floodway that allows floodwater to flow

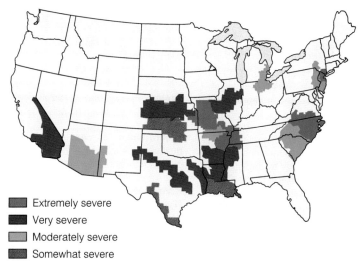

■ Extremely severe
■ Very severe
■ Moderately severe
■ Somewhat severe

Figure 14-25 Generalized map of *flood-prone areas* in the United States. State boundaries are shown in black. More detailed state and local maps are used to show the likelihood and severity of floods within these general areas. Flood frequency data and maps of flood-prone areas do not tell us when floods will occur, but they give a general idea of how often and where floods might occur, based on an area's history. (Data from U.S. Geological Survey)

through a community with minimal damage. These *prevention,* or *precautionary,* approaches are based on thousands of years of experience that can be summed up in one idea: *Sooner or later the river (or the ocean) always wins.*

14-6 TYPES, EFFECTS, AND SOURCES OF WATER POLLUTION

What Are the Major Types and Effects of Water Pollutants? **Water pollution** is any chemical, biological, or physical change in water quality that has a harmful effect on living organisms or makes water unsuitable for desired uses. Table 14-1 lists major classes of water pollutants along with their major human sources and harmful effects.

How Do We Measure Water Quality? Scientists use a number of methods to measure water quality. One involves measuring the number of colonies of *coliform bacteria* present in a water sample. The World Health Organization (WHO) recommends a coliform bacteria count of 0 colonies per 100 milliliters for drinking water, and the U.S. Environmental Protection Agency (EPA) recommends a maximum level for swimming water of 200 colonies per 100 milliliters.

Water pollution from oxygen-demanding wastes and from plant nutrients (Table 14-1) can be determined by measuring the level of *dissolved oxygen.* Another method is to measure the *biological oxygen demand (BOD)* to determine the quantity of oxygen-demanding wastes in water.

Chemical analysis is widely used to determine the presence and concentrations of inorganic and organic

---

Table 14-1 Major Categories of Water Pollutants

INFECTIOUS AGENTS

Examples: Bacteria, viruses, protozoa, and parasitic worms
Major Human Sources: Human and animal wastes
Harmful Effects: Disease

OXYGEN-DEMANDING WASTES

Examples: Organic waste such as animal manure and plant debris that can be decomposed by aerobic (oxygen-requiring) bacteria
Major Human Sources: Sewage, animal feedlots, paper mills, and food processing facilities
Harmful Effects: Large populations of bacteria decomposing these wastes can degrade water quality by depleting water of dissolved oxygen. This causes fish and other forms of oxygen-consuming aquatic life to die.

INORGANIC CHEMICALS

Examples: Water-soluble (1) acids, (2) compounds of toxic metals such as lead (Pb), arsenic (As), and selenium (Se), and (3) salts such as NaCl in ocean water and fluorides (F^-) found in some soils

Major Human Sources: Surface runoff, industrial effluents, and household cleansers
Harmful Effects: Can (1) make freshwater unusable for drinking or irrigation, (2) cause skin cancers and crippling spinal and neck damage (F^-), (3) damage the nervous system, liver, and kidneys (Pb and As), (4) harm fish and other aquatic life, (5) lower crop yields, and (6) accelerate corrosion of metals exposed to such water.

ORGANIC CHEMICALS

Examples: Oil, gasoline, plastics, pesticides, cleaning solvents, detergents
Major Human Sources: Industrial effluents, household cleansers, surface runoff from farms and yards
Harmful Effects: Can (1) threaten human health by causing nervous system damage (some pesticides), reproductive disorders (some solvents), and some cancers (gasoline, oil, and some solvents) and (2) harm fish and wildlife

PLANT NUTRIENTS

Examples: Water-soluble compounds containing nitrate

(NO_3^-), phosphate (PO_4^{3-}), and ammonium (NH_4^+) ions
Major Human Sources: Sewage, manure, and runoff of agricultural and urban fertilizers
Harmful Effects: Can cause excessive growth of algae and other aquatic plants, which die, decay, deplete water of dissolved oxygen, and kill fish. Drinking water with excessive levels of nitrates lowers the oxygen-carrying capacity of the blood and can kill unborn children and infants ("blue-baby syndrome").

SEDIMENT

Examples: Soil, silt
Major Human Sources: Land erosion
Harmful Effects: Can (1) cloud water and reduce photosynthesis, (2) disrupt aquatic food webs, (3) carry pesticides, bacteria, and other harmful substances, (4) settle out and destroy feeding and spawning grounds of fish, and (5) clog and fill lakes, artificial reservoirs, stream channels, and harbors.

RADIOACTIVE MATERIALS

Examples: Radioactive isotopes of iodine, radon,

uranium, cesium, and thorium
Major Human Sources: Nuclear power plants, mining and processing of uranium and other ores, nuclear weapons production, natural sources
Harmful Effects: Genetic mutations, miscarriages, birth defects, and certain cancers

HEAT (THERMAL POLLUTION)

Examples: Excessive heat
Major Human Sources: Water cooling of electric power plants and some types of industrial plants. Almost half of all water withdrawn in the United States each year is for cooling electric power plants.
Harmful Effects: Lowers dissolved oxygen levels and makes aquatic organisms more vulnerable to disease, parasites, and toxic chemicals. When a power plant first opens or shuts down for repair, fish and other organisms adapted to a particular temperature range can be killed by the abrupt change in water temperature—known as *thermal shock.*

---

chemicals that pollute water. Finally, we can use living organisms as *indicator species* to monitor water pollution. For example, scientists can remove aquatic plants such as cattails and analyze them to determine pollution in areas contaminated with fuels, solvents, and other organic chemicals.

What Are Point and Nonpoint Sources of Water Pollution? **Point sources** discharge pollutants at specific locations through pipes, ditches, or sewers into bodies of surface water (Figure 14-26). Examples include factories, sewage treatment plants (which remove some but not all pollutants), active and abandoned underground mines, and oil tankers.

Because point sources are at specific places, they are easy to identify, monitor, and regulate. Most developed countries control point source discharges of many harmful chemicals into aquatic systems. However, there is little control of such discharges in most developing countries.

Nonpoint sources cannot be traced to any single site of discharge (Figure 14-26). They are usually large land areas or airsheds that pollute water by runoff, subsurface flow, or deposition from the atmosphere. Examples include acid deposition (Figure 12-9, p. 264) and runoff of chemicals into surface water from crop-

lands, livestock feedlots, logged forests, urban streets, lawns, golf courses, and parking lots. There has been little progress in controlling nonpoint water pollution because of the difficulty and expense of identifying and controlling discharges from so many diffuse sources.

Is the Water Safe to Drink? Some *good news* is that according to the WHO the percentage of people in developing countries with access to clean drinking water increased from 30% in 1970 to 72% in 2000. The *bad news* is that 1.4 billion people—about one in five—in developing countries do not have access to clean drinking water.

Only 6 of China's 27 largest cities provide drinking water that meets government standards. In Russia, half of all tap water is unfit to drink. About 290 million Africans—about equal to the entire U.S. population—do not have access to safe drinking water.

The United Nations estimates it would cost about $23 billion a year over 8–10 years to bring low-cost safe water and sanitation to the 1.4 billion people who do not have access to clean drinking water. These expenditures could prevent many of the 3.4 million deaths (including 2 million children under age 5) and 3.4 billion cases of illness caused each year by drinking unsafe water.

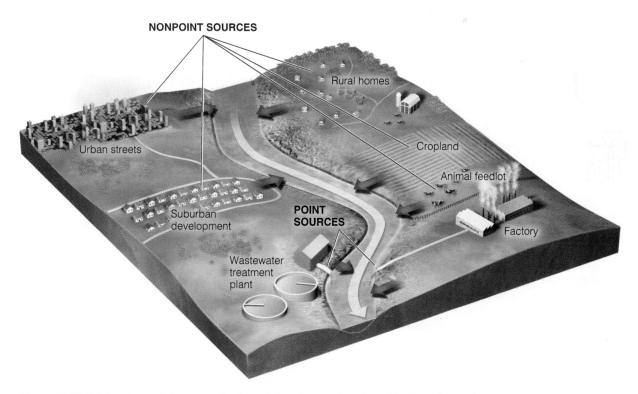

Figure 14-26 Point and nonpoint sources of water pollution. It is much easier to identify and control point sources than more dispersed nonpoint sources.

Currently, the world is spending only about $16 billion a year on clean water efforts. The $7-billion shortfall is about equal to what the world spends every 4 days for military purposes.

14-7 POLLUTION OF FRESHWATER STREAMS, LAKES, AND GROUNDWATER

What Are the Water Pollution Problems of Streams? Flowing streams, including large ones called *rivers*, can recover rapidly from degradable, oxygen-demanding wastes and excess heat through a combination of dilution and bacterial decay. This natural recovery process works as long as pollutants do not overload the streams and drought, damming, or water diversion for agriculture and industry do not reduce their flows. However, these natural dilution and biodegradation processes do not eliminate slowly degradable and nondegradable pollutants.

In a flowing stream, the breakdown of degradable wastes by bacteria depletes dissolved oxygen. This reduces or eliminates populations of organisms with high oxygen requirements until the stream is cleansed of wastes. The depth and width of the resulting *oxygen sag curve* (Figure 14-27), and thus the time and distance needed for a stream to recover, depend on the volume of incoming degradable wastes and the stream's volume, flow rate, temperature, and pH level (Figure 9-20, p. 192). Similar oxygen sag curves can be plotted when heated water from industrial and power plants is discharged into streams.

What Is the Good News about Stream Pollution? There is *good news* about stream pollution in parts of the world. Water pollution control laws enacted in the 1970s have greatly increased the number and quality of wastewater treatment plants in the United States and many other developed countries. Such laws require industries to reduce or eliminate point-source discharges into surface waters.

Individuals (Individuals Matter, p. 327) and environmental groups spurred these efforts. They have enabled the United States to hold the line against increased pollution of most of its streams by disease-causing agents and oxygen-demanding wastes. This is an impressive accomplishment given the rise in the country's economic activity, resource consumption, and population since passage of these laws.

One success story is that of the cleanup of Ohio's Cuyahoga River. It was so polluted that in 1959 and

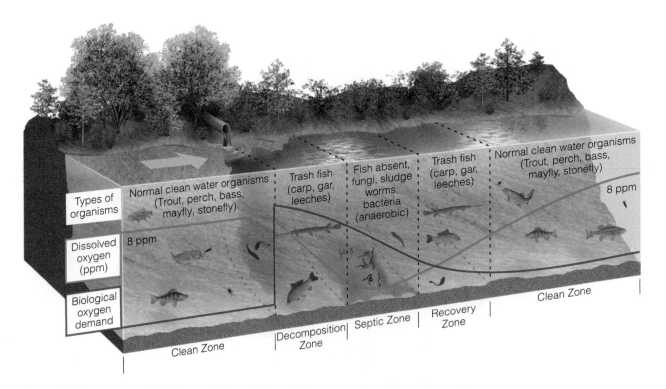

Figure 14-27 *Natural capital:* dilution and decay of degradable, oxygen-demanding wastes and heat in a stream, showing the oxygen sag curve (red) and the curve of oxygen demand (blue). Depending on flow rates and the amount of pollutants, streams recover from oxygen-demanding wastes and heat if they are given enough time and are not overloaded.

Rescuing a River

INDIVIDUALS MATTER

In the 1960s, Marion Stoddart moved to Groton, Massachusetts, on the Nashua River, then considered one of the nation's filthiest rivers. Dead fish bobbed on its waves and at times the water was red, green, or blue from pigments discharged by paper mills.

Instead of thinking nothing could be done, she committed herself to restoring the Nashua and establishing public parklands along its banks.

She did not start by filing lawsuits or organizing demonstrations. Instead she created a careful cleanup plan and approached state officials with it in 1962. They laughed, but she was not deterred and began practicing the most time-honored skill of politics: one-on-one persuasion. She identified the power brokers in the riverside communities and began to educate them, win them over, and get them to cooperate in cleaning up the river.

She also got the state to ban open dumping in the river. When promised federal matching funds promised for building a waste treatment plant failed to materialize, Stoddart gathered 13,000 signatures on a petition sent to President Richard Nixon. The funds arrived in a hurry.

Stoddart's next success was getting a federal grant to beautify the river. She hired high school dropouts to clear away mounds of debris. When the river cleanup was completed, she persuaded communities along the river to create some 2,400 hectares (6,000 acres) of riverside park and woodlands along both banks.

Now, four decades later, the Nashua is still clean. Several new water treatment plants have been built, and a citizen group founded by Stoddart keeps watch on water quality. The river supports many kinds of fish and other wildlife, and its waters are used for canoeing and other kinds of recreation. The project is testimony to what a committed individual can do to bring about change from the bottom up by getting people to work together.

For her efforts, the UN Environment Programme named Stoddart an outstanding worldwide worker for the environment. However, she might say the blue and canoeable Nashua itself is her best reward.

again in 1969 it caught fire and burned for several days as it flowed through Cleveland. The highly publicized image of this burning river prompted city, state, and federal officials to enact laws limiting the discharge of industrial wastes into the river and sewage systems and appropriate funds to upgrade sewage treatment fa-

cilities. Today the river has made a comeback and is widely used by boaters and anglers.

Another spectacular cleanup has occurred in Great Britain. In the 1950s, the Thames River was little more than a flowing anaerobic sewer. Now, after more than 40 years of effort and hundreds of millions of dollars spent by British taxpayers and private industry, the Thames has made a remarkable recovery. Commercial fishing is thriving and the number of fish species has increased 20-fold since 1960. In addition, many species of waterfowl and wading birds have returned to their former feeding grounds.

What Is the Bad News about Stream Pollution?
Despite progress in improving stream quality in most developed countries, large fish kills and drinking water contamination still occur. One cause of these disasters is accidental or deliberate releases of toxic inorganic and organic chemicals by industries or mines. Another source of harmful chemicals is malfunctioning sewage treatment plants. A third cause is nonpoint runoff of pesticides and excess plant nutrients from cropland and animal feedlots.

Available data indicate that stream pollution from discharges of sewage and industrial wastes is a serious and growing problem in most developing countries, where there is little waste treatment. Numerous streams in the former Soviet Union and in eastern European countries suffer from severe pollution. Industrial wastes and sewage pollute more than two-thirds of India's water resources. Untreated sewage and industrial wastes severely pollute 54 of the 78 streams monitored in China. In Latin America and Africa, most streams passing through urban or industrial areas suffer from severe pollution.

Why Are Lakes and Reservoirs More Vulnerable to Pollution than Most Streams? In lakes and reservoirs, dilution of pollutants often is less effective than in streams for two reasons. *First,* lakes and reservoirs often contain stratified layers (Figure 6-40, p. 135) that undergo little vertical mixing.

Second, they have little flow. For example, the flushing and changing of water in lakes and large artificial reservoirs can take from 1 to 100 years, compared with several days to several weeks for streams.

Consequently, lakes and reservoirs are more vulnerable than streams to contamination by runoff or discharge of plant nutrients, oil, pesticides, and toxic substances such as lead, mercury, and selenium. These contaminants can kill bottom life and fish and birds that feed on contaminated aquatic organisms. Many toxic chemicals and acids (Figure 12-9, p. 264) also enter lakes and reservoirs from the atmosphere.

As they pass through food webs in lakes, the concentrations of some chemicals can be biologically magnified. Examples include DDT (Figure 10-4, p. 205), PCBs (Figure 14-28), some radioactive isotopes, and some mercury compounds.

What Is Cultural Eutrophication? Lakes receive inputs of nutrients and silt eroded and running off from the surrounding land basin. **Eutrophication** is the name given to this natural nutrient enrichment of lakes. Over time, some lakes become more eutrophic (Figure 6-41, right, p. 136), but others do not because of differences in the surrounding drainage basin.

Near urban or agricultural areas, human activities can greatly accelerate the input of plant nutrients to a lake, which results in a process known as **cultural eutrophication.** Such a change is caused mostly by nitrate- and phosphate-containing effluents from various sources (Figure 14-29).

During hot weather or drought, this nutrient overload produces dense growths of organisms such as algae, cyanobacteria, water hyacinths, and duckweed. When the algae die, their decomposition by aerobic bacteria depletes dissolved oxygen in the sur-

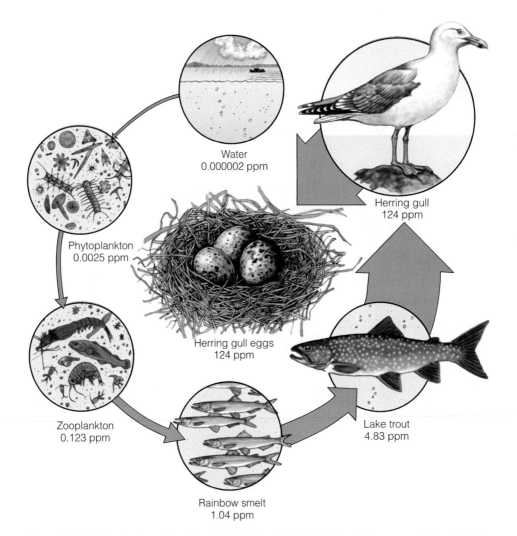

Water
0.000002 ppm

Phytoplankton
0.0025 ppm

Herring gull
124 ppm

Herring gull eggs
124 ppm

Zooplankton
0.123 ppm

Lake trout
4.83 ppm

Rainbow smelt
1.04 ppm

Figure 14-28 *Biological magnification* of PCBs (polychlorinated biphenyls) in an aquatic food chain in the Great Lakes. Most of the 209 different PCBs are insoluble in water, soluble in fats, and resistant to biological and chemical degradation—properties that result in their accumulation in the tissues of organisms and their biological amplification in food chains and webs. Although the long-term health effects on people exposed to low levels of PCBs are unknown, high doses of PCBs in laboratory animals produce liver and kidney damage, gastric disorders, birth defects, skin lesions, hormonal changes, smaller penis size, and tumors. Boys in Taiwan exposed to PCBs while in their mothers' wombs the womb developed abnormally small penises. In the United States, manufacture and use of PCBs have been banned since 1976. Before then, millions of metric tons of these long-lived chemicals were released into the environment. Many of them still exist in bottom sediments of lakes, streams, and oceans.

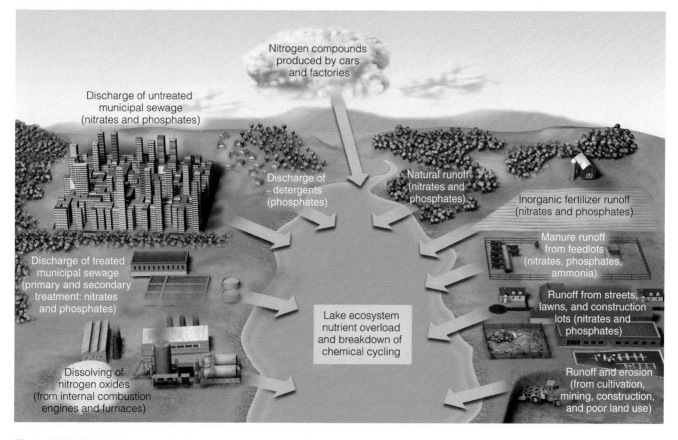

Figure 14-29 Principal sources of nutrient overload causing *cultural eutrophication* in lakes and coastal areas. The amount of nutrients from each source varies according to the types and amounts of human activities occurring in each airshed and watershed. The enlarged populations of algae and plants (stimulated by increased nutrient input) die. Then their decomposition by aerobic bacteria lowers levels of dissolved oxygen. This can kill fish and other aquatic life and reduce biodiversity and the aesthetic and recreational value of the lake.

face layer of water near the shore and in the bottom layer. This oxygen depletion can kill fish and other aerobic aquatic animals. If excess nutrients continue to flow into a lake, anaerobic bacteria take over and produce gaseous decomposition products such as smelly, highly toxic hydrogen sulfide and flammable methane gases.

According to the EPA, about one-third of the 100,000 medium to large lakes and about 85% of the large lakes near major population centers in the United States have some degree of cultural eutrophication. One-fourth of the lakes in China suffer from cultural eutrophication.

Solutions: How Can We Prevent or Reduce Cultural Eutrophication? There are several ways to *prevent* or *reduce* cultural eutrophication. They include advanced (but expensive) waste treatment to remove nitrates and phosphates before wastewater enters lakes, bans or limits on phosphates in household detergents and other cleaning agents, and soil conservation and land-use control to reduce nutrient runoff.

There are also several ways to *clean up* lakes suffering from cultural eutrophication. Examples are mechanically removing excess weeds, controlling undesirable plant growth with herbicides and algicides, and pumping air through lakes and reservoirs to avoid oxygen depletion (an expensive and energy-intensive method).

As usual, pollution prevention is more effective and usually cheaper in the long run than cleanup. If excessive inputs of limiting plant nutrients stop, a lake can usually return to its previous state.

Case Study: Pollution in the Great Lakes The five interconnected Great Lakes (Figure 14-30, p. 330) contain at least 95% of the fresh surface water in the United States and 20% of the world's fresh surface water. The Great Lakes basin is home for about 30% of the Canadian population and 14% of the U.S. population.

Despite their enormous size, these lakes are vulnerable to pollution from point and nonpoint sources because less than 1% of the water entering the Great Lakes flows out to the St. Lawrence River each year. In

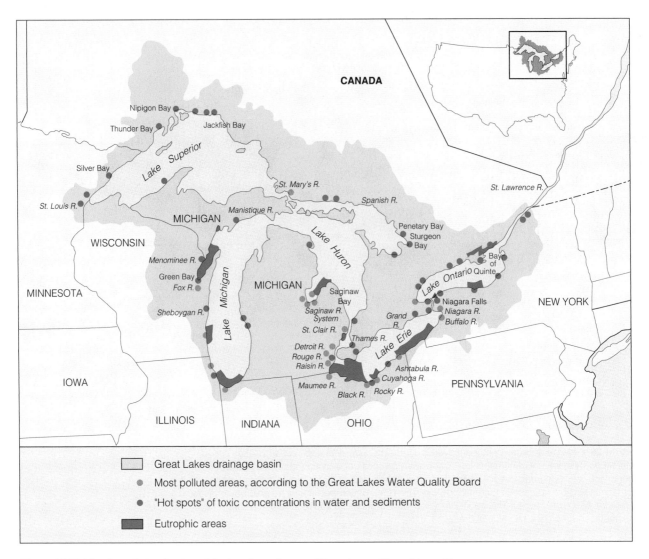

Figure 14-30 The *Great Lakes basin* and the locations of some of its water quality problems. The Great Lakes region is dotted with several hundred abandoned toxic waste sites listed by the EPA as Superfund sites to receive cleanup priority. (Data from Environmental Protection Agency)

addition to land runoff, these lakes receive large quantities of acids, pesticides, and other toxic chemicals by deposition from the atmosphere—often blown in from hundreds or thousands of kilometers away.

By the 1960s, many areas of the Great Lakes were suffering from severe cultural eutrophication, huge fish kills, and contamination from bacteria and a variety of toxic industrial wastes. The impact on Lake Erie was particularly intense because it is the shallowest of the Great Lakes and has the highest concentrations of people and industrial activity along its shores. Many bathing beaches had to be closed, and by 1970 the lake had lost nearly all its native fish.

Here is some *good news*. Since 1972, a $20 billion Great Lakes pollution control program has been carried out jointly by Canada and the United States. This

program has decreased algal blooms, increased dissolved oxygen levels and sport and commercial fishing catches, and allowed most swimming beaches to reopen. It has also significantly decreased levels of phosphates, coliform bacteria, and many toxic industrial chemicals.

These improvements occurred mainly because of new or upgraded sewage treatment plants, better treatment of industrial wastes, and bans on use of phosphate detergents, household cleaners, and water conditioners.

Here is some *bad news*. According to a 2000 survey by the EPA, 78% of the shoreline of the Great Lakes is not clean enough for swimming or for supplying drinking water. Nonpoint land runoff of pesticides and fertilizers from urban sprawl now surpasses industrial pollution as the greatest threat to the lakes. Sediments in 26 toxic hot spots (Figure 14-30) remain heavily polluted.

About half of the input of toxic compounds comes from atmospheric deposition of pesticides, mercury from coal-burning plants, and other toxic chemicals from as far away as Mexico and Russia. Toxic chemicals such as PCBs have built up in food chains and webs (Figure 14-28), contaminated many types of sport fish, and depleted populations of birds, river otters, and other animals feeding on contaminated fish. A recent survey by Wisconsin biologists found that one fish in four taken from the Great Lakes is unsafe for human consumption.

Finally, annual EPA funding for Great Lakes cleanup declined from $9 million in 1992 to $2 million in 2001. One result of this decreased funding is that the EPA staff charged with coordinating the Great Lakes cleanup dropped from 21 people in 1999 to one person in 2001.

Some environmentalists call for banning use of toxic chlorine compounds as bleach in the pulp and paper industry around the Great Lakes. They would also ban new incinerators (which can release toxic chemicals into the atmosphere) in the area and the discharge into the lakes of 70 toxic chemicals that threaten human health and wildlife. Officials of the industries involved strongly oppose such bans.

Why Is Groundwater Pollution Such a Serious Problem? According to many scientists, a serious threat to human health is the out-of-sight pollution of groundwater, a prime source of water for drinking and irrigation.

Studies indicate that groundwater pollution comes from numerous sources (Figure 14-31). People who dump or spill gasoline, oil, and paint thinners and other organic solvents onto the ground also contaminate groundwater. Although experts rate groundwater pollution as a low-risk ecological problem, they consider

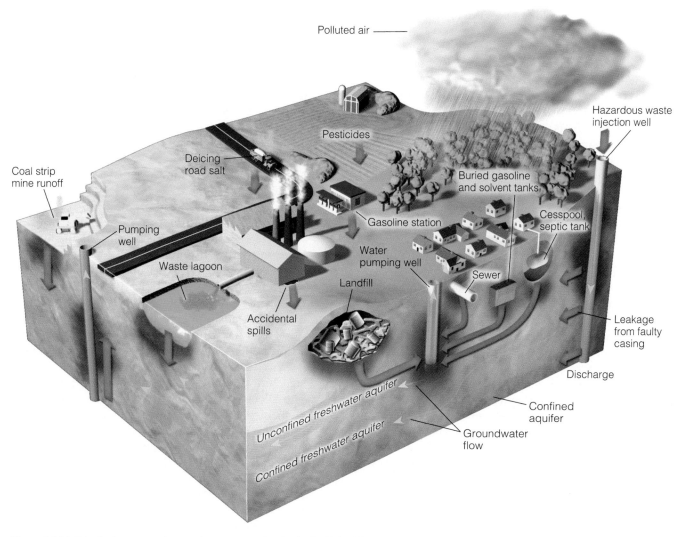

Figure 14-31 Principal sources of groundwater contamination in the United States.

pollutants in drinking water (much of it from groundwater) a high-risk health problem (Figure 10-15, left, p. 216).

When groundwater becomes contaminated, it cannot cleanse itself of *degradable wastes* as flowing surface water does (Figure 14-27). Groundwater flows so slowly (usually less than 0.3 meters meter or 1 foot per day) that contaminants are not diluted and dispersed effectively. Groundwater also has much smaller populations of decomposing bacteria and cold temperatures that slow down the chemical reactions that decompose wastes.

Thus it can take hundreds to thousands of years for contaminated groundwater to cleanse itself of *degradable* wastes. On a human time scale, *nondegradable wastes* (such as toxic lead, arsenic, and fluoride) are there permanently.

What Is the Extent of Groundwater Pollution?

The answer is that *we do not know* because few countries go to the great expense of locating, tracking, and testing aquifers. However, scientific studies in scattered parts of the world provide us with several pieces of *bad news*.

According to the EPA and the U.S. Geological Survey, one or more organic chemicals contaminate about 45% of *municipal* groundwater supplies in the United States. An EPA survey found that one-third of 26,000 industrial waste ponds and lagoons in the United States have no liners to prevent toxic liquid wastes from seeping into aquifers. One-third of these sites are within 1.6 kilometers (1 mile) of a drinking water well.

The General Accounting Office estimated in 2002 that at least 76,000 underground tanks storing gasoline, diesel fuel, home heating oil, and toxic solvents were leaking their contents into groundwater in the United States (Figure 14-31). During this century, scientists expect many of the millions of such tanks installed around the world in recent decades to corrode, leak, contaminate groundwater, and become a major global health problem. Determining the extent of a leak from a single underground tank can cost $25,000–250,000, and cleanup costs range from $10,000 to more than $250,000.

In coastal areas, excessive pumping of water from aquifers can lead to contamination of drinking water by saltwater intrusion (Figure 14-16, p. 316). Finally, toxic *arsenic (As)* contaminates drinking water through the drilling of tubewells into aquifers where soils and rock are naturally rich in arsenic. According to the WHO, more than 112 million people are drinking water with arsenic levels 5–100 times the WHO standard. They include an estimated 30 million people in parts of Bangladesh and 6 million people in India's state of West Bengal. According to estimates by the WHO, arsenic in drinking water will cause 200,000–270,00 premature deaths from cancer in Bangladesh alone. In the United States, standards for arsenic in drinking water are controversial (Case Study, below).

Case Study: How Much Arsenic Should We Allow in Drinking Water? Arsenic is a naturally occurring element in various rocks. This means it can be released into groundwater and surface water from the natural weathering of minerals in deep soils and rocks.

In addition, arsenic can be released into the air and water from coal burning, copper and lead smelting, municipal trash incinerators, wood-preserving treatments, leaching from landfills containing arsenic-laden ash produced by coal-burning power plants, and use of certain arsenic-containing pesticides.

According to a 1999 report by the U.S. National Academy of Sciences, long-term exposure to arsenic in drinking water can cause cancer of the skin, bladder, and lungs, possibly cause kidney and liver cancer, and cause skin lesions and hardening of the skin (keratosis). Some studies have also linked arsenic in drinking water to adult-onset diabetes, cardiovascular disease, anemia, and disorders of the immune, nervous, and reproductive systems. In 2001, researchers at the Dartmouth Medical School reported that arsenic is also a potent endocrine disrupter (Case Study, p. 208).

Since 1942 the acceptable level of arsenic in U.S. drinking water has been 50 parts per billion (50 ppb). This is five times the international standard of 10 ppb adopted in 1993 by the WHO and in 1998 by the 15-member European Union.

In 1962, the U.S. Public Health Service proposed lowering the U.S. drinking water standard for arsenic from 50 ppb to 10 ppb. After 37 years of scientific reviews, including a 1999 study by the National Academy of Sciences, the EPA proposed that the U.S. drinking water standard for arsenic be reduced to the international standard of 10 ppb.

According to WHO scientists, even the 10-ppb standard is not safe. Many scientists call for lowering the standard to 3–5 ppb. A 2001 study by the U.S. National Academy of Sciences found that routinely drinking water with arsenic levels of even 3 ppb poses a 1 in 1,000 risk of developing bladder or lung cancer.

According to the EPA, implementing the 10-ppb standard would make drinking water safer for at least 11 million Americans in areas with high arsenic levels in their groundwater (Figure 14-32). This would require about 3,000 communities to improve their drinking water treatment at an average annual cost of about $12 per person. Annual benefits of $140–198 million would be expected from reduction of arsenic-related bladder and liver cancers. In 2002, a team of scientists from several universities developed genetically modified plants that could help remove arsenic from naturally and artificially polluted soil and water.

Figure 14-32 Arsenic concentrations in groundwater of the United States in 1999. (Data from U.S. Geological Survey)

Mining, coal, wood treatment, and other arsenic-producing companies oppose the 10-ppb standard and say the estimated costs are too low. In addition, some small communities contend they cannot afford to implement the new standard. However, the federal government could subsidize some or all treatment costs for small systems.

In March 2001, President George W. Bush (under lobbying pressure from mining interests, coal companies, the wood products industry, and other arsenic producers) had the EPA withdraw implementation of the new EPA standard, arguing it would cost too much to implement. Health scientists and environmentalists were outraged and accused the Bush administration of caving in to the demands of industries whose activities release toxic levels of arsenic into drinking water supplies. In less than a month the Bush administration reversed its controversial decision and lowered the acceptable concentration of arsenic from 50 ppb to 10 ppb, to take effect in 2006.

Solutions: How Can We Protect Groundwater?

Contaminated aquifers are almost impossible to clean because of their enormous volume, inaccessibility, and slow movement. Pumping polluted groundwater to the surface, cleaning it up, and returning it to the aquifer is very expensive.

Thus *preventing contamination is the only effective way to protect groundwater resources.* There are several

ways to do this. One is to monitor aquifers near landfills and underground tanks. Another is to require leak detection systems for underground tanks used to store hazardous liquids.

We can ban or more strictly regulate disposal of hazardous wastes in deep injection wells and landfills. Also, we could store hazardous liquids above ground in tanks with systems that detect and collect leaking liquids.

14-8 OCEAN POLLUTION

How Much Pollution Can the Oceans Tolerate?

The *bad news* is that the oceans are the ultimate sinks for much of the waste matter we produce. The *good news* is that oceans can dilute, disperse, and degrade large amounts of raw sewage, sewage sludge, oil, and some types of degradable industrial waste, especially in deep-water areas. More *good news* is that some forms of marine life have been more resilient than originally expected.

This has led some scientists to suggest it is safer to dump sewage sludge and most other hazardous wastes into the deep ocean than to bury them on land or burn them in incinerators. Other scientists disagree, pointing out we know less about the deep ocean than we do about outer space or the moon. They add that dumping waste in the ocean would delay urgently needed pollution prevention and promote further degradation of this vital part of the earth's life-support system.

How Do Pollutants Affect Coastal Areas? Areas along coasts—especially wetlands and estuaries, coral reefs, and mangrove swamps—bear the brunt of our enormous inputs of pollutants and wastes into the ocean (Figure 14-33). This is not surprising because about 40% of the world's population lives on or within 100 kilometers (62 miles) of the coast and 14 of the world's 15 largest metropolitan areas (each with 10 million people or more) are near coastal waters (Figure 11-20, p. 239). Also, coastal populations are growing more rapidly than the global population.

In most coastal developing countries (and in some coastal developed countries), municipal sewage and industrial wastes are dumped into the sea without treatment. For example, about 85% of the sewage from large cities along the Mediterranean Sea (with a coastal population of 200 million people during tourist season) is discharged into the sea untreated. This causes widespread beach pollution and shellfish contamination.

Recent studies of some U.S. coastal waters have found vast colonies of human viruses from raw sewage, effluents from sewage treatment plants (which do not remove viruses), and leaking septic tanks. According to one study, about one-fourth of the people using coastal beaches in the United States develop ear infections, sore throats and eyes, respiratory disease, or gastrointestinal disease.

Runoff of sewage and agricultural wastes into coastal waters and acid deposition from the atmo-

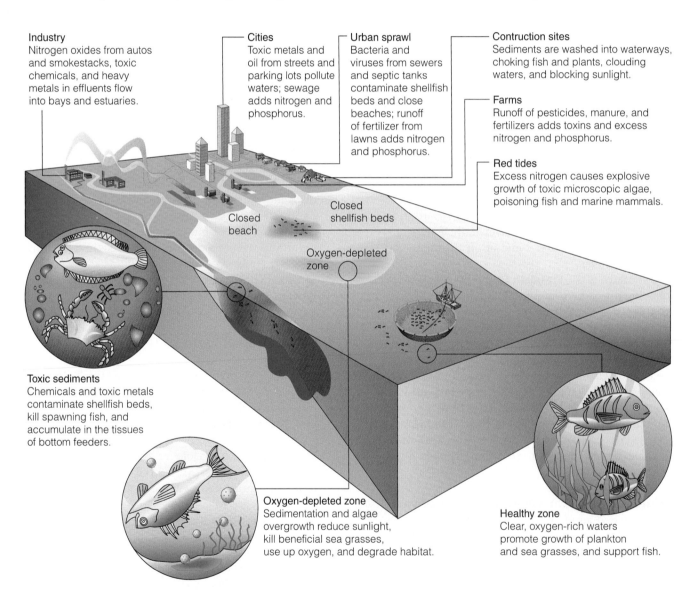

Industry
Nitrogen oxides from autos and smokestacks, toxic chemicals, and heavy metals in effluents flow into bays and estuaries.

Cities
Toxic metals and oil from streets and parking lots pollute waters; sewage adds nitrogen and phosphorus.

Urban sprawl
Bacteria and viruses from sewers and septic tanks contaminate shellfish beds and close beaches; runoff of fertilizer from lawns adds nitrogen and phosphorus.

Contruction sites
Sediments are washed into waterways, choking fish and plants, clouding waters, and blocking sunlight.

Farms
Runoff of pesticides, manure, and fertilizers adds toxins and excess nitrogen and phosphorus.

Red tides
Excess nitrogen causes explosive growth of toxic microscopic algae, poisoning fish and marine mammals.

Closed beach

Closed shellfish beds

Oxygen-depleted zone

Toxic sediments
Chemicals and toxic metals contaminate shellfish beds, kill spawning fish, and accumulate in the tissues of bottom feeders.

Oxygen-depleted zone
Sedimentation and algae overgrowth reduce sunlight, kill beneficial sea grasses, use up oxygen, and degrade habitat.

Healthy zone
Clear, oxygen-rich waters promote growth of plankton and sea grasses, and support fish.

Figure 14-33 How residential areas, factories, and farms contribute to the pollution of coastal waters and bays.

Figure 14-34 A large zone of oxygen-depleted water (less than 2 ppm dissolved oxygen) forms for half of the year in the Gulf of Mexico as a result of oxygen-depleting alga blooms. It is created mostly by huge inputs of nitrate (NO_3^-) and phosphate (PO_4^{3-}) plant nutrients from the massive Mississippi River basin. In 2000, the area of the zone (shown here) was the largest ever—22,000 square kilometers (8,500 square miles). This is roughly equivalent to the area of the state of New Hampshire or the Central American country of Belize.

sphere (Figure 12-9, p. 264) introduce large quantities of nitrate (NO_3^-) and phosphate (PO_4^{3-}) plant nutrients, which can cause explosive growth of harmful algae. These *harmful algal blooms* (HABs) are called red, brown, or green tides, depending on their color. They can release waterborne and airborne toxins that damage fisheries, kill some fish-eating birds, reduce tourism, and poison seafood.

Each year some 61 large *oxygen-depleted zones* (sometimes inaccurately called dead zones) form in the world's coastal waters. They result from excessive nonpoint inputs of fertilizers and animal wastes from land runoff and deposition of nitrogen compounds from the atmosphere. In these zones, much of the aquatic life dies or moves elsewhere. The biggest such zone in U.S. waters and the third largest in the world forms every summer in a narrow stretch of the Gulf of Mexico (Figure 14-34).

Some *promising news* is that scientists in Asia and other parts of the world are experimenting with using certain types of fine clay to help control algal blooms. With this approach, siltlike particles of certain clays are flung into the water. The idea is to find clay with particles fine and heavy enough to stick to the algae and remove them by weighing them down like microanchors. More tests are needed to determine the effectiveness of the method and to be sure that the clay particles do not harm other aquatic organisms.

Case Study: The Chesapeake Bay The largest estuary in the United States, the Chesapeake Bay, is in trouble because of human activities. Between 1940 and 2002, the number of people living in the Chesapeake Bay area grew from 3.7 million to 17 million, and may soon reach 18 million.

The estuary receives wastes from point and non-point sources scattered throughout a huge drainage basin that includes 9 large rivers and 141 smaller streams and creeks in parts of six states (Figure 14-35). The bay has become a huge pollution sink because it is quite shallow and only 1% of the waste entering it is flushed into the Atlantic Ocean.

Phosphate and nitrate levels have risen sharply in many parts of the bay, causing algae blooms and oxygen depletion (Figure 14-35). Studies have shown that point sources, primarily sewage treatment plants, account for about 60% by weight of the phosphates. Non-point sources, mostly runoff from urban, suburban, and agricultural land and deposition from the atmosphere, account for about 60% by weight of the nitrates.

Large quantities of pesticides run off cropland and urban lawns into the bay. Also, industries discharge large amounts of toxic wastes (often in violation of their discharge permits) into the Chesapeake. Commercial harvests of oysters, crabs, and several important fish have fallen sharply since 1960 because of a combination of overfishing, pollution, and disease.

In 1983, the United States implemented the country's most ambitious attempt at *integrated coastal management,* the Chesapeake Bay Program. Between 1985 and 2000, phosphorus levels declined 27%, nitrogen

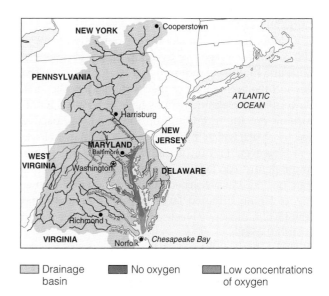

Figure 14-35 *Chesapeake Bay,* the largest estuary in the United States, is severely degraded as a result of water pollution from point and nonpoint sources in six states and from deposition of air pollutants.

levels dropped 16%, and grasses growing on the bay's floor have made a comeback. This is a significant achievement given the increasing population in the watershed and the fact that nearly 40% of the nitrogen inputs come from the atmosphere.

There is still a long way to go and state and federal funding for the project has fallen sharply. But the Chesapeake Bay Program shows what can be done when diverse interested parties work together to achieve goals that benefit both wildlife and people.

What Pollutants Do We Dump into the Ocean?
Dumping industrial waste off U.S. coasts has stopped, although it still occurs in a number of other developed countries and some developing countries. However, barges and ships still legally dump large quantities of **dredge spoils** (materials, often laden with toxic metals, scraped from the bottoms of harbors and rivers to maintain shipping channels) at 110 sites off the Atlantic, Pacific, and Gulf coasts of the United States.

In addition, many countries dump into the ocean large quantities of *sewage sludge*. It is a gooey mixture of toxic chemicals, infectious agents, and settled solids removed from wastewater at sewage treatment plants. Since 1992, the United States has banned this practice.

Fifty countries with at least 80% of the world's merchant fleet have agreed not to dump sewage and garbage at sea, but this agreement is difficult to enforce and violations are common. Most ship owners save money by dumping wastes at sea and if caught risk only small fines. Each year as many as 2 million seabirds and more than 100,000 marine mammals (including whales, seals, dolphins, and sea lions) die when they ingest or become entangled in fishing nets, ropes, and other debris dumped into the sea or discarded on beaches.

Under the London Dumping Convention of 1972, 100 countries agreed not to dump highly toxic pollutants and high-level radioactive wastes in the open sea beyond the boundaries of their national jurisdictions. Since 1983, these same nations have observed a moratorium on the dumping of low-level radioactive wastes at sea, which in 1994 became a permanent ban.

However, France, Great Britain, Russia, China, and Belgium, can legally exempt themselves from this ban. In 1992, it was learned that for decades the former Soviet Union dumped large quantities of high- and low-level radioactive wastes into the Arctic Ocean and its tributaries.

What Are the Effects of Oil on Ocean Ecosystems?
Crude petroleum (oil as it comes out of the ground) and *refined petroleum* (fuel oil, gasoline, and other processed petroleum products) are accidentally or deliberately released into the environment from a number of sources.

Tanker accidents and blowouts at offshore drilling rigs (when oil escapes under high pressure from a borehole in the ocean floor) get most of the publicity because of their high visibility. However, more oil is released into the ocean during normal operation of offshore wells, from washing oil tankers and releasing the oily water, and from oil pipeline and storage tank leaks.

Natural oil seeps also release large amounts of oil into the ocean at some sites, but most ocean oil pollution comes from activities on land. Almost half (some experts estimate 90%) of the oil reaching the oceans is waste oil dumped, spilled, or leaked onto the land or into sewers by cities, industries, and people changing their own motor oil. The EPA estimates that just 4 liters (1 gallon) of used motor oil has the potential of contaminating up to 4 million liters (1 million gallons) of drinking water.,

The effects of oil on ocean ecosystems depend on a number of factors: type of oil (crude or refined), type of aquatic system, amount released, distance of release from shore, time of year, weather conditions, average water temperature, and ocean currents.

Volatile organic hydrocarbons in oil immediately kill a number of aquatic organisms, especially in their vulnerable larval forms. Some other chemicals form tarlike globs that float on the surface and coat the feathers of birds (especially diving birds) and the fur of marine mammals. This oil coating destroys their natural insulation and buoyancy, causing many of them to drown or die of exposure from loss of body heat.

Heavy oil components that sink to the ocean floor or wash into estuaries can smother bottom-dwelling organisms such as crabs, oysters, mussels, and clams or make them unfit for human consumption. Some oil spills have killed reef corals.

Research shows that most (but not all) forms of marine life recover from exposure to large amounts of *crude oil* within 3 years. But recovery from exposure to *refined oil*, especially in estuaries, can take 10 years or longer. The effects of spills in cold waters and in shallow enclosed gulfs and bays generally last longer. Estuaries and salt marshes suffer the most and longest lasting damage.

In 1989, the Exxon Valdez tanker went off course, hit rocks, and released large amounts of oil into Alaska's Prince William Sound. Twelve years later in 2001 the National Marine Fisheries Service found that small remaining oil patches on about 28 beaches were still releasing toxins that harm wildlife. Exxon Mobil claim that the sound is entirely healed from the spill and that any lingering oil pollution comes from sources other than the Exxon Valdez supertanker, such as abandoned mines, natural oil seeps, and fuel spills from fishing boats.

Oil slicks that wash onto beaches can have a serious economic impact on coastal residents, who lose income from fishing and tourist activities. Oil-polluted beaches washed by strong waves or currents become clean after about a year, but beaches in sheltered areas remain contaminated for several years. Despite their localized harmful effects, experts rate oil spills as a low-risk ecological problem (Figure 10-15, left, p. 216).

How Well Can We Clean Up Oil Spills?   If they are not too large, oils oil spills can be partially cleaned up by mechanical, chemical, fire, and natural methods. *Mechanical methods* include using *floating booms* to contain the oil spill or keep it from reaching sensitive areas, *skimmer boats* to vacuum up some of the oil into collection barges, and *absorbent devices* (such as large mesh pillows filled with feathers or hair) to soak up oil on beaches or in waters too shallow for skimmer boats.

Chemical methods include using *coagulating agents* to cause floating oil to clump together for easier pickup or to sink to the bottom (where it usually does less harm) and *dispersing agents* to break up oil slicks. However, these agents can damage some types of organisms.

Fire can burn off floating oil, but crude oil is hard to ignite, and this approach produces air pollution. In time, the natural action of wind and waves mixes or emulsifies oil with water (like emulsified salad dressing), and bacteria biodegrade some of the oil.

Scientists estimate that current methods can recover no more than 12–15% of the oil from a major spill. This explains why preventing oil pollution is the most effective and in the long run the least costly approach.

Evidence for this approach comes from a number of oil tanker spills. One is the large 1989 spill from the *Exxon Valdez* oil tanker in Alaska's Prince William Sound that ended up costing $6 billion, including cleanup costs and damage fines. Another is the massive oil spill when the oil tanker *Prestige* sank off the coast of Spain in 2002. The ship is expected to leak about twice as much oil as the *Exxon Valdez* for years, possibly until 2006. Both of these spills would probably not have happened if the tankers had been equipped with double hulls instead of single hulls.

Since the 1989 *Exxon Valdez* spill environmentalists have been pushing for a much faster shift to safer double-hull tankers. But oil companies were able to delay the global phaseout of single-hull tankers for 25 years, until 2015, saying it would cost too much money. In 2003, two-thirds of the world's oil tankers still had the older and more vulnerable single-hull design. Oil spills would also be reduced if governments required more vigilant inspection of tankers entering and leaving ports. Establishing a global network of ecologically sensitive zones where tanker traffic is prohibited can reduce the harmful ecological and economic effects of oil spills.

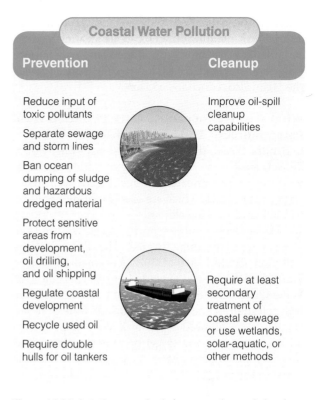

Coastal Water Pollution	
Prevention	**Cleanup**
Reduce input of toxic pollutants	Improve oil-spill cleanup capabilities
Separate sewage and storm lines	
Ban ocean dumping of sludge and hazardous dredged material	
Protect sensitive areas from development, oil drilling, and oil shipping	
Regulate coastal development	Require at least secondary treatment of coastal sewage or use wetlands, solar-aquatic, or other methods
Recycle used oil	
Require double hulls for oil tankers	

Figure 14-36 Solutions: methods for preventing and cleaning up excessive pollution of coastal waters.

Solutions: How Can We Protect Coastal Waters?   Analysts have suggested various ways to prevent and reduce excessive pollution of coastal waters (Figure 14-36). The key to protecting oceans is to reduce the flow of pollution from the land and from streams emptying into the ocean (Figure 14-33). Thus ocean pollution control must be linked with land land-use and air pollution policies because an estimated 33% of all pollutants entering the ocean worldwide comes from air emissions from land-based sources.

14-9 SOLUTIONS: PREVENTING AND REDUCING SURFACE WATER POLLUTION

What Can We Do about Water Pollution from Nonpoint Sources?   A number of methods can reduce nonpoint water pollution, most of it from agriculture. Farmers can reduce fertilizer running off into surface waters and leaching into aquifers by using slow-release fertilizer and using none on steeply sloped land. They can also plant buffer zones of vegetation between cultivated fields and nearby surface water.

Pesticide runoff can be reduced by applying pesticides only when needed and relying more on biological control of pests. Ways to control runoff and infiltration of manure from animal feedlots include improving manure control, planting buffers, and not locating feedlots and animal waste on steeply sloped land near surface water and in flood zones. Reforesting critical watersheds can also reduce soil erosion.

What Can We Do about Water Pollution from Point Sources?: The Legal Approach According to Sandra Postel, director of the Global Water Policy Project, most cities in developing countries discharge 80–90% of their untreated sewage directly into rivers, streams, and lakes, which are used for drinking water, bathing, and washing clothes.

Developed countries purify most wastes from point sources to some degree. The Federal Water Pollution Control Act of 1972 (renamed the Clean Water Act when it was amended in 1977) and the 1987 Water Quality Act form the basis of U.S. efforts to control pollution of the country's surface waters. It sets standards for allowed levels of key water pollutants and requires polluters to get permits specifying how much of various pollutants they can discharge into aquatic systems.

In 1995, the EPA proposed a *discharge trading policy* to use market forces to reduce water pollution (as has been done with sulfur dioxide for air pollution control, Case Study, p. 274). In 2003, the EPA proposed putting this policy into effect. It allows water pollution sources to pollute at higher levels than allowed in their permits by buying credits from permit holders with pollution levels below their allowed levels. The EPA contends that the trading system will encourage quicker cleanups than relying on conventional water pollution regulations.

Some environmentalists support discharge trading. However, they warn that such a system is no better than the caps set for total pollution levels in various areas and call for careful scrutiny of the cap levels. The program proposed by the EPA sets no caps. They also warn that discharge trading could allow pollutants to build up to dangerous levels in areas where credits are bought and should not be used as a way to control any toxic pollutants. In addition, they call for gradually lowering the caps to encourage prevention of water pollution and development of better technology for controlling water pollution. This strategy is not a part of the proposed EPA discharge trading system.

Here is some *good news*. The Clean Water Act of 1972 led to a number of improvements in U.S. water quality between 1972 and 2000. The percentage of U.S. stream lengths tested that are fishable and swimmable increased from 36% to 61% and the amount of topsoil lost through agricultural runoff was cut by about 1.1 billion metric tons (1 billion tons) annually. Also, the proportion of the U.S. population served by sewage treatment plants increased from 32% to 74% and annual wetland losses decreased by 83%.

There is also some *bad news*. In 2000, the EPA found that 45% of the country's lakes, 39% of the stream lengths (up from 26% in 1984), and 51% of the estuaries (up from 44% in 1998) it surveyed were too polluted for swimming or fishing. The number of polluted streams, lakes, and estuaries could be much higher because only 19% of the country's stream lengths, 43% of its lake and reservoir area, and 36% of its estuaries have been tested for water quality.

Runoff of animal wastes from hogs, poultry, and cattle pollutes 70% of U.S. rivers. Large quantities of toxic industrial wastes are illegally dumped into U.S. rivers each year. Fish caught in more than 1,400 different waterways and 28% of the nation's lakes are unsafe to eat because of high levels of pesticides and other toxic substances.

A 2002 study by the U.S. Geological Survey found that 80% of 139 streams in 30 states studied were contaminated with antibiotics, steroids, synthetic hormones, and other organic chemicals commonly found in wastewater.

A 2002 study of EPA data by the U.S. Public Interest Research Group found that between 1999 and 2001 about 81% of wastewater treatment plants and chemical and industrial facilities in the United States polluted waterways at least once a year beyond what their federal permits allowed.

Finally, according to a 2001 study by the National Academy of Sciences, the area of wetlands in the United States is continuing to fall, despite a government goal of "no net loss" in the area and function of wetlands.

Should We Strengthen or Weaken the U.S. Clean Water Act? Some environmentalists and a 2001 report by the EPA's inspector general call for the Clean Water Act to be strengthened. Suggested improvements include increased funding and authority to control nonpoint sources of pollution, increased monitoring of state programs to see that pollution permits are not allowed to expire, and strengthened programs to prevent and control toxic water pollution. Other suggestions include providing more funding and authority for integrated watershed and airshed planning to protect groundwater and surface water from contamination, and expanding the rights of citizens to bring lawsuits to ensure that water pollution laws are enforced. The National Academy of Sciences also calls for halting the loss of wetlands, higher standards for wetland restoration, and creating new wetlands before filling any natural wetlands.

Many people oppose these proposals, contending that the Clean Water Act's regulations and government wetlands regulations are already too restrictive and costly. Farmers and developers see the law as a curb on their rights as property owners to fill in wetlands. They also believe they should be compensated for any property value losses resulting from federal regulations protecting wetlands.

State and local officials want more discretion in testing for and meeting water quality standards. They argue that in many communities it is unnecessary and too expensive to test for all the water pollutants required by federal law.

What Can We Do about Water Pollution from Point Sources? The Technological Approach
As population, urbanization, and industrialization grow, the volume of wastewater needing treatment will increase rapidly. In rural and suburban areas with suitable soils, sewage from each house can be discharged into a **septic tank** (Figure 14-37). About 25% of all homes in the United States are served by septic tanks.

In U.S. urban areas, most waterborne wastes from homes, businesses, factories, and storm runoff flow through a network of sewer pipes to wastewater treatment plants. Some cities have separate lines for runoff of storm water. However, 1,200 U.S. cities have combined the lines for these two systems because it is cheaper. When rains cause combined sewer systems to overflow, they discharge untreated sewage directly into surface waters.

Sewage reaching a treatment plant typically undergoes one or both of three two levels of purification. One is **primary sewage** treatment. It is a *mechanical* process that uses screens to filter out debris such as sticks, stones, and rags and allows suspended solids to settle out as sludge in a settling tank (Figure 14-38, p. 340). By itself, primary treatment removes about 60% of the suspended solids and 30% of the oxygen-demanding organic wastes from sewage but removes no phosphates, nitrates, salts, radioisotopes, or pesticides.

A second level is called **secondary sewage** treatment. It is a *biological* process in which aerobic bacteria remove up to 90% of biodegradable, oxygen-demanding organic wastes (Figure 14-38, p. 340).

A combination of primary and secondary treatment (Figure 14-38) removes about 97% by weight of the suspended solids, 95–97% of the oxygen-demanding organic wastes, 70% of most toxic metal compounds and nonpersistent synthetic organic chemicals, 70% of the phosphorus (mostly as phosphates), 50% of the nitrogen (mostly as nitrates), and 5% of dissolved salts. This process removes only a tiny fraction of long-lived radioactive isotopes and persistent organic substances such as some pesticides.

The *good news* is that because of the Clean Water Act, most U.S. cities have combined primary and secondary sewage treatment plants. Some *bad news* is that according to the EPA, at least two-thirds of these plants have at times violated water pollution regulations. Also, 500 cities have failed to meet federal standards for sewage treatment plants, and 34 East Coast

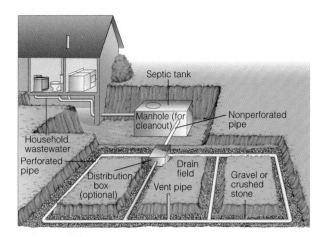

Figure 14-37 Solutions: *septic tank system* used for disposal of domestic sewage and wastewater in rural and suburban areas. This system traps greases and large solids and discharges the remaining wastes over a large drainage field. As these wastes percolate downward, the soil filters out some potential pollutants, and soil bacteria decompose biodegradable materials. To be effective, septic tank systems must be properly installed in soils with adequate drainage, not placed too close together or too near well sites, and pumped out when the settling tank becomes full.

cities simply screen out large floating objects from their sewage before discharging it into coastal waters.

A third level of cleanup is **advanced sewage treatment.** It is a series of specialized chemical and physical processes that remove specific pollutants left in the water after primary and secondary treatment. Advanced treatment is rarely used because such plants typically cost twice as much to build and four times as much to operate as secondary plants.

Before discharge, water from primary, secondary, or advanced treatment undergoes *bleaching* to remove water coloration and *disinfection* to kill disease-carrying bacteria and some but not all viruses. The usual method for doing this is *chlorination*. However, chlorine can react with organic materials in water to form small amounts of chlorinated hydrocarbons. Some of these chemicals can cause cancers in test animals and may damage the human nervous, immune, and endocrine systems (Case Study, p. 208).

Use of other disinfectants, such as ozone and ultraviolet light, is increasing. But they cost more and their effects do not last as long as chlorination.

What Should We Do with Sewage Sludge?
Sewage treatment produces a toxic and gooey *sludge*. In the United States, about 9% by weight of this sludge is converted to compost for use as a soil conditioner and 36% is applied to farmland, forests, golf courses, cemeteries, parkland, highway medians, and degraded land as fertilizer. The remaining 55% is dumped in conventional landfills (where it can contaminate groundwater) or incinerated (which can pollute the air with toxic

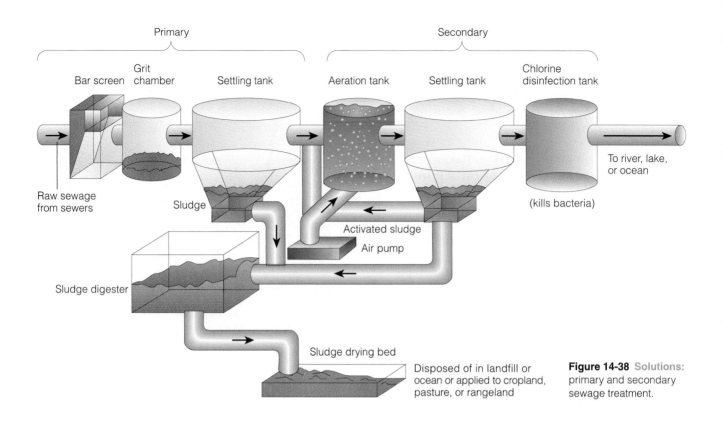

Primary | Secondary

Bar screen | Grit chamber | Settling tank | Aeration tank | Settling tank | Chlorine disinfection tank

Raw sewage from sewers

Sludge

Activated sludge

Air pump

Sludge digester

Sludge drying bed

To river, lake, or ocean

(kills bacteria)

Disposed of in landfill or ocean or applied to cropland, pasture, or rangeland

Figure 14-38 Solutions: primary and secondary sewage treatment.

chemicals and produces a toxic ash usually buried in landfills that the EPA says will leak eventually).

From an environmental standpoint, it is desirable to recycle the plant nutrients in sewage sludge to soil on land to grow crops, trees, or other plants. As long as harmful bacteria and toxic chemicals are not present, sludge is considered a safe way to fertilize land used for food crops or livestock. However, removing bacteria (usually by heating), toxic metals, and organic chemicals is expensive and rarely done in the United States.

According to a 2002 report by the National Academy of Sciences, the EPA is using outdated science to set standards for using sewage sludge as a fertilizer in the United States. The report also warned about public health risks from using sewage sludge as a commercial fertilizer.

A growing number of health problems and lawsuits are related to the use of sludge to fertilize crops in the United States. To protect consumers and avoid lawsuits, some food packers (such as Del Monte and Heinz) have banned produce grown on farms using sludge as a fertilizer.

How Can We Improve Sewage Treatment? Environmental scientist Peter Montague calls for redesigning the sewage treatment system to prevent toxic and hazardous waste from reaching sewage treatment plants and thus from getting into sludge and the

water discharged from such plants. He suggests several ways to do this. One is to require industries and businesses to remove toxic and hazardous wastes from water sent to municipal sewage treatment plants. Another is to encourage industries to reduce or eliminate toxic chemical use and waste.

We could also encourage use of less harmful household chemicals and have harmful chemicals picked up and disposed of safely on a regular basis as part of garbage collection services. Finally, we could have households, apartment buildings, and offices eliminate sewage outputs by switching to modern *composting toilet systems* that are installed, maintained, and managed by professionals. Such systems would be cheaper than current sewage systems because they do not require vast systems of underground pipes connected to centralized sewage treatment plants. They would also conserve large amounts of water.

Solutions: How Can We Treat Sewage by Working with Nature? Some individuals and communities are purifying contaminated water by working with nature. Ecologist John Todd designs, builds, and operates innovative ecological wastewater treatment systems called *living machines*. They look like aquatic botanical gardens and are powered by the sun in greenhouses or outdoors, depending on the climate.

This ecological purification process begins when sewage flows into a passive solar greenhouse or outdoor sites containing rows of large open tanks populated by an increasingly complex series of organisms. In the first set of tanks, algae and microorganisms decompose organic wastes, with sunlight speeding up the decomposition process. Water hyacinths, cattails, bulrushes, and other aquatic plants growing in the tanks take up the resulting nutrients.

After flowing though several of these natural purification tanks, the water passes through an artificial marsh of sand, gravel, and bulrush plants to filter out algae and remaining organic waste. Some of the plants also absorb (sequester) toxic metals such as lead and mercury and secrete natural antibiotic compounds that kill pathogens.

Next the water flows into aquarium tanks. There snails and zooplankton consume microorganisms and are in turn consumed by crayfish, tilapia, and other fish that can be eaten or sold as bait. After 10 days, the clear water flows into a second artificial marsh for final filtering and cleansing.

The water can be made pure enough to drink by using ultraviolet light or passing the water through an ozone generator, usually immersed out of sight in an attractive pond or wetland habitat. Selling the ornamental plants, trees, and baitfish produced as by-products of such living machines helps reduce costs. Operating costs are about the same as for a conventional sewage treatment plant.

Mark Nelson has developed a small, low-tech, and inexpensive artificial wetland system to treat raw sewage from hotels, restaurants, and homes in developing countries.

This *wastewater garden* system removes 99.9% of fecal coliform bacteria and more than 80% of the nitrates and phosphates from incoming sewage that in most developing countries is often dumped untreated into the ocean or into shallow holes in the ground. The water flowing out of such systems can be used to irrigate gardens or fields or to flush toilets and thus help save water.

Solutions: How Can We Use Wetlands to Treat Sewage?   Waste treatment is one of the important ecological services provided by wetlands. Some communities work with nature by using nearby natural wetlands to treat sewage, and others create artificial wetlands for such purposes. More than 150 cities and towns in the United States use natural and artificial wetlands to treat sewage as a low-tech, low-cost alternative to expensive waste treatment plants.

Some communities have created artificial wetlands to treat their water, as the residents of Arcata, California, did, led by Humboldt State University Professors Robert Gearheart and George Allen.

In this coastal town of 17,000, some 63 hectares (155 acres) of wetlands has been created between the town and the adjacent Humboldt Bay. The marshes, developed on land that was once a dump, act as an inexpensive natural waste treatment plant. The project cost less than half the estimated cost of a conventional treatment plant.

Here is how it works. First, sewage goes to sedimentation tanks, where the solids settle out as sludge that is then removed and processed for use as fertilizer. Next, the liquid is pumped into oxidation ponds, where remaining wastes are broken down by bacteria. After a month or so, the water is released into the artificial marshes, where plants and bacteria carry out further filtration and cleansing.

Although the water is clean enough for direct discharge into the bay, state law requires it first be chlorinated. The town chlorinates the water and then dechlorinates it before sending it into the bay, where oyster beds thrive.

The marshes and lagoons also serve as an Audubon Society bird sanctuary and provide habitats for thousands of otters, seabirds, and marine animals. The town celebrates its natural sewage treatment system with an annual "Flush with Pride" festival.

How Is Drinking Water Purified?   Treatment of water for drinking by city dwellers is much like wastewater treatment. Areas that depend on surface water usually store it in a reservoir for several days. This improves taste and clarity by increasing dissolved oxygen content and allowing suspended matter to settle.

Next the water is pumped to a purification plant and treated to meet government drinking water standards. Before disinfection, the water is usually run through sand filters and activated charcoal. In areas with very pure groundwater sources, little treatment is necessary.

Ways to purify drinking water can be simple. In tropical countries without centralized water treatment systems, the WHO is urging people to purify their drinking water by exposing a plastic bottle filled with contaminated water to the sun, which kills most bacteria.

In Bangladesh, women receive strips of cloth for filtering cholera-producing bacteria from drinking water. Villages where women use such strips to strain water have reduced cholera cases by 50%.

How Is the Quality of Drinking Water Protected?   About 54 countries, most of them in North America and Europe, have safe drinking water standards. The U.S. Safe Drinking Water Act of 1974 requires the EPA to establish national drinking water standards, called *maximum contaminant levels,* for any pollutants that may have adverse effects on human health.

Privately owned wells are not required to meet federal drinking water standards for two reasons. *First,* it costs at least $1,000 costs to test each well and owners would need to retest their water every few years. *Second,* some homeowners oppose mandatory testing and compliance.

Figuring how many people in the United States get sick or die each year from drinking contaminated water is difficult. Estimates vary and include 1 million illnesses and about 900 deaths according to a 1997 report by the Centers for Disease Control and Prevention and 239,000 illnesses and 50 deaths according to a 1999 EPA study.

Should the U.S. Safe Drinking Water Act Be Strengthened or Weakened? Health scientists and environmentalists call for strengthening the U.S. Safe Drinking Water Act in several ways. One is to combine many of the drinking water treatment systems that serve fewer than 3,300 people each with nearby larger systems. Another is to strengthen and enforce public notification requirements about violations of drinking water standards. They also call for banning all lead in new plumbing pipes, faucets, and fixtures (current law allows fixtures with up to 10% lead to be sold as lead free). According to the Natural Resources Defense Council (NRDC), such improvements would cost about $30 a year per U.S. household.

However, Congress is being pressured by water-polluting industries to weaken the Safe Drinking Water Act. *One* proposal is to eliminate national tests of drinking water and public notification requirements about violations of drinking water standards.

A *second* proposal is to allow states to give drinking water systems a permanent right to violate the standard for a given contaminant if the provider claims it cannot afford to comply. Another suggestion is to eliminate the requirement that water systems use affordable, feasible technology to remove cancer-causing contaminants.

Finally, there are suggestions to greatly reduce the EPA budget for enforcing the Clean Water Act clean and safe water laws.

Is Bottled Water the Answer? Despite some problems, experts say the United States has some of the world's cleanest drinking water. Yet about half of all Americans worry about getting sick from tap water contaminants, and about 60% drink bottled water or install expensive water purification systems.

Studies indicate that many people drinking bottled water are being cheated and in some cases may end up drinking water dirtier than the much cheaper water they can get from their taps.

Tests indicate that bacteria contaminate an estimated one-third of the bottled water purchased in the United States. To be safe, consumers purchasing bottled water should determine whether the bottling company belongs to the International Bottled Water Association (IBWA) and adheres to its testing requirements.* Some companies pay $2,500 annually to obtain more stringent certification by the National Sanitation Foundation, an independent agency that tests for 200 chemical and biological contaminants.

Use of bottled water also causes some environmental problems. They include the 1.4 million metric tons (1.5 million tons) of plastic bottles thrown away globally each year and toxic gases and liquids released during the manufacture of plastic water bottles. In addition, greenhouse gases are emitted during the shipping and delivery of bottled water and some aquifers are depleted to supply such water.

Before drinking expensive bottled water and buying costly home water purifiers, health officials suggest that consumers have their water tested by local health authorities or private labs (not companies trying to sell water purification equipment). The goals are to identify what contaminants (if any) must be removed and determine the type of purification needed to remove such contaminants. Independent experts contend that unless tests show otherwise, for most urban and suburban Americans served by large municipal drinking water systems, home water treatment systems are not worth the expense and maintenance hassles.

Buyers should check out companies selling water purification equipment and be wary of claims that the EPA has approved a treatment device. Although the EPA does *register* such devices, it neither tests nor approves them.

14-10 SOLUTIONS: ACHIEVING A MORE SUSTAINABLE WATER FUTURE

How Can We Use and Manage Water More Sustainably? Sustainable water use is based on the commonsense principle stated in an old Inca proverb: "The frog does not drink up the pond in which it lives." Figure 14-39 lists ways to implement this principle.

The challenge in developing such a *blue revolution* is to implement a mix of strategies built around irrigating crops more efficiently, using water-saving technologies in industries and homes, improving and integrating management of water basins and groundwater supplies, and integrating policies for managing water resources and reducing air pollution.

---

*Check for the IBWA seal of approval on the bottle or contact the International Bottled Water Association (1700 Diagonal Road, Suite 650, Alexandria, VA 22314; telephone: 703-683-5213; information hotline: 1-800-WATER-11; website: www.bottledwater. org) for a member list.

Accomplishing such a revolution in water use and management will be difficult and controversial. However, water experts contend that not developing such strategies will eventually lead to economic and health problems and increased environmental degradation and loss of biodiversity. It will also heighten tensions and perhaps lead to armed conflicts or economic competition over shared water supplies (p. 305).

How Can We Reduce Water Pollution? Individuals Matter   As with air pollution, it is encouraging that since 1970 most of the world's developed countries have enacted laws and regulations that have significantly reduced point sources of water pollution. Most of these improvements were the result of political pressure by individuals and organized groups of individuals on elected officials. However, little has been done to reduce water pollution in most developing countries.

To environmentalists the next step is to increase efforts to reduce and prevent water pollution by asking the question:. *"How can we not produce water pollutants in the first place?"*

Figure 14-40 lists ways to do this over the next several decades. A key lesson of ecology is *everything is connected.* This requires integrating government policies for preventing and reducing water pollution with policies for air pollution, agriculture, biodiversity protection, energy, solid and hazardous wastes, climate change, land use, poverty, and population control. Otherwise we run the risk of transferring a pollution problem from one part of the environment to another and to future generations.

Like the shift to *controlling air pollution* between 1970 and 2000, this new shift to *preventing water pollution* will not take place without political pressure on elected officials by individual citizens and groups of such citizens. See the website material for this chapter for some actions you can take to help reduce water pollution.

It is a hard truth to swallow, but nature does not care if we live or die. We cannot survive without the oceans, for example, but they can do just fine without us.

ROGER ROSENBLATT

REVIEW QUESTIONS

1. Define the boldfaced terms in this chapter.

2. Explain why there is a danger of water wars in the Middle East.

3. List nine unique properties of water and explain the importance of each property.

4. What percentage of the earth's total volume of water is easily available for use by people?

5. Distinguish among *surface runoff, reliable runoff, watershed, groundwater, zone of saturation, water table, aquifer,* and *natural recharge.* What is the connection between groundwater and surface water in the hydrological cycle? How can water in aquifers be depleted?

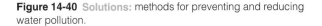

- Prevent groundwater contamination
- Greatly reduce nonpoint runoff
- Reuse treated wastewater for irrigation
- Find substitutes for toxic pollutants
- Work with nature to treat sewage
- Practice four R's of resource use (Refuse, Reduce, Reuse, Recycle)
- Reduce resource waste
- Reduce air pollution
- Reduce poverty
- Reduce birth rates

Figure 14-40 Solutions: methods for preventing and reducing water pollution.

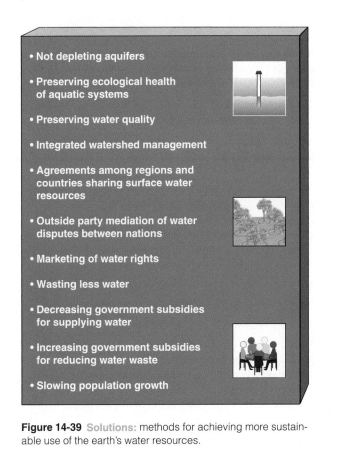

- Not depleting aquifers
- Preserving ecological health of aquatic systems
- Preserving water quality
- Integrated watershed management
- Agreements among regions and countries sharing surface water resources
- Outside party mediation of water disputes between nations
- Marketing of water rights
- Wasting less water
- Decreasing government subsidies for supplying water
- Increasing government subsidies for reducing water waste
- Slowing population growth

Figure 14-39 Solutions: methods for achieving more sustainable use of the earth's water resources.

6. Since 1900, how much has the total withdrawal and per capita withdrawal of water by humans increased? About what percentage of the world's reliable surface runoff do we use? About what percentage of the water we withdraw each year is used for **(a)** irrigation, **(b)** industry, and **(c)** residences and cities?

7. What are the major water uses and problems of **(a)** the eastern United States and **(b)** the western United States?

8. List four causes of water scarcity. Distinguish between *water stress* and *water scarcity*. About how many people live in countries that are water stressed or water scarce? How many people are likely to be living in such countries by 2025?

9. List six ways to increase the supply of fresh water in a particular area.

10. List the major advantages and disadvantages of building large dams and reservoirs to supply fresh water. List the advantages and disadvantages of China's Three Gorges dam project.

11. List six ecological services that rivers provide.

12. Describe the effects of the Aral Sea water transfer project in central Asia.

13. Describe the controversy over the California Water Transfer Project.

14. List the advantages and disadvantages of supplying more water by withdrawing groundwater. What is *saltwater intrusion*, and what harm does it cause?

15. Summarize the advantages and disadvantages of withdrawing groundwater from the Ogallala Aquifer in the United States.

16. Why is groundwater depletion such an urgent problem? List nine ways to prevent or slow groundwater depletion.

17. List the advantages and disadvantages of increasing supplies of fresh water by desalination of salt water. How useful are seeding clouds and towing icebergs in increasing to increase water supplies in water-short areas?.

18. About what percentage of the water used by people throughout the world is wasted? To what percentage could we reduce such waste? List two major causes of water waste.

19. List ways to reduce water waste in **(a)** irrigation and **(b)** industry, homes, and businesses. List four advantages of *drip irrigation*, and explain how it could help bring about a revolution in improving water efficiency. What is *Xeriscaping?*

20. What is a *floodplain?* What three major ecological and economic services do floodplains provide? List four reasons why so many people live on floodplains.

21. List the major benefits and disadvantages of floods. List four ways in which humans increase the damages from floods. Describe the nature and causes of the flooding problems in Bangladesh. List five ways to reduce the risks of flooding.

22. What is *water pollution?* What are eight types of water pollutants? What are the major sources and effects of each type?

23. Distinguish between *point* and *nonpoint sources of water pollution,* and give two examples of each type. Which type is easier to control? Why?

24. Summarize the good and bad news about the safety of drinking water in the world.

25. What are the major water pollution problems of streams? Describe how streams can handle some loads of biodegradable wastes. What are the limitations of this approach?.

26. Summarize the good and bad news about stream pollution in parts of the world.

27. List two reasons why dilution of pollution often is less effective in lakes than in streams.

28. Distinguish between *eutrophication* and *cultural eutrophication*. What are the major causes of cultural eutrophication? List three methods for preventing cultural eutrophication and three methods for cleaning up cultural eutrophication.

29. Summarize the good and bad news about attempts to reduce water pollution in the Great Lakes.

30. List five major sources of groundwater contamination. List three reasons why groundwater pollution is such a serious problem.

31. List three examples indicating the seriousness of groundwater pollution. List four ways to prevent groundwater contamination.

32. What causes arsenic pollution in drinking water? Discuss the controversy over reducing the allowed levels of arsenic found in drinking water in the United States.

33. List the major pollution problems of the oceans. Why are most of these problems found in coastal areas?

34. Summarize the major pollution problems of the Chesapeake Bay in the United States and the progress made in dealing with these problems.

35. What are the major sources of oil pollution in the world's oceans? Summarize the major effects of oil on ocean ecosystems. Describe mechanical and chemical methods for cleaning up oil spills. Discuss the limitations of cleaning up oil pollution. List three ways to reduce the chances and effects of spills by oil tankers.

36. List seven ways to prevent and three ways to reduce pollution of coastal waters.

37. List five ways to help prevent water pollution from nonpoint sources.

38. Explain how the United States and most developed countries have reduced water pollution from point sources by enacting laws. List four pieces of *good news* and five pieces of *bad news* about such efforts.

39. List five ways in which environmentalists believe water pollution control laws in the United States should be strengthened. List three reasons why there is opposition to such changes.

40. Distinguish among *septic tanks, primary sewage treatment, secondary sewage treatment,* and *advanced sewage treatment* as ways to reduce water pollution. List ways to deal with the sludge produced by waste treatment methods. Discuss the controversy over using sewage sludge to fertilize crops. List four ways to prevent toxic and hazardous wastes from reaching sewage treatment plants.

41. Describe the use of *living machines, wetlands,* and *wastewater gardens* to treat sewage based on working with nature.

42. How is drinking water purified? How is the quality of drinking water protected in the United States? How successful have these efforts been? List three suggestions made by environmentalists and health scientists for strengthening the U.S. Safe Drinking Water Act and four suggestions made by opponents for weakening the act. List the pros and cons of drinking bottled water.

43. List 11 ways to use the world's water more sustainably and 4 four disadvantages of not implementing such strategies.

44. List 10 ways to put more emphasis on preventing and reducing water pollution.

CRITICAL THINKING

1. How do human activities increase the harmful effects of prolonged drought? How can we reduce these effects?

2. Do you believe the projected benefits of China's Three Gorges dam and reservoir project on the Yangtze River will outweigh its potential drawbacks? Explain. What are the alternatives?

3. What role does population growth play in water supply problems?

4. What ecological and economic lessons can we learn from the Aral Sea tragedy?

5. Should the price of water be raised sharply to include more of its environmental costs and to encourage water conservation? Explain. What harmful and beneficial effects might this have on **(a)** business and jobs, **(b)** crop production, **(c)** your lifestyle and the lifestyles of any children or grandchildren you might have, **(d)** the poor, and **(e)** the environment?

6. List five major ways to conserve water for personal use (see website material for this chapter). Which, if any, of these practices do you now use or intend to use?

7. How do human activities contribute to flooding and flood damage? How can these effects be reduced?

8. Is each of the eight categories of pollutants listed in Table 14-1 (p. 324) most likely to originate from **(a)** point sources or **(b)** nonpoint sources?

9. Do you believe the drinking standard for arsenic in drinking water in the United States (or the country where you live) should be set at **(a)** 10 ppb or **(b)** less than 3 ppb? Explain. If you lived in a community affected by such a standard, how much more per year would you be willing to pay on your water bill to implement the standard?

10. Are you for or against banning all dumping of wastes and untreated sewage into the ocean? Explain. If so, where would you put the wastes instead? What exceptions would you permit, and why? How would you enforce such regulations?

11. Congratulations! You are in charge of managing the world's water resources. What are the three most important things you would do?

12. Congratulations! You are in charge of sharply reducing water pollution throughout the world. What are the three most important things you would do?

PROJECTS

1. In your community,
   a. What are the major sources of the water supply?
   b. How is water use divided among agricultural, industrial, power plant cooling, and public uses?
   c. Who are the biggest consumers of water?
   d. What has happened to water prices (adjusted for inflation) during the past 20 years? Are they too low to encourage water conservation and reuse?
   e. What water supply problems are projected?
   f. How is water being wasted?

2. Develop a water conservation plan for your school and submit it to school officials.

3. Consult with local officials to identify any floodplain areas in your community. Develop a map showing these areas and the types of activities (such as housing, manufacturing, roads, and recreational use) found on these lands. Evaluate management of such floodplains in your community and come up with suggestions for improvement.

4. In your community,
   a. What are the principal nonpoint sources of contamination of surface water and groundwater?
   b. What is the source of drinking water?
   c. How is drinking water treated?
   d. How many times during each of the past 5 years have levels of tested contaminants in the drinking water violated federal standards? Were violations reported to the public?
   e. Has pollution led to fishing bans or warnings not to eat fish from any lakes or streams in your region?
   f. Is groundwater contamination a problem? If so, where, and what has been done about the problem?
   g. Is there a vulnerable aquifer or critical recharge zone that needs protection to ensure the quality of groundwater? Is your local government aware of this? What action (if any) has it taken?

5. Arrange a class or individual tour of a sewage treatment plant in your community. Compare the processes it uses with those shown in Figure 14-38 (p. 340). What happens to the sludge produced by this plant? What improvements, if any, would you suggest for this plant?

6. Use library research, the Internet, and user interviews to evaluate the relative effectiveness and costs of home water purification devices. Determine the type or types of water pollutants each device removes and the effectiveness of this process.

7. Make a concept map of this chapter's major ideas, using the section heads and subheads and the key terms (in boldface type). See material on the website for this book about how to prepare concept maps.

INTERNET STUDY RESOURCES AND RESOURCES FOR FURTHER READING AND RESEARCH

The website for this book contains helpful study aids and many ideas for further reading and research. Log on to

http://biology.brookscole.com/miller 10

and click on the Chapter-by-Chapter area. Choose Chapter 14 and select a resource:

■ Flash Cards allows you to test your mastery of the Terms and Concepts to Remember for this chapter.

■ Tutorial Quizzes provides a multiple-choice practice quiz.

■ Student Guide to InfoTrac will lead you to Critical Thinking Projects that use InfoTrac College Edition as a research tool.

■ References lists the major books and articles consulted in writing this chapter.

■ Hypercontents takes you to an extensive list of sites with news, research, and images related to individual sections of the chapter.

INFOTRAC COLLEGE EDITION

Improve your skills with InfoTrac College Edition, a searchable online database of articles from more than 700 periodicals. Log on to

http://www.infotrac-college.com

or access InfoTrac through the website for this book. Try to find the following articles:

1. University of California at Irvine. 2002. Satellite data on underground aquifer levels can be vital to water resource management, UC Irvine hydrologist finds; Use of NASA technology seen as breakthrough to understanding water availability, movement. *Ascribe Higher Education News Service* (June 10). *Keywords:* "satellite," "aquifer," and "levels." A new satellite now allows hydrologists to assess more accurately the condition of the world's aquifers. This has important implications for water management.

2. Rabalais, N. N., R. E. Turner, and D. Scavia. 2002. Beyond science into policy: Gulf of Mexico hypoxia and the Mississippi River. *BioScience* 52: 129. *Keywords:* "Gulf of Mexico" and "hypoxia." Hypoxia (reduced oxygen levels) and cultural eutrophication are a serious problem in the northern Gulf of Mexico. Is there a solution?

15 SOLID AND HAZARDOUS WASTE

Pollution Control

Waste Treatment

There Is No "Away": Love Canal

Between 1942 and 1953, Hooker Chemicals and Plastics (owned by OxyChem since 1968) sealed chemical wastes containing at least 200 different chemicals into steel drums and dumped them into an old canal excavation (called Love Canal after its builder, William Love) near Niagara Falls, New York.

In 1953, Hooker Chemicals filled the canal, covered it with clay and topsoil, and sold it to the Niagara Falls school board for $1. The company inserted a disclaimer in the deed denying legal liability for any injury caused by the wastes. In 1957, Hooker warned the school board not to disturb the clay cap because of the possible danger from toxic wastes.

By 1959, an elementary school, playing fields, and 949 homes had been built in the 10-square-block Love Canal area (Figure 15-1). Some of the roads and sewer lines crisscrossing the dump site disrupted the clay cap covering the wastes. In the 1960s, an expressway was built at one end of the dump. It blocked groundwater from migrating to the Niagara River and allowed contaminated groundwater and rainwater to build up and overflow the disrupted cap.

Residents began complaining to city officials in 1976 about chemical smells and chemical burns their children received playing in the canal area, but their concerns were ignored. In 1977, chemicals began leaking from the badly corroded steel drums into storm sewers, gardens, basements of homes next to the canal, and the school playground.

In 1978, after media publicity and pressure from residents led by Lois Gibbs (a mother galvanized into action as she watched her children come down with one illness after another; Guest Essay, p. 352), the state acted. It closed the school and arranged for the 239 homes closest to the dump to be evacuated, purchased, and destroyed.

Two years later, after protests from families still living fairly close to the landfill, President Jimmy Carter declared Love Canal a federal disaster area, had the remaining families relocated, and offered federal funds to buy 564 more homes. Residents of all but 72 of the homes moved out. Some residents who remained claim that the other residents, environmentalists, and the media exaggerated the problem.

After more than 15 years of court cases, OxyChem agreed in 1994 to pay a $98 million settlement to New York State and take responsibility for all future treatment of wastes and wastewater at the Love Canal site. In 1999, the company also agreed to reimburse the federal government and New York State $7.1 million for the Love Canal cleanup.

Because of the difficulty in linking exposure to a variety of chemicals to specific health effects (Section 10-2, p. 204), the long-term health effects of exposure to hazardous chemicals on Love Canal residents remain unknown and controversial. However, for the rest of their lives the evacuated families will worry about the possible effects of the chemicals on themselves and their children and grandchildren.

The dumpsite has been covered with a new clay cap and surrounded by a drainage system that pumps leaking wastes to a new treatment plant. In June 1990, the U.S. Environmental Protection Agency (EPA) allowed state officials to begin selling 234 of the remaining houses in the area (renamed Black Creek Village). Most of the houses have been sold. Buyers must sign an agreement stating that New York State and the federal government make no guarantees or representations about the safety of living in these homes.

The Love Canal incident is a vivid reminder of three lessons from nature. *First*, we can never really throw anything away. *Second*, wastes often do not stay put. *Third*, preventing pollution is much safer and cheaper than trying to clean it up.

New York State Department of Environmental Conservation

Figure 15-1 The Love Canal housing development near Niagara Falls, New York, was built near a hazardous waste dump site. The photo shows the area when it was abandoned in 1980. In 1990, the EPA allowed people to buy some of the remaining houses and move back into the area.

Solid wastes are only raw materials we're too stupid to use.
ARTHUR C. CLARKE

This chapter addresses the following questions:

- What are solid waste and hazardous waste, and how much of each type do we produce?

- What can we do to reduce, reuse, and recycle solid waste and hazardous waste?

- What are we doing to recycle paper and plastics?

- What are the advantages and disadvantages of burning or burying wastes?

- What can we do to reduce exposure to lead, mercury, hazardous chlorine compounds, and dioxins?

- How is hazardous waste regulated in the United States?

- How can we make the transition to a more sustainable low-waste society?

15-1 WASTING RESOURCES

What Is Solid Waste, and How Much Is Produced? Solid waste is any unwanted or discarded material that is not a liquid or a gas. The United States, with only 4.6% of the world's population, produces about 33% of the world's solid waste. About 98.5% of this solid waste comes from mining, oil and natural gas production, agriculture, sewage sludge, and industrial activities (Figure 15-2).

The remaining 1.5% of solid waste is **municipal solid waste (MSW)** from homes and businesses in or near urban areas. The amount of MSW, often called *garbage*, produced in the United States comes to about 200 million metric tons (220 million tons) per year—al-

most twice as much as in 1970. Each year this is enough waste to fill a bumper-to-bumper convoy of garbage trucks encircling the globe almost eight times.

The annual MSW amounts to an average of 770 kilograms (1,700 pounds) per person in the United States, the world's highest per capita solid waste production. About 54% of this MSW is dumped in landfills, 30% is recycled or composted, and 16% is burned in incinerators.

What Does It Mean to Live in a High-Waste Society? According to architect and environmental designer William McDonough, the industrial revolution that has been taking place for about 275 years has a number of consequences. It puts huge amounts of toxic material into the air, water, and soil. It puts potentially valuable resources in landfills or other holes all over the planet, where they are too difficult or expensive to retrieve. It has spurred thousands of complex government regulations, mainly designed to keep most people from being poisoned or harmed too quickly instead of keeping people and natural systems safe.

It creates prosperity by cutting down and digging up natural resources and then burying or burning them when it is through with them. It depletes and degrades the earth's natural capital (Figure 1-2, p. 3, and top half of back cover) and erodes biodiversity and human cultural diversity. Finally, it counts these consequences as economic progress because they raise the gross national income (p. 22).

Solid waste is an indicator of how wasteful a society is. Here are a few of the solid wastes consumers throw away in the high-waste economy found in the United States:

- Enough aluminum to rebuild the country's entire commercial airline fleet every 3 months

- Enough tires each year to encircle the planet almost three times

- About 18 billion disposable diapers per year, which if linked end to end would reach to the moon and back seven times

- About 2 billion disposable razors, 30 million cell phones, 18 million computers, and 8 million television sets each year. By 2005, Americans will throw away an estimated 130 million cell phones and 50 million computers per year.

- Discarded carpet each year that would cover 7,800 square kilometers (3,000 square miles)—more than enough to carpet the state of Delaware

- About 1.7 billion metric tons (3.7 trillion pounds) of construction waste per year—an average of 6 metric tons (13,200 pounds) per American

- About 2.5 million nonreturnable plastic bottles every hour

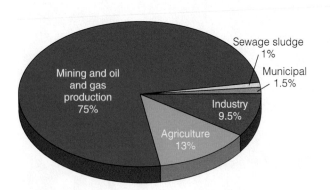

Figure 15-2 Sources of the estimated 11 billion metric tons (12 billion tons) of solid waste produced each year in the United States. Mining, oil and gas production, agricultural, and industrial activities produce 55 times as much solid waste as household activities. (Data from U.S. Environmental Protection Agency and U.S. Bureau of Mines)

- About 670,000 metric tons (1.5 billion pounds) of edible food per year

- Enough office paper each year to build a wall 3.5 meters (11 feet) high across the country from New York City to San Francisco, California

- Some 186 billion pieces of junk mail (an average of 660 per American) each year, about 45% of which are thrown in the trash unopened. This includes 17 billion catalogs, very few of them using recycled paper.

According to the EPA, *electronic waste* (*e-waste*) consisting of discarded TV sets, cell phones, computers, and other electronic devices is the fastest growing solid waste problem in the United States and the world. This waste is also a source of toxic and hazardous wastes (such as compounds containing lead, mercury, cadmium, and bromine) that can contaminate the air, surface water, groundwater, and soil.

What Is Hazardous Waste, and How Much Is Produced?   In the United States, **hazardous waste** is legally defined as any discarded solid or liquid material that meets one or more of four criteria. *First,* it contains one or more of 39 toxic, carcinogenic, mutagenic, or teratogenic compounds (Section 10-3, p. 208) at levels that exceed established limits (many solvents, pesticides, and paint strippers). *Second,* it catches fire easily (gasoline, paints, and solvents). *Third,* it is reactive or unstable enough to explode or release toxic fumes (acids, bases, ammonia, chlorine bleach). *Fourth,* it is capable of corroding metal containers such as tanks, drums, and barrels (industrial cleaning agents and oven and drain cleaners).

This official definition of hazardous wastes (mandated by the U.S. Congress) does *not* include the following materials: radioactive wastes, hazardous and toxic materials discarded by households (Figure 15-3), mining wastes, oil- and gas-drilling wastes (routinely discharged into surface waters or dumped into unlined pits and landfills), liquid waste containing organic hydrocarbon compounds (80% of all liquid hazardous waste), cement kiln dust produced when liquid hazardous wastes are burned in a cement kiln, and wastes from the thousands of small businesses and factories that generate less than 100 kilograms (220 pounds) of hazardous waste per month.

As a result, *hazardous waste laws do not regulate 95% of the country's hazardous waste.* Most other countries, especially developing countries, regulate even less, if any, of the hazardous wastes they produce.

Including all categories, the EPA estimates that the United States produces about 75% of the world's hazardous waste. This amounts to at least 5.5 billion metric tons (12 trillion pounds) of hazardous waste per year—an average of 19 metric tons (42,000 pounds) per American.

What Is the Threat from Release of Toxic Chemicals from Industrial Plants Because of Accidents or Terrorism?   Managers of industrial plants that manufacture and use chemicals work hard to prevent accidental release of

What Harmful Chemicals Are in Your Home?

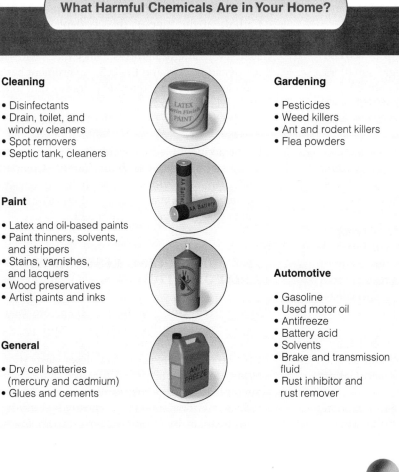

Cleaning

- Disinfectants
- Drain, toilet, and window cleaners
- Spot removers
- Septic tank, cleaners

Paint

- Latex and oil-based paints
- Paint thinners, solvents, and strippers
- Stains, varnishes, and lacquers
- Wood preservatives
- Artist paints and inks

General

- Dry cell batteries (mercury and cadmium)
- Glues and cements

Gardening

- Pesticides
- Weed killers
- Ant and rodent killers
- Flea powders

Automotive

- Gasoline
- Used motor oil
- Antifreeze
- Battery acid
- Solvents
- Brake and transmission fluid
- Rust inhibitor and rust remover

Figure 15-3 What harmful chemicals are in your home? Congress has exempted disposal of these materials from regulatory oversight.

chemicals that can harm workers or nearby residents. But accidents can happen, as thousands of people living near a pesticide manufacturing plant in Bhopal, India, learned in 1984 (Case Study, below).

The 2001 act of terrorism on New York City's World Trade Center Towers and the Pentagon has heightened concerns about terrorist acts against such plants. Roughly 20,000 industrial plants in the United States contain large quantities of hazardous chemicals. Analysts view such plants as easy targets for bombing or acts of sabotage by terrorists.

A 1999 government study found that chemical plant security ranged from "fair to very poor." Several hundred chemical plants have filed risk management plans indicating that a worst-case terrorist act at their sites could spread toxic chemicals as far away as 23 kilometers (14 miles) or more.

Case Study: A Black Day in Bhopal, India December 2, 1984, will long be a black day for India because on that date the world's worst industrial accident occurred at a Union Carbide pesticide plant in Bhopal, India.

Some 36 metric tons (40 tons) of highly toxic methyl isocyanate (MIC) gas, used to produce carbamate pesticides, erupted from an underground storage tank. Investigation found that water leaking into the tank through faulty valves and corroded pipes caused an explosive chemical reaction. Once in the atmosphere, some of the toxic MIC was converted to more deadly hydrogen cyanide gas.

The toxic cloud of gas settled over about 78 square kilometers (30 square miles), exposing up to 600,000 people. Many of these people were illegal squatters living near the plant because they had no other place to go. Since the safety sirens at the plant had been turned off to save money, the deadly cloud spread through Bhopal without warning.

Eyes, mouths, and lungs burned. Some victims tried to flee through the narrow streets but many were trampled. Others drowned in their own bodily fluids from exposure to the toxic gas. Many of the old, infirm, and very young died in their sleep.

According to Indian officials, at least 8,000 people died within a few days after the accident. The International Campaign for Justice in Bhopal estimates that by 2001 about 20,000 people had died as a result of the accident. An international team of medical specialists estimated in 1996 that 50,000–60,000 people sustained permanent injuries such as blindness, lung damage, and neurological problems. The International Campaign for Justice in Bhopal puts the number of people suffering from chronic illnesses from the accident at 120,000–150,000 people.

The accident caused an estimated at $4.1 billion in economic damages. Indian officials claim that Union Carbide probably could have prevented the tragedy by spending no more than $1 million to upgrade plant equipment and improve safety.

According to an investigation by India's Central Bureau of Investigation (CBI), corporate managers of Union Carbide in the United States made a decision to save money by cutting back on maintenance and safety and alarm systems because the plant had proven to be a financial disappointment for the company. The CBI found that on the night of the disaster six safety measures designed to prevent a leak of toxic materials were inadequate, shut down, or malfunctioning and the alarm systems had been turned off.

After the accident, Union Carbide reduced the corporation's liability risks for compensating victims by selling off a portion of its assets and giving much of the profits to its shareholders in the form of special dividends. In 1994, Union Carbide sold its holdings in India and later was taken over by Dow Chemical.

In 1989, Union Carbide agreed to pay an out-of-court settlement of $470 million to compensate the victims without admitting any guilt or negligence concerning the accident. The company also spent $100 million to build a hospital for the victims. By 2002, most victims with injuries had received $600 in compensation, and families of victims who died had received less than $3,000. About $272 billion of the settlement remains in the hands of the Indian government.

In 1992, the Court of the Chief Judicial Magistrate for Bhopal charged Warren Anderson, CEO of Union Carbide at the time of the accident, with "culpable homicide" (the equivalent of manslaughter) and issued a warrant for his arrest. Since then he has refused to appear in court and the U.S. government has not responded to India's request to extradite him to India to stand trial.

On December 2, 1999—the 15th anniversary of the disaster—survivors and families of people killed by the accident filed a lawsuit in a New York U.S. district court charging the Union Carbide Company and Anderson with violating international law and the fundamental human rights of the victims and survivors of the Bhopal plant accident.

15-2 PRODUCING LESS WASTE AND POLLUTION

What Are Our Options? There are two major ways to deal with the solid and hazardous waste we create. One is *waste management*. This is a *high-waste approach* (Figure 3-15, p. 61) that views waste production as an unavoidable product of economic growth. It attempts to manage the resulting wastes in ways that reduce environmental harm, mostly by burying them, burning them, or shipping them off to another state or

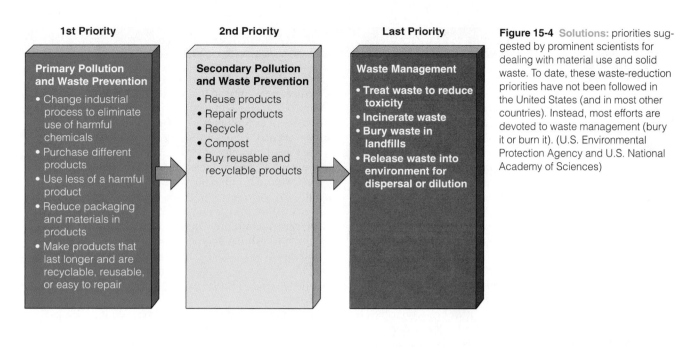

1st Priority

Primary Pollution and Waste Prevention

- Change industrial process to eliminate use of harmful chemicals
- Purchase different products
- Use less of a harmful product
- Reduce packaging and materials in products
- Make products that last longer and are recyclable, reusable, or easy to repair

2nd Priority

Secondary Pollution and Waste Prevention

- Reuse products
- Repair products
- Recycle
- Compost
- Buy reusable and recyclable products

Last Priority

Waste Management

- Treat waste to reduce toxicity
- Incinerate waste
- Bury waste in landfills
- Release waste into environment for dispersal or dilution

Figure 15-4 Solutions: priorities suggested by prominent scientists for dealing with material use and solid waste. To date, these waste-reduction priorities have not been followed in the United States (and in most other countries). Instead, most efforts are devoted to waste management (bury it or burn it). (U.S. Environmental Protection Agency and U.S. National Academy of Sciences)

country. In effect, it transfers solid and hazardous waste from one part of the environment to another.

The second approach is *waste and pollution prevention*, a *low-waste approach* that recognizes there is no "away." It also views most solid and hazardous waste either as potential resources (that we should be recycling, composting, or reusing) or as harmful substances that we should not be using in the first place (Figure 3-16, p. 61, and Figures 15-4 and 15-5). This approach focuses on discouraging waste production and encouraging waste reduction and prevention (Guest Essays, pp. 352 and 354).

What Would It Be Like to Live in a Low-Waste Society? Currently, the order of priorities for dealing with solid and hazardous wastes in the United States (and in most other countries) is the reverse of the order suggested by prominent scientists in Figures 15-4 and 15-5. It does not have to be that way. Some scientists and economists estimate that 60–80% of the solid and hazardous waste produced can be eliminated by a combination of *reduction, reuse, recycling* (including composting), and *redesign* of manufacturing processes and buildings, creating a low-waste society.

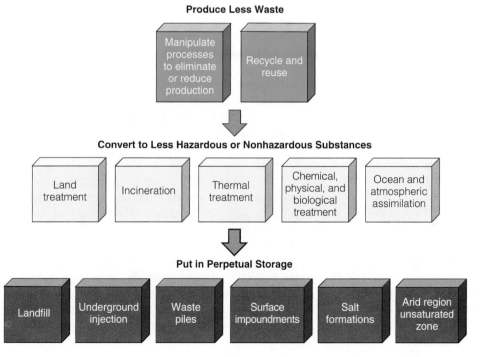

Produce Less Waste

Manipulate processes to eliminate or reduce production

Recycle and reuse

Convert to Less Hazardous or Nonhazardous Substances

Land treatment

Incineration

Thermal treatment

Chemical, physical, and biological treatment

Ocean and atmospheric assimilation

Put in Perpetual Storage

Landfill

Underground injection

Waste piles

Surface impoundments

Salt formations

Arid region unsaturated zone

Figure 15-5 Solutions: priorities suggested by prominent scientists for dealing with hazardous waste. To date, these priorities have not been followed in the United States (and in most other countries). (U.S. National Academy of Sciences)

We Have Been Asking the Wrong Questions About Wastes

Lois Marie Gibbs

GUEST ESSAY

In 1977, Lois Marie Gibbs was a young mother with two children living near the Love Canal toxic dumpsite. She had never engaged in any sort of political action until her children began experiencing unexplained illnesses and she learned that toxic chemicals were oozing from the dump site into many of the area's yards and basements. Then she organized her neighborhood and became the president and major strategist for the Love Canal Homeowners Association. This grassroots political action brought hazardous waste issues to national prominence and spurred passage of the federal Superfund legislation to help clean up abandoned hazardous waste sites. Lois Gibbs then moved to Washington, D.C., and formed Citizens' Clearinghouse for Hazardous Wastes (renamed the Center for Health, Environment, and Justice), an organization that has helped more than 7,000 community organizations protect themselves from hazardous wastes. Her story is told in her autobiography, Love Canal: My Story *(SUNY Press, 1982). She was also the subject of a CBS movie,* Lois Gibbs: The Love Canal, *which aired in 1982. Her latest book is* Dying from Dioxin *(Boston: South End Press, 1995).*

Just about everyone knows our environment is in danger. One of the most serious threats is the massive amount of waste we put into the air, water, and ground every year. All across the United States and around the world are thousands of places that have been, and continue to be, polluted by toxic chemicals, radioactive waste, and just plain garbage.

For generations, the main question people have asked is, "Where do we put all this waste? It's got to go somewhere." That is the wrong question, as has been shown by a series of experiments in waste disposal and by the simple fact that there is no "away" in "throwaway."

We tried dumping our waste in the oceans. That does not work. We tried injecting it into deep, underground wells. That does not work. We've been trying to build landfills that do not leak, but according to the EPA all landfills eventually leak. We've been trying to get rid of waste by burning it in high-tech incinerators; that only produces different types of pollution, such as air pollution and toxic ash. Even recycling, which is a very good thing to do, suffers from the same problem as all the other methods: It addresses waste *after* it has been produced.

For many years, people have been assuming, "It's got to go somewhere," but now many people, especially young people, are starting to ask why. Why do we produce so much waste? Why do we need products and services that have so many toxic by-products? Why can't industry change the way it makes things so that it stops producing so much waste?

When you start asking these questions, you start getting answers that lead to *pollution prevention* and *waste reduction* instead of *pollution control* and *waste management.* People, young and old, who care about waste reduction and pollution prevention are asking why so many goods are wrapped in excessive, throwaway packaging. They are challenging companies that sell pesticides, cleaning fluids, batteries, and other hazardous products to either

According to environmental designer William McDonough, in a low-waste society citizens, engineers, scientists, government leaders, and industrial leaders would work together to design a world with little pollution, waste, and poverty. *Factories* would emit no pollutants to the air or water or emit effluents so clean they can be used as drinking water. *Buildings* would produce more energy than they consume, and purify their own wastewater and air.

Transportation and agricultural systems would improve the quality of life by not harming humans and destroying and degrading much of the earth's natural capital (Figure 1-2, p. 3, and top half of back cover). *Toxic or hazardous materials* would be recycled or reused within industrial cycles or used as raw materials for new products instead of being buried, burned, or discarded.

Emphasis would be on producing *biodegradable products* (such as containers, clothes, cleaning products, and other items) that at the end of their useful life

can safely decompose and become *biological nutrients* in soil—thus mimicking what happens in nature (Figure 8-14, right, p. 171).

Solutions: How Can We Reduce Waste and Pollution? Here are some ways to reduce resource use, waste, and pollution. *First,* consume less. Before buying anything, ask questions such as: Do I really need this? Can I buy it secondhand? Can I borrow or lease it?

Second, redesign manufacturing processes and products to use less material and energy (Solutions, p. 353). *Third,* redesign manufacturing processes to produce less waste and pollution. Most toxic organic solvents can be recycled within plants or replaced with water-based or citrus-based solvents (Individuals Matter, p. 302). Hydrogen peroxide can be used instead of toxic chlorine to bleach paper and other materials. A CO_2-based process can replace dry cleaning with toxic organic solvents such as perchloroethylene (PERC).

remove the toxins from those products or take them back for recovery or recycling rather than disposing of them in the environment. They are demanding alternatives to throwaway materials in general.

Since 1988, hundreds of student groups have contacted my organization to get help and advice in taking these effective types of actions. Oregon students took legal action to get rid of cups and plates made from bleached paper because the paper contains the deadly poison dioxin. As a result the school systems switched to reusable cups, plates, and utensils.

Dozens of student groups have joined with local grassroots organizations to get toxic-waste sites cleaned up or to stop construction of new toxic-waste sites, radioactive-waste sites, or waste incinerators.

Waste issues are not simply environmental issues; they are also tied up with our economy, which is geared to producing and then disposing of waste. *Somebody* is making money from every scrap of waste and has a vested interest in keeping things the way they are.

Waste issues are also issues of *justice* and *fairness*. There is a lot of debate between industry officials and government risk analysts and regulators about so-called acceptable risk [Section 10-5, p. 215]. These individuals decide the degree of people's exposure to toxic chemicals but rarely ask the people who will actually be exposed how they feel about it.

Risk analysts often say, "There's only a one in a million chance of increased death from this toxic chemical." That may be true, but suppose I took a pistol and went to the edge of your neighborhood and began shooting

into it. There's probably only a one in a million chance that I'd hit somebody, but would you issue me a license to do that?

As long as we do not stand up for our rights and demand that "bullets" in the form of hazardous chemicals not be "fired" in our neighborhoods, we are giving environmental regulators and waste producers a license to kill a certain number of us without our even being consulted.

From my personal experience, I know that decisions made to dump wastes at Love Canal [p. 347] and in thousands of other places were *not* made purely on the basis of the best available scientific knowledge. The same holds true for decisions about how to manage the wastes we produce today and how to produce less waste.

We live in a world that is shaped by decisions based on money and power. If you really want to understand what's behind any given environmental issue, the first question you should ask is, "Who stands to profit from this?" Then ask, "Who is going to pay the price?" You can then identify both sides of the issue and decide whether you want to be part of the problem or part of the solution.

Critical Thinking

1. Do you believe we should put primary emphasis on pollution prevention and waste reduction? Explain.

2. What changes would you be willing to make in your own lifestyle to prevent pollution and reduce waste?

Fourth, *develop products that are easy to repair, reuse, remanufacture, compost, or recycle.* For example, a new Xerox photocopier, with every part reusable or recyclable for easy remanufacturing, should eventually save the company $1 billion in manufacturing costs. In 2000, remanufacturing by Xerox allowed the company to recycle 80% of its nonhazardous waste and keep 144 million metric tons (158 million tons) of e-waste out of landfills.

Fifth, *design products to last longer.* Today's tires have an average life of 97,000 kilometers (60,000 miles), but researchers believe this use could be extended to at least 160,000 kilometers (100,000 miles).

Sixth, *eliminate or reduce unnecessary packaging.* Here are some key questions for designers, manufacturers, and consumers to ask about packaging: Is it necessary? Can it use fewer materials? Can it be reused? Are the resources that went into it renewable? Does it contain the highest feasible amount of consumer-discarded (postconsumer) recycled material? Can it be

biodegraded into harmless nutrients that are recycled in the earth's natural chemical cycles? Is it designed to be recycled easily? Can it be incinerated without producing harmful air pollutants or a toxic ash? Can it be buried and decomposed in a landfill without producing chemicals that can contaminate groundwater?

Finally, *use trash taxes* to reduce waste, and use the revenues to reduce taxes on income and wealth, as a number of European countries have done. A related *pay-as-you-throw* system reduces solid waste and encourages recycling by basing garbage collection charges on the amount of waste a household or business generates for disposal.

Solutions: How Can We Do More with Less? Increasing Resource Productivity   During the last few decades, a *design revolution* has allowed businesses to use less material and energy per unit of goods and services. This has been done mostly by finding substitutes for products that use less material and redesigning

Cleaner Production: A New Environmentalism for the 21st Century

Peter Montague

Peter Montague is director of the Environmental Research Foundation in Washington, D.C., which studies and informs the public about environmental problems and the technologies and policies that might help solve them. He has served as project administrator of a hazardous waste research program at Princeton University and has taught courses in environmental impact analysis at the University of New Mexico. He is the coauthor of two books on toxic heavy metals in the natural environment and editor of Rachel's Environment and Health Biweekly, *an informative newsletter on environmental problems and solutions.*

Environmentalism as we have known it for the last 33 years is dead. The environmentalism of the 1970s advocated strict numerical controls on releases into the environment of *dangerous wastes* (any unwanted or uncontrolled materials that can harm living things or disrupt ecosystems).

However, after several decades of effort by government regulatory agencies and concerned citizens (the environmental movement), most dangerous chemicals are not regulated in any way. Even the few that are covered by regulations have not been adequately controlled.

In short, the *pollution management* approach to environmental protection has failed; *pollution prevention* is our only hope. An ounce of prevention really is worth a pound of cure.

Here is the situation facing environmentalists today:

- *All waste disposal—landfilling, incineration, deep-well injection—is polluting because "disposal" means dispersal into the environment.* Once wastes are created, they cannot be contained or controlled because of the scientific laws of matter conservation [p. 56] and energy conservation [p. 59]. The old environmentalism failed to recognize this important truth and thus squandered enormous resources trying to achieve the impossible.

- *The inevitable result of our reliance on waste treatment and disposal systems has been an unrelenting buildup of toxic synthetic materials in humans and other forms of life worldwide.* For example, breast milk of women in industrialized countries is so contaminated with pesticides and industrial hydrocarbons that if human milk were bottled and sold commercially, the U.S. Food and Drug Administration could ban it as unsafe for human consumption. If a whale beaches itself on U.S. shores and dies, its body must be treated as "hazardous waste" because whales contain concentrations of PCBs [polychlorinated biphenyls; Figure 14-28, p. 328] legally defined as hazardous.

or improving products so they take less material and energy to produce.

Paper documents such as product catalogs, phone directories, technical reference manuals, and parts directories can be accessed on CD-ROMs, DVD-ROMs, or at various Internet sites—saving millions of dollars and tons of paper. All U.S. phone books could be put on three CD-ROMs and a single DVD-ROM could hold all the world's phone numbers.

A skyscraper built today includes about 35% less steel than the same building built in the 1960s because of the use of lighter-weight but higher-strength steel. The weight of cars has been reduced by about 25% by using such steel and replacing many steel parts with lightweight plastics and composite materials. This has increased fuel efficiency without compromising performance and safety. Similar changes have reduced the average weight per unit of appliances such as stoves, washers, dryers, air conditioners, TV sets, and computers.

Since the mid-1970s, the thickness of plastic grocery bags has been reduced by 70% (without sacrificing strength), plastic milk jugs weigh 40% less, aluminum drink cans contain one-third less aluminum, steel cans are 60% lighter, disposable diapers contain 50% less paper pulp, and plastic frozen food bags weigh 89% less.

These improvements in resource productivity are important, but according to some analysts they can be greatly increased through a new *resource productivity revolution*. In their 1999 book *Natural Capitalism*, Paul Hawken (Guest Essay, p. 4), Amory Lovins, and Hunter Lovins contend we have the knowledge and technology to greatly increase resource productivity by getting 75–90% more work or service from each unit of material resources we use.

To these analysts, the only major impediments to such an economic and ecological revolution are laws, policies, taxes, and subsidies that continue to reward inefficient resource use and fail to reward efficient resource use.

15-3 SOLUTIONS: CLEANER PRODUCTION AND SELLING SERVICES INSTEAD OF THINGS

What Is the Ecoindustrial Revolution, and What Are Its Benefits? There are growing signs that a new *ecoindustrial revolution* will take place over

- *The ability of humans and other life forms to adapt to changes in their environment is limited by the genetic code each form of life inherits.* Continued contamination at a rate hundreds of times faster than we can adapt will subject humans to increasingly widespread sickness and degradation of the species and could ultimately lead to extinction.

- *Damage to humans (and other life forms) is abundantly documented.* Birds, fish, and humans in industrialized countries are enduring steadily rising levels of cancer, genetic mutations, and damage to their nervous, immune, and hormonal systems [Case Study, p. 208] as a result of pollution.

If we will but look, the handwriting is on the wall everywhere.

To deal with these problems, industrial societies must abandon their reliance on waste treatment and disposal and on the regulatory system of numerical standards created to manage the damage that results from relying on waste disposal instead of waste prevention. We must quickly move the industrialized and industrializing countries to new technical approaches accompanied by new industrial goals: *clean production* or zero-discharge systems.

The concept of clean production involves industrial systems that avoid or eliminate dangerous wastes and dangerous products and minimize the use and waste of raw materials, water, and energy. Goods manufactured in a clean production process must not damage natural ecosystems throughout their entire life cycle.

Clean production does not rely on *end-of-pipe* pollution controls such as filters or scrubbers [Figure 12-19, p. 275] or chemical, physical, or biological treatment. The concept of clean production does not include measures that pretend to reduce the volume of waste by incineration or concentration, use dilution to mask the hazard, or transfer pollutants from one environmental medium to another.

A new industrial pattern, and thus a new environmentalism, is emerging. Human survival and life quality depend on our willingness to make and pay for the changes needed to shift to this cleaner form of industrial production.

Critical Thinking

1. Some environmentalists point to the successes of the *pollution management* approach to environmental protection practiced during the past 33 years and do not agree it has failed. What is your position? Explain.

2. List three undesirable economic, health, consumption, and lifestyle changes you might experience as a consequence of putting much greater emphasis on pollution prevention.

---

the next 50 years as a way to help achieve industrial, economic, and environmental sustainability. The goal of this rapidly developing concept of *cleaner production* (Guest Essay, above), or *industrial ecology,* is to make industrial manufacturing processes more sustainable by redesigning them to mimic how nature sustains itself (Figure 8-14, p. 171). One way to do this is to recycle and reuse most chemicals (Figure 3-16, p. 61).

Another way is have industries interact in complex *resource exchange webs* where the wastes of one manufacturer become raw materials for another—similar to food webs in natural ecosystems (Figure 4-18, p. 77). An example of this industrial exchange ecosystem concept exists in Kalundborg, Denmark, where an electric power plant and a number of nearby industries, farms, and homes are working together to save money and reduce their outputs of waste and pollution. They do this by exchanging their waste outputs and thus converting them into resources or raw materials for use in another industry (Figure 15-6, p. 356). Today more than 25 ecoindustrial parks similar to the one in Kalundborg are being developed around the world.

In addition to eliminating most waste and pollution, these industrial forms of *biomimicry* provide many economic benefits for businesses. They reduce the costs of controlling pollution and complying with pollution regulations. They also improve the health and safety of workers by reducing exposure to toxic and hazardous material and thus reduce company health-care insurance costs. Future legal liability for toxic and hazardous wastes also disappears.

Biomimicry also stimulates companies to come up with new, environmentally beneficial chemicals, processes, and products that can be sold worldwide. It also gives companies a better image among consumers based on results rather than on public relations campaigns.

In 1975, the Minnesota Mining and Manufacturing Company (3M), which makes 60,000 different products in 100 manufacturing plants, began a Pollution Prevention Pays (3P) program. It redesigned equipment and processes, used fewer hazardous raw materials, identified hazardous chemical outputs (and recycled or sold them as raw materials to other companies), and began making more nonpolluting products.

By 1998, 3M's overall waste production was down by one-third, its air pollutant emissions per unit of production were 70% lower, and the company had

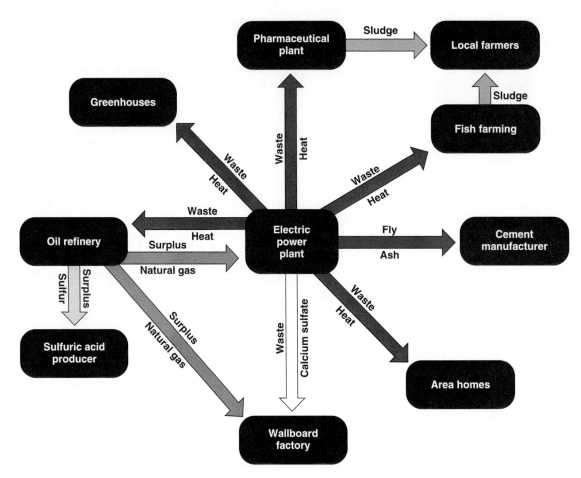

Figure 15-6 Solutions: *industrial ecosystem* in Kalundborg, Denmark, reduces waste production by mimicking a natural food web. The wastes of one business become the raw materials for another business.

saved more than $750 million in waste disposal and material costs.

Since 1990, a growing number of companies have adopted similar pollution prevention programs. For example, between 1992 and 2002 Xerox saved $2 billion through reuse, recycling, and elimination of hazardous materials from its products.

What Is a Service Flow Economy, and What Are Its Advantages? In the mid-1980s, German chemist Michael Braungart and Swiss industry analyst Walter Stahel independently proposed a new economic model that would provide profits while greatly reducing resource use and waste. Their idea involves shifting from our current *material flow economy* (Figure 3-15, p. 61) to a *service flow economy* over the next few decades. Instead of buying most goods outright, customers would lease or rent the *services* such goods provide.

With such a *service flow* or *product stewardship economy*, a product produced by a manufacturer remains as an asset that yields more profit if it

- Uses the minimum amount of materials

- Lasts as long as possible

- Is easy to maintain, repair, remanufacture, reuse, or recycle

- Provides customers with the services they want instead of trying to keep selling them newer models of outmoded products

Here are some examples illustrating that such an economic shift based on the concept of *eco-leasing* is under way. Since 1992, the Xerox Corporation has been leasing most of its copy machines as part of its mission to provide *document services* instead of selling photocopiers. When the service contract expires, Xerox takes the machine back for reuse or remanufacture and has a goal of sending no material to landfills or incinerators. To save money, machines are designed to use recycled paper, have few parts, be energy efficient, and emit as little noise, heat, ozone, and copier chemical waste as possible.

Ray Anderson

Ray Anderson is CEO of Interface, a company based in Atlanta, Georgia, that makes carpet tiles. The company is the world's largest commercial carpet manufacturer with 26 factories in six countries, customers in 110 countries, and more than $1 billion in annual sales.

Anderson changed the way he viewed the world and his business after reading Paul Hawken's book *The Ecology of Commerce* (Guest Essay, p. 4). In 1994, he announced plans to develop the nation's first totally sustainable green corporation.

He has implemented hundreds of projects with the goals of zero waste, greatly reduced energy use, and eventually zero use of fossil fuels by relying on renewable solar energy. By 1999, the company had reduced resource waste by almost 30% and reduced energy waste enough to save $100 million. One of Interface's factories in California runs on solar cells to produce the world's first solar-made carpet.

To achieve the goal of zero waste, Anderson plans to stop selling carpet and lease it as a way to control recycling. For a monthly fee, the company will install, clean, and inspect the carpet on a monthly basis, repair worn carpet tiles overnight, and recycle worn-out tiles into new carpeting. As Anderson puts it, "We want to harvest yesterday's carpets and recycle them with zero scrap going to the landfill and zero emissions into the ecosystem—and run the whole thing on sunlight."

DuPont and several other chemical companies have developed processes to remove the nylon and plastic PVC fibers in carpet and recycle it into other lower-quality products (downcycling). Interface has gone further and developed a new polymer material, called Solenium, that when worn out can be completely recycled back into new carpet tiles (more desirable closed-loop recycling), does not mildew, is highly stain resistant, and is easily cleaned with water. Making this material takes fewer steps, uses up to 40% less raw material and energy, and produces 99.7% less waste than making normal carpet. The material lasts about four times longer than conventional carpet.

The company also has plans to install and lease a raised-floor system that goes beneath its carpet tiles and integrate this with cooling and heating services provided by other service companies.

Anderson is one of a growing number of business leaders committed to finding a more economically and ecologically sustainable way to do business while still making a profit for stockholders. Between 1993 and 1998, the company's revenues doubled and profits tripled, mostly because the company saved $130 million in material costs with an investment of less than $40 million. Anderson says he is having a blast.

Ray Anderson, CEO of Interface, Inc.

Ray Anderson, CEO of a large carpet tile company, plans to lease rather than sell carpet (Individuals Matter, above). For years, 160 firms, called *chauffagistes,* have been providing 10 million buildings in metropolitan France with heat. These firms provide *warmth services* by contracting to keep a client's space within a specified temperature during certain hours at a designated cost.

Carrier, the world's leading maker of air conditioning equipment, now sells leases to provide its customers with *cooling services.* Carrier also teams up with other service providers to install super-efficient windows and more efficient lighting and make other energy-efficiency upgrades that reduce the cooling needs of its customers. Carrier makes money by having to install less or even no air conditioning equipment to provide cooling services for its customers.

Dow and several other chemical companies are doing a booming business in leasing organic solvents (mostly used to remove grease from surfaces), photographic developing chemicals, and dyes and pigments. In this *chemical service* business, the company delivers the chemicals, helps the client set up a recovery system, takes away the recovered chemicals, and delivers new chemicals as needed.

15-4 REUSE

What Are the Advantages of Refillable Containers? The *good news* is that reuse is a form of waste reduction that extends resource supplies and reduces energy use and pollution even more than recycling.

The *bad news* is that we have increasingly substituted throwaway tissues for reusable handkerchiefs, disposable paper towels and napkins for reusable cloth ones, throwaway beverage containers for refillable ones, and throwaway paper plates and cups and plastic utensils for reusable plates, cups, and silverware.

Two examples of reuse are refillable glass beverage bottles and refillable soft drink bottles made of polyethylene terephthalate (PET) plastic. Unlike throwaway and recyclable cans and bottles, refillable beverage bottles create local jobs related to their collection and refilling and thus help stimulate local economies. Moreover, studies by Coca-Cola and PepsiCo of Canada show that their soft drinks in 0.5-liter (16-ounce) bottles cost one-third less in refillable bottles than in throwaway bottles.

Denmark and Canada's Prince Edward Island have led the way by banning all beverage containers that cannot be reused. To encourage use of refillable glass bottles, Ecuador has a refundable beverage container deposit fee that is 50% of the cost of the drink. In Finland, 95% of the soft drink, beer, wine, and spirits containers are refillable, and in Germany, 73% are refillable.

Other examples of reusable items are metal or plastic lunch boxes and plastic containers for storing lunch box items and refrigerator leftovers (instead of using throwaway plastic wrap and aluminum foil).

Cloth bags can be used to carry groceries and other items instead of paper or plastic bags. Plastic and paper bags are both environmentally harmful, and the question of which is the more damaging has no clear-cut answer. To encourage people to bring reusable bags, stores in the Netherlands charge for paper or plastic bags. In 2002, use of plastic shopping bags dropped by 90% after the government imposed a 15¢ tax on each bag.

We can use shipping pallets made of recycled plastic waste instead of throwaway wood pallets. In 1991, Toyota shifted entirely to reusable shipping containers. A similar move by the Xerox Corporation saves the company $2–5 million per year.

Another example of reuse involves *tool libraries* (such as those in Berkeley, California, and Takoma Park, Maryland) where people can check out a variety of power and hand tools.

Finally, in the not-too-distant future we may be able to use e-paper over and over. It is a flexible and cordless computer screen being developed by Xerox that looks like a sheet of paper and uses no energy for storing or viewing writing or images. It can be electronically written and rewritten at least a million times—making it equivalent to more than a million sheets of paper.

15-5 RECYCLING

What Are the Two Types of Recycling? Recycling has a number of benefits to people and the environment (Figure 15-7). There are two types of recycling for materials such as glass, metals, paper, and plastics. One is *closed-loop recycling.* It occurs when wastes discarded by consumers (*postconsumer wastes*) are recycled to produce new products of the same type (such as newspaper into newspaper and aluminum cans into aluminum cans). This reduces pollution and use of virgin resources and saves energy.

The second type is *downcycling* in which waste materials are converted into different and usually lower-quality products.

Environmentalists urge us not to be lulled by labels claiming that paper and plastic bags or other items are recyclable. Just about anything is recyclable but only two things count. First is whether an item is actually recycled (ideally by closed-loop recycling). The second is whether we complete the recycling loop by buying products using the maximum feasible content of postconsumer recycled materials.

How Much Municipal Solid Waste Is Recycled or Composted in the United States? Currently, about 30% of the municipal solid waste (MSW) in the United States is recycled or composted, and in 2002 the EPA announced a goal of increasing the rate to at least 35% by 2005. This is a good start but pilot studies in several U.S. and European communities show that 60–80% of all MSW could be recycled and composted.

How Useful Is Composting? Biodegradable organic waste such as paper, food scraps, and lawn waste can be broken down by composting—a way to produce plant nutrients that can be recycled into the soil. Biodegradable wastes make up about 35% by weight of the MSW output in the United States. However, only about 5% of the MSW in the United States is composted (compared to 17% in France and 10% in Switzerland). Some cities in Austria, Belgium, Denmark, Germany, Luxembourg, and Switzerland recover and compost more than 85% of their biodegradable wastes.

Individuals can compost biodegradable wastes in backyard bins or indoor containers. They can also be collected and composted in centralized community facilities, as is done in many western European countries.

The resulting compost can be used as an organic soil fertilizer, topsoil, or landfill cover. Compost can also be used to help restore eroded soil on hillsides and along highways, strip-mined land, overgrazed areas, and eroded cropland.

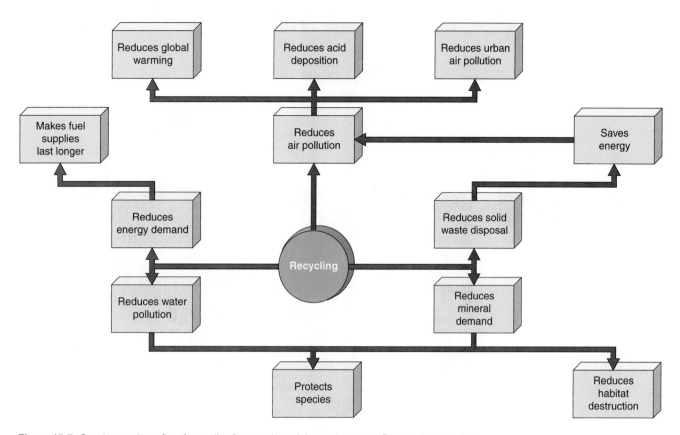

Figure 15-7 *Good news: benefits of recycling* for people and the environment. Despite its many benefits, recycling is still an output approach that deals with wastes after they are produced instead of a way to reduce the overall flow of resources.

To be successful, a large-scale composting program must overcome siting problems (few people want to live near a giant compost pile or plant) and control odors. It must also exclude toxic materials that can contaminate the compost and make it unsafe for fertilizing crops and lawns.

Is Centralized Recycling of Mixed Solid Waste the Answer? Large-scale recycling can be accomplished by collecting mixed urban waste and transporting it to centralized *materials-recovery facilities (MRFs)*. There, machines shred and automatically separate the mixed waste to recover valuable materials for sale to manufacturers as raw materials (Figure 15-8, p. 360). The remaining paper, plastics, and other combustible wastes are recycled or burned to produce steam or electricity to run the recovery plant or to sell to nearby industries or homes. Ash from the incinerator is buried in a landfill.

More than 225 MRFs operate in the United States. However, there are drawbacks. Such plants are expensive to build, operate, and maintain (which is why some have been shut down). They can emit toxic air pollutants if not operated properly and they produce a toxic ash that must be disposed of safely.

Finally, MRFs must have a large input of garbage to make them financially successful. Thus their owners have a vested interest in increasing *throughput* of matter and energy resources to produce more trash, the reverse of what prominent scientists believe we should be doing (Figure 15-4).

Is Separating Solid Wastes for Recycling the Answer? To many solid waste experts, it makes more sense economically and environmentally for households and businesses to separate their trash into recyclable categories (such as glass, paper, metals, certain types of plastics, and compostable materials). Then compartmentalized city collection trucks, private haulers, or volunteer recycling organizations pick up the segregated wastes and sell them to scrap dealers, compost plants, and manufacturers. Another alternative (especially in less populated areas) is to establish a network of drop-off centers, buyback centers, and deposit-refund programs in which people deliver and sell or donate separated recyclable materials.

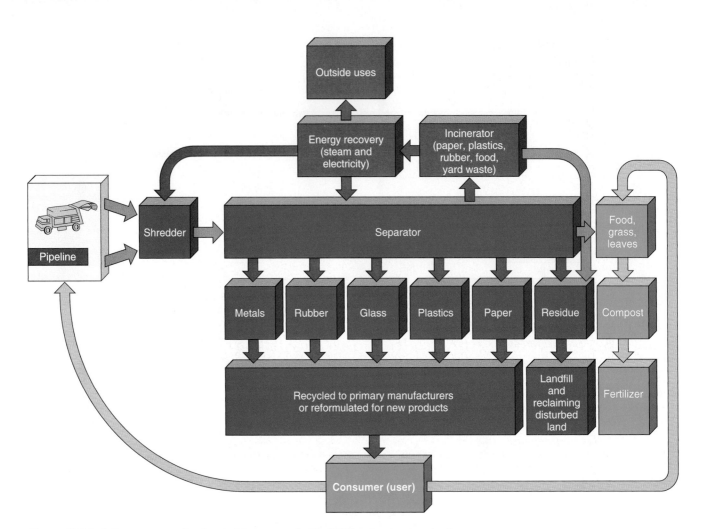

Figure 15-8 Solutions: a generalized *materials-recovery facility (MRF)* sorts mixed wastes for recycling and burning to produce energy. Because such plants need high volumes of trash to be economical, they discourage reuse and waste reduction.

The *source separation* approach has several advantages compared to the centralized approach. It produces much less air and water pollution and has low startup costs and moderate operating costs.

It also saves more energy and provides more jobs per unit of material than MRFs, landfills, and incinerators and yields cleaner and usually more valuable recyclables.

Finally, it educates people about the need for waste reduction, reuse, and recycling.

Case Study: How Much Wastepaper Is Being Recycled? Paper (especially newspaper and cardboard) is one of the easiest materials to recycle. Recycling paper involves removing its ink, glue, and coating and then reconverting it to pulp that is pressed into new paper. A variety of high-quality recycled papers are available to meet all types of printing demands at competitive prices (including the paper used in this book).

In 2000, the United States recycled about 49% of its wastepaper (up from 25% in 1989) and 70% of its corrugated cardboard containers. Despite a 49% recycling rate, the amount of paper thrown away each year in the United States is more than all of the paper consumed in China (where the paper recycling rate is only 27%). At least ten other countries recycle 50–97% of their wastepaper and paperboard, with a global recycling rate of 43%.

Chlorine (Cl_2) and chlorine compounds (such as chlorine dioxide, ClO_2) are used to bleach about 40% of the world's pulp for making paper. These compounds are corrosive to processing equipment, hazardous for workers, hard to recover and reuse, and extremely harmful when released to the environment. A growing number of paper mills (mostly in Europe) are replacing chlorine-based bleaching chemicals with oxygen-based chemicals such as hydrogen peroxide (H_2O_2) or ozone (O_3).

Buying recycled paper products can save trees and energy and reduce pollution, but it does not necessarily reduce solid waste. Only paper products made from *postconsumer waste*—waste intercepted on its way from consumer to the landfill or incinerator—does that.

Most recycled paper is made from *preconsumer waste:* scraps and cuttings recovered from paper and printing plants. Because paper manufacturers have always recycled this waste, it has never contributed to landfill problems. Now this paper is labeled "recycled" as a marketing ploy, giving the false impression that people who buy such products (often at higher prices) are helping reduce solid waste. Most so-called "recycled" paper has no more than 50% recycled fibers, with only 10% from postconsumer waste. Environmentalists propose that governments require companies to use labels giving postconsumer recycled content and indicating whether the paper was bleached with chlorine or by a chlorine-free process.

Some *good news* is that in 2002 Staples, a major office supply company, pledged to phase out purchases of paper products from endangered forests, and achieve an average of 30% postconsumer recycled content in all paper products sold by the company. The company will provide annual reports on progress toward reaching these goals. The company adopted these goals in response to a two-year grassroots campaign by The Paper Campaign, a coalition of dozens of citizens groups dedicated to protecting forests and increasing the use of recycled paper.

Case Study: Is It Feasible to Recycle Plastics? Currently, only about 7% by weight of the plastic wastes and 10% of plastic containers in the United States are recycled. There are three reasons why so little plastic is recycled. *First,* many plastics are difficult to isolate from other wastes. This happens because plastics occur in many different and often difficult-to-identify forms of resins (polymer molecules used to make different types of plastics) and sometimes are composites or laminated layers of different resins. Most plastics also contain stabilizers and other chemicals that must be removed before recycling.

Second, recovering individual plastic resins does not yield much material because only small amounts of any given resin are used per product. *Third,* the price of oil (adjusted for inflation) used to produce petrochemicals for making plastic resins is so low that the cost of virgin plastic resins is about 40% lower than that of recycled resins. An exception is PET, used mostly in plastic drink bottles.

These problems explain why mandating that plastic products contain a certain amount of recycled plastic resins is unlikely to work. It could also hinder the use of plastics in reducing the resource content and weight of many widely used items such as plastic bags, bottles, and other containers.

Does Recycling Make Economic Sense? The answer is yes and no, depending on different ways of looking at the economic and environmental benefits and costs of recycling. Critics contend that recycling has become almost a religion that is above criticism regardless of how much it costs communities. These critics say recycling does not make sense if it costs more to recycle materials than to send them to a landfill or incinerator. They also point out that recycling is often not needed to save landfill space because many areas in the United States are not running out of landfill space.

Finally, critics concede that recycling may make economic sense for valuable and easy-to-recycle materials (such as aluminum, paper, and steel) but not for cheap or plentiful resources (such as glass from silica) and most plastics (which are expensive to recycle).

However, recycling proponents argue that recycling programs should not be judged on whether they pay for themselves any more than conventional garbage disposal systems based on land burial or incineration are. They also point out that the primary benefit of recycling is not reducing the use of landfills and incinerators but the other important benefits it provides for people and the environment (Figure 15-7).

They also point to studies showing that the net economic, health, and environmental benefits of recycling (Figure 15-7) far outweigh the costs. Also, they remind us that the recycling industry is an important part of the U.S. economy. It employs about 1.1 million people and has $236 billion in annual income—much larger than either the mining or waste management and disposal industries.

Cities that make money by recycling and have higher recycling rates tend to use two major strategies. *First,* they have a *single pickup system* (for both materials to be recycled and garbage that cannot be recycled) instead of a more expensive dual collection system.

Second, they use a *pay-as-you-throw system* that charges households and businesses on the basis of how much trash they throw away and charges nothing for items to be recycled or composted. San Francisco, California, uses such a system to recycle almost 50% of its MSW.

Why Don't We Have More Recycling and Reuse? Four factors hinder recycling and reuse. *First* is failure to include the harmful environmental and health costs of raw materials in the market prices of consumer items (full-cost pricing, p. 24). *Second,* there are more government tax breaks and subsidies for resource-extracting industries than for recycling and reuse industries.

Third, tipping fees or charges for depositing wastes in landfills in the United States are low. For example, such fees average $41 per metric ton ($37 per ton) in the United States compared to fees of up to $220–330 per metric ton ($200–300 per ton) in Europe. When tipping fees rise to $110 per metric ton ($100 per ton) or more, industries save money by recycling their solid waste. *Fourth,* there is a lack of large steady markets for recycled materials.

Solutions: How Can We Encourage Recycling and Reuse?   Analysts suggest a number of ways for overcoming the obstacles to recycling and reuse. These include taxing virgin resources and phasing out subsidies for extracting them. At the same time, taxes on recycled materials based on postconsumer waste content could be lowered or eliminated and subsidies for reuse and postconsumer waste recycling would be provided.

Other approaches include greatly increasing use of the *pay-as-you-throw* system and encouraging or requiring government purchases of recycled products to help increase demand and lower prices. We should also view landfilling and incineration of solid wastes as last resorts to be used only for wastes that cannot be reused, composted, or recycled (Figure 15-4).

We can pass laws requiring companies to take back and recycle or reuse packaging discarded by consumers. Globally, at least 29 countries (including Japan and 20 European nations) have such "take-back" laws, which also encourage reduction of packaging.

We can also require manufacturers to take back and recycle or reuse appliances, computers and other electronic equipment, and motor vehicles at the end of their useful lives. This is required for car manufacturers and appliance makers in European Union countries and for major appliances in Japan. Soon laws in such countries are expected to require manufacturers to take back used computers and other electronic devices for recycling, reuse, or remanufacture.

Finally, we can require labels on all products listing preconsumer and postconsumer recycled content. This can help consumers make informed choices about the environmental consequences of the products they buy.

15-6 DETOXIFYING, BURNING, BURYING, AND EXPORTING WASTES

How Can We Use Biological Methods to Detoxify Hazardous Waste?   In Denmark, all hazardous and toxic waste from industries and households is de-

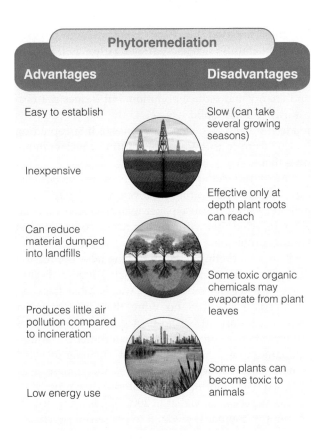

Figure 15-9 **Trade-offs:** advantages and disadvantages of using *phytoremediation* to remove or detoxify hazardous waste.

livered to 21 transfer stations throughout the country. The waste is then transferred to a large treatment facility. There about 75% of the waste is detoxified by biological and chemical methods and the rest is buried in a carefully designed and monitored landfill.

Some scientists and engineers consider biological treatment of hazardous waste, or *bioremediation,* to be the wave of the future for cleaning up some types of toxic and hazardous waste. In this process, microorganisms (usually natural or genetically engineered bacteria) and enzymes help destroy toxic or hazardous substances or convert them to harmless compounds.

Another biological way to treat hazardous wastes is *phytoremediation.* It involves using natural or genetically engineered plants to absorb, filter, and remove contaminants from polluted soil and water. Figure 15-9 lists advantages and disadvantages of phytoremediation.

Is Detoxifying Hazardous Waste with a Plasma Torch the Answer?   Plasma—known as a fourth state of matter—is an ionized gas consisting of electrically conductive ions and electrons. In a *plasma torch* a sustained electric arc is created by passage of electrical current through a gas to generate very high temperatures.

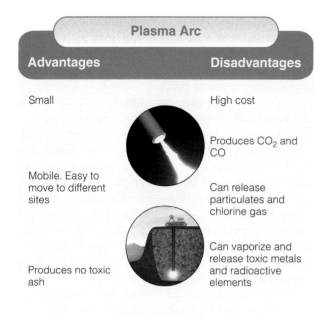

Plasma Arc

Advantages	Disadvantages
Small	High cost
	Produces CO_2 and CO
Mobile. Easy to move to different sites	Can release particulates and chlorine gas
	Can vaporize and release toxic metals and radioactive elements
Produces no toxic ash	

Figure 15-10 Trade-offs: advantages and disadvantages of using a *plasma arc torch* to detoxify hazardous wastes.

Such devices can consist of a plasma torch, reaction chamber (or furnace), and scrubber or other air pollution control devices to remove unwanted by-products. High temperatures from the torch can decompose liquid or solid hazardous organic material into ions and atoms (pyrolysis) that can be converted into simple molecules, cleaned up, and released as a gas. They can also convert hazardous inorganic matter into a molten glassy material (vitrification) that encapsulates toxic metals and keeps them from leaching into groundwater.

Figure 15-10 lists the advantages and disadvantages of using a plasma arc torch to detoxify hazardous waste.

Is Burning Solid and Hazardous Waste the Answer? In the United States, about 16% of the mixed trash in municipal solid waste is combusted in about 170 *mass-burn incinerators* (Figure 15-11). About 80% of the country's hazardous waste is burned in 172 commercial incinerators, cement kilns, and lightweight

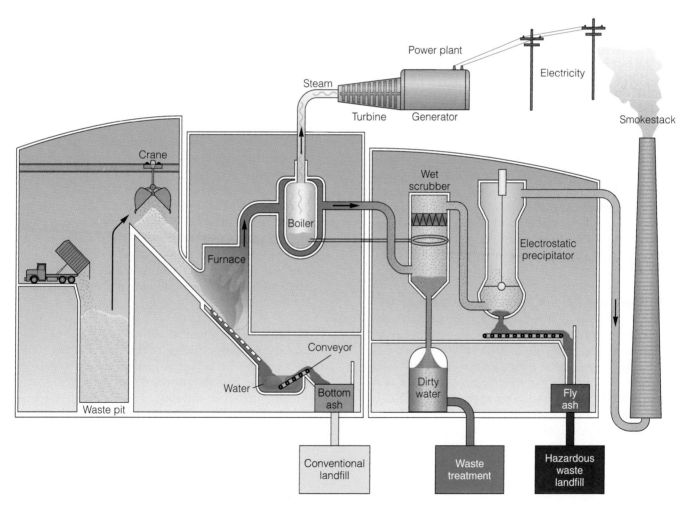

Figure 15-11 Solutions: *waste-to-energy incinerator* with pollution controls that burns mixed solid waste and recovers some of the energy to produce steam used for heating or producing electricity. (Adapted from EPA, *Let's Reduce and Recycle*)

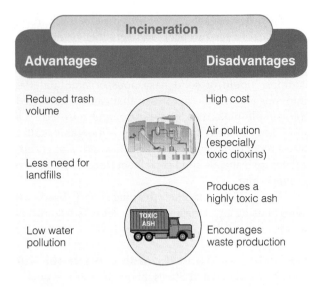

Incineration

Advantages

Reduced trash volume

Less need for landfills

Low water pollution

Disadvantages

High cost

Air pollution (especially toxic dioxins)

Produces a highly toxic ash

Encourages waste production

TOXIC ASH

Figure 15-12 Trade-offs: advantages and disadvantages of incinerating solid and hazardous waste.

aggregate kilns. The other 20% is combusted in industrial boilers and other types of industrial furnaces. Figure 15-12 lists the advantages and disadvantages of using incinerators to burn solid and hazardous waste.

Since 1985, incineration of wastes in some parts of the world has decreased because of high costs, health threats from air pollution, and intense citizen opposition. More than 280 new incinerator projects have been delayed or canceled in the United States since 1985. By 2006 an estimated 125 incinerators in the United States will be closed because of their inability to meet new and stricter air pollution standards.

Sweden banned the construction of new incinerators in 1985. Rhode Island and West Virginia banned solid waste incineration in 1992. In 1999 the Philippines became the first country to ban all waste incineration, followed by Costa Rica.

Is Land Disposal of Solid Waste the Answer?
About 54% by weight of the MSW in the United States is buried in sanitary landfills (compared to 90% in the United Kingdom, 80% in Canada, 15% in Japan, and 12% in Switzerland). In a **sanitary landfill,** solid wastes are spread out in thin layers, compacted, and covered daily with a fresh layer of clay or plastic foam.

Modern state-of-the-art landfills on geologically suitable sites are lined with clay and plastic before being filled with garbage (Figure 15-13). The bottom is covered with a second impermeable liner, usually made of several layers of clay, thick plastic, and sand. This liner collects *leachate* (rainwater contaminated as it percolates through the solid waste) and is intended to prevent its leakage into groundwater.

Collected leachate is pumped from the bottom of the landfill, stored in tanks, and sent to a regular sewage treatment plant or an on-site treatment plant. When full, the landfill is covered with clay, sand, gravel, and topsoil to prevent water from seeping in. Several wells are drilled around the landfill to monitor any leakage of leachate into nearby groundwater.

These new landfills are equipped with a connected network of vent pipes to collect landfill gas (consisting mostly of two greenhouse gases, methane and carbon dioxide; Table 13-1, p. 282) released by the underground (anaerobic) decomposition of wastes. The methane is filtered out and burned in small gas turbines to produce steam or electricity for nearby facilities or sold to utilities for use as a fuel. Figure 15-14 (p. 366) lists the advantages and disadvantages of using sanitary landfills to dispose of solid waste.

Thousands of older and abandoned landfills in the United States (and elsewhere) do not have such systems and will emit methane and carbon dioxide, both potent greenhouse gases, for decades.

Contamination of groundwater and nearby surface water by leachate from unlined and lined older landfills is also a serious problem. Some 86% of older U.S. landfills studied have contaminated groundwater, and a fifth of all Superfund hazardous waste sites are former municipal landfills that will cost billions of dollars to clean up.

According to G. Fred Lee (an experienced landfill consultant) and Ann Christy (a researcher at Ohio State University), the best solution to the leachate problem is to apply clean water to landfills continuously and collect and treat the resulting leachate in carefully designed and monitored facilities. They contend that after 10–20 years of such washing, little potential for groundwater pollution should remain. This wetting would also break down many wastes in 5–10 years instead of 100 years or longer in dry landfills. In addition, this approach would provide more room for more trash in the same landfill and allow old landfills to be dug out and used again.

Is Land Disposal of Hazardous Waste the Answer? Hazardous waste in the United States is disposed of on land in deep underground wells, state-of-the-art landfills, aboveground storage facilities, and surface impoundments such as ponds, pits, or lagoons.

In *deep-well disposal,* liquid hazardous wastes are pumped under pressure through a pipe into dry, porous geologic formations or zones of rock far beneath aquifers tapped for drinking and irrigation water (Figure 14-31, p. 331). Theoretically, these liquids soak into the porous rock material and are isolated from overlying groundwater by essentially impermeable layers of rock.

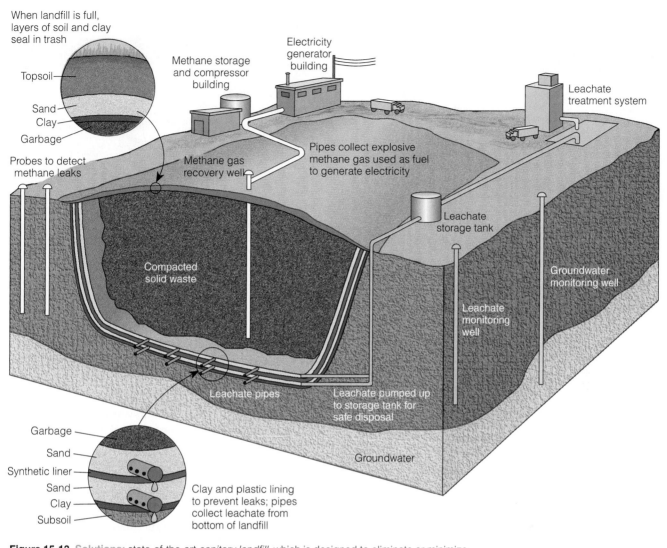

When landfill is full, layers of soil and clay seal in trash

Topsoil
Sand
Clay
Garbage

Probes to detect methane leaks

Methane storage and compressor building

Electricity generator building

Methane gas recovery well

Pipes collect explosive methane gas used as fuel to generate electricity

Leachate treatment system

Leachate storage tank

Compacted solid waste

Groundwater monitoring well

Leachate monitoring well

Leachate pipes

Leachate pumped up to storage tank for safe disposal

Groundwater

Garbage
Sand
Synthetic liner
Sand
Clay
Subsoil

Clay and plastic lining to prevent leaks; pipes collect leachate from bottom of landfill

Figure 15-13 Solutions: state-of-the-art *sanitary landfill,* which is designed to eliminate or minimize environmental problems that plague older landfills. Even such state-of-the-art landfills are expected to leak eventually, passing both the effects of contamination and cleanup costs on to future generations. Since 1997, only this state-of-the art type of landfill can operate in the United States. As a result, many older and small landfills have been closed and replaced with larger local and regional modern landfills.

Figure 15-15 (p. 366) lists the advantages and disadvantages of deep-well disposal of liquid hazardous wastes. Many scientists believe current regulations for deep-well disposal are inadequate and should be improved.

Surface impoundments are excavated depressions such as ponds, pits, or lagoons into which liquid hazardous wastes are drained and stored (Figure 14-31, p. 331). As water evaporates, the waste settles and becomes more concentrated. Figure 15-16 (p. 366) lists the advantages and advantages of this method. EPA studies found that 70% of these storage basins in the United States have no liners, and as many as 90% may threaten groundwater. According to the EPA, all liners

are likely to leak eventually and can contaminate groundwater.

Liquid and solid hazardous waste can also be put into drums or other containers and buried in carefully designed and monitored *secure hazardous waste landfills* (Figure 15-17, p. 367). Sweden goes further and buries its concentrated hazardous wastes in underground vaults made of reinforced concrete. By contrast, in the United Kingdom, most hazardous wastes are mixed with household garbage and stored in hundreds of conventional landfills throughout the country.

Hazardous wastes also can be stored in carefully designed *aboveground buildings*. These two-story buildings are built of reinforced concrete to prevent damage

Sanitary Landfills

Advantages	Disadvantages
No open burning	Noise and traffic
Little odor	Dust
Low groundwater pollution if sited properly	Air pollution from toxic gases and volatile organic compounds
Can be built quickly	Releases greenhouse gases (methane and CO_2)
Low operating costs	Groundwater contamination
Can handle large amounts of waste	Slow decomposition of wastes
Filled land can be used for other purposes	Encourages waste production
No shortage of landfill space in many areas	Eventually leaks and can contaminate groundwater

Figure 15-14 Trade-offs: advantages and disadvantages of using sanitary landfills to dispose of solid waste.

Deep Underground Wells

Advantages	Disadvantages
Safe method if sites are chosen carefully	Leaks or spills at surface
	Leaks from corrosion of well casing
Wastes can be retrieved if problems develop	Existing fractures or earthquakes can allow wastes to escape into groundwater
Easy to do	
Low cost	Encourages waste production

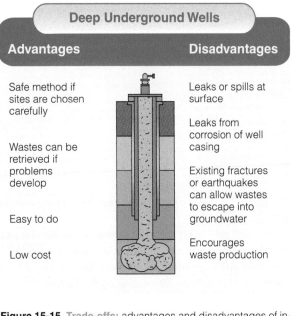

Figure 15-15 Trade-offs: advantages and disadvantages of injecting liquid hazardous wastes into deep underground wells.

by storms and hurricanes and to help contain any leakage. They also use fans and filters to create a negative air pressure to prevent the release of toxic gases and have automatic backup power systems in case of an electrical failure.

Surface Impoundments

Advantages	Disadvantages
Low construction costs	Groundwater contamination from leaking liners (or no lining)
Low operating costs	Air pollution from volatile organic compounds
Can be built quickly	
Wastes can be retrieved if necessary	Overflow from flooding
	Disruption and leakage from earthquakes
Can store wastes indefinitely with secure double liners	Promotes waste production

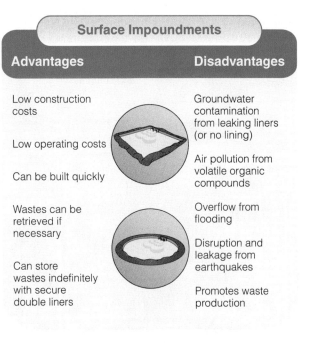

Figure 15-16 Trade-offs: advantages and disadvantages of storing liquid hazardous wastes in surface impoundments.

The first floor contains no wastes but has inspection walkways so people can easily check for leaks from the upper story. Any leakage is collected, treated, solidified, and returned to the storage building. Earthquakes could damage such structures, but this is a potential problem for all methods for storing hazardous wastes.

Each year there are more than 500,000 shipments of hazardous wastes (mostly to landfills and incinerators in trucks or by train) in the United States. On average, trucks and trains carrying hazardous materials (those to be used as raw materials as well as wastes) are involved in about 13,000 accidents per year in the United States. In a typical year, these accidents kill about 100 people, cause more than 10,000 injuries, and lead to evacuation of more than 500,000 people. Most communities do not have the equipment and trained personnel needed to deal with hazardous waste spills.

Is Exporting Hazardous Waste the Answer? Some hazardous waste producers in the United States and several other industrialized countries have been getting rid of some of these wastes by legally (or illegally) shipping them to other countries, especially developing countries.

Waste disposal firms can charge high prices for picking up hazardous wastes. If they can then dispose of them (legally or illegally) at low costs, they pocket huge profits. According to a 2001 study by the UN Food and Agriculture Organization, more than 450,000 met-

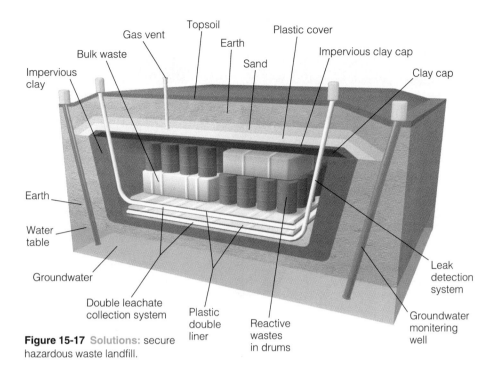

Figure 15-17 Solutions: secure hazardous waste landfill.

Labels on figure: Gas vent, Topsoil, Plastic cover, Bulk waste, Earth, Impervious clay cap, Sand, Impervious clay, Clay cap, Earth, Water table, Groundwater, Double leachate collection system, Plastic double liner, Reactive wastes in drums, Leak detection system, Groundwater monitering well

ric tons (500,000 tons) of banned or expired pesticides are seriously threatening the health of millions of people and the environment in many developing countries.

In 1989, countries met and drew up the *Basel Convention on Hazardous Waste*. It requires exporters to get approval from the recipient nation before a shipment of hazardous wastes can be sent. In 1995, the Basel Convention was strengthened to ban all hazardous waste exports from developed countries to developing countries.

If enforced, this ban on hazardous waste exports will help. However, it would not end illegal trade in these wastes because the potential profits are much too great. To most environmental scientists, the real solution to the hazardous waste problem is to produce as little as possible in the first place (Figure 15-5).

15-7 CASE STUDIES: LEAD, MERCURY, CHLORINE, AND DIOXINS

What Is the Threat from Lead? Because it is a chemical element, lead (Pb) does not break down in the environment. Lead is a potent neurotoxin that can harm the nervous system, especially in young children. Each year, 12,000–16,000 American children under age 9 are treated for acute lead poisoning, and about 200 die. About 30% of the survivors suffer from palsy, partial paralysis, blindness, and mental retardation.

Lead can also cause damage at levels far below those that cause acute lead poisoning, especially in children and unborn fetuses. Research indicates that children under age 6 and unborn fetuses with fairly low blood levels of lead are especially vulnerable to nervous system impairment, a lowered IQ (by 4–7 points), a shortened attention span, hyperactivity, hearing damage, and various behavior disorders.

It is *good news* that between 1976 and 2000 the percentage of U.S. children ages 1–5 with lead levels above the current blood level standard dropped from 85% to 2.2%, preventing at least 9 million childhood lead poisonings. The primary reason is that government regulations banned leaded gasoline in 1976 (with phaseout to be completed by 1986) and lead-based paints in 1970 (but illegal use continued until about 1978).

The *bad news* is that even with the encouraging drop in average blood levels of lead, the U.S. Centers for Disease Control and Prevention estimate that at least 800,000 U.S. children still have unsafe levels of lead, caused by exposure to a number of sources. Examples, include particles of lead compounds in soil and street and house dust and paint in older houses. Lead can also leach from service lines and pipes and faucets containing lead. More bad news is that a 1993 study by the National Academy of Sciences and numerous other studies indicate there is no safe level of lead in children's blood. This means several million children in the United States may be suffering from reduced mental capacity and other harmful effects of lead poisoning. Health scientists have proposed a number of ways to help protect children from lead poisoning (Figure 15-18, p. 368).

Taking most of these actions will cost an estimated $50 billion in the United States. However, health officials say the alternative is to keep poisoning and mentally handicapping millions of children.

Although the threat from lead has been reduced in the United States, this is not the case in many developing countries. The World Health Organization (WHO) estimates that 130–200 million children around the world are at risk from lead poisoning, and 15–18 million children in developing countries have permanent brain damage because of lead poisoning (mostly from use of leaded gasoline).

What Is the Threat from Mercury? Mercury (Hg) is a heavy metal that has been widely used in thermometers, dental fillings, fluorescent lights, mercury

light switches, and other electrical equipment. Because it is a chemical element, mercury (like lead) does not break down in the environment. According to the federal Agency for Toxic Substances and Disease Registry, long-term exposure to mercury in organic or inorganic form "can permanently demage the brain, kidneys, and developing fetuses."

Once released into the environment from natural or human sources, it can travel thousands of kilometers and in the process contaminate soil, water systems, and the atmosphere. As it moves through the atmosphere and other parts of the environment, elemental mercury often is converted to toxic inorganic and organic mercury compounds (Figure 15-19).

Humans are exposed to mercury in two ways. One is by inhaling vaporized elemental mercury (Hg) or particulates of inorganic mercury (Hg^{2+}) salts (such as HgS and $HgCl_2$). The other is eating fish contaminated with highly toxic methyl mercury (CH_3Hg). The greatest risk is brain damage from exposure to low levels of methyl mercury by fetuses and young children whose nervous systems are still developing.

Although it is released naturally from rocks, soil, and volcanoes, 50–75% of mercury emissions are believed to come from human activities. According to the EPA, the three major human-based sources of mercury in the United States are coal-burning power plants (31% of emissions), municipal waste incinerators (19%), and medical waste incinerators (11%).

These human-related sources emit elemental mercury (Hg) vapor and particles of inorganic mercury (Hg^{2+}) salts into the atmosphere. These nondegradable pollutants can be carried long distances by winds and deposited onto land and aquatic systems, mostly by rainfall. Once in an aquatic system, mercury enters into a complex cycle in which one form is converted to another (Figure 15-19).

Some *good news* is that according to the EPA, most Americans are not at risk from mercury exposure. Exceptions are pregnant women and children less than 6 years old who eat certain types of fish, and people who consume unusually large quantities of certain types of fish. According to a 2000 report by the National

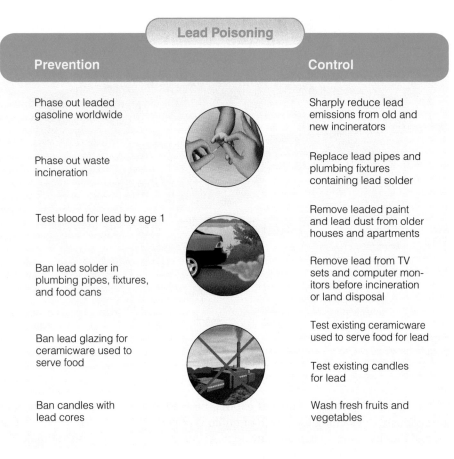

Figure 15-18 Solutions: ways to help protect children from lead poisoning.

Academy of Sciences, 60,000 pregnant women a year in the United States are putting their fetuses at risk of brain damage because of methyl mercury in fish they eat during pregnancy. Figure 15-20 (p. 370) list ways to prevent or control human exposure to mercury.

What Should We Do about Chlorine? Modern society depends heavily on chlorine (Cl_2) and chlorine-containing compounds. However, some of the approximately 11,000 chlorine-containing organic compounds we produce are persistent, accumulate in body fat, and are harmful to human health (according to animal and other toxicity studies; Section 10-2, p. 204).

The three largest uses of chlorine are in plastics (mostly polyvinyl chloride or PVC) and solvents and for paper and pulp bleaching. Together, they account for about 80% of all chlorine use. Most of these uses have substitutes and could be phased out within 10–20 years.

About 5% of the chlorine produced in the United States is used to purify drinking water (1%) and wastewater from sewage treatment plants (4%). Much of this could be replaced gradually with ozone and other nonchlorine purification processes. However, drinking

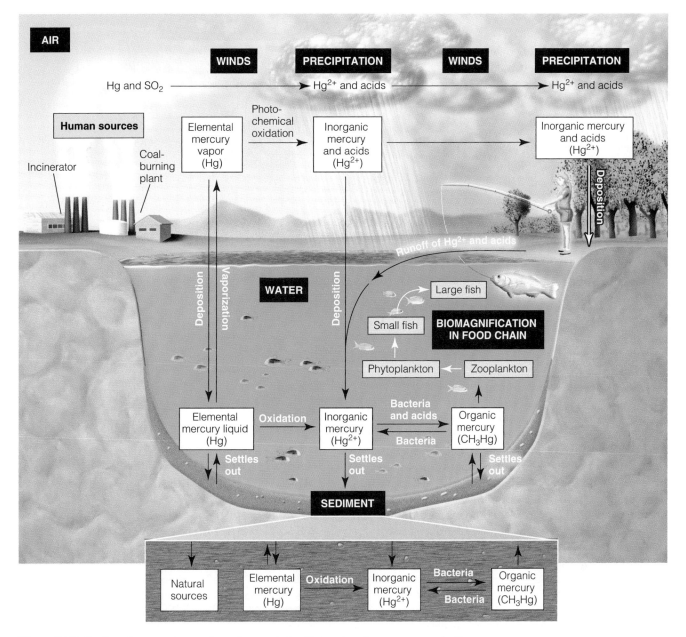

Figure 15-19 Cycling of mercury in aquatic environments in which various forms are converted from one form to another. The most toxic form to humans is methyl mercury (CH$_3$Hg), which can be biologically magnified in aquatic food chains. Some mercury is also released back into the atmosphere as mercury vapor.

water may still need to be purified by chlorination because it kills harmful microorganisms in water lines.

The remaining 35% of chlorine is used to manufacture a variety of organic and inorganic chemicals, many of them for small and specialized uses. Each of these chemicals could be evaluated to determine which are essential (many pharmaceuticals) and which might be replaced by affordable and environmentally acceptable substitutes.

If given proper financial incentives and 10–20 years to develop and phase in substitutes, chemical

producers could profit from sales of chlorine substitutes and protect themselves from adverse publicity, lawsuits, and cleanup costs.

How Dangerous Are Dioxins? *Dioxins* are a family of more than 75 different chlorinated hydrocarbon compounds. They form (along with toxic furans) as unwanted by-products in high-temperature chemical reactions involving chlorine and hydrocarbons.

Worldwide, incineration of municipal and medical wastes accounts for about 70% of dioxin and furan releases to the atmosphere. Other sources include burning wood in fireplaces, coal-fired power plants, metal smelting and refining facilities, wood pulp and paper mills, and sludge from municipal wastewater treatment plants.

About 30 dioxin compounds are considered to have significant toxicity. One dioxin compound, TCDD, is the most toxic and the most widely studied. Dioxins such as TCDD are persistent chemicals that linger in the environment for decades, especially in soil and human fat tissue. Food accounts for about 90% of the human exposure to dioxins.

Concern about the harmful health effects of exposure to low levels of dioxins on humans and wildlife is growing. A 2001 draft report of an EPA-sponsored comprehensive review by more than 100 scientists around the world came to three major conclusions.

First, TCDD is a definite human carcinogen, and other dioxin compounds are likely human carcinogens, especially for people who eat large amounts of fatty meats and dairy products. *Second*, the most powerful effect of exposure to low levels of dioxin on humans is disruption of the reproductive, endocrine, and immune systems (Case Study, p. 208) and their harmful effects on developing fetuses. *Third*, very small levels of dioxin in the environment can cause serious damage to certain wildlife species.

Industries producing dioxins say the dangers of long-term exposure of humans to low levels of dioxins are overestimated. They cite a 2000 study by Kyle Steenland, a researcher with the National Institute for Occupational Safety and Health, who found no significant risk of cancer for the general public exposed to low levels of toxic dioxins.

Because it will take decades to resolve these issues, some environmental scientists call for stricter regulations to reduce dioxin emissions now as a *precautionary strategy*. These scientists recommend two ways to sharply reduce emissions of dioxins in the United States (and elsewhere). One is to use nonchlorine methods to bleach paper (as is done in several European countries). The other is to eliminate chlorinated hydrocarbon compounds that produce dioxins from hazardous wastes burned in incinerators, iron ore sintering plants, and cement kilns.

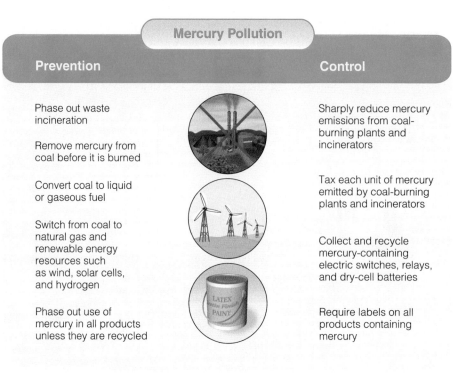

Figure 15-20 Solutions: ways to prevent or control inputs of mercury into the environment from human activities—mostly coal-burning plants and incinerators.

15-8 HAZARDOUS WASTE REGULATION IN THE UNITED STATES

What Is the Resource Conservation and Recovery Act? In 1976, the U.S. Congress passed the *Resource Conservation and Recovery Act* (*RCRA*, pronounced "RICK-ra") and amended it in 1984. This law has three major requirements. *First*, the EPA is to identify hazardous wastes and set standards for their management by states. *Second*, firms that store, treat, or dispose of more than 100 kilograms (220 pounds) of hazardous wastes per month must have a permit stating how such wastes are to be managed. *Third*, permit holders must use a *cradle-to-grave* system to keep track of waste they transfer from a point of origin to an approved off-site disposal facility.

What Is the Superfund Act? In 1980, the U.S. Congress passed the *Comprehensive Environmental Response, Compensation, and Liability Act,* commonly known as the *Superfund* program. Through taxes on chemical raw materials, this law (plus later amendments) has provided a trust fund to achieve three goals.

The *first* goal is to identify abandoned hazardous waste dump sites (Spotlight, right) and underground tanks leaking toxic chemicals. The *second* goal is to protect and if necessary clean up groundwater near such sites and clean up the sites. When they can be found, responsible parties must pay for the cleanup.

The *third* goal is to put the worst sites that represent an immediate and severe threat to human health on a *National Priorities List (NPL)*. These sites are to be cleaned up using the most cost-effective method.

Since 1981, about 1,980 sites have been placed on the National Priorities List for cleanup because they pose a real or potential threat to nearby populations. Emergency cleanup has been carried out at almost all sites, and wastes on more than half the sites have been contained and stabilized to prevent leakage.

Between 1981 and 2000, 750 sites were cleaned up and removed from the list at a cost of more than $300 billion. The 1,238 remaining sites are in various stages of cleanup, including 56% still awaiting a cleanup plan.

California's Santa Clara County, the birthplace of the nation's semiconductor industry, contains more Superfund priority sites than any other county in the nation. This is not surprising because the semiconductor industry is one of the world's most chemical-intensive industries—a single plant uses 500–1,000 different chemicals many of them toxic.

The former U.S. Office of Technology Assessment and the Waste Management Research Institute estimate the Superfund list could eventually include at least 10,000 priority sites, with cleanup costs of up to $1 trillion, not counting legal fees. Other studies project 2,000 sites with estimated cleanup costs of $200–300 billion over the next 30 years.

These estimated costs are only for cleanup to prevent future damage and do not include the health and ecological costs associated with such wastes. Some environmentalists and economists cite this as compelling evidence that preventing pollution costs much less than cleaning it up.

What Are Brownfields? *Brownfields* are abandoned industrial and commercial sites that in most cases are contaminated. Examples include factories, junkyards, older landfills, and gas stations. Some 450,000–600,000 brownfield sites exist in the United States, many in economically distressed inner cities.

Many of these sites could be cleaned up and reborn as parks, nature reserves, athletic fields, and neighborhoods. But first old oil and grease, industrial solvents, toxic metals, and other contaminants must be removed from their soil and groundwater.

Such efforts have been hampered by concerns of urban planners, developers, and lenders about legal liability for past contamination of such sites. However, Congress and almost half of the states have passed laws limiting the financial liability for developers of brownfield sites and their lenders.

Using Honeybees to Detect Toxic Pollutants

SPOTLIGHT

Honeybees are being used to detect the presence of toxic and radioactive chemicals in concentrations as low as several parts per trillion. On their forays from a hive, bees pick up water, nectar, pollen, suspended particulate matter, volatile organic compounds, and radioactive material found in the air near the sites they visit.

The bees then bring these materials back to the hive. There they fan the air vigorously with their wings to regulate the hive's temperature. This action releases and circulates pollutants they picked up into the air inside the hive.

Scientists have put portable hives, each containing 7,000–15,000 bees, near known or suspected hazardous waste sites. A small copper tube attached to the side of each hive pumps air out. Portable equipment is used to analyze the air for toxic and radioactive materials.

To find out where the bees have gone to forage for food, a botanist uses a microscope to examine pollen grains to determine what kinds of plants they come from. These data can be correlated with the plants found in an area up to 1.6 kilometers (1 mile) from each hive. This allows scientists to develop maps of toxicity levels and hot spots on large tracts of land.

This approach is much cheaper than setting up a number of air pollution monitors around mines, hazardous waste dumps, and other sources of toxic and radioactive pollutants. These *biological indicator* species have helped locate and track toxic pollutants and radioactive material at more than 30 sites across the United States.

Critical Thinking

Because honeybees can pick up toxic pollutants anywhere they go, should honey from all beehives be tested for such pollutants before it is placed on the market? Explain.

As a result, by 2002, more than 40,000 former brownfield sites had been redeveloped as part of urban revitalization and many other projects are under way. Here are two examples of brownfield projects. In Dallas, Texas, a brownfield site consisting of a 100-year-old city dump, abandoned grain silos, an aging coal-burning power plant, and a railroad maintenance building has been cleaned up and converted to a $420 million downtown site for a sports and entertainment arena and a complex of apartments, offices and stores.

Environmental Justice for All

Robert D. Bullard

Robert D. Bullard is professor of sociology and director of the Environmental Justice Resource Center at Clark Atlanta University.
For more than a decade, he has conducted research in the areas of urban land use, housing, community development, industrial facility siting, and environmental justice. He is the author of four books and more than three dozen articles, monographs, and scholarly papers that address concerns about environmental justice. His book Dumping in Dixie: Race, Class, and Environmental Quality, *2nd ed. (Westview Press, 1994) has become a standard text in the field. Other books are* Confronting Environmental Racism *(South End Press, 1993) and* Unequal Protection: Environmental Justice and Communities of Color *(Sierra Club Books, 1994).*

Despite widespread media coverage and volumes written on the U.S. environmental movement, environmentalism and social justice have seldom been linked. Nevertheless, an environmental revolution has been taking shape in the United States that combines the environmental and social justice movements into one framework.

People of color (African-Americans, Latinos, Asians, Pacific Islanders, and Native Americans), working-class people, and poor people in the United States suffer disproportionately from industrial toxins, dirty air and drinking water, unsafe work conditions, and the location of noxious facilities such as municipal landfills, incinerators, and toxic-waste dumps.

The *environmental justice* movement attempts to dismantle exclusionary zoning ordinances, discriminatory land-use practices, differential enforcement of environmental regulations, disparate siting of risky technologies, and the dumping of toxic waste on the poor and people of color in the United States and in developing countries.

Despite the government's attempts to level the playing field, all communities are not created equal when it comes to resolving environmental and public health concerns. More than 300,000 farm workers (more than 90% of whom are people of color) and their children are poisoned by pesticides sprayed on crops in the United States. Some 3–4 million children (many of them African-Americans or Latinos living in the inner city) are poisoned by lead-based paint in old buildings, lead-soldered pipes and water mains, lead-tainted soil contaminated by industry, and air pollutants from smelters.

All communities do not bear the same burden or reap the same benefits from industrial expansion. Nationally, 60% of African-Americans and 50% of Latinos live in communities with at least one uncontrolled toxic-waste

Developers in Philadelphia, Pennsylvania, plan to rehabilitate an 18-kilometer-long (11-mile-long) abandoned industrial park along the Delaware River, just north of downtown.

15-9 SOLUTIONS: ACHIEVING A LOW-WASTE SOCIETY

What Is the Role of Grassroots Action? Bottom-Up Change In the United States, local citizens have worked together to prevent hundreds of incinerators, landfills, or treatment plants for hazardous and radioactive wastes from being built in or near their communities. Opposition has grown as numerous studies have shown that such facilities have traditionally been located in communities populated mostly by African Americans, Asian Americans, Latinos, and poor whites. This practice has been cited as an example of *environmental injustice* (Guest Essay, above).

Most members of such groups recognize that health risks from incinerators and landfills, when averaged over the entire country, are quite low. However, they also know the risks for the people near these facilities are much higher. These people, not the rest of the population, are the ones whose health, lives, and property values are being threatened.

Manufacturers and waste industry officials point out that something must be done with the toxic and hazardous wastes produced to provide people with certain goods and services. They contend that if local citizens adopt a "not in my backyard" (NIMBY) approach, the waste still ends up in someone's backyard.

Many citizens do not accept this argument. To them, the best way to deal with most toxic or hazardous wastes is to produce much less of them, as suggested by the National Academy of Sciences (Figure 15-5 and Guest Essays, pp. 352 and 354). For such materials, their goal is "not in anyone's backyard" (NIABY) or "not on planet Earth" (NOPE) by emphasizing pollution prevention and use of the precautionary principle.

What Can Be Done at the International Level? The POPs Treaty In 2001, the UN Commission on Human Rights declared that being able to live free of pollution is a basic human right. There is a long way to go in converting this ideal into reality, but important progress has been made.

Between 1989 and 1994, an international treaty to limit transfer of hazardous waste from one country to another was developed (p. 367). In 2000, delegates from 122 countries completed a global treaty to control 12 *persistent organic pollutants* (POPs), which will go into effect when ratified by 50 countries.

site. Three of the five largest hazardous-waste landfills are located in communities that are predominantly African-American or Latino.

Environmental justice does not stop at the U.S. border. Environmental injustices exist from the *favelas* of Rio de Janeiro, Brazil, to the shantytowns of Johannesburg, South Africa. Members of the environmental justice movement are also questioning the wasteful and unsustainable development models being exported to the developing world.

Grassroots leaders are demanding justice. Residents of communities such as West Dallas and Texarkana (Texas), West Harlem (New York), Rosebud (South Dakota), Kettleman City (California), and Sunrise, Lions, and Wallace (Louisiana) see their struggle for environmental justice as a life-and-death matter. Unfortunately, their stories of environmental injustice are not broadcast into the nation's living rooms during the nightly news, nor are they splashed across the front pages of national newspapers and magazines. To a large extent, the communities that are the victims of environmental injustice remain invisible to the larger society.

The environmental justice movement is led, planned, and to a large extent funded by people who are not part of the established environmental community or the "Big 10" environmental organizations. Most environmental justice groups are small and operate with resources generated from the local community.

For too long these groups and their leaders have been invisible and their stories muted. This is changing as these grassroots groups are forcing their issues onto the nation's environmental agenda.

The United States has a long way to go in achieving environmental justice for all its citizens. The membership of decision-making boards and commissions still does not reflect the racial, ethnic, and cultural diversity of the country, and token inclusion of people of color on boards and commissions does not necessarily mean that their voices will be heard or their cultures respected. The ultimate goal of any inclusion strategy should be to democratize the decision-making process and empower disenfranchised people to speak and do for themselves.

Critical Thinking

1. How would you define environmental injustice? Can you identify any examples of environmental injustice in your community?

2. Have you been a victim of environmental injustice? Compare your answers with those of other members of your class.

These widely used toxic chemicals are insoluble in water and soluble in fat. This means that in the fatty tissues of humans and other organisms feeding at high trophic levels in food webs, they can become concentrated at levels hundreds of thousand times higher than in the general environment (Figure 10-4, p. 205, and Figure 14-28, p. 328). These persistent pollutants can also be transported long distances by wind and water.

The list of 12 chemicals, called the *dirty dozen,* includes DDT, eight other chlorine-containing persistent pesticides, PCBs, dioxins, and furans. The goals of the treaty are to ban or phase out use of these chemicals and detoxify or isolate stockpiles of such chemicals in warehouses and dumps. About 25 countries can continue using DDT to combat malaria until safer alternatives are available. Developed nations will provide developing nations about $150 million per year to help them switch to safer alternatives for the 12 POPs.

In 2000, the Swedish Parliament enacted a law that would ban by 2020 all chemicals that are persistent and can bioaccumulate in living tissue. This law requires an industry to show that a chemical it uses is safe rather than having the government show it is dangerous.

How Can We Make the Transition to a Low-Waste Society?
According to the famous physicist Albert Einstein, "A clever person solves a problem, a wise person avoids it." To prevent pollution and reduce waste, many environmental scientists urge us to understand and live by four key principles:

- Everything is connected.
- There is no "away" for the wastes we produce.
- Dilution is not always the solution to pollution.
- The best and cheapest way to deal with waste and pollution is to produce less of them and then reuse and recycle most of the materials we use (Figures 15-4 and 15-5).

Here is some *good news.* There is growing interest and use of increased *resource productivity* (p. 353), *cleaner production* (p. 354 and Guest Essay, p. 354), *industrial ecosystems* (p. 355 and Figure 15-6), and *service flow* businesses (p. 356).

In addition, at least 24 countries have *eco-labeling programs* that certify a product or service as having met specified environmental standards. The first national eco-labeling, Germany's *Blue Angel* program, began in 1978. It now awards its seal of approval to more than 3,900 products and services.

Such revolutions start off slowly but can accelerate rapidly as their economic, ecological, and health advantages become more apparent to investors, business leaders, elected officials, and citizens. See the

website material for this chapter for some actions you can take to reduce your production of solid waste and hazardous waste.

The key to addressing the challenge of toxics use and wastes rests on a fairly straightforward principle: harness the innovation and technical ingenuity that has characterized the chemicals industry from its beginning and channel these qualities in a new direction that seeks to detoxify our economy.

ANNE PLATT McGINN

REVIEW QUESTIONS

1. Define the boldfaced terms in this chapter.

2. Summarize what happened at Love Canal and the lessons learned from this environmental problem.

3. Distinguish between *solid waste* and *municipal solid waste.* What are the major sources of solid waste in the United States? What happens to municipal solid waste in the United States?

4. List seven characteristics of a high-waste society. List five examples of solid waste thrown away in the United States.

5. What is *hazardous waste?* What percentage of the overall hazardous waste produced in the United States is regulated by law? What percentage of the world's hazardous waste does the United States produce?

6. Summarize the threat from release of toxic chemicals from industrial plants because of accidents or terrorism. Describe the 1984 accident at the Union Carbide pesticide plant in Bhopal, India.

7. Distinguish between the *high-waste* and *low-waste* approaches to solid and hazardous waste management. List six characteristics of a low-waste society.

8. According to the U.S. National Academy of Sciences, what should be the five goals of solid and hazardous waste management in order of their importance as listed in Figures 15-4 and 15-5? According to some scientists, what percentage of solid and hazardous waste produced could be eliminated through a combination of waste reduction, reuse, and recycling (including composting)?

9. What does Lois Gibbs (p. 352) think we should do with solid and hazardous waste, and what major questions does she believe we should ask about such wastes?

10. List seven ways to reduce waste and pollution.

11. What is *resource productivity?* List five ways in which resource productivity has been improved. By how much do some experts think we can improve resource productivity?

12. What is the *ecoindustrial revolution* based on *cleaner production?* Describe the resource exchange system used in Denmark and the pollution prevention program implemented by the Minnesota Mining and Manufacturing

(3M) company in the United States. List five economic benefits of cleaner production.

13. Explain why Peter Montague (p. 354) believes the regulation and standards approach used to control pollution during the past 33 years has largely failed. Describe what he believes should replace this approach.

14. What is a *service flow economy?* What are four economic advantages of such an economy for businesses and consumers? List four examples of how a service economy is being implemented. Describe how Ray Anderson (Individuals Matter, p. 357) is developing a carpet tile service economy business.

15. What is *reuse?* What are three advantages of using this approach for waste reduction? List five examples of reuse.

16. Distinguish between *closed-loop recycling* and *downcycling.* About what percentage of the municipal solid waste produced in the United States is recycled and composted? What percentage do experts believe could be recycled and composted? What is *compost,* and how is it used as a way to deal with solid waste?

17. Distinguish between the *centralized recycling* of mixed solid waste and *consumer separation* of solid waste. List the advantages and disadvantages of each approach to recycling.

18. Describe how paper is recycled. Distinguish between *preconsumer* and *postconsumer paper waste.*

19. List three reasons why so few plastics are recycled.

20. List the advantages and disadvantages of recycling. What four factors hinder recycling and reuse? List nine ways to encourage more recycling and reuse.

21. Describe Denmark's hazardous waste detoxification program.

22. Distinguish between *bioremediation* and *phytoremediation.* List the advantages and disadvantages of using phytoremediation for detoxifying hazardous waste.

23. List the advantages and disadvantages of using a *plasma torch* to detoxify hazardous waste.

24. Describe the major components of a *mass-burn incinerator.* List the advantages and disadvantages of dealing with solid and hazardous waste by burning it in incinerators.

25. Describe the major components of a *sanitary landfill.* List the advantages and disadvantages of dealing with solid and hazardous waste by burying it in sanitary landfills.

26. List the advantages and disadvantages of storing hazardous wastes in **(a)** deep underground wells, **(b)** surface impoundments, **(c)** secure landfills, and **(d)** aboveground buildings.

27. Describe what is being done at the international level about the exporting of hazardous wastes from one country to another.

28. Describe the hazards of lead exposure, and list five ways to prevent and eight ways to control lead poisoning of children.

29. Describe the hazards of mercury exposure. List five ways to prevent and four ways to control inputs of mercury into the environment from human sources.

30. What are three major problems with many chlorine-containing compounds? What are the three major uses of chlorine? How can we reduce exposure to harmful chlorine-containing compounds?

31. What are *dioxins?* How are they produced, what harm can they cause, and what are two ways to reduce exposure to these hazardous chemicals?

32. How is the Resource Conservation and Recovery Act used to deal with the problem of hazardous wastes in the United States?

33. What is the Superfund Act?

34. What are *brownfields,* and what has been done to help redevelop such sites in the United States? Give an example of a reclaimed brownfield site.

35. What is *environmental injustice?* How does Robert Bullard (p. 372) describe the environmental justice movement?

36. Describe international efforts to control use of 12 persistent organic pollutants (POPs).

37. List four principles that can be used as guidelines for making the transition to a lower-waste society.

CRITICAL THINKING

1. Explain why you support or oppose requiring that **(a)** all beverage containers be reusable, **(b)** all households and businesses put recyclable materials into separate containers for curbside pickup, **(c)** garbage-collecting systems implement the pay-as-you-throw approach, and **(d)** consumers pay for plastic or paper bags at grocery and other stores to encourage the use of reusable shopping bags.

2. Use the second law of thermodynamics (p. 59) to explain why a *properly designed* source-separation recycling program takes less energy and produces less pollution than a centralized program that collects mixed waste over a large area and hauls it to a centralized facility where workers or machinery separate the wastes for recycling.

3. Are you for or against bringing about an *ecoindustrial revolution?* Explain. Do you believe it will be possible to phase in such a revolution in the country where you live over the next two to three decades? Explain.

4. Explain why some businesses participating in an exchange and chemical-cycling network (Figure 15-6, p. 356) might produce large amounts of waste for use as resources within the network rather than redesigning their manufacturing processes to reduce waste production. Is this acceptable? Explain.

5. Are you for or against shifting to a *service flow economy?* Explain. Do you believe it will be possible to shift to a *service flow economy* in the country where you live over the next two to three decades? Explain.

6. Would you oppose having a hazardous waste landfill, waste treatment plant, deep-injection well, or incinerator in your community? Explain. If you oppose these disposal facilities, how do you believe the hazardous waste generated in your community and your state should be managed?

7. Give your reasons for agreeing or disagreeing with each of the following proposals for dealing with hazardous waste:
   a. Reduce the production of hazardous waste and encourage recycling and reuse of hazardous materials by charging producers a tax or fee for each unit of waste generated.
   b. Ban all land disposal and incineration of hazardous waste to encourage recycling, reuse, and waste treatment and to protect air, water, and soil from contamination.
   c. Provide low-interest loans, tax breaks, and other financial incentives to encourage industries producing hazardous waste to reduce, recycle, reuse, treat, and decompose such waste.
   d. Ban the shipment of hazardous waste from one country to another.

8. Congratulations! You are in charge of bringing about cleaner production, resource productivity, and a service flow economic revolution throughout the world over the next 20 years. List the three most important components of your strategy for implementing each of these shifts.

PROJECTS

1. For 1 week, keep a list of the solid waste you throw away. What percentage of this waste consists of materials that could be recycled, reused, or burned for energy? What percentage of the items could you have done without in the first place? Tally and compare the results for your entire class.

2. What percentage of the municipal solid waste in your community is **(a)** landfilled, **(b)** incinerated, **(c)** composted, and **(d)** recycled? What technology is used in local landfills and incinerators? What leakage and pollution problems have local landfills or incinerators had? Does your community have a recycling program? Is it voluntary or mandatory? Does it have curbside collection? Drop-off centers? Buyback centers?

3. What hazardous wastes are produced **(a)** at your school and **(b)** in your community? What happens to these wastes?

4. Make a concept map of this chapter's major ideas, using the section heads and subheads and the key terms (in boldface). Look on the website for this book for information about making concept maps.

INTERNET STUDY RESOURCES AND RESOURCES FOR FURTHER READING AND RESEARCH

The website for this book contains helpful study aids and many ideas for further reading and research. Log on to

http://biology.brookscole.com/miller 10

and click on the Chapter-by-Chapter area. Choose Chapter 15 and select a resource:

■ Flash Cards allows you to test your mastery of the Terms and Concepts to Remember for this chapter.

■ Tutorial Quizzes provides a multiple-choice practice quiz.

■ Student Guide to InfoTrac will lead you to Critical Thinking Projects that use InfoTrac College Edition as a research tool.

■ References lists the major books and articles consulted in writing this chapter.

■ Hypercontents takes you to an extensive list of sites with news, research, and images related to individual sections of the chapter.

INFOTRAC COLLEGE EDITION

Improve your skills with InfoTrac College Edition, a searchable online database of articles from more than 700 periodicals. Log on to

http://www.infotrac-college.com

or access InfoTrac through the website for this book. Try to find the following articles:

1. Isaacs, L. 2001. Rounding up household hazardous waste: Local governments have developed several ways to keep household chemicals and poisonous materials from harming the environment. *American City and County* 116: 24. *Keyword:* "phytoremediation." Dealing with hazardous wastes generated in the home is becoming more of a problem than ever. This article looks at some of the ways that household hazardous waste is being managed by a variety of municipalities.

2. Evans, L. D. 2002. The dirt on phytoremediation. *Journal of Soil and Water Conservation* 57: 13. *Keyword:* "household hazardous wastes." Using plants to clean up certain hazardous wastes has been done for some time, but studies show that plants can contribute even more to the cleanup effort than previously realized.

16 FOOD RESOURCES

Perennial Crops on the Kansas Prairie

When you think about farms in Kansas, you probably picture seemingly endless fields of wheat or corn plowed up and planted each year. By 2040, the picture might change, thanks to pioneering research at the nonprofit Land Institute near Salina, Kansas.

The institute, headed by plant geneticist Wes Jackson, is experimenting with an ecological approach to agriculture on the midwestern prairie. This approach relies on planting a mixture of different crops in the same area, a technique called *polyculture*. This involves planting a mix of perennial grasses (Figure 16-1, left), legumes (a source of nitrogen fertilizer, Figure 16-1, right), sunflowers, grain crops, and plants that provide natural insecticides in the same field.

The goal is to grow food mimicking many of the natural conditions of the prairie without losing fertile grassland soil (Figure 9-18, top right, p. 191). Institute researchers believe that *perennial polyculture* can be blended with *modern monoculture* to reduce its harmful environmental effects.

Because these plants are perennials, there is no need to plow up and prepare the soil each year to replant them. This takes much less labor than conventional monoculture or diversified organic farms that grow annual crops. It also reduces soil erosion because the unplowed soil is not exposed to wind and rain, and it reduces the need for irrigation because the deep roots of such perennials retain more water than annuals. There is also less pollution from chemical fertilizers and pesticides.

Thirty-three years of research by the institute have shown that various mixtures of perennials grown in parts of the midwestern prairie could be used as important sources of food. One such mix of perennial crops includes *eastern gamma grass* (a warm-season grass that is a relative of corn with three times as much protein as corn and twice as much as wheat; Figure 16-1, left), *mammoth wildrye* (a cool-season grass that is distantly related to rye, wheat, and barley), *Illinois bundleflower* (a wild nitrogen-producing legume that can enrich the soil and whose seeds can serve as livestock feed; Figure 16-1, right), and *Maximilian sunflower* (which produces seeds with as much protein as soybeans).

These discoveries may help us come closer to producing and distributing enough food to meet everyone's basic nutritional needs without degrading the soil, water, air, and biodiversity that support all food production.

Figure 16-1 Solutions: the Land Institute in Salina, Kansas, is a farm, a prairie laboratory, and a school dedicated to changing the way we grow food. It advocates growing a diverse mixture (polyculture) of edible perennial plants to supplement traditional annual monoculture crops. Two of these perennial crops are eastern gamma grass (top, left) and the Illinois bundleflower (bottom, right).

There are two spiritual dangers in not owning a farm. One is the danger of supposing that breakfast comes from the grocery, and the other that heat comes from the furnace.

ALDO LEOPOLD

This chapter addresses the following questions:

- How is the world's food produced?

- How are green revolution and traditional methods used to raise crops?

- How much has food production increased, how serious is malnutrition, and what are the environmental effects of producing food?

- How can we increase production of crops, meat, and fish and shellfish?

- How do government policies affect food production?

- What are pesticides and what are the advantages and disadvantages of using chemicals to kill insects and weeds?

- What alternatives are there to using conventional pesticides, and what are the advantages and disadvantages of each alternative?

- How can we design and shift to more sustainable agricultural systems?

16-1 HOW IS FOOD PRODUCED?

What Three Systems Provide Us with Food? Some Good and Bad News Historically, humans have depended on three systems for their food supply. *Croplands* mostly produce grains and provide about 76% of the world's food. *Rangelands* produce meat, mostly from grazing livestock, and supply about 17% of the world's food. *Oceanic fisheries* supply about 7% of the world's food.

Some *good news* is that since 1950 there has been a staggering increase in global food production from all three systems. This phenomenal growth in food productivity occurred because of technological advances such as increased use of tractors and farm machinery and high-tech fishing boats and gear; inorganic chemical fertilizers; irrigation; pesticides; high-yield varieties of wheat, rice, and corn; densely populated feedlots and enclosed pens for raising cattle, pigs, and chickens; and aquaculture ponds and ocean cages for raising some types of fish and shellfish.

To feed the world's 9.3 billion people projected by 2050, we must produce and equitably distribute more food than has been produced since agriculture began about 10,000 years ago and do this in an environmentally sustainable manner. Some analysts believe we can continue expanding the use of industrialized agriculture to produce the necessary food.

Some *bad news* is that other analysts contend that environmental degradation, pollution, lack of water for irrigation, overgrazing by livestock, overfishing, and loss of vital ecological services (Figure 1-2, p. 3, and top half of back cover) may limit future food production. A key problem is that human activities continue to take over or degrade more of the planet's *net primary productivity* (Figure 4-22, p. 80), which supports all life. This chapter analyzes the advantages and disadvantages of the world's current crop, meat, and fish production systems and how these systems can be made more sustainable.

What Plants and Animals Feed the World? The earth has perhaps 30,000 plant species with parts that people can eat. However, only 15 plant and 8 terrestrial animal species supply an estimated 90% of our global intake of calories. Just three grain crops—*wheat, rice,* and *corn*—provide more than half the calories people consume. These three grains, and most other food crops, are *annuals,* whose seeds must be replanted each year.

Two-thirds of the world's people survive primarily on traditional grains (mainly rice, wheat, and corn), mostly because they cannot afford meat. As incomes rise, people consume more grain, but indirectly in the form of meat (mostly beef, pork, and chicken), eggs, milk, cheese, and other products of grain-eating domesticated livestock.

Fish and shellfish are an important source of food for about 1 billion people, mostly in Asia and in coastal areas of developing countries. But on a global scale fish and shellfish supply only 7% of the world's food, less than 6% of the protein in the human diet, and 1% of the energy in the human diet.

What Are the Major Types of Food Production? All crop production involves replacing species-rich late successional communities such as mature grasslands and forests with an early successional community (Figure 7-12, p. 153) consisting of a single crop (*monoculture,* Figure 6-18, p. 119) or a mixture of crops (*polyculture*).

The two major types of agricultural systems are industrialized and traditional. **Industrialized agriculture,** or **high-input agriculture,** uses large amounts of fossil fuel energy, water, commercial fertilizers, and pesticides to produce huge quantities of single crops (monocultures) or livestock animals for sale. Practiced

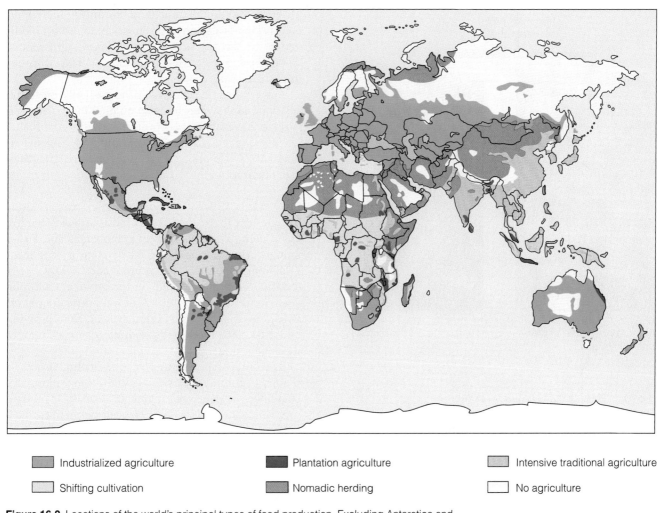

	Industrialized agriculture		Plantation agriculture		Intensive traditional agriculture
	Shifting cultivation		Nomadic herding		No agriculture

Figure 16-2 Locations of the world's principal types of food production. Excluding Antarctica and Greenland, agricultural systems cover almost one-third of the earth's land surface and account for an annual output of food worth about $1.3 trillion.

on about 25% of all cropland, mostly in developed countries (Figure 16-2), high-input industrialized agriculture has spread since the mid-1960s to some developing countries.

Plantation agriculture is a form of industrialized agriculture used primarily in tropical developing countries. It involves growing cash crops (such as bananas, coffee, soybeans, sugarcane, cocoa, and vegetables) on large monoculture plantations, mostly for export and sale in developed countries.

An increasing amount of livestock production in developed countries is industrialized. Large numbers of cattle are brought to densely populated *feedlots*, where they are fattened up for about 4 months before slaughter. Most pigs and chickens in developed countries spend their entire lives in densely populated pens and cages and are fed mostly grain grown on cropland.

Traditional agriculture consists of two main types, which together are practiced by about 2.7 billion people (43% of the world's people) in developing countries and provide about 20% of the world's food supply. **Traditional subsistence agriculture** typically uses mostly human labor and draft animals to produce only enough crops or livestock for a farm family's survival. Examples of this very low-input type of agriculture include numerous forms of shifting cultivation in tropical forests (Figure 1-13, p. 13) and nomadic livestock herding (Figure 16-2).

In **traditional intensive agriculture,** farmers increase their inputs of human and draft labor, fertilizer, and water to get a higher yield per area of cultivated land. They produce enough food to feed their families and to sell for income. Croplands, like natural ecosystems, provide ecological and economic services (Figure 16-3, p. 380).

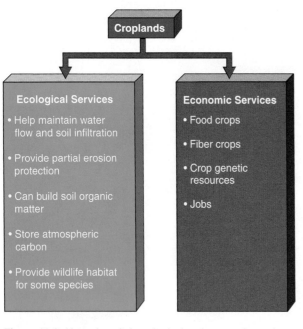

Figure 16-3 *Natural capital:* ecological and economic services provided by croplands.

16-2 PRODUCING FOOD BY GREEN REVOLUTION AND TRADITIONAL TECHNIQUES

How Have Green Revolutions Increased Food Production? High-Input Monocultures in Action Farmers can produce more food by farming more land or getting higher yields per unit of area from existing cropland. Since 1950, most of the increase in global food production has come from increased yields per unit of area of cropland in a process called the **green revolution.**

This green revolution involves three steps. *First,* develop and plant monocultures (Figure 6-18, p. 119) of selectively bred or genetically engineered high-yield varieties of key crops such as rice, wheat, and corn. *Second,* produce high yields of these crops by using large inputs of fertilizer, pesticides, and water. *Third,* increase the number of crops grown per year on a plot of land through *multiple cropping.*

This high-input approach dramatically increased crop yields in most developed countries between 1950 and 1970 in what is called the *first green revolution* (Figure 16-4).

A *second green revolution* has been taking place since 1967 (Figure 16-4) by introducing fast-growing dwarf varieties of rice and wheat, specially bred for tropical and subtropical climates, into several developing countries. With enough fertile soil and fertilizer, water, and pesticides, yields of these new plants (Figure 16-5) can be two to five times those of traditional wheat and rice varieties. The fast growth also allows farmers to grow two or even three crops a year (multi-

ple cropping) on the same land. Producing more food on less land is also an important way to protect biodiversity by saving large areas of forests, grasslands, wetlands, and easily eroded mountain terrain from being used to grow food.

These yield increases depend not only on fertile soil and ample water but also on high inputs of fossil fuels to run machinery, produce and apply inorganic fertilizers and pesticides, and pump water for irrigation. All told, high-input green revolution agriculture uses about 8% of the world's oil output.

Case Study: Food Production in the United States In the United States industrialized farming has become *agribusiness* as big companies and larger family-owned farms have taken control of almost three-fourths of U.S. food production. Only about 650,000 Americans (2% of the population) are full-time farmers. However, about 9% of the population is involved in the U.S. agricultural system, from growing and processing food to distributing it and selling it at the supermarket.

In terms of total annual sales, agriculture is bigger than the automotive, steel, and housing industries combined. It generates about 18% of the country's gross national product and 19% of all jobs in the private sector, employing more people than any other industry.

Here are three pieces of *good news* about the highly productive industrialized agricultural system in the United States. *First,* with only 0.3% of the world's farm labor force, U.S. farms produce about 17% of the world's grain (most consumed by U.S. livestock) and nearly half of the world's grain exports.

Second, since 1950, U.S. farmers have used green revolution techniques to more than double the yields of key crops without cultivating more land. Indeed, the amount of land used to grow crops in the United States decreased by 23% between 1950 and 2000. Such increases in the yield per hectare of key crops have kept large areas of forests, grasslands, wetlands, and easily erodible land from being converted to farmland.

Third, the country's agricultural system has become increasingly efficient. While the U.S. output of crops, meat, and dairy products has been increasing steadily since 1975, the major inputs of labor and resources—with the exception of pesticides—to produce each unit of that output have fallen steadily since 1950.

This industrialization of agriculture has been made possible by the availability of cheap energy, most of it from oil. Agriculture consumes about 17% of all commercial energy in the United States each year (Figure 16-6, p. 382). Some *good news* is that the input of energy needed to produce a unit of food has fallen considerably and most plant crops in the United States provide more food energy than the energy used to grow them.

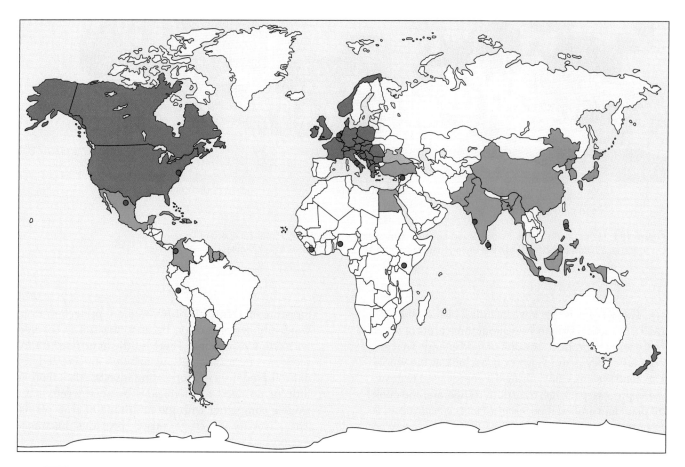

First green revolution
(developed countries)

Second green revolution
(developing countries)

● Major international agricultural
research centers and seed banks

Figure 16-4 Countries whose crop yields per unit of land area increased during the two green revolutions. The first (blue) took place in developed countries between 1950 and 1970; the second (green) has occurred since 1967 in developing countries with enough rainfall or irrigation capacity. Several agricultural research centers and gene or seed banks (red dots) play a key role in developing high-yield crop varieties.

Figure 16-5 Solutions: A high-yield, semidwarf variety of rice called IR-8 (left), developed as part of the second green revolution. It was produced by crossbreeding two parent strains of rice: PETA from Indonesia (center) and DGWG from China (right). The shorter and stiffer stalks of the new variety allow the plants to support larger heads of grain without toppling over and increase the benefit of applying more fertilizer.

International Rice Research Institute, Manila

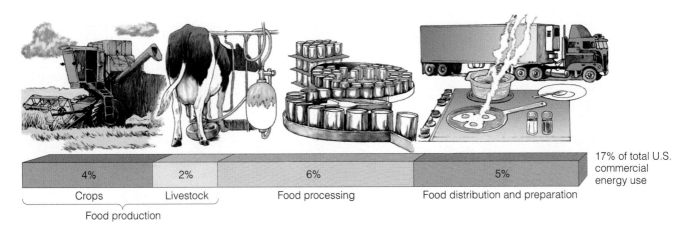

4%	2%	6%	5%	17% of total U.S. commercial energy use
Crops	Livestock	Food processing	Food distribution and preparation	

Food production

Figure 16-6 In the United States, industrialized agriculture uses about 17% of all commercial energy. On average, a piece of food eaten in the United States has traveled 2,100 kilometers (1,300 miles).

The *bad news* is that if we include livestock, the U.S. food production system uses about three units of fossil fuel energy to produce one unit of food energy. That energy efficiency is much lower if we look at the whole U.S. food system. Considering the energy used to grow, store, process, package, transport, refrigerate, and cook all plant and animal food, *about 10 units of nonrenewable fossil fuel energy are needed to put 1 unit of food energy on the table.* By comparison, every unit of energy from human labor in traditional subsistence farming provides at least 1 unit of food energy and up to 10 units of food energy using traditional intensive farming.

What Growing Techniques Are Used in Traditional Agriculture? Low-Input Agrodiversity in Action Traditional farmers in developing countries grow about 20% of the world's food on about 75% of its cultivated land. Many traditional farmers simultaneously grow several crops on the same plot, a practice known as **interplanting.** Such crop diversity reduces the chance of losing most or all of the year's food supply to pests, bad weather, and other misfortunes.

Interplanting strategies vary. One type is **polyvarietal cultivation,** which involves planting a plot with several varieties of the same crop. Another is **intercropping.** It involves growing two or more different crops at the same time on a plot (for example, a carbohydrate-rich grain that uses soil nitrogen and a protein-rich legume that puts it back). A third type is **agroforestry,** or **alley cropping,** in which crops and trees are grown together (Figure 9-27c, p. 198).

A fourth type is **polyculture.** It is a more complex form of intercropping in which many different plants maturing at various times are planted together. Low-input *polyculture* has a number of advantages. There is less need for fertilizer and water because root systems at different depths in the soil capture nutrients and

moisture efficiently. It provides more protection from wind and water erosion because the soil is covered with crops year-round. There is little or no need for insecticides because multiple habitats are created for natural predators of crop-eating insects. Also, there is little or no need for herbicides because weeds have trouble competing with the multitude of crop plants. The diversity of crops raised provides insurance against bad weather.

In addition, recent ecological research found that on average, low-input polyculture produces higher yields per hectare of land than high-input monoculture. For example, a 2001 study by ecologists Peter Reich and David Tilman found that carefully controlled polyculture plots with 16 different species of plants consistently outproduced plots with 9, 4, or only 1 type of plant species. Wes Jackson is using polyculture to grow perennial crops on prairie land in the United States (p. 377).

16-3 FOOD PRODUCTION, NUTRITION, AND ENVIRONMENTAL EFFECTS

How Much Has Food Production Increased? Figure 16-7 illustrates the success of using high-input monoculture farming to produce food and ward off sharp rises in hunger, malnutrition, and food prices. Here are three pieces of *good news* about food production. *First,* world grain production almost tripled (Figure 16-7, left), and per capita grain production rose by about 36% (Figure 16-7, right) between 1950 and 1990.

Second, the relative per capita food production in most parts of world increased between 1961 and 2002 (Figure 16-8). To maintain good health and resist disease, we need fairly large amounts of *macronutrients*

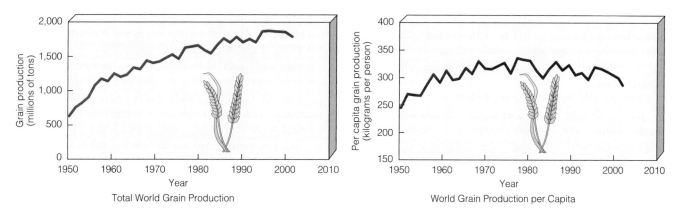

Figure 16-7 *Good news:* total worldwide grain production of wheat, corn, and rice (left), and per capita grain production (right), 1950–2002. In order, the world's three largest grain-producing countries in 2002 were China, the United States, and India. (Data from U.S. Department of Agriculture, Worldwatch Institute, and UN Food and Agriculture Organization)

(such as protein, carbohydrates, and fats) and smaller amounts of *micronutrients* consisting of various vitamins (such as A, C, and E) and minerals (such as iron, iodine, and calcium).

People who cannot grow or buy enough food to meet their basic energy needs suffer from **undernutrition.** Chronically undernourished children are likely to suffer from mental retardation and stunted growth. They also tend to be much more susceptible to infectious diseases (such as measles and diarrhea), which kill one child in four in developing countries.

Many of the world's poor can afford to live only on a low-protein, high-carbohydrate diet consisting only of grains such as wheat, rice, or corn. Many of

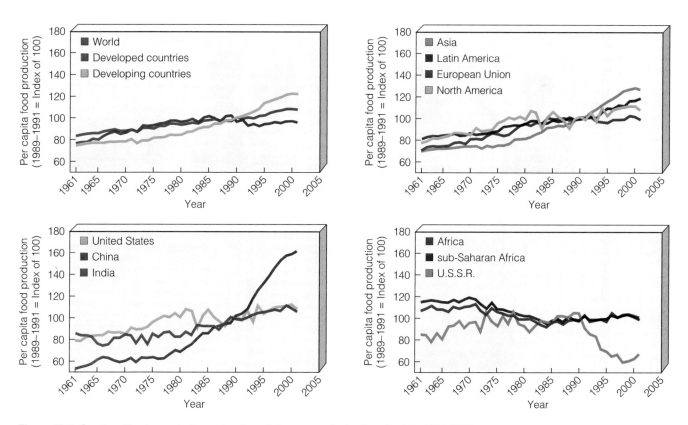

Figure 16-8 *Good and bad news:* indexes showing relative per capita food production, 1961–2001, using 1989–1991 as the index base. (Data from UN Food and Agriculture Organization)

them suffer from **malnutrition:** deficiencies of protein and other key nutrients. Many of the world's desperately poor people, especially children (Figure 1-5, p. 6), suffer from both chronic undernutrition and chronic malnutrition.

Chronically undernourished and malnourished people are disease prone and adults are too weak to work productively or think clearly. As a result, their children also tend to be underfed, malnourished, and susceptible to disease. If these children survive to adulthood, many are locked in a malnutrition and poverty cycle (Figure 16-9) that can continue for generations.

How Serious Are Undernutrition and Malnutrition? Here is some *good news.* According to the UN Food and Agriculture Organization (FAO), the average daily food intake in calories per person in the world and in developing countries rose sharply between 1961 and 2000, and is projected to continue rising through 2030 (Figure 16-10). In addition, the estimated number of chronically undernourished or malnourished people fell from 918 million in 1970 to 815 million in 2001. About 774 million of these people were in developing countries (especially in Asia and Africa, and 41 million were in developed countries.

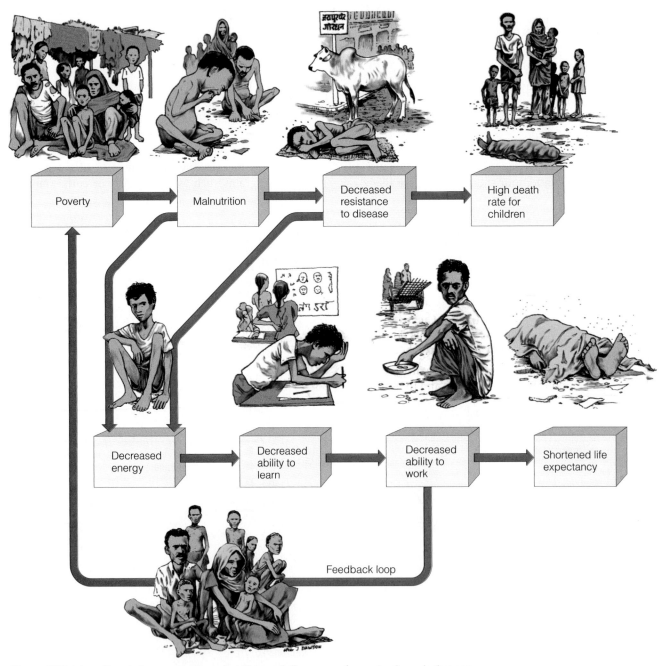

Figure 16-9 Interactions between poverty, malnutrition, and disease can form a tragic cycle that can perpetuate such conditions in succeeding generations of families.

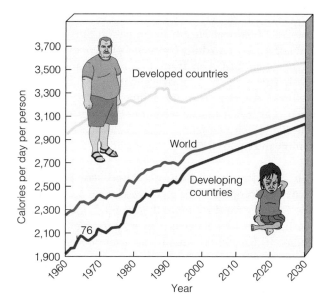

Figure 16-10 The average daily food intake in calories per person in the world, developing countries, and developed countries: 1961–2000 and projected increases to 2030. The average adult male needs about 2,500 calories per day for good health. (Data from UN Food and Agriculture Organization)

However, according to the FAO, World Bank, and WHO there is also some *bad news*. About one of every six people in developing countries (including about one of every three children below age 5) is chronically undernourished or malnourished. The 1996 World Food Summit goal of cutting the number of undernourished and malnourished people to 400 million by 2015 will probably not be achieved before 2030—15 years later than the original goal.

Finally, according to 2002 data from the WHO and the Food and Agriculture Organization, in 2001 at least 9 million people—6 million of them children under age 5—died prematurely from a combination of several causes. They were undernutrition and malnutrition, infectious diseases caused by drinking contaminated water, and increased susceptibility to normally nonfatal infectious diseases (such as measles and diarrhea, Figure 10-10, p. 211) because of their weakened condition from malnutrition. This means that each day at least 25,000 people die prematurely because of malnutrition and infectious disease caused mostly by poverty (Figure 10-16, p. 217).

Solutions: What Can We Do to Save Children from Malnutrition and Disease? According to the WHO, seven of ten deaths of children under age 5 in developing countries can be attributed to one or a combination of five main causes: malnutrition, pneumonia, diarrhea, measles, and malaria.

Some *good news* is that studies by the United Nations Children's Fund (UNICEF) indicate that one-half to two-thirds of childhood deaths from nutrition-

related causes could be prevented at an average annual cost of only $5–10 per child, or only 10–19¢ per week.

This life-saving program would involve the following measures:

- Immunizing children against childhood diseases such as measles

- Encouraging breast-feeding (except for mothers with AIDS)

- Preventing dehydration from diarrhea by giving infants a mixture of sugar and salt in a glass of water

- Preventing blindness by giving children a vitamin A capsule twice a year at a cost of about 75¢ per child or fortifying common foods with vitamin A and other micronutrients at a cost of about 10¢ per child annually

- Providing family planning services to help mothers space births at least 2 years apart

- Increasing education for women, with emphasis on nutrition, drinking water sterilization, and child care

How Serious Are Micronutrient Deficiencies?
According to the WHO, about 2 billion people—about one out of three people—suffer from a deficiency of one or more vitamins and minerals. The most widespread micronutrient deficiencies in developing countries involve *vitamin A, iron,* and *iodine.*

According to the WHO, 120–140 million children in developing countries are deficient in vitamin A (found in dairy products and green and yellow vegetables). This puts them at risk for blindness and premature death because even mild vitamin A deficiency reduces children's resistance to infectious diseases such as diarrhea and measles. Globally about 250,000 children under age 6 go blind each year from a lack of vitamin A and up to 80% of them die within a year.

Other nutritional deficiency diseases are caused by the lack of minerals. Too little *iron* (a component of hemoglobin that transports oxygen in the blood) causes anemia. Iron is found in legumes, green leafy vegetables, eggs, and meat. Iron deficiency causes fatigue, makes infection more likely, and increases a woman's chances of dying in childbirth. It also increases an infant's chances of dying of infection during its first year of life and cripples efforts to improve primary school education because developing brains need adequate iron to learn.

According to a 1999 survey by the WHO, one of every three people, mostly women and children in tropical developing countries, suffers from iron deficiency. The most severe problem is in India, where more than 80% of all pregnant women are anemic.

Elemental *iodine* is essential for proper functioning of the thyroid gland, which produces a hormone that controls the body's rate of metabolism. Iodine is found in seafood and crops grown in iodine-rich soils. Chronic lack of iodine can cause stunted growth, mental retardation, and goiter (an abnormal enlargement of the thyroid gland that can lead to deafness).

Some *good news* is that WHO and UNICEF programs to get countries to add iodine to salt slashed the percentage of the world's people with iodine deficiency from 29% to 12% between 1994 and 1999. Some *bad news* is that this still leaves at least 740 million people in developing countries with too little iodine in their diet. According to estimates by the UN Educational, Scientific, and Cultural Organization (UNESCO), about 26 million children suffer brain damage each year from lack of iodine and 600 million people—mostly in South and Southeast Asia—suffer from goiter.

The FAO estimates that spending $24 billion a year would be enough to end hunger and malnutrition within 10–15 years. According to the FAO, this expenditure would yield at least $120 billion per year in benefits as a result of longer, healthier, and more productive lives for several hundred million people freed from malnutrition and poverty.

How Serious Is Overnutrition? Overnutrition occurs when food energy intake exceeds energy use and causes excess body fat. Overnourished people are classified as *overweight* if they are roughly 4.5–14 kilograms (10–30 pounds) over a healthy body weight and *obese* if they are more than 14 kilograms (30 pounds) over a healthy weight. Too many calories (Figure 16-10, top curve), too little exercise, or both can cause overnutrition.

People who are underfed and underweight and those who are overfed and overweight face similar health problems: *lower life expectancy, greater susceptibility to disease and illness,* and *lower productivity and life quality.* Overnutrition is the second leading cause of preventable deaths after smoking, mostly from heart disease, cancer, stroke, and diabetes. An estimated 300,000 Americans die prematurely each year as a result of being overweight or obese—the second greatest health risk after smoking (Figure 10-1, p. 203). This death toll is roughly equivalent to two fully loaded jumbo (400-passenger) jets crashing accidentally every day with no survivors.

A study of thousands of Chinese villagers indicates that the healthiest diet for humans is largely vegetarian, with only 10–15% of calories coming from fat. This is in contrast to the typical meat-based diet, in which 40% of the calories come from fat.

About 14% or one out of seven adults in developed countries suffers from overnutrition. The situation is much worse in the United States. According to the 2002 National Health Examination Survey, about 65% of American adults (ages 20–74) were overweight or obese in 2000—up from 45% in 1961 and 47% in 1981. This is the highest overnutrition rate of any developed country. The $37 billion Americans spend each year trying to lose weight is 1.5 times more than the $24 billion per year needed to eliminate undernutrition and malnutrition in the world.

Do We Produce Enough Food to Feed the World's People? The *good news* is that according to the FAO we produce more than enough food to meet the basic nutritional needs of every person on the earth today. If distributed equally, the grain currently produced worldwide is enough to give everyone a meatless subsistence diet. According to food analysts, there should also be more food for more people in the future (Figure 16-10).

The *bad news* for the one out of eight people not getting enough to eat is that food is not distributed equally among the world's people because of differences in soil, climate, political and economic power, and average per capita income throughout the world.

Most agricultural experts agree that *the principal cause of hunger and malnutrition is and will continue to be poverty*, which prevents poor people from growing or buying enough food regardless of how much is available. For example, according to the United Nations, in the 1990s nearly 80% of all chronically undernourished and malnourished children lived in countries with food surpluses.

What Are the Environmental Effects of Producing Food? Agriculture has significant harmful effects on air, soil, water, and biodiversity (Figure 16-11). Some analysts believe these harmful environmental effects can be overcome and will not limit future food production.

Other analysts disagree. For example, according to Norman Myers (Guest Essay, p. 102), the future ability to produce more food will be limited by a combination of environmental factors. They include *soil erosion* (Figure 9-21, p. 193), *desertification* (Figure 9-23, p. 195), *salinization and waterlogging of irrigated lands* (Figure 9-25, p. 196), *water deficits and droughts* (Figure 14-8, p. 311), *loss of wild species* that provide the genetic resources for improved forms of foods, and the *effects of global warming* in some parts of the world (Figure 13-12, p. 291 and Figure 13-13, p. 291).

According to a 2002 study by the UN Department for Economic and Social Affairs, close to 30% of the world's cropland has been degraded by soil erosion, salinity, and chemical pollution, and 17% has been seriously degraded.

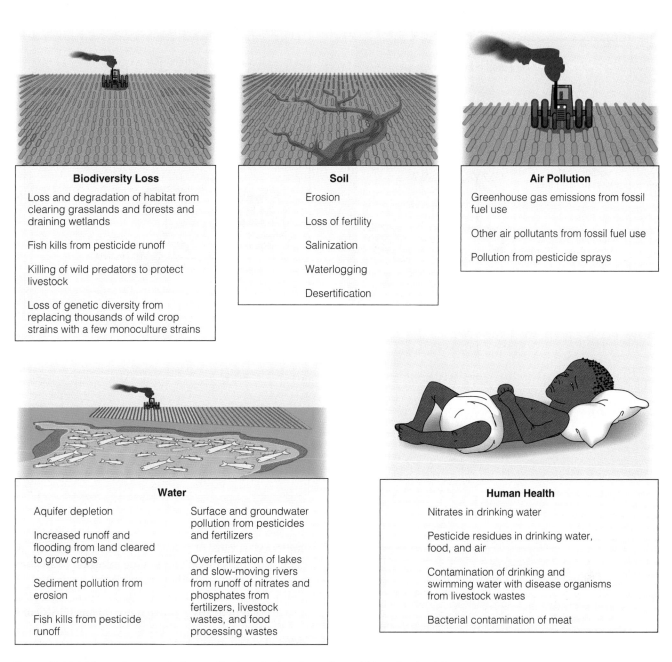

Figure 16-11 Major environmental effects of food production. According to UN studies, land degradation reduced cumulative food production worldwide by about 13% on cropland and 4% on pastureland between 1950 and 2000.

16-4 INCREASING WORLD CROP PRODUCTION

Is Increasing Crop Yields the Answer? Agricultural experts expect most future increases in food yields per hectare to result from improved crossbred strains of plants and from expansion of green revolution technology to new parts of the world.

Traditional crossbreeding is a slow process that typically takes 15 years or more to produce a commer-

cially valuable new variety. Also, it can combine traits only from species that are close to one another genetically. In addition, crossbred varieties are useful for only about 5–10 years before pests and diseases reduce their effectiveness.

Is Genetic Engineering the Answer? Scientists are working to create new green revolutions—actually *gene revolutions*—by using genetic engineering and other forms of biotechnology to develop genetically

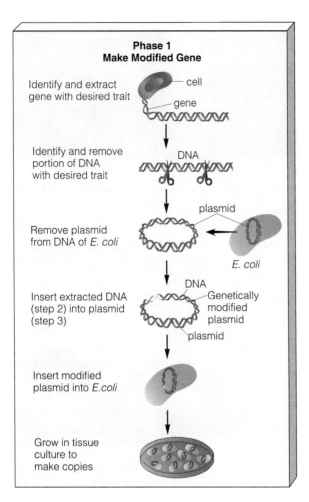

Phase 1
Make Modified Gene

Identify and extract gene with desired trait — cell / gene

Identify and remove portion of DNA with desired trait — DNA

Remove plasmid from DNA of *E. coli* — plasmid / *E. coli*

Insert extracted DNA (step 2) into plasmid (step 3) — DNA / Genetically modified plasmid / plasmid

Insert modified plasmid into *E.coli*

Grow in tissue culture to make copies

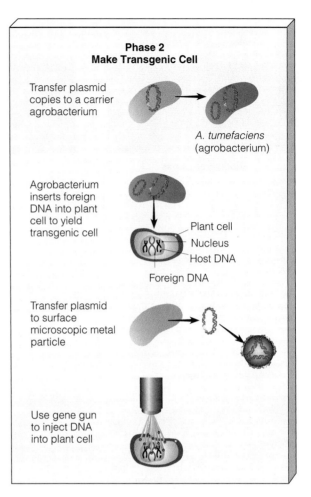

Phase 2
Make Transgenic Cell

Transfer plasmid copies to a carrier agrobacterium — *A. tumefaciens* (agrobacterium)

Agrobacterium inserts foreign DNA into plant cell to yield transgenic cell — Plant cell / Nucleus / Host DNA / Foreign DNA

Transfer plasmid to surface microscopic metal particle

Use gene gun to inject DNA into plant cell

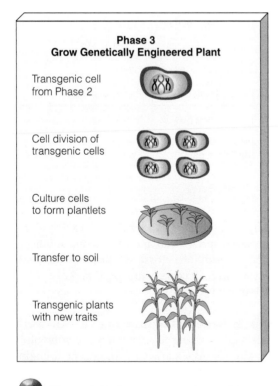

Phase 3
Grow Genetically Engineered Plant

Transgenic cell from Phase 2

Cell division of transgenic cells

Culture cells to form plantlets

Transfer to soil

Transgenic plants with new traits

Figure 16-12 Steps in genetically modifying a plant.

improved strains of crops. **Genetic engineering,** or **gene splicing,** of food crops is the insertion of an alien gene into a commercially valuable plant (or animal) to give it new beneficial genetic traits. Such organisms are called **genetically modified organisms (GMOs).**

Compared to traditional crossbreeding, gene splicing takes about half as much time to develop a new crop or animal variety, cuts costs, and allows the insertion of genes from almost any other organism into crop or animal cells. Figure 16-12 outlines the steps involved in developing a genetically modified, or transgenic, plant. Scientists are also using *advanced tissue culture techniques* to produce only the desired parts of a plant such as its oils or fruits.

Ready or not, the world is entering the *age of genetic engineering.* The United States is the world's largest producer of genetically modified crops. For example, the U.S. Department of Agriculture (USDA) estimated that 75% of soybeans, 71% of cotton, and 34% of corn planted by U.S. farmers in 2001 were GMOs. Nearly two-thirds of the food products on U.S. supermarket shelves contain genetically engineered crops, and the proportion is increasing rapidly.

However, there is growing controversy over the use of *genetically modified food (GMF).* Such food is seen as a potentially sustainable way to solve world food problems by its producers and investors and called potentially dangerous "Frankenfood" by its critics. Figure 16-13 summarizes the projected advantages and disadvantages of this new technology.

Critics recognize the potential benefits of genetically modified crops. However, they warn we know far too little about the potential harm to human health and ecosystems from widespread use of such crops. In addition, genetically modified organisms cannot be recalled if they cause harmful genetic and ecological effects. They call for more controlled field experiments, more research and long-term safety testing to better understand the risks, and stricter regulation of this new and rapidly growing technology.

Many analysts and consumer advocates believe governments should require mandatory labeling of genetically modified foods. This would provide consumers with information to help them make more informed choices about the foods they buy. Such labeling is required in Japan, Europe, South Korea, Canada, Australia, and New Zealand and is favored by 81% of Americans polled in 1999.

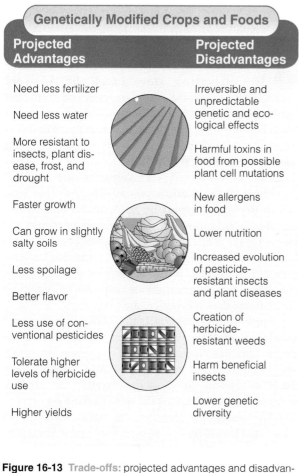

Figure 16-13 Trade-offs: projected advantages and disadvantages of genetically modified crops and foods.

Industry representatives oppose mandatory labeling because they claim that genetically modified foods are not substantially different from conventional foods. Also, they fear that labeling such foods would hurt sales by arousing unwarranted suspicion.

There is also concern about the use of genetic engineering to produce bioengineered pathogens (mostly bacteria and viruses) that could be used to devastate a country's crops and livestock animals during warfare or as an act of terrorism. According to a 2002 report by the National Academy of Sciences, conventional and bioengineered pathogens for harming crops or livestock are widely available and fairly easy to produce and pose a major threat to U.S. agriculture. According to the study, a bioterrorism attack on U.S. agriculture is highly unlikely to result in famine or malnutrition, but it could harm people, disrupt the economy, and cause widespread public confusion and fear.

Can We Continue Expanding the Green Revolution? Many analysts believe it is possible to produce enough food to feed the 9.3 billion people projected by 2050. This would be accomplished through new advances in gene-splicing technology and by spreading the use of new and existing high-yield green revolution techniques throughout most of the world.

In 2002, for example, the International Rice Institute announced it was working on two new varieties of genetically modified rice. One, called *aerobic rice,* could be grown with much less water than conventional rice strains. The other, called *dream rice,* is supposed to provide more iron, vitamin A, and lysine than conventional rice varieties. If dream rice lives up to its potential, it will be especially helpful in reducing chronic malnutrition among the poor. In addition, in 2002 scientists reported developing a form of genetically engineered rice that can withstand drought, salt water, and cold temperatures by borrowing a gene from *Escherichia coli (E. coli)* bacteria.

Other analysts point to several factors that have limited the success of the green and gene revolutions to date and may continue to do so. *First,* without huge amounts of fertilizer and water, most green revolution crop varieties produce yields that are no higher (and are sometimes lower) than those from traditional strains. This is why the second green revolution has not spread to many arid and semiarid areas such as much of Africa and Australia (Figure 16-4).

Second, green revolution and genetically engineered crop strains and their high inputs of water, fertilizer, and pesticides cost too much for most subsistence farmers in developing countries.

Third, grain yields per hectare are still increasing in many parts of the world but at a much slower rate. For example, grain yields rose about 2.1% a year between 1950 and 1990 but dropped to 1.1% per year between 1990 and 2000.

Fourth, crop yields in some areas may start dropping as soil erodes and loses fertility, irrigated soil becomes salty and waterlogged (Figure 9-24, p. 196), underground and surface water supplies become depleted and polluted with pesticides and nitrates from fertilizers, and populations of rapidly breeding pests develop genetic immunity to widely used pesticides. We do not know how close we are to such environmental limits in various parts of the world.

Fifth, increased loss of biodiversity can limit the genetic raw material needed for future green and gene revolutions.

Finally, according to Indian economist Vandana Shiva, overall gains in crop yields from new green and gene revolution varieties may be much lower than claimed. The yields are based on comparisons between the output per hectare of old and new *monoculture* varieties rather than between the even higher yields per hectare for *polyculture* cropping systems and the new monoculture varieties that often replace polyculture crops.

Will People Try New Foods? Some analysts recommend greatly increased cultivation of less widely known plants to supplement or replace such staples as wheat, rice, and corn. One of many possibilities is the *winged bean,* a protein-rich legume now common only in New Guinea and Southeast Asia. This fast-growing plant is a good source of protein and has so many edible parts it has been called a supermarket on a stalk. It also needs little fertilizer because of nitrogen-fixing nodules in its roots.

Insects—called *microlivestock*—are also important potential sources of protein, vitamins, and minerals in many parts of the world. There are about 1,500 edible insect species. Some are important food items in many parts of the world. Examples include black ant larvae (served in tacos in Mexico), giant waterbugs (crushed into vegetable dip in Thailand), *Mopani,* or emperor moth caterpillars (eaten in South Africa), cockroaches (eaten by Kalahari desert dwellers), lightly toasted butterflies (a favorite food in Bali), and fried ants (sold on the streets of Bogota, Colombia). Most of these insects are 58–78% protein by weight—three to four times as protein rich as beef, fish, or eggs. Two problems are getting farmers to take the financial risk of cultivating new types of food crops and convincing consumers to try new foods.

Some plant scientists believe we should rely more on polycultures of perennial crops, which are better adapted to regional soil and climate conditions than most annual crops (p. 377). Using perennials would also eliminate the need to till soil and replant seeds each year, greatly reduce energy use, save water, and reduce soil erosion and water pollution from eroded sediment. However, large seed companies that make their money selling farmers seeds each year for annual crops generally oppose this idea.

Is Irrigating More Land the Answer? About 40% of the world's food production comes from the 18% of the world's cropland that is irrigated. Irrigated land produces about 70% of the grain harvest in China, 50% in India, and 15% in the United States.

Some *good news* is that between 1950 and 2002, the world's irrigated area tripled, with most of the growth occurring from 1950 to 1978. Some *bad news* is that since 1978, the amount of irrigated land per person has been falling and is projected to fall much more between 2000 and 2050. One reason for this downward trend is that the world population has grown faster than irrigated agriculture since 1978. Other factors are aquifer depletion (p. 315), inefficient use of irrigation water (p. 318), soil salinization (Figure 9-24, p. 196) on irrigated cropland, and disruption of water supplies in some food-growing areas from global warming (Figure 13-12, p. 291). In addition, the majority of the world's farmers do not have enough money to irrigate their crops.

According to analysts, there are three key methods for using water more sustainably in crop production. One is to increase irrigation efficiency (p. 318). Another is to shift to crops that need less water. For example, we could depend less on water-thirsty rice and sugarcane and more on wheat and sorghum. The third is to withdraw water from aquifers no faster than they are replenished.

Is Cultivating More Land the Answer? Theoretically, the world's cropland could be more than doubled by clearing tropical forests and irrigating arid land (Figure 16-14). The problem is that much of this is *marginal land* with poor soil fertility (Figure 9-18, top left and bottom left, p. 191), steep slopes, or both. Thus cultivation of such land is unlikely to be sustainable.

Much of the world's potentially cultivable land lies in dry areas, especially in Australia and Africa. Large-scale irrigation in these areas would require expensive dam projects with a mixture of beneficial and harmful impacts (Figure 14-9, p. 312) and use large inputs of fossil fuel to pump water long distances. This could also deplete groundwater supplies by removing water faster than it is replenished.

Furthermore, these potential increases in cropland would not offset the projected loss of almost 30% of today's cultivated cropland caused by erosion, overgrazing, waterlogging, salinization, and urbanization. Such expansion in cropland would also reduce wildlife habitats and thus the world's biodiversity.

Thus *many analysts believe that significant expansion of cropland is unlikely over the next few decades.* For example, the International Food Policy Institute and FAO estimate that 80% of the projected increase in food pro-

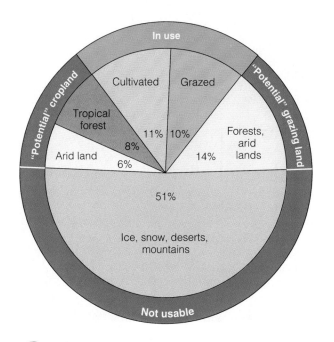

In use

Cultivated   Grazed

"Potential" cropland

"Potential" grazing land

Tropical
forest

11% 10%

8%

Arid land

6%

Forests,
arid
lands

14%

51%

Ice, snow, deserts,
mountains

Not usable

Figure 16-14 Classification of the earth's land. Theoretically, we could double the amount of cropland by clearing tropical forests and irrigating arid lands. However, converting these lands into cropland would destroy valuable forest resources, reduce the earth's biodiversity, affect water quality and quantity, and cause other serious environmental problems, usually without being cost effective.

duction between 1993 and 2020 will come from increased yields per hectare and only 20% from expansion of cropland.

Can We Grow More Food in Urban Areas? Currently, about 800 million urban gardens provide about 15% of the world's food. Food experts believe that people in urban areas could live more sustainably and save money by growing more of their food. Such food could be grown in empty lots, in backyards, on rooftops and balconies, and by raising fish in tanks and sewage lagoons. A study by the UN Center for Human Settlements estimated that up to 50% of the total area in many cities in developing countries is vacant public land that could be used to produce food.

16-5 PRODUCING MORE MEAT

What Are Rangeland and Pasture? About 40% of the earth's ice-free land is **rangeland** that is too dry, too steeply sloped, or too infertile to grow crops. This type of land supplies forage or vegetation for grazing (grass-eating) and browsing (shrub-eating) animals. About 5 billion cattle, sheep, and goats graze on about 42% of the world's rangeland. Much of the rest is too dry, cold, or remote from population centers to support large numbers of livestock. About 29% of the total

U.S. land area is rangeland, most of it short-grass prairie in the arid and semiarid western half of the country. Livestock also graze in **pastures:** managed grasslands or enclosed meadows usually planted with domesticated grasses or other forage.

Most rangeland grasses have deep and complex root systems (Figure 9-18, top right, p. 191) that help anchor the plants. These roots also extract underground water so plants can withstand drought, and store nutrients so plants can grow again after a drought or fire.

Blades of rangeland grass grow from the base, not the tip. Thus, as long as only its upper half—called its *metabolic reserve*—is eaten and its lower half remains, rangeland grass is a renewable resource that can be grazed again and again. The exposed metabolic reserve of a grass plant is where photosynthesis takes place to provide food for the deep roots of rangeland grasses. If all or most of the lower half of the plant is eaten, the plant is weakened and can die.

How Is Meat Produced, and What Are Its Environmental Consequences? Meat and meat products are good sources of high-quality protein. Between 1950 and 2002, world meat production increased more than fivefold, and per capita meat production more than doubled. Globally, animal products provide about 15% of the energy and 30% of the protein in the human diet.

About 80% of the world's cattle, sheep, and goats are raised on rangeland by open grazing or nomadic herding (Figure 16-2). Some analysts expect most future increases in meat production to come from densely populated *feedlots,* where animals are fattened for slaughter by feeding on grain grown on cropland or meal produced from fish. Feedlots account for about 43% of the world's beef production and more than half of the world's poultry and pork.

The *good news* is that this industrialized approach increases meat productivity. The *bad news* is that it also has a number of harmful environmental effects. It creates foul odors, and causes water pollution when lagoons storing animal wastes collapse or are flooded. Nitrates from animal wastes can also contaminate drinking water wells.

Meat production also increases pressure on the world's grain supply because livestock and fish raised for food consume about 37% of the world's grain production (70% in the United States). In addition, meat production requires increased inputs of fossil fuel (Figure 16-6).

Producing meat can also endanger wildlife species. According to a 2002 report by the National Public Lands Grazing Campaign, livestock grazing has contributed to population declines of 22% of the country's threatened and endangered species.

Finally, meat production uses more than half of the water withdrawn worldwide from rivers and aquifers each year. Most of this water is used to irrigate crops fed to livestock and wash away manure from crowded livestock pens and feedlots.

How Can We Produce Meat More Sustainably? Livestock and fish vary widely in the efficiency with which they convert grain into animal protein (Figure 16-15). A more sustainable form of meat production would involve shifting from less grain-efficient forms of animal protein, such as beef or pork, to more grain-efficient ones, such as poultry or farmed fish.

Some environmentalists have called for reducing livestock production (especially cattle) to decrease its environmental effects and to feed more people. This would decrease the environmental impact of livestock production, but it would not free up much land or grain to feed more of the world's hungry people.

Cattle and sheep that graze on rangeland use a resource (grass) that humans cannot eat, and most of this land is not suitable for growing crops. Moreover, because of poverty, insufficient economic aid, and the nature of global economic and food distribution systems, very little if any additional grain grown on land used to raise livestock or livestock feed would reach the world's hungry people.

What Are the Effects of Overgrazing and Undergrazing? Overgrazing can limit livestock production. **Overgrazing** occurs when too many animals graze for too long and exceed the carrying capacity of a grassland area. Excessive numbers of domestic livestock feeding for too long in a particular area causes most overgrazing.

Such overgrazing lowers the net primary productivity of grassland vegetation (Figure 4-22, p. 80) and reduces grass cover. It also and exposes the soil to erosion by water and wind (Figure 16-16), compacts the

Figure 16-16 Rangeland: overgrazed (left) and lightly grazed (right). About 14% of U.S. topsoil loss is directly associated with livestock grazing.

soil (which diminishes its capacity to hold water), and is a major cause of desertification (Figure 9-23, p. 195).

Some grassland can suffer from **undergrazing,** where absence of grazing for long periods (at least 5 years) can reduce the net primary productivity of grassland vegetation and grass cover. Moderate grazing of such areas removes accumulations of standing dead material and stimulates new biomass production.

What Is the Condition of the World's Rangelands? *Range condition* usually is classified as *excellent* (containing more than 75% of its potential forage production), *good* (51–75%), *fair* (26–50%), or *poor* (0–25%). Limited data from surveys in various countries indicate that overgrazing by livestock has caused as much as 20% of the world's rangeland to lose productivity, mostly by desertification (Figure 9-22, p. 195).

Most of the rangeland in the United States is in the West. About 60% is privately owned and the rest is public land managed by the Bureau of Land Management (BLM) and the U.S. Forest Service. Only about 2% of the 120 million cattle and 10% of the 20 million sheep raised in the United States graze on public rangelands.

According to a 2000 survey by the Bureau of Land Management, about 64% of the nonarctic U.S. public rangeland it managed was in unsatisfactory (fair or poor) condition, compared with 84% in 1936. This represents a great improvement, but there is still a long way to go.

According to some wildlife and rangeland experts, estimates of rangeland condition do not take

Kilograms of grain needed per kilogram of body weight

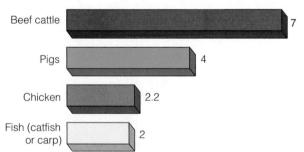

Beef cattle — 7
Pigs — 4
Chicken — 2.2
Fish (catfish or carp) — 2

Figure 16-15 Efficiency of converting grain into animal protein. Data in kilograms of grain per kilogram of body weight added. (Data from U.S. Department of Agriculture)

into account severe damage to heavily grazed thin strips of lush vegetation along streams called **riparian zones.** These ecologically important areas help keep streams from drying out during droughts by storing and releasing water slowly from spring runoff and summer storms. They also provide habitats, food, water, and shade for wildlife in the arid and semiarid western lands.

Studies indicate that 65–75% of the wildlife in the western United States totally depends on riparian habitats. According to a 1999 study in the *Journal of Soil and Water Conservation,* livestock grazing has damaged approximately 80% of stream and riparian ecosystems in the United States.

Some *good news* is that we can restore riparian areas by using fencing to restrict access to degraded areas and developing off-stream-watering sites for livestock. Sometimes protected areas can recover in a few years.

How Can Rangelands Be Managed More Sustainably to Produce More Meat? The primary goal of sustainable rangeland management is to maximize livestock productivity without overgrazing or undergrazing rangeland vegetation. There are two major ways to do this. One is to control the number, types, and distribution of livestock grazing on land. The other is to restore and improve degraded rangeland.

The most widely used method for more sustainable rangeland management is to control the number of grazing animals and the duration of their grazing in a given area so the carrying capacity of the area is not exceeded. However, determining the carrying capacity of a range site is difficult and costly. In addition, carrying capacity varies with factors such as climatic conditions (especially drought), soil type, invasions by new plant or animal species, kinds of grazing animals, and intensity of grazing.

Livestock tend to aggregate around natural water sources and stock ponds. As a result, areas around water sources tend to be overgrazed and other areas can be undergrazed. Managers can use several strategies to prevent this and help promote more uniform use of rangeland. Damaged rangeland and riparian zones can be fenced off and livestock can be moved from one grazing area to another. Ranchers can also provide supplemental feed at selected sites and locate water holes and tanks and salt blocks in strategic places.

A more expensive and less widely used method of rangeland management involves suppressing the growth of unwanted plants by herbicide spraying, mechanical removal, or controlled burning. A cheaper way to discourage unwanted vegetation is controlled, short-term trampling by large numbers of livestock.

16-6 CATCHING AND RAISING MORE FISH AND SHELLFISH

How Are Fish and Shellfish Harvested? The world's third major food-producing system consists of **fisheries:** concentrations of particular aquatic species suitable for commercial harvesting in a given ocean area or inland body of water. Some commercially important marine species of fish and shellfish are shown in Figure 16-17. Fish and shellfish supply about 7% of the global food supply and are the primary source of

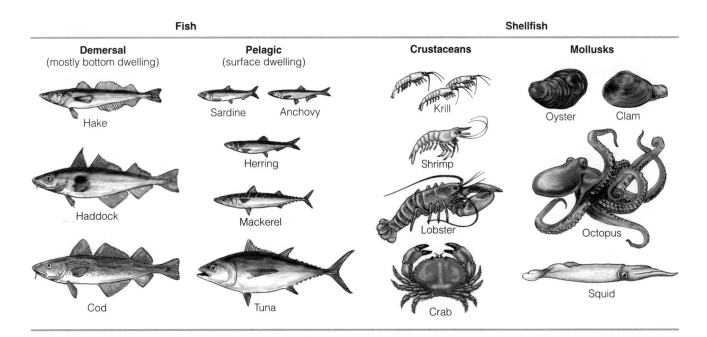

Figure 16-17 Some major types of commercially harvested marine fish and shellfish.

animal protein for about 1 billion people, mostly in developing countries.

The world's commercial fishing industry is dominated by industrial fishing fleets using satellite positioning equipment, sonar, huge nets, spotter planes, and factory ships that can process and freeze their catches. About 55% of the annual commercial catch of fish and shellfish comes from the ocean using harvesting methods shown in Figure 16-18.

The rest of the annual catch comes from using aquaculture to raise marine and freshwater fish in ponds and underwater cages (33%) and from inland freshwater fishing from lakes, rivers, reservoirs, and ponds (12%). About one-third of the world fish harvest (mostly small surface-dwelling species such as an-

chovy, herring, and menhaden) is used as animal feed, fish meal, and oils.

Can We Harvest More Fish and Shellfish? The *good news* is that between 1950 and 2000, the annual commercial fish catch (marine plus freshwater harvest) increased almost fivefold (Figure 16-19, left), and the per capita seafood catch more than doubled (Figure 16-19, right).

The *bad news* is that the commercial fish catch has dropped and leveled off since 1982 (Figure 16-19, left). Also, the per capita commercial fish catch has been falling since 1992 (Figure 16-19, right) and may continue to decline because of overfishing, pollution, habitat loss, and population growth.

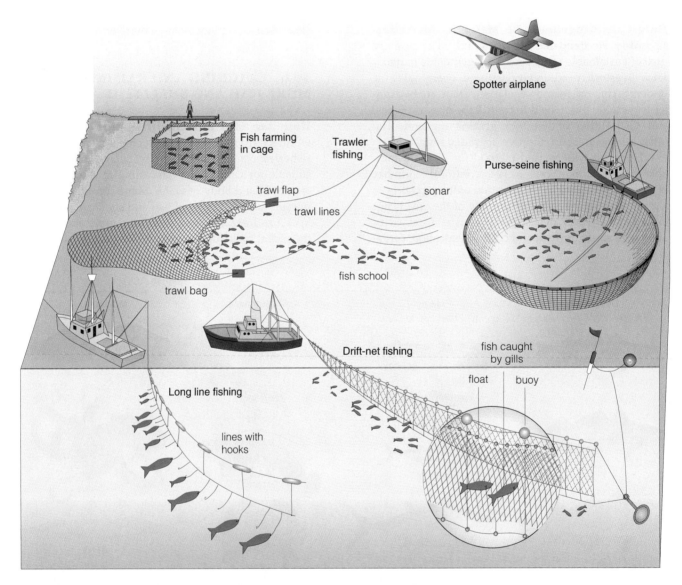

Figure 16-18 Major commercial fishing methods used to harvest various marine species.

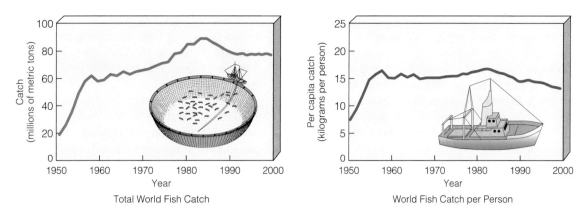

Figure 16-19 World fish catch (left) and world fish catch per person (right), 1950–2000. Worldwide per capita fish catch did not rise much between 1968 and 1989 and has dropped since then. The total catch and per capita catches since 1990 are probably about 10% lower than shown here because of the discovery in 2002 that since 1990 China had apparently been inflating its fish catches. (Data from UN Food and Agriculture Organization and Worldwatch Institute)

Connections: How Are Overfishing and Habitat Degradation Affecting Fish Harvests? Fish are renewable resources as long as the annual harvest leaves enough breeding stock to renew the species for the next year. Ideally, an annual **sustainable yield**—the size of the annual catch that could be harvested indefinitely without a decline in the population of a species—should be established for each species to avoid depleting the stock.

However, determining sustainable yields is difficult because it is hard to estimate mobile aquatic populations. Also, sustainable yields shift from year to year because of changes in climate, pollution, and other factors. In addition, sustainably harvesting the entire annual surplus of one species may severely reduce the population of other species that rely on it for food.

Overfishing is the taking of so many fish that too little breeding stock is left to maintain numbers; that is, overfishing is a harvest of a species that exceeds its sustainable yield. Prolonged overfishing leads to **commercial extinction,** when the population of a species declines to the point at which it is no longer profitable to hunt for them. Fishing fleets then move to a new species or a new region, hoping that the overfished species will eventually recover.

High levels of *bycatch* (the nontarget fish that are caught in nets and then thrown back into the sea) also deplete fisheries. Nearly one-fourth of the annual global fish catch is bycatch that depletes marine biodiversity and does not provide food for people.

Here are several pieces of *bad news* about the world's fish stocks. *First,* a 2001 report by the UN Food and Agriculture Organization found that about 75% of the world's 200 commercially valuable marine fish species are either overfished or fished to their estimated sustainable yield. The primary cause of this

depletion of fish stocks is too many fishing boats pursuing too few fish—another example of the tragedy of the commons (p. 7). According to the Ocean Conservancy, "we are spending the principal of our marine fish resources rather than living off the interest they provide."

Second, studies by the U.S. National Fish and Wildlife Foundation show that 14 major commercial fish species in U.S. waters (accounting for one-fifth of the world's annual catch and half of all U.S. stocks) are severely depleted (Figure 16-20). Some *good news* is that since 1996 some 25 species are no longer overfished and 70 others continue to recover.

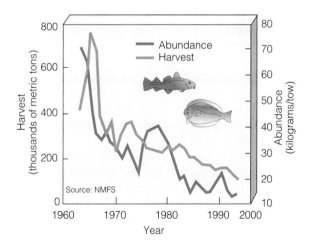

Figure 16-20 The harvest of groundfishes (yellowtail flounder, haddock, and cod) in the Georges Bank off the coast of New England in the North Atlantic, once one of the world's most productive fishing grounds, has declined sharply since 1965. Stocks dropped to such low levels that since December 1994 the National Marine Fisheries Services has banned fishing of these species in the Georges Bank. (Data from U.S. National Marine Fisheries Service)

Third, degradation, destruction, and pollution of wetlands, estuaries, coral reefs, salt marshes, and mangroves threaten populations of fish and shellfish. *Fourth,* global warming that is projected over the next 50–100 years (Figure 13-8, p. 286) is a threat to the global fish catch. Warmer seawater can degrade or destroy highly productive coral reefs and enhance the harmful effects of habitat degradation and pollution on fish populations.

Finally, a 2002 analysis by the WorldFish Center and the International Food Policy Institute warned that the world's growing population and overfishing mean that 1 billion people in developing countries will face shortages of fish, their most important source of protein, by 2020.

Should Governments Continue Subsidizing Fishing Fleets? Because of overfishing and the overcapacity of the fishing fleet, it costs the global fishing industry about $120 billion a year to catch $70 billion worth of fish. Government subsidies such as fuel tax exemptions, price controls, low-interest loans, and grants for fishing gear make up most of the $50 billion annual deficit of the industry. Without such subsidies, the world's fishing industry would be bankrupt and the number of fish caught would approach their sustainable yield.

Phasing out these subsidies would cause a loss of jobs for some fishers and fish processors in coastal communities. But to fishery biologists the alternative is worse. Continuing to subsidize excess fishing allows fishers to keep their jobs a little longer while making less and less money until the fishery collapses. Then all jobs are gone, and fishing communities suffer even more—another example of the tragedy of the commons (p. 7) in action.

What Is Aquaculture? Aquaculture, in which fish and shellfish are raised for food, supplies about 33% of the world's commercial fish harvest. Some *good news* is that aquaculture production increased fivefold between 1984 and 2001.

Developing countries account for about 85% of the world's aquaculture production. China is the world leader in aquaculture (producing about 68% of the world's output), followed by India, Japan, and the United States.

There are two basic types of aquaculture. One is **fish farming.** It involves cultivating fish in a controlled environment (often a coastal or inland pond, lake, reservoir, or rice paddy) and harvesting them when they reach the desired size.

The other is **fish ranching.** It involves holding anadromous species such as salmon (that live part of their lives in fresh water and part in salt water) in cap-

tivity for the first few years of their lives, usually in fenced-in areas or floating cages in coastal lagoons and estuaries (Figure 16-18). Then the fish are released, and adults are harvested when they return to spawn (Figure 16-21).

Species cultivated in developing countries (mostly in inland aquaculture ponds, lakes, reservoirs, and rice paddies) include carp (especially in China and India), catfish (in the United States), tilapia, milkfish, clams, and oysters. These herbivorous species feed on phytoplankton and other aquatic plants and thus eat low on the food chain.

In developed countries and some rapidly developing countries in Asia, aquaculture is used mostly to stock lakes and streams with game fish or to raise expensive fish and shellfish (such as oysters, catfish, crayfish, rainbow trout, shrimp, and salmon). Most of these species are exported to Japan, Europe, and North America. Aquaculture now produces 90% of all oysters, 40% of all salmon (75% in the United States), 50% of internationally traded shrimp and prawns, and 65% of freshwater fish sold in the global marketplace. Many of these species are fed grain or fish meal.

Is Aquaculture the Answer? Figure 16-22 (p. 398) lists major advantages and disadvantages of aquaculture. Aquaculture may soon overtake cattle ranching as a source of animal protein. Future trends in cattle and aquaculture and in our diets are likely to depend on the efficiency with which cattle and fish convert grain to protein (Figure 16-15).

Fish have a large cost advantage by requiring about 2 kilograms of grain to add 1 kilogram of live weight, whereas cattle require about 7 kilograms of grain to add 1 kilogram of live weight. Water scarcity is another factor. It takes about 1,000 kilograms of water to produce 1 kilogram of grain, whereas raising fish in an aquaculture pond takes much less water per kilogram of fish.

Some analysts project that freshwater and saltwater aquaculture production could provide 41% of the world's seafood by 2020. Other analysts warn that the harmful environmental effects of aquaculture (Figure 16-22) could limit future production.

But some aquaculture proponents point to new trends and research that may help decrease the harmful environmental effects of aquaculture. In his 2002 book *Ecological Aquaculture,* aquaculture expert Barry Costa-Pierce called for a number of changes in aquaculture practices to reduce harmful environmental impacts and make them more sustainable.

One reform is to reduce use of fishmeal as a feed so aquaculture programs do not consume more protein than they produce. Others are to reduce pollution by improving management of aquaculture wastes and

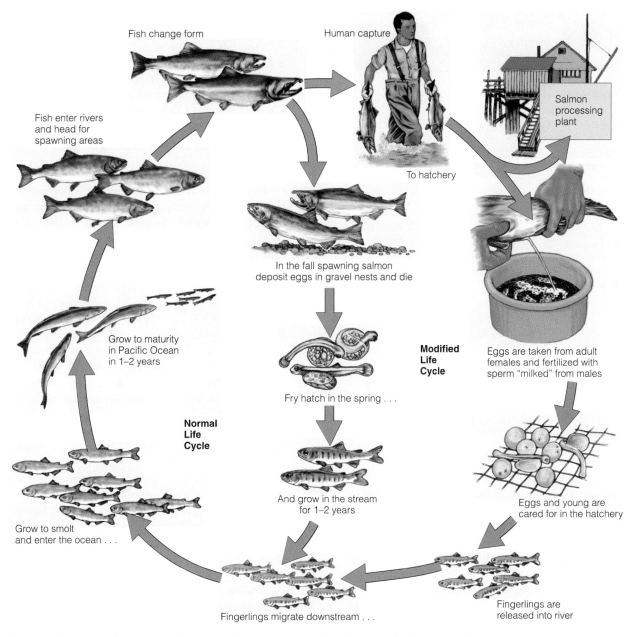

Figure 16-21 Normal life cycle of wild salmon (left) and human-modified life cycle of hatchery-raised salmon (right). Salmon spend part of their lives in fresh water and part in salt water.

Fish change form

Human capture

To hatchery

Salmon processing plant

Fish enter rivers and head for spawning areas

In the fall spawning salmon deposit eggs in gravel nests and die

Modified Life Cycle

Eggs are taken from adult females and fertilized with sperm "milked" from males

Fry hatch in the spring . . .

Grow to maturity in Pacific Ocean in 1–2 years

Normal Life Cycle

And grow in the stream for 1–2 years

Eggs and young are cared for in the hatchery

Grow to smolt and enter the ocean . . .

Fingerlings migrate downstream . . .

Fingerlings are released into river

eliminating the use of chemicals that harm human and ecosystem health. To increase efficiency, he calls for establishing better controls to reduce escape of aquaculture species into the wild and for recovery of escaped individuals. Finally, environmental labeling programs would certify that fish and shellfish were raised in a sustainable manner.

Other improvements have also been suggested. One is to restrict location of fish farms to reduce loss of mangrove forests and other threatened coastal envi-

ronments. Another suggestion is to increase production of herbivorous aquaculture fish species (such as carp, tilapia, and shellfish) that need little or no grain or fishmeal in their diets.

However, even under the most optimistic projections, increasing both the wild catch and aquaculture will not increase world food supplies significantly. The reason is that currently fish and shellfish supply only about 1% of the calories and 7% of the protein in the human diet.

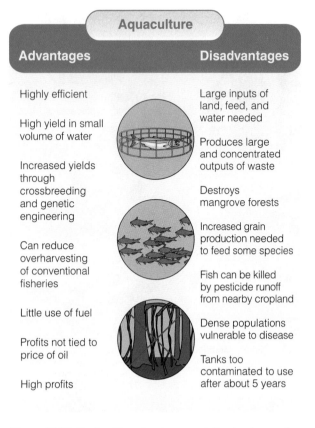

Aquaculture

Advantages	Disadvantages
Highly efficient	Large inputs of land, feed, and water needed
High yield in small volume of water	Produces large and concentrated outputs of waste
Increased yields through crossbreeding and genetic engineering	Destroys mangrove forests
Can reduce overharvesting of conventional fisheries	Increased grain production needed to feed some species
	Fish can be killed by pesticide runoff from nearby cropland
Little use of fuel	Dense populations vulnerable to disease
Profits not tied to price of oil	Tanks too contaminated to use after about 5 years
High profits	

Figure 16-22 Trade-offs: advantages and disadvantages of aquaculture.

Solutions: How Can We Manage Fisheries More Sustainably? Analysts suggest a number of measures for managing global fisheries more sustainably and protecting marine biodiversity. One is to *improve fishery regulations.* This involves setting and enforcing fishery quotas well below their estimated maximum sustained yields, requiring selective gear that avoids catching unwanted or undersized fish, and improving monitoring and enforcement of fishing regulations.

A second method is to use *economic approaches.* One is to sharply reduce or eliminate government fishing subsidies. Another is to impose fees for harvesting fish and shellfish from publicly owned and managed offshore waters. The money could be used for government fishery management, as Australia does. A third method involves setting up programs to certify sustainable fisheries, as the Marine Stewardship Council has done for six fisheries.

There is also a need to *reduce bycatch levels.* Ways to do this include using wider-mesh nets to allow smaller species and smaller individuals of the targeted species to escape, having observers on fishing vessels, and licensing boats to catch several species instead of only one target species. In addition, laws can be enacted that prohibit throwing edible and marketable fish back

to sea. Namibia and Norway have done this, with the law enforced by onboard observers.

Another strategy to achieve sustainable fisheries is to *establish protected areas.* Ways to do this include closing fisheries during certain seasons and establishing no-fishing or protected marine areas to allow depleted fish species to recover. Also, integrated coastal management programs can be used to promote both sustainable fishing and the ecological health of marine ecosystems.

Another strategy is to use labels that allow consumers to identify fish that have been harvested sustainably. See the website material for this chapter for lists of seafood species that consumers should consider not buying and those that are safe for now, as compiled in 2000 by the National Audubon Society and the Monterey Bay Aquarium.

16-7 GOVERNMENT AGRICULTURAL POLICY

How Do Government Agricultural Policies Affect Food Production? Agriculture is a financially risky business. Whether farmers have a good year or a bad year depends on factors over which they have little control: weather, crop prices, crop pests and diseases, interest rates, and the global market. Because of the need for reliable food supplies despite fluctuations in these factors, most governments provide various forms of assistance to farmers and consumers.

Governments use three main approaches to do this. One is to *keep food prices artificially low.* This makes consumers happy but means farmers may not be able to make a living.

Another is to *give farmers subsidies to keep them in business and encourage them to increase food production.* Globally, government price supports and other subsidies for agriculture total more than $500 billion per year (including $100 billion per year in the United States). If government subsidies are too generous and the weather is good, farmers may produce more food than can be sold. The resulting surplus depresses food prices, which reduces the financial incentive for farmers in developing countries to increase domestic food production.

A third approach is to *eliminate most or all price controls and subsidies and let farmers and fishers respond to market demand without government interference.* However, some analysts urge that any phaseout of farm and fishery subsidies should be coupled with increased aid for the poor and the lower middle class, who would suffer the most from any increase in food prices. Many environmentalists say that instead of eliminating all subsidies we should use them to reward farmers and ranchers who protect the soil, conserve water, reforest degraded land, protect and restore wetlands, conserve wildlife, and practice more sustainable agriculture and fishing.

16-8 PROTECTING FOOD RESOURCES: USING CONVENTIONAL CHEMICAL PESTICIDES TO CONTROL PESTS

What Is a Pest?   A **pest** is any species that competes with us for food, invades lawns and gardens, destroys wood in houses, spreads disease, or is simply a nuisance. Worldwide, only about 100 species of plants (which we call weeds), animals (mostly insects), fungi, and microbes (which can infect crop plants and livestock animals) cause about 90% of the damage to the crops we grow.

Most pest species are *r-selected* or *opportunist species* (Figure 8-10, p. 167). Typically, such species reproduce rapidly, spread quickly when a disturbance opens up a new habitat or niche (as in the early stages of ecological succession, Figure 7-12, p. 153), and can temporarily take over a biological community.

Insects cause much of the damage to the crops we grow. This is not surprising because insects make up about 75% of the earth's known species, reproduce rapidly, and through natural selection can rapidly develop protective traits such as genetic resistance to pesticides. As we try to keep a small number of insect pests from eating some of the food we grow, we should also remember that our lives depend on the many important ecological roles that insects play (p. 64).

In natural ecosystems and many polyculture agroecosystems (p. 377), *natural enemies* (predators, parasites, and disease organisms) control the populations of 50–90% of pest species as part of the earth's ecosystem services (Figure 1-2, p. 3, and top half of back cover) and help keep any one species from taking over for very long.

When we change agriculture from polyculture to monoculture (Figure 6-18, p. 119) and spray fields with huge amounts of pesticides, we upset many of these natural population checks and balances. Then we must devise ways to protect our monoculture crops, tree plantations, and lawns from insects and other pests that nature once controlled at no charge.

What Are Pesticides?   To help control pest organisms, we have developed a variety of **pesticides** (or *biocides*): chemicals to kill organisms we consider undesirable. Common types of pesticides include *insecticides* (insect killers), *herbicides* (weed killers), *fungicides* (fungus killers), *nematocides* (roundworm killers), and *rodenticides* (rat and mouse killers).

We did not invent the use of chemicals to repel or kill other species; plants have been producing chemicals to ward off or poison herbivores that feed on them for about 225 million years. This is a never-ending, ever-changing process: herbivores overcome various plant defenses through natural selection; then the plants use natural selection to develop new defenses.

As the human population grew and agriculture spread, people began looking for ways to protect their crops, mostly by using chemicals to kill or repel insect pests. Sulfur was used as an insecticide well before 500 B.C.; by the 1400s, people were applying toxic compounds of arsenic, lead, and mercury to crops as insecticides. Farmers abandoned this approach in the late 1920s when the increasing number of human poisonings and fatalities prompted a search for less toxic substitutes.

In the 1600s, farmers used nicotine sulfate, extracted from tobacco leaves, as an insecticide. In the mid-1800s, two more natural pesticides were introduced: *pyrethrum* (obtained from the heads of chrysanthemum flowers) and *rotenone* (extracted from the roots of various tropical forest legumes). These *first-generation pesticides* were mainly natural chemicals borrowed from plants that had been defending themselves from insects for eons.

A major pest control revolution began in 1939, when entomologist Paul Müller discovered that DDT (dichlorodiphenyltrichloroethane), a chemical known since 1874, was a potent insecticide. DDT, the first of the so-called *second-generation pesticides,* soon became the world's most used pesticide, and Müller received the Nobel Prize in 1948 for his discovery. Since then, chemists have made hundreds of other second-generation synthetic chemical pesticides.

How Are Pesticides Used Today?   Since 1950, pesticide use has risen more than 50-fold, and most of today's pesticides are more than ten times as toxic as those used in the 1950s. Worldwide, about 2.3 million metric tons (2.5 million tons) of second-generation pesticides—worth about $44 billion—are used yearly. About 75% of these chemicals are used in developed countries, but use in developing countries is soaring.

Since 1980, pesticide use for growing crops has leveled off in the United States. About 25% of pesticide use in the United States is for ridding houses, gardens, lawns, parks, playing fields, swimming pools, and golf courses of pests. According to the U.S. Environmental Protection Agency (EPA), the average lawn in the United States is doused with ten times more synthetic pesticides per hectare than U.S. cropland.

The EPA estimates that 84% of U.S. homes use pesticide products such as bait boxes, pest strips, bug bombs, flea collars, and pesticide pet shampoos and

INDIVIDUALS MATTER

Rachel Carson

Rachel Carson began her professional career as a biologist for the Bureau of U.S. Fisheries (later to become the U.S. Fish and Wildlife Service). In that capacity, she carried out research on oceanography and marine biology and wrote articles about the oceans and topics related to the environment. In 1949, she became editor-in-chief of the bureau's publications.

In 1951, she wrote *The Sea Around Us*, which described in easily understandable terms the natural history of oceans and how humans were harming them. This book was on the best-seller list for 86 weeks, sold more than 2 million copies, was translated into 32 languages, and won a National Book Award.

During the late 1940s and throughout the 1950s, DDT and related compounds were increasingly used to kill insects that ate food crops, attacked trees, bothered people, and transmitted diseases such as malaria.

In 1958, DDT was sprayed to control mosquitoes near the home and private bird sanctuary of Olga Huckins, a good friend of Carson. After the spraying, Huckins witnessed the agonizing deaths of several of her birds. In distress she asked Carson whether she could find someone to investigate the effects of pesticides on birds and other wildlife.

Carson decided to look into the issue herself and found that almost no independent research on the environmental effects of pesticides existed. As a well-trained scientist, she surveyed the scientific literature, became convinced that pesticides could harm wildlife and humans, and methodically built a case against the widespread use of pesticides.

In 1962, she published her findings in popular form in *Silent Spring*, an allusion to the silencing of "robins, catbirds, doves, jays, wrens, and scores of other bird voices" because of their exposure to pesticides. She pointed out that "for the first time in the history of the world, every human being is now subjected to dangerous chemicals, from the moment of conception until death."

Carson's book was read by many scientists, politicians, and policy makers and was embraced by the public. However, the chemical industry viewed the book as a serious threat to booming pesticide sales and mounted a $250,000 cam-paign to discredit Carson. A parade of critical reviewers and industry scientists claimed her book was full of inaccuracies, made selective use of research findings, and failed to give a balanced account of the benefits of pesticides.

Some critics even claimed that as a woman she was incapable of understanding the highly scientific and technical subject of pesticides. Others charged that she was a hysterical woman and a radical nature lover trying to scare the American public in order to sell books.

During this period of intense controversy Carson was suffering from terminal cancer, but she was able to defend her research and strongly counter her critics. She died in 1964, about 18 months after the publication of *Silent Spring*, without knowing that many historians consider her work an important contribution to the emerging modern environmental movement in the United States.

Rachel Carson (1907–64)

weed killers for lawns and gardens. Each year, more than 250,000 people in the United States become ill because of household pesticide use. Such pesticides are a major source of accidental poisonings and deaths for children under age 5.

Some pesticides, called *broad-spectrum agents,* are toxic to many species; others, called *selective* or *narrow-spectrum agents,* are effective against a narrowly defined group of organisms. Pesticides vary in their *persistence,* the length of time they remain deadly in the environment . In 1962, biologist Rachel Carson warned against relying on synthetic organic chemicals to kill insects and other species we deem pests (Individuals Matter, above).

What Is the Case for Pesticides? Proponents of conventional chemical pesticides contend that their benefits outweigh their harmful effects. Conventional pesticides have a number of important benefits.

First, they save human lives. Since 1945, DDT and other chlorinated hydrocarbon and organophosphate insecticides probably have prevented the premature deaths of at least 7 million people from insect-transmitted diseases such as malaria (carried by the *Anopheles* mosquito; Figure 10-13, p. 214), bubonic plague (rat fleas), typhus (body lice and fleas), and West Nile virus (mosquito).

Second, they increase food supplies and lower food costs. According to the UN Food and Agriculture

Organization, about 55% of the world's potential human food supply is lost to pests before (35%) or after (20%) harvest. Pests before and after harvest destroy an estimated 37% of the potential U.S. food supply; insects cause 13% of these losses, plant pathogens 12%, and weeds 12%. As a result, food production for humans and livestock worth at least $65 billion a year is lost to pests. Without pesticides, these losses would be worse, and food prices would rise.

Third, pesticides increase profits for farmers. Pesticide companies estimate that every $1 spent on pesticides leads to an increase in U.S. crop yields worth approximately $4 (but studies have shown this benefit drops to about $2 if the harmful effects of pesticides are included).

Fourth, chemical pesticides work faster and better than alternatives. Pesticides control most pests quickly at a reasonable cost, have a long shelf life, are easily shipped and applied, and are safe when handled properly. When genetic resistance occurs, farmers can use stronger doses or switch to other pesticides.

Fifth, when pesticides are used properly, their health risks are considered insignificant compared with their benefits. According to Elizabeth Whelan, director of the American Council on Science and Health (ACSH), which presents the position of the pesticide industry, "The reality is that pesticides, when used in the approved regulatory manner, pose no risk to either farm workers or consumers." According to the EPA the worst-case scenario is that synthetic pesticides in food cause 0.5–1% of all cancer-related deaths in the United States—or 3,000–6,000 premature deaths per year—far less than the estimated number of lives saved each year by pesticides.

Sixth, newer pesticides are safer and more effective than many older pesticides. Greater use is being made of botanicals and microbotanicals, derived originally from plants, which are safer to users and less damaging to the environment. Genetic engineering (Figure 16-12) is also being used to develop pest-resistant crop strains and genetically altered crops that produce pesticides.

Finally, many new pesticides are used at very low rates per unit area compared with older products. For example, application amounts per hectare for many new herbicides are 1/100 the rates for older ones, and genetically engineered crops could reduce the use of toxic insecticides.

Scientists continue to search for the ideal pest-killing chemical, which would have these qualities:

- *Kill only the target pest.*
- *Harm no other species.*
- *Disappear or break down into something harmless after doing its job.*

- *Avoid genetic resistance in target organisms.*
- *Be more cost effective than doing nothing.*

The search continues, but so far no known natural or synthetic pesticide chemical meets all or even most of these criteria.

What Is the Case against Pesticides? Opponents of widespread pesticide use believe their harmful effects outweigh their benefits. They cite several major problems with the use of conventional pesticides.

A major problem is that their widespread use *accelerates the development of genetic resistance to pesticides.* Insects breed rapidly, and within 5–10 years (much sooner in tropical areas) they can develop immunity to pesticides through directional natural selection (Figure 5-5, left, p. 97) and come back stronger than before. Weeds and plant disease organisms also develop genetic resistance, but more slowly. Since 1945, about 1,000 species of insects, mites, weed species, plant diseases, and rodents (mostly rats) have developed genetic resistance to one or more pesticides (Figure 16-23).

Because of genetic resistance, many insecticides (such as DDT) no longer protect people from insect-transmitted diseases in some parts of the world. This has led to the resurgence of such diseases as malaria (Figure 10-12, p. 214). Genetic resistance can also put farmers on a *pesticide treadmill,* whereby they pay more and more for a pest control program that often becomes less and less effective.

Another problem is that *broad-spectrum insecticides kill natural predators and parasites that help control the populations of pest species.* Wiping out natural predators

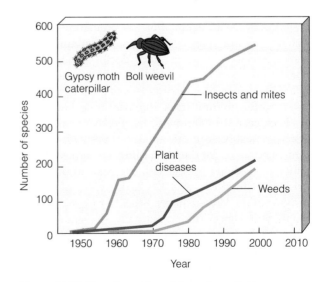

Figure 16-23 Rise of *genetic resistance to pesticides,* 1945–98. (Data from U.S. Department of Agriculture and the Worldwatch Institute)

can unleash new pests whose populations their predators had previously held in check and cause other unexpected effects (Connections, p. 170). Of the 300 most destructive insect pests in the United States currently, 100 were secondary pests that became major pests after widespread use of insecticides.

Also, *pesticides do not stay put.* According to the U.S. Department of Agriculture (USDA), no more than 2% (and often less than 0.1%) of the insecticide applied to crops by aerial or ground spraying reaches the target pests. Also, less than 5% of herbicide applied to crops reaches the target weeds.

Pesticides that miss their target pests can end up in the air, surface water, groundwater, bottom sediments, food, and nontarget organisms, including humans and wildlife (Figure 10-4, p. 205). Crops that have been genetically altered to release small amounts of pesticides directly to pests can help overcome this problem. But this can promote genetic resistance to such pesticides.

Some pesticides harm wildlife. According to the USDA and the U.S. Fish and Wildlife Service, each year pesticides applied to cropland in the United States wipe out about 20% of U.S. honeybee colonies and damage another 15%. This costs farmers at least $200 million per year from reduced pollination of vital crops. Pesticides also kill more than 67 million birds and 6–14 million fish, and menace about 20% of the endangered and threatened species in the United States.

Finally, *some pesticides can threaten human health.* The WHO and the UN Environment Programme (UNEP) estimate that pesticides seriously poison 3 million agricultural workers in developing countries (at least 300,000 in the United States) each year. This causes an estimated 18,000 deaths (about 25 in the United States)—an average of 490 premature deaths each day. Health officials believe the actual number of pesticide-related illnesses and deaths among the world's farm workers probably is greatly underestimated because of poor records, lack of doctors and disease reporting in rural areas, and inaccurate diagnoses.

Each year 110,000 Americans, mostly children, get sick from misuse or unsafe storage of pesticides in the home, and about 20 die. According to studies by the National Academy of Sciences, exposure to pesticide residues in food causes 4,000–20,000 cases of cancer per year in the United States. Because roughly 50% of people with cancer die prematurely, this amounts to about 2,000–10,000 premature deaths per year in the United States from exposure to legally allowed pesticide residues in foods. This is higher than the EPA estimate of 3,000–6,000 premature deaths per year.

Some scientists are becoming increasingly concerned about possible genetic mutations, birth defects, nervous system disorders (especially behavioral disorders), and effects on the immune and endocrine systems from long-term exposure to low levels of various pesticides (Case Study, p. 208).

Case Study: How Successful Have Pesticides Been in Reducing Crop Losses in the United States? Studies indicate that pesticides have not been as effective in reducing crop losses to pests in the United States as agricultural experts had hoped, mostly because of genetic resistance and reductions in natural predators.

David Pimentel, an expert in insect ecology, has evaluated data from more than 300 agricultural scientists and economists and come to three major conclusions.

First, although the use of synthetic pesticides has increased 33-fold since 1942, more of the U.S. food supply is lost to pests today (an estimated 37%) than in the 1940s (31%). Losses attributed to insects almost doubled (from 7% to 13%) despite a 10-fold increase in the use of synthetic insecticides.

Second, the estimated environmental, health, and social costs of pesticide use in the United States range from $4 billion to $10 billion per year. The International Food Policy Research Institute puts the estimate much higher, at $100–200 billion per year, or $5–10 in damages for every dollar spent on pesticides.

Third, alternative pest control practices (p. 403) could halve the use of chemical pesticides on 40 major U.S. crops without reducing crop yields.

Numerous studies and experience show that pesticide use can be reduced sharply without reducing yields, and in some cases yields increase. Sweden has cut pesticide use in half with almost no decrease in crop yields. Campbell Soup uses no pesticides on tomatoes it grows in Mexico, and yields have not dropped. After a 65% cut in pesticide use on rice in Indonesia, yields increased by 15%.

How Are Pesticides Regulated in the United States? The Federal Insecticide, Fungicide, and Rodenticide Act (FIFRA) was established by Congress in 1947 and amended in 1972. It requires EPA approval for use of all commercial pesticides. Pesticide companies must evaluate the biologically active ingredients in their products for toxicity to animals and, by extrapolation, to humans (p. 206).

EPA officials then review such data from pesticide companies to determine whether the pesticide can be registered for use. When a pesticide is approved for use on fruits or vegetables, the EPA sets a *tolerance level* specifying the amount of toxic pesticide residue that can legally remain on the crop when the consumer eats it.

What Goes Around Can Come Around

U.S. pesticide companies can make and export to other countries pesticides that have been banned or severely restricted—or never even approved—in the United States. Between 1997 and 2000, U.S. exports of such pesticides (most to developing countries) averaged more than 24 metric tons (26 tons) per day—or 1 metric ton (1.1 ton) per hour. Other industrial countries also export banned and unapproved pesticides.

But what goes around can come around. In what environmentalists call a *circle of poison*, residues of some of these banned or unap-

proved chemicals exported to other countries can return to the exporting countries on imported food. Persistent pesticides such as DDT can also be carried by winds from other countries to the United States.

Environmentalists have urged the U.S. Congress—without success—to ban such exports. Supporters of pesticide exports argue that such sales increase economic growth and provide jobs and that banned pesticides are exported only with the consent of the importing countries. They also contend that if the United States did not export pesticides, other countries would.

In 1998, more than 50 countries met to finalize an international

treaty that requires exporting countries to have informed consent from importing countries for exports of 22 pesticides and 5 industrial chemicals. In 2000, more than 100 countries developed an international agreement to ban or phase out the use of 12 especially hazardous persistent organic pollutants (POPs)—9 of them persistent chlorinated hydrocarbon pesticides (p. 372).

Critical Thinking

Should U.S. companies be allowed to export pesticides that have been banned, severely restricted, or not approved for use in the United States? Explain.

How Well Is the Public Protected from Exposure to Pesticides in the United States? Here are two pieces of *good news. First,* the EPA banned or severely restricted the use of 56 active pesticide ingredients between 1972 and 2002. The banned chemicals include most chlorinated hydrocarbon insecticides, several carbamates and organophosphates, and the systemic herbicide 2,4,5-T (Silvex).

Second, the 1996 Food Quality Protection Act (FQPA) increased public protection from pesticides. It requires manufacturers to demonstrate the safety of active ingredients in new pesticide products for infants and children. The EPA must also consider exposure to more than one pesticide when setting pesticide tolerance levels.

Environmentalists believe U.S. pesticide laws should be strengthened further to help prevent contamination of groundwater by pesticides, improve the safety of farm workers who are exposed to high levels of pesticides, and allow citizens to sue the EPA for not enforcing the law. Pesticide manufacturers strongly oppose such changes.

Here is some *bad news.* Banned or unregistered pesticides may be manufactured in the United States and exported to other countries (Connections, above).

According to scientific literature reviewed by the EPA, approximately 165 of the active ingredients approved for use in U.S. pesticide products are known or suspected human carcinogens. By 2002, use of only 41 of these chemicals had been banned by the EPA or discontinued voluntarily by manufacturers.

Here is more *bad news.* According to studies by the National Academy of Sciences, federal laws regulating pesticide use in the United States are inadequate and poorly enforced by the EPA, Food and Drug Administration (FDA), and USDA. Another study by the National Academy of Sciences found that up to 98% of the potential risk of developing cancer from pesticide residues on food grown in the United States would be eliminated if EPA standards were as strict for pre-1972 pesticides as they are for later ones.

Representatives from pesticide companies dispute these findings. Indeed, the food industry denies that eating food that has been grown using pesticides for the past 50 years has ever harmed anyone in the United States. The industry also claims that the benefits of pesticides (p. 400) far outweigh their disadvantages (p. 401).

Pesticide control laws in the United States could be improved. But most other countries (especially developing countries) have not made nearly as much progress as the United States in regulating pesticides.

16-9 PROTECTING FOOD RESOURCES: ALTERNATIVES TO CONVENTIONAL CHEMICAL PESTICIDES

What Are Other Ways to Control Pests? Many scientists believe we should greatly increase the use of biological, ecological, and other alternative methods for controlling pests and diseases that affect crops and human health. A number of methods are available.

One is the use of certain *cultivation practices.* Examples are rotating the types of crops planted in a field each year, adjusting planting times so major insect pests either starve or get eaten by their natural predators, and growing crops in areas where their major pests do not exist. Other strategies involve planting trap crops to lure pests away from the main crop and planting rows of hedges or trees around fields to hinder insect invasions and provide habitats for their natural enemies (with the added benefit of reduced soil erosion; Figure 9-27d, p. 198). Also, farmers can increase the use of polyculture (p. 377), which uses plant diversity to reduce losses to pests.

Genetic engineering (Figure 16-12) *can be used to speed up the development of pest- and disease-resistant crop strains* (Figure 16-24). However, there is controversy over whether the projected advantages of the increasing use of genetically modified plants and foods outweigh their projected disadvantages (Figure 16-13).

We can increase the use of *biological pest control.* It involves importing natural predators (Figure 16-25), parasites, and disease-causing bacteria and viruses to help regulate pest populations. This approach focuses on selected target species, is nontoxic to other species, and minimizes genetic resistance. Also, it can save large amounts of money—about $25 for every $1 invested in controlling 70 pests in the United States.

However, biological control has some disadvantages. Biological agents cannot always be mass produced. They are often slower acting and more difficult

Figure 16-25 Solutions: biological pest control. An adult convergent ladybug (right) is consuming an aphid (left).

to apply than conventional pesticides and the agents can sometimes multiply and become pests themselves. In addition, the biological agents must be protected from pesticides sprayed in nearby fields.

Another strategy is to use *insect birth control.* This involves raising males of insect pest species in the laboratory and sterilizing them by exposure to radiation or chemicals. Then the sterile males are released into an infested area to mate unsuccessfully with fertile wild females. This method has been used to control the screwworm fly, a major livestock pest from the southeastern United States (Figure 16-26), and the Mediterranean fruit fly (medfly) during a 1990 outbreak in California.

Problems include high costs, difficulties in knowing the mating time and behavior of each target insect, and the large number of sterile males needed. In addition, there are few species for which this strategy works, and sterile males must be released continually to prevent pest resurgence.

Sex attractants can also be used. These chemicals, called *pheromones,* can lure pests into traps or attract their natural predators into crop fields (usually the more effective approach). More than 50 companies worldwide sell about 250 pheromones to control pests.

Some advantages are that these chemicals attract only one species, work in trace amounts, have little chance of causing genetic resistance, and are not harmful to nontarget species. However, it is costly and time consuming to identify, isolate, and produce the specific sex attractant for each pest or predator.

Figure 16-24 Solutions: results of one example of using *genetic engineering* to reduce pest damage. Both tomato plants were exposed to destructive caterpillars. The normal plant's leaves are almost gone (left), whereas the genetically altered plant (right) shows little damage.

U. S. Department of Agriculture

Figure 16-26 Infestation of a steer by screwworm fly larvae in Texas. An adult steer can be killed in 10 days by thousands of maggots feeding on a single wound.

We can also use *hormones that disrupt an insect's normal life cycle* (Figure 16-27), *causing the insect to fail to reach maturity and reproduce.* Insect hormones have the same advantages as sex attractants. However, they take weeks to kill an insect, often are ineffective with large infestations of insects, and sometimes break down before they can act. They must be applied at exactly the right time in the target insect's life cycle and can sometimes affect the target's predators and other nonpest species. They are also difficult and costly to produce.

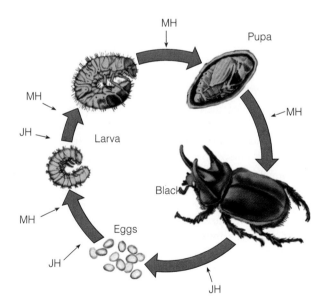

Figure 16-27 For normal insect growth, development, and reproduction to occur, certain juvenile hormones (JH) and molting hormones (MH) must be present at genetically determined stages in the insect's life cycle. If applied at the proper time, synthetic hormones disrupt the life cycles of insect pests and help control their populations.

Some farmers have been able to control some insect pests by *spraying them with hot water.* This has worked well on cotton, alfalfa, and potato fields and in citrus groves in Florida. The cost is roughly equal to that of using chemical pesticides.

Another strategy is to *expose foods to high-energy gamma radiation.* Such *food irradiation* extends food shelf life and kills insects and parasitic worms (such as trichinae in pork). It also kills harmful bacteria. Examples are *salmonella* (which infects at least 51,000 Americans and kills 2,000 each year) and *E. coli* (which infects more than 20,000 Americans and kills about 250 each year). According to the U.S. FDA and the WHO, more than 2,000 studies show that foods exposed to low doses of gamma radiation are safe for human consumption.

Critics of irradiating food argue that it forms trace amounts of certain chemicals which have caused cancer in laboratory animals. They also point out that the long-term health effects of eating irradiated food are unknown and consumers do not want old and possibly less nutritious food made to appear fresh and healthy by irradiation. They believe that all irradiated food should be labeled so consumers can make informed choices. Another problem is that poorly protected facilities for food irradiation contain radioactive isotopes that terrorists could steal and use to make dirty nuclear weapons.

Is Integrated Pest Management the Answer?
An increasing number of pest control experts and farmers believe the best way to control crop pests is a carefully designed **integrated pest management (IPM)** program. In this approach, each crop and its pests are evaluated as parts of an ecological system. Then a control program is developed that includes cultivation and biological and chemical methods applied in proper sequence and with the proper timing.

The overall aim of IPM is not to eradicate pest populations but to reduce crop damage to an economically tolerable level. Fields are monitored carefully to determine when an economically damaging level of pests has been reached. When this happens farmers first use biological methods (natural predators, parasites, and disease organisms) and cultivation controls, including vacuuming up harmful bugs. Small amounts of insecticides (mostly based on natural insecticides produced by plants) are applied only as a last resort. Also, different chemicals are used to slow development of genetic resistance and to avoid killing predators of pest species.

In 1986, the Indonesian government banned the use of 57 of the 66 pesticides used on rice and phased out pesticide subsidies over a 2-year period. It also launched a nationwide education program to help farmers switch to IPM. The results were dramatic:

Between 1987 and 1992, pesticide use dropped by 65%, rice production rose by 15%, and more than 250,000 farmers were trained in IPM techniques. Norway and Sweden have imposed substantial taxes on conventional pesticides, with the goal of cutting current pesticide use by 25–50%.

The experiences of various countries show that a well-designed IPM program can reduce pesticide use and pest control costs by 50–90%, decrease preharvest pest-induced crop losses by 50%, and improve crop yields. It can also reduce inputs of fertilizer and irrigation water and slow the development of genetic resistance because pests are assaulted less often and with lower doses of pesticides.

Thus IPM is an important form of *pollution prevention* that reduces risks to wildlife and human health. Consumers Union estimates that if all U.S. farmers practiced IPM by 2020, public health risks from pesticides would drop by 75%.

Despite its promise, IPM, like any other form of pest control, has some disadvantages. It requires expert knowledge about each pest situation and is slower acting than conventional pesticides. In addition, methods developed for a crop in one area might not apply to areas with even slightly different growing conditions. Also, initial costs may be higher, although long-term costs typically are lower than those of using conventional pesticides.

Widespread use of IPM is hindered by government subsidies of conventional chemical pesticides and opposition from agricultural chemical companies, whose pesticide sales would drop sharply. In addition, farmers get most of their information about pest control from pesticide salespeople (and in the United States from USDA county farm agents). Most of these advisers do not have adequate training in IPM.

A 1996 study by the National Academy of Sciences recommended that the United States shift from chemically based approaches to ecologically based pest management approaches. A growing number of scientists urge the USDA to promote IPM in the United States. Ways to do this include adding a 2% sales tax on pesticides and using the revenue to fund IPM research and education. We could also set up a federally supported IPM demonstration project on at least one farm in every county, and train USDA field personnel and county farm agents in IPM so they can help farmers use this alternative.

Farmers who use IPM or other approved alternatives to pesticides could receive federal and state subsidies, and perhaps government-backed crop insurance. As effective IPM methods are developed for major pest species, the government could gradually phase out subsidies to farmers who depend almost entirely on conventional chemical pesticides.

16-10 SOLUTIONS: MORE SUSTAINABLE AGRICULTURE

What Is More Sustainable Agriculture? As we have seen in this chapter, two major ways to increase the world's food production are to increase crop yields per hectare and to increase the amount of land used to grow crops.

The total area of cropland is unlikely to expand because of a lack of affordable and environmentally sustainable land. In addition, increasing the yields per area of existing cropland may be limited in some areas because of a lack of water for irrigation, reduced genetic diversity needed to develop new crop strains, slowing of increases in yields per hectare, and the environmental effects of food production, which degrade existing cropland (Figure 16-11).

If these projections are correct, there are three main ways to reduce hunger and malnutrition and the harmful environmental effects of agriculture. *One* is to slow population growth (p. 238). *Another* is to reduce poverty (p. 28) so people can grow or buy enough food for their survival and good health.

The third is to develop and phase in systems of more **sustainable agriculture** or **low-input agriculture** (also called **organic farming**) over the next three decades. Currently, organic farming is used on less than 1% of the world's cropland (0.2% in the United States) but on 6–10% of the cropland in many European countries. However, this type of farming is growing rapidly. In 2002, it was a $25 billion global market and a $12 billion market in the United States.

The U.S. Department of Agriculture has established several rules for organic food production. *First,* crops are grown on land that has not been fertilized with sewage sludge or chemical fertilizers. Soil health is managed primarily through use of organic fertilizers, crop rotation, and planting cover crops.

Second, organic farmers do not use chemical insecticides. They treat pests and plant diseases primarily with natural insect predators, traps, and insect repellents. They also do not use chemical herbicides. Instead, weeds are controlled mostly by mulching and hand weeding or mechanical cultivation.

Third, genetically engineered crops and livestock animals are not used. *Fourth,* animals used to provide meat and eggs must be raised on pure organic feed and receive no growth hormones or antibiotics. They must also be raised under living conditions that allow for exercise, freedom of movement, and reduction of stress. *Finally,* organic food cannot be irradiated.

Figure 16-28 lists the major components of more sustainable, low-input agriculture. In 2002, agricultural scientists Paul Mader and David Dubois reported the results of a 21-year study comparing organic and

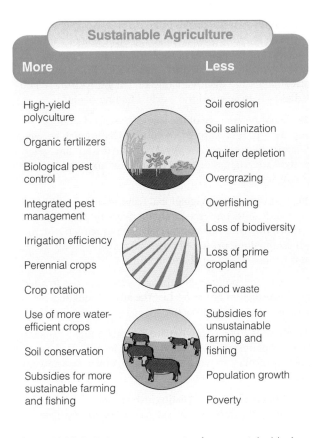

Sustainable Agriculture

More	Less
High-yield polyculture	Soil erosion
Organic fertilizers	Soil salinization
Biological pest control	Aquifer depletion
Integrated pest management	Overgrazing
Irrigation efficiency	Overfishing
Perennial crops	Loss of biodiversity
Crop rotation	Loss of prime cropland
Use of more water-efficient crops	Food waste
Soil conservation	Subsidies for unsustainable farming and fishing
Subsidies for more sustainable farming and fishing	Population growth
	Poverty

Figure 16-28 Solutions: components of more sustainable, low-throughput agriculture.

conventional farming. Their results and those from other studies have shown that for most crops low-input organic farming has a number of advantages over conventional high-input farming. They include use of up to 56% less energy per unit of yield, and improved soil health and fertility. Organic farming also provides more habitats for wild plant and animal species and generally is more profitable for the farmer than high-input farming.

Most proponents of more sustainable agriculture are not opposed to high-yield agriculture. Instead, they see it as vital for protecting the earth's biodiversity by reducing the need to cultivate new and often marginal land. They call for using environmentally sustainable forms of both high-yield polyculture and high-yield monoculture for growing crops.

How Can We Make the Transition to More Sustainable Agriculture? A growing number of agricultural analysts say that over the next 30 years we need to make a transition from unsustainable and environmentally harmful agriculture (Figure 16-11) to more sustainable forms of agriculture (Figure 16-28).

Analysts suggest four major strategies to help farmers make the transition to more sustainable agri-

culture. *First,* greatly increase research on sustainable agriculture and improving human nutrition. *Second,* set up demonstration projects throughout each country so farmers can see how sustainable agricultural systems work. *Third,* increase agricultural aid to developing countries, with emphasis on developing more sustainable, low-input agriculture. *Fourth,* establish training programs in sustainable agriculture for farmers and government agricultural officials and encourage the creation of college curricula in sustainable agriculture and human nutrition.

Phasing in more sustainable agriculture involves applying the four principles of sustainability (Figure 8-14, p. 171) to producing food. The goal is to feed the world's people while sustaining and restoring the earth's natural capital (Figure 1-2, p. 3, and top half of back cover). See the website material for this chapter for some actions you can take to help promote more sustainable agriculture.

We need to build an agriculture as sustainable as the nature we destroy.

WES JACKSON

REVIEW QUESTIONS

1. Define the boldfaced terms in this chapter.

2. What are *perennial crops*? What advantages do they have over conventional annual crops?

3. What three systems supply most of our food? What three crops provide most of the world's food?

4. Distinguish among *industrialized agriculture, plantation agriculture, traditional subsistence agriculture,* and *traditional intensive agriculture.*

5. What is a *green revolution,* and what three steps does it involve? Distinguish between the first and second green revolutions.

6. Explain how producing more food on less land can help protect biodiversity.

7. Describe the nature and importance of the agricultural industry in the United States. List three pieces of good news about the U.S. agricultural industry. How energy efficient is the entire industrialized agriculture system in the United States?

8. Distinguish among *interplanting, polyvarietal cultivation, intercropping, agroforestry,* and *polyculture.* List six advantages of low-input polyculture.

9. List three pieces of *good* news about world food production since 1961.

10. Distinguish between *undernutrition* and *malnutrition.* List two pieces of good news and three pieces of bad news about global undernutrition and malnutrition. About how many chronically undernourished and malnourished people are there in the world? About how many people die prematurely each year from undernutrition, malnutrition, or diseases worsened by malnutrition?

11. List six major ways to reduce sickness and premature death of children from malnutrition and disease.

12. What are the effects of deficiencies of vitamin A, iron, and iodine? How serious are these nutrition problems?

13. What is *overnutrition* and how serious is this problem?

14. What is the primary cause of hunger in the world?

15. List the major harmful environmental effects of producing food.

16. Describe how genetically improved crop strains are developed by *genetic engineering.* List the advantages and disadvantages of growing more food by genetically modifying crops and foods.

17. List some good news about the possibility of increasing crop yields through improved green revolutions. List six factors that could limit greatly increased food production through green revolution and genetic engineering techniques.

18. Describe two types of new foods. What are two problems in making a switch to new types of foods?

19. What percentage of the world's food is produced on irrigated cropland? What is the good news and bad news about irrigation since 1950? List six factors that have contributed to the decrease in irrigated land per person since 1978.

20. What is the potential for cultivating more land for growing crops in tropical forests and arid areas? What are the problems with doing this?

21. How could we grow more food in urban areas?

22. Distinguish between *rangeland* and *pasture.* Explain how rangeland grass can be a renewable resource for livestock and wild herbivores.

23. How is meat produced? List four problems with using the industrialized approach to increase meat productivity.

24. Distinguish between *overgrazing* and *undergrazing.* What are three major effects of overgrazing? Describe the general condition of rangelands throughout the world and in the United States. What are *riparian areas,* and why are they important? Describe three ways to manage rangelands more sustainably for meat production.

25. What are *fisheries?* What is *aquaculture?* What percentage of the global food supply comes from fish and shellfish? What percentage of the world's fish and shellfish comes from the ocean and what percentage comes from aquaculture? Distinguish among *trawling, purse-seine, longlining,* and *drift-net* methods for harvesting fish.

26. List the good news and bad news about trends in the world's total fish catch and per capita fish catch since 1950.

27. What is the *sustainable yield* of a fishery? Distinguish between *overfishing* and *commercial extinction* of a fish species. List two pieces of *good news* and four pieces of *bad news* about the world's fish stocks. What are the advantages and disadvantages of using government subsidies to help support the world's fishing fleets?

28. What percentage of the world's fish and shellfish supply is produced by aquaculture? Distinguish between *fish farming* and *fish ranching.* What are the advantages and disadvantages of aquaculture? List six ways to improve aquaculture.

29. List ways to manage global fisheries more sustainably and protect marine biodiversity by **(a)** fishery regulations, **(b)** economic approaches, **(c)** reducing bycatch, **(d)** establishing protected areas, **(e)** consumer education, and **(f)** increased use of aquaculture.

30. List three policy types that governments use to affect food production or prices. List the advantages and disadvantages of each approach.

31. What is a *pest* and what are typical biological characteristics of pest species? How are species we consider pests controlled in nature? What happens to this natural pest control service when we simplify ecosystems?

32. What is a *pesticide?* Distinguish among *insecticides, herbicides, fungicides, nematocides,* and *rodenticides.* Distinguish between *first-generation* and *second-generation* chemical pesticides and give an example of each. Distinguish between *broad-spectrum* and *narrow-spectrum* pesticides. Describe the increased use of synthetic chemical pesticides in the world and in the United States since 1950.

33. Who was Rachel Carson? Describe her efforts to warn the public of possible dangers from overuse of synthetic chemical pesticides.

34. List seven benefits of relying on synthetic chemical pesticides. List five traits of the ideal pest-killing chemical.

35. Summarize the harmful effects of synthetic chemical pesticides in terms of **(a)** development of genetic resistance among pests, **(b)** killing of natural predators and parasites that help control pest species, **(c)** creation of new pest species, **(d)** migration of pesticides into the environment, **(e)** effects on wildlife, and **(f)** effects on the health of pesticide and farm workers and the general public.

36. According to studies, how successful have pesticides been in reducing crop losses in the United States?

37. With respect to pesticides, what is the circle of poison? How might it affect you?

38. How are pesticides regulated in the United States? List two pieces of good news and six pieces of bad news about the effectiveness of pesticide control laws in the United States. How have pesticide companies reacted to the bad news about pesticides?

39. List six cultivation practices that can be used to help control pests. List the advantages and disadvantages of using the following pest control methods: **(a)** genetic engineering to speed up the development of pest- and disease-resistant crop strains, **(b)** biological control, **(c)** insect birth control through sterilization, **(d)** insect sex attractants, **(e)** insect hormones (pheromones), **(f)** food irradiation, and **(g)** integrated pest management (IPM). List five ways to increase the use of integrated pest management.

40. What is *sustainable* or *low-input agriculture,* and what are its major components? How is organic food produced? List four advantages of organic farming over conventional crop production. List four ways to help farmers make the transition to more sustainable agriculture.

CRITICAL THINKING

1. Summarize the major economic and ecological advantages and limitations of each of the following proposals for increasing world food supplies and reducing hunger over the next 30 years: **(a)** cultivating more land by clearing tropical forests and irrigating arid lands, **(b)** catching more fish in the open sea, **(c)** producing more fish and shellfish with aquaculture, and **(d)** increasing the yield per area of cropland.

2. Why should it matter to people in developed countries that many people in developing countries are malnourished and hungry? What are the three most important actions you believe should be taken to reduce hunger and malnutrition **(a)** in the country where you live and **(b)** in the world?

3. Some people argue that starving people could get enough food by eating nonconventional plants and insects; others point out that most starving people do not know what plants and insects are safe to eat and cannot take a chance on experimenting when even the slightest illness could kill them. If you had no money to grow or buy food, would you collect and eat protein-rich grasshoppers, moths, beetles, or other insects?

4. What could happen to energy-intensive agriculture in the United States and other industrialized countries if world oil prices rose sharply? How might this affect your life?

5. Some have suggested that some rangelands could be used to raise wild grazing animals for meat instead of conventional livestock. Others consider it unethical to raise and kill wild herbivores for food. What do you think? Explain.

6. Should **(a)** all price supports and other government subsidies paid to farmers be eliminated, and **(b)** governments phase in agricultural tax breaks and subsidies to encourage farmers to switch to more sustainable farming? Explain your answers. Try to consult one or more farmers in answering these questions.

7. What are the economic and ecological advantages and disadvantages of relying more on **(a)** a small number of genetic varieties of major crops and livestock, **(b)** genetically modified food, and **(c)** perennial food crops?

Explain why you support or oppose each of these approaches.

8. Do you believe government subsidies for the fishing industry should be eliminated or sharply reduced? Explain. How would you feel about eliminating such subsidies if your livelihood depended on fishing?

9. Suppose you live near a coastal area and a company wants to use a fairly large area of coastal marshland for an aquaculture operation. If you were an elected local official, would you support or oppose such a project? Explain. What safeguards or regulations would you impose on the operation?

10. Overall, do you think the benefits of synthetic chemical pesticides outweigh their disadvantages? Explain your position.

11. If increased mosquito populations threatened you with malaria or West Nile fever, would you spray DDT in your yard and inside your home to reduce the risk? Explain. What are the alternatives?

12. Congratulations! You are in charge of the world. List the three most important features of your **(a)** agricultural policy and **(b)** global pest control strategy.

PROJECTS

1. If possible, visit both a conventional industrialized farm and an organic or low-input farm. Compare **(a)** soil erosion and other forms of land degradation, **(b)** use and costs of energy, **(c)** use and costs of pesticides and inorganic fertilizer, **(d)** use and costs of natural pest control and organic fertilizer, **(e)** yields per hectare for the same crops, and **(f)** overall profit per hectare for the same crops.

2. Evaluate cattle grazing on private and public rangeland and pastures in your local area. Try to document any harmful environmental impacts. Have the economic benefits to the community outweighed any harmful environmental effects?

3. Try to gather data evaluating the harmful environmental effects of nearby agriculture on your local community. What is being done to reduce these effects?

4. Use health and other local government records to estimate how many people in your community suffer from undernutrition or malnutrition. Has this problem increased or decreased since 1980? What are the basic causes of this hunger problem, and what is being done to alleviate it? Share the results of your study with local officials, and present your own plan for improving efforts to reduce hunger in your community.

5. How are bugs and weeds controlled in **(a)** your yard and garden, **(b)** the grounds of your school, and **(c)** public school grounds, parks, and playgrounds in your community?

6. Make a survey of all pesticides used in or around your home. Compare the results for your entire class.

7. Make a concept map of this chapter's major ideas, using the section heads and subheads and the key terms (in boldface type). Look on the website for this book for information about making concept maps.

INTERNET STUDY RESOURCES AND RESOURCES FOR FURTHER READING AND RESEARCH

www.

The website for this book contains helpful study aids and many ideas for further reading and research. Log on to

http://biology.brookscole.com/miller 10

and click on the Chapter-by-Chapter area. Choose Chapter 16 and select a resource:

- Flash Cards allows you to test your mastery of the Terms and Concepts to Remember for this chapter.

- Tutorial Quizzes provides a multiple-choice practice quiz.

- Student Guide to InfoTrac will lead you to Critical Thinking Projects that use InfoTrac College Edition as a research tool.

- References lists the major books and articles consulted in writing this chapter.

- Hypercontents takes you to an extensive list of sites with news, research, and images related to individual sections of the chapter.

INFOTRAC COLLEGE EDITION

Improve your skills with InfoTrac College Edition, a searchable online database of articles from more than 700 periodicals. Log on to

http://www.infotrac-college.com

or access InfoTrac through the website for this book. Try to find the following articles:

1. Snell, M. B. 2001. Against the grain. *Sierra* 86: 30. *Keywords:* "transgenic crops" and "developing countries." Genetically engineered food crops may seem like a boon to the hungry populations in developing countries, but many of these countries are concerned about the overall environmental effects of these foods.

2. Davis, S. 2002. Insect communication discovery could lead to new pesticides. (Brief article) *Pesticide and Toxic Chemical News* 30: 9. *Keywords:* "insect communication" and "pesticides." A new generation of pheromone-blocking pesticides may mean more and environmentally safer target-specific insecticides.

17 SUSTAINING BIODIVERSITY: THE ECOSYSTEM APPROACH

Biodiversity

Who's Afraid of the Big Gray Wolf?

At one time, the gray wolf (Figure 17-1) ranged over most of North America. This changed between 1850 and 1900 when an estimated 2 million wolves were shot, trapped, and poisoned by ranchers, hunters, and government employees. The idea was to make the West and the Great Plains safe for livestock and for big game animals prized by hunters.

It worked. When Congress passed the U.S. Endangered Species Act in 1973, only about 400–500 gray wolves remained in the lower 48 states, primarily in Minnesota and Michigan. In 1974, the U.S. Fish and Wildlife Service (USFWS) listed the gray wolf as endangered in all 48 lower states except Minnesota. Alaska was not included because it had 6,000–8,000 gray wolves.

With such protection, the gray wolf population increased to about 3,000 in eight states (Arizona, Idaho, Michigan, Minnesota, Montana, New Mexico, Wisconsin, and Wyoming). In 2000, this recovery prompted the USFWS to propose moving the gray wolf from the endangered to the threatened category in these states.

Ecologists recognize the important role this keystone predator species once played in parts of the West and the Great Plains. These wolves culled herds of bison, elk, caribou, and mule deer, and kept down coyote populations. They also provided uneaten meat for scavengers such as ravens, bald eagles, ermines, and foxes.

In recent years, herds of elk, moose, and antelope have expanded. Their larger numbers have devastated some of the area's vegetation, increased erosion, and threatened the niches of other wildlife species. Reintroducing a keystone species such as the gray wolf into a terrestrial ecosystem is one way to help sustain its biodiversity and prevent environmental degradation.

In 1987, the USFWS proposed reintroducing gray wolves into the Yellowstone ecosystem. This idea brought angry protests. Some objections came from ranchers who feared

the wolves would attack their cattle and sheep; one enraged rancher said that the idea was "like reintroducing smallpox." Other protests came from hunters who feared the wolves would kill too many big game animals, and from mining and logging companies who worried the government would halt their operations on wolf-populated federal lands.

Since 1995, federal wildlife officials have caught gray wolves in Canada and relocated them in Yellowstone National Park and northern Idaho. By 2002, the population of these wolves had grown by natural reproduction to around 270. These wolves plus others that have migrated from Canada mean that about 700 gray wolves now roam the northern Rockies.

In 2002, the U.S. Fish and Wildlife Service declared that biologically the gray wolf has recovered in Wyoming, Montana, and Idaho and proposed changing the status of this species from endangered to threatened in these states. This means that the wolves can be hunted or trapped. The new status also means that ranchers can kill wolves that attack their livestock.

Environmentalists contend that eliminating federal protection of the gray wolf is premature. But Fish and Wildlife officials say their job under the Endangered Species Act is to protect the gray wolf from extinction and that this has been achieved.

Population growth, economic development, and poverty are exerting increasing pressure on the world's forests, grasslands, parks, wilderness, oceans, rivers, and other storehouses of biodiversity. Biodiversity researchers and protectors urge us to use these renewable natural resources more sustainably, as discussed in this chapter.

Figure 17-1 The *gray wolf* is a threatened species in the lower 48 states. Ranchers, hunters, miners, and loggers have vigorously opposed efforts to return this keystone species to its former habitat in the Yellowstone National Park area. However, wolves were reintroduced beginning in 1995 and now number around 270.

This chapter addresses the following questions:

- How have human activities affected the earth's biodiversity?

- How is land used in the world and in the United States?

- What are the major types of public lands in the United States, and how are they used?

- Why are forest resources important, and how are they used and managed in the world and in the United States?

- Is tropical deforestation a serious problem? How can we help sustain tropical forests?

- What problems do parks face, and how should we manage them?

- How should we establish, design, protect, and manage terrestrial nature reserves?

- How can we protect and sustain aquatic systems?

- What is ecological restoration, and why is it important?

17-1 HUMAN IMPACTS ON BIODIVERSITY

How Have Human Activities Affected Global Biodiversity? Figure 17-2 lists factors that tend to increase or decrease biodiversity. The *bad news* is that many human activities decrease biodiversity (Figure 17-3). According to biodiversity expert Edward O. Wilson, "The natural world is everywhere disappearing before our eyes—cut to pieces, mowed down, plowed under, gobbled up, replaced by human artifacts."

Here are examples of how human activities have decreased and degraded the earth's biodiversity. In terms of *land use,* humans have taken over, disturbed, or degraded 40–50% of the earth's land surface (Figure 8-12, p. 169), especially by filling in wetlands and converting grasslands and forests to crop fields and urban areas. Also, logging and land conversion have reduced global forest cover by at least 20%, and possibly as much as 50%.

In the United States, a 1999 study by the U.S. Geological Survey found that 95–98% of the virgin forests in the lower 48 states has been destroyed since 1620 and 90% of Hawaii's dry forests has disappeared. In addition, 98% of tallgrass prairie in the Midwest and Great Plains has disappeared, and 99%

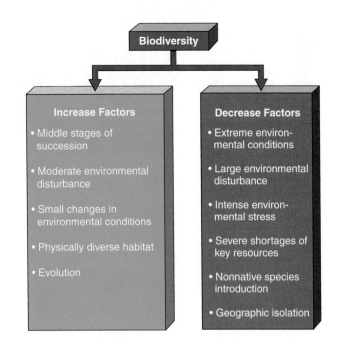

Figure 17-2 Factors that tend to increase or decrease the earth's biodiversity.

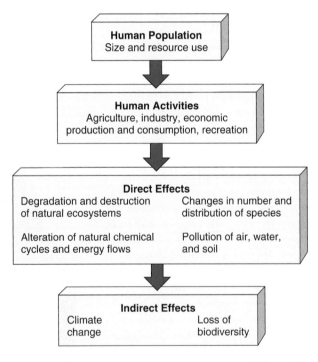

Figure 17-3 Major connections between human activities and the earth's biodiversity.

of California's native grassland and 85% of its redwood forests are gone.

We are also affecting the *net primary productivity* that supports all life. Humans use, waste, or destroy about 27% of the earth's total potential net primary productivity and 40% of the net primary productivity of the planet's terrestrial ecosystems (Figure 4-22, p. 80)

Human activities are also degrading the world's *aquatic biodiversity*. About half of the world's wetlands and half of the wetlands in the United States were lost during the last century. According to a 2002 study by the U.S. Commission on Ocean Policy, each year the United States loses about 162,000 hectares (400,000 acres) of coastal wetlands that provide essential spawning, feeding, and nursery areas for three-fourths of U.S. commercial fish catches.

An estimated 27% of the world's diverse coral reefs (Figure 6-26, p. 125, and Figure 6-35, p. 132) have been severely damaged. By 2050 another 70% may be severely damaged or eliminated.

About 75% of the world's 200 commercially valuable marine fish species are either overfished or fished to their estimated sustainable yield. According to a 2002 preliminary report by the U.S. Commission on Ocean Policy, about 40% of U.S. commercial fish stocks are depleted or overfished.

Human activities also contribute to the *premature extinction of species*. Biologists estimate that the current global extinction rate of species is at least 100 times and probably 1,000 to 10,000 times what it was before humans existed. The rate of premature extinction is expected to increase during this century as more people use more of the world's resources.

A 2000 joint study by the World Conservation Union and Conservation International and a 1999 study by the World Wildlife Fund found that human activities threaten a significant percentage of various types of species with premature extinction. This includes 34% of the world's fish species (51% of freshwater species), 24% of mammals, 20% of reptiles, 14% of plants, and 11% of birds.

A 2000 survey by the Nature Conservancy and the Association for Biodiversity Information found that about 33% of 21,000 animal and plant species in the United States are vulnerable to premature extinction.

These threats to the world's biodiversity are projected to increase sharply by 2018 (Figure 17-4).

How Can We Reduce Biodiversity Loss? Some *good news* is that we know how to slow down this biological impoverishment. Figure 17-5 (p. 414) outlines the goals, strategies, and tactics for preserving and restoring the ecosystems and aquatic systems that provide habitats and resources for the world's species (as discussed in this chapter) and preventing the premature extinction of species (as discussed in Chapter 18).

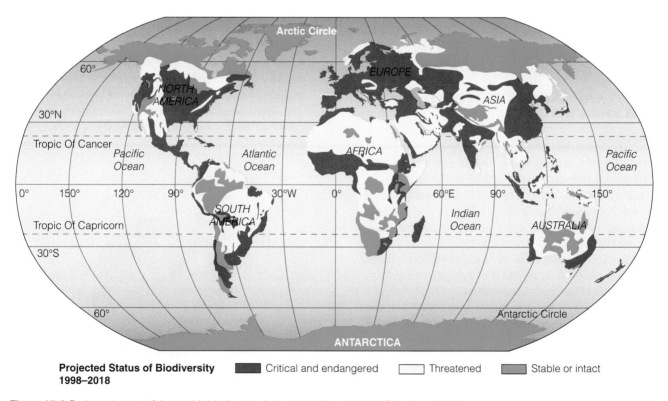

Projected Status of Biodiversity 1998–2018 ■ Critical and endangered □ Threatened ■ Stable or intact

Figure 17-4 Projected status of the earth's biodiversity between 1998 and 2018. (Data from World Resources Institute, World Conservation Monitoring Center, and Conservation International)

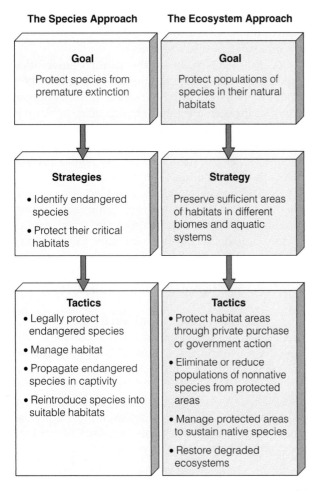

The Species Approach

Goal

Protect species from premature extinction

↓

Strategies

- Identify endangered species
- Protect their critical habitats

↓

Tactics

- Legally protect endangered species
- Manage habitat
- Propagate endangered species in captivity
- Reintroduce species into suitable habitats

The Ecosystem Approach

Goal

Protect populations of species in their natural habitats

↓

Strategy

Preserve sufficient areas of habitats in different biomes and aquatic systems

↓

Tactics

- Protect habitat areas through private purchase or government action
- Eliminate or reduce populations of nonnative species from protected areas
- Manage protected areas to sustain native species
- Restore degraded ecosystems

Figure 17-5 Solutions: goals, strategies, and tactics for protecting biodiversity.

17-2 LAND USE IN THE WORLD AND THE UNITED STATES

How Is Land Used? Figure 17-6 shows the major categories of land use in the world and Figure 17-7 shows land use and ownership in the United States. Human activities have disturbed about two-thirds of the earth's inhabitable land area (Figure 8-12, p. 169) and a significant portion of the land in the United States (Figure 17-8).

No nation has set aside as much of its land—about 42%—for public use, resource extraction, enjoyment, and wildlife as has the United States (Figure 17-7, right). The federal government manages roughly 35% of the country's land that belongs to every American. About 73% of this federal public land is in Alaska. Another 22% is in the western states (where 60% of all land is public land).

What Are the Major Types of U.S. Public Lands? Federal public lands are classified as *multiple-use lands, moderately restricted-use lands,* and *restricted-use lands.*

Multiple-Use Lands

- The *National Forest System,* managed by the U.S. Forest Service, consists of 155 forests (Figure 17-9, p. 416) and 22 grasslands. These forests are used for logging, mining, livestock grazing, farming, oil and gas extraction, recreation, sport hunting, sport and commercial fishing, and conservation of watershed, soil, and wildlife resources. Off-road vehicles usually are restricted to designated routes.

- The Bureau of Land Management (BLM) manages the *National Resource Lands* in the western states and Alaska. It emphasizes providing a secure domestic supply of energy and strategic minerals and preserving rangelands for livestock grazing under a permit system.

Moderately Restricted-Use Lands

- The USFWS manages the 524 *National Wildlife Refuges* (Figure 17-9). Most refuges protect habitats and breeding areas for waterfowl and big game to provide a harvestable supply for hunters; a few protect endangered species from extinction. Permitted activities include sport hunting, trapping, sport and commercial fishing, oil and gas development, mining, logging, grazing, some military activities, and farming—as long as the Department of the Interior finds such uses compatible with the purposes of each unit.

Restricted-Use Lands

- The *National Park System* consists of 385 units managed by the National Park Service (NPS). These units include 55 major parks (mostly in the West; Figure 17-9) and 324 national recreation areas, monuments, memorials, battlefields, historic sites, parkways, trails, rivers,

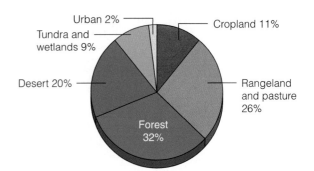

Figure 17-6 How the world's land is used. Excluding uninhabitable areas of rock, ice, desert, and steep mountain terrain, only about 27% of the planet's land area remains undisturbed by human activities (Figure 8-12, p. 169). (Data from UN Food and Agriculture Organization)

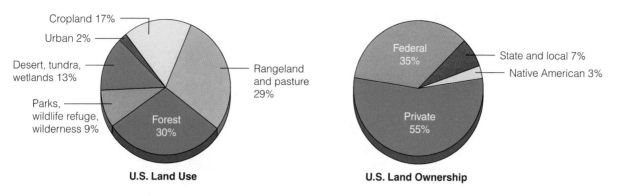

Figure 17-7 How land is used (left) and owned (right) in the United States. (Data from U.S. Department of Agriculture)

seashores, and lakeshores. Only camping, hiking, sport fishing, and boating can take place in the national parks. Motor vehicles must stay on park roads, although some parks permit use of off-road vehicles. In national recreation areas, the NPS permits these same activities, plus sport hunting, mining, and oil and gas drilling.

- The 630 roadless areas of the *National Wilderness Preservation System* lie within the national parks, na-

tional wildlife refuges, and national forests. Management of these areas is the responsibility of the NPS (42%), Forest Service (33%), USFWS (20%), and BLM (5%). These areas are open only for recreational activities such as hiking, sport fishing, camping, nonmotorized boating, and, in some areas, sport hunting and horseback riding. Roads, motorized vehicles, logging, livestock grazing, mining, commercial activities, and buildings are banned, except when they predate the wilderness designation.

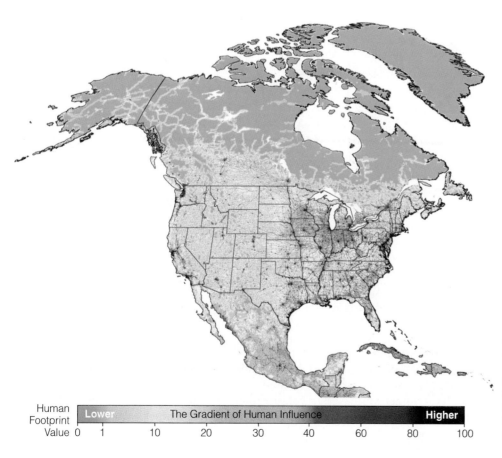

Figure 17-8 The human ecological footprint in North America. Colors represent the percentage of each area influenced by human activities. This is a portion of the diagram in Figure 8-12 (p. 169) showing the human footprint on all of the earth's land surface. [Wildlife Conservation Society in collaboration with the Center for International Earth Science Information Network (CIESIN)]

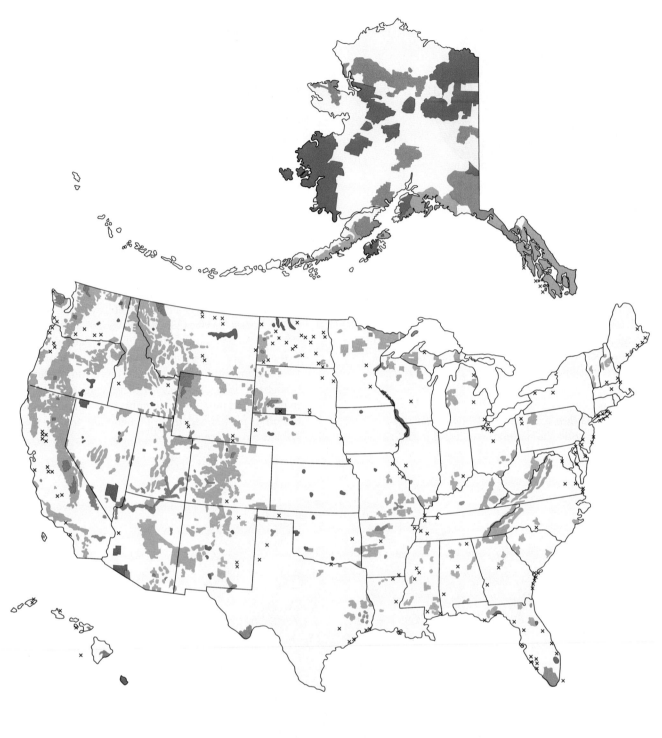

■ National parks and preserves   ▢ National forests   ■ (and Xs) National wildlife refuges

Figure 17-9 National forests, national parks, and wildlife refuges managed by the U.S. federal government. U.S. citizens jointly own these and other public lands. (Data from U.S. Geological Survey)

How Should U.S. Public Lands Be Managed?
Federal public lands contain an estimated 20% of the country's oil reserves, 30% of its natural gas reserves, 40% of its coal reserves, 40% of its commercial forests, and large amounts of hard-rock minerals.

Since the 1800s there has been controversy over how public lands should be used and managed, mostly because of the valuable resources they contain. Most conservation biologists and environmental economists and many free-market economists believe the following four principles should govern use of public land:

- Protecting biodiversity, wildlife habitats, and the ecological functioning of public land ecosystems should be the primary goal.

- No one should receive government subsidies or tax breaks for using or extracting resources on public lands.

- The American people deserve fair compensation for extraction of any resources from their property.

- All users or extractors of resources on public lands should be responsible for any environmental damage they cause.

Aldo Leopold's land-use ethic (Individuals Matter, p. 39) is the basis for most of these guiding principles.

There is strong and effective opposition to these ideas. Economists, developers, and resource extractors tend to view public lands in terms of their usefulness in providing mineral, timber, and other resources and their ability to increase short-term economic growth.

They have succeeded in blocking implementation of the four principles just listed. For example, in recent years, the government has given more than $1 billion a year in subsidies to privately owned mining, logging, and grazing interests using U.S. public lands.

Some developers and resource extractors go further and have mounted a campaign to get the U.S. Congress to pass laws that would

- Sell public lands or their resources to corporations or individuals, at less than their fair market value.

- Slash federal funding for regulatory administration of public lands.

- Cut all old-growth forests in the national forests and replace them with tree plantations.

- Open all national parks, national wildlife refuges, and wilderness areas to oil drilling, mining, off-road vehicles, and commercial development.

- Do away with the National Park Service and launch a 20-year construction program of new concessions and theme parks run by private firms in the former national parks.

- Continue mining on public lands under the provisions of the 1872 Mining Law, which allows mining interests to pay no royalties to taxpayers for hard-rock minerals they remove and not be responsible for environmental damage they cause.

- Repeal the Endangered Species Act or modify it to allow economic factors to override protection of endangered and threatened species.

- Redefine government-protected wetlands so about half of them would no longer be protected.

- Prevent individuals or groups from legally challenging uses of public land for private financial gain.

Since 2001 the Bush administration and the U.S. Congress have opened large tracts of public land (mostly in the West and Alaska) to logging, drilling for oil and natural gas, mining, off-road vehicles, and nuclear waste storage.

Case Study: Livestock and U.S. Public Rangeland About 25,000 U.S. ranchers hold permits to graze less than 3% of the country's cattle and 10% of the sheep on BLM and Forest Service rangelands in 16 western states. Permit holders pay the federal government a grazing fee for this privilege.

Since 1981, Congress has set grazing fees on public rangeland at one-fourth to one-tenth the going rate for comparable private land. This means that taxpayers give the roughly 1% of U.S. ranchers with federal grazing permits subsidies amounting to $100–500 million a year, depending on how the costs of federal rangeland management are calculated.

The public subsidy does not end with low grazing fees. Other government costs include water pipelines, stock ponds, weed control, livestock predator control (about 96,000 wild animals were killed in 1999 by the USDA predator control program), clearing of undesirable vegetation, grass planting, erosion, and biodiversity loss. When these costs are added, a 2002 study by a Bureau of Land Management economist found that taxpayers are giving 25,000 ranchers an annual subsidy of about $500 million to $1 billion (an average of $20,000–40,000 per rancher) to produce less than 3% of the country's beef and 10% of its mutton (from sheep). It is not surprising that politically influential permit holders have fought hard to block any change in this system.

Ranchers with permits say that grazing fees on public lands should be low because most private rangeland is more productive than public rangeland. Environmentalists contend this varies with different tracts of land, and even when true, current grazing fees on public rangelands are much too low. Ranchers also argue that lower grazing fees help keep the price of beef and mutton down for consumers. Environmentalists counter that there is little evidence to support this claim.

Here are three positions on the use of public rangelands (mostly in the West). *One* is to phase out commercial grazing of livestock on western public lands over the next 10–15 years. Proponents contend that water-poor western rangeland (Figure 14-7, p. 310) is not a good place to raise cattle and sheep, which need a lot of water. Also, these livestock animals

The Eco-Rancher

Wyoming rancher Jack Turnell is one of a new breed of cowpuncher who gets along with environmentalists. He talks about riparian ecology and biodiversity as fluently as he does about cattle: "I guess I have learned how to bridge the gap between the environmentalists, the bureaucracies, and the ranching industry."

Turnell grazes cattle on his 32,000-hectare (80,000-acre) ranch south of Cody, Wyoming, and 16,000 hectares (40,000 acres) of Forest Service land on which he has grazing rights. For the first decade after he took over the ranch, he raised cows the conventional way. Since then, he has made some changes.

Turnell disagrees with the proposals by some environmentalists to raise grazing fees and remove sheep and cattle from public rangeland. He believes if ranchers are kicked off the public range, ranches like his will be sold to developers and chopped up into vacation sites and homes ("ranchettes"). This will destroy the range for wildlife and livestock alike.

At the same time, he believes there are more ecologically sustainable ways to operate ranches. To demonstrate this, Turnell began rotating his cows away from the riparian areas, and gave up most uses of fertilizers and pesticides. He also

crossed his Hereford and Angus cows with a French breed that tends to congregate less around water. He makes most of his ranching decisions in consultation with range and wildlife scientists, and uses photographs to monitor changes in range condition.

The results have been impressive. Willows and other plant life line riparian areas on the ranch and Forest Service land. This has provided lush habitat for an expanding population of wildlife, including pronghorn antelope, deer, moose, elk, bear, and mountain lions. In addition, this eco-rancher makes more money because the higher-quality grass puts more meat on his cattle.

Jack Turnell is a rancher who is trying to practice more environmentally sustainable ranching.

can overgraze rangeland (Figure 16-16, left, p. 392) and degrade riparian areas, and they threaten or endanger at least 175 plant and animal species.

A *second* position is to get government out of managing public rangeland. Instead, allow it to be managed by grazing boards made up of ranchers exempt from most environmental laws or sold to private ranching or development interests.

A *third* position is to encourage progressive ranchers (Individuals Matter, above), federal land managers, and environmentalists to work together to develop more sustainable ways to manage public rangelands. This would also help keep ranchers using public and private lands in business so rangeland is not lost to houses, condos, small "ranchettes," and tourist attractions.

Although their numbers are still small, some coalitions of environmentalists, federal range managers, and ranchers are working together to develop more sustainable ways of using public rangeland. Here are three ways analysts suggest doing this. *First,* ban or strictly limit grazing on riparian areas. *Second,* ban or strictly limit grazing on any rangeland where it endangers habitats of threatened and endangered species, causes desertification, or is otherwise ecologically unsustainable.

Third, allow individuals and environmental groups to purchase grazing permits and not use the land for grazing. In 2000, the U.S. Supreme Court upheld this practice.

So far, western ranchers with grazing permits have had enough political power to prevent government officials and government land management agencies from implementing most of these measures.

17-3 MANAGING AND SUSTAINING FORESTS

What Are the Major Types of Forests? Forests with 50% or more tree cover occupy about 32% of the earth's land surface (excluding Greenland and Antarctica) and provide numerous ecological and economic services (Figure 17-10). About 47% of the world's forests are tropical, 33% boreal, 11% temperate, and 9% subtropical (Figure 6-11, p. 113).

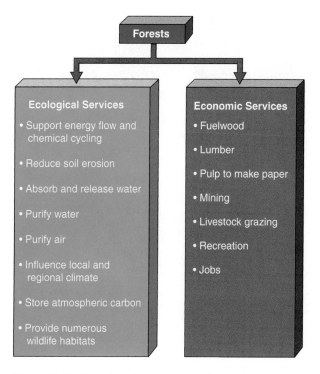

Figure 17-10 *Natural capital:* major ecological and economic services provided by forests.

Ecological Services
- Support energy flow and chemical cycling
- Reduce soil erosion
- Absorb and release water
- Purify water
- Purify air
- Influence local and regional climate
- Store atmospheric carbon
- Provide numerous wildlife habitats

Economic Services
- Fuelwood
- Lumber
- Pulp to make paper
- Mining
- Livestock grazing
- Recreation
- Jobs

Forest managers and ecologists also classify forests into three major types based on their age and structure. *One* type is **old-growth** or **frontier forests.** They are uncut forests or regenerated forests that have not been seriously disturbed by human activities or natural disasters for at least several hundred years. Old-growth forests provide ecological niches for a multitude of wildlife species (Figure 6-21, p. 121).

A *second* type is **second-growth forests.** They are stands of trees resulting from secondary ecological succession (Figure 7-12, p. 153) after the trees in an area have been removed by *human activities* (such as clear-cutting for timber or conversion to cropland) or *natural forces* (such as fire, hurricanes, or volcanic eruption).

Tree plantations, also called **tree farms,** are a *third* type. They are managed tracts with uniformly aged trees of one species that are harvested by clear-cutting as soon as they become commercially valuable. Then they are replanted and clear-cut again on regular cycles (Figure 17-11).

These plantations, created mostly by clearing old-growth and second-growth forests, now occupy about 5% of the world's tree cover area (62% of them in Asia, mostly in China and India) and produce about 10% of the world's wood. During the past 8,000 years we have cut down many of the world's frontier forests and replaced most of them with secondary forests and tree plantations (Figure 17-12, p. 420).

What Are the Major Types of Forest Management? The total volume of wood produced by a particular stand of forest varies as it goes through different stages of growth and ecological succession (Figure 17-13, p. 420). Producing fuelwood or fiber for paper production requires harvesting the trees on a short rotation cycle (Figure 17-11), well before the volume of wood peaks (point A in Figure 17-13). Harvesting of

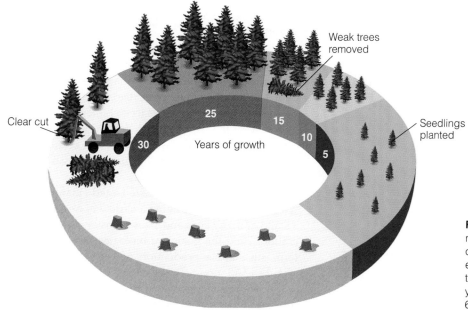

Clear cut

Weak trees removed

Seedlings planted

30 25 15 10 5

Years of growth

Figure 17-11 Short (25- to 30-year) rotation cycle of cutting and regrowth of a monoculture tree plantation in modern industrial forestry. In tropical countries, where trees can grow more rapidly year-round, the rotation cycle can be 6–10 years.

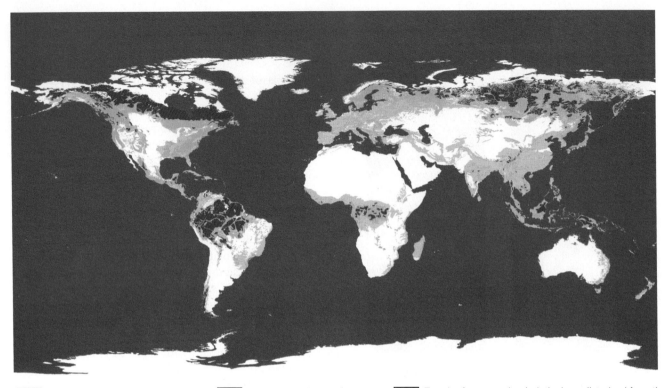

■ Frontier forests 8,000 years ago   ■ Current non-frontier forests   ■ Frontier forests today (relatively undisturbed forest)

Figure 17-12 Major changes in frontier and nonfrontier forests during the past 8,000 years. (Global Forest Watch, World Resources Institute. Reproduced with permission)

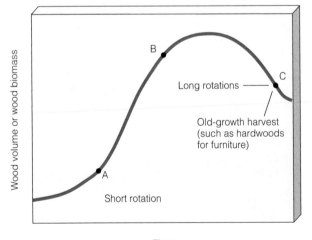

Figure 17-13 Changes in wood volume over various growth and harvest cycles in forest management. Forest management occurs over a cycle of decisions and events called a *rotation* (Figure 17-11). The most important steps in a rotation include taking an inventory of the site, developing a forest management plan, building roads into the site, preparing the site for harvest, harvesting timber, and regenerating and managing the site until the next harvest.

pulpwood plantations typically occurs in 6- to 10-year rotations in the tropics and 20- to 30-year rotations in temperate regions (Figure 17-11).

Harvesting at point B in Figure 17-13 gives the maximum yield of wood per unit of time. If the goal is high-quality wood for fine furniture or veneer, managers use longer rotations. This allows the development of larger older-growth trees whose rate of growth has leveled off (point C in Figure 17-13).

There are two forest management systems. *One* is **even-aged management,** which involves maintaining trees in a given stand at about the same age and size. In this approach, sometimes called *industrial forestry,* a simplified *tree plantation* replaces a biologically diverse old-growth or second-growth forest. The plantation consists of one or two fast-growing and economically desirable species that can be harvested every 6–100 years, depending on the species.

A *second* type is **uneven-aged management,** which involves maintaining a variety of tree species in a stand at many ages and sizes to foster natural regeneration. Here the goals are biological diversity, long-term sustainable production of high-quality timber (Figure 17-13, point C), selective cutting of individual mature or intermediate-aged trees, and multiple use of the forest for timber, wildlife, watershed protection, and recreation.

The rate of economic return plays a major role in determining whether forest owners use short-term, even-aged management or longer-term, uneven-aged management. Before the 1960s, forest owners figured a rate of return on their investment of 2–3%. Because the trees grew about as fast as the money invested in them, owners could afford to wait decades before they harvested them, normally using uneven-aged management.

Since the 1960s, many forest owners have been basing their management decisions on returns of almost 10%, reflecting what they can often earn by putting their money into other investments. This means they can make more money by clear-cutting diverse uneven-aged forests, investing the profits in something else, growing new even-aged stands of trees as quickly as possible, cutting them down, reinvesting the money, and repeating this process until the soil is exhausted.

How Are Trees Harvested? The first step in forest management is to build roads for access and timber removal. Even with careful design, logging roads have a number of harmful effects (Figure 17-14). Examples are increased erosion and sediment runoff into waterways, habitat fragmentation, and biodiversity loss. Logging roads also expose forests to invasion by nonnative pests, diseases, and wildlife species, and open up once-inaccessible forests to farmers, miners, ranchers, hunters, and off-road vehicle users. In addition, logging roads on public lands in the United States disqualify the land for protection as wilderness.

Once loggers can reach a forest, they use various methods to harvest the trees (Figure 17-15, p. 422). With **selective cutting,** intermediate-aged or mature trees in an uneven-aged forest are cut singly or in small groups (Figure 17-15a). Selective cutting reduces crowding, encourages growth of younger trees, maintains an uneven-aged stand of trees of different species, and allows natural regeneration from surrounding trees. It can also help protect the site from soil erosion and wind damage, can be used to remove diseased trees, and allows a forest to be used for multiple purposes.

Loggers use a form of selective cutting called *high grading* to selectively cut trees in many tropical forests. It involves cutting and removing only the largest and best specimens of the most desirable species. Studies have shown that for every large tree cut down, 16 or 17 other trees are damaged or pulled down because a network of vines usually connects the trees in tropical forest canopies. This reduction of the forest canopy (Figure 6-21, p. 121) causes the forest floor to become warmer, drier, and more flammable and increases erosion of the forest's thin and usually nutrient-poor soil (Figure 9-18, bottom left, p. 191).

Some tree species grow best in full or moderate sunlight in medium to large clearings. Three major methods are used to harvest such species. *One* is **shelterwood cutting,** which removes all mature trees in two or three cuttings over a period of about 10 years (Figure 17-15b).

Another is **seed-tree cutting.** In this case, loggers harvest nearly all a stand's trees in one cutting but leave a few uniformly distributed seed-producing trees to regenerate the stand (Figure 17-15c).

The *third* approach is **clear-cutting,** which removes all trees from an area in a single cutting (Figure 17-15d). Figure 17-16 (p. 423) lists the advantages and disadvantages of clear-cutting.

A clear-cutting variation that can provide a sustainable timber yield without widespread destruction is **strip cutting** (Figure 17-15e). It involves clear-cutting a strip of trees along the contour of the land, with the corridor narrow enough to allow natural regeneration within a few years. After regeneration, loggers cut another strip above the first, and so on. This allows clear-cutting of a forest in narrow strips over several decades with minimal damage.

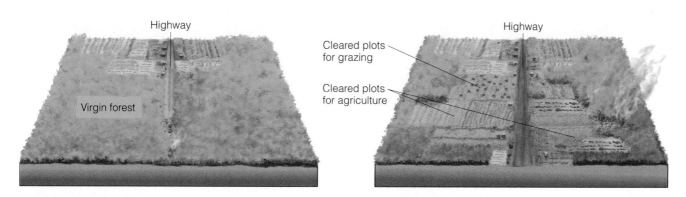

Figure 17-14 Building roads into previously inaccessible forests paves the way to fragmentation, destruction, and degradation.

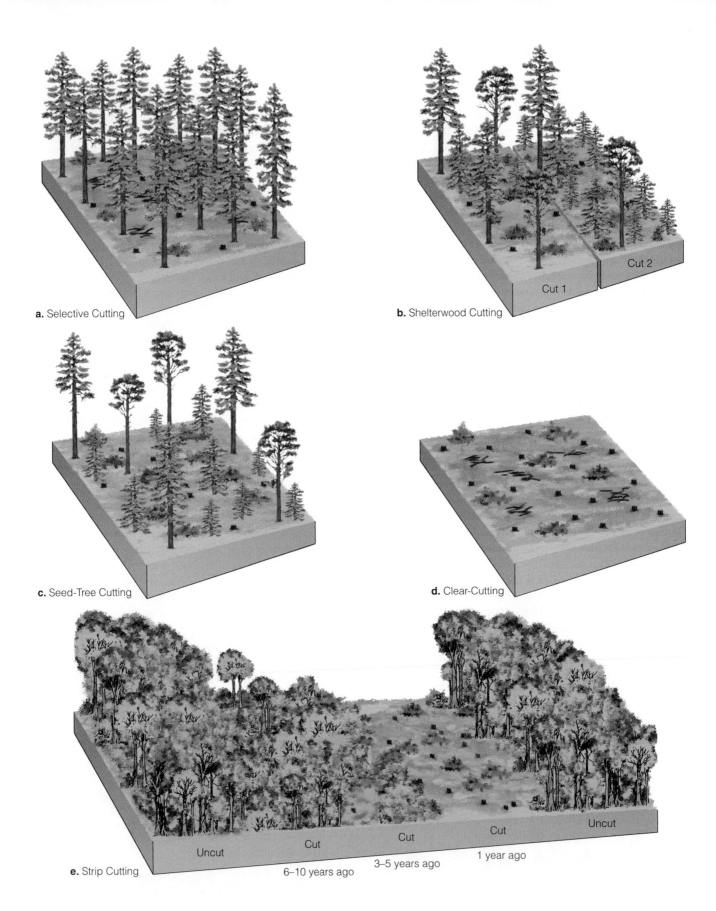

a. Selective Cutting

b. Shelterwood Cutting

Cut 1

Cut 2

c. Seed-Tree Cutting

d. Clear-Cutting

e. Strip Cutting

Uncut

Cut

6–10 years ago

Cut

3–5 years ago

Cut

1 year ago

Uncut

Figure 17-15 Tree harvesting methods.

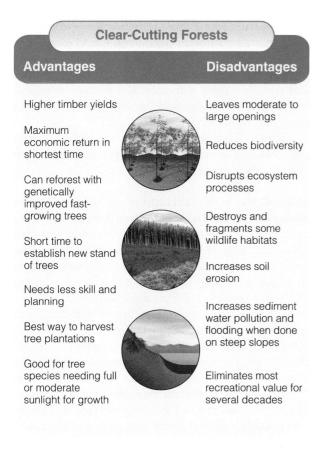

Clear-Cutting Forests

Advantages	Disadvantages
Higher timber yields	Leaves moderate to large openings
Maximum economic return in shortest time	Reduces biodiversity
Can reforest with genetically improved fast-growing trees	Disrupts ecosystem processes
Short time to establish new stand of trees	Destroys and fragments some wildlife habitats
Needs less skill and planning	Increases soil erosion
Best way to harvest tree plantations	Increases sediment water pollution and flooding when done on steep slopes
Good for tree species needing full or moderate sunlight for growth	Eliminates most recreational value for several decades

Figure 17-16 Trade-offs: advantages and disadvantages of clear-cutting forests.

What Is Happening to the World's Forests?
Forests are renewable resources as long as the rate of cutting and degradation does not exceed the rate of regrowth. Global estimates of forest cover change are difficult to make because of lack of satellite and radar data, unmonitored land-use change, and different definitions of what constitutes a forest.

Here are two pieces of *bad news. First,* surveys by the World Resources Institute (WRI) indicate that over the past 8,000 years human activities have reduced the earth's original forest cover by 20–50% (Figure 17-12). Of the remaining natural forests, about 22% are relatively undisturbed and remain intact, while the rest have been highly altered and fragmented by human activities (Figure 8-12, p. 169). *Second,* surveys by the UN Food and Agricultural Organization (FAO) and the World Resources Institute indicate that the global rate of forest cover loss during the 1990s was between 0.2% and 0.5% a year, and at least another 0.1–0.3% of the world's forests were degraded. About 95% of these losses took place in the tropics. According to conservation biologists, even at the lowest estimated rate a huge area of the world's forests is being lost and degraded each year.

There is also some *good news. First,* the total area of many temperate forests in North America and Europe has increased slightly because of reforestation from secondary ecological succession on cleared forest areas and abandoned croplands. *Second,* some of the cut areas of tropical forest have increased tree cover from regrowth and planting of tree plantations. However, ecologists do not believe that tree plantations (with their much lower biodiversity) should be counted as forest any more than croplands should be counted as grassland.

What Is the Economic Value of the Earth's Ecological Services? Do the economic advantages of clear-cutting remaining frontier forests and older secondary forests for timber and development outweigh the economic disadvantages of losing or decreasing the important ecological services they provide? The answer depends on the accounting system we use. Currently, forests are valued mostly for their economic services (Figure 17-10, right).

However, suppose we shifted to *full-cost pricing* (p. 24) by estimating and taking into account the ecological services provided by forests. This would favor not clear-cutting most remaining frontier forests and older second-growth forests because their long-term ecological services (Figure 17-10, left) would be much more valuable than their short-term economic services. With this more realistic accounting system, it would be much more profitable to use these and other forests more sustainably by living off of the ecological income they provide.

In 1997, a team of 13 ecologists, economists, and geographers attempted to estimate the monetary worth of the earth's natural ecological services (Figure 1-2, p. 3, and top half of back cover), including those of forests. According to this crude appraisal led by ecological economist Robert Costanza of the University of Maryland, the economic value of income from the earth's natural capital is at least $36 trillion per year. This is fairly close to the annual gross world product of about $48 trillion in 2002. To provide an annual natural income of $36 trillion per year, the world's natural capital would have a value of at least $500 trillion—an average of about $82,000 for each person on earth.

In this conservative estimate, the world's forests provided us with ecological services worth about $4.7 trillion per year—equal in value to about one-tenth of the annual gross world product.

To make these estimates, the researchers divided the earth's surface into 16 biomes (Figure 6-11, p. 113) and aquatic life zones (they omitted deserts and tundra because of a lack of data). Then they agreed on a list of 17 goods and services provided by nature (Figure 1-2, p. 3, and top half of back cover) and sifted through more than 100 studies that attempted to put a dollar value on such services in the 16 different types of ecosystems.

In 2002, a team of scientists led by zoologist Andrew Balmford estimated that living off the ecological income of a global network of nature reserves would provide goods and services worth at least $400 trillion more each year than converting such wild ecosystems to cropland, housing, or other human uses. To come up with this estimate, the researchers reviewed more than 300 case studies and compared the difference in the economic benefits provided by relatively intact ecosystems with those by versions of such systems converted to human uses. If these estimates are correct, the economic benefits of preserving and living off the natural income provided by wild ecosystems are 100 times greater than destroying or degrading them for other human uses.

This study also found that when ecosystems ranging from tropical forests to coral reefs are converted from their wild state to human-managed systems, about half of the economic value of their ecological services is lost. As a result, these researchers estimated that habitat destruction and degradation costs the world about $250 billion a year in new losses of ecological goods and services provided for us by nature at no cost. And because most of these conversions are permanent on a human time scale, these losses continue every year into the future, in addition to each year's new losses.

The study estimates that by spending about $45 billion a year to conserve natural habitat on the land and in the oceans, we would see a net return on the goods and services produced by these habitats of $400–520 billion per year. This expenditure is about seven times the $6.5 billion we currently spend to sustain natural areas around the world—about half of which is provided by the United States.

Ecological economist Robert Costanza says: "We have been cooking the books for a long time by leaving out the worth of nature." Biologist David Suzuki warns: "Our economic system has been constructed under the premise that natural services are free. We can't afford that luxury any more."

If these practices are uneconomical, why haven't we changed our accounting system to reflect these losses? According to ecological economists there are three major reasons. *First,* economic savings provided by conserving nature benefit everyone now and in the future, whereas profits made by exploiting nature are immediate and benefit a relatively small group of individuals. This can be summed up in a simple equation: STG = LTG or *short-term greed (STG)* eventually leads to *long-term ecological and economic greed (LTG).*

Second, many current government subsidies and tax incentives support destruction and degradation of wild ecosystems for short-term economic gain and mask the true costs of these activities. Researchers estimate that current earth-degrading subsidies amount to $1–2 billion per year. Conservationists support phasing out these environmentally harmful subsidies and replacing them with subsidies and tax incentives to protect wild systems and live off the ecological and economic income they provide.

Finally, most people are unaware of the value of the ecological services and income provided by nature and take them for granted.

Should We Make Estimates of the Economic Value of the Earth's Ecological Services? Some analysts believe estimates of the economic worth of ecological services are misleading and dangerous because they put a dollar value on ecosystem services that have an infinite value, being irreplaceable. In the 1970s, economist E. F. Schumacher warned that "to undertake to measure the immeasurable is absurd" and is a "pretense that everything has a price."

The researchers who made the estimates admit that the figures rely on many assumptions and omissions and could easily be too low by a factor of 10 to 1 million or more. For example, their calculations included only estimates of the ecosystem services themselves, not the natural capital that generates them, and omitted the value of nonrenewable minerals and fuels. They also recognize that as the supply of ecosystem services declines their value will rise sharply, and such services can be viewed as having an infinite value.

However, they contend their estimates are much more accurate than the *very* low or *zero* value the market usually assigns to these ecosystem services. They hope such estimates will call people's attention to the facts that the earth's ecosystem services are essential for all humans and their economies, their economic value is huge, and they are a source of ecological income as long as they are used sustainably.

How Can Forests Be Managed More Sustainably? Instead of increased use of conventional short-rotation industrial forestry (Figure 17-11), biodiversity researchers and a growing number of foresters call for more sustainable forest management. Figure 17-17 lists ways to do this.

There is disagreement over how much of the world's remaining old-growth forests should be protected from human activities. Some conservation biologists call for protecting as much as possible of these forests so they can serve as centers of biodiversity and future evolution.

Others point out that forests are dynamic not static. Through natural ecological succession they undergo renewal, change, and deterioration from both human and natural factors.

- Grow more timber on long rotations

- Rely more on selective cutting and strip cutting

- No clear-cutting, seed-tree, or shelterwood cutting on steeply sloped land

- No fragmentation of remaining large blocks of forest

- Sharply reduce road building into uncut forest areas

- Leave most standing dead trees and fallen timber for wildlife habitat and nutrient recycling

- Certify timber grown by sustainable methods

- Include ecological services of trees and forests in estimating economic value

Figure 17-17 Solutions: ways to manage forests more sustainably.

Some ecologists suggest using two factors to determine how much and which remaining forest to protect. *First,* protect enough forest so that the rate of forest loss and degradation by human and natural factors in a particular area is balanced by the rate of forest renewal. *Second,* place primary emphasis on identifying and protecting forests areas based on their state of decline, degree of biodiversity, and threats from nearby human activities.

Solutions: How Can We Certify Sustainably Grown Timber?   Collins Pine owns and manages a large area of productive timberland in northeastern California. Since 1940 the company has used selective cutting to help maintain ecological, economic, and social sustainability of its timberland.

Since 1993, Scientific Certification Systems (SCS) has evaluated the company's timber production. SCS is part of the nonprofit Forest Stewardship Council (FSC). It was formed in 1993 to develop a list of environmentally sound practices for use in certifying timber and products made from such timber.

Each year, SCS evaluates Collins's landholdings to ensure that cutting has not exceeded long-term forest regeneration, roads and harvesting systems have not caused unreasonable ecological damage, soils are not damaged, downed wood (boles) and standing dead trees (snags) are left to provide wildlife habitat, and the company is a good employer and a good steward of its land and water resources.

Another successful example of sustainable forestry certification involves the Menominee nation. Since

1890, they have been selectively harvesting trees of mixed species and ages from their tribal reservation land near Green Bay, Wisconsin—the state's single largest tract of virgin forest.

Each year, the Rainforest Alliance's Smart Wood Program evaluates whether the Menominee harvest lumber from their tribal forest in an environmentally and socially responsible manner. Thousands of people flock to the reservation each year to look at a sustainable forest operation that honors wildlife and biodiversity while providing jobs and income for the local economy.

In 2001, the World Wildlife Fund called for adoption of the sustainable management principles of the FSC by the world's five largest companies that harvest and process timber and buy wood products. By doing so, these five companies could halt logging of old-growth forests and still meet the world's industrial wood and wood fiber needs on one-fifth of the world's forests.

How Do Pathogens and Insects Affect Forests? Insects and tree diseases damage and kill trees in natural forests and tree plantations. A number of insect species have caused serious damage to certain tree species in the United States. One type consists of *bark beetles* such as the Asian long-horned beetle. They bore channels through the layer beneath the bark of spruce, fir, birch, and pine trees.

Other examples are the *spruce budworm* and *gypsy moth* larvae accidentally introduced from Europe in 1869 and now established in 16 states. They can kill trees by eliminating foliage needed to carry out photosynthesis and produce food.

There are several ways to reduce the impact of tree diseases and of insects on forests. They include preserving biodiversity and banning imported timber that might introduce harmful new pathogens or insect pests. We can also selectively remove infected and infested trees or clear-cut infected and infested areas and burn all debris. Other approaches are to treat diseased trees with antibiotics and develop tree species that are disease-resistant. We can also control insect pests by applying conventional pesticides and using integrated pest management (p. 405).

How Do Fires Affect Forest Ecosystems?   Intermittent natural fires set by lightning are an important part of the ecological cycle of some types of forests and grasslands. Such fires maintain the vegetation of many ecosystems at a certain stage of ecological succession (Figure 7-12, p. 153). In such ecosystems, occasional fires burn away much of the low-lying vegetation and small trees, add soil nutrients, and allow growth of new vegetation.

Figure 17-18 Surface fires (left) usually burn undergrowth and leaf litter on a forest floor and can help prevent more destructive crown fires (right) by removing flammable ground material. Sometimes carefully controlled surface fires are deliberately set to prevent buildup of flammable ground material in forests.

Surface fire   **Crown fire**

Several types of fires can affect forest ecosystems. Some, called *surface fires* (Figure 17-18, left), usually burn only undergrowth and leaf litter on the forest floor. These fires can kill seedlings and small trees but spare most mature trees and allow most wild animals to escape.

Occasional surface fires have a number of ecological benefits. They burn away flammable ground material and help prevent more destructive fires. They also release valuable mineral nutrients (tied up in slowly decomposing litter and undergrowth), increase the activity of underground nitrogen-fixing bacteria, stimulate the germination of certain tree seeds (such as those of the giant sequoia, lodgepole pine, and jack pine), and help control pathogens and insects. In addition, some wildlife species such as deer, moose, elk, muskrat, woodcock, and quail depend on occasional surface fires to maintain their habitats and provide food in the form of vegetation that sprouts after fires.

Some extremely hot fires, called *crown fires* (Figure 17-18, right), may start on the ground but eventually burn whole trees and leap from treetop to treetop. They usually occur in forests which have had no surface fires for several decades. This allows dead wood, leaves, and other flammable ground litter to build up. These rapidly burning fires can destroy most vegetation, kill wildlife, increase soil erosion, and burn or damage human structures in their paths.

Sometimes surface fires go underground and burn partially decayed leaves or peat. Such *ground fires* are most common in northern peat bogs. They may smolder for days or weeks and are difficult to detect and extinguish.

How Can We Reduce Forest Damage from Fire?

Four approaches used to help protect forest resources from fire are *prevention, prescribed burning* (setting controlled ground fires to prevent buildup of flammable material), *presuppression* (early detection and control of fires), and *suppression* (fighting fires once they have started). Ways to prevent forest fires include requiring burning permits, closing all or parts of a forest to travel and camping during periods of drought and high fire danger, and educating the public.

The Smokey Bear educational campaign of the Forest Service and the National Advertising Council, for example, has prevented countless forest fires in the United States. It has also saved many lives and prevented billions of dollars in losses of trees, wildlife, and human structures.

This educational program has convinced most members of the public that all forest fires are bad and should be prevented or put out. According to ecologists, however, trying to prevent all forest fires can increase the likelihood of highly destructive crown fires (Figure 17-18, right) because it allows large quantities of highly flammable underbrush and undergrowth and smaller trees to accumulate in some forests. This can convert normally harmless surface fires into fires intense enough to destroy the larger fire-resistant species needed for forest regeneration and threaten nearby human structures.

Ecologists propose two principles for the fire management of forests:

■ Set carefully controlled prescribed ground fires in some forests. The goal is to keep down flammable ground litter in fire-adapted forests as part of the natural ecological cycle of succession and regeneration (Figure 7-12, p. 153).

■ Allow many fires in national parks, national forests, and wilderness areas to burn as long as they do not threaten human structures and life.

Should Logging Companies Be Allowed to Thin U.S. National Forests to Help Prevent Crown Fire Damage? Logging companies say they can help prevent crown fires if Congress allows them to clear-cut more trees or selectively remove medium and large trees in national forests and other public lands. In 2003, with urging from the Bush administration and logging companies, Congress passed a law allowing logging companies to thin most national forests to help prevent large fires.

The law, called the Healthy Forests Initiative, puts no limits on the size of the trees that can be cut and prohibits environmental reviews and appeals by citizens to reduce the logging of old-growth trees. It also eliminates a 20-year-old regulation that requires the Forest Service to help prevent wildlife species from becoming endangered.

According to biologists there are several reasons why this program would increase—not decrease—the chances of severe forest fires. *First*, the most dangerous fuel is the unwanted waste material or "slash" left behind after a forest is cut. Indeed, many of the worst fires in U.S. history (including some of those during the 1990s) burned through cleared forest areas.

Second, U.S. Forest Service data show that the era of large fire losses did not begin in the 1990s when there was a decline in logging. Instead, it started in the 1980s when there was greatly increased timber harvesting in the national forests.

Third, removing the most fire-resistant medium and large trees and replacing them with dense growths of young trees and rapidly growing underbrush increases the chances of large forest fires.

Fourth, a 2002 study by the Fire Sciences Laboratory in Missoula, Montana (the country's premier laboratory for studying forest fire behavior) found that protecting houses or other buildings from fires—even in the most tinder-dry forest—requires thinning only within a zone of 46–61 meters (150–200 feet) around such buildings.

Fifth, in September 2002 about two dozen of the nation's top fire ecologists sent a letter to President Bush stating that existing scientific research shows that logging mature trees will likely worsen the fire problem. Instead, they called for thinning undergrowth and using prescribed burning to reduce fire risks in national forests.

Critics of the Healthy Forests Initiative law say that saving forests from burning by clearing them or removing their biggest and most fire-resistant trees is not saving them at all. They charge that *salvage logging* in national forests and other public lands is a way to put forest policy in the hands of the timber industry. The law would bypass environmental reviews and appeals by citizens, undermine scientific analysis in the planning process, and revoke protection of endangered species in national forests. Also, such a thinning program would increase road building and commercial logging in national forests and in roadless areas eligible for inclusion in the national wilderness system.

According to Bush administration officials, the new rules will simplify and streamline the process for managing federal forests. This will save hundreds of millions of dollars that could be applied to restoring watersheds and providing better recreational opportunities. These officials deny claims by environmentalists that the new rules will weaken forest protection and contend that they will allow even broader public participation. Environmentalists see no evidence for this claim.

How Do Air Pollution and Climate Change Threaten Forests? Forests at high elevations and those downwind from urban and industrial centers are exposed to a variety of air pollutants that can harm trees (especially conifers) and make them more vulnerable to drought, diseases, and insects (Figure 12-12, p. 267). The solution is to reduce emissions of the offending pollutants from coal-burning power plants, industrial plants, and motor vehicles (Section 12-7, p. 273).

In coming decades, an even greater threat to forests (especially temperate and boreal forests) may come from regional climate changes brought about by global warming. In areas getting less precipitation as a result of global warming, some tree species would die out, and the threat of forest fires would increase (Figure 13-14, p. 292). Ways to deal with projected global warming are discussed in Section 13-6 (p. 292).

17-4 FOREST RESOURCES AND MANAGEMENT IN THE UNITED STATES

What Is the Status of Forests in the United States? Forests cover about 30% of U.S. land area (Figure 17-7, left), provide habitats for more than 80% of the country's wildlife species, and supply about two-thirds of the nation's total water runoff.

There is some *good news* about U.S. forests. *First*, forests (including tree plantations) in the United States cover more area than they did in 1920, when the country's population was around 100 million. Many of the old-growth or frontier forests that were cleared or partially cleared between 1620 and 1960 have grown back naturally through secondary ecological succession as fairly diverse second-growth (and in some cases third-growth) forest.

Second, the United States was the world's first country to set aside large areas of forest in protected areas. By 2000, protected forest made up about 40% of the country's total forest area, most in the national forests (Figure 17-9).

There is also some *bad news* about U.S. forests. *First*, most of the existing old-growth forests in the lower 48 states were cut between 1620 and 1965. *Second*, since the mid-1960s, an increasing area of the nation's remaining old-growth and fairly diverse second-growth forests has been clear-cut and replaced with biologically simplified tree plantations.

According to biodiversity researchers, replacing diverse old-growth and secondary forests with tree plantations reduces overall forest biodiversity and disrupts ecosystem processes such as energy flow and chemical cycling. However, proponents argue that tree plantations help preserve overall forest biodiversity by reducing the pressure to clear-cut more diverse old-growth and second-growth forests.

Why Should We Care about National Forests?
The 156 national forests (Figure 17-9) managed by the U.S. Forest Service provide a number of economic, ecological, and recreational benefits. From an *economic standpoint,* they contain about 19% of the country's forest area and supply about 3% of the nation's softwood timber (down from 15% in the 1980s but expected to rise again under Bush administration plans). They also serve as grazing lands for more than 3 million cattle and sheep each year and provide about $4 billion worth of minerals, oil, and natural gas per year.

These forests contain a network of more than 612,000 kilometers (380,000 miles) of roads, equal in area to the country's entire interstate highway system. Taxpayers pay for building and maintaining these mostly narrow dirt roads that give private logging companies access to timber stands.

From an *ecological standpoint* the national forests provide habitats for 17% of the federally listed threatened and endangered species and hundreds of other wild species and fish and contain about one-third of the country's protected wilderness area. They also are principal habitats for thousands of pollinator species that contribute $4–7 billion per year to U.S. agriculture.

In addition, they provide some of the country's cleanest drinking water (worth about $3.7 billion per year) for more than 60 million Americans in more than 3,400 communities. This single ecological service is worth more than the annual value of the timber harvested from these lands.

These forests also have great *recreational value.* They receive more visits for recreation, hunting, and fishing than any other federal public lands.

Case Study: How Should U.S. National Forests Be Managed? The Forest Service is required by law to manage national forests according to two principles. One is the principle of *sustained yield,* which states that potentially renewable tree resources should not be harvested or used faster than they are replenished. The other is the principle of *multiple use,* which says that each of these forests should be managed for a variety of simultaneous uses such as sustainable timber har-

vesting, recreation, livestock grazing, watershed protection, and wildlife.

For decades there has been intense controversy over the use of forest resources in the national forests. Timber companies push to cut as much of this timber as possible at low prices. Biodiversity experts and environmentalists call for sharply reducing or eliminating tree harvesting in public forests and using more sustainable forest management practices (Figure 17-17) for timber cutting in national forests. They also favor having timber companies pay more for trees they harvest from national forests. Finally, they favor managing national forests primarily to provide recreation and to sustain their biodiversity, water resources, and other ecological services.

Between 1930 and 1988, timber harvesting from national forests increased sharply, mostly because of three factors. *One* was intense lobbying of Congress by timber company interests. A *second* factor is a law that allows the Forest Service to keep most of the money it makes on timber sales. Thus timber cutting is a key way for the Forest Service to increase its budget.

Third, a 1908 law gives counties within the boundaries of national forests 25% of the gross receipts from timber sales. This encourages county governments to push for increased timber harvesting in these forests.

By law, the Forest Service must sell timber for no less than the cost of reforesting the cleared land. However, this price does not include the government-subsidized cost of building and maintaining access roads for timber removal by logging companies. Usually, the companies also buy the timber for less than they would pay a private landowner for an equivalent amount of timber.

The Forest Service's timber-cutting program loses money because revenue from timber sales does not cover the costs of road building, timber sale preparation, administration, and other overhead costs. Because of such government subsidies, timber sales from U.S. federal lands have turned a profit for taxpayers in only 3 of the last 100 years. According to the Congressional Budget Office, eliminating losses from timber sales in the national forest would save taxpayers about $1.6 billion over the next 10 years.

Figure 17-19 lists advantages and disadvantages of logging in national forests. According to a 2000 study by the accounting firm Econorthwest, recreation, hunting, and fishing in national forests generate about 2.9 million jobs and add about $234 billion to the national economy each year. By contrast, logging, mining, grazing, and other extractive uses in the national forests add about 407,000 jobs and $23 billion to the economy per year.

In 2001, timber company officials were pleased that the Bush administration proposed a 40% increase in timber sales in national forests and reversed de-

Logging in US National Forests

Advantages	Disadvantages
Helps meet country's timber needs	Provides only 3% of timber needs
Cut areas grow back	Ample private forest land to meet timber needs
	Has little effect on timber and paper prices
Keeps lumber and paper prices down	Damages nearby rivers and fisheries
	Recreation in national forests provides more local jobs and income for local communities than logging
Provides jobs in nearby communities	
Promotes economic growth in nearby communities	Decreases recreational opportunities

Figure 17-19 Trade-offs: advantages and disadvantages of allowing logging in U.S. national forests.

cisions by the Clinton administration to block road construction in roadless areas of national forests. The administration also eliminated requirements that national forests must be managed to protect wildlife and for ecological sustainability. Finally, it pushed for laws allowing timber companies to do salvage logging without having to obey most environmental laws.

How Can We Cut Fewer Trees by Making Paper from Tree-Free Fibers? Producing paper is the world's fastest growing use of wood. One way to reduce the pressure to harvest trees for paper production is to make paper by using fiber that does not come from trees. *Tree-free fibers* for making paper come from two sources: *agricultural residues* left over from crops (such as wheat, rice, and sugar) and *fast-growing crops* (such as kenaf and industrial hemp).

Currently, tree-free paper fibers account for about 7% of the world's fiber supply for paper, with 97% of it used in developing countries and less than 1% in the United States. China uses tree-free pulp, such as rice straw and other agricultural wastes left after harvest, to make 60% of its paper.

Most of the small amount of tree-free paper produced in the United States is made from the fibers of a rapidly growing woody annual plant called *kenaf* (pro-

nounced "kuh-NAHF"; see photo on p. xxi). Compared to pulpwood, kenaf needs less herbicide because it grows faster than most weeds and reduces insecticide use because its outer fibrous covering is nearly insect-proof. Growing kenaf does not deplete soil nitrogen because it is a nitrogen fixer. In addition, breaking down the kenaf fibers takes less energy and fewer chemicals, and thus produces less toxic wastewater than using conventional trees.

Kenaf paper currently costs three to five times more than virgin or recycled paper stock. However, prices probably would come down if demand increases and greater production allows lower unit costs.

However, biologists urge caution against the widespread planting of kenaf or industrial hemp. They fear that such rapidly growing plants could take over some ecosystems, crowd out other plant species, and disrupt food webs. Also, large plantations of such plants could compete with land needed to provide food for the world's growing population.

17-5 TROPICAL DEFORESTATION

How Fast Are Tropical Forests Being Cleared and Degraded? Tropical forests cover about 6% of the earth's land area (roughly the area of the lower 48 states of the United States). Tropical forests come in several varieties, including rain forests (which receive rainfall almost daily), deciduous forests with one or two dry seasons each year, dry deciduous forests, and forests on hills and mountains.

Climatic and biological data suggest that mature tropical forests once covered at least twice as much area as they do today, with most of the destruction occurring since 1950. Satellite scans and ground-level surveys used to estimate forest destruction indicate that large areas of tropical forests are being cut rapidly in parts of South America (especially Brazil), Africa, and Asia. Haiti has lost 99% of its original forest cover, the Philippines 97%, and Madagascar 84%.

Brazil has about half the world's remaining tropical rain forest in the vast Amazon basin, which is half the size of the continental United States. This important center of biodiversity is home for up to 30% of the world's plant and animal species.

In 1970, deforestation affected only 1% of the area of the Amazon basin. By 2000, almost 15% had been deforested. According to a 2001 study by Penn State researcher James Alcock, without immediate and aggressive action to change current forest destruction and degradation practices, Brazil's original Amazon rain forests will largely disappear within 40–50 years.

The current rates of tropical deforestation and degradation are debated because of several factors. *One* is difficulty in interpreting satellite images. *Another* is that some countries hide or exaggerate deforestation rates for political and economic reasons. A *third* factor is that governments and international agencies define forest, deforestation, and forest degradation in different ways.

For these reasons, estimates of total global tropical forest loss vary from 50,000 square kilometers (19,300 square miles) to 170,000 square kilometers (65,600 square miles) per year. This amounts to a loss of 0.3–1% per year—forestry researchers at the World Resources Institute put the estimated annual loss at 0.9–1.0%. Scientists estimate that each year serious degradation and fragmentation affect an equivalent area of these forests. Thus tropical forests are being cut and degraded at a rate of 0.6–2% per year—high enough to lose or degrade half of the world's remaining tropical forests in 25–83 years.

Why Should We Care about Tropical Forests?

Most biologists believe that cutting and degrading most remaining old-growth tropical forests is a serious global environmental problem because of the important ecological and economic services they provide (Figure 17-10). For example, tropical forest plants provide chemicals used as blueprints for making most of the world's prescription drugs (Figure 17-20).

Solutions: The Incredible Neem Tree

Suppose a single plant existed that could quickly reforest bare land, supply fuelwood and lumber in dry areas, provide natural alternatives to toxic pesticides, be used to treat numerous diseases, and help control human population growth? There is: the *neem tree,* a broadleaf evergreen member of the mahogany family.

This remarkable tropical species, native to India and Burma, is ideal for reforestation because it can grow to maturity in only 5–7 years. It grows well in poor soil in semiarid lands such as those in Africa, providing abundant fuelwood, lumber, and lamp oil.

It also contains various natural pesticides. Chemicals from its leaves and seeds can repel or kill more than 200 insect species, including termites, gypsy moths, locusts, boll weevils, and cockroaches.

Extracts from neem seeds and leaves (Figure 17-20) can fight bacterial, viral, and fungal infections. Villagers call the tree a "village pharmacy" because its chemicals can relieve so many different health problems. People also use the tree's twigs as an antiseptic toothbrush and the oil from its seeds to make toothpaste and soap.

That is not all. Neem-seed oil evidently acts as a strong spermicide and may help in the development of a much-needed male birth control pill. According to a

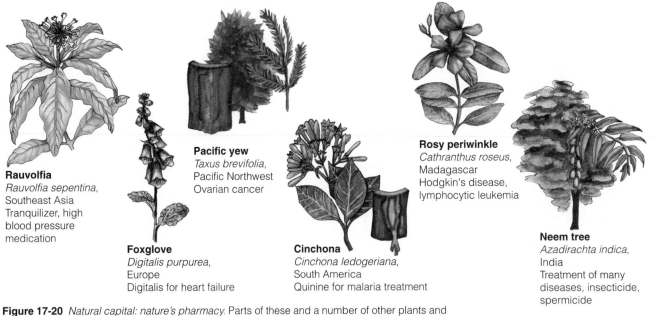

Rauvolfia
Rauvolfia sepentina,
Southeast Asia
Tranquilizer, high
blood pressure
medication

Pacific yew
Taxus brevifolia,
Pacific Northwest
Ovarian cancer

Foxglove
Digitalis purpurea,
Europe
Digitalis for heart failure

Cinchona
Cinchona ledogeriana,
South America
Quinine for malaria treatment

Rosy periwinkle
Cathranthus roseus,
Madagascar
Hodgkin's disease,
lymphocytic leukemia

Neem tree
Azadirachta indica,
India
Treatment of many
diseases, insecticide,
spermicide

Figure 17-20 *Natural capital: nature's pharmacy.* Parts of these and a number of other plants and animals (many of them found in tropical forests) are used to treat a variety of human ailments and diseases. About 70% of the 3,000 plants identified by the National Cancer Institute as sources of cancer-fighting chemicals come from tropical forests. Despite their economic and health potential, fewer than 1% of the estimated 125,000 flowering plant species in tropical forests (and a mere 1,100 of the world's 260,000 known plant species) have been examined for their medicinal properties. Many of these tropical plant species are likely to become extinct before we can study them.

study by the U.S. National Academy of Sciences, the neem tree "may eventually benefit every person on the planet."

Despite its numerous advantages, ecologists caution against widespread planting of neem trees outside its native range. As a nonnative species, it could take over and displace native species because of its rapid growth and resistance to pests.

What Causes Tropical Deforestation and Degradation? Tropical deforestation results from a number of interconnected causes (Figure 17-21). These factors are related to population growth, poverty, and government policies that encourage deforestation.

Population growth and poverty combine to drive subsistence farmers and the landless poor to tropical forests, where they try to grow enough food to survive. Government subsidies can accelerate deforestation by making timber or other tropical forest resources cheap relative to their full ecological value. Governments in Indonesia, Mexico, and Brazil also encourage the poor to colonize tropical forests by giving them title to land they clear. In addition, international lending agencies encourage developing countries to borrow huge sums of money from developed countries to finance projects such as roads, mines, logging operations, oil drilling, and dams in tropical forests.

The depletion and degradation of a tropical forest follows a typical sequence. First, a road is cut deep into the forest interior for logging and settlement (Figure 17-14) and hunters are hired to kill wild animals to provide logging and other work crews with bushmeat.

Bromeliad

Primary Causes:
Rapid population growth
Poverty
Exploitive government policies
Exports to developed countries
Failure to include ecological services in evaluating forest resources

Toucan

Scarlet macaw

Golden lion marmoset

Orchid

Secondary Causes:
Roads
Logging
Unsustainable peasant farming
Cash crops
Cattle ranching
Tree plantations
Flooding from dams
Mining
Oil drilling
Fires

Blue morpho butterfly

Figure 17-21 Major interconnected causes of the destruction and degradation of tropical forests. These factors ultimately are related to population growth, poverty, and government policies that encourage deforestation.

Loggers use selective cutting to remove the best timber (high grading). This topples many other trees because of their shallow roots and the network of vines connecting trees in the forest's canopy. Timber exports to developed countries contribute significantly to tropical forest depletion and degradation. But domestic use accounts for more than 80% of the trees cut in developing countries.

After the best timber has been removed, timber companies often sell the land to ranchers. Within a few years they typically overgraze it and sell it to settlers who have migrated to the forest hoping to grow enough food to survive. Then they move their land-degrading ranching operations to another area of the forest.

The settlers cut most of the remaining trees, burn the debris after it has dried for about a year, and plant crops using slash-and-burn agriculture (Figure 1-13, p. 13). They can also endanger species by hunting them for bushmeat.

After a few years of crop growing and rain erosion, the nutrient-poor tropical soil (Figure 9-18, bottom left, p. 191) is depleted of nutrients. Then the settlers move on, hoping for a few years of good crops on newly cleared land.

In some areas large sections of tropical forest are cleared for raising cash crops (such as sugarcane, bananas, pineapples, strawberries, and coffee)—mostly for export to developed countries. Tropical forests are also cleared for mining and oil drilling and to build dams on rivers that flood large areas of the forest.

Healthy rain forests do not burn. But increased logging, settlements, grazing, and farming along roads built in these forests fragment and dry out areas of forests. This means they are easily ignited by lightning and easily burned by farmers and ranchers.

For example, between 1997 and 1999 huge areas of forest burned in Indonesia, Malaysia, Brazil, Guatemala, Nicaragua, and Mexico. Farmers started some of these fires to prepare fields for planting or cattle grazing. Corporations cleared forests and burned the remaining plant residues to establish pulp, palm oil, and rubber plantations. The resulting highly polluted air sickened tens of millions of people, killed hundreds, caused billions of dollars in damage, and released large amounts of CO_2 into the atmosphere.

Finally, if forest destruction occurs over a large enough area, the climate may become so dry that secondary ecological succession will lead to tropical grassland instead of a tropical forest.

Solutions: How Can We Reduce Deforestation and Degradation of Tropical Forests? Analysts have suggested various ways to protect tropical forests and use them more sustainably (Figure 17-22). One promising method is to help new settlers in tropical forests learn how to practice small-scale sustainable agriculture and forestry. The Lacandon Maya Indians of Chiapas, Mexico, for example, use a multilayered system of agroforestry to cultivate as many as 75 crop species on 1-hectare (2.5-acre) plots for up to 7 years. After that, they plant a new plot to allow regeneration of the soil in the original plot.

In the lush rain forests of Peru's Palcazú Valley, Yaneshé Indians use strip cutting (Figure 17-15e) to harvest tropical trees for lumber. Tribe members also act as consultants to help other forest dwellers set up similar systems.

Another approach is to use debt-for-nature swaps, conservation easements, and conservation concessions to make it financially profitable for countries to protect tropical forests. In a *debt-for-nature swap,* participating countries act as custodians for protected forest reserves in return for foreign aid or debt relief. Since the first debt-for-nature swap in 1987 governments and private groups have carried out such swaps in 30 countries

Figure 17-22 Solutions: ways to protect tropical forests and use them more sustainably.

Kenya's Green Belt Movement

In Kenya, Wangari Maathai founded the Green Belt Movement in 1977. The goals of this highly regarded women's self-help group are to establish tree nurseries, raise seedlings, and plant and protect a tree for each of Kenya's 30 million people. By 2002, the 100,000 members of this grassroots group had planted and protected more than 20 million trees.

The success of this project has sparked the creation of similar programs in more than 30 other African countries. This inspiring leader has said,

I don't really know why I care so much. I just have something inside me that tells me that there is a problem and I have to do something about it. And I'm sure it's the same

voice that is speaking to everyone on this planet, at least everybody who seems to be concerned about the fate of the world, the fate of this planet.

Wangari Maathai was the first Kenyan woman to earn a Ph.D. (in anatomy) and to head an academic department (veterinary medicine) at the University of Nairobi. She organized the internationally acclaimed Green Belt Movement in 1977. For her work in protecting the environment she has received many honors, including the Goldman Prize, the Right Livelihood Award, the UN Africa Prize for Leadership, and the Golden Ark Award.

In a *conservation easement,* a private organization, country, or group of countries compensates other countries for protecting selected forest areas. With a *conservation concession,* a nongovernmental conservation organization protects land from logging by leasing it from a government.

Another important tool is using an international system for evaluating and certifying timber produced by sustainable methods (p. 425). Loggers can also use gentler methods for harvesting trees. For example, cutting canopy vines (lianas) before felling a tree can reduce damage to neighboring trees by 20–40%, and using the least obstructed paths to remove the logs can halve the damage to other trees. Finally, governments and individuals can mount efforts to reforest and rehabilitate degraded tropical forests and watersheds (Individuals Matter, above).

17-6 MANAGING AND SUSTAINING NATIONAL PARKS

How Popular Is the Idea of National Parks?
Today, more than 1,100 national parks larger than 10 square kilometers (4 square miles) are located in more than 120 countries. These parks cover a total area equal to that of the U.S. states of Alaska, Texas, and California combined.

The U.S. national park system, established in 1912, has 55 national parks (sometimes called the country's *crown jewels*), most of them in the West (Figure 17-9). State, county, and city parks supplement these national parks. Most state parks are located near urban

areas and have about twice as many visitors per year as the national parks.

How Are Parks Being Threatened? Parks everywhere are under pressure from external and internal threats. According to a 1999 study by the World Bank and the World Wildlife Fund, only 1% of the parks and wildlife reserves in developing countries receive protection. The other 99% are mostly *paper parks* that exist in name only and have no protection.

Local people invade most of these parks in search of wood, cropland, game animals, and other natural products for their daily survival. Loggers, miners, and wildlife poachers (who kill animals to obtain and sell items such as rhino horns, elephant tusks, and furs) also invade many of these parks. Park services in developing countries typically have too little money and too few personnel to fight these invasions, either by force or by education.

Another problem is that most national parks are too small to sustain many large animal species. Also, many parks suffer from invasions by nonnative species that can reduce the populations of native species and cause ecological disruption.

Popularity is one of the biggest problems of national and state parks in the United States and other developed countries. Because of increased numbers of roads, cars, and affluent people, annual recreational visits to major U.S. national parks increased more than fourfold between 1950 and 2000, and visits to state parks rose sevenfold.

Many of the national parks in the United States provide spectacular scenery, solitude, and enjoyable recreational experiences for visitors. However, because of a lack of park funds, visitors to some of the most heavily used parks find closed campgrounds, uncollected garbage, debris on trails and beaches, dirty toilets, and fewer nature lectures and tours by park rangers. During the summer, users entering the most popular U.S. national and state parks often face hour-long backups and experience noise, congestion, and stress instead of peaceful solitude.

U.S. Park Service rangers spend an increasing amount of their time on law enforcement instead of conservation, management, and education. Many overworked and underpaid rangers are leaving for better paying jobs.

Many parks suffer damage from the migration or deliberate introduction of nonnative species. European wild boars (imported to North Carolina in 1912 for hunting) threaten vegetation in part of the Great Smoky Mountains National Park. The Brazilian pepper tree has invaded Florida's Everglades National Park. Nonnative mountain goats in Washington's Olympic National Park trample native vegetation and accelerate soil erosion. While some nonnative species have moved into parks, some native species of animals and plants (including many threatened or endangered species) are killed or removed illegally in almost half of U.S. national parks.

Nearby human activities that threaten wildlife and recreational values in many national parks include mining, logging, livestock grazing, coal-burning power plants, water diversion, and urban development. Polluted air, drifting hundreds of kilometers, kills ancient trees in California's Sequoia National Park and often blots out the awesome views at Arizona's Grand Canyon. Car smog damages many plant species in the heavily used Great Smoky Mountains National Park in Tennessee and North Carolina. Indeed, a 2002 study found that the park had higher ozone levels during parts of the year than any U.S. urban area except Los Angeles, California. According to the National Park Service, air pollution affects scenic views in national parks more than 90% of the time.

That is not all. Mountains of trash wash ashore daily at Padre Island National Seashore in Texas. Water use in Las Vegas, Nevada (Spotlight, p. 321) threatens to shut down geysers in the Death Valley National Monument. Visitors to Sequoia National Park and Kentucky's Mammoth Cave complain of raw sewage flowing through a parking lot. Worn-out sewage treatment plants in Yellowstone National Park dump untreated waste into Yellowstone Lake and other park waterways. Unless a massive ecological restoration project is successful, Florida's Everglades National Park may dry up because of diversion of water for urban areas and agriculture.

Solutions: How Can Management of U.S. National Parks Be Improved? The National Park Service has two goals that increasingly conflict. One is to preserve nature in the parks. The other is to make nature in the parks more available to the public. Analysts have made a number of suggestions for sustaining and expanding the national park system in the United States (Figure 17-23).

The 2001 report of the National Park System Advisory Board made many similar recommendations. It also called for increased emphasis on biodiversity preservation, research devoted to understanding national parks and using them as living laboratories, science-based management, and increased efforts to educate park visitors about nature.

Private concessionaires provide campgrounds, restaurants, hotels, and other services for park visitors. Some analysts call for requiring concessionaires to compete for contracts and pay franchise fees equal to 22% of their gross (not net) receipts. Currently concessionaires in national parks pay the government an average of only about 6–7% of their gross receipts in franchise fees; many large concessionaires with long-term contracts pay as little as 0.75% of their gross receipts. Another suggestion is to allow concessionaires to lease but not own facilities inside parks.

- Integrate plans for managing parks and nearby federal lands

- Add new parkland near threatened parks

- Buy private land inside parks

- Locate visitor parking outside parks and use shuttle buses for entering and touring heavily used parks

- Increase funds for park maintenance and repairs

- Survey wildlife in parks

- Raise entry fees for visitors and use funds for park management and maintenance

- Limit number of visitors to crowded park areas

- Increase number and pay of park rangers

- Encourage volunteers to give visitor lectures and tours

- Seek private donations for park maintenance and repairs

Figure 17-23 Solutions: suggestions for sustaining and expanding the national park system in the United States. (Wilderness Society and National Parks and Conservation Association)

17-7 ESTABLISHING, DESIGNING, AND MANAGING NATURE RESERVES

How Much of the Earth's Land Should We Protect from Human Exploitation? Most ecologists and conservation biologists believe the best way to preserve biodiversity is through a worldwide network of protected areas. Currently, more than 17,000 nature reserves, parks, wildlife refuges, wilderness, and other areas provide strict or partial protection for about 10% of the world's land area. However, many existing reserves are too small to protect their native wild species. They also receive too little protection to prevent illegal and unsustainable exploitation of their plant and animal resources.

Conservation biologists call for strict protection of at least 20% of the earth's land area in a global system of biodiversity reserves that includes multiple examples of all the earth's biomes. Doing this will take action and funding by national governments, private groups (Solutions, right), and cooperative ventures involving governments, businesses, and private conservation groups.

Some *good news* is that in 2002 Canada announced plans to create ten huge new national parks and five marine conservation areas between 2002 and 2007. This will almost double the area occupied by the country's current 39 national parks.

Most developers and resource extractors oppose protecting even the current 10% of the earth's remaining undisturbed ecosystems. They contend that most of these areas contain valuable resources that would add to economic growth. Ecologists and conservation biologists disagree. They view protected areas as islands of biodiversity that help sustain all life and economies and serve as centers of future evolution (Guest Essay, p. 102).

Case Study: What Has Costa Rica Done to Protect Its Land from Degradation? Once tropical forests completely covered Costa Rica, which is smaller in area than West Virginia. Between 1963 and 1983, however, politically powerful ranching families cleared much of the country's forests to graze cattle. They exported most of the beef produced to the United States and western Europe.

Despite such widespread forest loss (which continues today), tiny Costa Rica is a superpower of biodiversity, with an estimated 500,000 plant and animal species. A single park in Costa Rica is home for more bird species than in all of North America.

In the mid-1970s, Costa Rica established a system of reserves and 28 national parks that by 2002 included about 25% of its land—6% of it in reserves for indigenous peoples. Costa Rica now has a larger proportion

The Nature Conservancy

Private groups play an important role in establishing wildlife refuges and other reserves to protect biological diversity. For example, since 1951, the Nature Conservancy has preserved more than 44,500 square kilometers (17,200 square miles) of vital wildlife habitats in the United States and 243,000 square kilometers (94,000 square miles) throughout Canada, Latin America, the Caribbean, Asia, and the Pacific Ocean.

The Nature Conservancy has more than 1 million members throughout the world. It has one of the lowest overhead rates of any nonprofit organization: 85% of all contributions go directly to its conservation programs.

This organization began in 1951 when an association of professional ecologists wanted to use their scientific knowledge to conserve natural areas. Since then, this science-based organization has used the most sophisticated scientific knowledge available to identify and rank "hot spot" sites that are unique and ecologically significant and whose biodiversity or existence is threatened by development or other human activities.

After identifying sites, the organization uses a variety of techniques to see they receive legal protection. The group also uses science-based management plans to maintain or restore the ecological health of each site and provide long-term stewardship.

This organization uses private and corporate donations to maintain a fund for buying ecologically important pieces of land or wetlands threatened by development. If it cannot buy land for habitat protection, the conservancy helps landowners obtain tax benefits in exchange for accepting legal restrictions or conservation easements preventing development. Other techniques include long-term management agreements and debt-for-nature swaps. Landowners also receive sizable tax deductions by donating their land to the Nature Conservancy in exchange for lifetime occupancy rights.

Through such efforts, this organization has created the world's largest system of private natural areas and wildlife sanctuaries, using the guiding principle of *land conservation through private action*.

Critical Thinking

Do you favor this private approach to protecting biodiversity over the government approach to protecting public lands and endangered species? Explain. Would you be in favor of selling or giving some public lands to private groups such as the Nature Conservancy? Explain.

of land devoted to biodiversity conservation than any other country.

The country's parks and reserves are consolidated into eight *megareserves* designed to sustain about 80% of the country's biodiversity (Figure 17-24). Each reserve contains a protected inner core surrounded by buffer zones that local and indigenous people use for sustainable logging, food growing, cattle grazing, hunting, fishing, and eco-tourism.

One reason for this accomplishment in biodiversity protection was the establishment of the Organization of Tropical Studies in 1963. It is a consortium of more than 50 U.S. and Costa Rican universities with the goal of promoting research and education in tropical ecology. The resulting infusion of several thousand scientists helped Costa Ricans appreciate their country's great biodiversity. The program also led to establishment in 1989 of the National Biodiversity Institute (INBio), a private nonprofit organization set up to survey and catalog the country's biodiversity.

This biodiversity conservation strategy has paid off. Today, the $1 billion a year tourism business (almost two-thirds of it from eco-tourists) is the country's largest source of income.

Some *bad news* is that legal and illegal deforestation threatens this plan because of the country's population growth, poverty (which affects 10% of its people), and lack of government regulations. According to a 2001 government-commissioned study, about 34% of the country's timber is harvested illegally. As a result of legal and illegal timber cutting, this small country still loses roughly 400 square kilometers (150 square miles) of primary forest per year—four times the rate of loss in Brazil.

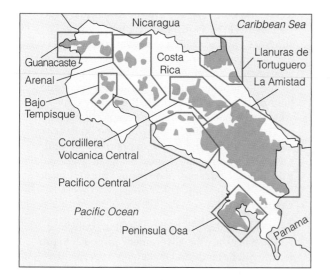

Figure 17-24 Solutions: Costa Rica has consolidated its parks and reserves into eight *megareserves* designed to sustain about 80% of the country's rich biodiversity.

In addition, without careful government control, the 1 million tourists visiting Costa Rica each year can degrade some of the protected areas. Increased tourism can also stimulate the building of too many hotels, resorts, and other potentially harmful forms of development.

What Principles Should Be Used to Establish and Manage Nature Reserves? According to most ecologists and conservation biologists, the selection, design, and management of biodiversity reserves should be guided by the three ecological principles. *First,* ecosystems are rarely at a stable point (the "balance of nature" concept) and thus cannot be locked up and protected from human disturbances. Instead, they are mostly in an ever-changing *nonequilibrium state* because of disturbances caused by natural processes and human activities. *Second,* ecosystems and biological communities experiencing fairly frequent but moderate disturbances have the greatest diversity of species. This is known as the *intermediate disturbance hypothesis* (Figure 7-13, p. 154). *Third,* we should view most reserves as *habitat islands* surrounded by a sea of developed and fragmented land.

Conservation biologists use the theory of island biogeography (Figure 7-5, p. 143) to help them locate areas in greatest danger of losing species diversity and estimate the size of a nature reserve needed to prevent it from losing species. They also use the theory to evaluate how closely a series of small wildlife reserves should be spaced to allow immigration from one preserve to another if a species in one reserve becomes locally extinct. In addition, they use the theory to assess the size and number of protected corridors needed to connect various reserves and encourage the spread of protected species between them.

Experience shows that two social principles are also useful in establishing, managing, and protecting reserves. *First,* include local people in the planning and design of a reserve. *Second,* create user-friendly reserves that allow local people to use parts of a reserve or a buffer zone surrounding a reserve for sustainable timber cutting, livestock grazing, growing crops, hunting, and fishing (Figure 17-25). Also, train local people as guides and local wildlife experts and hire them to help restore degraded areas. This turns local people into partners in protecting a reserve from unsustainable uses instead of destroyers.

Should Reserves Be as Large as Possible? There is general agreement that large reserves are the best way to maintain viable populations of large, wide-ranging species (such as panthers, elephants, and grizzly bears), sustain more species, and provide greater habitat diversity than small reserves. Large reserves also minimize the area of outside edges exposed to natural disturbances (such as fires and hurricanes), invading species, and human disturbances from nearby developed areas.

However, research indicates that in some locales, several well-placed, medium-sized, and isolated reserves may better protect a wider variety of habitats and preserve more biodiversity than a single large reserve of the same area. Also, such reserves are less likely to be simultaneously devastated by a single event such as a flood, fire, disease, or invasion by a nonnative species than a single large reserve. A mixture of both large and small reserves (Figure 17-24) may be the best way to protect a variety of species and communities against a number of different threats.

Should Corridors Connect Reserves? Establishing protected habitat corridors between reserves can help support more species and allow migration of vertebrates that need large ranges. They also permit migration of individuals and populations when environmental conditions in a reserve deteriorate and help preserve animals that must make seasonal migrations to obtain food. Corridors may also enable some species to shift their ranges if global warming makes their current ranges uninhabitable.

On the other hand, corridors can threaten once-isolated populations by allowing the movement of pest species, disease, fire, and exotic species between reserves. They also increase exposure of migrating species to natural predators, human hunters, and pollution. In addition, corridors can be costly to acquire, protect, and manage.

What Are Biosphere Reserves? In 1971, the UN Educational, Scientific, and Cultural Organization (UNESCO) created the Man and the Biosphere (MAB) Programme to improve the relationship between people and the environment. A major goal of the program is to establish at least one (and ideally five or more) *biosphere reserves* in each of the earth's 193 biogeographical zones.

Ideally, each reserve should be large enough to prevent gradual species loss and combine both conservation and sustainable use of natural resources. Today there are more than 425 biosphere reserves in 95 countries.

Each reserve must receive a nomination from its national government and meet certain size requirements. It must contain three zones (Figure 17-25). *First* is a *core area*. It contains an important ecosystem that the government legally protects from all human activities except nondestructive research and monitoring.

Second is a *buffer zone* that surrounds and protects the core area. In this zone, emphasis is on nondestruc-

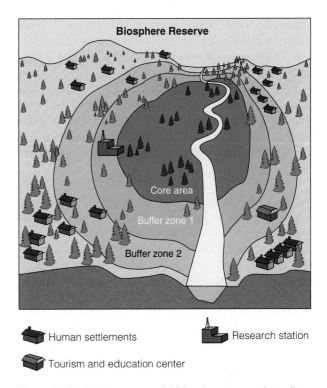

Human settlements

Research station

Tourism and education center

Figure 17-25 Solutions: a model *biosphere reserve*. In traditional parks and wildlife reserves, well-defined boundaries keep people out and wildlife in. By contrast, biosphere reserves recognize people's needs for access to sustainable use of various resources in parts of the reserve.

tive research, education, and recreation. Local people can also carry out sustainable logging, agriculture, livestock grazing, hunting, and fishing in this buffer zone, as long as such activities do not harm the core.

Third, a second *buffer*, or *transition*, *zone*, surrounds the inner buffer. In this zone local people can engage in more intensive but sustainable forestry, grazing, hunting, fishing, agriculture, and recreation than in the inner buffer zone.

So far, most biosphere reserves fall short of the ideal and receive too little funding for their protection and management. An international fund to help countries protect and manage biosphere reserves would cost about $100 million per year—about what the world's nations spend on weapons every 90 minutes.

What Areas Should Receive Top Priority for Establishing Reserves? In reality, few countries are physically, politically, or financially able to set aside and protect large biodiversity reserves. Ecologists suggest using two approaches to deal with this problem and protect as much of the earth's remaining biodiversity as possible.

One is a *prevention strategy* designed to reduce the future loss of biodiversity. It focuses international efforts on establishing a variety of large and small reserves in the world's most biodiverse countries (Figure 17-26) and in threatened species-rich areas within such countries.

The *second* is an *emergency action* strategy that identifies and quickly protects *biodiversity hot spots* (Figure 17-27). They are areas especially rich in plant and animal species that are found nowhere else and are in great danger of extinction or serious ecological disruption.

The 25 hot spots identified so far (Figure 17-27) cover 1.4% of the land area of the earth (about the same land area as the U.S. states of Alaska and Texas combined) and consist mostly of tropical forests. They contain at least 60% of the earth's terrestrial biodiversity and are the only locations for more than one-third of the planet's known terrestrial plant and animal species (including 55% of all primate species and 22% of all carnivore species).

Currently, many of the 1.1 billion people who live in or near these hot spots suffer from extreme poverty and malnutrition. In 19 of the 25 hot spots the human population is growing more rapidly than world population. In addition, 10 of the 25 hot spots are located in water-short regions,

According to conservation biologists, spending about $500 million annually over a 5-year period could go far in safeguarding these endangered centers of biodiversity. This expenditure for protecting at least a third of the world's threatened species is equivalent to about what the world's nations spend on weapons every 8 hours. According to Norman Myers (Guest Essay, p. 102): "I can think of no other biodiversity initiative that could achieve so much at a comparatively small cost, as the hot spots strategy."

What Is Wilderness?   One way to protect undeveloped lands from human exploitation is by legally setting them aside as wilderness. According to the U.S. Wilderness Act of 1964, *wilderness* consists of areas "of undeveloped land affected primarily by the forces of nature, where man is a visitor who does not remain." U.S. President Theodore Roosevelt summarized what we should do with wilderness: "Leave it as it is. You cannot improve it."

The U.S. Wilderness Society estimates that a wilderness area should contain at least 4,000 square kilometers (1,500 square miles); otherwise it can be affected by air, water, and noise pollution from nearby human activities. Figure 17-28 (p. 440) shows remaining large wild areas of the earth's surface with the least influence from human activities. Conservation biologists call for these last wild areas to be protected from harmful human activities.

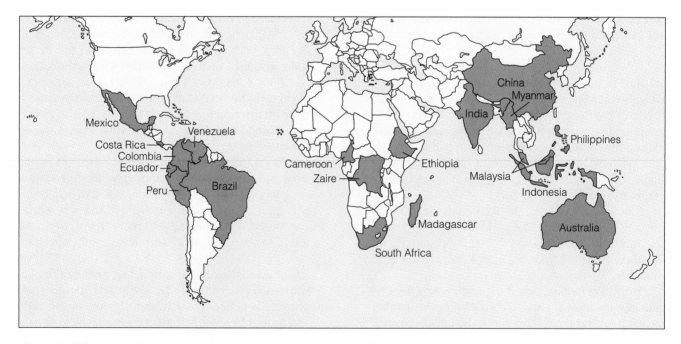

Figure 17-26 *Natural capital:* the earth's 19 most biodiverse countries. Conservation biologists believe efforts to preserve repositories of biodiversity as protected wilderness areas and other nature reserves should be concentrated in these 19 countries, which contain at least 60% of the world's known terrestrial species. Protecting the world's biodiversity will take financial and scientific help from developed countries. (Data from Conservation International and World Wildlife Fund)

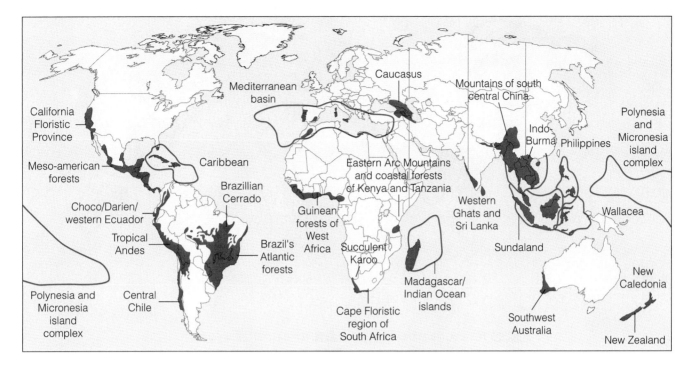

Figure 17-27 Twenty-five *hot spots* identified by ecologists as important but endangered centers of biodiversity that contain a large number of endemic plant and animal species found nowhere else. (Data from Conservation International)

Why Preserve Wilderness? According to wilderness supporters, there are several reasons why we need wild places. *One* is that they are areas where people can experience the beauty of nature and observe natural biological diversity. A *second* is that such areas can enhance the mental and physical health of visitors by allowing them to get away from noise, stress, development, and large numbers of people. Wilderness preservationist John Muir advised us,

> *Climb the mountains and get their good tidings. Nature's peace will flow into you as the sunshine into the trees. The winds will blow their freshness into you, and the storms their energy, while cares will drop off like autumn leaves.*

According to the late renowned conservationist David Brower: *"Without wilderness the world is a cage."*

Even those who never use wilderness areas may want to know they are there, a feeling expressed by novelist Wallace Stegner:

> *Save a piece of country . . . and it does not matter in the slightest that only a few people every year will go into it. This is precisely its value . . . . We simply need that wild country available to us, even if we never do more than drive to its edge and look in. For it can be a means of reassuring ourselves of our sanity as creatures, a part of the geography of hope.*

Some critics oppose protecting wilderness for its scenic and recreational value for a small number of people. They believe this is an outmoded concept that keeps some areas of the planet from being economically useful to humans.

To most biologists, however, the most important reasons for protecting wilderness and other areas from exploitation and degradation are to *preserve the biodiversity they contribute* as a vital part of the earth's natural capital (Figure 1-2, p. 3, and top half of back cover) and to *protect them as centers for evolution* in response to mostly unpredictable changes in environmental conditions (Guest Essay, p. 102). In other words, wilderness is a biodiversity and wilderness bank and an eco-insurance policy.

Some analysts also believe wilderness should be preserved because the wild species it contains have an ethical right to exist (or struggle to exist) and play their roles in the earth's ongoing saga of biological evolution and ecological processes, without human interference. In other words, wild species have *intrinsic value*, regardless of their usefulness to us.

Wilderness advocates call for protecting more wilderness in the world's species-rich countries (Figure 17-26), most endangered hot spots (Figure 17-27), and remaining areas that have been the least influenced by human activities (Figure 17-28).

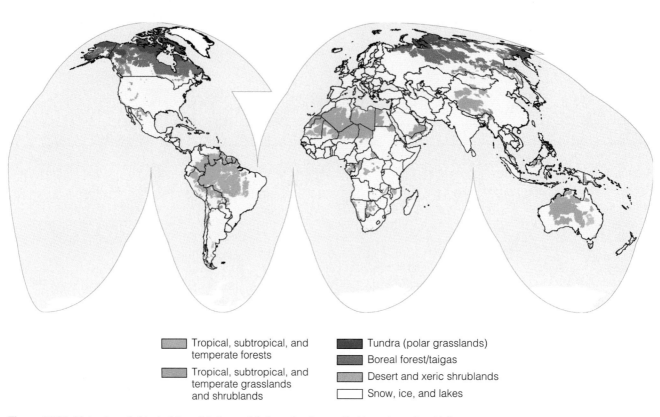

▨ Tropical, subtropical, and temperate forests	■ Tundra (polar grasslands)
▨ Tropical, subtropical, and temperate grasslands and shrublands	▨ Boreal forest/taigas
	▨ Desert and xeric shrublands
	☐ Snow, ice, and lakes

Figure 17-28 *Natural capital:* last of the wild, the world's large land areas that have been least influenced by human activities. [Wildlife Conservation Society in collaboration with the Center for International Earth Science Information Network (CIESIN)]

Case Study: How Much Wilderness Should Be Protected in the United States? In the United States, preservationists have been trying to save wild areas from development since 1900. Overall, they have fought a losing battle. Not until 1964 did Congress pass the Wilderness Act. It allowed the government to protect undeveloped tracts of public land from development as part of the National Wilderness Preservation System.

To wilderness supporters, the *good news* is that the area of protected wilderness in the United States increased tenfold between 1970 and 2000. The *bad news* is that only about 4.6% of U.S. land is protected as wilderness (almost three-fourths of it in Alaska) and only 1.8% of the land area of the lower 48 states is protected (most of it in the West). According to a 1999 study by the World Conservation Union (IUCN), the United States ranks 42nd among nations in terms of terrestrial area protected as wilderness, and Canada is in 36th place.

More *bad news* is that only 4 of the 413 wilderness areas in the lower 48 states are larger than 4,000 square kilometers (1,500 square miles). Also, the system includes only 81 of the country's 233 distinct ecosystems.

In addition, most wilderness areas in the lower 48 states are threatened habitat islands in a sea of development.

Almost 400,000 square kilometers (150,000 square miles) in scattered blocks of public lands could qualify for designation as wilderness (about 60% of it in the national forests). Conservation biologists urge wilderness protection for all of this land. But officials of timber, mining, and other resource extraction industries lobby elected officials to build roads in these areas to make them ineligible for wilderness protection.

Some wilderness advocates also call for creating *wilderness recovery areas.* They would do this by closing and obliterating nonessential roads in large areas of public lands, restoring wildlife habitats, allowing natural fires to burn, and reintroducing key species that have been driven from such areas.

Some ecologists and conservation biologists call for development of *The Wildlands Project (TWP)* to establish a network of protected wildlands and ecosystems throughout as much of the United States as possible. The goal of this science-based approach is to reconnect many of the existing fragmented and threatened pieces of wild nature for use by wildlife and people. Accomplishing this would require cooperative efforts among government agencies, scientists, conservation groups, and private owners and users.

There is strong opposition to expansion of wilderness areas, establishment of wilderness recovery areas, and the Wildlands Project. Such opposition comes mainly from timber, mining, ranching, energy, and other interests who want to extract resources from these and most other public lands or convert them to private ownership (p. 417).

How Should We Manage Wilderness? To protect the most popular areas from damage, wilderness managers designate sites where camping is allowed and limit the number of people using these sites at any one time. They also use wilderness rangers to patrol vulnerable areas and enlist volunteers to pick up trash discarded by thoughtless users who do not follow the *leave no trace (LNT)* wilderness ethic.

Environmental historian and wilderness expert Roderick Nash suggests dividing wilderness areas into three categories:

- Easily accessible, popular areas that have trails, bridges, hikers' huts, outhouses, assigned campsites, and extensive ranger patrols

- Large, remote wilderness areas used only by people who get a permit by demonstrating their wilderness skills

- Undisturbed biologically unique areas with no human entry allowed

17-8 PROTECTING AND SUSTAINING AQUATIC SYSTEMS

What Do We Know About the Earth's Aquatic Biodiversity? Although we live on a water planet (Figure 6-27, p. 126), we know fairly little about the earth's aquatic systems. Indeed, less than 5% of the earth's global ocean has been explored and mapped with the same level of detail as the surface of the moon and Mars. According to aquatic scientists, scientific investigation of poorly understood aquatic systems is a poorly funded *research frontier* whose study could result in immense ecological and economic benefits.

What Are Some General Patterns of Marine Biodiversity? Despite our lack of knowledge about overall marine biodiversity, scientists have established three general patterns of marine biodiversity.

First, the greatest marine biodiversity occurs in coral reefs (Figure 6-35, p. 132) and on the deep-sea floor. *Second,* biodiversity is higher near coasts than in the open sea because of the greater variety of producers, habitats, and nursery areas in coastal areas.

Third, biodiversity is higher in the benthic (bottom) region of the ocean than in the pelagic (surface) region because of the greater variety of habitats and food sources on the ocean bottom.

Why Is It Difficult to Protect Marine Biodiversity? Protecting marine biodiversity is difficult for several reasons. A major threat comes from coastal development and the accompanying massive inputs of sediment and other wastes from land. They harm shore-hugging species and threaten biologically diverse and highly productive coastal ecosystems such as coral reefs, marshes, and mangrove forest swamps.

Another problem is that much of the damage to the oceans and other bodies of water is not visible to most people. Also, many people view the seas as an inexhaustible resource that can absorb an almost infinite amount of waste and pollution.

Another problem is that most of the world's ocean area lies outside the legal jurisdiction of any country. Thus it is an open-access resource, subject to overexploitation because of the tragedy of the commons (p. 7).

Finally, no effective international agreements protect biodiversity in the open seas. Such agreements are difficult to develop, monitor, and enforce.

How Can We Protect and Sustain Marine Systems? Ways to sustain the biodiversity of marine systems include

- Establishing protected areas

- Using integrated coastal management

- Regulating and preventing ocean pollution (Section 14-8, p. 333)

- Protecting endangered and threatened species (Chapter 18)

What Is the Role of International Agreements and Protected Marine Sanctuaries? Individual governments have used several international treaties, agreements, and actions to protect living marine resources in parts of the world. There is some *good news. First,* under the United Nations Law of the Sea, all coastal nations have sovereignty over the waters and seabed up to 19 kilometers (12 miles) offshore and almost total jurisdiction over their Exclusive Economic Zone (EEZ), which extends 320 kilometers (200 miles) offshore. Taken together, the nations of the world have jurisdiction over 36% of the ocean surface and 90% of the world's fish stocks.

Second, the United Nations Environment Programme has spearheaded efforts to develop 12 regional agreements to protect large marine areas shared by several countries. About 90 of the world's 350 biosphere reserves (p. 437) include coastal or marine habitats that receive partial protection.

Third, New Zealand has 16 "no take" marine reserves, which prohibit all fishing. Other such reserves have been established in Australia, Belize, the Galapagos Islands, and the Caribbean. The United States has designated 12 marine sanctuaries.

Fourth, the National Center for Ecological Analysis and Synthesis reported in 2001 that scientific studies show that within fully protected marine reserves fish populations double, fish size grows by 30%, fish reproduction triples, and species diversity is 23% higher. Furthermore, this improvement happens within 2–4 years after strict protection begins and lasts for decades.

Finally, since 1986 the World Conservation Union (IUCN) has helped establish a global system of *marine protected areas (MPAs),* mostly at the national level. The 1,300 existing MPAs help protect about 0.2% of the earth's total ocean area.

There is some *bad news. First,* less than 0.01% of the world's ocean area consists of fully protected marine reserves. In the United States, the total area of fully protected marine habitat is only about 130 square kilometers (50 square miles).

Second, many existing marine sanctuaries and MPAs allow extractive activities that are prohibited in marine reserves. *Finally,* stresses from nearby coastal areas can disrupt protected marine reserves unless they are also protected as part of an integrated coastal management plan.

In 1997, an international group of marine scientists called for governments to increase fully protected marine reserves to 20% of the ocean's surface by 2020.

What Is the Role of Integrated Coastal Management? *Integrated coastal management* is a community-based effort to develop and use coastal resources more sustainably. The overall aim is for groups competing for the use of coastal resources to identify shared problems and goals. Then they attempt to develop workable and cost-effective solutions that preserve biodiversity and environmental quality while meeting economic and social needs.

Ideally, the overall goal is to zone the land and sea portions of an entire coastal area. Such zoning would include some fully protected marine reserves where no exploitive human activities are allowed and other zones where different kinds and levels of human activities are permitted. Australia's huge Great Barrier Reef Marine Park is managed in this way. Currently, more than 100 integrated coastal management programs are being developed throughout the world.

In the United States, 90 coastal counties are working to establish coastal management systems, but

fewer than 20 of these plans have been implemented. Since the early 1980s, people have also worked to develop and implement an integrated coastal management plan for the Chesapeake Bay (p. 335, and Figure 14-35, p. 335).

How Are Wetlands Protected in the United States? Coastal wetlands (Figure 6-32, p. 129) and inland wetlands are reservoirs of aquatic biodiversity that provide many important ecological and economic services. These systems are under intense pressure from development and pollution.

In the United States, a federal permit is required to fill or to deposit dredged or fill material into wetlands occupying more than 1.2 hectares (3 acres). Some *good news* is that according to the U.S. Fish and Wildlife Service, this law has helped cut the average annual wetland loss by 80% since 1969.

The *bad news* is that attempts to weaken this law by using unscientific criteria to classify areas as wetlands have continued. Also, only about 8% of remaining inland wetlands are under federal protection and federal, state, and local wetland protection is weak.

How Can Wetlands Be Sustained and Restored? Ecologists and environmentalists call for several strategies to protect and sustain wetlands. *One* is to enact and enforce laws to protect existing wetlands from destruction and degradation. A *second* is to use comprehensive land-use planning to steer developers, farmers, and resource extractors away from existing wetlands. *Third,* biologists call for requiring creation and evaluation of a new wetland before destroying any existing wetland. *Finally,* we can restore degraded wetlands. Many developers, farmers, and resource extractors oppose such wetland protection.

17-9 ECOLOGICAL RESTORATION

How Can We Rehabilitate and Restore Damaged Ecosystems? Some *bad news* is that almost every place on the earth has been degraded at least to some degree by human activities (Figure 8-12, p. 169, and Figure 17-8). The *good news* is that much of the environmental damage we have inflicted on nature is at least partially reversible through ecological restoration.

By studying how natural ecosystems recover, scientists are learning how to speed up repair operations using the following approaches:

■ *Restoration:* trying to return a particular degraded habitat or ecosystem to a condition as similar as possible to its predegraded state. Three difficulties are lack of knowledge about the previous composition of a system, need to deal with a moving target, and changes in climate, soil, and species composition that may make it impossible to restore an area to an earlier state.

- *Rehabilitation:* any attempt to restore at least some of a degraded system's natural species and ecosystem functions. Examples include removing pollutants and replanting areas such as mining sites, landfills, and clear-cut forests to reduce soil erosion.

- *Replacement:* replacing a degraded ecosystem with another type of ecosystem. For example, a productive pasture or tree farm may replace a degraded forest.

- *Creating artificial ecosystems:* using ecological principles to create human-designed ecosystems for specific purposes. Examples are the ecological wastewater treatment systems developed by John Todd (p. 340) and the artificial wetlands developed to treat sewage in Arcata, California (Solutions, p. 341).

Four basic steps are used in ecological restoration and rehabilitation. *First,* identify what caused the degradation (such as pollution, farming, overgrazing, mining, or invading species). *Second,* eliminate or sharply reduce these factors. Examples include removing toxic soil pollutants, adding nutrients to depleted soil, adding new topsoil, and eliminating disruptive nonnative species.

Third, protect the area from further degradation and from the disruptive effects of fires. *Fourth,* monitor restoration efforts, assess success, and modify strategies as needed.

Case Study: Ecological Restoration of a Tropical Dry Forest in Costa Rica Costa Rica is the site of one of the world's largest *ecological restoration* projects. In the lowlands of the country's Guanacaste National Park (Figure 17-24), a small tropical dry deciduous forest has been burned, degraded, and fragmented by large-scale conversion to cattle ranches and farms.

Now it is being restored and relinked to the rain forest on adjacent mountain slopes. The goal is to eliminate damaging nonnative grass and cattle and reestablish a tropical dry forest ecosystem over the next 100–300 years.

Ecologists have identified two keys to restoring the resulting grasslands to tropical dry forest. One is to exclude fires and cattle grazing from much of the park. The other is to enhance seed dispersal from remaining small patches of native woodland.

The strategy seems to be working. Areas of the park covered with monoculture expanses of nonnative grasses 10–15 years ago are now healthy secondary-growth dry forests of native species. This project illustrates two lessons about restoration. *First,* the causes of degradation must be known. *Second,* enough remnants of native plant species must be present for seed dispersal and regrowth.

Daniel Janzen, professor of biology at the University of Pennsylvania and a leader in the field of restoration ecology, has helped galvanize international support and has raised more than $10 million for this restoration project.

He recognizes that ecological restoration and protection of the park will fail unless the people in the surrounding area believe they will benefit from such efforts. Janzen's vision is to make the nearly 40,000 people who live near the park an essential part of the restoration of the degraded forest, a concept he calls *biocultural restoration.*

By actively participating in the project, local residents reap educational, economic, and environmental benefits. Local farmers make money by sowing large areas with tree seeds and planting seedlings started in Janzen's lab. Local grade school, high school, and university students and citizens' groups study the ecology of the park and go on field trips to the park. The park's location near the Pan American Highway makes it an ideal area for eco-tourism, which stimulates the local economy.

The project also serves as a training ground in tropical forest restoration for scientists from all over the world. Research scientists working on the project give guest classroom lectures and lead some of the field trips.

Janzen recognizes that in a few decades today's children will be running the park and the local political system. If they understand the importance of their environment, they are more likely to protect and sustain its biological resources. He believes that education, awareness, and involvement—not guards and fences—are the best ways to restore degraded ecosystems and protect largely intact ecosystems from unsustainable use.

Is Ecological Restoration the Best Approach? Some analysts worry that environmental restoration could encourage continuing environmental destruction and degradation by suggesting any ecological harm we do can be undone. However, ecologists point out that preventing ecosystem damage in the first place is cheaper and more effective than any form of ecological restoration.

Another concern is government policies that allow developers to destroy one ecosystem or wetland if they protect, restore, or "create" a similar one of roughly the same size. This trade-off, or *mitigation,* approach is preferable to wanton destruction of ecosystems. However, according to ecological restoration expert John Berger, "The purpose of ecological restoration is to repair previous damage, not legitimize further destruction."

What Is the Next Step? Individuals Matter In 2002, Edward O. Wilson, considered to be one of the world's foremost experts on biodiversity, published a book called *The Future of Life* (Knopf, New York). In this book, he proposed the following priorities for protecting most of the world's remaining ecosystems and species:

■ *Take immediate action to preserve the world's 25 biological hot spots (Figure 17-27).*

■ *Keep intact the world's remaining frontier forests, which are the earth's last remaining true wilderness areas (Figure 17-12).*

■ *Cease logging of old-growth (frontier) forests everywhere.*

■ *Concentrate on protecting and restoring everywhere the world's lakes and river systems, which are the most threatened ecosystems of all.*

■ *Determine the world's marine hot spots and assign them the same priority for immediate action as for those on land.*

■ *Complete the mapping of the world's terrestrial and aquatic biodiversity so we know what we have and can make conservation efforts more precise and cost-effective.*

■ *Ensure that the full range of the earth's terrestrial and aquatic ecosystems are included in a global conservation strategy.*

■ *Make conservation profitable.* This involves finding ways to raise the income of people who live in or near nature reserves so they can become partners in their protection and sustainable use. It also requires providing financial help from private and government sources to governments that protect their forests and other nature reserves. Also, it is important to help governments understand that eco-tourism, bioprospecting, and eventually greenhouse gas (carbon) trading credits for protecting remaining wild land can yield more income than logging or clearing such land for agriculture or ranching.

This strategy for protecting the earth's precious biodiversity will not be implemented without political pressure on elected officials by individual citizens and groups of such citizens. See the website material for this chapter for some actions you can take to help sustain the earth's biodiversity.

We abuse land because we regard it as a commodity belonging to us. When we see land as a community to which we belong, we may begin to use it with love and respect.

ALDO LEOPOLD

REVIEW QUESTIONS

1. Define the boldfaced terms in this chapter.

2. Evaluate the reintroduction of the gray wolf as a keystone predator species in the Yellowstone ecosystem.

3. List major factors that tend to increase or decrease biodiversity. List examples of how human activities have decreased and degraded the earth's biodiversity in terms of **(a)** land use, **(b)** net primary productivity, **(c)** aquatic systems, and **(d)** species extinctions.

4. Describe two major ways to reduce biodiversity loss and the strategies and tactics used for each method.

5. What are the major uses of land in **(a)** the world and **(b)** the United States?

6. What percentage of the land in the United States does the federal government own and manage as public lands? Describe the five major types of public lands in the United States and list the major uses allowed on each type.

7. List four principles that most biologists and some economists believe should govern the use of public land in the United States. Compare these principles with the views of users of mineral and other resources about the use, ownership, and management of the country's public land.

8. List the advantages and disadvantages of providing government subsidies to ranchers holding permits to graze livestock on public lands in the United States. Describe major positions on the use of public rangelands. List four ways suggested by some ranchers, environmentalists, and federal range managers for using public rangeland more sustainably. Describe how Jack Turnell has managed rangeland he uses.

9. List **(a)** eight important ecological services provided by forests and **(b)** seven important economic benefits of forests.

10. Distinguish among *old-growth forests, second-growth forests,* and *tree plantations.*

11. Describe the rotation cycle for harvesting and managing a forest. Distinguish between *even-aged* and *uneven-aged management* of a forest, and list the advantages and disadvantages of each type.

12. Describe five major ways for harvesting trees. List the advantages and disadvantages of **(a)** selective cutting and **(b)** clear-cutting.

13. Why is it difficult to make estimates of change in forest cover? List two pieces of bad news and two pieces of good news about trends in the world's forest cover.

14. Describe efforts by ecologists, economists, and geographers to estimate the monetary worth of the earth's natural ecological services. What price did they put on the annual value of the ecological services of the world's **(a)** forests, **(b)** natural capital, and **(c)** how much the economic benefits of conserving the earth's wild ecosystems exceed the economic value of destroying or degrading them for human use? List three reasons why we have not used full-cost accounting to evaluate the earth's natural capital. Describe the controversy over estimating the economic worth of ecosystem services.

15. List eight ways to use forests more sustainably. Describe the benefits of certifying that timber has been produced sustainably.

16. List six ways to reduce the harmful impacts of tree diseases and of insects on forests.

17. Distinguish among *surface, crown,* and *ground* forest fires. List four approaches used to protect forest resources from fire. How do some plants and animals benefit from forest fires? List the advantages and disadvantages of reducing the chances of crown fires on public lands by **(a)** setting prescribed fires, **(b)** allowing most natural fires to burn, and **(c)** cutting trees to thin out forests.

18. Summarize the threats to forests from **(a)** air pollution and **(b)** global warming.

19. Describe two pieces of *good news* and two pieces of *bad news* about the extent and condition of forests in the United States.

20. List nine services provided by U.S. national forests. What are the principles of *sustained yield* and *multiple use* that are supposed to guide the use of national forests in the United States?

21. What are the advantages and disadvantages of giving government subsidies to companies cutting timber in national forests? List the advantages and disadvantages of making **(a)** timber cutting and **(b)** recreation the primary use of national forests.

22. What are some advantages of making paper from tree-free sources such as rice straw and kenaf fibers?

23. How rapidly are tropical forests being cleared and degraded?

24. Explain why we should care about preserving tropical forests and using them sustainably. Describe the many uses of the tropical *neem tree,* and list the advantages and disadvantages of widespread planting of neem trees.

25. List three underlying causes of the items listed as direct causes of tropical deforestation and degradation in Figure 17-21, p. 431).

26. List nine ways to prevent or reduce tropical deforestation and degradation and three ways to restore degraded tropical forests. What are *debt-for-nature swaps, conservation easements,* and *conservation concessions?*

27. Describe **(a)** ways to achieve more sustainable farming and logging in tropical forests and **(b)** Kenya's Green Belt Movement.

28. What are the major threats to national parks in the United States and in other countries? List 11 ways to

improve national park management in the United States.

29. According to scientists, what is the minimum percentage of the earth's land area that we should strictly protect from harmful human activities?

30. Describe **(a)** efforts by Costa Rica to establish reserves to protect its biodiversity and **(b)** the role of the Nature Conservancy in establishing nature reserves in the United States and other parts of the world.

31. List three scientific principles and two social principles for guiding the selection, design, and management of nature reserves.

32. List the advantages and disadvantages of **(a)** large reserves and **(b)** establishing corridors between reserves.

33. What is a *biosphere reserve?* Describe the three zones of such a reserve.

34. Explain the importance of protecting biodiversity in **(a)** the world's centers of biodiversity, **(b)** biodiversity hot spots, and **(c)** the earth's remaining wildlands.

35. What is *wilderness,* and why is it important? List the advantages and disadvantages of protecting more wilderness. List three proposed categories for wilderness areas.

36. Describe three general patterns governing where we find marine biodiversity.

37. List five factors that make it difficult to protect marine biodiversity.

38. List four major ways to protect and sustain marine biodiversity.

39. List five pieces of *good news* and three pieces of *bad news* about establishing marine sanctuaries to help protect aquatic biodiversity.

40. What is *integrated coastal management?* Give two examples of its implementation in the United States.

41. Describe efforts to protect wetlands in the United States. List four methods for protecting and restoring wetlands.

42. Distinguish among ecosystem *restoration, rehabilitation,* and *replacement.* List four basic steps in carrying out ecological restoration or rehabilitation.

43. Describe efforts to restore a degraded tropical dry forest in Costa Rica. What is *biocultural restoration?*

44. What are two concerns some people have about ecological restoration?

45. List eight ways suggested by biodiversity expert Edward O. Wilson for protecting most of the world's remaining ecosystems and species.

CRITICAL THINKING

1. Do you agree or disagree with the program to reintroduce populations of the gray wolf in the Yellowstone ecosystem? Explain. Do you favor reintroducing grizzly bears to Yellowstone or other public lands in the western United States? Explain.

2. Explain why you agree or disagree with **(a)** the four principles that biologists and some economists have suggested for using public land in the United States (p. 417) and **(b)** the nine suggestions made by developers and resource extractors for managing and using U.S. public land (p. 417).

3. Explain why you agree or disagree with each of the proposals for providing more sustainable use of **(a)** public rangeland in the United States listed on pp. 417–418 and **(b)** forests throughout the world listed in Figure 17-17, p. 425.

4. Suppose you inherited a timber company that had been managed sustainably for many decades by using selective cutting. If you switched to clear-cutting most of the land, you could make millions in a short time, sell the clear-cut land, and probably invest the money elsewhere at a much higher rate of return than long-term sustainable timber harvesting. What would you do with your inheritance? Explain.

5. Should the U.S. government continue providing private companies that harvest timber from U.S. national forests with subsidies for reforestation and for building and maintaining access roads? Explain.

6. Explain why you agree or disagree with each of the proposals listed in Figure 17-22 (p. 432) for protecting the world's tropical forests. Should developed countries provide most of the money to preserve remaining tropical forests in developing countries? Explain.

7. List five actions you could take to help preserve some of the world's tropical forests. Which, if any, of these actions do you plan to carry out?

8. Explain why you agree or disagree with each of the proposals listed in Figure 17-23 (p. 434) concerning the U.S. national park system.

9. An ecologist recommends a controlled burn of vegetation in a heavily forested nearby state park to help reduce the chance of serious damage from a major forest fire. Many local citizens oppose the burn. If you were an elected official, would you support the proposed controlled burn? If you approved the burn, what restrictions would you impose for ecological and human health reasons?

10. Are you in favor of establishing more wilderness areas in the United States, especially in the lower 48 states (or in the country where you live)? Explain. What might be some drawbacks?

11. Congratulations! You are in charge of protecting and sustaining the world's terrestrial and aquatic biodiversity. List the three most important features of your policies for using and managing **(a)** forests, **(b)** parks, **(c)** wilderness and other protected biodiversity reserves, and **(d)** the world's oceans.

PROJECTS

1. Obtain a topographic map of the region where you live and use it to identify local, state, and federally owned lands in the form of parks, rangeland, forests, and wilderness areas. Identify the government agency responsible for managing each of these areas, and evaluate how well these agencies are preserving the natural resources on this land on your behalf.

2. If possible, try to visit **(a)** a diverse old-growth forest, **(b)** an area that has been recently clear-cut, and **(c)** an area that was clear-cut 5–10 years ago. Compare the biodiversity, soil erosion, and signs of rapid water runoff in each of the three areas.

3. Evaluate timber harvesting on private and public lands in your local area. What are the most widely used harvesting methods? Try to document any harmful environmental impacts. Have the economic benefits to the community outweighed any harmful environmental effects?

4. Use the library and Internet to find one example of a successful ecological restoration project not discussed in this chapter and one that failed. For each example, describe the strategy used, the ecological principles involved, and why the project succeeded or failed.

5. Survey the condition of a nearby wetland, coastal area, river, or stream. Has its condition improved or deteriorated during the last 10 years? What local, state, or national efforts are being used to protect this aquatic system? Develop a plan for protecting this system.

6. Make a concept map of this chapter's major ideas, using the section heads and subheads and the key terms (in boldface). Look on the website for this book for information about making concept maps.

INTERNET STUDY RESOURCES AND RESOURCES FOR FURTHER READING AND RESEARCH

The website for this book contains helpful study aids and many ideas for further reading and research. Log on to

http://biology.brookscole.com/miller 10

and click on the Chapter-by-Chapter area. Choose Chapter 17 and select a resource:

■ Flash Cards allows you to test your mastery of the Terms and Concepts to Remember for this chapter.

■ Tutorial Quizzes provides a multiple-choice practice quiz.

■ Student Guide to InfoTrac will lead you to Critical Thinking Projects that use InfoTrac College Edition as a research tool.

■ References lists the major books and articles consulted in writing this chapter.

- Hypercontents takes you to an extensive list of sites with news, research, and images related to individual sections of the chapter.

INFOTRAC COLLEGE EDITION

Improve your skills with InfoTrac College Edition, a searchable online database of articles from more than 700 periodicals. Log on to

http://www.infotrac-college.com

or access InfoTrac through the website for this book. Try to find the following articles:

1. Sanderson, Eric, et al. 2002. The human footprint and the last of the wild. *BioScience* 52: 891. *Keywords:* "ecological footprint" and "human footprint." Techniques and maps showing the degree of human influence on various parts of the earth's land surface.

2. Helmuth, L. 1999. Can this swamp be saved? *Science News* 155: 252. *Keywords:* "swamp" and "Florida Everglades." The restoration of the Florida Everglades is a huge undertaking. It took a long time to mess it up; will we be able to fix it?

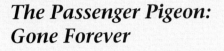

The Passenger Pigeon: Gone Forever

In the early 1800s, bird expert Alexander Wilson watched a single migrating flock of an estimated 2 million passenger pigeons darken the sky for more than 4 hours. By 1914, the passenger pigeon (Figure 18-1) had disappeared forever.

How could a species that was once the most common bird in North America become extinct in only a few decades? The answer is *humans.* The main reasons for the extinction of this species were uncontrolled commercial hunting and loss of the bird's habitat and food supply as forests were cleared to make room for farms and cities.

Passenger pigeons were good to eat, their feathers made good pillows, and their bones were widely used for fertilizer. They were easy to kill because they flew in gigantic flocks and nested in long, narrow colonies.

Commercial hunters would capture one pigeon alive, sew its eyes shut, and tie it to a perch called a stool. Soon a curious flock would land beside this "stool pigeon." Then the birds would be shot or ensnared by nets that might trap more than 1,000 birds at once.

Beginning in 1858, passenger pigeon hunting became a big business. Shotguns, traps, artillery, and even dynamite were used. Burning grass or sulfur below their roosts sometimes suffocated birds. Shooting galleries used live birds as targets. In 1878, one professional pigeon trapper made $60,000 by killing 3 million birds at their nesting grounds near Petoskey, Michigan.

By the early 1880s, only a few thousand birds remained. At that point, recovery of the species was doomed because the females laid only one egg per nest. On March 24, 1900, a young boy in Ohio shot the last known wild passenger pigeon. The last passenger pigeon on earth, a hen named Martha after Martha Washington, died in the Cincinnati Zoo in 1914. Her stuffed body is now

on view at the National Museum of Natural History in Washington, D.C.

Eventually, all species become extinct or evolve into new species. However, biologists estimate that every day 2–200 species become prematurely extinct primarily because of human activities. Studies indicate that this rate of loss of biodiversity is expected to increase (Figure 17-4, p. 413) as the human population grows, consumes more resources, disturbs more of the earth's land (Figure 8-12, p. 169), and uses more of the earth's net plant productivity that supports all species.

Figure 18-1 Passenger pigeons, extinct in the wild since 1900. The last known passenger pigeon died in the Cincinnati Zoo in 1914.

The last word in ignorance is the person who says of an animal or plant: "What good is it?" . . . If the land mechanism as a whole is good, then every part of it is good, whether we understand it or not. . . . Harmony with land is like harmony with a friend; you cannot cherish his right hand and chop off his left.

ALDO LEOPOLD

This chapter addresses the following questions:

- Are human activities causing an extinction crisis?
- Why should we care about species extinction?
- What human activities endanger wildlife?
- How can we prevent premature extinction of species?
- How can we manage game animals more sustainably?

18-1 SPECIES EXTINCTION

What Are Three Types of Species Extinction?

Biologists distinguish among three levels of species extinction. *First,* there is *local extinction.* It occurs when a species is no longer found in an area it once inhabited but is still found elsewhere in the world. Most local extinctions involve losses of one or more populations of species.

The *second* type is *ecological extinction.* It occurs when so few members of a species are left that it can no longer play its ecological roles in the biological communities where it is found.

The *third* type is *biological extinction,* when a species is no longer found anywhere on the earth (Figures 18-1 and 18-2). Biological extinction is forever.

What Are Endangered and Threatened Species?

Biologists classify species heading toward biological extinction as either *endangered* or *threatened* (Figure 18-3, pp. 450–451). An **endangered species** has so few individual survivors that the species could soon become extinct over all or most of its natural range. Unless it is protected, an endangered species makes the transition to critically endangered, then to the living dead, and finally to oblivion.

A **threatened,** or **vulnerable, species** is still abundant in its natural range but because of declining numbers is likely to become endangered in the near future. Endangered and threatened species are ecological smoke alarms.

Some species have characteristics that make them more vulnerable than others to ecological and biological extinction (Figure 18-4, p. 452).

Case Study: Why Should We Care about Bats?

Worldwide, there are 950 known species of bats—the only mammals that can fly. However, bats have two traits that make them vulnerable to extinction. *First,* they reproduce slowly. *Second,* many bat species live in huge colonies in caves and abandoned mines, which people sometimes block. This prevents them from leaving to get food and can disturb their hibernation.

Bats play important ecological roles. About 70% of all bat species feed on crop-damaging nocturnal insects and other insect pest species such as mosquitoes. This makes them the primary control agents for such insects.

In some tropical forests and on many tropical islands, *pollen-eating bats* pollinate flowers, and *fruit-eating bats* distribute plants throughout tropical forests by excreting undigested seeds.

As keystone species, such bats are vital for maintaining plant biodiversity and for regenerating large areas of tropical forest cleared by human activities. If you enjoy bananas, cashews, dates, figs, avocados, or mangos, you can thank bats.

Passenger pigeon   Great Auk   Dodo   Dusky seaside sparrow   Aepyornis (Madagascar)

Figure 18-2 Some animal species that have become prematurely extinct largely because of human activities, mostly habitat destruction and overhunting.

Grizzly bear
(threatened)

Kirkland's warbler

White top pitcher plant

Arabian oryx
(Middle East)

African elephant
(Africa)

Mojave desert tortoise
(threatened)

Swallowtail butterfly

Humpback chub

Golden lion tamarin
(Brazil)

Siberian tiger
(Siberia)

West Virginia spring
salamander

Giant panda
(China)

Whooping crane

Knowlton cactus

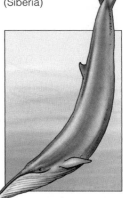

Blue whale

Mountain gorilla
(Africa)

Pine barrens
tree frog (male)

Swamp pink

Hawksbill sea turtle

El Segunda blue butterfly

Figure 18-3 *Natural capital:* species that are endangered or threatened with premature extinction largely because of human activities. Almost 30,000 of the world's species and 1,200 of those in the United States are officially listed as being in danger of becoming extinct. Most biologists believe the actual number of species at risk is much larger.

Florida manatee

Northern spotted owl (threatened)

Gray wolf

Florida panther

Bannerman's turaco (Africa)

Devil's hole pupfish

Snow leopard (Central Asia)

Symphonia (Madagascar)

Black-footed ferret

Utah prairie dog

Ghost bat (Australia)

California condor

Black lace cactus

Black rhinoceros (Africa)

Oahu tree snail

In the United States, researchers sometimes attach radio transmitters to bats to help monitor the environmental health of commercially managed forests. This can help timber companies make decisions about when and how to harvest trees while simultaneously protecting wildlife habitats.

Many people mistakenly view bats as fearsome, filthy, aggressive, rabies-carrying bloodsuckers. However, most bat species are harmless to people, livestock, and crops. In the United States, only 10 people have died of bat-transmitted disease in four decades of recordkeeping; more Americans die each year from falling coconuts.

Because of unwarranted fears of bats and lack of knowledge about their vital ecological roles, several bat species have been driven to extinction. Currently, 26% of the world's bat species, including the ghost bat (Figure 18-3), are listed as endangered or threatened. Conservation biologists urge us to view bats as valuable allies, not as enemies.

How Do Biologists Estimate Extinction Rates?
Recall that the number and diversity of the earth's species at any given time is determined by the interplay between extinction and speciation (Section 5-4, p. 100). Evolutionary biologists estimate that 99.9% of

Characteristic	Examples
Low reproductive rate (K-strategist)	Blue whale, giant panda, rhinoceros
Specialized niche	Blue whale, giant panda, Everglades kite
Narrow distribution	Many island species, elephant seal, desert pupfish
Feeds at high trophic level	Bengal tiger, bald eagle, grizzly bear
Fixed migratory patterns	Blue whale, whooping crane, sea turtles
Rare	Many island species, African violet, some orchids
Commercially valuable	Snow leopard, tiger, elephant, rhinoceros, rare plants and birds
Large territories	California condor, grizzly bear, Florida panther

Figure 18-4 Characteristics of species that are prone to ecological and biological extinction.

all the species that have ever existed are now extinct because of a combination of background extinction, mass extinctions, and mass depletions taking place over thousands to millions of years. Biologists also talk of an *extinction spasm,* where large numbers of species are lost over a period of a few centuries or at most 1,000 years.

Mass extinctions, mass depletions, and extinction spasms temporarily reduce biodiversity. But they also create opportunities for the evolution of new species to fill new or vacated ecological niches under the changed environmental conditions. According to fossil evidence, it takes at least 5 million years for such *adaptive radiations* to rebuild biological diversity after a large loss of species.

One critical question is: How have human activities affected natural extinction rates? Another is: How are we likely to affect such rates in the near future (50–100 years)? Answering these questions is difficult and controversial. One reason is that the extinction of a species typically takes a long time and is not easy to document. Another reason is that we have identified only about 1.4–1.8 million of the world's estimated 5–100 million species (a best guess is 12–14 million species). Also, we know little about most of the species we have identified.

Scientists use several methods to estimate extinction rates. One way is to observe how the number of

species present increases with the size of an area. This *species–area relationship* suggests that on average, a 90% loss of habitat causes the extinction of about 50% of the species living in that habitat. For example, if we assume about 50% of the world's existing terrestrial species live in tropical forests and about one-third of the remaining tropical forests will be cut or burned during the next few decades, the species–area relationship suggests about 1 million species in these forests will become extinct during this period.

Scientists also observe how species diversity changes at different latitudes (Figure 7-3, p. 142). They also use models to estimate the risk that a particular population of a species will become endangered or extinct within a certain time. This includes models such as the *theory of island biogeography* (Figure 7-5, p. 143). Scientists have also developed models for estimating the *minimum area of suitable habitat* needed to ensure the survival of a population in a region for a specified time.

Such estimates of extinction rates vary because of differing assumptions about the earth's total number of species (as mentioned above), the proportion of these species that are found in tropical forests (50–80%), and the rate at which tropical forests are being cleared (0.6–2% per year).

What Effects Are Human Activities Having on Extinction Rates? Before we came on the scene the estimated extinction rate was roughly one species per million species annually. This amounted to an annual extinction rate of about 0.0001% per year.

Using the methods just described, biologists conservatively estimate that the current rate of extinction is at least 1,000 to 10,000 times the rate before humans existed. This amounts to an annual extinction rate of at least 0.1% to 1% per year. This exponential rate of loss may seem small, but recall that exponential change carried out over a period of 100 years or more amounts to a huge change in numbers (Figure 1-1, p. 1).

So how many species are we losing prematurely each year? This depends on how many species are on the earth. Assuming that the extinction rate is 0.1%, each year we are losing 5,000 species per year if there are 5 million species, 14,000 if there are 14 million species, and 100,000 if there are 100 million species.

Most biologists would consider the premature loss of 1 million species over 100–200 years an extinction crisis or spasm that if kept up would lead to a mass depletion or even a mass extinction. At an extinction rate of 0.1% a year we would lose 1 million species in 200 years with 5 million species, 71 years with 14 million species, and 10 years with 100 million species. How many years would it take to lose 1 million species for each of these three species estimates if the extinction rate is 1% a year?

If 1% a year extinction rate is correct, researchers Edward O. Wilson and Stuart Primm estimate that 20% of the world's current animal and plant species could be gone by 2030 and 50% by the end of this century. In the words of biodiversity expert Norman Myers (Guest Essay, p. 102), "Within just a few human generations, we shall—in the absence of greatly expanded conservation efforts—impoverish the biosphere to an extent that will persist for at least 200,000 human generations or 20 times longer than the period since humans emerged as a species."

Most biologists consider extinction rates of 0.1%–1% to be conservative estimates for several reasons. *First,* both the exponential rate of species loss and the extent of biodiversity loss are likely to increase during the next 50–100 years because of the projected exponential growth of the world's human population and per capita resource use (Figure 17-4, p. 413).

Second, current and projected extinction rates are much higher than the global average in parts of the world that are especially rich in biodiversity and thus contain the majority of the world's species. Conservation biologists estimate that such biologically rich areas could lose 25–50% of their estimated species within a few decades. They urge us to focus our efforts on slowing the much higher rates of extinction in the world's most biodiverse countries (Figure 17-26, p. 438) and in its hot spots (Figure 17-27, p. 439) as the best way to protect much of the earth's biodiversity from being lost prematurely.

Third, we are eliminating, degrading, and simplifying many biologically diverse environments (such as tropical forests, tropical coral reefs, wetlands, and estuaries) that serve as potential colonization sites for the emergence of new species. Thus, in addition to increasing the rate of extinction, we may also be limiting long-term recovery of biodiversity by reducing the rate of speciation for some types of species. In other words, we are also creating a *speciation crisis* (Guest Essay, p. 102).

Philip Levin, Donald Levin, and other biologists also argue that the increasing fragmentation and disturbance of habitats throughout the world may increase the speciation rate for rapidly reproducing opportunist species such as rodents, cockroaches (Spotlight, p. 99) and other insects, and weeds. Thus, the real long-term threat to biodiversity from current human activities may not be a permanent decline in the number of species but a long-term erosion in the earth's variety of species and habitats.

According to biologist Jennifer Hughes and her colleagues at Stanford University, the loss of *local populations* of key species may be a better indicator of biodiversity loss than species extinction. The reason is that these local populations provide most of the ecological services (Figure 1-2, p. 3, and top half of back cover) of an area. These researchers estimate that the loss of local populations is tens of thousands of times greater than the current estimated rate of species loss.

Some people, most of them not biologists, say the current estimated extinction rates are too high and are based on inadequate data and models. Researchers agree that their estimates of extinction rates are based on inadequate data and sampling. They are also continually striving to get better data and improve the models they use to estimate extinction rates.

However, they point to clear evidence that human activities have increased the rate of species extinction and that this rate is likely to rise. According to these biologists, arguing over the numbers and waiting to get better data and models should not be used as excuses for inaction. They call for us to implement a *precautionary strategy* now to help prevent a significant decrease in the earth's genetic, species, ecological, and functional diversity.

To these biologists, we are not heeding Aldo Leopold's (Individuals Matter, p. 39) warning about preserving biodiversity as we tinker with the earth: "To keep every cog and wheel is the first precaution of intelligent tinkering."

18-2 WHY SHOULD WE CARE ABOUT SPECIES EXTINCTION?

Why Preserve Wild Species and Ecosystems? So what is all the fuss about? If all species eventually become extinct, why should we worry about losing a few more because of our activities? Does it matter that the passenger pigeon (Figure 18-1), the great auk (Figure 18-2), the green sea turtle (photo on front cover), the 80–100 remaining Florida panthers (Figure 18-3), or some unknown plant or insect in a tropical forest becomes prematurely extinct because of our activities?

If new species evolve to take the place of ones lost through extinction spasms, mass depletions, or mass extinctions, why should we care if we speed up the extinction rate over the next 50–100 years? The answer is that it will take at least 5 million years for speciation to rebuild the biodiversity we destroy during this century.

Because ecosystems are constantly changing in response to changing environmental conditions, why should we try to preserve ecosystem diversity? Does it matter that tropical forests, grasslands, wetlands, coral reefs, and other systems making up the earth's ecological diversity are being destroyed or degraded by human activities?

Conservation biologists and ecologists contend we should act to preserve the earth's overall biodiversity because its genes, species, ecosystems, and ecological processes have two types of value. One is *instrumental value* because of their usefulness to us. The other is *intrinsic value* because they exist, regardless of whether they have any usefulness to us.

The earth's biodiversity has several types of instrumental values. One is the *ecological services* they provide (Figure 1-2, p. 3, and top half of back cover). Another is the *economic services* they supply such as development of food crops by crossbreeding (Figure 16-5, p. 381) and genetic engineering (Figure 16-12, p. 388), fuelwood and lumber, paper, and medicine (Figure 17-20, p. 430).

Another is the *genetic information* in species. This allows them to adapt to changing environmental conditions and to form new species. This information is also used in genetic engineering to produce new types of crops and foods and edible vaccines for viral diseases such as hepatitis B. Carelessly eliminating many of the species making up the world's vast genetic library is like burning books before we read them.

Finally, the earth's wild plants and animals and natural ecosystems provide us with *recreational pleasure.* For example, each year Americans spend more than three times more on watching wildlife than on watching movies or professional sporting events.

Wildlife tourism, or *eco-tourism,* generates at least $500 billion per year worldwide, and perhaps twice as much. Conservation biologist Michael Soulé estimates that one male lion living to age 7 generates $515,000 in tourist dollars in Kenya but only $1,000 if killed for its skin. Similarly, over a lifetime of 60 years, a Kenyan elephant is worth about $1 million in eco-tourist revenue—many times more than its tusks are worth when sold illegally for their ivory.

Ideally, eco-tourism should not cause ecological damage. In addition, it should provide income for local people to motivate them to preserve wildlife and funds for the purchase and maintenance of wildlife preserves and conservation programs. Much eco-tourism does not meet these standards, and excessive and unregulated eco-tourism can destroy or degrade fragile areas and promote premature species extinction. The website for this chapter lists some guidelines for evaluating eco-tours.

What Is the Intrinsic Value of Biodiversity? Some people believe all wild species and ecosystems have *intrinsic value.* Each species is said to have an inherent value and a right to exist that are unrelated to its usefulness to humans.

According to this view, we have two ethical responsibilities related to the effects of our activities on the earth's natural capital. One is to protect species from becoming prematurely extinct. The other is to prevent the degradation of the world's ecosystems and its overall biodiversity.

Some people distinguish between the survival rights of plants and those of animals, mostly for practical reasons. Poet Alan Watts once said he was a vegetarian "because cows scream louder than carrots."

Other people distinguish among various types of species. For example, they might think little about getting rid of the world's mosquitoes, cockroaches (Spotlight, p. 99), rats, or disease-causing bacteria.

Some proponents go further and assert that each individual organism, not just each species, has a right to survive without human interference. Others apply this to individuals of some species but not to those of other species. Unless they are strict vegetarians, for example, some people see no harm in having others kill domesticated animals in slaughterhouses to provide them with meat, leather, and other products. However, these same people might deplore the killing of wild animals such as deer, squirrels, or rabbits.

Others emphasize the importance of preserving the whole spectrum of biodiversity by protecting entire ecosystems rather than individual species or organisms, as discussed in Chapter 17.

Biologists caution us not to focus conservation efforts primarily on protecting relatively big organisms—the plants and animals we can see and are familiar with. They remind us that the true foundation of the earth's ecosystems and ecological processes are the invisible bacteria and the algae, fungi, and other microorganisms that decompose the bodies of larger organisms and recycle the nutrients needed by all life (Connections, p. 65).

18-3 EXTINCTION THREATS FROM HABITAT LOSS AND DEGRADATION

What Is the Role of Habitat Loss and Degradation? Figure 18-5 shows the underlying and secondary causes of the endangerment and premature extinction of wild species. According to biodiversity researchers, the greatest threat to wild species is habitat loss (Figure 18-6, p. 456), degradation, and fragmentation.

Deforestation of tropical forests (p. 429) is the greatest eliminator of species. This is followed by destruction of coral reefs and wetlands, plowing of grasslands (Figure 6-18, p. 119), and pollution of aquatic systems (streams, lakes, and oceans). Globally, temperate biomes have been affected more by habitat disturbance, degradation, and fragmentation than tropical biomes because of widespread development in temperate developed countries over the past 200 years (Figure 8-12, p. 169).

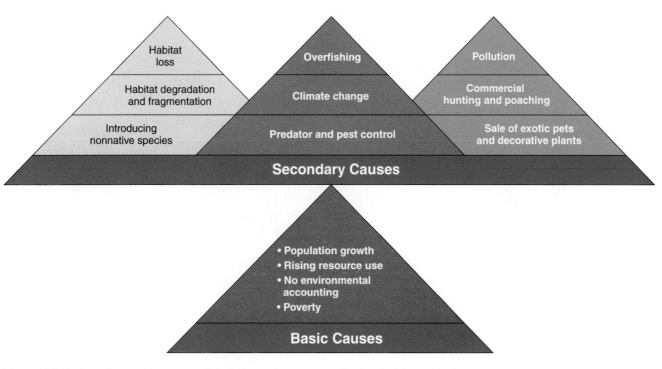

Figure 18-5 Basic and secondary causes of depletion and premature extinction of wild species. The two biggest secondary causes of wildlife depletion and premature extinction are habitat loss, fragmentation, and degradation and deliberate or accidental introduction of nonnative species into ecosystems.

According to the Nature Conservancy, the major types of habitat disturbance threatening endangered species in the United States are, in order of importance: agriculture, commercial development, water development, outdoor recreation (including off-road vehicles), livestock grazing, and pollution.

Island species, many of them *endemic species* found nowhere else on earth, are especially vulnerable to extinction. Scientists use the theory of island biogeography (Figure 7-5, p. 143) to project the number and percentage of species that would become extinct when habitats on islands are destroyed, degraded, or fragmented. They have also applied the model to the protection of national parks, tropical rain forests, lakes, and nature reserves—many of which can be viewed as *habitat islands* in an inhospitable sea of human-altered habitat.

What Is the Role of Habitat Fragmentation?
Habitat fragmentation occurs when a large, continuous area of habitat is reduced in area and divided into a patchwork of isolated areas or fragments. Habitat fragmentation causes three main problems. *First,* it decreases the sustainable population size for many species when an existing population is divided into two or more isolated subpopulations.

Second, it causes an increase in surface area or edge. This makes some species more vulnerable to predators,

competition from nonnative and pest species, wind, and fire. *Finally,* it creates barriers that limit the ability of some species to disperse and colonize new areas, find enough to eat, and find mates.

Certain types of species are especially vulnerable to local and regional extinction because of habitat fragmentation. They include species that are rare, that need to roam unhindered over a large area, and that cannot rebuild their population because of a low reproductive capacity. Also included are species with specialized niches (habitat or resource needs) and ones that are sought by people for furs, food, medicines, or other uses.

Case Study: How Do Human Activities Affect Bird Species? Approximately 70% of the world's 9,800 known bird species are declining in numbers (58%) or are threatened with extinction (12%), mostly because of habitat loss and fragmentation. A 2002 National Audubon Society study found that 25% of U.S. bird species are declining in numbers or at risk of disappearing. Figure 18-7 (p. 457) shows the ten most threatened songbird species in the United States according to a 2001 study by the National Audubon Society.

Conservation biologists view this decline of bird species as an early warning of the greater loss of biodiversity to come. Birds are excellent *environmental*

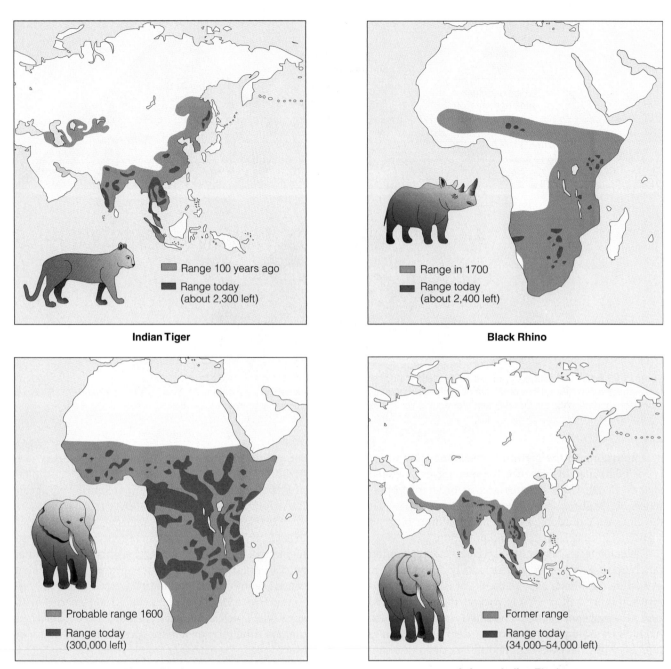

Indian Tiger

Range 100 years ago

Range today
(about 2,300 left)

Black Rhino

Range in 1700

Range today
(about 2,400 left)

African Elephant

Probable range 1600

Range today
(300,000 left)

Asian or Indian Elephant

Former range

Range today
(34,000–54,000 left)

Figure 18-6 Reductions in the ranges of four wildlife species, mostly the result of habitat loss and hunting. What will happen to these and millions of other species when the world's human population doubles and per capita resource consumption rises sharply in the next few decades? (Data from International Union for the Conservation of Nature and World Wildlife Fund)

indicators because they live in every climate and biome, respond quickly to environmental changes in their habitats, and are easy to track and count.

Besides serving as indicator species, birds play important ecological roles. They include helping control populations of rodents and insects (which decimate many tree species), pollinating a variety of flowering plants, spreading plants throughout their habitats (by consuming plant seeds and excreting them in their droppings), and scavenging dead animals.

| Cerulean warbler | Sprague's pipit | Bichnell's thrush | Black-capped vireo | Golden-cheeked warbler |

| Florida scrub jay | California gnatcatcher | Kirtland's warbler | Henslow's sparrow | Bachman's warbler |

Figure 18-7 *Natural capital:* the ten most threatened species of U.S. songbirds according to a 2002 study by the National Audubon Society. Most of these species are threatened because of habitat loss and fragmentation.

18-4 EXTINCTION THREATS FROM NONNATIVE SPECIES

What Harm Do Nonnative Species Cause in the United States? After habitat loss and degradation, the deliberate or accidental introduction of nonnative species into ecosystems is the biggest cause of animal and plant extinctions. The nonnative invaders arrive from other continents as stowaways on aircraft, in the ballast water of tankers and cargo ships, and as hitchhikers on imported products such as wooden packing crates.

Figure 18-8 (p. 458) shows some of the estimated 50,000 nonnative species that have been deliberately or accidentally introduced into the United States. Here are three pieces of *bad news* about such nonnative species. *First,* they cause at least $137 billion per year in damages and pest control costs—an average loss of $16 million per hour—according to a 2000 study by David Pimentel.

Second, they threaten 49% of the more than 1,200 endangered and threatened species in the United States (and 95% of those in Hawaii), according to the U.S. Fish and Wildlife Service. *Third,* they are blamed for about 68% of fish extinctions in the United States between 1900 and 2000.

What Is the Role of Deliberately Introduced Species? Deliberate introduction of nonnative species can be beneficial or harmful depending on the species and where they are introduced. We depend heavily on nonnative organisms for ecosystem services, food, shelter, medicine, and aesthetic enjoyment.

The problem is that some introduced species have no natural predators, competitors, parasites, or pathogens to control their numbers in their new habitats. This can enable them to reduce or wipe out populations of many native species and trigger ecological disruptions. One example of a deliberately introduced plant species is the *kudzu* ("CUD-zoo") *vine,* which

Deliberately Introduced Species

Purple looselife

European starling

African honeybee
("Killer bee")

Nutria

Salt cedar
(Tamarisk)

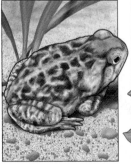

Marine toad
(Giant toad)

Water hyacinth

Japanese beetle

Hydrilla

European wild boar
(Feral pig)

Accidentally Introduced Species

Sea lamprey
(attached to lake trout)

Argentina fire ant

Brown tree snake

Eurasian muffe

Common pigeon
(Rock dove)

Formosan termite

Zebra mussel

Asian long-horned beetle

Asian tiger mosquito

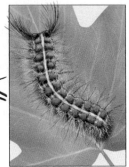

Gypsy moth larvae

Figure 18-8 (facing page) Some nonnative species that have been deliberately or accidentally introduced into the United States.

grows rampant in the southeastern United States (see Case Study below).

Deliberately introduced animal species have also caused ecological and economic damage. One case is the estimated 1 million *European wild (feral) boars* (Figure 18-8) found in parts of Florida and other states. They hog food from endangered animals, root up farm fields, and cause traffic accidents. Game and wildlife officials have had little success in controlling their numbers with hunting and trapping and say there is no way to stop them. Another example is the estimated 30 million *feral cats* and 41 million *outdoor pet cats* introduced into the United States; they kill about 568 million birds per year.

Case Study: Deliberate Introduction of the Kudzu Vine In the 1930s, the *kudzu vine* was imported from Japan and planted in the southeastern United States to help control soil erosion. It does control erosion. But it is so prolific and difficult to kill that it engulfs hillsides, trees, abandoned houses and cars, stream banks, patches of forest, and anything else in its path (Figure 18-9).

This vine, sometimes called "the vine that ate the South," has spread throughout much of the southern United States. It could spread as far north as the Great Lakes by 2040 if projected global warming occurs.

Kudzu is considered a menace in the United States. But Asians use a powdered kudzu starch in beverages, gourmet confections, and herbal remedies for a range of diseases. A Japanese firm has built a large kudzu farm and processing plant in Alabama and ships the extracted starch to Japan.

Although kudzu can engulf and kill trees it could eventually help save trees from loggers. Research at the Georgia Institute of Technology indicates that kudzu may be used as a source of tree-free paper.

What Is the Role of Accidentally Introduced Species? In the late 1930s, the extremely aggressive *Argentina fire ant* (Figure 18-8) was introduced accidentally into the United States in Mobile, Alabama. The ants may have arrived on shiploads of lumber or coffee imported from South America or by hitching a ride in the soil-containing ballast water of cargo ships.

Without natural predators, fire ants have spread rapidly by land and water (they can float) throughout the South, from Texas to Florida and as far north as Tennessee and North Carolina. They are also found in Puerto Rico and recently invaded California and New Mexico.

Figure 18-9 Kudzu taking over a house and a truck. This vine can grow 0.3 meter (1 foot) per day and is now found from east Texas to Florida and as far north as southeastern Pennsylvania and Illinois. Kudzu was deliberately introduced into the United States for erosion control, but it cannot be stopped by being dug up or burned. Grazing by goats and repeated doses of herbicides can destroy it, but goats and herbicides also destroy other plants, and herbicides can contaminate water supplies. Recently, scientists have found a common fungus (*Myrothecium verrucaria*) that can kill kudzu within a few hours, apparently without harming other plants.

Wherever fire ants have gone, they have sharply reduced or wiped out up to 90% of native ant populations. Their extremely painful stings have killed deer fawns, birds, livestock, pets, and at least 80 people allergic to their venom. These ants have invaded cars and caused accidents by attacking drivers, made crop fields unplowable, disrupted phone service and electrical power, caused some fires by chewing through underground cables, and cost the United States an estimated $600 million per year.

Widespread pesticide spraying in the 1950s and 1960s temporarily reduced fire ant populations. In the end, however, this chemical warfare hastened the advance of the rapidly multiplying fire ant in two ways. *First*, it reduced populations of many native ant species. *Second*, it promoted development of genetic resistance to heavily used pesticides in the rapidly multiplying fire ants through directional natural selection (Figure 5-5, left, p. 97).

Researchers at the U.S. Department of Agriculture are experimenting with use of biological controls such as a tiny parasitic Brazilian fly and a pathogen imported from South America to reduce fire ant plantations. Before widespread use of these biological control agents, researchers must be sure they will not

The Termite from Hell

Forget killer bees and fire ants. The homeowner's nightmare is the Formosan termite (Figure 18-8). It is the most voracious, aggressive, and prolific of more than 2,000 known termite species.

These termites invaded the Hawaiian Islands by 1900. They probably arrived on the U.S. mainland during or soon after World War II in wooden packing materials on military cargo ships that docked in southern ports such as New Orleans, Louisiana, and Houston, Texas.

Formosan termites consume wood nine times faster than domestic termites. Their huge colonies can contain up to 73 million insects—73 times as many as a nest of native termites.

Domestic termite colonies have to be in contact with soil, which can be treated around the outside of a building to reduce infestation. However, Formosan termites can establish a colony in an attic. This makes applying pesticides around the perimeter of a building virtually worthless in fighting these pests.

Over the past decade, the Formosan termite has caused more damage in New Orleans than hurricanes, floods, and tornadoes combined. Infestations affect as many as 90% of the houses and one-third of the oak trees in the city. The famous French Quarter has one of the world's most concentrated infestations.

Once confined to Louisiana, these termites have invaded at least a dozen other states, including Alabama, Florida, Mississippi, North and South Carolina, Texas, and California. They cause at least $1.1 billion in damage each year.

Most pesticides do not work on these termites. In New Orleans, the U.S. Department of Agriculture is using a variety of techniques all at once in an attempt to control the species in a heavily infested 15-block area of the French Quarter. They hope to develop other techniques for dealing with these invaders elsewhere.

Critical Thinking

What important ecological roles do termites play in nature? If the Formosan termite and other termite species could be eradicated (a highly unlikely possibility), would you favor doing this? Explain.

Japanese ship carrying tires to a Houston, Texas, recapping plant. Since then, they have spread to 25 states. Aggressive biters, these mosquitoes can transmit 17 potentially fatal tropical viruses, including dengue fever, yellow fever, and forms of encephalitis. University of Kentucky scientists have flown a radar-equipped plane over the Ohio River Valley to find the Asian tiger mosquito. Radar cannot detect the mosquitoes, but it can spot hidden mounds of scrap tires, where the insects breed. Another harmful invader is the *Formosan termite* (Case Study, left).

Solutions: How Can We Reduce the Threat from Nonnative Species? Once a nonnative species gets established in an ecosystem, its wholesale removal is almost impossible—somewhat like trying to get smoke back into a chimney or trying to unscramble an egg. Thus the best way to limit the harmful impacts of nonnative species is to prevent them from being introduced and becoming established.

There are four major ways to do this. *First,* identify major characteristics that allow species to become successful invaders and the types of ecosystems that are vulnerable to invaders (Figure 18-10). Such information can be used to screen out potentially harmful invaders.

Second, inspect types of goods coming into a country that are likely to contain invader species. *Third,* identify major harmful invader species and pass international laws banning their transfer from one country to another (as is now done for endangered species).

Finally, require ships to discharge their ballast (bilge) water and replace it with salt water at sea before entering ports, or require them to sterilize such water or pump nitrogen into it (Individuals Matter, p. 461).

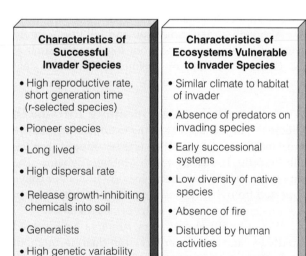

Characteristics of Successful Invader Species	Characteristics of Ecosystems Vulnerable to Invader Species
• High reproductive rate, short generation time (r-selected species) • Pioneer species • Long lived • High dispersal rate • Release growth-inhibiting chemicals into soil • Generalists • High genetic variability	• Similar climate to habitat of invader • Absence of predators on invading species • Early successional systems • Low diversity of native species • Absence of fire • Disturbed by human activities

Figure 18-10 Some general characteristics of successful invader species and ecosystems vulnerable to invading species.

cause problems for native ant species or become pests themselves.

Another accidentally introduced species is the *Asian tiger mosquito* (Figure 18-8), which can breed in scrap tires. In 1985 they arrived in the United States in a

Killing Invader Species and Saving Shipping Companies Money

INDIVIDUALS MATTER

A large cargo ship typically has a dozen or more ballast tanks below deck. Each tank is the size of a high school gymnasium and holds millions of gallons of water.

When a ship takes on cargo and leaves a port it sucks water into its ballast tanks to keep it low in the water, submerge its rudder, and help maintain stability. The enormous amount of water sucked into the ship also contains large numbers of fish, crabs, clams, and other species found in the port's local waters.

When the ship's cargo is removed at its destination its ballast water is released until the ship is loaded again. This dumps millions of foreign organisms into rivers and bays. Thus, cargo ships moving about 80% of the goods traded internationally play a major role in the release of nonnative aquatic organisms into various parts of the world.

In 2002, researchers Mario Tamburri and Kerstin Wasson found that pumping nitrogen gas into ballast tanks while a ship is at sea virtually eliminates dissolved oxygen in the ballast water. This saves the shipping industry money by reducing corrosion of a ship's steel compartments. In addition, within three days it kills most fish, crabs, clams, and other potential invader species lurking in the ballast tanks.

18-5 EXTINCTION THREATS FROM HUNTING AND POACHING

What Is the Role of Commercial Hunting and Poaching? The international trade in wild plants and animals is roughly a $20 billion a year business. According to a 2002 study by the World Life Conservation Society, the value of illegal international trade in endangered and threatened species or their parts is put at $4–5 billion per year. Organized crime has moved into illegal wildlife smuggling because of the huge profits involved. Smuggling wildlife (including many endangered species) is the third largest illegal cross-border smuggling activity after arms and drugs.

The demand for illegal wildlife comes mostly from wealthy collectors and other consumers in Asia, the Middle East, North America, and Europe. At least two-thirds of all live animals illegally smuggled around the world die in transit.

Worldwide, some 622 animal and plant species face extinction, mostly because of illegal trade. To poachers, a *live mountain gorilla* is worth $150,000, a

gyrfalcon $120,000, a *panda pelt* $100,000 (with only about 1,000 pandas thought to remain in the wild), a *chimpanzee* $50,000, an *Imperial Amazon macaw* $30,000, and an endangered *golden lion tamarin* $20,000. A *rhinoceros horn* is worth as much as $28,600 per kilogram ($13,000 per pound) because of its use in dagger handles in the Middle East and as a fever reducer and alleged aphrodisiac in China and other parts of Asia.

In much of West and Central Africa, wildlife in the form of *bushmeat* is an important source of protein for many local people (Figure 18-11). In Africa's Congo basin, the demand for bushmeat has replaced habitat loss as the biggest threat to the animals whose meat is eaten or sold. The bushmeat trade is also increasing in Asia, the Caribbean, and Central and South America.

In 1950, an estimated 100,000 tigers existed in the world. Despite international protection, today less than 7,500 tigers remain in the wild (about 4,000 of them in India), mostly because of habitat loss and poaching for their furs and bones. Bengal tigers are at risk because a tiger fur sells for $100,000 in Tokyo. With the body parts of a single tiger worth $5,000–20,000, it is not surprising that illegal hunting has skyrocketed, especially in India. Without emergency action, few or no tigers may be left in the wild within 20 years.

As commercially valuable species become endangered, the demand for them on the black market soars. This increases their chances of premature extinction from poaching. Most poachers are not caught. Even when caught, they find the money to be made far outweighs the risk of fines and the much smaller risk of imprisonment.

Figure 18-11 *Bushmeat*, such as this gorilla head, is consumed as a source of protein by local people in parts of West Africa and sold in the national and international marketplace. You can find bushmeat on the menu in Cameroon and the Congo in West Africa as well as in Paris, France, and Brussels, Belgium. Much of this is provided through organized and illegal poaching.

© Karl Ammann, Biosynergy Institute

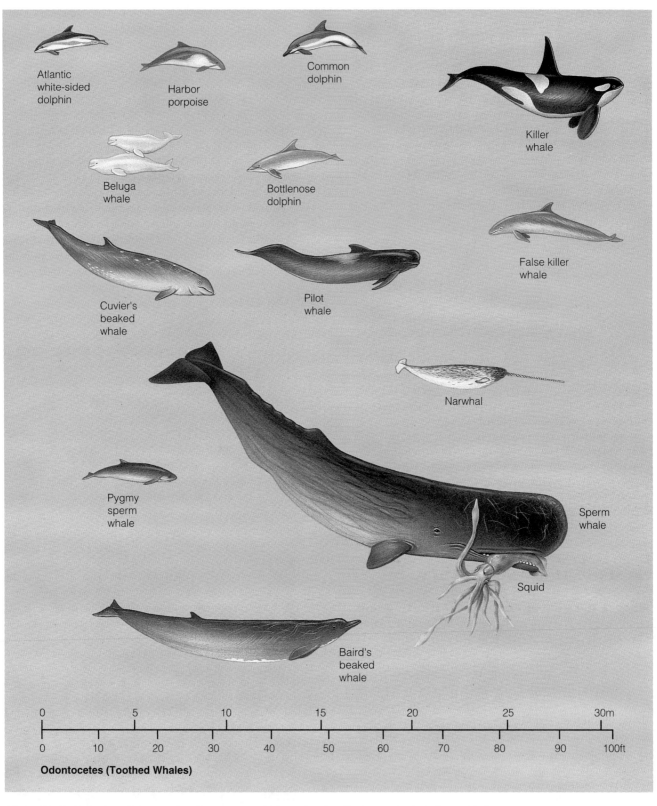

Figure 18-12 *Natural capital:* examples of cetaceans, which can be classified as toothed whales and baleen whales.

Case Study: Should Commercial Whaling Be Resumed? *Cetaceans* are an order of mostly marine mammals ranging in size from the 0.9-meter (3-foot) porpoise to the giant 15- to 30-meter (50- to 100-foot) blue whale. They are divided into two major groups: *toothed whales* and *baleen whales* (Figure 18-12).

Toothed whales, such as the porpoise, sperm whale, and killer whale (orca), bite and chew their food and

Humpback whale

Bowhead whale

Right whale

Minke whale

Blue whale

Feeding on krill

Fin whale

Sei whale

Gray whale

Mysticetes (Baleen Whales)

feed mostly on squid, octopus, and other marine animals. *Baleen whales*, such as the blue, gray, humpback, and finback, are filter feeders. Instead of teeth, they have several hundred horny plates made of baleen, or whalebone, that hang down from the upper jaw. These plates filter plankton from the seawater, especially tiny shrimplike krill (Figure 4-18, p. 77). Baleen whales are the more abundant of the two cetacean groups.

Near Extinction of the Blue Whale

CASE STUDY

The biologically endangered blue whale (Figure 18-12) is the world's largest animal. Fully grown, it is more than 30 meters (100 feet) long—longer than three train boxcars—and weighs more than 25 elephants. The adult has a heart the size of a Volkswagen Beetle car, and some of its arteries are so big that a child could swim through them.

Blue whales spend about 8 months of the year in Antarctic waters. There they find an abundant supply of krill (Figure 4-18, p. 77), which they filter daily by the trillions from seawater. During the winter, they migrate to warmer waters, where their young are born.

Before commercial whaling began, an estimated 200,000 blue whales roamed the Antarctic Ocean. Today, the species has been hunted to near biological extinction for its oil, meat, and bone.

A combination of prolonged overharvesting and certain natural characteristics of blue whales caused its decline. Their huge size made them easy to spot. They were caught in large numbers because they grouped together in their Antarctic feeding grounds. They also take 25 years to mature sexually and have only one offspring every 2–5 years. This low reproductive rate makes it difficult for the species to recover once its population falls beneath a certain threshold.

Blue whales have not been hunted commercially since 1964 and have been classified as an endangered species since 1975. Despite this protection, some marine biologists fear that too few blue whales remain for the species to recover and avoid extinction. Others believe that with continued protection they will make a slow comeback.

Their extinction could also be hastened by melting polar ice because of climate change. This melting reduces populations of krill, which are a main food source for blue whales (Figure 4-18, p. 77).

Critical Thinking

Opponents of commercial whaling contend that resuming commercial whaling for some whale species such as minke, pilot, and gray could lead to illegal harvesting of blue whales. Japan contends that excess population of minkes in Antarctic waters is threatening the blue whale population by consuming much of the krill they eat (Figure 4-18, p. 77). What scientific evidence would you require to resolve this issue?

Whales are fairly easy to kill because of their large size and their need to come to the surface to breathe. Mass slaughter has become efficient with the use of fast ships, harpoon guns, and inflation lances (which pump dead whales full of air and make them float).

Whale harvesting, mostly in international waters, has followed the classic pattern of a tragedy of the commons (p. 7), with whalers killing an estimated 1.5 million whales between 1925 and 1975. This overharvesting reduced the populations of 8 of the 11 major species to the point at which it no longer paid to hunt and kill them (*commercial extinction*), and drove some commercially prized species such as the giant blue whale to the brink of biological extinction (Case Study, above).

In 1946, the International Convention for the Regulation of Whaling established the International Whaling Commission (IWC) to regulate the whaling industry by setting annual quotas to prevent overharvesting and commercial extinction. However, IWC quotas often were based on inadequate data or ignored by whaling countries. Without any powers of enforcement, the IWC has not been able to stop the decline of most commercially hunted whale species to the point at which they were commercially extinct.

In 1970, the United States stopped all commercial whaling and banned all imports of whale products. Under intense pressure from environmentalists, the U.S. government, and governments of many nonwhaling countries in the IWC, the IWC has imposed a moratorium on commercial whaling since 1986. As a result, the estimated number of whales killed commercially worldwide dropped from 42,480 in 1970 to about 1,200 in 2002.

Despite the ban, IWC members Japan and Norway have continued to hunt certain whale species, and Iceland resumed hunting whales in 2002. Japan, Norway, Iceland, Russia, and a growing number of small tropical island countries (which Japan has brought into the IWC to support its position) continue working to overthrow the IWC ban on commercial whaling and reverse the international ban on buying and selling whale products.

Proponents of lifting the ban on commercial whaling argue that whaling should be allowed because it has long been a traditional part of the economies and cultures of countries such as Japan, Iceland, and Norway. They also contend that the ban is based on emotion, not updated scientific estimates of whale populations. According to IWC estimates, the population of minke whales (which Japan and Norway continue to hunt) now numbers about 1 million (with an estimated 760,000 in Antarctic waters) and that of pilot whales about 1.4 million. They see no scientific reason for not resuming controlled hunting of these species, along with sperm, Bryde's, and gray whales (in the eastern Pacific).

Environmentalists disagree for several reasons. Some argue that whales are peaceful, intelligent, sensitive, and highly social mammals that pose no threat to humans and should be protected for ethical reasons. Also, some question the estimates of minke, Bryde's, gray, and pilot whale populations, noting the inaccuracy of past IWC estimates of whale populations. Finally, most fear that opening the door to any commercial whaling may eventually lead to widespread harvesting of most whale species by weakening current international disapproval and legal sanctions against commercial whaling.

18-6 OTHER EXTINCTION THREATS

What Is the Role of Predator and Pest Control? People try to exterminate species that compete with them for food and game animals. For example, U.S. fruit farmers exterminated the Carolina parakeet around 1914 because it fed on fruit crops. The species was easy prey because when one member of a flock was shot, the rest of the birds hovered over its body, making themselves easy targets.

African farmers kill large numbers of elephants to keep them from trampling and eating food crops. Many ranchers, farmers, and hunters in the United States support the killing of coyotes, wolves, and other species that can prey on livestock and on species prized by game hunters.

Since 1929, U.S. ranchers and government agencies have poisoned 99% of North America's prairie dogs because horses and cattle sometimes step into the burrows and break their legs. This has nearly wiped out the endangered black-footed ferret (Figure 18-3; about 600 left in the wild), which preyed on the prairie dog.

What Is the Role of the Market for Exotic Pets and Decorative Plants? The global legal and illegal trade in wild species for use as pets is a huge and very profitable business. However, for every live animal captured and sold in the pet market, an estimated 50 other animals are killed.

About 25 million U.S. households have exotic birds as pets, 85% of them imported. More than 60 bird species, mostly parrots, are endangered or threatened because of this wild bird trade. According to the U.S. Fish and Wildlife Service, collectors of exotic birds may pay $10,000 for a threatened hyacinth macaw smuggled out of Brazil; however, during its lifetime, a single macaw left in the wild might yield as much as $165,000 in tourist income. A 1992 study suggested that keeping a pet bird indoors for more than 10 years doubles a person's chances of getting lung cancer from inhaling tiny particles of bird dander.

Other wild species whose populations are depleted because of the pet trade include amphibians, reptiles, mammals, and tropical fish (taken mostly from the coral reefs of Indonesia and the Philippines). Divers catch tropical fish by using plastic squeeze bottles of cyanide to stun them. For each fish caught alive, many more die. In addition, the cyanide solution kills the coral animals that create the reef, which is a center for marine biodiversity.

Some exotic plants, especially orchids and cacti, are endangered because they are gathered (often illegally) and sold to collectors to decorate houses, offices, and landscapes. A collector may pay $5,000 for a single rare orchid, and a single rare mature crested saguaro cactus can earn cactus rustlers as much as $15,000.

What Are the Roles of Climate Change and Pollution? Most natural climate changes in the past have taken place over long periods of time (Figure 13-2, p. 281), which gave species more time to adapt and evolve to survive in these changing environmental conditions. But human activities such as greenhouse gas emissions and deforestation are projected to bring about rapid climate change during this century (Figure 13-9, p. 287).

Projected climate changes may benefit some species (such as those adapted to a warmer climate) but can have harmful effects on other species. According to a 2000 study by the World Wildlife Fund, global warming could increase extinction by altering one-third of the world's wildlife habitats by 2100. This includes 70% of the habitat in high-altitude arctic and boreal biomes.

Another problem is that some species may not have enough time to adapt or migrate (Figure 13-14, p. 292). However, species that can thrive in a warmer climate may expand their populations and range.

Pollution threatens populations and species in a number of ways. A major extinction threat is from the unintended effects of pesticides. According to the U.S. Fish and Wildlife Service, each year in the United States pesticides kill about 20% of the country's beneficial honeybee colonies and kill more than 67 million birds and 6–14 million fish. They also threaten about 20% of the country's endangered and threatened species.

18-7 PROTECTING WILD SPECIES FROM DEPLETION AND EXTINCTION: THE RESEARCH AND LEGAL APPROACH

How Can Bioinformatics Help Protect Biodiversity? To protect biodiversity, we need basic information about the biology and ecology of wild species. Examples of such information include the name, description, and distribution of each species. For each species we also need information about the status of its populations, habitat requirements, and interactions with other species.

Bioinformatics is the applied science of managing, analyzing, and communicating biological information. It involves building computer databases to organize and store useful biological information, developing computer tools to find, visualize, and analyze the information, and communicating the information, especially using the Internet.

One example is *Species 2000,* an Internet-based global research project designed to provide information about all the earth's known species as the baseline data set for studies of global biodiversity. Users worldwide will be able to verify the scientific name, status, and classification of any known species via the Species Locator and have online access to data on any species drawn from an array of participating databases.

How Can International Treaties Help Protect Endangered Species? Several international treaties and conventions help protect endangered or threatened wild species. One of the most far reaching is the 1975 *Convention on International Trade in Endangered Species (CITES).* This treaty, now signed by 160 countries, lists some 900 species that cannot be commercially traded as live specimens or wildlife products because they are in danger of extinction. The treaty also restricts international trade of 29,000 other species because they are at risk of becoming threatened.

The *good news* is that CITES has helped reduce international trade in many threatened animals, including elephants, crocodiles, and chimpanzees. The *bad news* is that the effects of this treaty are limited because enforcement is difficult and spotty and convicted violators often pay only small fines. Also, member countries can exempt themselves from protecting any listed species, and much of the highly profitable illegal trade in wildlife and wildlife products goes on in countries that have not signed the treaty.

The *Convention on Biological Diversity (CBD),* ratified by 172 countries, legally binds signatory governments to reversing the global decline of biological diversity. However, its implementation has proceeded slowly because some key countries (such as the United States) have not ratified the treaty, and it has no severe penalties or other enforcement mechanisms.

How Can National Laws Help Protect Endangered Species? The United States controls imports and exports of endangered wildlife and wildlife products through two laws. One is the *Lacey Act of 1900.* It prohibits transporting live or dead wild animals or their parts across state borders without a federal permit. The other is the *Endangered Species Act of 1973 (ESA),* which was amended in 1982 and 1988. It makes it illegal for Americans to import or trade in any product made from an endangered or threatened species unless it is used for an approved scientific purpose or to enhance the survival of the species.

The ESA authorizes the National Marine Fisheries Service (NMFS) to identify and list endangered and threatened ocean species and the USFWS to identify and list all other endangered and threatened species. These species cannot be hunted, killed, collected, or injured in the United States.

Any decision by either agency to add or remove a species from the list must be based on biological information only, not on economic or political considerations. However, economic factors can be used in deciding whether and how to protect endangered habitat and in developing recovery plans for listed species. The act also forbids federal agencies to carry out, fund, or authorize projects that would jeopardize an endangered or threatened species or destroy or modify the critical habitat it needs to survive. On private lands, fines and even jail sentences can be imposed to ensure protection of the habitats of endangered species.

Between 1973 and 2002, the number of U.S. species on the official endangered and threatened list increased from 92 to about 1,260 species (about 59% of them plants and 41% animals). According to a 2000 study by the Nature Conservancy, about 33% of the country's species are at risk of extinction, and 15% of them are at high risk. This amounts to about 30,000 species, compared to the roughly 1,260 species currently protected under the ESA. The study found that many of the country's rarest and most imperiled species are concentrated in a few hot spots (Figure 18-13).

The ESA generally requires the secretary of the interior to designate and protect the critical habitat needed for the survival and recovery of each listed species. So far, only about 125 designated critical habitats have been established.

Getting listed is only half the battle. Next, the USFWS or the NMFS is supposed to prepare a plan to help the species recover. The *good news* is that final recovery plans had been developed and approved for

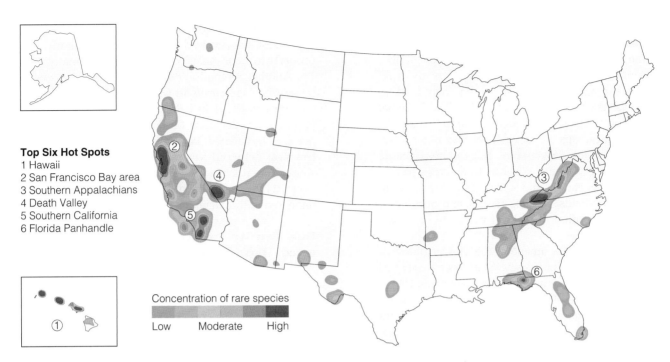

Top Six Hot Spots
1 Hawaii
2 San Francisco Bay area
3 Southern Appalachians
4 Death Valley
5 Southern California
6 Florida Panhandle

Concentration of rare species

Low   Moderate   High

Figure 18-13 *Natural capital:* biodiversity hot spots in the United States. This map shows areas that contain the largest concentrations of rare and potentially endangered species. (Data from State Natural Heritage Programs, the Nature Conservancy, and Association for Biodiversity Information)

about 77% of the endangered or threatened U.S. species by 2002. The *bad news* is that about half of those plans exist only on paper.

The ESA requires all commercial shipments of wildlife and wildlife products to enter or leave the country through one of nine designated ports. Few illegal shipments are confiscated (Figure 18-14) because the 60 USFWS inspectors can examine only about one-fourth of at least 90,000 shipments that enter and leave the United States each year. Even if caught, many violators are not prosecuted, and convicted violators often pay only a small fine.

Steve Hillebrand/U.S. Fish and Wildlife Service

Figure 18-14 Confiscated products made from endangered species. Because of a scarcity of funds and inspectors, probably no more than one-tenth of the illegal wildlife trade in the United States is discovered. The situation is even worse in most other countries.

How Can Private Landowners Be Encouraged to Protect Endangered Species?

One problem is that the ESA has encouraged some developers, timber companies, and other private landowners to avoid government regulation by managing their land to reduce its use by endangered species. The National Association for Homebuilders, for example, has published practical tips for developers and other landowners to avoid ESA issues. Suggestions include planting crops, plowing fields between crops to prevent native vegetation and endangered species from occupying the fields, clearing forests, and burning or managing vegetation to make it unsuitable for local endangered species.

In 1982, Congress amended the ESA to allow the secretary of the interior to use *habitat conservation plans (HCPs)*. They are designed to strike a compromise between the interests of private landowners and the interests of endangered and threatened species without reducing the recovery chances of a protected species. With an HCP, landowners, developers, or loggers are allowed to destroy some critical habitat or kill all or part of an endangered or threatened species population on private land in exchange for taking steps to protect that species.

Such protective measures might include setting aside a part of the species' habitat as a preserve, paying to relocate the species to another suitable habitat, or paying money to have the government buy suitable habitat elsewhere. Once the plan is approved it cannot be changed, even if new data show that the plan is inadequate to protect a species and help it recover.

Some wildlife conservationists support this approach because it can head off use of evasive techniques and reduce political pressure to weaken or eliminate the ESA. However, there is growing concern that such plans are being developed without enough scientific evaluation of their effects on a species' recovery. Two suggestions for improving HCPs are to develop better scientific standards for such plans and to have the plans reviewed by a scientific advisory committee. Another suggestion is to require private landowners to make more compensating efforts as an insurance policy when there are not enough data to assess the impact of the plan on the affected species.

In 1999, the USFWS approved two new approaches for encouraging private landowners to protect threatened or endangered species. One consists of *safe harbor agreements* in which landowners voluntarily agree to take specified steps to restore, improve, or maintain habitat for threatened or endangered species located on their land. In return, landowners get technical help. They also receive government assurances that the natural resources involved will not face future re-

strictions once the agreement is over, and that after the agreement has expired landowners can return the property to its original condition without penalty.

Another method is the use of *voluntary candidate conservation agreements* in which landowners agree to take specific steps to help conserve a species whose population is declining but is not yet listed as endangered or threatened. Participating landowners receive technical help and assurances that no additional resource use restrictions will be imposed on the land covered by the agreement if the species is listed as endangered or threatened in the future.

How Can We Protect Threatened and Endangered Marine Species?

The Endangered Species Act has also been used to protect a number of endangered and threatened marine reptiles and mammals (especially whales, seals, sea lions, and turtles; Figure 18-15).

Three of eight major sea turtle species (Figure 18-16, p. 470) are endangered (Kemp's ridley, leatherbacks, and hawksbills), and the rest are threatened. Two major threats to the world's endangered or threatened sea turtles are loss or degradation of beach habitat (where they come ashore to lay their eggs) and legal and illegal taking of their eggs. Other threats include unintentional capture and drowning by commercial fishing boats (especially shrimp trawlers) and increased use of the turtles as sources of food, medicinal ingredients, tortoiseshell (for jewelry), and leather from their flippers. In China, for example, some sea turtles sell for up to $1,500.

Two major problems hinder protecting marine biodiversity by protecting endangered species. One is lack of knowledge about marine species. The other is the difficulty in monitoring and enforcing treaties to protect marine species, especially in the open ocean.

Should the Endangered Species Act Be Weakened?

Opponents of the ESA contend it has not worked and has caused severe economic losses by hindering development on private land. Since 1995, efforts to weaken the ESA have included the following suggested changes:

- Make protection of endangered species on private land voluntary.

- Have the government compensate landowners if it forces them to stop using part of their land to protect endangered species.

- Make it harder and more expensive to list newly endangered species by requiring government wildlife officials to navigate through a series of hearings and peer review panels.

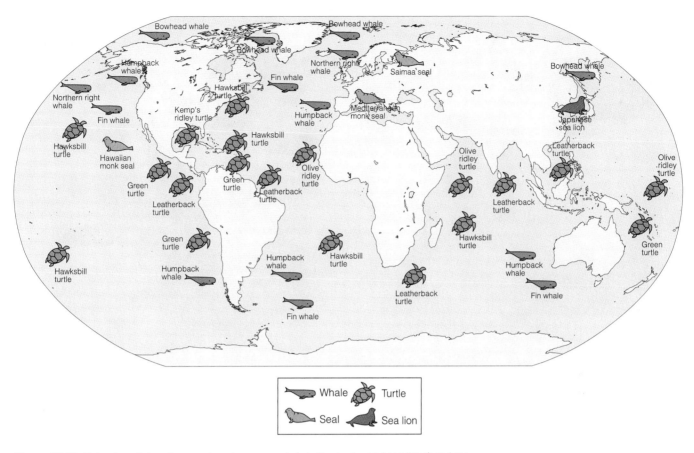

Figure 18-15 *Natural capital:* endangered marine mammals (whales, seals, and sea lions) and reptiles (turtles). Many marine fish, seabirds (Figure 5-7, p. 100), and invertebrate species are also threatened.

- Allow the secretary of the interior to permit a listed species to become extinct without trying to save it and to determine whether a species should be listed.

- Allow the secretary of the interior to give any state, county, or landowner permanent exemption from the law, with no requirement for public notification or comment.

- Prohibit the public from commenting on or bringing lawsuits to change poorly designed HCPs.

Should the Endangered Species Act Be Strengthened? Most conservation biologists and wildlife scientists contend that the ESA has not been a failure. They also cite several pieces of information to refute the charge that the ESA has caused severe economic losses.

One is that since 1979, only about 0.05% of the almost 200,000 projects evaluated by the USFWS have been blocked or canceled as a result of the ESA. The act specifically allows for economic concerns. By law, a decision to list a species must be based solely on science. However, once a species is listed, economic considerations can be weighed against species protection in protecting critical habitat and designing and implementing recovery plans.

Furthermore, the act authorizes a special cabinet-level panel, nicknamed the "God Squad," to exempt any federal project from having to comply with the act if the economic costs are too high.

Finally, the act allows the government to issue permits and exemptions to landowners with listed species living on their property and use habitat conservation plans, safe harbor agreements, and candidate conservation agreements to bargain with private landowners.

A study by the U.S. National Academy of Sciences recommended three major changes to make the ESA more scientifically sound and effective. *First,* greatly increase the meager funding for implementing the act. *Second,* develop recovery plans more quickly. *Third,* when a species is first listed, establish a core survival habitat as a temporary emergency measure that could support the species for 25–50 years.

Most biologists and wildlife conservationists believe the United States should develop a new system

Figure 18-16 *Natural capital:* major species of sea turtles that have roamed the seas for 150 million years, showing their relative adult sizes. Three of these species (Kemp's ridley, leatherbacks, and hawksbills) are endangered, and the rest are threatened as a result of human activities.

to protect and sustain its biological diversity and ecosystem function based on three principles:

- Find out what species and ecosystems the country has.

- Locate and protect the most endangered ecosystems (Figure 18-13) and species.

- Give private landowners who agree to help protect specific endangered species and ecosystems financial incentives (tax breaks and write-offs), technical help, and assurances of no additional requirements in the future (safe harbor and candidate conservation agreements).

What Has the Endangered Species Act Accomplished? Critics of the ESA call it an expensive failure because only a few species have been removed from the endangered list. Most biologists strongly disagree that the act has been a failure, for several reasons.

First, species are listed only when they are in serious danger of extinction. This is like setting up a poorly funded hospital emergency room that takes only the most desperate cases, often with little hope for recovery, and saying it should be shut down because it has not saved enough patients.

Second, it takes decades for most species to become endangered or threatened. Thus it usually takes decades to bring a species in critical condition back to the point where it can be removed from the list. Expecting the ESA (which has been in existence only since 1973) to quickly repair the biological depletion of many decades is unrealistic.

Third, the most important measure of the law's success is that the conditions of almost 40% of the listed species are stable or improving. A hospital emergency room taking only the most desperate cases and then stabilizing or improving the condition of 40% of its patients would be considered an astounding success.

Finally, the federal endangered species budget was only $126 million in 2002—about one-third the cost of one C-17 transport plane or 44¢ a year per U.S. citizen.

To most biologists, it is amazing that so much has been accomplished in stabilizing or improving the condition of almost 40% of the listed species on a shoestring budget.

Should We Try to Protect All Endangered and Threatened Species? Because of limited funds, information, and trained personnel, only a few endangered and threatened species can be saved. Some analysts suggest concentrating the limited funds available for preserving threatened and endangered wildlife on species that have the best chance for survival, play the most important or keystone ecological roles in an ecosystem, and are potentially useful for agriculture, medicine, or industry.

Recent research suggests we should not evaluate the health of an ecosystem on the basis of sheer numbers of species. Instead, we should focus on which species play keystone roles and are tolerant to environmental change such as acid deposition, climate change, and toxins. In addition to protecting keystone species, biologists call for protection of *keystone resources*. Examples are salt licks and mineral pools that provide essential minerals for wildlife, deep pools in streams and springs that serve as refuges for fish and other aquatic species during dry periods, and hollow tree trunks used as breeding sites and homes for many bird and mammal species.

Others oppose selective protection of species on ethical grounds or contend we do not have enough biological information to make such evaluations. Proponents argue that in effect we are already deciding by default which species to save, and the selective approach is more effective and a better use of limited funds than the current one. What do you think?

18-8 PROTECTING WILD SPECIES FROM DEPLETION AND EXTINCTION: THE SANCTUARY APPROACH

How Can Wildlife Refuges and Other Protected Areas Help Protect Endangered Species? In 1903, President Theodore Roosevelt established the first U.S. federal wildlife refuge at Pelican Island, Florida. Since then the National Wildlife Refuge System has grown to 524 refuges (Figure 17-9, p. 416). Some 34 million Americans visit these refuges each year to hunt, fish, hike, or watch birds and other wildlife.

More than three-fourths of the refuges are wetlands for protecting migratory waterfowl. About 20% of the U.S. endangered and threatened species have habitats in the refuge system, and some refuges have been set aside for specific endangered species. These have helped Florida's key deer, the brown pelican, and the trumpeter swan to recover.

Conservation biologists call for setting aside more refuges to help protect endangered plants. They also urge Congress and state legislatures to allow abandoned military lands that contain significant wildlife habitat to become national or state wildlife refuges.

According to a General Accounting Office study, activities considered harmful to wildlife occur in nearly 60% of the nation's wildlife refuges. According to a 2001 Audubon Society study, the U.S. refuge system is in a state of crisis. Pollutants and invasive species are degrading many refuges, and the system has a $1.6 billion backlog of unmet operations and maintenance needs. A 2002 study by the National Wildlife Refuge Association found that invasions by nonnative species are wreaking havoc on wildlife refuges across the country and causing more than $100 billion in damages each year.

Research shows that fully protected marine reserves can help build up populations of depleted fish and other aquatic species (p. 442). However, only a tiny fraction of the world's oceans have been set aside as fully protected marine reserves.

Can Gene Banks, Botanical Gardens, and Farms Help Save Most Endangered Plant Species? *Gene* or *seed banks* preserve genetic information and endangered plant species by storing their seeds in refrigerated, low-humidity environments. Most of the world's 50 seed banks have focused on storing seeds of the approximately 100 plant species that provide about 90% of the food consumed by humans. However, some banks are devoting more attention to storing seeds for a wider range of species that may be threatened with extinction or a loss of genetic diversity.

Scientists urge the establishment of many more such banks, especially in developing countries. However, some species cannot be preserved in gene banks, and maintaining the banks is very expensive.

The world's 1,600 *botanical gardens* and *arboreta* contain about 4 million living plants, representing about 80,000 species or approximately 30% of the world's known plant species. The world's largest botanical garden is the Royal Botanical Gardens of England at Kew. It contains an estimated 25,000 species of living plants, or about 10% of the world's total. About 2,700 of these species are listed as threatened.

Botanical gardens are increasingly focusing on cultivation of rare and endangered plant species. In the United States, the Center for Plant Conservation coordinates efforts by 28 botanical gardens to store nearly 600 endangered U.S. plant species and propagate and reintroduce some of them into the wild.

Botanical gardens also help educate an estimated 150 million visitors a year about the need for plant conservation. However, these sanctuaries have too little storage capacity and too little funding to preserve most of the world's rare and threatened plants.

We can take pressure off some endangered or threatened species by raising them on *farms* for commercial sale. One example is the use of farms in Florida to raise alligators for their meat and hides. Another example is butterfly farms in Papua New

Guinea, where many butterfly species are threatened by habitat destruction and fragmentation, commercial overexploitation, and environmental degradation.

Can Zoos and Aquariums Help Protect Most Endangered Animal Species? Zoos, aquariums, game parks, and animal research centers are being used to preserve some individuals of critically endangered animal species, with the long-term goal of reintroducing the species into protected wild habitats.

Two techniques for preserving endangered terrestrial species are egg pulling and captive breeding. *Egg pulling* involves collecting wild eggs laid by critically endangered bird species and then hatching them in zoos or research centers. In *captive breeding,* some or all of the wild individuals of a critically endangered species are captured for breeding in captivity, with the aim of reintroducing the offspring into the wild.

Other techniques for increasing the populations of captive species include artificial insemination, surgical implantation of eggs of one species into a surrogate mother of another species (embryo transfer), use of incubators, and cross-fostering, in which the young of a rare species are raised by parents of a similar species. Scientists also use computer databases of the family lineages of species in zoos and DNA analysis to match individuals for mating and to prevent genetic erosion through inbreeding.

Recently there has been increased use of *genetic cloning* to help revive endangered species. Proponents urge zoos and wildlife managers to collect and freeze cells of endangered species for cloning. However, several conservation groups fear this could divert attention and very limited funds from what endangered species need most: protected habitats and in some cases protection from poaching.

Captive breeding programs at zoos in Phoenix, Arizona, and San Diego and Los Angeles, California, temporarily saved the nearly extinct Arabian oryx (Figure 18-3). This large antelope once lived throughout the Middle East. By 1960, it had been hunted nearly to extinction in the wild (with only about 48 left) by people riding in jeeps and helicopters and wielding rifles and machine guns. Between 1980 and 1996, the return of oryx bred in captivity to protected habitats in the Middle East allowed the population to reach about 400. However, since 1996, poachers have been capturing these animals for sale to private zoos. As a result, the wild population has been reduced to about 100, and the species is again facing extinction.

Before release, many captive-raised animals and birds need extensive training to help them learn how to find food and shelter, avoid predators, and interact in social groups. Because released populations often are small and thus vulnerable, scientists try to estimate the minimum viable population needed for an introduced species to survive.

After more than two decades of captive breeding efforts, only a handful of endangered species have been returned to the wild. Examples shown in Figure 18-3 include the black-footed ferret, California condor, Arabian oryx, and golden lion tamarin. However, most reintroductions fail because of lack of suitable habitat, inability of individuals bred in captivity to survive in the wild, or renewed overhunting or capture of some returned species (such as the Arabian oryx).

Lack of space and money limit efforts to maintain populations of endangered animal species in zoos and research centers. The captive population of each species must number 100–500 individuals to avoid extinction through accident, disease, or loss of genetic diversity through inbreeding. Recent genetic research indicates that 10,000 or more individuals are needed for an endangered species to maintain its capacity for biological evolution.

Today's zoos and research centers have space to preserve healthy and sustainable populations of only about 925 of the 2,000 large vertebrate species that could vanish from the planet. According to one estimate, using all the space in the 201 accredited U.S. zoos for captive breeding could sustain only about 100 large animal species on a long-term basis.

Public aquariums that exhibit unusual and attractive fish and some marine animals such as seals and dolphins help educate the public about the need to protect such species. In the United States, more than 35 million people visit aquariums each year. However, unlike some zoos, public aquariums have not served as effective gene banks for endangered marine species, especially marine mammals that need large volumes of water.

Instead of seeing zoos and aquariums as sanctuaries, some critics see most of them as prisons for once wild animals. They also contend that zoos and aquariums foster the notion that we do not need to preserve large numbers of wild species in their natural habitats.

Some zoos display many of their animals in natural settings and use this to educate visitors about the need to preserve natural habitats for wildlife throughout the world. However, in some zoos elephants and other animals that roam free in habitat areas during the day are chained up in a barn at night.

Some people criticize zoos and aquariums for putting on shows with animals wearing clothes, riding bicycles, or performing tricks. They contend that this fosters the idea that the animals are there primarily to entertain us by doing things people do and in the process raising money for their keepers.

Regardless of their benefits and drawbacks, conservation biologists point out that zoos, aquariums, and botanical gardens are not a biologically or economically feasible solution for most of the world's current endangered species and the much larger number expected to become endangered over the next few decades.

18-9 WILDLIFE MANAGEMENT

How Can Wildlife Populations Be Managed?
Wildlife management involves manipulating wildlife populations (especially game species) and their habitats for their welfare and for human benefit. The wildlife management approach manages game species for sustained yield. This is done by using laws to regulate hunting and fishing, establishing harvest quotas, developing population management plans, improving wildlife habitat, and using international treaties to protect migrating game species such as waterfowl.

In the United States, funds for state game management programs come from the sale of hunting and fishing licenses and from federal taxes on hunting and fishing equipment. Two-thirds of the states also have provisions on state income tax returns that allow individuals to contribute money to state wildlife programs.

The first step in wildlife management is to decide which species are to be managed in a particular area. This is controversial. Ecologists and conservation biologists emphasize preserving biodiversity, wildlife conservationists are concerned about endangered species, bird-watchers want the greatest diversity of bird species, and hunters want large populations of game species.

In the United States and other developed countries, most wildlife management is devoted to producing surpluses of game animals and birds for hunters. Since the passage of the Wildlife Restoration Act in 1937, there has been a large increase in the number of game animals (such as white-tailed deer, wild turkeys, Rocky Mountain elk, and pronghorn antelope) sought by many sport hunters.

After goals have been set, the wildlife manager must develop a management plan. Ideally, this plan is based on principles of *ecological succession* (Figure 7-11, p. 152, and Figure 7-12, p. 153) and *wildlife population dynamics* (Figure 8-3, p. 162, and Figure 8-7, p. 164). Managers must also have an understanding of the cover, food, water, space, and other habitat needs of each species to be managed. In addition, they should know the estimated *maximum sustained yield (MSY)* of a population, in which harvested individuals are not removed faster than they can be replaced through reproduction. A wildlife manager must also consider the number of potential hunters, their likely success

rates, and the regulations for preventing excessive hunting.

This information is difficult, expensive, and time consuming to obtain. It involves much educated guesswork and trial and error, which is why wildlife management is as much an art as a science. Management plans must also be sensitive to political pressures from conflicting groups and to budget constraints.

How Useful Is Sport Hunting in Managing Wildlife Populations? Most developed countries use sport hunting laws to manage populations of game animals. Licensed hunters are allowed to hunt only during certain times of the year to protect animals during their mating season. Limits are also set on the size, number, and sex of animals that can be killed, and on the number of hunters allowed in a given area.

Close control of sport hunting is difficult. Accurate data on game populations may not exist and may cost too much to get. People in communities near hunting areas, who benefit from money spent by hunters, may seek to have hunting quotas raised.

Sport hunting is also controversial. Proponents argue that without carefully regulated sport hunting, deer and other large game animals exceed the carrying capacity of their habitats and destroy vegetation they and other species need. For example, between 1900 and 2002, the estimated population of white-tailed deer in the United States increased from 500,000 to 25–30 million. Many of these deer invade subdivisions and eat shrubs and home gardens, raid farmers' fields and orchards, threaten rare plants and animals in some areas, and help spread Lyme disease (carried by deer ticks) to humans. As a result, deer in the United States cause at least $1.1 billion in losses per year. Deer are also involved in more than 500,000 vehicle collisions each year. These collisions kill more Americans than any other incidents with wild animals and injure thousands. Wildlife biologist William Porter estimates that without hunting, the U.S. deer population would be five times its current level and cause extensive ecological damage.

Deer are not the only problem. There are other wild species whose populations have grown because of hunting and trapping restrictions and lack of natural predators. They include *bear* populations in 20 states, *beavers* (which can chop down trees around waterways and threaten drinking water supplies), *elk* herds in Colorado, *golden eagles* (which can prey on sheep in the West), *cormorant birds* (which consume large numbers of catfish grown in aquaculture ponds in Alabama, Mississippi, and Louisiana), protected *sea lions* (which in the Northwest consume large numbers of endangered salmon and steelhead fish), *moose* (which in parts of the Northeast are destroying vegetation along some riverbanks), *nonmigratory Canada geese* (which have become pests in many areas), and *mountain lions* in parts of the West. As wildlife populations increase and more people move into areas inhabited by wildlife, the ethical dilemma over increased use of sport hunting and trapping will become more acute.

According to its proponents, sport hunting also provides recreational pleasure for millions of people (15 million in the United States), and stimulates local economies. It also provides money through sales of hunting licenses and taxes on firearms and ammunition (more than $1.7 billion since 1937) that is used to buy, restore, and maintain wildlife habitats and to support wildlife research in the United States. Environmental groups such as the Sierra Club and Defenders of Wildlife support carefully controlled sport hunting as a way to preserve biological diversity by helping prevent depletion of native species of plants and animals.

Some individuals and groups, including the Humane Society, oppose sport hunting. They argue that it inflicts unnecessary pain and suffering on wild animals, most of which are not killed to supply food that humans need for survival. They also contend that game managers often create a surplus of game animals by deliberately eliminating their natural predators (such as wolves). Then the managers claim that hunters should harvest the surplus to prevent habitat degradation or starvation of the game. Instead of eliminating natural predators, say opponents, wildlife managers should reintroduce them in some areas to reduce the need for sport hunting (p. 411).

Supporters of hunting point out that populations of many game species (such as deer) are so large that predators such as wolves cannot control them, and introducing predators can lead to the loss of nearby livestock. However, critics of hunting contend that the argument for hunting deer (which make up only 2% of the 200 million animals that hunters kill in the United States each year) is a smokescreen to allow killing of many other game species that do not threaten vegetation. What do you think?

How Can Populations of Migratory Waterfowl Be Managed? Migratory birds—including ducks, geese, swans, and many songbirds—journey north and south from one habitat to another each year, usually to find food, suitable climate, and other conditions necessary for reproduction. Such bird species use many different routes called *flyways*, but only about 15 are considered major routes (Figure 18-17). Some countries along such flyways have entered into agreements and treaties to protect crucial habitats needed by such species, both along their migration routes and at each end of their journeys.

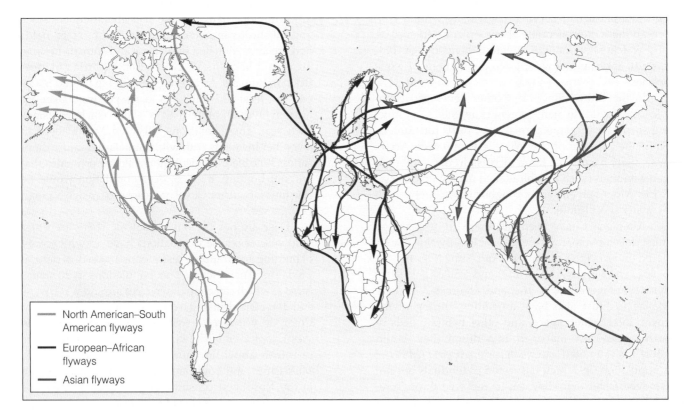

Legend:
- North American–South American flyways
- European–African flyways
- Asian flyways

Figure 18-17 Major *flyways* used by migratory birds, mostly waterfowl. Each route has a number of subroutes.

Wildlife officials manage waterfowl by regulating hunting, protecting existing habitats, and developing new habitats (including artificial nesting sites, ponds, and nesting islands). More than 75% of the federal wildlife refuges in the United States are wetlands used by migratory birds. Local and state agencies and private conservation groups such as Ducks Unlimited, the Audubon Society, and the Nature Conservancy have also established waterfowl refuges.

Since 1934, the Migratory Bird Hunting and Conservation Stamp Act has required waterfowl hunters to buy a duck stamp each season they hunt. Revenue from these sales goes into a fund to buy land and easements for the benefit of waterfowl.

Protecting these and other species from premature extinction and preserving the earth's biodiversity is a difficult, controversial, and challenging responsibility. See the website for this chapter for some actions you can take to help protect wildlife and preserve biodiversity.

We know what to do. Perhaps we will act in time.
EDWARD O. WILSON

REVIEW QUESTIONS

1. Define the boldfaced terms in this chapter.

2. What factors led to the extinction of the passenger pigeon in the United States?

3. Distinguish among *local*, *ecological*, and *biological* extinction of a species.

4. Distinguish between *endangered* and *threatened* species. List eight characteristics that make species vulnerable to biological extinction.

5. What are some important ecological roles of bats, and why are they vulnerable to extinction?

6. Give two reasons why it is so difficult to estimate how human activities are affecting extinction rates.

7. List four methods scientists use to estimate how human activities are affecting extinction rates. What is the estimated range of annual extinction rates caused by human activities based on using these methods? How do these rates compare with the estimated annual extinction rate before humans appeared?

8. List three reasons why most biologists believe their current estimates of extinction rates from human activities are probably too low.

9. Explain why we should care about species extinction. Distinguish between *instrumental* and *intrinsic* values of wildlife.

10. What are four underlying causes of the population reduction and extinction of wild species? Describe how each of the following factors contributes to the premature extinction of species, and give an example of a species af-

fected by each factor: **(a)** habitat loss and degradation, **(b)** habitat fragmentation, **(c)** deliberately introduced nonnative species, **(d)** accidentally introduced nonnative species, **(e)** commercial hunting and illegal hunting (poaching), **(f)** predator and pest control, **(g)** the legal and illegal market for exotic pets and decorative plants, and **(h)** climate change and pollution.

11. What percentage of the world's known bird species are **(a)** declining in numbers and **(b)** threatened with extinction? List three reasons why birds are excellent indicators of environmental conditions.

12. Give two examples of **(a)** deliberately introduced harmful species and **(b)** accidentally introduced harmful species.

13. What are major characteristics of **(a)** successful invader species and **(b)** ecosystems vulnerable to invader species?

14. List four ways to reduce the threat from nonnative species. Explain how pumping nitrogen into the ballast water of cargo ships may help reduce the transfer of nonnative species into aquatic systems throughout much of the world.

15. Distinguish between *toothed whales* and *baleen whales*. Explain how commercial whaling has been an example of the tragedy of the commons.

16. Describe factors leading to the near extinction of the blue whale.

17. What is *bioinformatics*, and how can it be used to help protect biodiversity?

18. List the benefits and limitations of protecting species using **(a)** the Convention on International Trade in Endangered Species (CITES) and **(b)** the Endangered Species Act (ESA) in the United States.

19. Distinguish among *habitat conservation plans*, *safe harbor agreements*, and *candidate conservation plans* used as ways to help implement the ESA.

20. List four threats to marine sea turtles.

21. Give reasons why you believe the ESA has been a failure or a success. List measures that would strengthen and weaken the ESA in the United States. List the major accomplishments of the Endangered Species Act.

22. List three guidelines that could be used to decide which species to protect.

23. Summarize the advantages and disadvantages of using the following to help protect endangered species: **(a)** wildlife refuges, **(b)** gene banks and botanical gardens, **(c)** zoos and animal research centers, and **(d)** aquariums. Distinguish between *egg pulling* and *captive breeding*.

24. What is *wildlife management*, and what are the major steps in managing wildlife? What are the major limitations of wildlife management?

25. Describe the advantages and disadvantages of using sport hunting to help manage wildlife populations.

26. Describe methods used to manage and protect populations of migratory waterfowl.

CRITICAL THINKING

1. Discuss your gut-level reaction to the following statement: "Eventually all species become extinct. Thus it does not really matter that the passenger pigeon is extinct and that the blue whale, the whooping crane, the California condor, and the world's remaining rhinoceros and tiger species are endangered mostly because of human activities." Be honest about your reaction, and give arguments for your position.

2. **(a)** Do you accept the ethical position that each species has the inherent right to survive without human interference, regardless of whether it serves any useful purpose for humans? Explain. Would you extend this right to *Anopheles* mosquito species, which transmit malaria, and to infectious bacteria? **(b)** Do you believe each individual of an animal species has an inherent right to survive? Explain. Would you extend such rights to individual plants and microorganisms and to tigers that have killed people? Explain.

3. Explain why you agree or disagree with **(a)** using animals for research, **(b)** keeping animals captive in a zoo, and **(c)** killing surplus animals produced by a captive-breeding program in a zoo when no suitable habitat is available for their release.

4. Your lawn and house are invaded by fire ants, which can cause painful bites. What would you do?

5. Which of the following statements best describes your feelings toward wildlife? **(a)** As long as it stays in its space, wildlife is OK. **(b)** As long as I do not need its space, wildlife is OK. **(c)** I have the right to use wildlife habitat to meet my own needs. **(d)** When you have seen one redwood tree, fox, elephant, or some other form of wildlife you have seen them all, so lock up a few of each species in a zoo or wildlife park and do not worry about protecting the rest. **(e)** Wildlife should be protected.

6. How do you feel about bats? What would you do if you found bats flying around your yard at night?

7. List your three favorite wild species. Examine why they are your favorites. Are they cute and cuddly looking, like the giant panda and the koala? Do they have humanlike qualities, like apes or penguins that walk upright? Are they large, like elephants or blue whales? Are they beautiful, like tigers and monarch butterflies? Are any of them plants? Are any of them species such as bats, sharks, snakes, or spiders that most people are afraid of? Are any of them microorganisms that help keep you alive? Reflect on what your choice of favorite species tells you about your attitudes toward most wildlife.

8. Environmental groups in a heavily forested state want to restrict logging in some areas to save the habitat of an endangered squirrel. Timber company officials argue that the well-being of one type of squirrel is not as important as the well-being of the many families affected

if the restriction causes them to lay off hundreds of workers. If you had the power to decide this issue, what would you do and why? Can you come up with a compromise?

9. Explain why you agree or disagree with each of the proposals made on pp. 468–469 to **(a)** weaken the ESA and **(b)** strengthen the act.

10. Explain why some developers and extractors of resources on public land oppose the development of databases such as Species 2000 and national databases of the species found in a particular country.

11. If you moved to the suburbs and deer invaded your yard and ate your shrubs and vegetables, what would you do?

12. Should commercial whaling of selected species with fairly high population levels be resumed? Explain. How would you establish, monitor, and enforce annual commercial whaling quotas?

13. Congratulations! You are in charge of preventing the premature extinction of the world's existing species from human activities. What would be the three major components of your program to accomplish this goal?

PROJECTS

1. Make a log of your own consumption of all products for a single day. Relate your level and types of consumption to the decline of wildlife species.

2. Identify examples of habitat destruction or degradation in your community that have had harmful effects on the populations of various wild plant and animal species. Develop a management plan for rehabilitating these habitats and species.

3. Choose a particular endangered animal or plant species and use the library or the Internet to find out what is being done to protect it from extinction. Develop a protection plan for this species.

4. Make a concept map of this chapter's major ideas, using the section heads and subheads and the key terms (in boldface). Look on the website for this book for information about making concept maps.

INTERNET STUDY RESOURCES AND RESOURCES FOR FURTHER READING AND RESEARCH

The website for this book contains helpful study aids and many ideas for further reading and research. Log on to

http://biology.brookscole.com/miller 10

and click on the Chapter-by-Chapter area. Choose Chapter 18 and select a resource:

■ Flash Cards allows you to test your mastery of the Terms and Concepts to Remember for this chapter.

- Tutorial Quizzes provides a multiple-choice practice quiz.

- Student Guide to InfoTrac will lead you to Critical Thinking Projects that use InfoTrac College Edition as a research tool.

- References lists the major books and articles consulted in writing this chapter.

- Hypercontents takes you to an extensive list of sites with news, research, and images related to individual sections of the chapter.

INFOTRAC COLLEGE EDITION

Improve your skills with InfoTrac College Edition, a searchable online database of articles from more than 700 periodicals. Log on to

http://www.infotrac-college.com

or access InfoTrac through the website for this book. Try to find the following articles:

1. Levin, D. A. 2002. Hybridization and extinction: In protecting rare species, conservationists should consider the dangers of interbreeding, which compound the better known threats to wildlife. *American Scientist* 90: 254. *Keywords:* "hybridization" and "extinction." An excellent introduction to the science of population genetics and species conservation.

2. Seneen, S. 2002. Going, going . . . exotic species are decimating America's native wildlife. *E Magazine* 13: 34. *Keywords:* "exotic species" and "native wildlife." Human aesthetics when it comes to wildlife may be having profound effects on populations of native species.

Bitter Lessons from Chernobyl

Chernobyl is a chilling word recognized around the globe as the site of a major nuclear disaster (Figure 19-1). On April 26, 1986, a series of explosions took place in one of the reactors in a nuclear power plant in Ukraine—then part of the Soviet Union. The explosions blew the massive roof off the reactor building and flung radioactive debris and dust high into the atmosphere. A huge radioactive cloud spread over much of Belarus, Russia, Ukraine, and other parts of Europe and eventually encircled the planet.

According to various UN studies, here are some consequences of this disaster, caused by poor reactor design and human error. By 2001, an estimated 8,000 people had died prematurely from radiation-related diseases because of the accident. The Ukrainian Health Ministry says that 125,000 people have died and 3.5 million people have become ill because of the accident. Almost 400,000 people had to leave their homes. Most were not evacuated until at least 10 days after the accident.

In 2003—seventeen years after the accident—an area of the former Soviet Union about the size of the state of Florida remains highly contaminated with radioactivity. Despite the danger, between 100,000 and 200,000 people either illegally remained or have returned to live inside this highly radioactive zone.

The accident exposed more than half a million people to dangerous levels of radioactivity and has caused several thousand cases of thyroid cancer. The total cost of the accident will reach at least $358 billion. This is many times more than the value of all nuclear electricity ever generated in the former Soviet Union.

The environmental refugees evacuated from the Chernobyl region had to leave their possessions behind. They also had to say good-bye to lush green wheat fields and blossoming apple trees, land their families had farmed for generations, cows and goats that would be shot because the grass they ate was radioactive, and their cats and dogs poisoned with radioactivity. They will not be able to return without exposing themselves to potentially harmful doses of ionizing radiation.

Chernobyl taught us that *a major nuclear accident anywhere is a nuclear accident everywhere.*

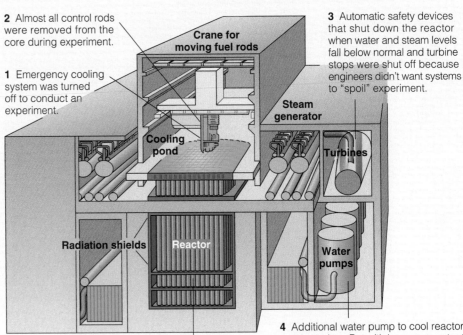

Figure 19-1 Major events leading to the Chernobyl nuclear power plant accident on April 26, 1986, in the former Soviet Union. The accident happened because engineers turned off most of the reactor's automatic safety and warning systems (to keep them from interfering with an unauthorized safety experiment). Also, the safety design of the reactor was inadequate (there was no secondary containment shell, as in Western-style reactors), and a design flaw led to unstable operation at low power. After the reactor exploded, crews exposed themselves to lethal levels of radiation to put out fires and encase the shattered reactor in a hastily constructed concrete tomb. This 19-story concrete tomb is sagging and full of holes that allow water to seep in and radioactive dust to drift out. Building a new tomb for the reactor will cost at least $1.5 billion—money the Ukrainian government does not have.

2 Almost all control rods were removed from the core during experiment.

1 Emergency cooling system was turned off to conduct an experiment.

3 Automatic safety devices that shut down the reactor when water and steam levels fall below normal and turbine stops were shut off because engineers didn't want systems to "spoil" experiment.

Crane for moving fuel rods

Cooling pond

Steam generator

Turbines

Radiation shields

Reactor

Water pumps

4 Additional water pump to cool reactor was turned on. But with low power output and extra drain on system, water didn't actually reach reactor.

5 Reactor power output was lowered too much, making it too difficult to control.

Typical citizens of advanced industrialized nations each consume as much energy in six months as typical citizens in developing countries consume during their entire life.

MAURICE STRONG

This chapter addresses the following questions:

- How should we evaluate energy alternatives?
- What are the advantages and disadvantages of oil?
- What are the advantages and disadvantages of natural gas?
- What are the advantages and disadvantages of coal?
- What are the advantages and disadvantages of conventional nuclear fission, breeder nuclear fission, and nuclear fusion?

19-1 EVALUATING ENERGY RESOURCES

What Types of Energy Do We Use? *Some 99% of the energy used to heat the earth and all of our buildings comes directly from the sun.* Without this direct input of essentially inexhaustible solar energy, the earth's average temperature would be $-240°C$ ($-400°F$), and life as we know it would not exist.

This direct input of solar energy also produces other *indirect forms of renewable solar energy.* Examples are wind, falling and flowing water (hydropower), and biomass (solar energy converted to chemical energy stored in chemical bonds of organic compounds in trees and other plants).

Commercial energy sold in the marketplace makes up the remaining 1% of the energy we use to supplement the earth's direct input of solar energy. Most commercial energy comes from extracting and burning *nonrenewable mineral resources* obtained from the earth's crust, primarily carbon-containing fossil fuels (oil, natural gas, and coal; Figure 19-2).

What Types of Energy Does the World Depend On? Over the past 60,000 years, major cultural changes (Section 1-7, p. 11) and technological advances have greatly increased energy use per person (Figure 19-3). As a result of such advances, about 82% of

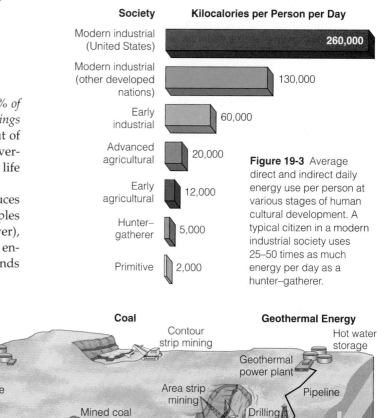

Figure 19-3 Average direct and indirect daily energy use per person at various stages of human cultural development. A typical citizen in a modern industrial society uses 25–50 times as much energy per day as a hunter–gatherer.

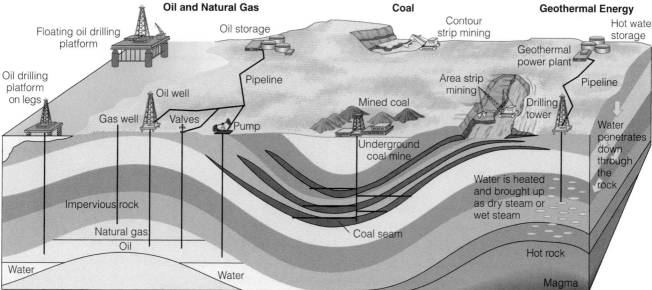

Figure 19-2 *Natural capital:* important nonrenewable energy resources that can be removed from the earth's crust are coal, oil, natural gas, and some forms of geothermal energy. Nonrenewable uranium ore is also extracted from the earth's crust and then processed to increase its concentration of uranium-235, which can be used as a fuel in nuclear reactors to produce electricity.

the commercial energy consumed in the world comes from *nonrenewable* energy resources (76% from fossil fuels and 6% from nuclear power; Figure 19-4, left).

Figure 19-5 shows the world's increase in consumption of energy by fuel type between 1970 and 2000, with projections to 2020. In developing countries, the main source of energy for heating and cooking for roughly half the world's population is renewable energy from *biomass* (mostly fuelwood and charcoal made from fuelwood)—as long as wood supplies are not harvested faster than they are replenished. Some *bad news* is that many of the world's people in developing countries face a *fuelwood shortag*e that is expected to get worse because of unsustainable harvesting of fuelwood. Many of these people (especially women and children) die prematurely from breathing particles emitted by burning wood indoors in open fires and poorly designed primitive stoves.

What Types of Energy Does the United States Depend On? The United States is the world's largest energy user. With only 4.6% of the population, in 2002 it used 26% of the world's commercial energy. This is more energy than the combined total used by the next four largest energy-consuming countries—China, Russia, Japan, and Germany. In contrast, India, with 16% of the world's people, uses about 3% of the world's commercial energy.

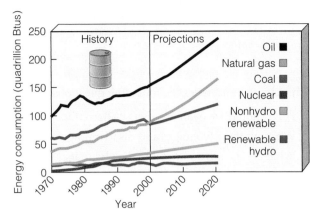

Figure 19-5 Global energy consumption by fuel type, 1970–2000, with projections to 2020. (A Btu is a British thermal unit, a standard measure of heat for value comparison of various fuels.) (Data from U.S. Department of Energy)

About 93% of the commercial energy used in the United States comes from *nonrenewable* energy resources (85% from fossil fuels and 8% from nuclear power; Figure 19-4, right).

Figure 19-6 shows energy consumption by fuel in the United States from 1970 to 2000, with projections to 2020. Note that the main projected trends between 2000 and 2020 are increased use of oil and natural gas and a leveling off of coal use.

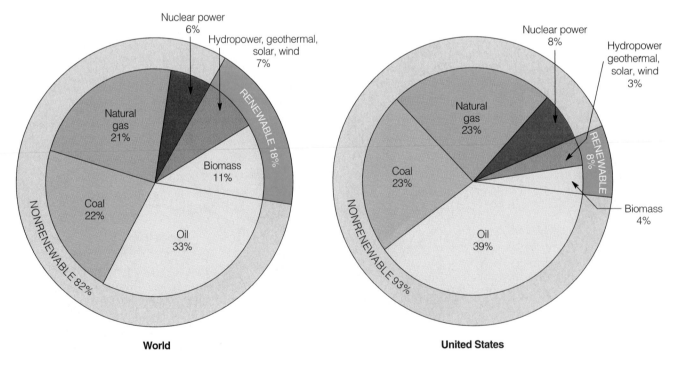

Figure 19-4 Commercial energy use by source for the world (left) and the United States (right) in 2001. Commercial energy amounts to only 1% of the energy used in the world; the other 99% is direct solar energy received from the sun and is not sold in the marketplace. (Data from U.S. Department of Energy, British Petroleum, and Worldwatch Institute)

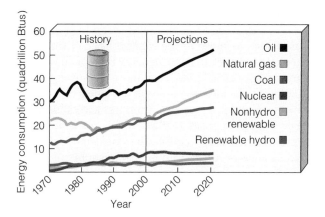

Figure 19-6 Energy consumption by fuel in the United States, 1970–2000, with projections to 2020. (Data from U.S. Department of Energy, *Annual Energy Review, 2002*)

An important environmental, economic, and political issue is what energy resources the United States might be using by 2050 and 2100. Figure 19-7 shows shifts in use of various sources of energy in the United States since 1800 and one scenario showing projected changes to a hydrogen–solar energy age by 2100.

According to the U.S. Department of Energy and the Environmental Protection Agency, burning fossil fuels causes more than 80% of U.S. air pollution and 80% of U.S. carbon dioxide emissions. Many energy experts contend that the need to use cleaner and less climate-disrupting (noncarbon) energy resources—not the depletion of fossil fuels—is the driving force behind the projected global transition to a solar–hydrogen energy age in the United States and other parts of the world before the end of this century (Figure 19-7).

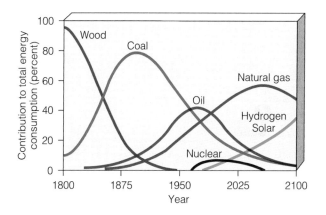

Figure 19-7 Shifts in the use of commercial energy resources in the United States since 1800, with projected changes to 2100. Shifts from wood to coal and then from coal to oil and natural gas have each taken about 50 years. The projected shift to hydrogen–solar energy in 2100 is only one of many possible scenarios depending on a variety of assumptions. (Data from U.S. Department of Energy)

How Can We Decide Which Energy Resources to Use? There is intense scientific, economic, and political controversy over which energy resources we should rely on now and in the future. As discussed in this chapter and the one that follows, each energy resource has advantages and disadvantages that must be carefully evaluated.

Energy policies need to be developed with the future in mind because experience shows that it usually takes at least 50 years and huge investments to phase in new energy alternatives to the point where they provide 10–20% of total energy use. Making projections such as those in Figure 19-7 involves trying to answer the following questions for *each* energy alternative:

- How much of the energy source will be available in the near future (the next 15–25 years) and the long term (the next 25–50 years)?

- What is this source's net energy yield?

- How much will it cost to develop, phase in, and use this energy resource?

- What government research and development subsidies and tax breaks will be provided for each energy resource to spur its development?

- How will dependence on each energy resource affect national and global economic and military security?

- How vulnerable is each source of energy to terrorism?

- How will extracting, transporting, and using the energy resource affect the environment, human health, and the earth's climate? Should these harmful costs be included in the market prices of each energy resource through a combination of taxes and phasing out environmentally harmful subsidies (full-cost pricing, p. 24)?

What Is Net Energy? The Only Energy That Really Counts It takes energy to get energy. For example, before oil is useful to us it must be found, pumped up from beneath the ground or ocean floor, transferred to a refinery and converted to useful fuels (such as gasoline, diesel fuel, and heating oil), transported to users, and burned in furnaces and cars. Each of these steps uses high-quality energy. The second law of thermodynamics (p. 59) tells us that some of the high-quality energy (Figure 3-10, p. 55) we use in each step is wasted and degraded to lower-quality energy.

The usable amount of *high-quality energy* available from a given quantity of an energy resource is its **net energy.** It is the total amount of energy available from an energy resource minus the energy needed to find, extract, process, and get that energy to consumers. It is calculated by estimating the total energy available from the resource over its lifetime minus the amount of energy *used* (the first law of thermodynamics), *automatically wasted* (the second law of thermodynamics), and *unnecessarily wasted* in finding, processing, concentrating, and transporting the useful energy to users.

Net energy is like your net spendable income (your wages minus taxes and job-related expenses). For example, suppose that for every 10 units of energy in oil in the ground we have to use and waste 8 units of energy to find, extract, process, and transport the oil to users. Then we have only 2 units of *useful energy* available from every 10 units of energy in the oil.

We can express net energy as the ratio of useful energy produced to the useful energy used to produce it. In the example just given, the *net energy ratio* would be 10/8, or 1.25. The higher the ratio, the greater the net energy. When the ratio is less than 1, there is a net energy loss.

Figure 19-8 shows estimated net energy ratios for various types of space heating, high-temperature heat for industrial processes, and transportation. Currently, oil has a high net energy ratio because much of it comes from large, accessible, and cheap-to-extract deposits such as those in the Middle East. When those sources are depleted, the net energy ratio of oil will decline and prices will rise. Then more money and high-quality fossil fuel energy will be needed to find, process, and deliver new oil from less accessible sites. They include deposits that are small and widely dispersed, buried deep in the earth's crust, and located in remote areas or further and further offshore.

Conventional nuclear energy has a low net energy ratio because large amounts of energy are needed to extract and process uranium ore, convert it into a usable nuclear fuel, build and operate nuclear power plants, dismantle the highly radioactive plants after their 15–60 years of useful life, and store the resulting highly radioactive wastes safely for 10,000–240,000 years depending on the types of radioisotopes they contain. Each of these steps in what is called the *nuclear fuel cycle* uses energy and costs money.

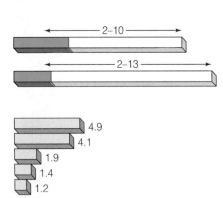

Space Heating

Passive solar	5.8
Natural gas	4.9
Oil	4.5
Active solar	1.9
Coal gasification	1.5
Electric resistance heating (coal-fired plant)	0.4
Electric resistance heating (natural-gas-fired plant)	0.4
Electric resistance heating (nuclear plant)	0.3

High-Temperature Industrial Heat

Surface-mined coal	28.2
Underground-mined coal	25.8
Natural gas	4.9
Oil	4.7
Coal gasification	1.5
Direct solar (highly concentrated by mirrors, heliostats, or other devices)	0.9

Electricity

Photovoltaic (solar) cells	2–10
Geothermal	2–13

Transportation

Natural gas	4.9
Gasoline (refined crude oil)	4.1
Biofuel (ethyl alcohol)	1.9
Coal liquefaction	1.4
Oil shale	1.2

Figure 19-8 *Net energy ratios* for various energy systems over their estimated lifetimes. The higher the net energy ratio, the greater the net energy available. (Data from U.S. Department of Energy and Colorado Energy Research Institute, *Net Energy Analysis*, 1976; and Howard T. Odum and Elisabeth C. Odum, *Energy Basis for Man and Nature*, 3rd ed., New York: McGraw-Hill, 1981)

What Is Crude Oil, and How Is It Extracted and Processed?

Petroleum, or **crude oil** (oil as it comes out of the ground), is a thick liquid consisting of hundreds of combustible hydrocarbons along with small amounts of sulfur, oxygen, and nitrogen impurities. This fossil fuel was produced by the decomposition of dead organic matter from plants (primarily plankton) and animals that were buried under lake and ocean sediments 2–140 million years ago. The resulting high temperatures and pressures over millions of years converted this dead organic matter to oil as part of the carbon cycle (Figure 4-24, p. 82).

Deposits of crude oil and natural gas often are trapped together under a dome deep within the earth's crust on land or under the seafloor (Figure 19-2). The crude oil is dispersed in pores and cracks in underground rock formations, somewhat like water saturating a sponge. To extract the oil, a well is drilled into the deposit. Then oil that is drawn by gravity out of the rock pores and into the bottom of the well is pumped to the surface.

On average, producers get only about 35% of the oil out of an oil deposit. They then abandon the well because the remaining *heavy crude oil* is too difficult or expensive to recover. As oil prices rise, it can become economical to remove about 10–25% of this remaining heavy oil by flushing the well with steam and water. But the net energy yield for such recovered oil is lower because it takes the energy in one-third of a barrel of refined oil to retrieve each barrel of heavy crude oil.

Drilling for oil causes only moderate damage to the earth's land because the wells occupy fairly little land area. However, oil drilling always involves some oil spills on land and in aquatic systems. In addition, harmful environmental effects are associated with the extraction, processing, and use of any nonrenewable resource from the earth's crust (Figure 9-9, p. 182).

According to oil producers, several improved extraction technologies can increase oil production without serious damage to environmentally sensitive areas. One is new equipment that allows oil and natural gas producers to drill deeper on land and the ocean bottom. In addition, oil producers can now use one drilling rig (derrick) to drill several gas or oil pockets at the same time. Another new technology allows oil or gas extraction from distances as far away as 8 kilometers (5 miles) by drilling at angles of 90 degrees or more (slant drilling).

After it is extracted, crude oil is transported to a *refinery* by pipeline, truck, or ship (oil tanker). There it is heated and distilled in gigantic columns to separate it into components with different boiling points (Figure 19-9). Some products of oil distillation, called **petrochemicals,** are used as raw materials in manu-

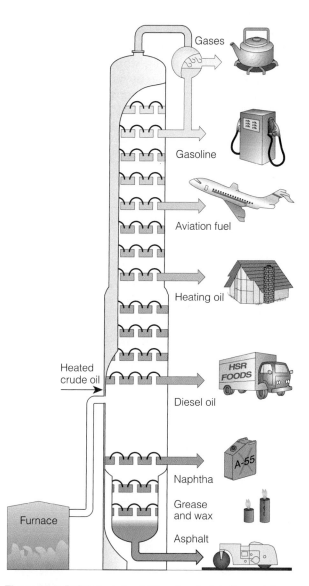

Figure 19-9 Refining crude oil. Based on their boiling points, components are removed at various levels in a giant distillation column. The most volatile components with the lowest boiling points are removed at the top of the column.

facturing industrial organic chemicals, pesticides, plastics, synthetic fibers, paints, medicines, and many other products.

Who Has the World's Oil Supplies?

Oil *reserves* are identified deposits from which oil can be extracted profitably at current prices with current technology. The 11 countries that make up the Organization of Petroleum Exporting Countries (OPEC)* have 78% of

---

*OPEC was formed in 1960 so developing countries with much of the world's known and projected oil supplies could get a higher price for this resource. Today its members are Algeria, Indonesia, Iran, Iraq, Kuwait, Libya, Nigeria, Qatar, Saudi Arabia, the United Arab Emirates, and Venezuela.

the world's estimated crude oil reserves. This explains why OPEC is expected to have long-term control over the supplies and prices of the world's conventional oil. Saudi Arabia, with 25%, has by far the largest proportion of the world's crude oil reserves, followed by Iraq (11%), the United Arab Emirates (9.3%), Kuwait (9.2%), and Iran (8.5%). Most of the world's remaining global crude oil reserves are found in Venezuela (7.4%), Africa (7.3%), the former Soviet Union (6.2%), Asia (4.2%, with 2% in China), the United States (2.9%), Mexico (2.6%), and western Europe (1.8%).

Case Study: How Much Oil Does the United States Have? Figure 19-10 shows the locations of the major known deposits of fossil fuels (oil, natural gas, and coal) in the United States and Canada and ocean areas where more crude oil and natural gas might be found. In 2001, about 25% of U.S. domestic

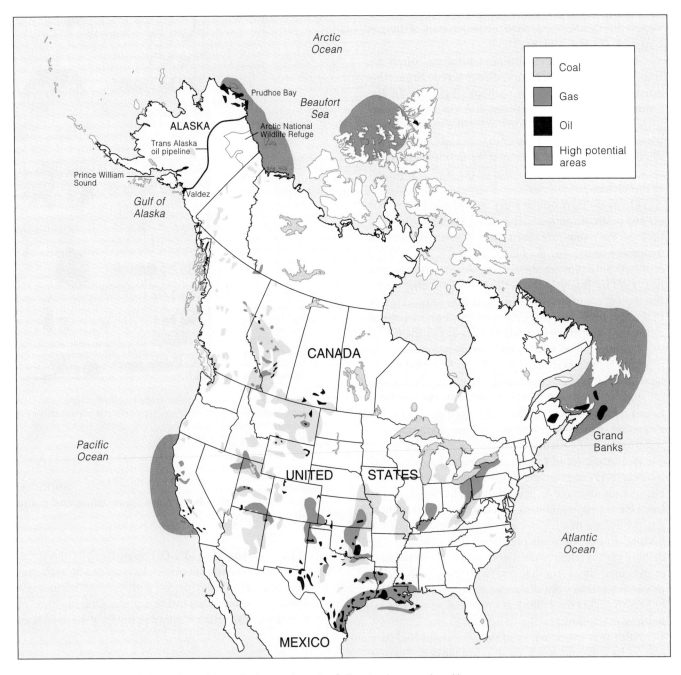

Figure 19-10 *Natural capital:* locations of the major known deposits of oil, natural gas, and coal in North America and offshore areas where more crude oil and natural gas might be found. Geologists do not expect to find very much new oil and natural gas in North America. (Data from Council on Environmental Quality and U.S. Geological Survey)

oil production came from offshore drilling (mostly off the coasts of Texas and Louisiana) and 17% from Alaska's North Slope.

The United States has only 2.9% of the world's oil reserves. However, it uses about 26% of the crude oil extracted worldwide each year (68% of it for transportation), mostly because oil is an abundant and convenient fuel to use and is cheap (Figure 19-11). Despite an upsurge in exploration and test drilling, U.S. oil extraction has declined since 1985, and most geologists do not expect a significant increase in domestic supplies (Figure 19-12).

In 2002, the United States imported about 55% of the oil it used (up from 36% in 1973 during the OPEC oil embargo). Reasons for this high dependence on imported oil are declining domestic oil reserves, higher production costs for domestic oil than for most sources of imported oil, and increased oil use. In 2001, the U.S. bill for oil imports was about $100 billion—an average of $11 million per hour. According to the Department of Energy (DOE), the United States could be importing at least 61% of the oil it uses by 2010 and 64% by 2020 (Figure 19-12).

How Long Will Oil Supplies Last? Production of the world's estimated oil reserves is expected to peak between 2010 and 2030 (Figure 19-13, top), and production of estimated U.S. reserves peaked in 1975 (Figure 19-12 and Figure 19-13, bottom).

Identified global reserves of oil should last about 53 years at the current usage rate, and 42 years if usage increases exponentially by about 2% per year. Undiscovered oil that is thought to exist might add another

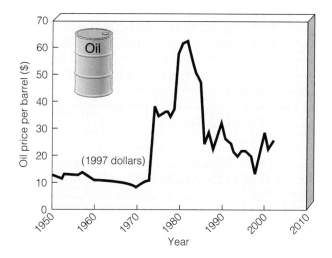

Figure 19-11 Inflation-adjusted price of oil, 1950–2002. When adjusted for inflation, oil costs about the same as it did in 1975. Although low oil prices have stimulated economic growth, they have discouraged improvements in energy efficiency and increased use of renewable energy resources. (Data from U.S. Department of Energy and Department of Commerce)

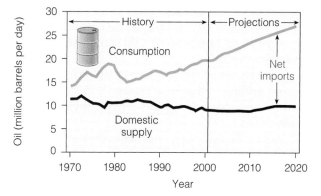

Figure 19-12 U.S. petroleum supply, consumption, and imports, 1970–2002, with projections to 2020. (U.S. Department of Energy)

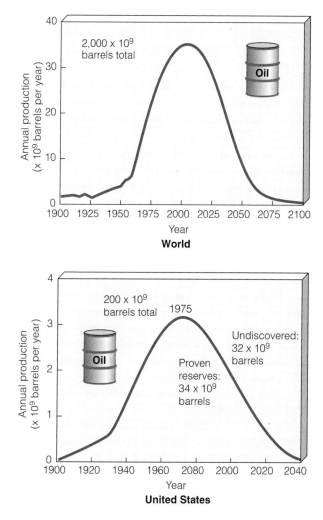

Figure 19-13 Petroleum production curves for the world (top) and the United States (bottom). (Data from U.S. Geological Survey)

20–40 years to global oil supplies, probably at higher prices. Thus *known and projected supplies of oil are expected to be 80% depleted within 42–93 years depending on the annual rate of use.*

U.S. oil reserves should last about 15–24 years (Figure 19-13, bottom) at current consumption rates and 10–15 years if consumption increases as projected. However, potential reserves might yield an additional 24 years of production. Thus *U.S. oil supplies are projected to be 80% depleted within 10–48 years, depending on the annual rate of use.*

Some analysts contend that rising oil prices (when oil consumption exceeds oil production) will stimulate exploration and lead to discovery of new reserves to meet future demand through the next century or longer. Other analysts argue that such projections ignore the consequences of the high (1–5% per year) exponential growth in oil consumption.

For example, suppose that we continue to use oil at the current rate with no increase in oil consumption (an unlikely assumption). Even under this conservative estimate

- Saudi Arabia, with the world's largest crude oil reserves, could supply world oil needs for about 10 years.

- The estimated reserves under Alaska's North Slope (the largest ever found in North America) would meet current world demand for only 6 months or U.S. demand for 3 years.

- The estimated reserves in Alaska's Arctic National Wildlife Refuge would meet the world's current oil demand for only 1–5 months or U.S. oil demand for 7–24 months (see Case Study at right).

- Estimated oil reserves on other federal lands on Alaska's North Slope would meet the current world oil demand for 10 days to 2 months or U.S. demand for 2–11 months.

In short, to keep using conventional oil at the *current rate,* we must discover global oil reserves equivalent to a new Saudi Arabian supply *every 10 years.*

According to the Bush administration, the United States can reduce its dependence on imported oil and have more control over global oil prices by increasing domestic oil supplies. Most analysts consider this unrealistic because the United States has only 2.9% of the world's oil reserves and uses 26% of the world's annual oil production. The country also produces most of its dwindling supply of oil at a high cost of $5–7.50 per barrel compared to production costs of less than $1.50 per barrel in Persian Gulf countries. Thus opening all of the U.S. coastal water, forests, and wild places to drilling would hardly put a dent in world oil prices or meet much of the growing U.S. demand for oil (Figures 19-6 and 19-12).

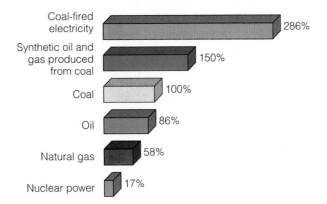

Figure 19-14 CO_2 emissions per unit of energy produced by various fuels, expressed as percentages of emissions produced by burning coal directly.

Burning any carbon-containing fossil fuel releases CO_2 into the atmosphere and thus can promote global warming. Figure 19-14 compares the relative amounts of CO_2 emitted per unit of energy by the major fossil fuels and nuclear power. Currently, burning oil mostly as gasoline and diesel fuel for transportation accounts for 43% of global CO_2 emissions.

Case Study: Should Oil and Gas Development Be Allowed in the Arctic National Wildlife Refuge? The Arctic National Wildlife Refuge (ANWR) on Alaska's North Slope (Figure 19-10) contains more than one-fifth of all land in the U.S. National Wildlife Refuge System (Figure 17-9, p. 416) and has been called the crown jewel of the system. The refuge's coastal plain, its most biologically productive part, is the only stretch of Alaska's arctic coastline not open to oil and gas development.

The Alaskan National Interest Lands Conservation Act of 1980 requires authorization from Congress before drilling or other development can take place on this coastal plain. For years, U.S. oil companies have been lobbying Congress to grant them permission to carry out exploratory drilling in the coastal plain because they believe this area might contain oil and natural gas deposits.

Environmentalists and conservationists strongly oppose drilling in this area. Figure 19-15 summarizes the opposing positions in this intense environmental controversy.

According to drilling opponents, improving motor vehicle fuel efficiency is a much faster, cheaper, cleaner, and more secure way to increase future oil supplies. For example, requiring new SUVs and light trucks used in the United States to have the same average fuel efficiency as new cars would save more oil in 10 years than would ever be produced from the ANWR.

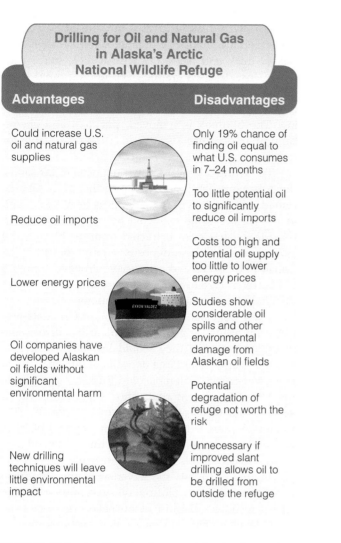

Drilling for Oil and Natural Gas in Alaska's Arctic National Wildlife Refuge

Advantages	Disadvantages
Could increase U.S. oil and natural gas supplies	Only 19% chance of finding oil equal to what U.S. consumes in 7–24 months
Reduce oil imports	Too little potential oil to significantly reduce oil imports
Lower energy prices	Costs too high and potential oil supply too little to lower energy prices
Oil companies have developed Alaskan oil fields without significant environmental harm	Studies show considerable oil spills and other environmental damage from Alaskan oil fields
New drilling techniques will leave little environmental impact	Potential degradation of refuge not worth the risk
	Unnecessary if improved slant drilling allows oil to be drilled from outside the refuge

Figure 19-15 Trade-offs: advantages and disadvantages of drilling for oil and natural gas in Alaska's Arctic National Wildlife Refuge (ANWR).

What Is the Future of Conventional Oil? A Brief History of the Age of Oil Figure 19-16 (p. 488) lists the advantages and disadvantages of using conventional crude oil as an energy resource. We are heavily dependent on conventional oil today because it has a high energy content and net energy yield, low cost (as long as supplies exceed demand), and is easy to transport within and between countries.

In 1859, the world's first commercial oil well was drilled in Titusville, Pennsylvania. This was the beginning of the *Age of Oil,* which now supplies most of the world's energy (Figure 19-4, left).

We are not running out of oil, but many energy experts expect that by 2059—some 200 years after oil was discovered—we will probably begin shifting to increased dependence on natural gas and possibly hydrogen by 2100 (Figure 19-7).

Here are some milestones in the Age of Oil:

- **1905:** Oil supplies 10% of U.S. energy.
- **1925:** United States produces 71% of the world's oil.
- **1930:** Because of an oil glut, oil sells for 10¢ a barrel.
- **1953:** U.S. oil companies account for about half of the world's oil production and the United States is the world's leading oil exporter.
- **1955:** United States has 20% of the world's estimated oil reserves.
- **1960:** OPEC formed so developing countries, with most of the world's known oil and projected oil reserves, can get a higher price for their oil.
- **1973:** United States uses 30% of the world's oil, imports 36% of this oil, and has only 5% of the world's proven oil reserves.
- **1973–1974:** OPEC reduces oil imports to the West and bans oil exports to the U.S. because of its support for Israel in the 18-day Yom Kippur War with Egypt and Syria. World oil prices rise sharply (Figure 19-11) and lead to double-digit inflation in the United States and many other countries and a global economic recession.
- **1975:** Production of estimated U.S. oil reserves peaks.
- **1979:** Iran's Islamic Revolution shuts down most of Iran's oil production and reduces world oil production.
- **1981:** Iran–Iraq war pushes global oil prices to a historic high (Figure 19-11).
- **1983:** Facing an oil glut, OPEC cuts its oil prices.
- **1985:** U.S. domestic oil production begins to decline and is not expected to increase enough to affect the global price of oil or to reduce U.S. dependence on oil imports (Figure 19-12).
- **August 1990–June 1991:** United States and its allies fight the Persian Gulf War to oust Iraqi invaders of Kuwait and to protect Western access to Saudi Arabian and Kuwaiti oil supplies.
- **2002:** OPEC has 67% of world oil reserves and produces 40% of the world's oil. U.S. has only 2.9% of oil reserves, uses 26% of the world's oil production, and imports 55% of its oil.
- **2010:** U.S. could be importing at least 61% of the oil it uses as consumption continues to exceed production (Figure 19-12).
- **2010–2030:** Production of oil from the world's estimated oil reserves is expected to peak (Figure 19-13, top). Oil prices expected to increase gradually as the demand for oil increasingly exceeds the supply—unless the world decreases demand by wasting less energy and shifting to other sources of energy.
- **2010–2048:** Domestic U.S. oil reserves projected to be 80% depleted.
- **2042–2083:** Gradual decline in dependence on oil.

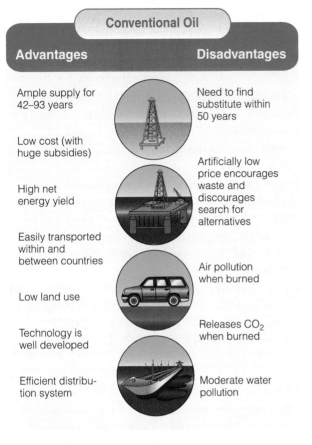

Conventional Oil

Advantages	Disadvantages
Ample supply for 42–93 years	Need to find substitute within 50 years
Low cost (with huge subsidies)	Artificially low price encourages waste and discourages search for alternatives
High net energy yield	
Easily transported within and between countries	Air pollution when burned
Low land use	Releases CO_2 when burned
Technology is well developed	
Efficient distribution system	Moderate water pollution

Figure 19-16 Trade-offs: advantages and disadvantages of using conventional oil as an energy resource.

In 1999, Mike Bowling, CEO of ARCO Oil, said, "We are embarked on the beginning of the last days of the Age of Oil." He went on to discuss the need for the world to shift from a carbon-based to a hydrogen-based energy economy during this century (Figure 19-7).

How Useful Are Heavy Oils from Oil Shale and Tar Sands? *Oil shale* is a fine-grained sedimentary rock (Figure 19-17, left) containing a solid combustible organic material called *kerogen*. This material can be distilled from oil shale by heating it in a large container to yield *shale oil* (Figure 19-17, right). Before the thick shale oil can be sent by pipeline to a refinery, it must be heated to increase its flow rate and processed to remove sulfur, nitrogen, and other impurities.

Some *good news* is that estimated potential global supplies of shale oil are about 240 times larger than estimated global supplies of conventional oil. The *bad news* is that most deposits of oil shale are of such a low grade that with current oil prices and technology it takes more energy and money to mine and convert the kerogen to crude oil than the resulting fuel is worth. However, as oil prices rise it may become economically feasible to exploit some reserves of oil shale.

Tar sand (or oil sand) is a mixture of clay, sand, water, and a combustible organic material called *bitumen* (a thick and heavy oil with a high sulfur content). Most deposits are too deep underground to be mined at a profit, but some deposits are close enough to the earth's surface to be removed by surface mining. The bitumen is removed, purified, and chemically upgraded into a synthetic crude oil suitable for refining.

The world's largest known deposits of tar sands, the Athabasca Tar Sands, lie in northeastern Alberta, Canada. About 10% of these deposits lie close enough to the surface to be extracted by surface mining. In Canada, oil has been extracted from tar sands since 1978. Currently, these deposits supply about 21% of Canada's oil needs and prices have dropped from $28 per barrel in 1978 to about $11 per barrel in 2001—less than half the current cost of conventional oil.

These deposits could supply all of Canada's projected oil needs for about 33 years at its current consumption rate, but they would last the world only about 2 years. Other large deposits of tar sands are in Utah, Venezuela, Colombia, and Russia. According to the U.S. Department of Energy, exploitation of tar sands could increase the global oil reserves by 50% within 25 years if the price of conventional oil rises above $30 per barrel and fivefold if the price of oil rises above $40 per barrel.

Figure 19-18 lists the advantages and disadvantages of using heavy oil from oil shale and tar sand as energy resources. Because of low net energy yields and high development and processing costs, neither of these resources is expected to provide much of the world's energy in the near future. However, as oil

Figure 19-17 Oil shale (left) and the shale oil (right) extracted from it. Big U.S. oil shale projects have been canceled because of excessive cost.

U.S. Department of Energy

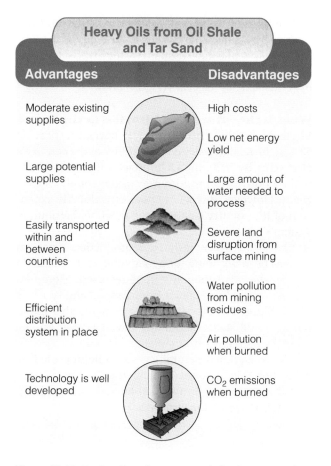

Heavy Oils from Oil Shale and Tar Sand

Advantages	Disadvantages
Moderate existing supplies	High costs
	Low net energy yield
Large potential supplies	
	Large amount of water needed to process
Easily transported within and between countries	Severe land disruption from surface mining
	Water pollution from mining residues
Efficient distribution system in place	
	Air pollution when burned
Technology is well developed	CO_2 emissions when burned

Figure 19-18 Trade-offs: advantages and disadvantages of using heavy oils from oil shale and tar sand as energy resources.

prices rise it may become economically feasible to exploit these two sources of heavy oil unless supplies of natural gas and other energy resources cost less to develop and produce fewer harmful environmental effects.

19-3 NATURAL GAS

What Is Natural Gas?   In its underground gaseous state, **natural gas** is a mixture of 50–90% by volume of methane (CH_4), the simplest hydrocarbon. It also contains smaller amounts of heavier gaseous hydrocarbons such as ethane (C_2H_6), propane (C_3H_8), and butane (C_4H_{10}), and small amounts of highly toxic hydrogen sulfide (H_2S), a by-product of naturally occurring sulfur in the earth.

Conventional natural gas usually lies above most reservoirs of crude oil (Figure 19-2). Like oil, the natural gas was formed from fossil deposits of plants (mostly phytoplankton) and animals buried on the seafloor for millions of years and subjected to high temperatures and pressures.

Unconventional natural gas is found by itself in other underground sources. One such source is *methane hydrate,* which is composed of small bubbles of natural gas trapped in ice crystals deep under the arctic permafrost and beneath deep-ocean sediments. So far it costs too much to get natural gas from such unconventional sources, but the extraction technology is being developed rapidly.

When a natural gas field is tapped, propane and butane gases are liquefied and removed as **liquefied petroleum gas (LPG).** LPG is stored in pressurized tanks for use mostly in rural areas that are not served by natural gas pipelines. The rest of the gas (mostly methane) is dried to remove water vapor. Then it is cleansed of poisonous hydrogen sulfide and other impurities and pumped into pressurized pipelines for distribution.

At a very low temperature of $-184°C$ ($-300°F$), natural gas can be converted to **liquefied natural gas (LNG).** This highly flammable liquid can then be shipped to other countries in refrigerated tanker ships.

Who Has the World's Natural Gas Supplies? Russia has about 31% of the world's natural gas reserves. Other countries with large known natural gas reserves are Iran (15%), Qatar (9%), Saudi Arabia (4%), the United Arab Emirates (4%), the United States (3%), Algeria (3%), and Venezuela (3%).

Geologists expect to find more natural gas, especially in unexplored developing countries. Most U.S. natural gas reserves are located in the same places as crude oil (Figures 19-2 and 19-10).

How Long Will Natural Gas Supplies Last? The long-term outlook for natural gas supplies is much better than for oil. *At the current consumption rate, known reserves and undiscovered, potential reserves of conventional natural gas in the world are expected to last 125 years and those in the United States for 65–80 years.*

Geologists estimate that *conventional* supplies of natural gas, plus *unconventional* supplies available at higher prices, will last at least 200 years at the current consumption rate and 80 years if usage rates rise 2% per year. Thus *global supplies of conventional and unconventional natural gas should last 205–325 years, depending on how rapidly natural gas is used.*

What Is the Future of Natural Gas?   Figure 19-19 (p. 490) lists the advantages and disadvantages of natural gas as an energy resource. Energy experts project greatly increased global use of natural gas during this century (Figure 19-5) because of its abundant supply, low production costs, and lower pollution and CO_2 per unit of energy than other fossil fuels (Figure 19-14).

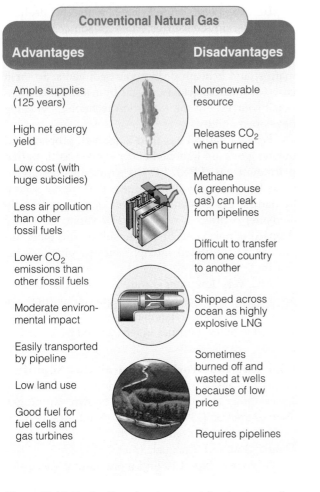

Conventional Natural Gas

Advantages	Disadvantages
Ample supplies (125 years)	Nonrenewable resource
High net energy yield	Releases CO_2 when burned
Low cost (with huge subsidies)	Methane (a greenhouse gas) can leak from pipelines
Less air pollution than other fossil fuels	
Lower CO_2 emissions than other fossil fuels	Difficult to transfer from one country to another
Moderate environmental impact	Shipped across ocean as highly explosive LNG
Easily transported by pipeline	Sometimes burned off and wasted at wells because of low price
Low land use	
Good fuel for fuel cells and gas turbines	Requires pipelines

Figure 19-19 Trade-offs: advantages and disadvantages of using conventional natural gas as an energy resource.

In *combined-cycle natural gas systems,* natural gas is burned in combustion turbines, which are essentially giant jet engines bolted to the ground. This system can produce electricity more efficiently than burning coal or oil or using nuclear power. It can also provide backup power for solar energy and wind power systems. In addition, these turbines produce much less CO_2 (Figure 19-14) and smog-causing nitrogen oxides per unit of energy than coal-burning power plants. Smaller combined-cycle natural gas units being developed could supply the heat and electricity needs of an apartment or office building.

In 2002, about 380,000 of the world's motor vehicles ran on liquefied or compressed natural gas. In 2002, Honda announced it would begin selling a CivicGX fueled by natural gas. The company has also teamed up with Toronto-based FuelMaker to mass-market a system that connects to a home's natural gas supply line and allows owners to fuel natural gas vehicles at home.

Because of its advantages over oil, coal, and nuclear energy, some analysts see natural gas as the best fuel to help make the transition to improved energy efficiency and greater use of solar energy and hydrogen over the next 50–100 years (Figure 19-7).

What Is the Future of Natural Gas in the United States? In 2001, natural gas was burned to provide 53% of the heat in U.S. homes and 16% of the country's electricity. By 2020, the DOE projects that natural gas will be burned to produce 32% of the country's electricity. However, this will require considerable expansion of the country's natural gas pipeline distribution system.

A major problem is that U.S. production of natural gas has been declining for a long time, and experts do not believe this situation will be reversed. More natural gas could be obtained from Canada and Alaska's North Slope. But the pipeline for bringing this gas to the lower 48 states—assuming that it is ever built—will take 7–10 years to complete.

More liquefied natural gas could be imported by ship. But this requires cooling the gas to −184°C (−300°F), shipping it in special tankers, and building special LNG receiving terminals.

19-4 COAL

What Is Coal, and How Is It Extracted and Processed? Coal is a solid fossil fuel formed in several stages as buried remains of land plants that lived 300–400 million years ago were subjected to intense heat and pressure over many millions of years (Figure 19-20). Coal contains small amounts of sulfur, released into the atmosphere as SO_2 when coal is burned. Trace amounts of mercury (Figure 15-19, p. 369) and radioactive materials are also released into the atmosphere when coal is burned.

Anthracite (which is about 98% carbon) is the most desirable type of coal because of its high heat content and low sulfur content (Figure 19-20). However, because it takes much longer to form, it is less common and therefore more expensive than other types of coal.

Some coal is extracted underground by miners working in tunnels and shafts (Figure 9-8, p. 181). Such mining is one of the world's most dangerous occupations because of accidents and black lung disease (caused by prolonged inhalation of coal dust particles). When coal lies close to the earth's surface, it is extracted by *area strip mining* (Figure 9-7c, p. 179) on flat terrain and *contour strip mining* (Figure 9-7d, p. 179) on hilly or mountainous terrain. In some cases, entire mountaintops are removed to expose seams of coal under the mountains.

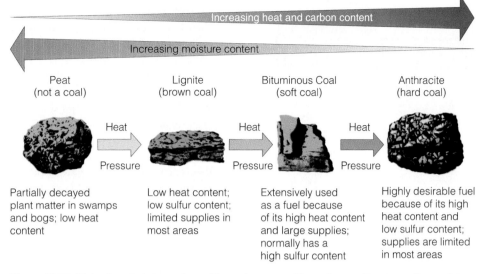

Increasing heat and carbon content

Increasing moisture content

| Peat (not a coal) | Lignite (brown coal) | Bituminous Coal (soft coal) | Anthracite (hard coal) |

Heat / Pressure → Heat / Pressure → Heat / Pressure →

Partially decayed plant matter in swamps and bogs; low heat content

Low heat content; low sulfur content; limited supplies in most areas

Extensively used as a fuel because of its high heat content and large supplies; normally has a high sulfur content

Highly desirable fuel because of its high heat content and low sulfur content; supplies are limited in most areas

Figure 19-20 *Natural capital:* stages in coal formation over millions of years. Peat is a soil material made of moist, partially decomposed organic matter. Lignite and bituminous coal are sedimentary rocks, whereas anthracite is a metamorphic rock (Figure 9-5, p. 178).

After coal is removed, it is transported (usually by train) to a processing plant, where it is broken up, crushed, and washed to remove impurities. After the coal is dried it is shipped (again usually by train) to users, mostly power plants and industrial plants.

How Is Coal Used, and Where Are the Largest Supplies? Coal provides about 22% of the world's commercial energy (23% in the United States). It is burned to generate 62% of the world's electricity (51% in the United States) and make 75% of its steel.

With current technology and prices the United States has 25% of global proven reserves. Russia has 16%, China 12%, and India 9%. Almost half of global coal consumption takes place in the United States (24%) and China (23%).

How Long Will Coal Supplies Last? Coal is the world's most abundant fossil fuel. *Identified* world reserves of coal should last at least 225 years at the current usage rate and 65 years if usage rises 2% per year. The world's *unidentified* coal reserves are projected to last about 900 years at the current consumption rate and 149 years if the usage rate increases 2% per year. Thus *identified and unidentified supplies of coal could last the world for 214–1,125 years, depending on the rate of usage.*

China, with 12% of the world's reserves, has enough coal to last 300 years at its current rate of consumption. Identified U.S. coal reserves should also last about 300 years at the current consumption rate, and unidentified U.S. coal resources could extend those supplies for perhaps another 100 years, at a higher cost.

What Is the Future of Coal? Figure 19-21 (p. 492) lists the advantages and disadvantages of using coal as an energy resource. Coal is very abundant. But it has the highest environmental impact of any fossil fuel in each of the following categories: land disturbance (Figure 19-2 and Figure 9-7, p. 179); air pollution (Figure 12-9, p. 264); CO_2 emissions (Figure 19-14), accounting for about 36% of the world's annual emissions; release of particles of toxic mercury when burned (Figure 15-19, p. 369); release of thousands of times more radioactive particles into the atmosphere per unit of energy produced than a normally operating nuclear power plant; and water pollution (Figure 9-10, p. 182). Each year in the United States alone, air pollutants from coal burning kill thousands of people (estimates range from 65,000 to 200,000), cause at least 50,000 cases of respiratory disease, and result in several billion dollars of property damage. Coal is also difficult and expensive to transport very far because it is heavy and bulky.

Energy costs from a new coal power plant in the United States are low (around 4¢ per kilowatt-hour*), mostly because of large government subsidies. However, according to a 2001 study by Stanford University researchers, this cost doubles to around 8¢ per kilowatt-hour if health and environmental costs are included. In contrast, the average cost of wind energy in the United States—including its environmental and health costs—is about 4¢ per kilowatt-hour.

---

*A *kilowatt-hour* (kWh), a basic unit of electricity, is equal to the energy consumed by a 100-watt bulb burning for 10 hours.

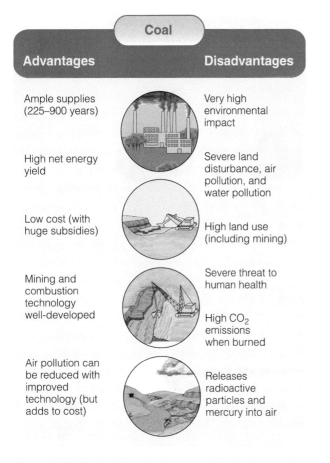

Coal

Advantages	Disadvantages
Ample supplies (225–900 years)	Very high environmental impact
High net energy yield	Severe land disturbance, air pollution, and water pollution
Low cost (with huge subsidies)	High land use (including mining)
Mining and combustion technology well-developed	Severe threat to human health
	High CO_2 emissions when burned
Air pollution can be reduced with improved technology (but adds to cost)	Releases radioactive particles and mercury into air

Figure 19-21 Trade-offs: advantages and disadvantages of using coal as an energy resource.

New ways have been developed to burn coal more cleanly and efficiently, such as *fluidized-bed combustion* and *coal gasification*. With large government subsidies, these technologies may be phased in over the next several decades. But they do little to reduce CO_2 emissions that are considered the major culprit in projected global warming and a major drawback of coal as an energy resource.

Most analysts project a leveling off of coal use over the next 40–50 years followed by a decline in its use (Figure 19-7). There are three major reasons for a shift away from coal in this century: its high CO_2 emissions (Figure 19-14), its harmful environmental and human health effects, and increased availability of less environmentally harmful ways to produce electricity (such as burning natural gas in combined-cycle gas turbines, wind energy, solar cells, and hydrogen).

What Are the Advantages and Disadvantages of Converting Solid Coal into Gaseous and Liquid Fuels? Solid coal can be converted into **synthetic natural gas (SNG)** by **coal gasification** (Figure 19-22) or into a liquid fuel such as methanol or synthetic gasoline by **coal liquefaction**. Figure 19-23 lists the advantages and disadvantages of using these *synfuels* produced from coal.

Unless it receives huge government subsidies, most analysts expect coal gasification to play only a minor role as an energy resource in the next 20–50 years.

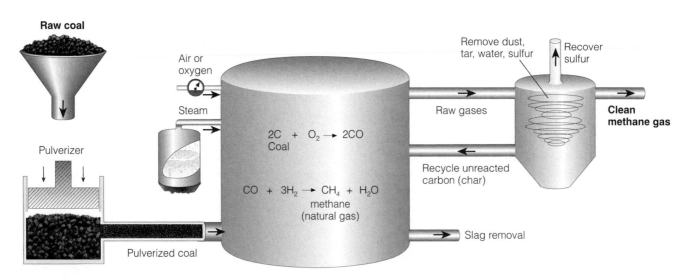

Figure 19-22 Solutions: coal gasification. Generalized view of one method for converting solid coal into synthetic natural gas (methane).

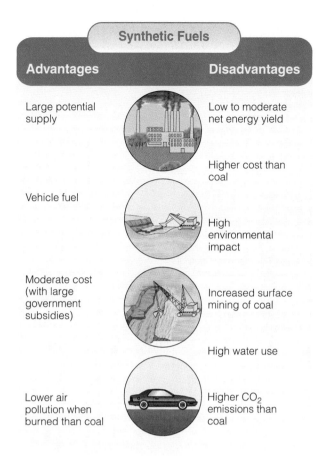

Synthetic Fuels

Advantages **Disadvantages**

Large potential supply — Low to moderate net energy yield

— Higher cost than coal

Vehicle fuel — High environmental impact

Moderate cost (with large government subsidies) — Increased surface mining of coal

— High water use

Lower air pollution when burned than coal — Higher CO_2 emissions than coal

Figure 19-23 Trade-offs: advantages and disadvantages of using synthetic natural gas (SNG) and other liquid synfuels produced from coal.

19-5 NUCLEAR ENERGY

How Does a Nuclear Fission Reactor Work? To evaluate the advantages and disadvantages of nuclear power, we need to know how a conventional nuclear power plant and its accompanying nuclear fuel cycle work. In a nuclear fission chain reaction, neutrons split the nuclei of atoms such as uranium-235 (Figure 3-11, p. 58) and plutonium-239 and release energy mostly as high-temperature heat as a result of a chain reaction (Figure 3-12, p. 58). In the reactor of a nuclear power plant, the rate of fission is controlled and the heat generated is used to produce high-pressure steam, which spins turbines that generate electricity.

Light-water reactors (LWRs) such as the one diagrammed in Figure 19-24 (p. 494) produce about 85% of the world's nuclear-generated electricity (100% in the United States). An LWR plant has the following key parts:

■ *Core,* containing 35,000–70,000 long, thin fuel rods, each packed with fuel pellets. Each pellet is about one-

third the size of a cigarette and contains the energy equivalent of 0.9 metric ton (1 ton) of coal. As the fuel is used up, the spent fuel rods are removed and replaced with new ones.

■ *Uranium oxide fuel,* consisting of about 97% non-fissionable uranium-238 and 3% fissionable uranium-235. To create a suitable fuel, the concentration of uranium-235 in the ore is increased (enriched) from 0.7% (its natural concentration in uranium ore) to 3% by removing some of the uranium-238.

■ *Control rods,* which are moved in and out of the reactor core to absorb neutrons and thus regulate the rate of fission and amount of power the reactor produces.

■ *Moderator,* which slows down the neutrons emitted by the fission process so the chain reaction (Figure 3-12, p. 58) can be kept going. This is a material such as liquid water (75% of the world's reactors, called *pressurized water reactors,* Figure 19-24), solid graphite (20% of reactors), or heavy water (deuterium oxide, D_2O; 5% of reactors). Graphite-moderated reactors (used in the ill-fated Chernobyl plant; Figure 19-1) can also produce fissionable plutonium-239 for nuclear weapons.

■ *Coolant,* usually water, which circulates through the reactor's core to remove heat (to keep fuel rods and other materials from melting) and produce steam for generating electricity.

■ *Containment vessel* to keep radioactive materials from escaping into the environment in case of an internal explosion or core meltdown within the reactor or an external threat to the reactor such as a plane crashing into the containment vessel. Containment vessels typically consist of a 1.2-meter (4-foot) steel-reinforced concrete wall along with a steel liner. A U.S. reactor survived a head-on test crash of a military jet without major damage. However, it is not known whether the typical containment shell would remain intact after a head-on crash of a large commercial airliner loaded with jet fuel.

■ *Water-filled pools* or *dry casks* for on-site storage of highly radioactive spent fuel rods removed when reactors are refueled. Most spent fuel rods are stored in 12-meter- (40-foot-) deep pools of boron-treated water to shield against radiation and to keep the fuel from heating up, catching fire, and releasing radioactive materials into the environment. Spent-fuel pools or casks are in a separate building not nearly as well protected as the reactor core and are much more vulnerable to a head-on crash from even a small plane and to a terrorist attack. The long-term goal is to transport spent fuel rods and other long-lived radioactive wastes to an underground facility where they must be stored safely for 10,000–240,000 years until their radioactivity falls to safe levels.

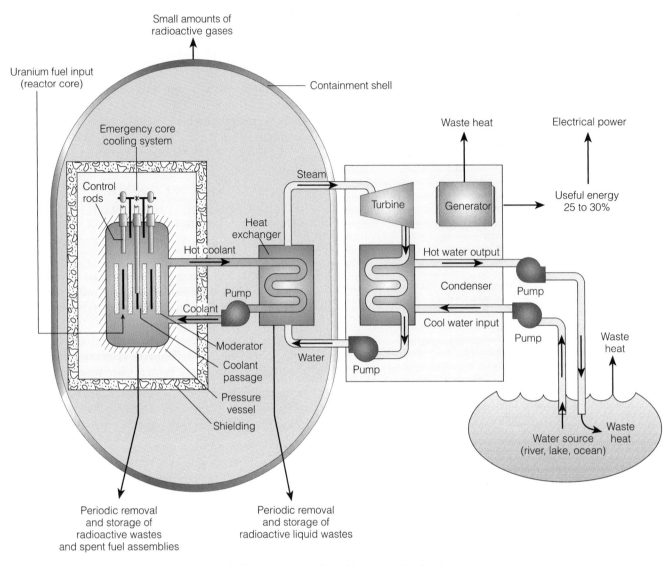

Figure 19-24 Light-water–moderated and –cooled nuclear power plant with a pressurized water reactor.

Nuclear power plants, each with one or more reactors, are only one part of the nuclear fuel cycle (Figure 19-25). Unlike other energy resources, nuclear energy results in the production of intensely radioactive materials that must be stored safely for 10,000–240,000 years.

In the *closed cycle* (Figure 19-25, dotted line), the fissionable isotopes uranium-235 and plutonium-239 are removed from spent fuel assemblies for reuse as nuclear fuel. Their removal means that the remaining radioactive wastes must be stored safely for about 10,000 years.

In the *open cycle* (solid line, Figure 19-25) these isotopes are not removed by reprocessing the nuclear wastes and are buried in an underground disposal fa-

cility. These wastes must be stored safely for about 240,000 years. Currently, these isotopes are rarely removed from spent fuel rods and other nuclear wastes because of high costs and the potential use of the removed isotopes in nuclear weapons.

In evaluating the safety and economic feasibility of nuclear power, energy experts and economists caution us to look at this entire cycle, not just the nuclear plant itself.

What Happened to Nuclear Power? Studies indicate that U.S. utility companies began developing nuclear power plants in the late 1950s for three reasons. *First,* the Atomic Energy Commission (which had the conflicting roles of promoting and regulating nuclear power) promised utility executives that nuclear power

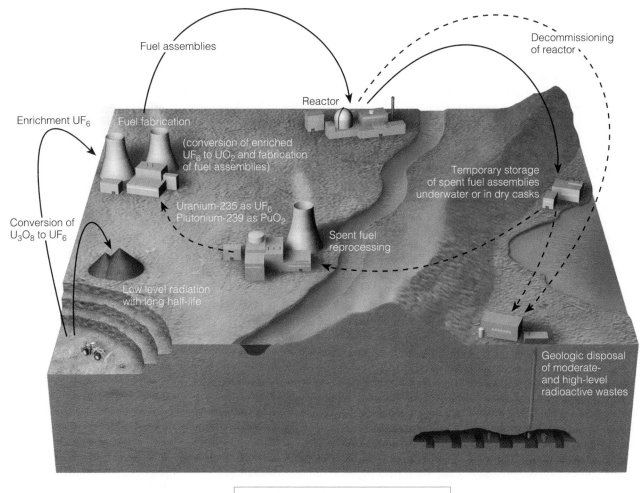

Fuel assemblies

Decommissioning
of reactor

Enrichment UF$_6$

Fuel fabrication

Reactor

(conversion of enriched
UF$_6$ to UO$_2$ and fabrication
of fuel assemblies)

Temporary storage
of spent fuel assemblies
underwater or in dry casks

Conversion of
U$_3$O$_8$ to UF$_6$

Uranium-235 as UF$_6$
Plutonium-239 as PuO$_2$

Spent fuel
reprocessing

Low level radiation
with long half-life

Geologic disposal
of moderate-
and high-level
radioactive wastes

Open fuel cycle today
Prospective "closed" end fuel cycle

Figure 19-25 The nuclear fuel cycle.

would produce electricity at a much lower cost than coal and other alternatives. Indeed, President Dwight D. Eisenhower declared in a 1953 speech that nuclear power would be "too cheap to meter."

Second, the government (taxpayers) paid about one-fourth of the cost of building the first group of commercial reactors and guaranteed there would be no cost overruns. *Third,* after insurance companies refused to insure nuclear power, Congress passed the Price-Anderson Act to protect the U.S. nuclear industry and utilities from significant liability to the general public in case of accidents.*

---

*This act limits the nuclear industry's liability in case of an accident to $9.5 billion and has the government (taxpayers) paying most of this. According to the U.S. Nuclear Regulatory Commission, a worst-case accident would cause more than $300 billion in damages.

In the 1950s, researchers predicted that by the year 2000, at least 1,800 nuclear power plants would supply 21% of the world's commercial energy (25% in the United States) and most of the world's electricity.

However, after more than 50 years of development, enormous government subsidies, and an investment of $2 trillion, these goals have not been met. Instead, by 2001, 436 commercial nuclear reactors in 32 countries were producing 6% of the world's commercial energy and 16% of its electricity. Since 1989, the growth in electricity production from nuclear power has essentially leveled off and is projected to decrease by 2020 (Figure 19-5).

Germany (which gets 31% of its electricity from nuclear power), Sweden (39%), Belgium (60%), the Netherlands, and Spain plan to phase out nuclear power over the next 20–30 years. Nuclear power is also losing some of its appeal in France, Japan, and China—all once solid supporters of its increased use. France and China (which has four reactors and is building nine more) have put a moratorium on building new nuclear plants, and France has cut its long-term target for building new reactors in half.

No new nuclear power plants have been ordered in the United States since 1978, and all 120 plants ordered since 1973 have been canceled. In 2001, there were 104 licensed and operating commercial nuclear power reactors in 31 states—most in the eastern half of the country. These reactors generated about 21% of the country's electricity and 8% of its total energy use. This percentage is expected to decline over the next two to three decades as existing plants wear out and are retired (Figure 19-6).

According to energy analysts and economists, there are several major reasons for the failure of nuclear power to grow as projected. They include multibillion-dollar construction cost overruns, higher operating costs and more malfunctions than expected, and poor management. Two other major setbacks have been public concerns about safety and stricter government safety regulations, especially after the accidents in 1986 at Chernobyl (p. 478) and in 1979 at Three Mile Island in Pennsylvania. Another problem is investor concerns about the economic feasibility of nuclear power. Also, concern has risen about the vulnerability of nuclear power plants to terrorist attack after destruction of the World Trade Center buildings in New York City on September 11, 2001.

What Are the Advantages and Disadvantages of Conventional Nuclear Power? Figure 19-26 lists the major advantages and disadvantages of nuclear power. Using nuclear power to produce electricity has some important advantages over coal-burning power plants (Figure 19-27).

How Safe Are Nuclear Power Plants and Other Nuclear Facilities? *Because of the built-in safety features, the risk of exposure to radioactivity from nuclear power plants in the United States and most other developed countries is extremely low.* However, a partial or complete meltdown or explosion is possible, as accidents at the Chernobyl (p. 478) nuclear power plant in Ukraine and the Three Mile Island plant in Pennsylvania have taught us.

The U.S. Nuclear Regulatory Commission (NRC) estimates there is a 15–45% chance of a complete core meltdown at a U.S. reactor during the next 20 years. The NRC also found that 39 U.S. reactors have an 80% chance of failure in the containment shell from a meltdown or an explosion of gases inside the containment structures.

Another concern is that 27 of 57 operating nuclear power plants in the United States failed mock, ground-based terrorist security tests made by the Nuclear Regulatory Commission prior to 1998. The NRC contends that the security weaknesses revealed by these tests have been corrected. But many nuclear power analysts are unconvinced and note that the government has stopped staging mock attacks on nuclear plants that reveal security shortcomings.

The 2001 destruction of New York City's two World Trade Center towers by terrorist attacks raised fears that a similar attack by a large plane loaded with fuel could break open a reactor's containment shell (Figure 19-24), set off a reactor meltdown, and create a major radioactive disaster. Nuclear officials believe that U.S. plants could survive such an attack. But a 2002 study by the Nuclear Control Institute found that the U.S. nuclear plants were not designed to withstand the crash of a large jet traveling at the impact speed of the two hijacked airliners that hit the World Trade

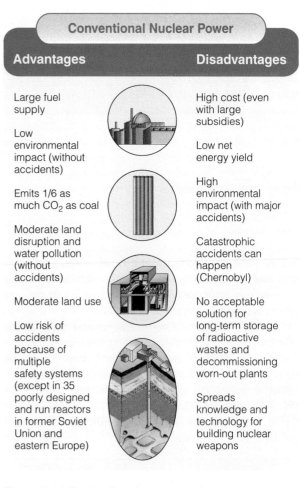

Figure 19-26 Trade-offs: advantages and disadvantages of using nuclear power to produce electricity. This evaluation includes the entire nuclear fuel cycle (Figure 19-25).

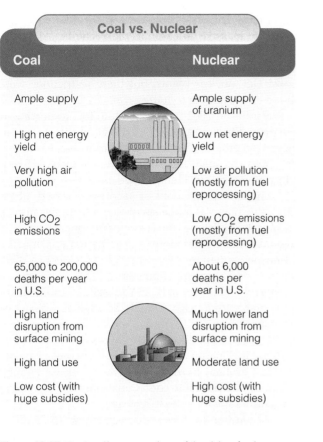

Coal vs. Nuclear

Coal	Nuclear
Ample supply	Ample supply of uranium
High net energy yield	Low net energy yield
Very high air pollution	Low air pollution (mostly from fuel reprocessing)
High CO$_2$ emissions	Low CO$_2$ emissions (mostly from fuel reprocessing)
65,000 to 200,000 deaths per year in U.S.	About 6,000 deaths per year in U.S.
High land disruption from surface mining	Much lower land disruption from surface mining
High land use	Moderate land use
Low cost (with huge subsidies)	High cost (with huge subsidies)

Figure 19-27 Trade-offs: comparison of the risks of using nuclear power and coal-burning plants to produce electricity.

Center. This is not surprising because in 1982 the U.S. Nuclear Regulatory Commission ruled that owners of nuclear power plants did not have to design the plants to survive threats such as kamikaze airliner crashes. According to the NRC, requiring such construction would make nuclear electricity too expensive to be competitive.

A 2002 report by Congressman Ed Markey of Massachusetts, based on analysis of NRC data, said that guards at U.S. nuclear plants are underpaid, undertrained, and incapable of repelling a serious ground attack by terrorists. It also said that 21 U.S. nuclear reactors are within 8 kilometers (5 miles) of an airport. In addition, a small or large aircraft or ground attack that cut off electrical power to a reactor could cause a core meltdown within about two hours without crashing into the reactor.

David Freeman was President Jimmy Carter's energy adviser, CEO of several major power companies, and is now energy adviser to the governor of California. In 2002 he warned, "The danger of penetration into a nuclear reactor—which is difficult but not impossible—is so horrendous that we've got to put out of our minds the building of any more nuclear power plants."

Throughout the world, nuclear scientists and government officials urge the shutdown of 35 poorly designed and poorly operated nuclear reactors in some republics of the former Soviet Union and in eastern Europe. This is unlikely without economic aid from the world's developed countries.

In the United States, there is widespread public distrust of the ability of the NRC and the Department of Energy (DOE) to enforce nuclear safety in commercial (NRC) and military (DOE) nuclear facilities. Congressional hearings in 1987 uncovered evidence that high-level NRC staff members destroyed documents and obstructed investigations of criminal wrongdoing by utilities, suggested ways utilities could evade commission regulations, and provided utilities and their contractors with advance notice of so-called surprise inspections. In 1996, George Galatis, a respected senior nuclear engineer, said, "I believe in nuclear power but after seeing the NRC in action I'm convinced a serious accident is not just likely, but inevitable. . . . They're asleep at the wheel."

The nuclear power industry claims that nuclear power plants in the United States have not killed anyone and cause far less environmental harm than coal-burning plants (Figure 19-27). However, according to U.S. National Academy of Sciences estimates, U.S. nuclear plants cause 6,000 premature deaths and 3,700 serious genetic defects each year. If correct, this annual death toll is much smaller than the 65,000–200,000 deaths per year caused by coal-burning plants in the United States. However, critics point out that the estimated annual deaths from both of these types of plants are unacceptable, given the availability of much less harmful alternatives such as combined-cycle natural gas turbines and wind power and eventually solar cells and hydrogen.

Despite these concerns, in 2002 the Bush administration—with strong lobbying from the nuclear power industry—was pushing to build 25–50 new nuclear power plants in the United States within the next two decades.

What Do We Do with Low-Level Radioactive Waste? Each part of the nuclear fuel cycle (Figure 19-25) produces low-level and high-level solid, liquid, and gaseous radioactive wastes with various half-lives (Table 3-1, p. 57). Wastes classified as *low-level radioactive wastes* give off small amounts of ionizing radiation and must be stored safely for 100–500 years before decaying to safe levels.

From the 1940s to 1970, most low-level radioactive waste produced in the United States (and most other countries) was put into steel drums and dumped into the ocean; the United Kingdom and Pakistan still dispose of their low-level radioactive wastes in this way.

Today, low-level waste materials from commercial nuclear power plants, hospitals, universities, industries, and other producers in the United States are put in steel drums and shipped to the two remaining regional landfills run by federal and state governments. Attempts to build new regional dumps for low-level radioactive waste using improved technology have met with fierce public opposition.

What Is High-Level Radioactive Waste? *High-level radioactive wastes* give off large amounts of ionizing radiation for a short time and small amounts for a long time. Such wastes must be stored safely for at least 10,000 years and about 240,000 years if plutonium-239 (with a half-life of 24,000 years) is not removed by reprocessing (Figure 19-25).

Case Study: How Safe Is High-Level Radioactive Waste Stored at U.S. Nuclear Power Plants? Most high-level radioactive wastes are spent fuel rods from commercial nuclear power plants. Some of these rods are stored in metal casks but most are stored in pools of boron-treated water at plant sites. A spent-fuel pool typically holds five to ten times more long-lived radioactivity than the radioactive core inside a plant's reactor.

Suppose that water drained out of a spent-fuel pool or that a dry storage cask ruptured—because of events such as earthquake, airplane impact, or terrorist act. Then the highly radioactive and thermally hot fuel would be exposed to air and steam. This would cause the zirconium outer cover of the fuel assemblies to catch fire and burn fiercely. The NRC acknowledges that such a fire could not be extinguished and would burn for days. This would release large amounts of radioactive materials into the atmosphere, contaminate large areas with radioactivity for many decades, and create economic and psychological havoc.

Unlike the reactor core with its thick concrete protective dome, spent-fuel pools have little protective cover. At about one-third of U.S. power plants the reactor is in one building and the spent-fuel pool is in a separate building with a thin corrugated metal roof and walls. At the rest of the plants, the spent-fuel pool is in a concrete building attached to the reactor building with the concrete offering much less protection than the much thicker concrete containment shell surrounding the reactor. The pools have back-up cooling systems to help prevent a fire, but these systems could malfunction or be destroyed by a terrorist attack. Dry casks stored in buildings outside a reactor are safer than storage in water pools. However, the casks can be ruptured by a severe attack.

A deliberate crash by a small airplane or an attack by a group of suicidal terrorists could drain a pool or rupture a dry cask containing spent fuel rods and cause a massive release of radioactivity. After the September 11 attacks in 2001, the Federal Aviation Administration banned small planes from flying over most U.S. nuclear plants, but these restrictions were lifted on November 6, 2001.

Studies in 2002 by the Institute for Resource and Security Studies and the Federation of American Scientists projected serious harm from possible terrorist attacks on spent-fuel storage areas at many U.S. nuclear power plants. Here are two examples.

Suppose an accident or sabotage released all radioactive material in the spent-fuel rods in the storage pool at the Millstone Unit 3 reactor in Connecticut. This would put five times more radioactive material into the atmosphere than the 1986 Chernobyl accident (p. 478). About 145,000 square kilometers (55,900 square miles)—more than the area of New York state—would be uninhabitable for at least 30 years because of radioactive contamination.

Or suppose all the fuel in two spent-fuel storage pools at the Sharon Harris Nuclear Plant (near Raleigh, North Carolina) burned. This would release enough radioactive material to contaminate an area larger than the entire state for at least 30 years.

According to these studies, about 161 million people—57% of the U.S. population—live within 121 kilometers (75 miles) of 131 aboveground spent-fuel storage sites in 39 states (most in the eastern half of the United States). In addition, these studies warned that decommissioned nuclear power plants might be even more vulnerable to sabotage because they store large amounts of spent fuel and have fewer security personnel than operating plants.

U.S. nuclear power officials consider such events to be highly unlikely worst-case scenarios and question some of the estimates. They also contend that nuclear power facilities are very safe from such attacks.

What Should We Do with High-Level Radioactive Waste? After more than 50 years of research, scientists still do not agree on whether there is a safe method for storing high-level radioactive waste. Some scientists believe the long-term safe storage or disposal of high-level radioactive wastes is technically possible. Others disagree, pointing out it is impossible to demonstrate that any method will work for the 10,000–240,000 years of fail-safe storage needed for such wastes. Following are five of the proposed methods and their possible drawbacks.

Bury it deep underground. This favored strategy is under study by all countries producing nuclear waste. In 2001, the U.S. National Academy of Sciences concluded that the geological repository option is the only scientifically credible long-term solution for safely isolating such wastes. However, according to an earlier 1990 report by the U.S. National Academy of Sciences, "Use of geological information to pretend to be able to make very accurate predictions of long-term site behavior is scientifically unsound."

Shoot it into space or into the sun. Costs would be very high, and a launch accident—like the explosion of the space shuttle *Challenger*—could disperse high-level radioactive wastes over large areas of the earth's surface. This strategy has been abandoned for now.

Bury it under the Antarctic ice sheet or the Greenland ice cap. The long-term stability of the ice sheets is not known. They could be destabilized by heat from the wastes, and retrieving the wastes would be difficult or impossible if the method failed. This strategy is prohibited by international law.

Dump it into descending subduction zones in the deep ocean (Figure 9-4, middle, p. 177). However, wastes eventually might be spewed out somewhere else by volcanic activity and containers might leak and contaminate the ocean before being carried downward. Also, retrieval would be impossible if the method did not work. This strategy is prohibited by international law.

Bury it in thick deposits of mud on the deep-ocean floor in areas that tests show have been geologically stable for 65 million years. The waste containers eventually would corrode and release their radioactive contents. This approach is prohibited by international law.

Change it into harmless, or less harmful, isotopes. Currently no way exists to do this. Even if a method was developed, costs probably would be very high, and the resulting toxic materials and low-level (but very long-lived) radioactive wastes would still need to be disposed of safely.

Case Study: What Should Be Done with High-Level Radioactive Wastes in the United States?

In 1985, the U.S. Department of Energy (DOE) announced plans to build a repository for underground storage of high-level radioactive wastes from commercial nuclear reactors on federal land in the Yucca Mountain desert region, 160 kilometers (100 miles) northwest of Las Vegas, Nevada.

By 2002, the federal government had spent about $7 billion evaluating the site as a permanent underground repository. The proposed facility (Figure 19-28) is expected to cost at least $58 billion to build (financed partly by a tax on nuclear power). It is scheduled to open by 2010.

In 2002, scientists working on the site released a report finding nothing to disqualify the site as a safe place to store high-level radioactive wastes. A number of scientists and energy analysts disagree with this

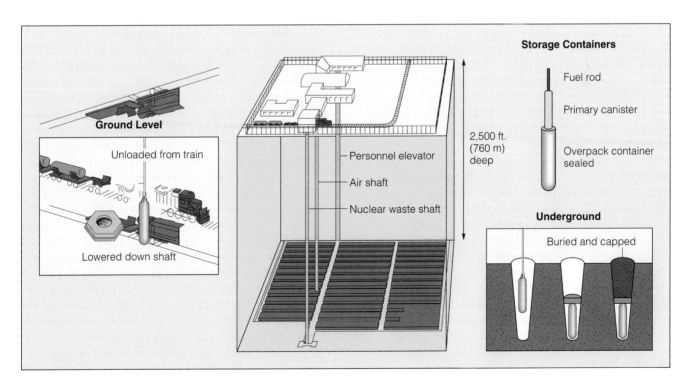

Figure 19-28 Solutions: general design for deep underground permanent storage of high-level radioactive wastes from commercial nuclear power plants in the United States. (U.S. Department of Energy)

assessment. For example, according to a December 2002 report by Congress's General Accounting Office, any decision to approve the site would be "premature" because 293 scientific questions about safety have not been answered.

Some scientists argue that it should never be allowed to open. They are concerned that rock fractures may allow water to leak into the site and corrode casks holding radioactive waste. They also point out a nearby active volcano and 32 active earthquake fault lines running through the site—an unusually high number. In 1998, Jerry Szymanski, formerly the DOE's top geologist at Yucca Mountain and now an outspoken opponent of the site, said that if water flooded the site it could cause an explosion so large that "Chernobyl would be small potatoes."

In January 2002 the Nuclear Waste Technical Review Board published its review of the site. According to this evaluation, it impossible to ensure that the wastes would remain safe for the thousands of years necessary to protect the environment and human health.

Despite such concerns, in January 2002, the U.S. energy secretary found that the site is scientifically sound and recommended to President Bush and Congress that highly radioactive waste from the nation's nuclear power plants be deposited under Nevada's Yucca Mountain. The secretary cited this as an important way to help protect wastes now stored at nuclear plants from possible terrorist attacks.

This decision raised a storm of protest from Nevada elected officials and citizens (80% of them opposed to the site) and others concerned about the safety of this approach. Opponents contend that *the Yucca Mountain waste site should not be opened because it can decrease national (homeland) security* for two major reasons. *First*, it would require about 56,000 shipments of wastes over much of the country—an average of five shipments a day for the estimated 30 years before the site is filled—in trucks and rail cars that are difficult to protect from a terrorist attack.* Nevada's Senator Harry Reid calls a truck carrying waste to the site "a dirty nuclear bomb on 18 wheels waiting to happen."

Second, it would not decrease the possibilities of sabotage of nuclear wastes stored in pools and casks at the country's nuclear plant sites because the plants will be producing new wastes about as fast as the old wastes are shipped out. By 2036, when the site may be filled, there will be about as much nuclear waste stored at nuclear plant sites as exists today.

---

*Unless the site has been closed down for national security reasons, you can go to the website, www.mapscience.org, type in any address, and view a map showing how close that area is to probable nuclear waste transportation routes to be used for at least 30 years.

The DOE and nuclear power proponents say the risks of an accident or sabotage of a nuclear waste shipment are negligible. Opponents believe the risks are underestimated, especially after the events of September 11, 2001.

In 2002, the U.S. National Academy of Sciences in collaboration with Harvard and University of Tokyo scientists urged the U.S. government to slow down and rethink the nuclear waste storage process. These scientists contend that storing spent-fuel rods in dry-storage casks in well-protected buildings at nuclear plant sites is an adequate solution for a number of decades in terms of safety and national security. This would buy time to carry out more research on this complex problem and to evaluate other sites and storage methods that might be more acceptable scientifically and politically.

Despite many objections from scientists and citizens, during the summer of 2002 Congress approved Yucca Mountain as the official site for storing the country's commercial nuclear wastes.

How Widespread Are Contaminated Radioactive Sites? In 1992, the EPA estimated that as many as 45,000 sites in the United States may be contaminated with radioactive materials—20,000 of them belonging to the DOE and the Department of Defense. Since 1992, the DOE has spent $60 billion on cleanup and estimates that the cleanup will cost taxpayers another $200 billion over the next 70 years (some analysts estimate $400–900 billion). According to a 2000 DOE report, more than two-thirds of 144 highly contaminated sites used to produce nuclear weapons will never be completely cleaned up and will have to be protected and monitored for centuries. Some critics question spending so much money on a problem ranked by scientific advisers to the EPA as a low-risk ecological problem and not among the top high-risk human health problems (Figure 10-15, p. 216).

The radioactive contamination situation in the United States pales in comparison to the legacy of nuclear waste and contamination in the republics of the former Soviet Union. Land in various parts of the former Soviet Union is dotted with areas severely contaminated by nuclear accidents, 25 operating nuclear power plants with flawed and unsafe designs, nuclear waste dump sites, radioactive waste-processing plants, and contaminated nuclear test sites. In addition, the government dumped nuclear wastes and retired nuclear-powered submarines into coastal waters.

A 1957 explosion of a nuclear waste storage tank at Mayak, a plutonium production facility in southern Russia, spewed 2.5 times as much radiation into the atmosphere as the Chernobyl accident. Because of radioactive contamination, no one can live within about 2,600 square kilometers (1,000 square miles) of the surrounding area. In addition, nearby Lake Karachay (which was a dumpsite for the facility's radioactive wastes between 1949 and 1967) is so radioactive that standing on its shores for about an hour would be fatal.

What Can We Do with Worn-Out Nuclear Plants?
After approximately 15–40 years of operation, a nuclear reactor becomes dangerously contaminated with radioactive materials, and many of its parts are worn out. Unless the plant's life can be extended by expensive renovation, it must be *decommissioned* or retired (the last step in the nuclear fuel cycle, Figure 19-25).

Scientists have proposed three ways to do this. *One* is to dismantle the plant and store its large volume of highly radioactive materials in high-level nuclear waste storage facilities (Figure 19-28), whose safety is questioned by many scientists.

A *second* approach is to put up a physical barrier around the plant and set up full-time security for 30–100 years before the plant is dismantled. A *third* option is to enclose the entire plant in a tomb that must last for several thousand years.

At least 228 large commercial reactors worldwide (20 in the United States) are scheduled for retirement by 2012. By 2033, all U.S. reactors will have to be retired, based on their original 40-year operating licenses.

Some utility companies and the Bush administration have proposed extending the life of current reactors to 50–60 years. Opponents contend that extending the life of reactors could increase the risk of nuclear accidents in aging reactors.

What Is the Connection between Nuclear Reactors and the Spread of Nuclear Weapons?
Currently, 60 countries—1 of every 3 in the world—have nuclear weapons (at least 10 nations) or the knowledge and ability to build them. Information and fuel needed to build these nuclear weapons have come mostly from the research and commercial nuclear reactors that the United States and 14 other countries have been giving away and selling in the international marketplace for decades.

Some *good news* is that between 1986 and 2001, the number of nuclear warheads held by the world's five largest nuclear powers declined by 55% and further declines are expected by 2012. The *bad news* is that this still leaves enough nuclear weapons to kill everyone in the world at least 30 times. This also greatly increases the amount of bomb-grade plutonium-239 removed from retired nuclear warheads that must be kept secure from use in nuclear weapons.

What Is the Threat from "Dirty" Radioactive Bombs?
There is growing concern, especially after the September 11, 2001, terrorist attacks in the United States, about threats from explosions of so-called *dirty radioactive bombs.* A dirty bomb consists of an explosive such as dynamite mixed with or wrapped around some form of radioactive material that is not difficult or expensive to obtain. Such a conventional bomb with a small amount of radioactive material (that often would fit in a coffee cup) is fairly easy to assemble.

Such materials can be stolen from thousands of loosely guarded and difficult to protect sources or bought on the black market. One source is hospitals using radioisotopes (such as cobalt-60) to treat cancer, diagnose various diseases, and sterilize some types of medical equipment. Other sources include university research labs, and industries using radioisotopes to detect leaks in underground pipes, irradiate food, examine mail and other materials, and detect flaws in pipe welds and boilers. Radioactive materials are also found in smoke detectors (americium-241).

Detonating a dirty bomb at street level or on a rooftop does not cause a nuclear blast. But it could have a number of harmful effects depending on the amount of explosive material used. For example, the blast and subsequent cancers could kill a dozen to 1,000 people in densely populated cities and spread radioactive material over several to hundreds of blocks.

The blast would contaminate buildings and soil in the affected area for up to ten times the half-life of the isotope used unless areas are cleaned up at great expense (billions of dollars). Cleanup would involve sandblasting or demolishing buildings, digging up asphalt and sidewalks, removing topsoil, and transporting the radioactive material to safe storage sites.

Affected areas would have to be evacuated for many years, perhaps decades, including shutdown of all business and government agency offices. Prolonged contamination would discourage businesses and agencies from reopening in such areas.

Finally, detonating a dirty bomb would cause intense psychological terror and panic throughout much of a country. As a result, terrorists would succeed in their primary objective.

Here are the results of two simulations by the Federation of American Scientists and the Center for Strategic and International Studies (a Washington think tank). In one scenario, a dirty bomb is detonated in Manhattan using a small piece of cobalt-60 stolen from a food irradiation plant. The explosion would spread contamination over about 300 city blocks, kill up to 1,000 people, and render much of New York City uninhabitable for decades.

In another scenario, terrorists explode a pea-sized piece of radioactive cesium-137 (fairly easy to obtain from medical equipment) on the mall in Washington, D.C., near the National Gallery of Art. The blast would kill about 20 people and contaminate for decades a

1.6-kilometer (1-mile) swath of land that includes the U.S. Capitol, Supreme Court, and Library of Congress buildings.

Since 1986, the NRC has recorded 1,700 incidents in the United States in which radioactive materials used by industrial, medical, or research facilities have been stolen or lost—an average of about 100 incidents per year. Since 1991, the International Atomic Energy Agency (IAEA) has detected 671 incidents of illicit trafficking in dirty-bomb materials.

Can Nuclear Power Reduce U.S. Dependence on Oil? Some proponents of nuclear power in the United States claim it will help reduce dependence on imported oil. However, other analysts point out that use of nuclear power has little effect on U.S. oil use because oil produced only 2.9% of the electricity in the United States in 2001. Also, the major use for oil is in transportation, which would not be affected by increasing the use of nuclear power plants to produce electricity.

Can We Afford Nuclear Power? Experience has shown that nuclear power is an expensive way to boil water to produce electricity, even when huge government subsidies partially shield it from free-market competition with other energy sources.

Costs rose dramatically in the 1970s and 1980s because of unanticipated safety problems and stricter regulations after the Three Mile Island and Chernobyl accidents. Estimates of the costs of producing nuclear power vary widely because of different variables used to make such calculations.

In 1995, the World Bank said that nuclear power is too costly and risky. *Forbes* business magazine has called the failure of the U.S. nuclear power program "the largest managerial disaster in U.S. business history, involving $1 trillion in wasted investment and $10 billion in direct losses to stockholders."

Some *good news* is that in recent years, the operating cost of many U.S. nuclear power plants has dropped, mostly because of less downtime. However, environmentalists and economists point out that the cost of nuclear power must be based on the entire nuclear fuel cycle (Figure 19-25), not merely the operating cost of individual plants. This includes mining and producing nuclear fuel, nuclear waste disposal, and decommissioning of worn-out plants. According to these analysts, when these costs are included the overall cost of nuclear power is very high (even with huge government subsidies) compared to many other energy alternatives.

Can We Develop New and Safer Types of Nuclear Reactors? The U.S. nuclear industry hopes to persuade the federal government and utility companies to build hundreds of smaller second-generation plants using standardized designs, which they claim are safer and can be built more quickly (in 3–6 years). These *advanced light-water reactors (ALWRs)* have built-in *passive safety features* designed to make explosions or the release of radioactive emissions almost impossible. However, according to *Nucleonics Week,* an important nuclear industry publication, "Experts are flatly unconvinced that safety has been achieved—or even substantially increased—by the new designs." In addition, these new designs do not eliminate the threats from the use of nuclear fuel in nuclear weapons and the expense and hazards of long-term radioactive waste storage and power plant decommissioning.

Is Breeder Nuclear Fission a Feasible Alternative? Some nuclear power proponents urge the development and widespread use of **breeder nuclear fission reactors,** which generate more nuclear fuel than they consume by converting nonfissionable uranium-238 into fissionable plutonium-239. Because breeders would use more than 99% of the uranium in ore deposits, the world's known uranium reserves would last at least 1,000 years, and perhaps several thousand years.

However, if the safety system of a breeder reactor fails, the reactor could lose some of its liquid sodium coolant, which ignites when exposed to air and reacts explosively if it comes into contact with water. This could cause a runaway fission chain reaction and perhaps a nuclear explosion powerful enough to blast open the containment building and release a cloud of highly radioactive gases and particulate matter. Leaks of flammable liquid sodium can also cause fires, which have happened with all experimental breeder reactors built so far.

In addition, existing experimental breeder reactors produce plutonium so slowly, it would take 100–200 years for them to produce enough to fuel a significant number of other breeder reactors. In 1994, the United States ended government-supported research for breeder technology after providing about $9 billion in research and development funding.

In December 1986, France opened a commercial-size breeder reactor. It was so expensive to build and operate that after spending $13 billion, the government spent another $2.75 billion to shut it down permanently in 1998. Because of this experience, other countries have abandoned their plans to build full-size commercial breeder reactors.

Is Nuclear Fusion a Feasible Alternative? Nuclear fusion in the sun provides 99% of the earth's energy. For decades, scientists have hoped that controlled nuclear fusion will provide an almost limitless source of high-temperature heat and electricity to supply most of the world's commercial energy. Research has focused on the D–T nuclear fusion reaction, in which two isotopes of hydrogen—deuterium (D) and tritium (T)—fuse at about 100 million°C (180 million°F; Figure 3-13, p. 59).

According to a 2001 Department of Energy task force chaired by Vice President Dick Cheney, fusion energy has a number of important advantages. It includes no emissions of conventional air pollutants and carbon dioxide. Because the hydrogen isotopes involved in the fusion reaction could be obtained from water, the fuel supply would be essentially unlimited. Some short-lived radioactive wastes would require burial and protection for about 100 years—much less than the 10,000–250,000 years needed to store such wastes produced by conventional nuclear fission.

There would be no risk of meltdown or release of large amounts of radioactive materials from a terrorist attack because only a small amount of fuel is present in the system at any time. There would also be little risk from additional proliferation of nuclear weapons because bomb-grade materials (such as enriched uranium-235 and plutonium-239) are not required for fusion energy.

Fusion power could also be used to destroy toxic wastes. Finally, fusion power plants could supply unlimited energy to produce electricity for ordinary use and to decompose water and produce the hydrogen gas needed to run a hydrogen economy by the end of this century (Figure 19-7).

This sounds great, but what is the catch? Why don't we already have fusion energy? The *bad news* is that after 50 years of research and huge expenditures of mostly government funds, controlled nuclear fusion is still in the laboratory stage. None of the approaches tested so far have produced more energy than they use.

In 1989, two chemists claimed to have achieved deuterium–deuterium (D–D) nuclear fusion at room temperature using a simple apparatus, but subsequent experiments have not substantiated their claim. Some scientists still hope to develop low-temperature nuclear fusion, but most nuclear physicists are skeptical that it can be done.

If researchers can eventually get more energy out of nuclear fusion than they put in, the next step would be to build a small fusion reactor and then scale it up to commercial size. This is an extremely difficult engineering problem. Also, the estimated cost of building and operating a commercial fusion reactor (even with huge government subsidies) is several times that of a comparable conventional fission reactor.

Proponents, including the Bush administration, contend that with greatly increased federal funding, a commercial nuclear fusion power plant might be built by 2030 or perhaps by 2020 (with emphasis on developing a new technique called muon-catalyzed fusion). However, many energy experts do not expect nuclear fusion to be a significant energy source until 2100, if then.

What Should Be the Future of Nuclear Power in the United States? Since 1948, nuclear energy (fission and fusion) has received about 58% of all federal energy research and development funds in the United States (compared to 22% for fossil fuels, 11% for renewable energy, and 8% for energy efficiency and conservation). Some analysts call for phasing out all or most government subsides and tax breaks for nuclear fission power and using such funds to subsidize and accelerate the development of promising newer energy technologies.

To these analysts, nuclear fission power is a complex, expensive, inflexible, and centralized way to produce electricity that is too vulnerable to terrorist attack. They believe it is a technology whose time has passed in a world where electricity will increasingly be provided by small, decentralized, easily expandable power plants such as natural gas turbines, wind turbines, arrays of solar cells, and fuel cells. According to investors and World Bank economic analysts, conventional nuclear power simply cannot compete in today's increasingly open, decentralized, and unregulated energy market.

Proponents of conventional nuclear power and nuclear fusion, including the Bush administration, disagree. They argue that governments should continue funding research and development and pilot plant testing of potentially safer and cheaper reactor designs along with breeder fission and nuclear fusion. They argue that we need to keep these nuclear options available for use in the future if natural gas turbines, improved energy efficiency, hydrogen-powered fuel cells, wind turbines, and other renewable energy options fail to keep up with electricity demands and reduce CO_2 emissions to acceptable levels.

The question is not whether there will be an energy revolution. It is already under way. The only questions are how rapidly it will unfold, whether it will move fast enough to prevent climate change from getting out of hand, and who will benefit most from the transition.

Lester R. Brown

REVIEW QUESTIONS

1. Define the boldfaced terms in this chapter.

2. Describe the causes and effects of the 1986 Chernobyl nuclear power plant accident in the Soviet Union.

3. What supplies 99% of the energy we use? What percentage of the remaining 1% of the energy we use comes from nonrenewable and from renewable energy in **(a)** the world, and **(b)** the United States?

4. Summarize the trends since 1950 in the global use of **(a)** coal, **(b)** oil, **(c)** natural gas, **(d)** nuclear power to produce electricity, and **(e)** wood to provide heat and cook food in developing countries.

5. How long does it usually take to phase in a new energy alternative to the point where it accounts for 10–20% of total energy use? What seven questions should we try to answer about each energy resource?

6. What is *net energy,* and why is it important in evaluating an energy resource? Why is the net energy for oil from the Middle East high and that for nuclear power low?

7. What is *petroleum,* or *crude oil?* How is oil extracted from the earth's crust?

8. What happens to crude oil at a refinery? What are *petrochemicals?*

9. Summarize the history of the Age of Oil.

10. Who has most of the world's oil reserves? What percentage of the world's oil reserves is found in the United States? What percentage of the world's oil does the United States use? What percentage of the oil used in the United States is imported?

11. How long are known and projected supplies of conventional oil expected to last in **(a)** the world and **(b)** the United States? List the advantages and disadvantages of drilling for oil and natural gas in Alaska's Arctic National Wildlife Refuge.

12. What are the advantages and disadvantages of using conventional oil as an energy resource?

13. What are the advantages and disadvantages of using heavy oil from shale oil and tar sand as energy resources?

14. What is *natural gas?* Who has most of the world's reserves of natural gas? What is *methane hydrate?* Distinguish between *liquefied petroleum gas (LPG)* and *liquefied natural gas (LNG).*

15. How long are known and projected supplies of natural gas expected to last in **(a)** the world and **(b)** the United States? What is a *combined-cycle natural gas turbine system,* and why is its use rising rapidly?

16. What are the advantages and disadvantages of using natural gas as an energy resource?

17. What is *coal,* and how is it formed? Distinguish among *lignite, bituminous,* and *anthracite* coal. How is coal extracted from the earth's crust? How is coal used? What four countries have the largest coal reserves?

18. How long are known and projected supplies of coal expected to last **(a)** in the world and **(b)** in the United States?

19. What are the advantages and disadvantages of using coal as an energy resource? What are the advantages and disadvantages of converting solid coal into gaseous and liquid fuels?

20. Describe how a *nuclear fission reactor* works. What are the seven major components of a *light-water nuclear reactor,* and what role does each play? What is the *nuclear fuel*

cycle? What is the difference between an open and a closed nuclear fuel cycle?

21. List three reasons why commercial nuclear power plants were developed in the United States after World War II. List seven factors that have contributed to the leveling off of the use of nuclear power plants to produce electricity.

22. List the major advantages and disadvantages of using conventional nuclear fission to produce electricity. Compare the advantages and disadvantages of using nuclear power and burning coal to produce electricity.

23. How safe are nuclear power plants?

24. What is being done with *low-level radioactive waste* produced by the nuclear fuel cycle?

25. Describe the threat from terrorist attacks on high-level radioactive wastes stored in pools of water or in dry casks at most nuclear power plants in the United States.

26. What are the options for dealing with *high-level radioactive waste?*

27. List the advantages and disadvantages of the proposed site for storing high-level nuclear wastes at Yucca Mountain in Nevada. How widespread are contaminated radioactive waste sites in **(a)** the United States and **(b)** the former Soviet Union?

28. What are the three options for retiring (decommissioning) worn-out nuclear power plants?

29. What is the relationship between the development of commercial nuclear power and the spread of nuclear weapons throughout much of the world?

30. Describe the threat from terrorists stealing or buying radioactive materials used in medicine, industry, and research facilities and using the materials to make dirty radioactive bombs.

31. How useful is nuclear power in reducing U.S. dependence on oil imports?

32. What are the advantages and disadvantages of second-generation nuclear reactor designs?

33. What are the advantages and disadvantages of using **(a)** breeder nuclear fission and **(b)** nuclear fusion as an energy resource? What are the advantages and disadvantages of continuing large-scale government subsidies for research and development of conventional nuclear power, breeder fission, and nuclear fusion?

CRITICAL THINKING

1. To continue using oil at the current rate (not the projected higher exponential increase in its annual use), we must discover and add to global oil reserves the equivalent of a new Saudi Arabian supply (the world's largest) *every 10 years.* Do you believe this is possible? If not, what effects might this have on your life and on the life of a child or grandchild you might have?

2. List five actions you can take to reduce your dependence on oil and resources such as gasoline and most plastics derived from oil (see website material for this chapter). Which of these things do you actually plan to do?

3. Explain why you are for or against continuing to increase oil imports in the United States (or the country where you live). What do you believe are the three best ways to reduce dependence on oil imports?

4. Explain why you agree or disagree with the following proposals by various energy analysts to help solve U.S. energy problems: **(a)** find and develop more domestic supplies of oil, **(b)** place a heavy federal tax on gasoline and imported oil to help reduce the waste of oil resources, **(c)** increase dependence on nuclear power, and **(d)** phase out all nuclear power plants by 2020.

5. Do you believe oil companies should be allowed to explore and remove oil and natural gas from the Arctic National Wildlife Refuge (ANWR)? Why or why not? Use the library or Internet to find out the political fate of this wildlife refuge.

6. What do you believe should be done with high-level radioactive wastes? Explain.

7. Would you favor having high-level nuclear waste transported by truck or train through the area where you live to a centralized underground storage site? Explain.

8. Explain why you agree or disagree with each of the following proposals made by the U.S. nuclear power industry: **(a)** Provide at least $100 billion in government subsidies to build a large number of better designed nuclear fission power plants to reduce dependence on imported oil and slow global warming. **(b)** Prevent the public from participating in hearings for licensing nuclear power plant level waste depository and on safety issues at the nation's nuclear reactors. **(c)** Restore government subsidies to develop a breeder nuclear fission reactor program. **(d)** Greatly increase federal subsidies for developing nuclear fusion.

9. If you had to choose, would you rather live next door to a coal-fired power plant or a nuclear plant? Explain.

10. Should the United States and other developed countries provide economic and technical aid for closing 35 poorly designed and poorly operated nuclear reactors in some republics of the former Soviet Union and in eastern Europe? Explain.

PROJECTS

1. Write a two-page scenario of what your life might be like without oil. Compare and discuss the scenarios developed by members of your class.

2. Use the library or the Internet to find information about the accident that took place at the Three Mile Island (TMI) nuclear power plant near Harrisburg, Pennsylvania, in 1979. According to the nuclear power industry, the TMI accident showed that its safety systems work because the accident caused no known deaths.

Other analysts dispute this claim and argue the accident was a wake-up call about the potential dangers of nuclear power plants that led to tighter and better safety regulations. Use the information you find to determine which of these positions you support, and defend your choice.

3. Make a concept map of this chapter's major ideas, using the section heads and subheads and the key terms (in boldface). Look on the website for this book for information about making concept maps.

INTERNET STUDY RESOURCES AND RESOURCES FOR FURTHER READING AND RESEARCH

The website for this book contains helpful study aids and many ideas for further reading and research. Log on to

http://biology.brookscole.com/miller 10

and click on the Chapter-by-Chapter area. Choose Chapter 19 and select a resource:

■ Flash Cards allows you to test your mastery of the Terms and Concepts to Remember for this chapter.

■ Tutorial Quizzes provides a multiple-choice practice quiz.

■ Student Guide to InfoTrac will lead you to Critical Thinking Projects that use InfoTrac College Edition as a research tool.

■ References lists the major books and articles consulted in writing this chapter.

■ Hypercontents takes you to an extensive list of sites with news, research, and images related to individual sections of the chapter.

INFOTRAC COLLEGE EDITION

Improve your skills with InfoTrac College Edition, a searchable online database of articles from more than 700 periodicals. Log on to

http://www.infotrac-college.com

or access InfoTrac through the website for this book. Try to find the following articles:

1. Geiselman, B. 2002. Yucca Mountain foes quake over plan; tremor won't dissuade energy officials. (Brief article) *Waste News* 8: 1. *Keywords:* "Yucca Mountain" and "earthquake." Nevada's proposed Yucca Mountain nuclear waste repository has been controversial for a number of years, but as the government prepares to make the site the official U.S. nuclear waste dump, the area's geologic stability is being questioned.

2. Carey, J. 2003. Taming the oil beast. *Business Week,* Feb. 24. *Keywords:* "Oil" and "energy policy." Propose six steps for a sensible national energy policy that is within our reach.

20 ENERGY EFFICIENCY AND RENEWABLE ENERGY

The Coming Energy-Efficiency and Renewable-Energy Revolution

Energy expert Amory Lovins (Guest Essay, p. 510) built a large, passively heated, superinsulated, partially earth-sheltered home and office in Snowmass, Colorado (Figure 20-1), where winter temperatures can drop to $-40°C$ ($-40°F$).

This structure also houses the research center for the Rocky Mountain Institute, an office used by 40 people. This office-home gets 99% of its space and water heating and 95% of its daytime lighting from the sun and uses one-tenth the usual amount of electricity for a structure of its size.

With today's superinsulating windows a house can have large numbers of windows without much heat loss in cold weather or heat gain in hot weather. Thinner insulation material now being developed will allow roofs and walls to be insulated far better than in today's best superinsulated houses.

A small but growing number of people in developed and developing countries are getting their electricity from *solar cells* that convert sunlight directly into electricity. They can be attached like shingles to a roof, used as a roof material, or applied to window glass as a coating. Solar-cell prices are high but are falling.

Many scientists and executives of oil and automobile companies believe we are in the beginning stages of a *solar–hydrogen revolution* to be phased in during this century as the Age of Oil (p. 487) begins winding down (Figure 19-7, p. 481). Electricity produced by large banks of solar cells or farms of wind turbines could be passed through water to make hydrogen gas (H_2). This clean-burning fuel could be used to fuel vehicles, industries, and buildings. Another solution is to burn hydrogen in energy-efficient *fuel cells* that produce electricity to run cars and appliances, heat water, and heat and cool buildings.

Burning hydrogen made by decomposing water produces water vapor and no carbon dioxide (CO_2). Thus shifting to hydrogen as our primary energy resource during this century would eliminate most of the world's air pollution. This would also greatly slow global warming—as long as the hydrogen is produced from water and not fossil fuels or other carbon-containing compounds.

Figure 20-1 The Rocky Mountain Institute in Colorado. This facility is a home and a center for the study of energy efficiency and sustainable use of energy and other resources. It is also an example of energy-efficient passive solar design.

Robert Millman/Rocky Mountain Institute

This chapter addresses the following questions:

- What are the advantages and disadvantages of improving energy efficiency?

- What are the advantages and disadvantages of using solar energy to heat buildings and water and produce electricity?

- What are the advantages and disadvantages of using flowing water (hydropower) to produce electricity?

- What are the advantages and disadvantages of using wind to produce electricity?

- What are the advantages and disadvantages of burning plant material (biomass) to heat buildings and water, produce electricity, and propel vehicles (biofuels)?

- What are the advantages and disadvantages of producing hydrogen gas and using it to produce electricity, heat buildings and water, and propel vehicles?

- What are the advantages and disadvantages of extracting heat from the earth's interior (geothermal energy)?

- What are the advantages and disadvantages of using smaller, decentralized micropower sources to heat buildings and water, produce electricity, and propel vehicles?

- How can we make a transition to a more sustainable energy future?

20-1 THE IMPORTANCE OF IMPROVING ENERGY EFFICIENCY

What Is Energy Efficiency? Doing More with Less **Energy efficiency** is the percentage of total energy input into an energy conversion device or system that does useful work and is not converted to low-quality, essentially useless heat. Improving the energy efficiency of a car motor, home heating system, or other energy conversion device involves using less energy to do more useful work.

How Much Energy Do We Waste? You may be surprised to learn that *84% of all commercial energy used in the United States is wasted* (Figure 20-2). About 41% of

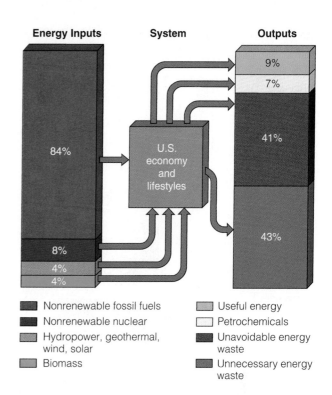

Figure 20-2 Flow of commercial energy through the U.S. economy. Note that only 16% of all commercial energy used in the United States ends up performing useful tasks or being converted to petrochemicals; the rest is either automatically and unavoidably wasted because of the second law of thermodynamics (41%) or wasted unnecessarily (43%).

this energy is wasted automatically because of the degradation of energy quality imposed by the second law of thermodynamics (p. 59). However, about 43% is wasted unnecessarily. This waste is caused mostly by using fuel-wasting motor vehicles, furnaces, and other devices and living and working in leaky, poorly insulated, poorly designed buildings (Guest Essay, p. 510).

According to the U.S. Department of Energy (DOE), *the United States unnecessarily wastes as much energy as two-thirds of the world's population consumes.* Improvements in energy efficiency since the OPEC oil embargo in 1973 have cut U.S. energy bills by $275 billion a year. But unnecessary energy waste still costs the United States about $300 billion per year—an average of $570,000 per minute. Reducing energy waste has a number of economic and environmental advantages (Figure 20-3, p. 508).

The United States (and most other developed countries) has come a long way in improving energy efficiency and saving money. But there is an exciting opportunity to do much better because the country still unnecessarily wastes almost half the energy it uses. Sharply reducing this energy waste will benefit the environment, people, and the economy.

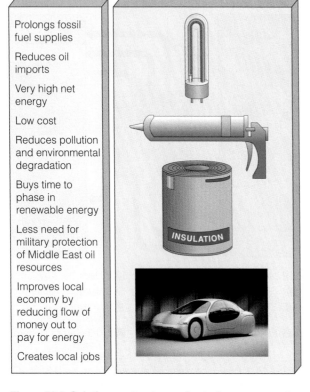

- Prolongs fossil fuel supplies
- Reduces oil imports
- Very high net energy
- Low cost
- Reduces pollution and environmental degradation
- Buys time to phase in renewable energy
- Less need for military protection of Middle East oil resources
- Improves local economy by reducing flow of money out to pay for energy
- Creates local jobs

INSULATION

Figure 20-3 Solutions: advantages of reducing energy waste. Global improvements in energy efficiency could save the world about $1 trillion per year—an average of $114 million per hour.

What Are the Energy Efficiencies of Common Devices?

The energy conversion devices we use vary in their energy efficiencies (Figure 20-4). We can save energy and money by buying more energy-efficient cars, lighting, heating systems, water heaters, air conditioners, and appliances. Some energy-efficient models may cost more initially, but in the long run they usually save money by having a lower **life cycle cost**: initial cost plus lifetime operating costs.

Here are three of the world's most energy-inefficient devices in widespread use today. *First* is the *incandescent light bulb,* which wastes 95% of its energy input of electricity. In other words, it is a *heat bulb. Second* is a motor vehicle with an *internal combustion engine,* which wastes 86–90% of the energy in its fuel. *Third* is a *nuclear power plant* producing electricity for space heating or water heating. Such a plant wastes about 86% of the energy in its nuclear fuel and probably 92% when the energy needed to deal with its radioactive wastes and to retire the plant is included. Energy experts call for us to replace these devices or greatly improve their energy efficiency over the next few decades.

Coal-burning power plants also are big energy wasters. About 34% of the energy in coal burned in a typical electric power plant is used to produce electricity. However, the remaining 66% ends up as waste heat that flows into the environment. As a result, U.S. coal-burning power plants throw away as much heat as all the energy used by Japan.

What Is Net Energy Efficiency?

Recall that the only energy that really counts is *net energy* (p. 481). The *net energy efficiency* of the entire energy delivery process for a space heater, water heater, or car is determined by the efficiency of each step in the energy conversion process.

Figure 20-5 shows the net energy efficiency for heating two well-insulated homes. One is heated with electricity produced at a nuclear power plant, transported by wire to the home, and converted to heat (electric resistance heating). The other is heated passively: direct solar energy enters through high-efficiency windows facing the sun and strikes heat-absorbing materials that store the heat for slow release.

This analysis shows that converting the high-quality energy in nuclear fuel to high-quality heat at several thousand degrees in the power plant, converting this heat to high-quality electricity, transmitting the electricity to users, and using the electricity to provide low-quality heat for warming a house to only about 20°C (68°F) is very wasteful of high-quality en-

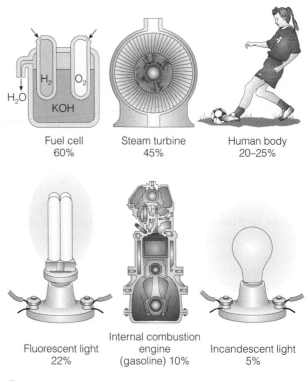

Fuel cell 60%　　Steam turbine 45%　　Human body 20–25%

Fluorescent light 22%　　Internal combustion engine (gasoline) 10%　　Incandescent light 5%

Figure 20-4 Energy efficiency of some common energy conversion devices.

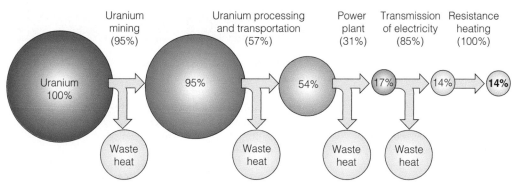

Uranium mining (95%) Uranium processing and transportation (57%) Power plant (31%) Transmission of electricity (85%) Resistance heating (100%)

Uranium 100% → 95% → 54% → 17% → 14% → **14%**

Waste heat Waste heat Waste heat Waste heat

Electricity from Nuclear Power Plant

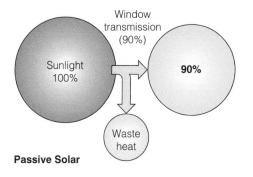

Window transmission (90%)

Sunlight 100% → **90%**

Waste heat

Passive Solar

Figure 20-5 Comparison of net energy efficiency for two types of space heating. The cumulative net efficiency is obtained by multiplying the percentage shown inside the circle for each step by the energy efficiency for that step (shown in parentheses). Because of the second law of thermodynamics, in most cases the greater the number of steps in an energy conversion process, the lower its net energy efficiency. About 86% of the energy used to provide space heating by electricity produced at a nuclear power plant is wasted. If the additional energy needed to deal with nuclear wastes and to retire highly radioactive nuclear plants after their useful life is included, then the net energy yield for a nuclear plant is only about 8% (or 92% waste). By contrast, with passive solar heating, only about 10% of incoming solar energy is wasted.

ergy (Figure 3-10, p. 55). Burning coal or any fossil fuel at a power plant to supply electricity and transmitting it long distances to heat water or space is also inefficient.

This example illustrates two general principles for saving energy. *First,* keep the number of steps in an energy conversion process as low as possible. Each time we convert energy from one form to another or transmit it, some useful energy is lost. *Second,* strive to have the highest possible energy efficiency for each step in an energy conversion process.

20-2 WAYS TO IMPROVE ENERGY EFFICIENCY

How Can We Use Waste Heat? Could we save energy by recycling energy? No. The second law of thermodynamics (p. 59) tells us that we cannot recycle energy. But we can slow the rate at which waste heat flows into the environment when high-quality energy is degraded.

For a house, the best way to do this is to insulate it thoroughly, eliminate air leaks (Figure 20-6), and equip

Figure 20-6 An infrared photo (thermogram) showing heat loss (red, white, and orange) around the windows, doors, roofs, and foundations of houses and stores in Plymouth, Michigan. Many homes and buildings in the United States (and in most other countries) are so full of leaks that their heat loss in cold weather and heat gain in hot weather are equivalent to having a large window-sized hole in the wall of the house.

Technology Is the Answer (But What Was the Question?)

Amory B. Lovins

Physicist and energy consultant Amory B. Lovins is one of the world's most respected experts on energy strategy. In 1989, he received the Delphi Prize for environmental work; in 1990, the Wall Street Journal *named him one of the 39 people most likely to change the course of business in the 1990s. He is research director at Rocky Mountain Institute, a nonprofit resource policy center that he and Hunter Lovins founded in Snowmass, Colorado, in 1982. He has served as a consultant to more than 200 utilities, private industries, and international organizations, and to many national, state, and local governments. He is active in energy affairs in more than 35 countries and has published several hundred papers and a dozen books on energy strategies and policies.*

It is fashionable to suppose that we're running out of energy and ask how we can get more of it. However, the more important questions are the following: How much energy do we need? What are the cheapest and least environmentally harmful ways to meet these needs?

How much energy it takes to make steel, run a car, or keep ourselves comfortable in our houses depends on how cleverly we use energy. For example, it is now cheaper to double the efficiency of most industrial electric motor drive systems than to fuel existing power plants to make electricity. Just this one saving can more than replace the entire U.S. nuclear power program. We know how to make lights five times as efficient as those currently in use and household appliances that give us the same work as now but use one-fifth as much energy (saving money in the process).

Within a decade automakers could have cars getting 64–128 kpl (150–300 mpg) on the road if consumers demanded such cars.

We know today how to make new buildings (and many old ones) so heat-tight (but still well ventilated) that they need essentially no outside energy to maintain comfort year-round, even in severe climates. In fact, I live and work in one [Figure 20-1].

These energy-saving measures are all cheaper than going out and getting more energy. However, the old view of the energy problem included a worse mistake than forgetting to ask how much energy we needed: It sought more energy, in any form, from any source, at any price, as if all kinds of energy were alike.

Just as there are different kinds of food, so there are many different forms of energy whose different prices and qualities suit them to different uses [Figure 3-10, p. 55]. After all, there is no demand for energy as such; nobody wants raw kilowatt-hours or barrels of sticky black goo. People instead want energy services: comfort, light, mobility, hot showers, cold beverages, and the ability to cook food and make cement. In developing energy resources we should start by asking, "What tasks do we want energy for, and what amount, type, and source of energy will do each task most cheaply?"

The real question is, "What is the cheapest way to do low-temperature heating and cooling?" The answer is weather-stripping, insulation, heat exchangers, greenhouses, superwindows (which have as much insulating value as the outside wall of a typical house), roof overhangs, trees, and so on. These measures generally cost the equivalent of buying electricity at about 0.5–2¢ per kilowatt-hour, the lowest-cost way by far to supply energy.

If we need more electricity, we should get it from the cheapest sources first. In approximate order of increasing price, these include

■ Converting to efficient lighting equipment. This would save the United States electricity equal to the output of 120 large power plants, plus $30 billion a year in fuel and maintenance costs.

■ Using more efficient electric motors to save up to half the energy used by motor systems. This would save electricity equal to the output of another 150 large power plants and repay the cost in about a year.

it with an air-to-air heat exchanger to prevent buildup of indoor air pollutants.

In office buildings and stores, waste heat from lights, computers, and other machines can be collected and distributed to reduce heating bills during cold weather. During hot weather, the collected heat can be vented outdoors to reduce cooling bills or to drive a type of air conditioner called an *absorption chiller.*

How Can We Save Energy and Money in Industry? *One way* some industries save energy and money is to use cogeneration, or combined heat and power (CHP) systems. In such a system two useful forms of energy (such as steam and electricity) are produced from the same fuel source. These systems have an efficiency of up to 80% (compared to about 30–40% for coal-fired boilers and nuclear power plants) and emit two-thirds less CO_2 per unit of energy produced than conventional coal-fired boilers.

Cogeneration has been widely used in western Europe for years. Its use in the United States (where it now produces only 9% of the country's electricity) and China is growing. In Germany, small cogeneration units that run on natural gas or liquefied petroleum

- Displacing the electricity now used for water heating and for space heating and cooling with good architecture, weatherization, insulation, and mostly passive solar techniques.

- Improving the energy efficiency of appliances, smelters, and the like.

Just these four measures can quadruple U.S. electrical efficiency, making it possible to run today's economy with no changes in lifestyles and using no power plants, whether old or new or fueled with oil, gas, coal, uranium, or solar energy. We would need only the present hydroelectric capacity, readily available small-scale hydroelectric projects, and a modest amount of wind power.

If we still wanted more electricity, the next cheapest sources would include cogenerating electricity and heat in industrial plants and power plants, using low-temperature heat engines run by industrial waste heat or by solar ponds, filling empty turbine bays and upgrading equipment in existing big dams, using modern wind turbines or small-scale hydroelectric turbines in good sites, using combined-cycle natural gas turbines, and using recently developed more efficient solar cells when their price is reduced by mass production.

Only after exhausting all these cheaper opportunities should we even consider building a new central power station of any kind—the slowest and costliest known way to get more electricity (or to save oil).

To emphasize the importance of starting with energy end uses rather than energy sources, consider a story from France. In the mid-1970s, energy conservation planners in the French government found that their biggest need for energy was to heat buildings and that even with good heat pumps, electricity would be the costliest way to do this. So they had a fight with their government-owned and -run utility company; they won, and electric heating was supposed to be discouraged or even phased out because it was so wasteful of money and fuel.

Meanwhile, down the street, the energy supply planners (who were far more numerous and influential in the French government) said, "Look at all that nasty imported oil coming into our country. We must replace that oil with some other source of energy. Voilà! Nuclear reactors can give us energy, so we'll build them all over the country." However, they paid little attention to who would use that extra energy and no attention to relative prices.

Thus, these two groups of the French energy establishment went on with their respective solutions to two different, indeed contradictory, French energy problems: *more energy of any kind* versus *the right kind to do each task in the most inexpensive way*. It was only in 1979 that these conflicting perceptions collided. The supply-side planners suddenly realized that the only thing they would be able to sell all that nuclear electricity for would be electric heating, which they had just agreed not to do.

Every industrial country is in this embarrassing position. Supply-oriented planners think the problem boils down to whether to build coal or nuclear power stations (or both). Energy-use planners realize that *no* kind of new power station can be an economic way to meet the needs for using electricity to provide low- and high-temperature heat and for the vehicular liquid fuels that are 92% of our energy problem.

So if we want to provide energy services at the lowest cost, we need to begin by determining what we need the energy for!

Critical Thinking

1. The author argues that building more nuclear, coal, or other electrical power plants to supply electricity for the United States is unnecessary and wasteful of energy and money. List your reasons for agreeing or disagreeing with this viewpoint.

2. Explain why you agree or disagree that increasing the supply of energy, instead of improving energy efficiency, is the wrong answer to our energy problems.

gas (LPG) supply restaurants, apartment buildings, and houses with all their energy. In 6–8 years, they pay for themselves in saved fuel and electricity.

Another way to save energy and money is to replace energy-wasting electric motors. Running electric motors (mostly in industry) consumes about one-fourth of the electricity produced in the United States. Most of these motors are inefficient because they run only at full speed with their output throttled to match the task. Each year a heavily used electric motor consumes 10 times its purchase cost in electricity—equivalent to using $200,000 worth of gasoline each year to fuel a $20,000 car. The costs of replacing such motors with new adjustable-speed drive motors would be paid back in about 1 year and save an amount of energy equal to that generated by 150 large (1,000-megawatt) power plants. A *third way* to save energy is to switch from low-efficiency incandescent lighting to higher-efficiency fluorescent lighting.

How Can We Save Energy in Transportation?
According to most energy analysts, the best way to save energy (especially oil) and money in transportation is to *increase the fuel efficiency of motor vehicles*.

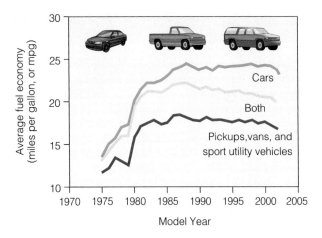

Figure 20-7 Average fuel economy of new vehicles sold in the United States, 1975–2003. (Data from U.S. Environmental Protection Agency and National Highway Traffic Safety Administration)

Some *good environmental news* is that between 1973 and 1985, the average fuel efficiency rose sharply for new cars sold in the United States and to a lesser degree for pickup trucks, minivans, and sport utility vehicles (SUVs) (Figure 20-7). This occurred because of the government-mandated *Corporate Average Fuel Economy (CAFE)* standards.

Some *bad environmental news* is that between 1985 and 2003, the average fuel efficiency for new motor vehicles sold in the United States leveled off or declined slightly (Figure 20-7). One reason for this is the increased popularity of energy-inefficient sport utility vehicles (SUVs), minivans, light trucks, and large autos that also produce about 20% of U.S. emissions of CO_2. Another reason is failure of elected officials to raise CAFE standards since 1985 because of opposition from automakers.

A 2001 study by the American Council for an Energy-Efficient Economy (ACEEE) analyzed the effects of increasing the fuel economy of new vehicles in the United States by just 5% a year for 10 years. Doing this would save 10–20 times more oil than the projected supply from the Arctic National Wildlife Refuge (p. 486) and more than three times the oil in the nation's current proven oil reserves.

Suppose that in 1991 the first Bush administration and Congress had established a policy of requiring that the average car in the United States get 14 kilometers per liter (kpl) [32 miles per gallon (mpg)] by 2001. According to energy analysts, doing this would have saved enough oil to eliminate all current oil imports to the United States from the Persian Gulf. A similar action by the second Bush administration and Congress could significantly reduce U.S. dependence on imported oil within 10 years.

Here are two pieces of *bad environmental news. First,* only 33 of the 934 cars, trucks, and vans on the market in the United States in 2003 got more than 13 kpl (30 mpg). *Second,* such models account for less than 1% of all car sales. One reason for this is that the inflation-adjusted price of gasoline today in the United States is low (Figure 20-8 and Connections, right). A second reason is that two-thirds of U.S. consumers prefer SUVs, pickup trucks, minivans, and other large, inefficient vehicles.

In Europe, where gasoline costs $0.80–1.30 per liter ($3–5 per gallon), subcompact cars are in much greater use—especially for urban trips. For example, Volkswagen has a four-seater subcompact car that gets 33 kpl (78 mpg) and has begun producing a smaller model that gets 100 kpl (235 mpg).

Are Hybrid-Electric Vehicles the Answer? There is rapidly growing interest in developing *superefficient cars* that could eventually get 34–128 kpl (80–300 mpg). One type of energy-efficient car uses a small *hybrid-electric internal combustion engine.* It runs on gasoline, diesel fuel, or natural gas and a small battery (recharged by the internal combustion engine) to provide the energy needed for acceleration and hill climbing (Figure 20-9).

Toyota introduced its first hybrid vehicle in 1997 and Honda and Nissan have been selling several mod-

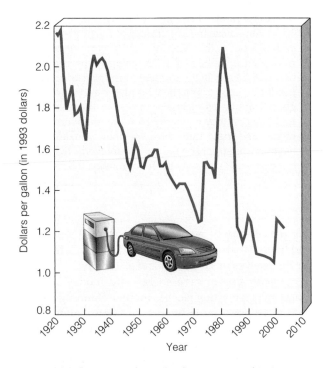

Figure 20-8 Real price of gasoline (in 1993 dollars) in the United States, 1920–2002. The 225 million motor vehicles in the United States use about 40% of the world's gasoline. Gasoline is one of the cheapest items American consumers buy and is less expensive than bottled water. (U.S. Department of Energy)

A **Combustion engine:**
Small, efficient internal combustion engine powers vehicle with low emissions.

B **Fuel tank:**
Liquid fuel such as gasoline, diesel, or ethanol runs small combustion engine.

C **Electric motor:**
Traction drive provides additional power, recovers braking energy to recharge battery.

D **Battery bank:**
High-density batteries power electric motor for increased power.

E **Regulator:**
Controls flow of power between electric motor and battery bank.

F **Transmission:**
Efficient 5-speed automatic transmission.

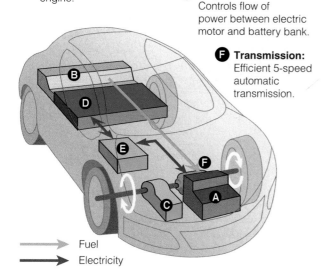

→ Fuel
→ Electricity

Figure 20-9 Solutions: general features of a car powered by a *hybrid gas–electric engine*. A small internal combustion engine recharges the batteries, thus reducing the need for heavy banks of batteries and solving the problem of the limited range of conventional electric cars. The bodies of future models of such cars probably will be made of lightweight composite plastics that offer more protection in crashes, do not need to be painted, do not rust, can be recycled, and have fewer parts than conventional cars. (Concept information from Daimler-Chrysler, Ford, Honda, and Toyota)

els of hybrid vehicles in the United States since 2000. Carmakers plan to introduce up to 20 hybrid models, including cars, trucks, SUVs, and vans, in the next 4–5 years. Sales are expected to reach 500,000 a year by 2006.

Fleets of diesel-hybrid buses have been in operation for several years in New York City and several other major cities around the world.

Are Fuel-Cell Cars the Answer? Another type of superefficient car is an electric vehicle that uses a *fuel cell* that burns hydrogen (H_2) fuel to produce the electricity (Figure 20-10, p. 514). In the cell, the hydrogen fuel (H_2) combines with oxygen (O_2) in the air to produce electrical energy to power the car and emits water vapor (H_2O) into the atmosphere. Most major automobile companies have developed prototype fuel-cell cars and plan to market a variety of such vehicles by 2010.

The Real Cost of Gasoline in the United States

CONNECTIONS

Economists and environmentalists point out that gasoline costs U.S. consumers much more than it appears. This is because most of the real cost of gasoline is not paid directly at the pump.

According to a 1998 study by the International Center for Technology Assessment, the hidden costs of gasoline to U.S. consumers are about $1.30–3.70 per liter ($5–14 per gallon), depending on how the costs are estimated. These hidden costs include government subsidies and tax breaks for oil companies and road builders, pollution control and cleanup, military protection of oil supplies in the Middle East, and environmental, health, and social costs. Such costs include increased medical bills and insurance premiums, time wasted in traffic jams, noise pollution, increased mortality from air and water pollution, urban sprawl, and harmful effects on wildlife species and habitats.

Economists point out that if these harmful costs were included as taxes in the market price of gasoline, we would have much more energy-efficient and less polluting cars. However, gasoline and car companies benefit financially by being able to pass these hidden costs on to consumers and future generations.

This is basically an education and political problem. Most consumers are unaware that they are paying these harmful costs and do not connect them with gasoline use. Also, politicians running on a platform of raising gasoline prices 3- to 11-fold in the United States would be committing political suicide.

Critical Thinking

Some economists have suggested that U.S. consumers might be willing to pay much more for gasoline if they understood they are already paying these hidden costs indirectly and the tax revenues from gasoline sales were used to reduce taxes on wages, income, and wealth and provide a safety net for low- and middle-class consumers (p. 27). Would you support or oppose such a taxshift proposal? Explain.

How Can Electric Bicycles and Scooters Reduce Energy Use and Waste? For urban trips, more people may begin using *electric bicycles* powered by a small electric motor. These bicycles, now being sold by several companies, cost $500–1,100 and travel at up to 32 kilometers per hour (kph) [20 miles per hour (mph)].

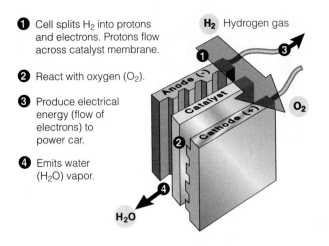

❶ Cell splits H_2 into protons and electrons. Protons flow across catalyst membrane.

❷ React with oxygen (O_2).

❸ Produce electrical energy (flow of electrons) to power car.

❹ Emits water (H_2O) vapor.

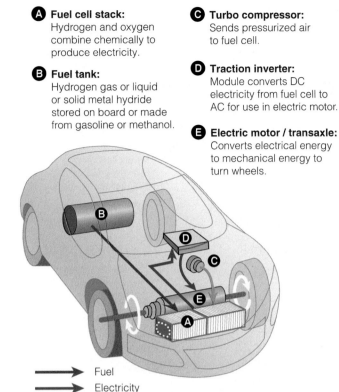

Ⓐ **Fuel cell stack:** Hydrogen and oxygen combine chemically to produce electricity.

Ⓑ **Fuel tank:** Hydrogen gas or liquid or solid metal hydride stored on board or made from gasoline or methanol.

Ⓒ **Turbo compressor:** Sends pressurized air to fuel cell.

Ⓓ **Traction inverter:** Module converts DC electricity from fuel cell to AC for use in electric motor.

Ⓔ **Electric motor / transaxle:** Converts electrical energy to mechanical energy to turn wheels.

→ Fuel

→ Electricity

Figure 20-10 Solutions: general features of an electric car powered by a *fuel cell* running on hydrogen gas. The hydrogen can be produced onboard the vehicle from gasoline or methanol or offboard for transfer to the vehicle from natural gas or by using renewable energy sources such as solar cells or wind turbines to decompose water into hydrogen and oxygen gas. Several automobile companies have developed prototypes and are working to get costs down. Prototype models are on the road now and an array fuel cell vehicles should be available to consumers by 2010. (Concept information from Daimler-Chrysler, Ford, Ballard, Toyota, and Honda)

They go about 48 kilometers (30 miles) without pedaling on a full electric charge and produce no pollution during operation (and only a small amount for the electricity used in recharging them).

In 2003, an electric bicycle powered by a small fuel cell became available for about $1,500–2000. A small metal container (similar to a propane gas container) that can be easily disconnected and refilled within seconds supplies the hydrogen.

Another alternative is an *electric scooter*. One model, the Nova Cruz Voloci, has a range of about 81 kilometers (50 miles), can travel at a top speed of 48 kph (30 mph), and costs about $2,500.

How Can We Design Buildings to Save Energy?
Atlanta's 13-story Georgia Power Company building uses 60% less energy than conventional office buildings of the same size. The largest surface of the building faces south to capture solar energy. Each floor extends out over the one below it. This blocks out the higher summer sun to reduce air conditioning costs but allows warming by the lower winter sun. Energy-efficient lights focus on desks rather than illuminating entire rooms.

If phased in over two to three decades, the Georgia Power model and other existing cost-effective commercial building technologies could reduce energy use by 75% in U.S. buildings and cut CO_2 emissions from buildings in half. This would save more than $130 bil-lion per year in energy bills—an average of $15 million an hour.

Another energy-efficient design is a *superinsulated house* (Figure 20-11). Such houses typically cost 5% more to build than conventional houses of the same

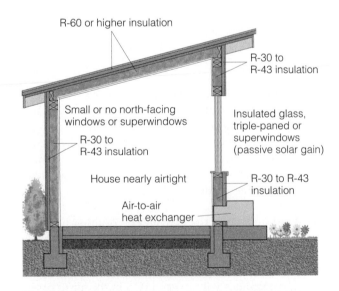

R-60 or higher insulation

R-30 to R-43 insulation

Small or no north-facing windows or superwindows

Insulated glass, triple-paned or superwindows (passive solar gain)

R-30 to R-43 insulation

House nearly airtight

R-30 to R-43 insulation

Air-to-air heat exchanger

Figure 20-11 Solutions: major features of a *superinsulated house*. Such a house is so heavily insulated and so airtight that it can be warmed by heat from direct sunlight, appliances, and human bodies, with little or no need for a backup heating system. An air-to-air heat exchanger prevents buildup of indoor air pollution.

size. But this extra cost is paid back by energy savings within about 5 years and can save a homeowner $50,000–100,000 over a 40-year period.

Since the mid-1980s there has been growing interest in building superinsulated houses called *straw-bale houses.* Their walls consist of compacted bales of certain types of straw, which are widely available at a low cost. Then the walls are covered with plaster or adobe. By 2002, more than 1,200 such homes had been built or were under construction in the United States (Guest Essay, p. 516). Making the walls from straw, an annually renewable agricultural residue often burned as a waste product, reduces the need for wood and thus slows deforestation. The main problem is getting banks and other moneylenders to recognize the potential of this and other unconventional types of housing and provide homeowners with construction loans.

Eco-roofs covered with plants have been used in Germany and in other parts of Europe for about 25 years. These plant-covered roof gardens (Figure 13-1, p. 280) provide good insulation, absorb storm water, and outlast conventional roofs.

How Can We Save Energy in Buildings? An important way to save energy is to *use the most energy-efficient ways to heat houses* (Figure 20-12). The most energy-efficient ways to heat space, in order, are a superinsulated house, passive solar heating, heat pumps in warm climates (but not in cold climates because at low temperatures they automatically switch to costly electric resistance heating), and high-efficiency (85–98%) natural gas furnaces. The most wasteful and expensive way is to use electric resistance heating with the electricity produced by a coal-fired or nuclear power plant.

The energy efficiency of existing houses and buildings can be improved significantly by adding insulation, plugging leaks, and installing energy-saving windows and lighting. About one-third of heated air in U.S. homes and buildings escapes through closed windows and holes and cracks (Figure 20-6)—equal to the energy in all the oil flowing through the Alaska pipeline every year. During hot weather these windows and cracks also let heat in, increasing the use of air conditioning. Between 1976 and 2002, the U.S. Department of Energy's Weatherization Assistance Program, which aids low-income Americans in making their homes more energy efficient, saved more than $1 billion in energy costs.

Replacing all windows in the United States with low-E (low-emissivity) windows would cut expensive losses from houses by two-thirds and reduce CO_2

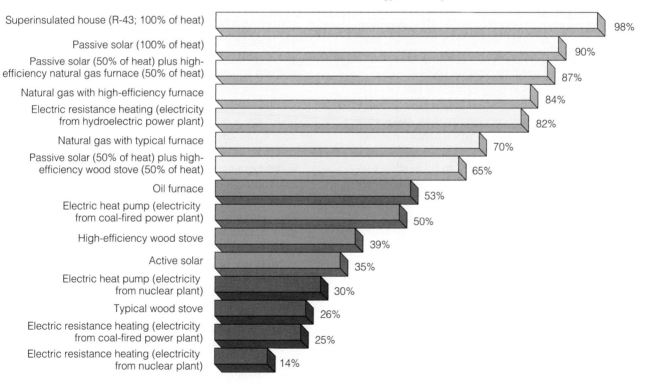

Net Energy Efficiency

Superinsulated house (R-43; 100% of heat)	98%
Passive solar (100% of heat)	90%
Passive solar (50% of heat) plus high-efficiency natural gas furnace (50% of heat)	87%
Natural gas with high-efficiency furnace	84%
Electric resistance heating (electricity from hydroelectric power plant)	82%
Natural gas with typical furnace	70%
Passive solar (50% of heat) plus high-efficiency wood stove (50% of heat)	65%
Oil furnace	53%
Electric heat pump (electricity from coal-fired power plant)	50%
High-efficiency wood stove	39%
Active solar	35%
Electric heat pump (electricity from nuclear plant)	30%
Typical wood stove	26%
Electric resistance heating (electricity from coal-fired power plant)	25%
Electric resistance heating (electricity from nuclear plant)	14%

Figure 20-12 Solutions: ways to heat an enclosed space such as a house, ranked by net energy efficiency. (Data from Howard T. Odum)

Living Lightly on the Earth at Round Mountain Organics

Nancy Wicks

Nancy Wicks is an eco-pioneer trying to live her ideals. She grew up in a small town in Iowa and did undergraduate studies in a village in Nepal. Both of these life experiences inspired her to live more sustainably by creating Round Mountain Organics, an organic garden at an altitude of 1.6 kilometers (8,500 feet) in the Rocky Mountains near Crested Butte, Colorado. Nancy lives in a passive solar strawbale house, which is powered by the wind and sun. She is a fanatic reuser, recycler, and composter. She received the "Sustainable Business of the Year 2000" award from the High Country Citizens' Alliance.

After studying in Nepal, where sustainability is a do or die situation, I have tried to incorporate as many sustainable practices into my life as possible at my house and organic garden business called Round Mountain Organics. This includes being a member of a buying coop (where buying in bulk not only saves resources but also saves money), reusing everything from plastic and paper bags to trays and pots for garden plants, and composting food waste (which saves money on trash bills and fertilizes the soil).

After moving onto the land that is now home to Round Mountain Organics, I spent 4 years planning and building an octagonal strawbale house with a stucco exterior—the first such house to be built in the county. I chose to build with straw for several reasons. *First*, straw is a natural building material and renewable resource. *Second*, there is a surplus of straw after harvesting grains such as wheat (which I used), oats, barley, and rice. *Third*, its insulation value of R-54 comes in handy when you live in an area where winter temperatures can dip to 40°F below zero. *Finally*, such houses are easy to build (with only 1 week needed to put up the strawbale walls). Since the 1980s strawbale houses have also been built in Arizona and New Mexico to beat the heat.

I used a passive solar design by orienting the house to the south to take advantage of Colorado's abundant sunshine. During the day the insulated window covers are drawn up and the sun shines onto flagstone tiles covering a cement slab that stores and releases heat slowly to keep the house comfortably warm or cool regardless of outside conditions. At night the window covers are let down to hold the heat in.

Because I live in one of the world's sunniest places, I decided not to get hooked up to the electrical grid and instead get my electricity from a small wind turbine and

---

emissions. Widely available superinsulating windows insulate as well as 8–12 sheets of glass. Although they cost 10–15% more than double-glazed windows, this cost is paid back rapidly by the energy they save. Even better windows will reach the market soon.

Simply wrapping a water heater in a $20 insulating jacket can save $45 a year and reduce CO_2 emissions. Leaky heating and cooling ducts in attics and unheated basements allow 20–30% of a home's heating and cooling energy to escape and draw unwanted moisture and heat into the home. Careful sealing can reduce this loss. Some designs for new homes keep the ducts inside the home's thermal envelope so that escaping hot or cool air leaks into the living space.

An energy-efficient way to heat hot water for washing and bathing is to use *tankless instant water heaters* (about the size of a small suitcase) fired by natural gas or LPG. These devices, widely used in many parts of Europe, heat water instantly as it flows through a small burner chamber, provide hot water only when it is needed, and use about 20% less energy than traditional water heaters.*

---

*They work great. I used them in a passive solar office and living space for 15 years. Models are available for $500–1,000 from companies such as Rinnai, Bosch, Takagi, and Envirotech.

A well-insulated, conventional natural gas or LPG water heater is fairly efficient. But all conventional natural gas and electric resistance heaters waste energy by keeping a large tank of water hot all day and night and can run out after a long shower or two.

Using electricity produced by any type of power plant is the most inefficient and expensive way to heat water for washing and bathing. A $425 electric water heater can cost $5,900 in energy over its 20-year life, compared to about $4,000 for a comparable natural gas water heater over the same period.

Cutting off lights, computers, TVs, and other appliances when they are not needed can make a big difference in energy use and bills. At 9 P.M. one weekday evening, major TV stations in Bangkok, Thailand, cooperated with the government in showing a dial that gave the city's current use of electricity. Viewers were asked to turn off unnecessary lights and appliances. They then watched the dial register a 735-megawatt drop in electricity use—a drop equal to the output of two medium-sized coal-burning power plants. This visual experience showed individuals that reducing their unnecessary electricity use could cut their bills and collectively close down power plants.

Setting higher energy-efficiency standards for new buildings is another way to save energy. Building codes

panels of solar cells. The electricity is stored in a bank of 12 batteries and an inverter converts the stored direct current (DC) electricity to ordinary 120-volt alternating current (AC).

If it's cloudy and not windy for a couple of days (which is rare), I fire up a small gas generator to charge the batteries. I also use some propane to provide hot water with a small on-demand water heater. Only a small pilot light stays lit until the hot water faucet is turned on. Then a large flame is ignited that the water is piped through. This way, I do not use energy to keep a tank of water hot around the clock.

I use many energy-saving devices. They include compact fluorescent light bulbs, an oversized pressure tank so the well pump does not have to kick on every time the faucet turns on, and a superinsulated energy-efficient DC refrigerator.

I use organic gardening to grow flowers, herbs, and vegetables for my own use and for sale to local residents and restaurants. I incorporate some pioneering organic gardening techniques such as Rudolph Steiner's biodynamics (developed in 1924) and Bill Mollison's permaculture (developed in 1978).

Insect pests are picked off by hand and beneficial insects such as ladybugs are used to eat harmful insects such as aphids. Compost, aged animal manure, and cover crops that are plowed in as green manure are used to add nutrients to the soil. Crop rotation is used so as not to deplete the soil of nutrients.

Cold frames (a type of mini-greenhouse) are used to extend the growing season to 150 days in the cold climate where there are only about 90 days without a killing frost. Recently I built a passively heated straw-bale greenhouse to extend the short growing season further and raise heat-loving plants such as tomatoes, cucumbers, and zucchinis. The chicken coop is in the northeast corner of the greenhouse, with the heat given off by the chickens helping to warm the greenhouse.

Now I am working on starting a nonprofit, Round Mountain Institute, Inc., to educate people on how they can live in harmony with the earth.

Critical Thinking

Would you like to live a lifestyle similar to that of Nancy Wicks? Explain. Why do you think more people do not try to live more sustainably, as she does? List three ways to help encourage people to adopt such a lifestyle.

could require that all new houses use 60–80% less energy than conventional houses of the same size, as has been done in Davis, California. Because of tough energy-efficiency standards, the average Swedish home consumes about one-third as much energy as the average American home of the same size.

Another way to save energy is to *buy the most energy-efficient appliances and lights.* Since 1978, the Department of Energy (DOE) has set federal energy-efficiency standards for more than 20 appliances used in the United States. A 2001 study by the National Academy of Sciences found that between 1978 and 2000, the $7 billion spent by the DOE on this program saved consumers more than $30 billion in energy costs and provided environmental benefits valued conservatively at $60–80 billion. The 2001 study also projected that the program will save U.S. consumers another $46 billion in energy costs between 2000 and 2020.* Programs like these exist in 43 other countries.

---

*Each year the American Council for an Energy-Efficient Economy (ACEEE) publishes a list of the most energy-efficient major appliances mass-produced for the U.S. market. A copy can be obtained from the council at 1001 Connecticut Avenue NW, Suite 801, Washington, DC 20036, or on its website at http://www.aceee.org/consumerguide/index.htm.

Energy-efficient lighting could save U.S. businesses and homes about $30 billion per year in electricity bills. Replacing a standard incandescent bulb with an energy-efficient compact fluorescent bulb (Figure 20-13, p. 518) saves about $48–70 per bulb over its 10-year life. Thus replacing 25 incandescent bulbs (in a house or building) with energy-efficient fluorescent bulbs saves $1,250–1,750. Students in Brown University's environmental studies program showed that the school could save more than $40,000 per year just by replacing the incandescent light bulbs in exit signs with compact fluorescent bulbs.

In 2001, researchers at the Lawrence Berkeley National Laboratory developed a very efficient high-intensity fluorescent table lamp. It uses two independently controllable and fully dimmable compact fluorescent bulbs, one directed downward and the other upward. The lamp can eliminate the need for overhead room lighting and also provide down lighting for reading and other tasks. It can also help decrease use of highly inefficient halogen bulbs, which can also start fires and increase air conditioning needs because of the intense heat they produce. In 2002, scientists at Sandia National Laboratories developed a new type of incandescent light bulb that raises the efficiency of such bulbs from 5% to about 60%.

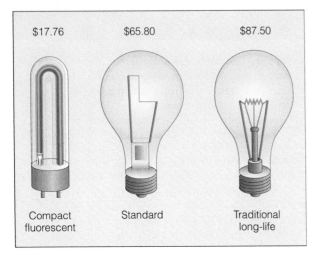

$17.76 $65.80 $87.50

Compact   Standard   Traditional
fluorescent   long-life

Figure 20-13 Solutions: cost of electricity for comparable light bulbs used for 10,000 hours. Because conventional incandescent bulbs are only 5% efficient and last only 1,500 hours, they waste enormous amounts of energy and money and add to the heat load of houses during hot weather. Socket-type fluorescent lights use one-fourth as much electricity as conventional bulbs. Although these bulbs cost $6–15 per bulb, they last up to 100,000 hours (60–70 times longer than conventional incandescent bulbs and 25 times longer than halogen bulbs), saving a lot of money (compared with less efficient incandescent and halogen bulbs) over their long life. Between 1998 and 2001, global sales of compact fluorescent bulbs rose from 45 million to 606 million per year—with 80% of them made in China. (Data from Electric Power Research Institute)

If all households in the United States used the most efficient frost-free refrigerator now available, 18 large (1,000-megawatt) power plants could close. Microwave ovens can cut electricity use for cooking by 25–50% (but not if used for defrosting food). Clothes dryers with moisture sensors cut energy use by 15%, and front-loading washers use 50% less energy than top-loading models but cost about the same. The SpinX dryer, which debuted in 2002, uses centrifugal force and minimal electricity to dry clothes. It can dry clothes in 2 minutes instead of 30 minutes in a conventional dryer.

Connections: How Can Using the Internet Save Energy and Paper and Reduce Global Warming? According to a 2000 report by researchers at the Center for Energy and Climate Solutions, increased use of the Internet for business and shopping transactions reduces energy use and decreases emissions of CO_2 and other air pollutants. For example, using the Internet for business transactions allows more employees to work at home and can reduce the need for retail, manufacturing, warehouse, and commercial office space.

Some online companies keep no merchandise in warehouses and have it shipped to customers directly from manufacturers. For example, Amazon.com uses 1/16 as much energy per unit of floor space to sell a book as a regular store. Internet use can also reduce the energy, paper, and materials used to produce, package, and market consumer items such as computer software and music CDs by downloading them.

Paper production (a highly energy- and resource-intensive industry) is expected to decrease as consumers use the Internet to download software and view magazines, newspapers, research articles, telephone directories, encyclopedias, and books. Paper use can also decrease as consumers and businesses send more e-mail and less conventional mail and more sellers replace paper catalogs with easily updated online catalogs.

Although this book is printed on paper, I used no paper in preparing it for the publisher. Instead, the manuscript was sent to the publisher electronically as attachments to e-mail messages. Copyediting of the manuscript was also done via e-mail attachments.

We can also envision a day not too far away when we can read and interact with most textbooks on websites maintained by book publishers or individual authors.

Why Are We Not Doing More to Reduce Energy Waste? With such an impressive array of benefits (Figure 20-3), why is there so little emphasis on improving energy efficiency by governments, businesses, and consumers? There are three major reasons. *First,* there is *a glut of low-cost oil* (Figure 19-11, p. 485) *and gasoline* (Figure 20-8). As long as energy is cheap, people are more likely to waste it and not make investments in improving energy efficiency.

Second, there is a *lack of sufficient government tax breaks and other economic incentives* for consumers and businesses to invest in improving energy efficiency. *Finally,* there is a *lack of information* about the availability of energy-saving devices and the amount of money such items can save consumers as revealed by *life cycle cost* analysis.

20-3 USING SOLAR ENERGY TO PROVIDE HEAT AND ELECTRICITY

What Are the Major Advantages and Disadvantages of Solar Energy? One of the four keys to sustainability (Figure 8-14, p. 171) based on learning from nature is to *rely mostly on renewable* solar energy. Figure 20-14 lists some of the advantages and disadvantages of making a shift to greatly increased use of direct solar energy and indirect forms of solar energy such as wind. Like fossil fuels and nuclear power (Chapter 19),

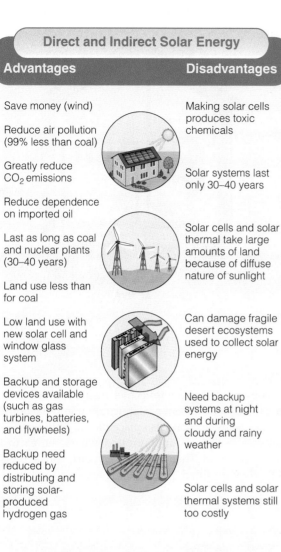

Direct and Indirect Solar Energy	
Advantages	**Disadvantages**
Save money (wind)	Making solar cells produces toxic chemicals
Reduce air pollution (99% less than coal)	
Greatly reduce CO_2 emissions	Solar systems last only 30–40 years
Reduce dependence on imported oil	
Last as long as coal and nuclear plants (30–40 years)	Solar cells and solar thermal take large amounts of land because of diffuse nature of sunlight
Land use less than for coal	
Low land use with new solar cell and window glass system	Can damage fragile desert ecosystems used to collect solar energy
Backup and storage devices available (such as gas turbines, batteries, and flywheels)	Need backup systems at night and during cloudy and rainy weather
Backup need reduced by distributing and storing solar-produced hydrogen gas	Solar cells and solar thermal systems still too costly

Figure 20-14 Trade-offs: major advantages and disadvantages of using direct and indirect solar energy systems to produce heat and electricity. Specific advantages and disadvantages of different direct and indirect solar and other renewable energy systems are discussed in this chapter.

each renewable energy alternative has a mix of advantages and disadvantages, as discussed in the remainder of this chapter.

Here are four pieces of *good news* about the increased use of renewable energy. *First,* in 2001 the European Union (EU) adopted nonbinding agreements for its member countries to get 12% of their total energy and 22% of their electricity from renewable energy by 2010.

Second, California gets about 12% of its electricity from renewable resources. *Third,* a 2001 joint study by the American Council for an Energy-Efficient Economy, the Tellus Institute, and the Union of Concerned Scientists showed how renewable energy could pro-

vide 20% of U.S. energy by 2020. *Finally,* according to Royal Dutch Shell International Petroleum, renewable energy could account for 50% of the world's energy production by 2050.

The *bad news* is that solar and wind power currently provide only about 1% of the world's commercial energy—mostly because they have received and continue to receive much lower government tax breaks, subsidies, and research and development funding than fossil fuels and nuclear power. This creates an uneven economic playing field.

How Can We Use Solar Energy to Heat Houses and Water? Buildings and water can be heated by solar energy using two methods: passive and active (Figure 20-15, p. 520). A **passive solar heating system** absorbs and stores heat from the sun directly within a structure (Figure 20-1, Figure 20-15, left, and Figure 20-16, p. 520).

Energy-efficient windows and attached greenhouses face the sun to collect solar energy by direct gain. Walls and floors of concrete, adobe, brick, stone, salt-treated timber, and water in metal or plastic containers store much of the collected solar energy as heat and release it slowly throughout the day and night. A small backup heating system such as a vented natural gas or propane heater may be used but is not necessary in many climates.

On a life cycle cost basis, good passive solar and superinsulated design is the cheapest way to heat a home or small building in regions where ample sunlight is available during the daytime. Such a system usually adds 5–10% to the construction cost, but the life cycle cost of operating such a house is 30–40% lower. The typical payback time for passive solar features is 3–7 years.

In an **active solar heating system,** collectors absorb solar energy, and a fan or a pump supplies part of a building's space-heating or water-heating needs (Figure 20-15, right). Several connected collectors are usually mounted on the roof with an unobstructed exposure to the sun. Some of the heat can be used directly. The rest can be stored in insulated tanks containing rocks, water, or a heat-absorbing chemical for release as needed. Active solar collectors can also supply hot water. Most analysts do not expect widespread use of active solar collectors for heating houses because of high costs, maintenance requirements, and unappealing appearance.

Figure 20-17 (p. 521) lists the major advantages and disadvantages of using passive or active solar energy for heating buildings. Passive solar cannot be used to heat existing homes and buildings not oriented to receive sunlight or whose access to sunlight is blocked by other buildings and structures.

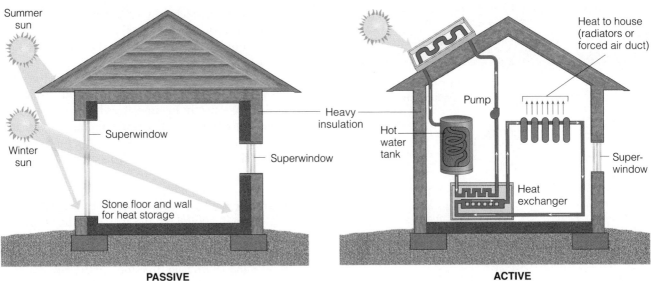

Figure 20-15 Solutions: passive and active solar heating for a home.

How Can We Cool Houses Naturally? Here are some ways to make a building cooler. Use superinsulation and superinsulating windows. Block the high summer sun with deciduous trees, window overhangs, or awnings (Figure 20-16, top left). Use windows and fans to take advantage of breezes and keep air moving. Suspend reflective insulating foil in an attic to block heat from radiating down into the house.

Homeowners can also place plastic *earth tubes* 3–6 meters (10–20 feet) underground where the earth is cool year-round. Then a tiny fan can pipe cool and partially dehumidified air into an energy-efficient house

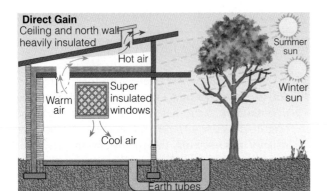

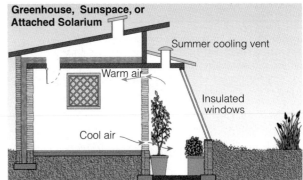

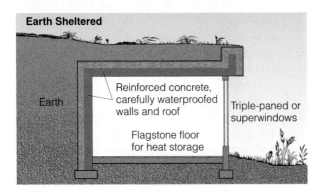

Figure 20-16 Solutions: three examples of *passive solar design* for houses.

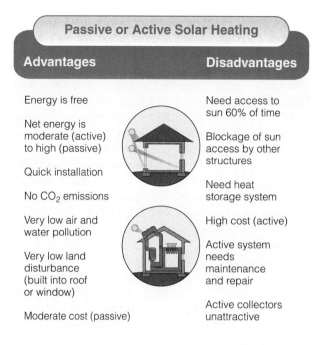

Passive or Active Solar Heating

Advantages	Disadvantages
Energy is free	Need access to sun 60% of time
Net energy is moderate (active) to high (passive)	Blockage of sun access by other structures
Quick installation	Need heat storage system
No CO_2 emissions	High cost (active)
Very low air and water pollution	Active system needs maintenance and repair
Very low land disturbance (built into roof or window)	Active collectors unattractive
Moderate cost (passive)	

Figure 20-17 Trade-offs: advantages and disadvantages of heating a house with passive or active solar energy.

(Figure 20-16, top left).* In dry climates building can be cooled by solar-powered evaporative air conditioners. However, they cost too much for residential use and do not work in humid climates.

How Can We Use Solar Energy to Generate High-Temperature Heat and Electricity? Several so-called *solar thermal systems* collect and transform radiant energy from the sun into high-temperature thermal energy (heat), which can be used directly or converted to electricity. One method uses a *central receiver system*, called a *power tower.* Huge arrays of computer-controlled mirrors called *heliostats* track the sun and focus sunlight on a central heat collection tower (Figure 20-18, top drawing). By 2006, Australia is planning to build a power tower in its sunny outback that will be more than twice the height of the world's tallest building.

Another approach is a *solar thermal plant* or *distributed receiver system*, in which sunlight is collected and focused on oil-filled pipes running through the middle of curved solar collectors (Figure 20-18, bottom drawing). This concentrated sunlight can generate temperatures high enough for industrial processes or for producing steam to run turbines and generate elec-

---

*They work. I used them in a passively heated and cooled office and home for 15 years. People allergic to pollen and molds should add an air purification system, but this is also necessary with a conventional cooling system.

tricity. At night or on cloudy days, high-efficiency combined-cycle natural gas turbines can supply backup electricity as needed.

A different type of distributed receiver system uses *parabolic dish collectors* (which look somewhat like TV satellite dishes) instead of parabolic troughs. These collectors can track the sun along two axes and generally are more efficient than troughs. A pilot plant is being built in northern Australia. The DOE projects that within 10–20 years, parabolic dishes with a natural gas turbine backup should be able to produce electrical power costing about the same as that from coal-burning plants.

Inexpensive *solar cookers* can focus and concentrate sunlight and cook food, especially in rural villages in sunny developing countries. They can be made by fitting an insulated box big enough to hold three or four pots with a transparent, removable top. Solar cookers reduce deforestation for fuelwood and the time and labor needed to collect firewood. They also reduce indoor air pollution from smoky fires.

Figure 20-18 lists the advantages and disadvantages of concentrating solar energy to produce high-temperature heat or electricity. Most analysts do not expect widespread use of such technologies over the next few decades for several reasons. One is their high costs, and another is lack of sufficient tax breaks and government research and development funding. Finally, there are much cheaper ways to produce electricity such as combined-cycle natural gas turbines and wind turbines.

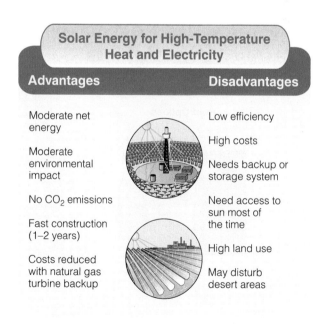

Solar Energy for High-Temperature Heat and Electricity

Advantages	Disadvantages
Moderate net energy	Low efficiency
Moderate environmental impact	High costs
No CO_2 emissions	Needs backup or storage system
Fast construction (1–2 years)	Need access to sun most of the time
Costs reduced with natural gas turbine backup	High land use
	May disturb desert areas

Figure 20-18 Trade-offs: advantages and disadvantages of using solar energy to generate high-temperature heat and electricity.

How Can We Produce Electricity with Solar Cells? Solar energy can be converted directly into electrical energy by **photovoltaic (PV) cells,** commonly called **solar cells** (Figure 20-19). A solar cell is a transparent wafer containing a *semiconductor* material with a thickness ranging from less than that of a human hair to that of a sheet of paper. Sunlight energizes and causes electrons in the semiconductor to flow, creating an electrical current.

A single solar cell produces only a tiny amount of electricity. Thus many cells are wired together in modular panels to produce the amount of electricity needed. The resulting direct current (DC) electricity can be stored in batteries and used directly. Another option is to convert the DC to conventional alternating current (AC) electricity by a separate inverter or an inverter built into the cells.

Traditional-looking solar-cell roof shingles and roofing material (developed in Japan) reduce the cost of solar-cell installations by saving on roof costs (Figure 20-19). Glass walls and windows of buildings can also have built-in solar cells. With this technology, the roof and glass walls and windows become a building's power plant.

Easily expandable banks of solar cells can be used to provide electricity in developing countries for 1.7 billion people in rural villages that have no electricity. They can also produce electricity at a small power plant, using combined-cycle natural gas turbines to provide backup power when the sun is not shining. Another possibility is to use arrays of solar cells to convert water to hydrogen gas that can be distributed to energy users by pipeline, as natural gas is. With financing from the World Bank, India (the world's number-one market for solar cells) is installing solar-cell systems in 38,000 villages, and Zimbabwe is bringing solar electricity to 2,500 villages.

Figure 20-20 lists the advantages and disadvantages of solar cells. By 2003, more than 1 million homes in the world (most of them in villages in developing countries and about 200,000 in the United States) were getting some or all of their electricity from solar cells.

Current costs of producing electricity from solar cells are high (about 30¢ per kilowatt-hour). But costs are expected to drop because of savings from mass production of solar cells and greatly increased research by major corporations and many governments in solar-cell design.

With a strong push from governments and private investors, by 2050 solar cells could provide up to 25% of the world's electricity (and at least 35% in the United States). If such projections are correct, the production, sale, and installation of solar cells could become one of the world's largest and fastest growing businesses.

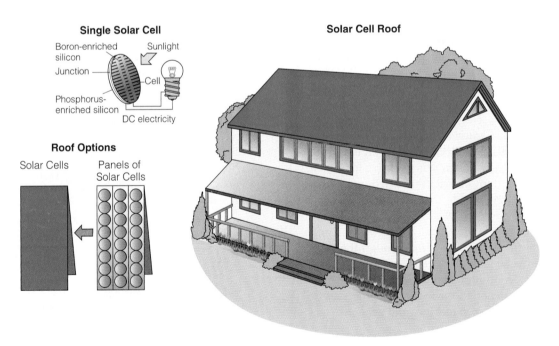

Single Solar Cell

Boron-enriched silicon
Sunlight
Junction
Cell
Phosphorus-enriched silicon
DC electricity

Roof Options

Solar Cells
Panels of Solar Cells

Solar Cell Roof

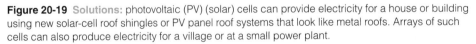

Figure 20-19 Solutions: photovoltaic (PV) (solar) cells can provide electricity for a house or building using new solar-cell roof shingles or PV panel roof systems that look like metal roofs. Arrays of such cells can also produce electricity for a village or at a small power plant.

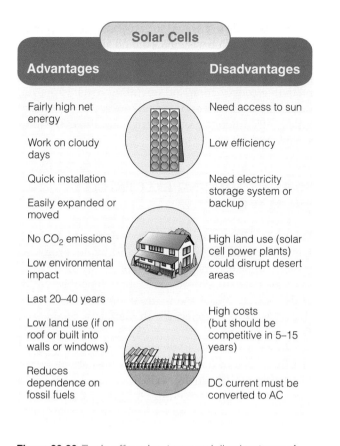

Solar Cells

Advantages	Disadvantages
Fairly high net energy	Need access to sun
Work on cloudy days	Low efficiency
Quick installation	Need electricity storage system or backup
Easily expanded or moved	
No CO_2 emissions	High land use (solar cell power plants) could disrupt desert areas
Low environmental impact	
Last 20–40 years	
Low land use (if on roof or built into walls or windows)	High costs (but should be competitive in 5–15 years)
Reduces dependence on fossil fuels	DC current must be converted to AC

Figure 20-20 Trade-offs: advantages and disadvantages of using solar cells to produce electricity.

20-4 PRODUCING ELECTRICITY FROM MOVING WATER AND FROM HEAT STORED IN WATER

How Can We Produce Electricity Using Hydropower Plants? Electricity can be produced from flowing water by three methods. *One* is large-scale hydropower, in which a high dam is built across a large river to create a reservoir (Figure 14-9, p. 312). Some of the water stored in the reservoir is allowed to flow through huge pipes at controlled rates, spinning turbines and producing electricity.

Another method is small-scale hydropower. In this case a low dam with no reservoir (or only a small one) is built across a small stream, and the stream's flow of water is used to spin turbines and produce electricity.

A *third* method is pumped-storage hydropower. First, pumps use surplus electricity from a conventional power plant to pump water from a lake or a reservoir to another reservoir at a higher elevation. When more electricity is needed, water in the upper reservoir is released, flows through turbines, and generates electricity on its return to the lower reservoir.

In 2001, hydropower supplied about 7% of the world's total commercial energy (2% in the United States), 20% of the world's electricity (6% in the United States but about 50% of the power used along the West Coast). It supplies 99% of the electricity in Norway, 75% in New Zealand, 50% in developing countries, and 25% in China.

Figure 20-21 lists the advantages and disadvantages of using large-scale hydropower plants to produce electricity. According to the United Nations, only about 13% of the world's exploitable potential for hydropower has been developed. Much of its untapped potential is in south Asia (especially China, p. 312), South America, and parts of the former Soviet Union.

Because of increasing concern about the harmful environmental and social consequences of large dams (Figure 14-9, p. 312), there has been growing pressure on the World Bank and other development agencies to stop funding new large-scale hydropower projects. According to a 2000 study by the World Commission on Dams, hydropower in tropical countries is a major emitter of greenhouse gases. This occurs because reservoirs that power the dams can trap rotting vegetation, which can emit greenhouse gases such as CO_2 and CH_4.

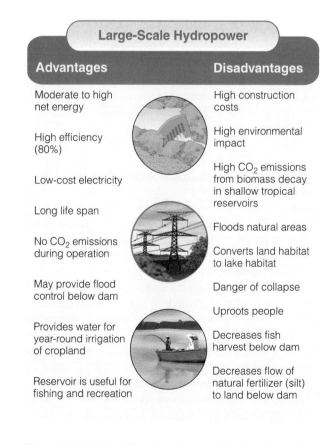

Large-Scale Hydropower

Advantages	Disadvantages
Moderate to high net energy	High construction costs
High efficiency (80%)	High environmental impact
Low-cost electricity	High CO_2 emissions from biomass decay in shallow tropical reservoirs
Long life span	
No CO_2 emissions during operation	Floods natural areas
	Converts land habitat to lake habitat
May provide flood control below dam	Danger of collapse
	Uproots people
Provides water for year-round irrigation of cropland	Decreases fish harvest below dam
	Decreases flow of natural fertilizer (silt) to land below dam
Reservoir is useful for fishing and recreation	

Figure 20-21 Trade-offs: advantages and disadvantages of using large dams and reservoirs to produce electricity.

Small-scale hydropower projects eliminate most of the harmful environmental effects of large-scale projects. However, their electrical output can vary with seasonal changes in stream flow.

Is Producing Electricity from Tides and Waves a Useful Option? Twice a day in high and low tides, water that flows into and out of coastal bays and estuaries can spin turbines to produce electricity. Two large tidal energy facilities are currently operating, one at the Rance estuary in France and the other in Canada's Bay of Fundy. Most analysts expect tidal power to make only a tiny contribution to world electricity supplies because of a lack of suitable sites and high construction costs.

The kinetic energy in ocean waves, created primarily by wind, is another potential source of electricity. Most analysts expect wave power to make little contribution to world electricity production, except in a few coastal areas with the right conditions (such as western England).

How Can We Produce Electricity from Heat Stored in Water? Japan and the United States have been evaluating the use of the large temperature differences between the cold deep waters and the sun-warmed surface waters of tropical oceans for producing electricity. If economically feasible, this would be done in *ocean thermal energy conversion (OTEC)* plants anchored to the bottom of tropical oceans in suitable sites. However, most energy analysts believe the large-scale extraction of energy from ocean thermal gradients may never compete economically with other energy alternatives.

Saline solar ponds, usually located near inland saline seas or lakes in areas with ample sunlight, can be used to produce electricity. Heat accumulated during the day in the denser bottom layer can be used to produce steam that spins turbines, generating electricity. A small experimental power plant on a saline solar pond on the Israeli shore of the Dead Sea operated for several years but was closed in 1989 because of high operating costs.

Freshwater solar ponds can be used to heat water and space. A shallow hole is dug and lined with concrete. A number of large black plastic bags, each filled with several centimeters of water, are placed in the hole and then covered with fiberglass insulation panels. The panels let sunlight in but keep most of the heat stored in the water during the daytime from being lost to the atmosphere. Typically, the water in the bags reaches its peak temperature in the afternoon. Then a computer turns on pumps that transfer hot water from the bags to large insulated tanks for distribution.

Saline and freshwater solar ponds use no energy storage and backup systems, emit no air pollution, and have a moderate net energy yield. Freshwater solar ponds can be built in almost any sunny area and have moderate construction and operating costs. However, saline and freshwater solar ponds are expected to make little contribution to global energy supplies in the foreseeable future.

20-5 PRODUCING ELECTRICITY FROM WIND

What Is the Status of Wind Power? In 2002, wind turbines (Figure 20-22) worldwide produced more than 32,000 megawatts of electricity, enough to meet the residential needs of 35 million people worldwide. About 73% of the world's wind power is produced in Europe, especially in Germany, Spain, and Denmark.

In 2002, the price of electricity produced by wind at prime sites in the United States was about 4¢ per kilowatt-hour (down from 38¢ per kilowatt-hour in the early 1980s). According to the Department of Energy, this is almost equal to the cost of electricity produced by new coal-fired power plants and half the cost if coal's health and environmental costs are included. This is also about half the cost of nuclear power if all nuclear fuel cycle costs are taken into account. Increased investments in wind power by governments and large corporations should reduce its costs further as a result of technological innovations and savings from mass production of wind turbines.

What Areas Have the Greatest Potential for Wind Power? In 2002, western European countries produced 2% of their electricity from wind (18% in Denmark). These countries expect to get at least 10% of their electricity from onshore and offshore wind turbines within 10 years. The German government plans to get 25% of its electricity from wind power by 2025 (up from 4.7% in 2002), much of it from building offshore wind farms in the Baltic and the North Sea.

Wind power is also being developed rapidly in India (the world's number-two market for wind energy), and China could easily double its wind-generating capacity.

Figure 20-23 shows the potential areas for use of wind power in the United States. Currently, wind power supplies less than 0.5% of America's electricity (enough to power more than 1.3 million homes) but 1.5% of California's electricity. However, the DOE calls the midwestern United States the "Saudi Arabia of wind." The Dakotas and Texas alone have enough wind resources to meet all the nation's electricity needs.

Wind Turbine

Gearbox
Electrical generator
Power cable

Wind Farm

Figure 20-22 Solutions: wind turbines, which can be used to produce electricity individually or in clusters, called wind farms. Since 1990, wind power has been the world's fastest growing source of energy.

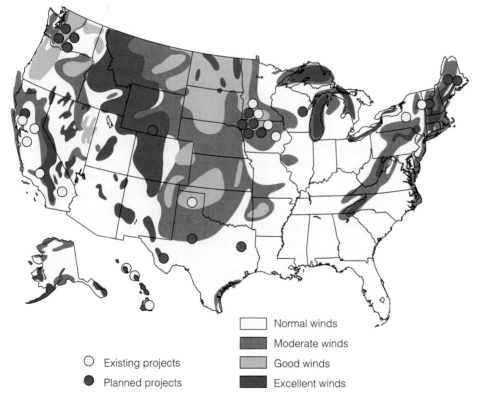

○ Existing projects
● Planned projects

☐ Normal winds
▨ Moderate winds
☐ Good winds
■ Excellent winds

Figure 20-23 Solutions: potential for use of wind power in the United States. In principle, exploiting the wind potential of just three states—North Dakota, South Dakota, and Texas—could provide all the power needs of the United States. (Data from U.S. Department of Energy)

According to the American Wind Energy Association, the United States could use wind power to produce 23% of its electricity by 2025.

A growing number of U.S. farmers and ranchers boost their income by leasing some of their cropland or ranchland for wind turbines while still growing crops or grazing cattle around the turbines. Currently, a U.S. farmer or rancher who leases 0.10 hectare (0.25 acre) of cropland or rangeland to the local utility as a site for a wind turbine can easily get $2,000 a year in royalties from providing the local community with electricity worth $100,000. Some are making more money by leasing their land for wind power production than by growing crops or raising cattle. This explains why many U.S. farmers and ranchers are joining environmentalists and wind industry executives in urging political leaders to increase government research and development and tax breaks for wind power.

What Are the Major Advantages and Disadvantages of Wind Power? Figure 20-24 lists the advantages and disadvantages of using wind to produce

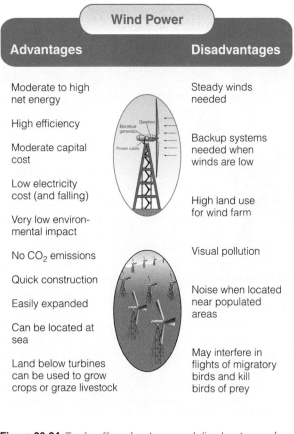

Figure 20-24 Trade-offs: advantages and disadvantages of using wind to produce electricity. Wind power experts project that by 2020 wind power could supply more than 10% of the world's electricity and 10–25% of the electricity used in the United States.

electricity. Some critics have alleged that wind turbines suck large numbers of birds into their wind stream. However, as long as wind farms are not located along bird migration routes most birds learn to fly around them. Also, studies have shown that much larger numbers of birds die when they are sucked into jet engines, killed by domesticated and feral cats, and crash into skyscrapers, plate glass windows, communications towers, and car windows.

Even larger numbers of birds, fish, and other forms of wildlife are killed by oil spills, air pollution, water pollution, and release of toxic wastes from use of fossil fuels such as coal and oil. The key questions are which types of energy resources lead to the lowest loss of wildlife and how we can minimize loss of wildlife from use of any energy resource.

In the long run, electricity from large wind farms in remote areas might be used to make hydrogen gas from water during off-peak periods—thus storing electricity from excess wind capacity in a useful fuel. The hydrogen could then be fed into a pipeline and storage system for fuel cells or gas turbines used to power cars, homes, and buildings.

Increasingly, many governments and corporations are recognizing that wind is a vast, climate-benign, renewable energy resource that can supply both electricity and hydrogen fuel at an affordable cost. If its current growth rate continues, wind power could produce 10% of the world's electricity by 2020.

20-6 PRODUCING ENERGY FROM BIOMASS

How Useful Is Burning Solid Biomass? Biomass consists of plant materials and animal wastes used as sources of energy. Biomass comes in many forms and can be burned directly as a solid fuel or converted into gaseous or liquid **biofuels** (Figure 20-25).

Most biomass is burned *directly* for heating, cooking, and industrial processes or *indirectly* to drive turbines and produce electricity. Burning wood and manure for heating and cooking supplies about 11% of the world's energy and about 30% of the energy used in developing countries. Almost 70% of the people living in developing countries heat their homes and cook their food by burning wood or charcoal. However, about 2.7 billion people in these countries cannot find or are too poor to buy enough fuelwood to meet their needs.

In the United States, biomass is used to supply about 4% of the country's commercial energy and 2% of its electricity (produced by about 350 biomass power plants). The U.S. government has a goal of increasing the use of biomass energy to 9% of the country's total commercial energy by 2010.

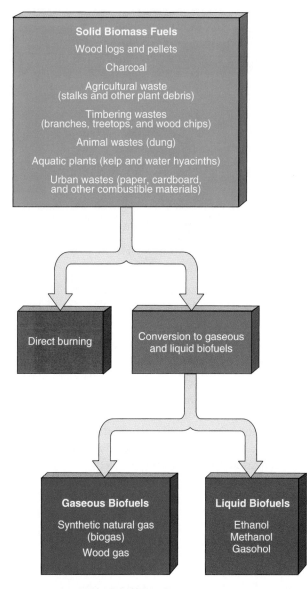

Figure 20-25 Principal types of biomass fuel.

that it makes more sense to use animal manure as a fertilizer and crop residues to feed livestock, retard soil erosion, and fertilize the soil.

Figure 20-26 lists the general advantages and disadvantages of burning solid biomass as a fuel. One problem is that burning biomass produces CO_2. However, if the rate of use of biomass does not exceed the rate at which it is replenished by new plant growth (which takes up CO_2), there is no net increase in CO_2 emissions.

Is Producing Gaseous and Liquid Fuels from Solid Biomass a Useful Option? Bacteria and various chemical processes can convert some forms of biomass into gaseous and liquid biofuels (Figure 20-25). Examples include *biogas* (a mixture of 60% methane and 40% CO_2), *liquid ethanol* (ethyl, or grain, alcohol), and *liquid methanol* (methyl, or wood alcohol).

In rural China, anaerobic bacteria in more than 6 million *biogas digesters* convert plant and animal wastes into methane fuel for heating and cooking. These simple devices can be built for about $50 including labor. After the biogas has been separated, the

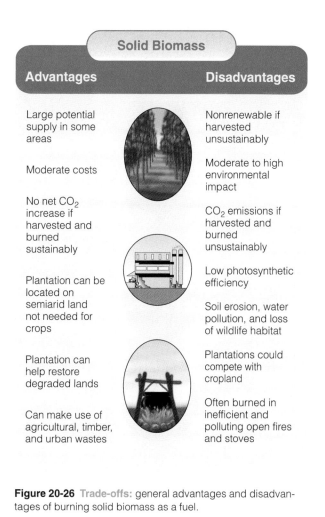

Figure 20-26 Trade-offs: general advantages and disadvantages of burning solid biomass as a fuel.

One way to produce biomass fuel is to plant, harvest, and burn large numbers of fast-growing trees (especially cottonwoods, poplars, sycamores, willows, and leucaenas), shrubs, perennial grasses (such as switchgrass), and water hyacinths in *biomass plantations.*

In agricultural areas, *crop residues* (such as sugarcane residues, rice husks, cotton stalks, and coconut shells) and *animal manure* can be collected and burned or converted into biofuels. According to a 1999 study by the Union of Concerned Scientists, energy crops and crop wastes from the Midwest alone could theoretically provide about 16% of the electricity used in the United States, without irrigation and without competing with food crops for land. Some ecologists argue

almost odorless solid residue is used as fertilizer on food crops or, if contaminated, on trees. When they work, biogas digesters are very efficient. However, they are also slow and unpredictable, a problem that could be corrected by developing more reliable models.

In the United States, livestock wastes are converted to biogas by bacteria by placing the wastes in a long, lined, insulated pit. A flexible liner stretching across the digester pit inflates like a balloon as it collects the biogas. The biogas may be used to heat the digester or nearby farm buildings or to produce electricity.

Some analysts believe liquid ethanol and methanol produced from biomass could replace gasoline and diesel fuel when oil becomes too scarce and expensive. *Ethanol* can be made from sugar and grain crops (sugarcane, sugar beets, sorghum, sunflowers, and corn) by fermentation and distillation. Gasoline mixed with 10–23% pure ethanol makes *gasohol*, which can be burned in conventional gasoline engines and is sold as super unleaded or ethanol-enriched gasoline. Figure 20-27 lists the advantages and disadvantages of using ethanol as a vehicle fuel.

Another alcohol, *methanol*, is made mostly from natural gas but also can be produced at a higher cost from wood, wood wastes, agricultural wastes (such as corncobs), sewage sludge, garbage, and coal. Some of

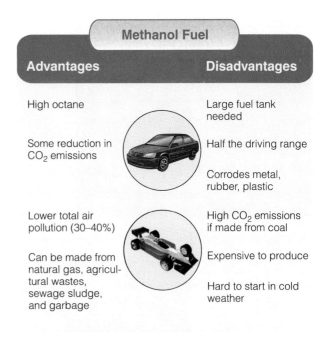

Figure 20-28 Trade-offs: general advantages and disadvantages of using methanol as a vehicle fuel.

the first generation of cars using hydrogen-powered fuel cells (Figure 20-10) will use reformers to convert carbon-containing natural gas, gasoline, or methanol to hydrogen. Figure 20-28 lists the advantages and disadvantages of using methanol as a vehicle fuel.

According to a 1997 analysis by David Pimentel and two other researchers, "Large-scale biofuel production is not an alternative to the current use of oil and is not even an advisable option to cover a significant fraction of it."

20-7 THE SOLAR–HYDROGEN REVOLUTION

What Can We Use to Replace Oil? Good-bye Oil and Smog, Hello Hydrogen When oil is gone (or when what is left costs too much to use), how will we fuel vehicles, industry, and buildings? Many scientists and executives of major oil companies and automobile companies say the fuel of the future is hydrogen gas (H_2)—envisioned in 1874 by science fiction writer Jules Verne in his book *The Mysterious Island*.

When hydrogen gas burns in air, it combines with oxygen gas in the air and produces nonpolluting water vapor ($2 H_2 + O_2 \longrightarrow 2 H_2O$).* Widespread use of this fuel would eliminate most of the air pollution problems we face today. It would also greatly reduce the

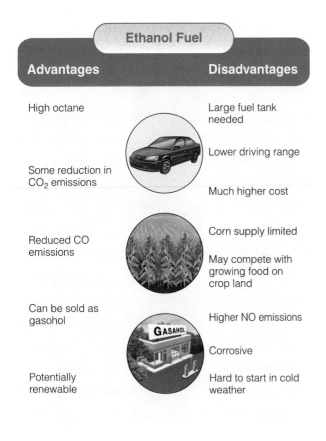

Figure 20-27 Trade-offs: general advantages and disadvantages of using ethanol as a vehicle fuel.

---

*Water vapor is a potent greenhouse gas. However, because there is already so much of it in the atmosphere, human additions of this gas are insignificant.

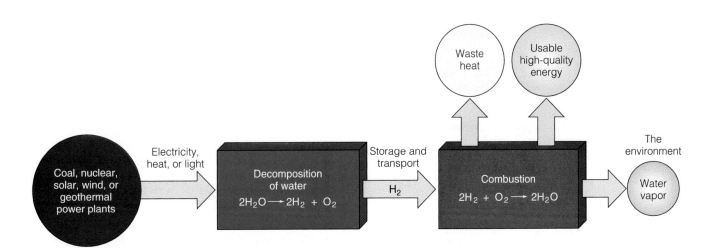

Figure 20-29 Solutions: hydrogen gas as an energy source. Producing hydrogen gas takes electricity, heat, or solar energy to decompose water, thus leading to a negative net energy yield. However, hydrogen is a clean-burning fuel that can replace oil, other fossil fuels, and nuclear energy. Using solar energy (probably solar cells and wind turbines) to produce hydrogen from water could eliminate most air pollution and greatly reduce the threat of global warming.

threat from global warming by emitting no CO_2—as long as the hydrogen is not produced from fossil fuels or other carbon-containing compounds.

The *bad news* is that although hydrogen is all around us it is chemically locked up in water and organic compounds such as methane and gasoline. The *good news* is that we can produce it from something we have plenty of: *water.* Water can be split by electricity (electrolysis) or high temperatures (thermolysis) into gaseous hydrogen and oxygen (Figure 20-29).

There are other ways to produce hydrogen. One is *reforming,* in which high temperatures and chemical

processes are used to separate hydrogen from carbon atoms in organic chemicals (hydrocarbons) found in conventional carbon-containing fuels such as natural gas, gasoline, or methanol. Gasification of coal (Figure 19-22, p. 492) or biomass can also produce it. Other sources are decomposition of sewage sludge and other wet biomass and some types of algae and bacteria (Spotlight, p. 530). These various sources of hydrogen could become, as some scientists put it, "tomorrow's oil." The resulting hydrogen could be used to provide most of the energy needed to run an economy (Figure 20-30).

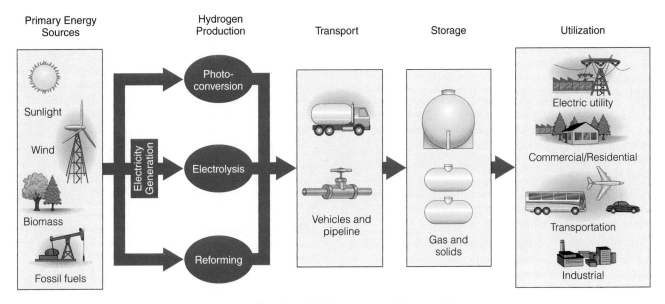

Figure 20-30 Solutions: hydrogen energy system. (Data from U.S. Department of Energy and the Worldwatch Institute)

SPOTLIGHT

Producing Hydrogen from Green Algae Found in Pond Scum

In a few decades we may be able to use large-scale cultures of green algae to produce hydrogen gas. This simple plant grows all over the world and is commonly found in pond scum.

When living in ordinary air and sunlight, green algae carry out photosynthesis like other plants and produce carbohydrates and oxygen gas. However, in 2000, Tasios Melis, a researcher at the University of California at Berkeley, found a way to make these algae produce bubbles of hydrogen rather than oxygen.

First, he grew cultures of hundreds of billions of the algae in the normal way with plenty of sunlight, nutrients, and water. Then he cut off their supply of two key nutrients: sulfur and oxygen. Within 20 hours, the plant cells underwent a metabolic change and switched from an oxygen-producing to a hydrogen-producing metabolism, allowing the researcher to collect hydrogen gas bubbling from the culture.

Melis believes he can increase the efficiency of this hydrogen-producing process tenfold. If so, sometime in the future a biological hydrogen plant might cycle a mixture of algae and water through a system of clear tubes exposed to sunlight to produce hydrogen. The gene responsible for producing the hydrogen might even be transferred to other plants to produce hydrogen.

Critical Thinking

What might be some ecological problems related to the widespread use of this method for producing hydrogen?

What Is the Catch? If you think using H_2 as the world's major energy source sounds too good to be true, you are right. Several problems must be solved to make hydrogen one of our primary energy resources, but scientists are making rapid progress in finding solutions to these challenges.

One problem is that it takes energy (and thus money) to produce this fuel. We could burn coal or synthetic natural gas (Figure 19-22, p. 492) to produce high-temperature heat or use electricity from coal-burning and conventional nuclear power plants to split water and produce hydrogen. However, this subjects us to the harmful environmental effects associated with using these fuels (Figure 19-21, p. 492; Figure 19-23, p. 493; Figure 19-26, p. 496; and Figure 19-27, p. 497). This also costs more than the hydrogen fuel is worth when environmental costs are included.

Producing hydrogen by coal gasification or from carbon-containing methane (natural gas), gasoline, or methanol (reforming) can be used. However, according to a 2002 article by physicist Marin Hoffer and a team of other scientists, this adds more CO_2 to the atmosphere per unit of heat generated than does burning these carbon-containing fuels directly. Thus doing this could accelerate global warming.

In 2003, environmentalists were pleased that President Bush talked about the benefits of a hydrogen economy and proposed spending $1.7 billion over the following two to five years to build partnerships with the private sector to develop new vehicle and fuel technologies and the infrastructure needed to make it practical and cost-effective for large numbers of Americans to choose fuel cell vehicles by 2020. Such a program would also dramatically improve America's energy security by significantly reducing the need for imported oil, reduce air pollution, and also reduce emissions of carbon dioxide. However, environmentalists were disappointed that President Bush's proposals relied mostly on generating the hydrogen from carbon-containing fuels, which produce carbon dioxide instead of placing more emphasis on using renewable energy resources to produce hydrogen fuel.

Most proponents of using hydrogen gas believe that if we are to get its very low pollution benefits, the energy to produce H_2 by decomposing water must come from low-polluting, renewable sources. The most likely sources are electricity generated by solar cells, wind farms, hydropower, and geothermal energy, or biological processes in bacteria and algae (Spotlight, left). The type of renewable energy used would vary in different parts of the world depending on its local and regional availability. Another possibility is to use electricity or heat produced by nuclear fusion power plants to decompose water if this energy alternative turns out to be technologically and economically feasible (p. 502).

If scientists and engineers can learn how to use various forms of direct and indirect solar energy to decompose water cheaply enough, they will set in motion a *solar–hydrogen revolution* over the next 50–100 years and change the world as much as the agricultural and industrial revolutions did. In effect, the world would shift from carbon-based *fossil fuel economies* (Figure 19-4, p. 480) to decarbonized *hydrogen economies*, powered increasingly by using solar energy (or perhaps nuclear fusion) to produce hydrogen gas from water (Figure 19-7, p. 481, and Figure 20-30). By using renewable solar energy, such an economy would follow the first of the four principles of sustainability based on observing how the earth sustains itself (Figure 8-13, p. 170, and Figure 8-14, p. 171).

Methane from natural gas may be used to produce hydrogen in the transition to a true renewable hydrogen system because of its large supply (p. 490) and lower production of air pollutants and CO_2 (Figure 19-14, p. 486) compared to other fossil fuels.

How Can We Store Hydrogen? Once hydrogen is produced we must have a way to store it for use as needed. One way is to store it in *compressed gas tanks* either above or below ground or aboard motor vehicles. Because of their large size and weight, such storage tanks have been more useful for buses and large trucks than for cars. However, in 2002, General Motors developed a lightweight, high-pressure, hydrogen storage tank that can be used on cars and can store enough hydrogen to provide a range of nearly 480 kilometers (300 miles) before refueling.

Another storage option is to convert hydrogen gas to more dense *liquid hydrogen*. This allows a larger quantity of hydrogen to be stored in stationary containers or aboard motor vehicles. However, the liquid hydrogen must be stored at very low temperatures, below $-250°C$ ($-420°F$). This is costly, takes a large input of energy (as much as 30% of the hydrogen's original fuel energy), and requires a large amount of insulation.

Hydrogen can also be stored in *solid metal hydride compounds*. When cooled, certain metals absorb and chemically bond the hydrogen in the metal's latticework of atoms. Heating the metal hydride compound releases the hydrogen gas as needed. This is a safe and efficient way to store hydrogen. But current metal hydrides are costly, heavy, and require energy to release the hydrogen.

Another possibility is to *absorb H_2 on activated charcoal or graphite nanofibers,* which when heated release hydrogen gas. Like hydrides, this is a safe and efficient way to store hydrogen, but an input of energy is needed to release the hydrogen.

Hydrogen gas can also be stored *inside tiny glass microspheres.* In 2002, scientists were also able to trap hydrogen gas in a framework of water molecules— called *clathrate hydrates.*

Some *good news* is that metal hydrides, charcoal powders, graphite nanofibers, and glass microspheres containing hydrogen will not explode or burn if a vehicle's tank is ruptured in an accident. Such tanks would be much safer than current gasoline tanks.

What Is the Role of Fuel Cells in the Solar–Hydrogen Revolution? In a *fuel cell* (Figures 20-4 and 20-10), hydrogen and oxygen gas combine to produce electrical current. Various versions of such cells can be used to power a car or bus and to meet the heating, cooling, and electrical needs of buildings.

Fuel cells have energy efficiencies of 65–95%. This is several times the efficiency of conventional gasoline-powered engines and electric cars and at least twice the efficiency of coal-burning and nuclear power plants. Fuel cells have no moving parts and are quiet. They emit only water and heat (and some CO_2 if the hydrogen is produced from carbon-containing substances such as gasoline, natural gas, propane, or methanol). Also, fuel cells are more reliable than the traditional electricity grid because they are not as susceptible to lightning strikes, fallen trees, and terrorist or military attacks.

Some fuel cells are tiny enough to fit into a cellular phone. Others are big enough to power a large building or factory. Smaller fuel cells can power bicycles, vacuum cleaners, laptop computers, lawn mowers, leaf blowers, and other devices.

With hydrogen-powered fuel cells, people would have their own personal power plant to run their lights, appliances, and car and to heat and cool their house. Here is some *good news.* A number of prototype fuel-cell systems for cars, buses, homes, and buildings are being tested and evaluated. Fleets of hydrogen-powered buses are running in various cities of the world.

In 1999, DaimlerChrysler, Royal Dutch Shell, and Norsk Hydro announced government-approved plans to turn the tiny country of Iceland into the world's first "hydrogen economy" by 2030–2040—the brainchild of chemist Bragi Árnason, known as "Professor Hydrogen." The country's abundant renewable geothermal energy, hydropower, and offshore winds will be used to produce hydrogen from seawater and the H_2 will be used to run its buses, passenger cars, fishing vessels, and factories. Royal Dutch Shell is already opening hydrogen filling stations in parts of Europe and plans to open a chain of such stations in Iceland. The first hydrogen filling station in the United States opened in Las Vegas, Nevada, in 2002.

The key problem with fuel cells so far is cost. For widespread use of fuel cells the price must be sharply reduced by improved technology and mass production. With greatly increased private and government-funded research and tax breaks, some analysts see this happening within 10 years. They envision fuel cells being used first by electric utilities, followed in order by midsize buildings, homes and small buildings, and motor vehicles. Some *good news* is that in 2002 researchers at the Lawrence Berkeley National Laboratory developed a solid oxide fuel cell (SOFC) that promises to generate electricity at one-tenth the cost of today's fuel cells and as cheaply as the most efficient gas turbine.

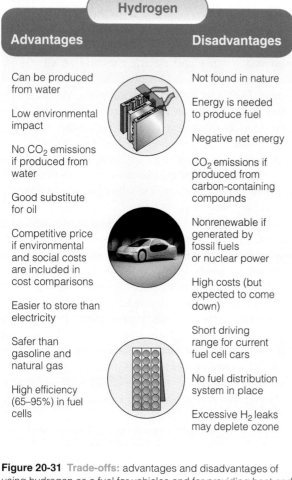

Hydrogen

Advantages	Disadvantages
Can be produced from water	Not found in nature
Low environmental impact	Energy is needed to produce fuel
No CO_2 emissions if produced from water	Negative net energy
Good substitute for oil	CO_2 emissions if produced from carbon-containing compounds
Competitive price if environmental and social costs are included in cost comparisons	Nonrenewable if generated by fossil fuels or nuclear power
Easier to store than electricity	High costs (but expected to come down)
Safer than gasoline and natural gas	Short driving range for current fuel cell cars
High efficiency (65–95%) in fuel cells	No fuel distribution system in place
	Excessive H_2 leaks may deplete ozone

Figure 20-31 Trade-offs: advantages and disadvantages of using hydrogen as a fuel for vehicles and for providing heat and electricity.

What Are the Advantages and Disadvantages of Hydrogen as an Energy Resource? Figure 20-31 lists the advantages and disadvantages of using hydrogen as an energy resource. The U.S. Department of Energy has a goal of hydrogen energy providing 10% of all U.S. energy consumption by 2025.

20-8 GEOTHERMAL ENERGY

How Can We Tap the Earth's Internal Heat? Heat contained in underground rocks and fluids is an important source of energy. Over millions of years, this **geothermal energy** from the earth's mantle (Figure 9-2, p. 174) has been transferred to three types of underground reservoirs. One contains *dry steam,* which consists of steam with no water droplets. Another is *wet steam,* which consists of a mixture of steam and water droplets. The third is *hot water* trapped in fractured or porous rock at various places in the earth's crust.

If such geothermal sites are close to the surface, wells can be drilled to extract the dry steam, wet steam

(Figure 20-32), or hot water. This thermal energy can be used to heat homes and buildings and to produce electricity. For example, geothermal energy is used to heat about 85% of Iceland's buildings.

Currently, about 22 countries (most of them in the developing world) are extracting energy from geothermal sites to produce about 1% of the world's electricity. Japan, with an abundance of geothermal energy, could get an estimated 30% of its electricity from this energy resource.

Geothermal electricity meets the electricity needs of 6 million Americans and supplies 6% of California's electricity. The world's largest operating geothermal system, called *The Geysers,* extracts energy from a dry steam reservoir north of San Francisco, California. But heat is being withdrawn from this geothermal site about 80 times faster than it is being replenished, converting this renewable resource to a nonrenewable source of energy. In 1999, Santa Monica, California, became the first city in the world to get all its electricity from geothermal energy.

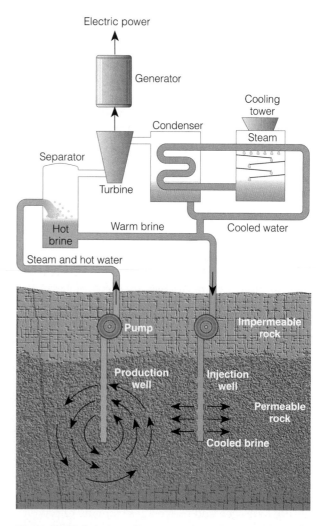

Figure 20-32 Solutions: Tapping the earth's internal heat or *geothermal energy* in the form of wet steam to produce electricity.

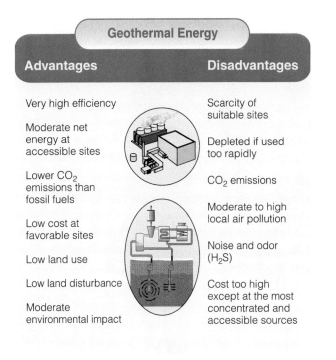

Geothermal Energy

Advantages	Disadvantages
Very high efficiency	Scarcity of suitable sites
Moderate net energy at accessible sites	Depleted if used too rapidly
Lower CO_2 emissions than fossil fuels	CO_2 emissions
Low cost at favorable sites	Moderate to high local air pollution
Low land use	Noise and odor (H_2S)
Low land disturbance	Cost too high except at the most concentrated and accessible sources
Moderate environmental impact	

Figure 20-33 Trade-offs: advantages and disadvantages of using *geothermal energy* for space heating and to produce electricity or high-temperature heat for industrial processes.

There are three other nearly nondepletable sources of geothermal energy. One is *molten rock* (magma). Another is *hot dry-rock zones,* where molten rock that has penetrated the earth's crust heats subsurface rock to high temperatures. A third source is low- to moderate-temperature *warm-rock reservoir deposits.* Heat from such deposits could be used to preheat water and run heat pumps for space heating and air conditioning. Hot dry-rock zones can be found almost anywhere about 8–10 kilometers (5–6 miles) below the earth's surface. Research is being carried out in several countries to see whether these zones can provide affordable geothermal energy.

Figure 20-33 lists the advantages and disadvantages of using geothermal energy. Currently, the cost of tapping geothermal energy is too high for all but the most concentrated and accessible sources. In 2000, the U.S. Department of Energy launched a program to have geothermal energy produce 10% of the electricity used in the western United States by 2020.

20-9 ENTERING THE AGE OF DECENTRALIZED MICROPOWER

What Is Micropower? According to the director of energy supply policy for the Edison Electric Institute, Chuck Linderman, the era of big central power plant systems (Figure 20-34) is over. Most energy analysts believe the chief feature of electricity production over the next few decades will be *decentralization* to dispersed, small-scale, **micropower systems** that generate

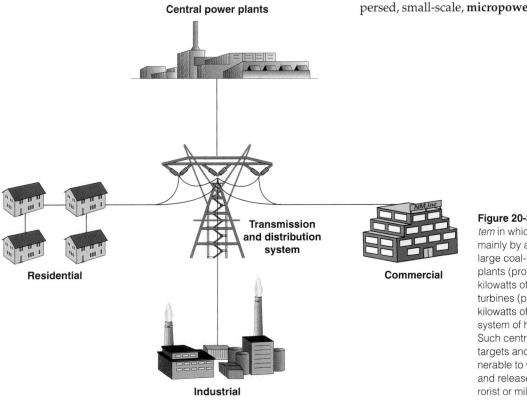

Central power plants

Transmission and distribution system

Residential

Commercial

Industrial

Figure 20-34 *Centralized power system* in which electricity produced mainly by a fairly small number of large coal-burning and nuclear power plants (producing 600,000 to 1 million kilowatts of power) and natural gas turbines (producing about 200,000 kilowatts of power) is distributed by a system of high-voltage wires to users. Such centralized systems are easy targets and make a country more vulnerable to widespread power outages and releases of radioactivity from terrorist or military attacks.

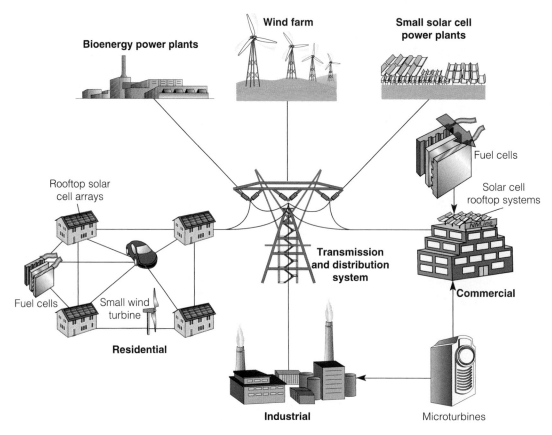

Figure 20-35 Solutions: *decentralized power system* in which electricity is produced by a large number of dispersed, small-scale *micropower systems* (producing 1–10,000 kilowatts of power). Some would produce power on site and others would feed the power they produce into a conventional electrical distribution system. Over the next few decades, many energy and financial analysts expect a shift to this type of power system.

1–10,000 kilowatts of power (Figure 20-35). This shift from centralized *macropower* to dispersed *micropower* is analogous to the computer industry's shift from large centralized mainframes to increasingly smaller, widely dispersed PCs, laptops, and handheld computers.

Examples of micropower systems include such energy-efficient systems as natural gas–burning *microturbines* for commercial buildings and residences (5–10,000 kilowatts), *wind turbines* (1–3,000 kilowatts), *Stirling engines* (0.1–100 kilowatts), *fuel cells* (1–10,000 kilowatts), and household *solar panels and solar roofs* (1–1,000 kilowatts). Figure 20-36 lists some of the advantages of decentralized micropower systems over traditional macropower systems.

20-10 SOLUTIONS: A SUSTAINABLE ENERGY STRATEGY

What Are the Best Energy Alternatives? Many scientists and many energy experts who have evaluated energy alternatives have reached three general conclusions:

- *There will be a gradual shift from centralized macropower systems* (Figure 20-34) *to smaller, decentralized micropower systems* (Figures 20-35).

- *The best alternatives are a combination of improved energy efficiency and using natural gas as a fuel to make the transition to increased use of a variety of small-scale, decentralized, locally available renewable energy resources and possibly nuclear fusion (if it proves feasible).*

- *Over the next 50 years, the choice is not between using nonrenewable fossil fuels and various types of renewable energy.* Because of their supplies and low prices, fossil fuels will continue to be used in large quantities. The challenge is to find ways to reduce the harmful environmental impacts of widespread fossil fuel use, with special emphasis on reducing air pollution and emissions of greenhouse gases as less harmful alternatives are phased in.

What Role Does Economics Play in Energy Resource Use? To most analysts the key to making a shift to more sustainable energy resources and societies

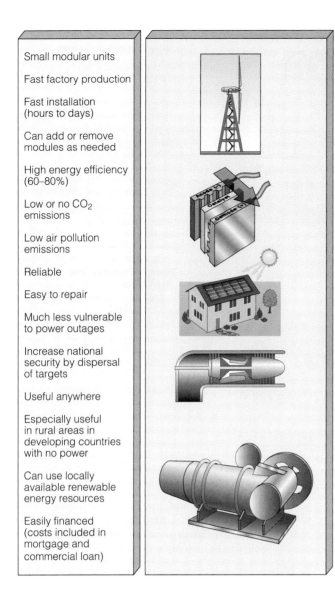

Figure 20-36 *Good news:* advantages of micropower systems.

- Small modular units
- Fast factory production
- Fast installation (hours to days)
- Can add or remove modules as needed
- High energy efficiency (60–80%)
- Low or no CO_2 emissions
- Low air pollution emissions
- Reliable
- Easy to repair
- Much less vulnerable to power outages
- Increase national security by dispersal of targets
- Useful anywhere
- Especially useful in rural areas in developing countries with no power
- Can use locally available renewable energy resources
- Easily financed (costs included in mortgage and commercial loan)

is not technology but economics and politics. Governments can use three basic economic and political strategies to help stimulate or dampen the short-term and long-term use of a particular energy resource.

One approach is to allow all energy resources to compete in a free market without any government interference. This is rarely politically feasible because of well-entrenched government intervention in the marketplace in the form of subsidies, taxes, and regulations. Furthermore, the free-market approach, with its emphasis on short-term profit, can inhibit development of new energy resources, which can rarely compete economically in their early stages without government support.

A *second* approach is to keep energy prices artificially low to encourage use of selected energy resources. This is done mostly by providing research and development subsidies and tax breaks and enacting regulations that help stimulate the development and use of energy resources receiving such support. For decades, this approach has been used to help stimulate the development and use of fossil fuels and nuclear power in the United States and most other developed countries. This follows the general economic rule that *you get more of what you reward.* So far, this approach has created an uneven economic playing field that encourages energy waste and rapid depletion of nonrenewable energy resources. It also discourages the development of energy alternatives such as energy efficiency and renewable energy that are getting much lower levels of subsidies and tax breaks than fossil fuels and nuclear power.

A *third* option is to keep energy prices artificially high to discourage use of a resource. Governments can raise the price of an energy resource by withdrawing existing tax breaks and other subsidies, enacting restrictive regulations, or adding taxes on its use. This increases government revenues, encourages improvements in energy efficiency, reduces dependence on imported energy, and decreases use of an energy resource that has a limited future supply.

Many economists favor *increasing taxes on fossil fuels* as a way to reduce air and water pollution and slow global warming. The tax revenues would be used to reduce income taxes on wages and profits (p. 27), improve energy efficiency, encourage use of renewable energy resources, and provide energy assistance to the poor and lower middle class. Some economists believe the public might accept these higher taxes if income and payroll taxes were lowered as gasoline or other fossil fuel taxes were raised (p. 27).

How Can We Develop a More Sustainable Energy Future? Figure 20-37 (p. 536) lists strategies for making the transition to a more sustainable energy future over the next few decades.

Energy experts estimate that implementing policies such as those shown in Figure 20-37 over the next 20–30 years could save money, create a net gain in jobs, reduce greenhouse gas emissions, and sharply reduce air pollution and water pollution. According to proponents, these policies would also increase national security in two ways. One is reduced dependence on imported oil. The other is decreased dependence on nuclear power and coal plants that are vulnerable to terrorist attacks (Figure 20-36).

Some *great news* is that we have the technology, creativity, and wealth to make the transition to a more sustainable energy future. The *challenging news* is that making this happen depends primarily on *politics*—which depends largely on pressure individuals put on elected officials. See this chapter's website for actions you can take to promote this transition.

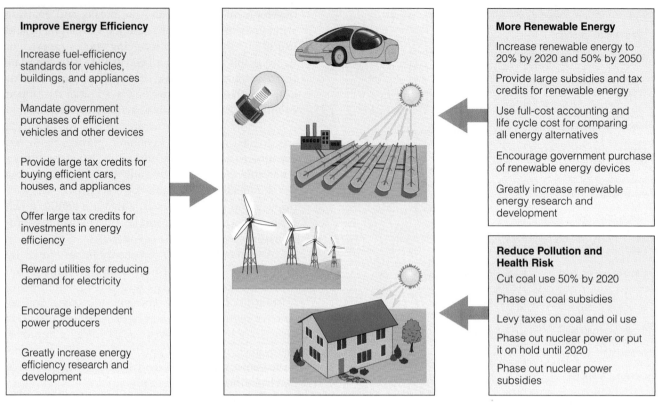

Improve Energy Efficiency

Increase fuel-efficiency standards for vehicles, buildings, and appliances

Mandate government purchases of efficient vehicles and other devices

Provide large tax credits for buying efficient cars, houses, and appliances

Offer large tax credits for investments in energy efficiency

Reward utilities for reducing demand for electricity

Encourage independent power producers

Greatly increase energy efficiency research and development

More Renewable Energy

Increase renewable energy to 20% by 2020 and 50% by 2050

Provide large subsidies and tax credits for renewable energy

Use full-cost accounting and life cycle cost for comparing all energy alternatives

Encourage government purchase of renewable energy devices

Greatly increase renewable energy research and development

Reduce Pollution and Health Risk

Cut coal use 50% by 2020

Phase out coal subsidies

Levy taxes on coal and oil use

Phase out nuclear power or put it on hold until 2020

Phase out nuclear power subsidies

Figure 20-37 Solutions: suggestions of various analysts to help make the transition to a more sustainable energy future.

A transition to renewable energy is inevitable, not because fossil fuel supplies will run out—large reserves of oil, coal, and gas remain in the world—but because the costs and risks of using these supplies will continue to increase relative to renewable energy.

MOHAMED EL-ASHRY

REVIEW QUESTIONS

1. Define the boldfaced terms in this chapter.

2. What is *energy efficiency?* How much of the energy used in the United States is wasted? What percentage is wasted because of the second law of thermodynamics, and what percentage is wasted unnecessarily? What is *life cycle cost?* What are three of the least efficient energy-using devices?

3. List the advantages of reducing energy waste.

4. What is *net energy,* and why is it important in evaluating an energy resource? Why is the net energy for oil from the Middle East high and that for nuclear power low? List two general principles for saving energy.

5. Explain why we cannot recycle energy. List three ways to slow down the flow of heat from a house. What are the sources of collectible waste heat in an office building, and what can be done with such heat in **(a)** winter and **(b)** summer?

6. What is *cogeneration,* and how efficient is it compared with producing electricity by a conventional coal-burning or nuclear power plant? List two other ways to save energy in industry.

7. What do most experts believe is the best way to save energy in transportation? List good and bad news about improvement of the average fuel efficiency for motor vehicles in the United States.

8. List the advantages and disadvantages of using **(a)** hybrid-electric cars and **(b)** fuel-cell cars. List the advantages of using *electric bicycles* and *electric scooters.*

9. Describe how we can save energy in homes by using **(a)** superinsulated houses, **(b)** strawbale houses, and **(c)** eco-roofs.

10. What are the four most efficient ways to heat a house? Describe ways to make an existing house more energy efficient. What are the most efficient and least efficient ways to heat water for washing and bathing? List advantages of switching from inefficient incandescent and halogen light bulbs to efficient compact fluorescent light bulbs.

11. Describe how using the Internet can save energy and help reduce CO_2 emissions.

12. List three reasons why there is little emphasis on saving energy in the United States, despite its important benefits.

13. What are the major advantages and disadvantages of relying more on direct and indirect renewable energy from the sun? List four pieces of good news and one piece of bad news about the use of direct and indirect renewable energy from the sun.

14. Distinguish between a *passive solar heating system* and an *active solar heating system,* and list the advantages and disadvantages of these systems for heating buildings.

15. Describe six ways to cool houses naturally.

16. Distinguish among the following solar systems used to generate high-temperature heat and electricity: **(a)** power tower, **(b)** solar thermal plant, **(c)** parabolic dish collection system, and **(d)** solar cookers. List the advantages and disadvantages of concentrating solar energy to produce high-temperature heat or electricity.

17. What is a *solar cell?* List the advantages and disadvantages of using solar cells to produce electricity.

18. Distinguish among *large-scale hydropower, small-scale hydropower,* and *pumped-storage hydropower* systems. List the advantages and disadvantages of using hydropower plants to produce electricity.

19. How can electricity be produced from tides and waves? What is the potential of these energy resources?

20. Describe the following systems for storing heat in water to produce electricity: **(a)** ocean thermal energy conversion (OTEC), **(b)** saline solar ponds, and **(c)** freshwater solar ponds. What is the potential of these energy resources?

21. What is a **(a)** wind turbine and **(b)** wind farm? List the advantages and disadvantages of using wind to produce electricity.

22. List the advantages and disadvantages of **(a)** burning solid biomass as a source of energy, **(b)** using ethanol as a vehicle fuel, and **(c)** using methanol as a vehicle fuel.

23. What is the *solar–hydrogen revolution?* What major impacts would such a revolution have on energy production, air pollution, and global warming? List five ways to produce hydrogen and six ways to store hydrogen. What is a *fuel cell,* and what are the advantages and disadvantages of using this technology? List the advantages and disadvantages of using hydrogen as a source of energy. Describe progress that is being made in bringing about a solar–hydrogen revolution.

24. What is *geothermal energy?* Describe three types of geothermal reservoirs. List the advantages and disadvantages of using geothermal energy to produce heat and electricity.

25. What is *micropower,* and what are its advantages over *macropower* electricity systems? Describe five types of micropower systems.

26. What three major conclusions have energy experts reached about possible future energy alternatives?

27. Summarize the three different economic approaches that can be used to stimulate or dampen the use of a particular energy resource. List the advantages and disadvantages of each approach.

28. List major ways various analysts have suggested to help make the transition to a more sustainable energy future by **(a)** improving energy efficiency (7 ways), **(b)** using more renewable energy (5 ways), and **(c)** reducing pollution and health risks from energy use (5 ways).

CRITICAL THINKING

1. A home builder installs electric baseboard heat and claims, "It is the cheapest and cleanest way to go." Apply your understanding of the second law of thermodynamics to evaluate this claim.

2. Someone tells you we can save energy by recycling it. How would you respond?

3. Should the Corporate Average Fuel Economy (CAFE) standards for motor vehicles used in the United States be increased, left at 1985 levels (the current situation), or eliminated? Explain. Should the CAFE standards for light trucks, vans, and sport utility vehicles be increased to the same level as for cars? Explain. List the positive and negative effects on your health and lifestyle if CAFE standards are **(a)** increased or **(b)** eliminated.

4. What are the five most important actions an individual can take to save energy at home and in transportation (see website for this chapter)? Which, if any, of these do you currently do? Which, if any, do you plan to do?

5. Congratulations! You have won $250,000 to build a house of your choice anywhere you want. What type of house would you build? Where would you locate it? What types of materials would you use? What types of materials would you *not* use? How would you heat and cool your house? How would you heat your water? Considering fuel and energy efficiency, what sort of lighting, stove, refrigerator, washer, and dryer would you use? Which of these appliances could you do without?

6. Explain why you agree or disagree with the following proposals by various energy analysts: **(a)** Federal subsidies for all energy alternatives should be eliminated so all energy choices can compete in a true free-market system. **(b)** All government tax breaks and other subsidies for conventional fuels (oil, natural gas, coal), synthetic natural gas and oil, and nuclear power (fission and fusion) should be phased out. They should be replaced with subsidies and tax breaks for improving energy efficiency and developing solar, wind, geothermal, hydrogen, and biomass energy alternatives. **(c)** Development of solar, wind, and hydrogen energy should be left to private enterprise and receive little or no help from the federal government, but nuclear energy and fossil fuels should continue to receive large federal subsidies.

7. Explain why you agree or disagree with the proposals suggested in Figure 20-37 (p. 536) as ways to promote a more sustainable energy future.

8. Congratulations! You are in charge of the world. List the five most important features of your energy policy.

PROJECTS

1. Make a study of energy use in your school and use the findings to develop an energy-efficiency improvement program. Present your plan to school officials.

2. Learn how easy it is to produce hydrogen gas from water using a battery, some wire for two electrodes, and a dish of water. Hook a wire to each of the poles of the battery, immerse the electrodes in the water, and observe bubbles of hydrogen gas being produced at the negative electrode and bubbles of oxygen at the positive electrode. Carefully add a small amount of battery acid to the water and notice that this increases the rate of hydrogen production.

3. Make a concept map of this chapter's major ideas, using the section heads and subheads and the key terms (in boldface). Look on the website for this book for information about making concept maps.

INTERNET STUDY RESOURCES AND RESOURCES FOR FURTHER READING AND RESEARCH

The website for this book contains helpful study aids and many ideas for further reading and research. Log on to

http://biology.brookscole.com/miller10

and click on the Chapter-by-Chapter area. Choose Chapter 20 and select a resource:

- Flash Cards allows you to test your mastery of the Terms and Concepts to Remember for this chapter.

- Tutorial Quizzes provides a multiple-choice practice quiz.

- Student Guide to InfoTrac will lead you to Critical Thinking Projects that use InfoTrac College Edition as a research tool.

- References lists the major books and articles consulted in writing this chapter.

- Hypercontents takes you to an extensive list of sites with news, research, and images related to individual sections of the chapter.

IINFOTRAC COLLEGE EDITION

Improve your skills with InfoTrac College Edition, a searchable online database of articles from more than 700 periodicals. Log on to

http://www.infotrac-college.com

or access InfoTrac through the website for this book. Try to find the following articles:

1. Nanney, J. 2002. How motor efficiencies affect system designs. *Plant Engineering* 56: 46. *Keywords:* "motor efficiencies" and "system designs." Electric motors are widely used in America's manufacturing plants. They are also notoriously inefficient. This article describes how revamping the motors in a manufacturing plant can help the "watts per part" of a company's bottom line.

2. Fairley, P. 2002. Wind power for pennies: a lightweight wind turbine is finally on the horizon and it just might be the breakthrough needed to give fossil fuels a run for their money. *Technology Review* 105: 40. *Keywords:* "wind power" and "fossil fuels." Using the wind as a way to generate electricity seems to be always just out of reach on a large scale, but advances in turbine design may make wind power a windfall for electricity users.

APPENDIX 1

UNITS OF MEASURE

LENGTH

Metric
1 kilometer (km) = 1,000 meters (m)
1 meter (m) = 100 centimeters (cm)
1 meter (m) = 1,000 millimeters (mm)
1 centimeter (cm) = 0.01 meter (m)
1 millimeter (mm) = 0.001 meter (m)

English
1 foot (ft) = 12 inches (in)
1 yard (yd) = 3 feet (ft)
1 mile (mi) = 5,280 feet (ft)
1 nautical mile = 1.15 miles

Metric–English
1 kilometer (km) = 0.621 mile (mi)
1 meter (m) = 39.4 inches (in)
1 inch (in) = 2.54 centimeters (cm)
1 foot (ft) = 0.305 meter (m)
1 yard (yd) = 0.914 meter (m)
1 nautical mile = 1.85 kilometers (km)

AREA

Metric
1 square kilometer (km^2) = 1,000,000 square meters (m^2)
1 square meter (m^2) = 1,000,000 square millimeters (mm^2)
1 hectare (ha) = 10,000 square meters (m^2)
1 hectare (ha) = 0.01 square kilometer (km^2)

English
1 square foot (ft^2) = 144 square inches (in^2)
1 square yard (yd^2) = 9 square feet (ft^2)
1 square mile (mi^2) = 27,880,000 square feet (ft^2)
1 acre (ac) = 43,560 square feet (ft^2)

Metric–English
1 hectare (ha) = 2.471 acres (ac)
1 square kilometer (km^2) = 0.386 square mile (mi^2)
1 square meter (m^2) = 1.196 square yards (yd^2)
1 square meter (m^2) = 10.76 square feet (ft^2)
1 square centimeter (cm^2) = 0.155 square inch (in^2)

VOLUME

Metric
1 cubic kilometer (km^3) = 1,000,000,000 cubic meters (m^3)
1 cubic meter (m^3) = 1,000,000 cubic centimeters (cm^3)
1 liter (L) = 1,000 milliliters (mL) = 1,000 cubic centimeters (cm^3)
1 milliliter (mL) = 0.001 liter (L)
1 milliliter (mL) = 1 cubic centimeter (cm^3)

English
1 gallon (gal) = 4 quarts (qt)
1 quart (qt) = 2 pints (pt)

Metric–English
1 liter (L) = 0.265 gallon (gal)
1 liter (L) = 1.06 quarts (qt)
1 liter (L) = 0.0353 cubic foot (ft^3)
1 cubic meter (m^3) = 35.3 cubic feet (ft^3)
1 cubic meter (m^3) = 1.30 cubic yards (yd^3)
1 cubic kilometer (km^3) = 0.24 cubic mile (mi^3)
1 barrel (bbl) = 159 liters (L)
1 barrel (bbl) = 42 U.S. gallons (gal)

MASS

Metric
1 kilogram (kg) = 1,000 grams (g)
1 gram (g) = 1,000 milligrams (mg)
1 gram (g) = 1,000,000 micrograms (μg)
1 milligram (mg) = 0.001 gram (g)
1 microgram (μg) = 0.000001 gram (g)
1 metric ton (mt) = 1,000 kilograms (kg)

English
1 ton (t) = 2,000 pounds (lb)
1 pound (lb) = 16 ounces (oz)

Metric–English
1 metric ton (mt) = 2,200 pounds (lb) = 1.1 tons (t)
1 kilogram (kg) = 2.20 pounds (lb)
1 pound (lb) = 454 grams (g)
1 gram (g) = 0.035 ounce (oz)

ENERGY AND POWER

Metric
1 kilojoule (kJ) = 1,000 joules (J)
1 kilocalorie (kcal) = 1,000 calories (cal)
1 calorie (cal) = 4,184 joules (J)

Metric–English
1 kilojoule (kJ) = 0.949 British thermal unit (Btu)
1 kilojoule (kJ) = 0.000278 kilowatt-hour (kW-h)
1 kilocalorie (kcal) = 3.97 British thermal units (Btu)
1 kilocalorie (kcal) = 0.00116 kilowatt-hour (kW-h)
1 kilowatt-hour (kW-h) = 860 kilocalories (kcal)
1 kilowatt-hour (kW-h) = 3,400 British thermal units (Btu)
1 quad (Q) = 1,050,000,000,000,000 kilojoules (kJ)
1 quad (Q) = 2,930,000,000,000 kilowatt-hours (kW-h)

TEMPERATURE CONVERSIONS

Fahrenheit($^\circ$F) to Celsius($^\circ$C):
$^\circ$C = ($^\circ$F − 32.0) ÷ 1.80
Celsius($^\circ$C) to Fahrenheit($^\circ$F):
$^\circ$F = ($^\circ$C × 1.80) + 32.0

APPENDIX 2

MAJOR EVENTS IN U.S. ENVIRONMENTAL HISTORY

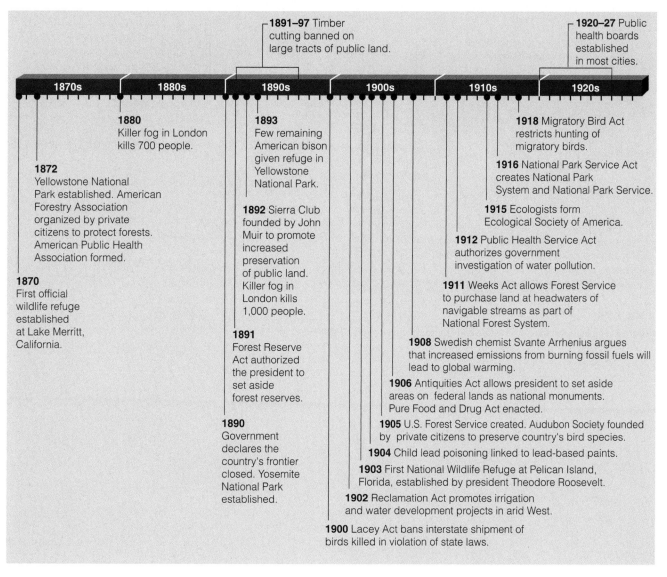

1870–1930

Figure 1 Examples of the increased role of the federal government in resource conservation and public health and establishment of key private environmental groups, 1870–1930.

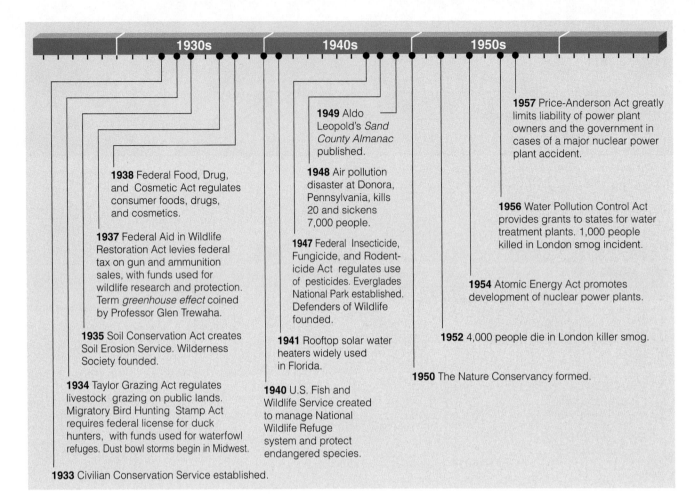

1930–1960

Figure 2 Some important conservation and environmental events, 1930–1960.

The following is the text from the timeline figure:

1930s **1940s** **1950s**

1938 Federal Food, Drug, and Cosmetic Act regulates consumer foods, drugs, and cosmetics.

1937 Federal Aid in Wildlife Restoration Act levies federal tax on gun and ammunition sales, with funds used for wildlife research and protection. Term *greenhouse effect* coined by Professor Glen Trewaha.

1935 Soil Conservation Act creates Soil Erosion Service. Wilderness Society founded.

1934 Taylor Grazing Act regulates livestock grazing on public lands. Migratory Bird Hunting Stamp Act requires federal license for duck hunters, with funds used for waterfowl refuges. Dust bowl storms begin in Midwest.

1933 Civilian Conservation Service established.

1949 Aldo Leopold's *Sand County Almanac* published.

1948 Air pollution disaster at Donora, Pennsylvania, kills 20 and sickens 7,000 people.

1947 Federal Insecticide, Fungicide, and Rodent-icide Act regulates use of pesticides. Everglades National Park established. Defenders of Wildlife founded.

1941 Rooftop solar water heaters widely used in Florida.

1940 U.S. Fish and Wildlife Service created to manage National Wildlife Refuge system and protect endangered species.

1957 Price-Anderson Act greatly limits liability of power plant owners and the government in cases of a major nuclear power plant accident.

1956 Water Pollution Control Act provides grants to states for water treatment plants. 1,000 people killed in London smog incident.

1954 Atomic Energy Act promotes development of nuclear power plants.

1952 4,000 people die in London killer smog.

1950 The Nature Conservancy formed.

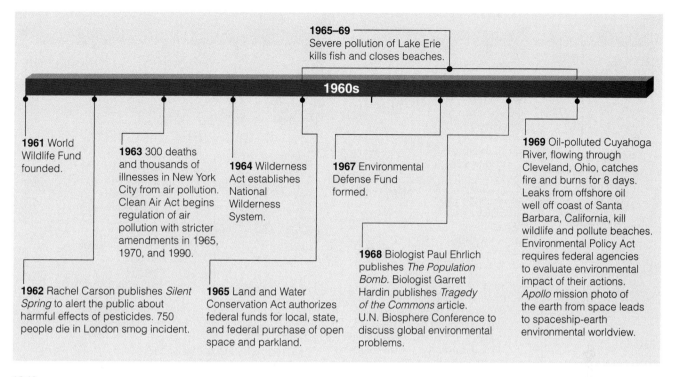

1965–69 Severe pollution of Lake Erie kills fish and closes beaches.

1960s

1961 World Wildlife Fund founded.

1963 300 deaths and thousands of illnesses in New York City from air pollution. Clean Air Act begins regulation of air pollution with stricter amendments in 1965, 1970, and 1990.

1964 Wilderness Act establishes National Wilderness System.

1967 Environmental Defense Fund formed.

1969 Oil-polluted Cuyahoga River, flowing through Cleveland, Ohio, catches fire and burns for 8 days. Leaks from offshore oil well off coast of Santa Barbara, California, kill wildlife and pollute beaches. Environmental Policy Act requires federal agencies to evaluate environmental impact of their actions. *Apollo* mission photo of the earth from space leads to spaceship-earth environmental worldview.

1962 Rachel Carson publishes *Silent Spring* to alert the public about harmful effects of pesticides. 750 people die in London smog incident.

1965 Land and Water Conservation Act authorizes federal funds for local, state, and federal purchase of open space and parkland.

1968 Biologist Paul Ehrlich publishes *The Population Bomb*. Biologist Garrett Hardin publishes *Tragedy of the Commons* article. U.N. Biosphere Conference to discuss global environmental problems.

1960s

Figure 3 Some important environmental events during the 1960s.

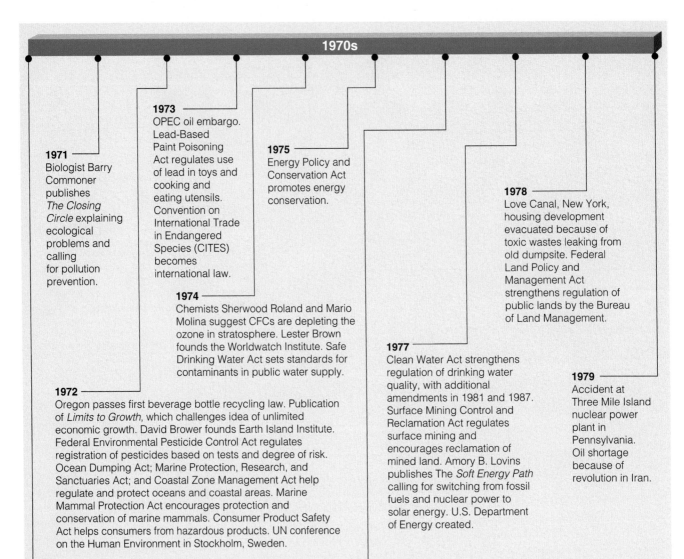

1970s

1971
Biologist Barry Commoner publishes *The Closing Circle* explaining ecological problems and calling for pollution prevention.

1973
OPEC oil embargo. Lead-Based Paint Poisoning Act regulates use of lead in toys and cooking and eating utensils. Convention on International Trade in Endangered Species (CITES) becomes international law.

1975
Energy Policy and Conservation Act promotes energy conservation.

1978
Love Canal, New York, housing development evacuated because of toxic wastes leaking from old dumpsite. Federal Land Policy and Management Act strengthens regulation of public lands by the Bureau of Land Management.

1974
Chemists Sherwood Roland and Mario Molina suggest CFCs are depleting the ozone in stratosphere. Lester Brown founds the Worldwatch Institute. Safe Drinking Water Act sets standards for contaminants in public water supply.

1977
Clean Water Act strengthens regulation of drinking water quality, with additional amendments in 1981 and 1987. Surface Mining Control and Reclamation Act regulates surface mining and encourages reclamation of mined land. Amory B. Lovins publishes The *Soft Energy Path* calling for switching from fossil fuels and nuclear power to solar energy. U.S. Department of Energy created.

1979
Accident at Three Mile Island nuclear power plant in Pennsylvania. Oil shortage because of revolution in Iran.

1972
Oregon passes first beverage bottle recycling law. Publication of *Limits to Growth*, which challenges idea of unlimited economic growth. David Brower founds Earth Island Institute. Federal Environmental Pesticide Control Act regulates registration of pesticides based on tests and degree of risk. Ocean Dumping Act; Marine Protection, Research, and Sanctuaries Act; and Coastal Zone Management Act help regulate and protect oceans and coastal areas. Marine Mammal Protection Act encourages protection and conservation of marine mammals. Consumer Product Safety Act helps consumers from hazardous products. UN conference on the Human Environment in Stockholm, Sweden.

1970
First Earth Day. EPA established by President Richard Nixon. Occupational Health and Safety Act promotes safe working conditions. Resources Recovery Act regulates waste disposal and encourages recycling and waste reduction. National Environmental Policy Act passed. Clean Air Act passed. Natural Resources Defense Council created.

1976
National Forest Management Act establishes guidelines for managing national forests.Toxic Control Substances Act regulates many toxic substances not regulated under other laws. Resource Conservation and Recovery Act requires tracking of hazardous waste and encourages recycling, resource recovery, and waste reduction. Noise Control Act regulates harmful noise levels. U.N. Conference on Human Settlements.

1970s

Figure 4 Some important environmental events during the 1970s, sometimes called the *environmental decade*.

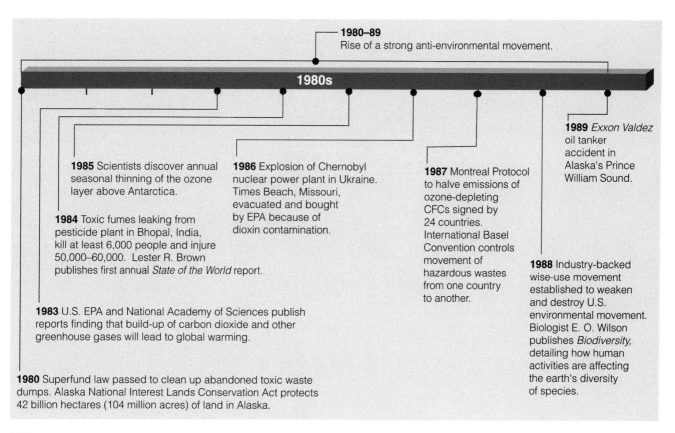

Figure 5 Some important environmental events during the 1980s.

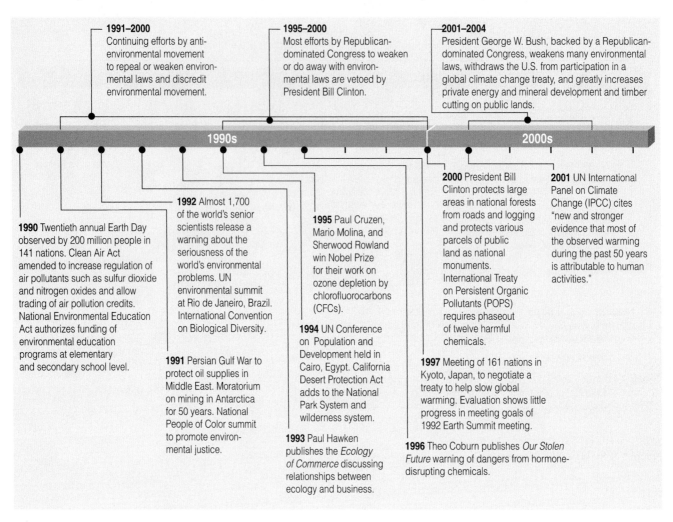

1991–2000
Continuing efforts by anti-environmental movement to repeal or weaken environmental laws and discredit environmental movement.

1995–2000
Most efforts by Republican-dominated Congress to weaken or do away with environmental laws are vetoed by President Bill Clinton.

2001–2004
President George W. Bush, backed by a Republican-dominated Congress, weakens many environmental laws, withdraws the U.S. from participation in a global climate change treaty, and greatly increases private energy and mineral development and timber cutting on public lands.

1990s

2000s

1990 Twentieth annual Earth Day observed by 200 million people in 141 nations. Clean Air Act amended to increase regulation of air pollutants such as sulfur dioxide and nitrogen oxides and allow trading of air pollution credits. National Environmental Education Act authorizes funding of environmental education programs at elementary and secondary school level.

1992 Almost 1,700 of the world's senior scientists release a warning about the seriousness of the world's environmental problems. UN environmental summit at Rio de Janeiro, Brazil. International Convention on Biological Diversity.

1991 Persian Gulf War to protect oil supplies in Middle East. Moratorium on mining in Antarctica for 50 years. National People of Color summit to promote environmental justice.

1995 Paul Cruzen, Mario Molina, and Sherwood Rowland win Nobel Prize for their work on ozone depletion by chlorofluorocarbons (CFCs).

1994 UN Conference on Population and Development held in Cairo, Egypt. California Desert Protection Act adds to the National Park System and wilderness system.

1993 Paul Hawken publishes the *Ecology of Commerce* discussing relationships between ecology and business.

2000 President Bill Clinton protects large areas in national forests from roads and logging and protects various parcels of public land as national monuments. International Treaty on Persistent Organic Pollutants (POPS) requires phaseout of twelve harmful chemicals.

1997 Meeting of 161 nations in Kyoto, Japan, to negotiate a treaty to help slow global warming. Evaluation shows little progress in meeting goals of 1992 Earth Summit meeting.

1996 Theo Coburn publishes *Our Stolen Future* warning of dangers from hormone-disrupting chemicals.

2001 UN International Panel on Climate Change (IPCC) cites "new and stronger evidence that most of the observed warming during the past 50 years is attributable to human activities."

1990–2004

Figure 6 Some important environmental events, 1990–2004.

GLOSSARY

abiotic   Nonliving. Compare *biotic*.

acclimation   Adjustment to slowly changing new conditions. Compare *threshold effect*.

acid   See *acid solution*.

acid deposition   The falling of acids and acid-forming compounds from the atmosphere to the earth's surface. Acid deposition is commonly known as *acid rain*, a term that refers only to wet deposition of droplets of acids and acid-forming compounds.

acid rain   See *acid deposition*.

acid solution   Any water solution that has more hydrogen ions (H^+) than hydroxide ions (OH^-); any water solution with a pH less than 7. Compare *basic solution, neutral solution*.

active solar heating system   System that uses solar collectors to capture energy from the sun and store it as heat for space heating and water heating. Liquid or air pumped through the collectors transfers the captured heat to a storage system such as an insulated water tank or rock bed. Pumps or fans then distribute the stored heat or hot water throughout a dwelling as needed. Compare *passive solar heating system*.

adaptation   Any genetically controlled structural, physiological, or behavioral characteristic that helps an organism survive and reproduce under a given set of environmental conditions. It usually results from a beneficial mutation. See *biological evolution, differential reproduction, mutation, natural selection*.

adaptive radiation   Process in which numerous new species evolve to fill vacant and new ecological niches in changed environments, usually after a mass extinction or mass depletion. Typically, this takes millions of years.

adaptive trait   See *adaptation*.

advanced sewage treatment   Specialized chemical and physical processes that reduce the amount of specific pollutants left in wastewater after primary and secondary sewage treatment. This type of treatment usually is expensive. See also *primary sewage treatment, secondary sewage treatment*.

aerobic respiration   Complex process that occurs in the cells of most living organisms, in which nutrient organic molecules such as glucose ($C_6H_{12}O_6$) combine with oxygen (O_2) and produce carbon dioxide (CO_2), water (H_2O), and energy. Compare *photosynthesis*.

age structure   Percentage of the population (or number of people of each sex) at each age level in a population.

agricultural revolution   Gradual shift from small, mobile hunting and gathering bands to settled agricultural communities in which people survived by learning how to breed and raise wild animals and to cultivate wild plants near where they lived. It began 10,000–12,000 years ago. Compare *environmental revolution, hunter–gatherers, industrial revolution, information and globalization revolution*.

agroforestry   Planting trees and crops together.

air pollution   One or more chemicals in high enough concentrations in the air to harm humans, other animals, vegetation, or materials. Excess heat and noise are also considered forms of air pollution. Such chemicals or physical conditions are called air pollutants. See *primary pollutant, secondary pollutant*.

albedo   Ability of a surface to reflect light.

alien species   See *nonnative species*.

allele   Slightly different molecular form found in a particular gene.

alley cropping   Planting of crops in strips with rows of trees or shrubs on each side.

alpha particle   Positively charged matter, consisting of two neutrons and two protons, that is emitted as a form of radioactivity from the nuclei of some radioisotopes. See also *beta particle, gamma rays*.

altitude   Height above sea level. Compare *latitude*.

anaerobic respiration   Form of cellular respiration in which some decomposers get the energy they need through the breakdown of glucose (or other nutrients) in the absence of oxygen. Compare *aerobic respiration*.

ancient forest   See *old-growth forest*.

animal manure   Dung and urine of animals used as a form of organic fertilizer. Compare *green manure*.

annual   Plant that grows, sets seed, and dies in one growing season. Compare *perennial*.

anthropocentric   Human-centered. Compare *biocentric*.

aquaculture   Growing and harvesting of fish and shellfish for human use in freshwater ponds, irrigation ditches, and lakes, or in cages or fenced-in areas of coastal lagoons and estuaries. See *fish farming, fish ranching*.

aquatic   Pertaining to water. Compare *terrestrial*.

aquatic life zone   Marine and freshwater portions of the biosphere. Examples include freshwater life zones (such as lakes and streams) and ocean or marine life zones (such as estuaries, coastlines, coral reefs, and the deep ocean).

aquifer   Porous, water-saturated layers of sand, gravel, or bedrock that can yield an economically significant amount of water.

arable land   Land that can be cultivated to grow crops.

area strip mining   Type of surface mining used where the terrain is flat. An earthmover strips away the overburden, and a power shovel digs a cut to remove the mineral deposit. After removal of the mineral, the trench is filled with overburden, and a new cut is made parallel to the previous one. The process is repeated over the entire site. Compare *dredging, mountaintop removal, open-pit mining, subsurface mining*.

arid   Dry. A desert or other area with an arid climate has little precipitation.

artificial selection   Process by which humans select one or more desirable genetic traits in the population of a plant or animal and then use *selective breeding* to end up with populations of the species containing large numbers of individuals with the desired traits. Compare *genetic engineering, natural selection*.

asexual reproduction   Reproduction in which a mother cell divides to produce two identical daughter cells that are clones of the mother cell. This type of reproduction is common in single-celled organisms. Compare *sexual reproduction*.

atmosphere   The whole mass of air surrounding the earth. See *stratosphere, troposphere*.

atom   Minute unit made of subatomic particles that is the basic building block of all chemical elements and thus all matter; the smallest unit of an element that can exist and still have the unique characteristics of that element. Compare *ion, molecule*.

atomic number Number of protons in the nucleus of an atom. Compare *mass number*.

autotroph See *producer*.

background extinction Normal extinction of various species as a result of changes in local environmental conditions. Compare *mass depletion, mass extinction*.

bacteria Prokaryotic, one-celled organisms. Some transmit diseases. Most act as decomposers and get the nutrients they need by breaking down complex organic compounds in the tissues of living or dead organisms into simpler inorganic nutrient compounds.

barrier islands Long, thin, low offshore islands of sediment that generally run parallel to the shore along some coasts.

basic solution Water solution with more hydroxide ions (OH^-) than hydrogen ions (H^+); water solution with a pH greater than 7. Compare *acid solution, neutral solution*.

benefit–cost analysis Estimates and comparison of short-term and long-term benefits (gains) and costs (losses) from an economic decision.

benthos Bottom-dwelling organisms. Compare *decomposer, nekton, plankton*.

beta particle Swiftly moving electron emitted by the nucleus of a radioactive isotope. See also *alpha particle, gamma rays*.

bioaccumulation An increase in the concentration of a chemical in specific organs or tissues at a level higher than would normally be expected. Compare *biomagnification*.

biocentric Life-centered. Compare *anthropocentric*.

biodegradable Capable of being broken down by decomposers.

biodegradable pollutant Material that can be broken down into simpler substances (elements and compounds) by bacteria or other decomposers. Paper and most organic wastes such as animal manure are biodegradable but can take decades to biodegrade in modern landfills. Compare *degradable pollutant, nondegradable pollutant, slowly degradable pollutant*.

biodiversity Variety of different species (*species diversity*), genetic variability among individuals within each species (*genetic diversity*), variety of ecosystems (*ecological diversity*), and functions such as energy flow and matter cycling needed for the survival of species and biological communities (*functional diversity*).

biofuel Gas or liquid fuel (such as ethyl alcohol) made from plant material (biomass).

biogeochemical cycle Natural processes that recycle nutrients in various chemical forms from the nonliving environment to living organisms and then back to the nonliving environment. Examples are the carbon, oxygen, nitrogen, phosphorus, sulfur, and hydrologic cycles.

bioinformatics Applied science of managing, analyzing, and communicating biological information.

biological amplification See *biomagnification*.

biological community See *community*.

biological diversity See *biodiversity*.

biological evolution Change in the genetic makeup of a population of a species in successive generations. If continued long enough, it can lead to the formation of a new species. Note that populations—not individuals—evolve. See also *adaptation, differential reproduction, natural selection, theory of evolution*.

biological pest control Control of pest populations by natural predators, parasites, or disease-causing bacteria and viruses (pathogens).

biomagnification Increase in concentration of DDT, PCBs, and other slowly degradable, fat-soluble chemicals in organisms at successively higher trophic levels of a food chain or web. Compare *bioaccumulation*.

biomass Organic matter produced by plants and other photosynthetic producers; total dry weight of all living organisms that can be supported at each trophic level in a food chain or web; dry weight of all organic matter in plants and animals in an ecosystem; plant materials and animal wastes used as fuel.

biome Terrestrial regions inhabited by certain types of life, especially vegetation. Examples are various types of deserts, grasslands, and forests.

biosphere Zone of earth where life is found. It consists of parts of the atmosphere (the troposphere), hydrosphere (mostly surface water and groundwater), and lithosphere (mostly soil and surface rocks and sediments on the bottoms of oceans and other bodies of water) where life is found. Sometimes called the *ecosphere*.

biotic Living organisms. Compare *abiotic*.

biotic potential Maximum rate at which the population of a given species can increase when there are no limits on its rate of growth. See *environmental resistance*.

birth rate See *crude birth rate*.

bitumen Gooey, black, high-sulfur, heavy oil extracted from tar sand and then upgraded to synthetic fuel oil. See *tar sand*.

breeder nuclear fission reactor Nuclear fission reactor that produces more nuclear fuel than it consumes by converting non-fissionable uranium-238 into fissionable plutonium-239.

broadleaf deciduous plants Plants such as oak and maple trees that survive drought and cold by shedding their leaves and becoming dormant. Compare *broadleaf evergreen plants, coniferous evergreen plants*.

broadleaf evergreen plants Plants that keep most of their broad leaves year-round. Examples are the trees found in the canopies of tropical rain forests. Compare *broadleaf deciduous plants, coniferous evergreen plants*.

buffer Substance that can react with hydrogen ions in a solution and thus hold the acidity or pH of a solution fairly constant. See *pH*.

calorie Unit of energy; amount of energy needed to raise the temperature of 1 gram of water 1°C (unit on Celsius temperature scale). See also *kilocalorie*.

cancer Group of more than 120 different diseases, one for each type of cell in the human body. Each type of cancer produces a tumor in which cells multiply uncontrollably and invade surrounding tissue.

capitalism See *capitalist market economic system*. Compare *pure command economic system, pure free-market economic system*.

capitalist market economic system Economic system built around controlling market prices of goods and services, global free trade, and maximizing profits for the owners or stockholders whose financial capital the company is using to do business. Compare *pure command economic system, pure free-market economic system*.

carbon cycle Cyclic movement of carbon in different chemical forms from the environment to organisms and then back to the environment.

carcinogen Chemicals, ionizing radiation, and viruses that cause or promote the development of cancer. See *cancer*. Compare *mutagen, teratogen*.

carnivore Animal that feeds on other animals. Compare *herbivore, omnivore*.

carrying capacity (K) Maximum population of a particular species that a given habitat can support over a given period of time.

cell Smallest living unit of an organism. Each cell is encased in an outer membrane or wall and contains genetic material (DNA) and other parts to perform its life function. Organisms such as bacteria consist of only one cell, but most of the organisms we are familiar with contain many cells. See *eukaryotic cell, prokaryotic cell*.

centrally planned economy See *pure command economic system*.

CFCs See *chlorofluorocarbons*.

chain reaction Multiple nuclear fissions, taking place within a certain mass of a fissionable isotope, that release an enormous amount of energy in a short time.

chemical One of the millions of different elements and compounds found naturally and synthesized by humans. See *compound, element*.

chemical change Interaction between chemicals in which there is a change in the chemical composition of the elements or compounds involved. Compare *nuclear change, physical change*.

chemical evolution Formation of the earth and its early crust and atmosphere, evolution of the biological molecules necessary for life, and evolution of systems of chemical reactions needed to produce the first living cells. These processes are believed to have occurred about 1 billion years before biological evolution. Compare *biological evolution*.

chemical formula Shorthand way to show the number of atoms (or ions) in the basic structural unit of a compound. Examples are H_2O, $NaCl$, and $C_6H_{12}O_6$.

chemical reaction See *chemical change*.

chemosynthesis Process in which certain organisms (mostly specialized bacteria) extract inorganic compounds from their environment and convert them into organic nutrient compounds without the presence of sunlight. Compare *photosynthesis*.

chlorinated hydrocarbon Organic compound made up of atoms of carbon, hydrogen, and chlorine. Examples are DDT and PCBs.

chlorofluorocarbons (CFCs) Organic compounds made up of atoms of carbon, chlorine, and fluorine. An example is Freon-12 (CCl_2F_2), used as a refrigerant in refrigerators and air conditioners and in making plastics such as Styrofoam. Gaseous CFCs can deplete the ozone layer when they slowly rise into the stratosphere and their chlorine atoms react with ozone molecules.

chromosome A grouping of various genes and associated proteins in plant and animal cells that carry certain types of genetic information. See *genes*.

clear-cutting Method of timber harvesting in which all trees in a forested area are removed in a single cutting. Compare, *seed-tree cutting, selective cutting, shelterwood cutting, strip cutting*.

climate Physical properties of the troposphere of an area based on analysis of its weather records over a long period (at least 30 years). The two main factors determining an area's climate are *temperature*, with its seasonal variations, and the amount and distribution of *precipitation*. Compare *weather*.

climax community See *mature community*.

coal Solid, combustible mixture of organic compounds with 30–98% carbon by weight, mixed with various amounts of water and small amounts of sulfur and nitrogen compounds. It forms in several stages as the remains of plants are subjected to heat and pressure over millions of years.

coal gasification Conversion of solid coal to synthetic natural gas (SNG).

coal liquefaction Conversion of solid coal to a liquid hydrocarbon fuel such as synthetic gasoline or methanol.

coastal wetland Land along a coastline, extending inland from an estuary, that is covered with salt water all or part of the year. Examples are marshes, bays, lagoons, tidal flats, and mangrove swamps. Compare *inland wetland*.

coastal zone Warm, nutrient-rich, shallow part of the ocean that extends from the high-tide mark on land to the edge of a shelflike extension of continental land masses known as the continental shelf. Compare *open sea*.

coevolution Evolution in which two or more species interact and exert selective pressures on each other that can lead each species to undergo various adaptations. See *evolution, natural selection*.

cogeneration Production of two useful forms of energy, such as high-temperature heat or steam and electricity, from the same fuel source.

commensalism An interaction between organisms of different species in which one type of organism benefits and the other type is neither helped nor harmed to any great degree. Compare *mutualism*.

commercial extinction Depletion of the population of a wild species used as a resource to a level at which it is no longer profitable to harvest the species.

commercial inorganic fertilizer Commercially prepared mixture of plant nutrients such as nitrates, phosphates, and potassium applied to the soil to restore fertility and increase crop yields. Compare *organic fertilizer*.

common-property resource Resource that people normally are free to use; each user can deplete or degrade the available supply. Most are renewable and owned by no one. Examples are clean air, fish in parts of the ocean not under the control of a coastal country, migratory birds, gases of the lower atmosphere, and the ozone content of the upper atmosphere (stratosphere). See *tragedy of the commons*.

community Populations of all species living and interacting in an area at a particular time.

competition Two or more individual organisms of a single species (*intraspecific competition*) or two or more individuals of different species (*interspecific competition*) attempting to use the same scarce resources in the same ecosystem.

compost Partially decomposed organic plant and animal matter used as a soil conditioner or fertilizer.

compound Combination of atoms, or oppositely charged ions, of two or more different elements held together by attractive forces called chemical bonds. Compare *element*.

concentration Amount of a chemical in a particular volume or weight of air, water, soil, or other medium.

coniferous evergreen plants Cone-bearing plants (such as spruces, pines, and firs) that keep some of their narrow, pointed leaves (needles) all year. Compare *broadleaf deciduous plants, broadleaf evergreen plants*.

coniferous trees Cone-bearing trees, mostly evergreens, that have needle-shaped or scalelike leaves. They produce wood known commercially as softwood. Compare *deciduous plants*.

consensus science Scientific data, models, theories, and laws that are widely accepted by scientists considered experts in the area of study. These results of science are very reliable. Compare *frontier science*.

conservation biologist Biologist who investigates human impacts on the diversity of life found on the earth (biodiversity) and develops practical plans for preserving such biodiversity. Compare *conservationist, ecologist, environmentalist, environmental scientist, preservationist, restorationist*.

conservation biology Multidisciplinary science created to deal with the crisis of maintaining the genes, species, communities, and ecosystems that make up earth's biological diversity. Its goals are to investigate human impacts on biodiversity and to develop practical approaches to preserving biodiversity.

conservationist Person concerned with using natural areas and wildlife in ways that sustain them for current and future generations of humans and other forms of life. Compare *conservation biologist, ecologist, environmentalist, environmental scientist, preservationist, restorationist*.

conservation-tillage farming Crop cultivation in which the soil is disturbed little (minimum-tillage farming) or not at all (no-till farming) to reduce soil erosion, lower labor costs, and save energy. Compare *conventional-tillage farming*.

constancy Ability of a living system, such as a population, to maintain a certain size. Compare *inertia, resilience*. See *homeostasis*.

consumer Organism that cannot synthesize the organic nutrients it needs and gets its organic nutrients by feeding on the tissues of producers or of other consumers; generally divided into *primary consumers* (herbivores), *secondary consumers* (carnivores), *tertiary (higher-level) consumers, omnivores*, and *detritivores* (decomposers and detritus feeders). In economics, one who uses economic goods.

contour farming Plowing and planting across the changing slope of land, rather than in straight lines, to help retain water and reduce soil erosion.

contour strip mining Form of surface mining used on hilly or mountainous terrain. A power shovel cuts a series of terraces into the side of a hill. An earthmover removes the overburden, and a power shovel extracts the coal, with the overburden from each new terrace dumped onto the one below. Compare *area strip mining, dredging, mountaintop removal, open-pit mining, subsurface mining*.

controlled burning Deliberately set, carefully controlled surface fires that reduce flammable litter and decrease the chances of damaging crown fires. See *ground fire, surface fire*.

conventional-tillage farming Crop cultivation method in which a planting surface is made by plowing land, breaking up the exposed soil, and then smoothing the surface. Compare *conservation-tillage farming*.

convergent plate boundary Area where earth's lithospheric plates are pushed together. See *subduction zone*. Compare *divergent plate boundary, transform fault*.

coral reef Formation produced by massive colonies containing billions of tiny coral animals, called polyps, that secrete a stony substance (calcium carbonate) around themselves for protection. When the corals die, their empty outer skeletons form layers and cause the reef to grow. They are found in the coastal zones of warm tropical and subtropical oceans.

core Inner zone of the earth. It consists of a solid inner core and a liquid outer core. Compare *crust, mantle.*

cost-benefit analysis See *benefit-cost analysis.*

critical mass Amount of fissionable nuclei needed to sustain a nuclear fission chain reaction.

crop rotation Planting a field, or an area of a field, with different crops from year to year to reduce soil nutrient depletion. A plant such as corn, tobacco, or cotton, which removes large amounts of nitrogen from the soil, is planted one year. The next year a legume such as soybeans, which adds nitrogen to the soil, is planted.

crown fire Extremely hot forest fire that burns ground vegetation and treetops. Compare *controlled burning, ground fire, surface fire.*

crude birth rate Annual number of live births per 1,000 people in the population of a geographic area at the midpoint of a given year. Compare *crude death rate.*

crude death rate Annual number of deaths per 1,000 people in the population of a geographic area at the midpoint of a given year. Compare *crude birth rate.*

crude oil Gooey liquid consisting mostly of hydrocarbon compounds and small amounts of compounds containing oxygen, sulfur, and nitrogen. Extracted from underground accumulations, it is sent to oil refineries, where it is converted to heating oil, diesel fuel, gasoline, tar, and other materials.

crust Solid outer zone of the earth. It consists of oceanic crust and continental crust. Compare *core, mantle.*

cultural eutrophication Overnourishment of aquatic ecosystems with plant nutrients (mostly nitrates and phosphates) because of human activities such as agriculture, urbanization, and discharges from industrial plants and sewage treatment plants. See *eutrophication.*

cyanobacteria Single-celled, prokaryotic, microscopic organisms. Before being reclassified as monera, they were called blue-green algae.

DDT Dichlorodiphenyltrichloroethane, a chlorinated hydrocarbon that has been widely used as a pesticide but is now banned in some countries.

death rate See *crude death rate.*

debt-for-nature swap Agreement in which a certain amount of foreign debt is canceled in exchange for local currency investments that will improve natural resource management or protect certain areas in the debtor country from harmful development.

deciduous plants Trees, such as oaks and maples, and other plants that survive during dry seasons or cold seasons by shedding their leaves. Compare *coniferous trees, succulent plants.*

decomposer Organism that digests parts of dead organisms and cast-off fragments and wastes of living organisms by breaking down the complex organic molecules in those materials into simpler inorganic compounds and then absorbing the soluble nutrients. Producers return most of these chemicals to the soil and water for reuse. Decomposers consist of various bacteria and fungi. Compare *consumer, detritivore, producer.*

deforestation Removal of trees from a forested area without adequate replanting.

degradable pollutant Potentially polluting chemical that is broken down completely or reduced to acceptable levels by natural physical, chemical, and biological processes. Compare *biodegradable pollutant, nondegradable pollutant, slowly degradable pollutant.*

degree of urbanization Percentage of the population in the world, or a country, living in areas with a population of more than 2,500 people (higher in some countries). Compare *urban growth.*

democracy Government by the people through their elected officials and appointed representatives. In a *constitutional democracy,* a constitution provides the basis of government authority and puts restraints on government power through free elections and freely expressed public opinion.

demographic transition Hypothesis that countries, as they become industrialized, have declines in death rates followed by declines in birth rates.

depletion time The time it takes to use a certain fraction, usually 80%, of the known or estimated supply of a nonrenewable resource at an assumed rate of use. Finding and extracting the remaining 20% usually costs more than it is worth.

desalination Purification of salt water or brackish (slightly salty) water by removal of dissolved salts.

desert Biome in which evaporation exceeds precipitation and the average amount of precipitation is less than 25 centimeters (10 inches) a year. Such areas have little vegetation or have widely spaced, mostly low vegetation. Compare *forest, grassland.*

desertification Conversion of rangeland, rain-fed cropland, or irrigated cropland to desertlike land, with a drop in agricultural productivity of 10% or more. It usually is caused by a combination of overgrazing, soil erosion, prolonged drought, and climate change.

detritivore Consumer organism that feeds on detritus, parts of dead organisms, and cast-off fragments and wastes of living organisms. The two principal types are *detritus feeders* and *decomposers.*

detritus Parts of dead organisms and cast-off fragments and wastes of living organisms.

detritus feeder Organism that extracts nutrients from fragments of dead organisms and their cast-off parts and organic wastes. Examples are earthworms, termites, and crabs. Compare *decomposer.*

deuterium (D; hydrogen-2) Isotope of the element hydrogen, with a nucleus containing one proton and one neutron and a mass number of 2.

developed country Country that is highly industrialized and has a high per capita GNI. Compare *developing country.*

developing country Country that has low to moderate industrialization and low to moderate per capita GNI. Most are located in Africa, Asia, and Latin America. Compare *developed country.*

dieback Sharp reduction in the population of a species when its numbers exceed the carrying capacity of its habitat. See *carrying capacity.*

differential reproduction Phenomenon in which individuals with adaptive genetic traits produce more living offspring than do individuals without such traits. See *natural selection.*

dioxins Family of 75 different chlorinated hydrocarbon compounds formed as unwanted by-products in chemical reactions involving chlorine and hydrocarbons, usually at high temperatures.

discount rate The economic value a resource will have in the future compared with its present value.

dissolved oxygen (DO) content Amount of oxygen gas (O_2) dissolved in a given volume of water at a particular temperature and pressure, often expressed as a concentration in parts of oxygen per million parts of water.

divergent plate boundary Area where earth's lithospheric plates move apart in opposite directions. Compare *convergent plate boundary, transform fault.*

DNA (deoxyribonucleic acid) Large molecules in the cells of organisms that carry genetic information in living organisms.

domesticated species Wild species tamed or genetically altered by crossbreeding for use by humans for food (cattle, sheep, and food crops), pets (dogs and cats), or enjoyment (animals in zoos and plants in gardens). Compare *wild species.*

dose The amount of a potentially harmful substance an individual ingests, inhales, or absorbs through the skin. Compare *response.* See *dose-response curve, median lethal dose.*

dose-response curve Plot of data showing effects of various doses of a toxic agent on a group of test organisms. See *dose, median lethal dose, response.*

doubling time The time it takes (usually in years) for the quantity of something growing exponentially to double. It can be calculated by dividing the annual percentage growth rate into 70.

drainage basin   See *watershed*.

dredge spoils   Materials scraped from the bottoms of harbors and streams to maintain shipping channels. High levels of toxic substances that have settled out of the water often contaminated these materials. See *dredging*.

dredging   Type of surface mining in which chain buckets and draglines scrape up sand, gravel, and other surface deposits covered with water. It is also used to remove sediment from streams and harbors to maintain shipping channels. See *dredge spoils*. Compare *area strip mining, contour strip mining, mountaintop removal, open-pit mining, subsurface mining*.

drift-net fishing   Catching fish in huge nets that drift in the water.

drought   Condition in which an area does not get enough water because of lower-than-normal precipitation or higher-than-normal temperatures that increase evaporation.

early successional plant species   Plant species found in the early stages of succession that grow close to the ground, can establish large populations quickly under harsh conditions, and have short lives. Compare *late successional plant species, mid-successional plant species*.

earthquake   Shaking of the ground resulting from the fracturing and displacement of rock, which produces a fault, or from subsequent movement along the fault.

ecological diversity   The variety of forests, deserts, grasslands, oceans, streams, lakes, and other biological communities interacting with one another and with their nonliving environment. See *biodiversity*. Compare *functional diversity, genetic diversity, species diversity*.

ecological efficiency   Percentage of energy transferred from one trophic level to another in a food chain or web.

ecological footprint   A measure of the ecological impact of the **(1)** consumption of food, wood products, and other resources, **(2)** use of buildings, roads, garbage dumps, and other things that consume land space, and **(3)** destruction of the forests needed to absorb the CO_2 produced by burning fossil fuels.

ecological niche   Total way of life or role of a species in an ecosystem. It includes all physical, chemical, and biological conditions a species needs to live and reproduce in an ecosystem. See *fundamental niche, realized niche*.

ecological restoration   Deliberate alteration of a degraded habitat or ecosystem to restore as much of its ecological structure and function as possible.

ecological succession   Process in which communities of plant and animal species in a particular area are replaced over time by a series of different and often more complex communities. See *primary succession, secondary succession*.

ecologist   Biological scientist who studies relationships between living organisms and their environment. Compare *conservation biologist, conservationist, environmentalist, environmental scientist, preservationist, restorationist*.

ecology   Study of the interactions of living organisms with one another and with their nonliving environment of matter and energy; study of the structure and functions of nature.

economic decision   Deciding what goods and services to produce, how to produce them, how much to produce, and how to distribute them to people.

economic depletion   Exhaustion of 80% of the estimated supply of a nonrenewable resource. Finding, extracting, and processing the remaining 20% usually costs more than it is worth; may also apply to the depletion of a renewable resource, such as a fish or tree species.

economic development   Improvement of living standards by economic growth. Compare *economic growth, environmentally sustainable economic development*.

economic growth   Increase in the capacity to provide people with goods and services produced by an economy; an increase in real GNI (GNP). Compare *economic development, environmentally sustainable economic development, sustainable economic development*.

economic resources   Natural resources, capital goods, and labor used in an economy to produce material goods and services. See *natural resources*.

economic system   Method that a group of people uses to choose what goods and services to produce, how to produce them, how much to produce, and how to distribute them to people. See *capitalist market economic system, pure command economic system, pure free-market economic system*.

economy   System of production, distribution, and consumption of economic goods.

ecosystem   Community of different species interacting with one another and with the chemical and physical factors making up its nonliving environment.

ecosystem services   Natural services or natural capital that support life on the earth and are essential to the quality of human life and the functioning of the world's economies. See *natural resources*.

edge effect   The existence of a greater number of species and a higher population density in a transition zone (ecotone) between two ecosystems than in either adjacent ecosystem. See *ecotone*.

electromagnetic radiation   Forms of kinetic energy traveling as electromagnetic waves. Examples are radio waves, TV waves, microwaves, infrared radiation, visible light, ultraviolet radiation, X rays, and gamma rays. Compare *ionizing radiation, nonionizing radiation*.

electron (e)   Tiny particle moving around outside the nucleus of an atom. Each electron has one unit of negative charge and almost no mass. Compare *neutron, proton*.

element   Chemical, such as hydrogen (H), iron (Fe), sodium (Na), carbon (C), nitrogen (N), or oxygen (O), whose distinctly different atoms serve as the basic building blocks of all matter. There are 92 naturally occurring elements. Another 23 have been made in laboratories. Two or more elements combine to form compounds that make up most of the world's matter. Compare *compound*.

endangered species   Wild species with so few individual survivors that the species could soon become extinct in all or most of its natural range. Compare *threatened species*.

endemic species   Species that is found in only one area. Such species are especially vulnerable to extinction.

energy   Capacity to do work by performing mechanical, physical, chemical, or electrical tasks or to cause a heat transfer between two objects at different temperatures.

energy efficiency   Percentage of the total energy input that does useful work and is not converted into low-quality, usually useless heat in an energy conversion system or process. See *energy quality, net energy*. Compare *material efficiency*.

energy productivity   See *energy efficiency*.

energy quality   Ability of a form of energy to do useful work. High-temperature heat and the chemical energy in fossil fuels and nuclear fuels are concentrated high-quality energy. Low-quality energy such as low-temperature heat is dispersed or diluted and cannot do much useful work. See *high-quality energy, low-quality energy*.

environment   All external conditions and factors, living and nonliving (chemicals and energy), that affect an organism or other specified system during its lifetime.

environmental degradation   Depletion or destruction of a potentially renewable resource such as soil, grassland, forest, or wildlife that is used faster than it is naturally replenished. If such use continues, the resource becomes nonrenewable (on a human time scale) or nonexistent (extinct). See also *sustainable yield*.

environmental ethics   Human beliefs about what is right or wrong environmental behavior.

environmentalist   Person who is concerned about the impact of people on environmental quality and believe that some human actions are degrading parts of the earth's life-support systems for humans and many other forms of life. Compare *conservation biologist, conservationist, ecologist, environmental scientist, preservationist, restorationist*.

environmental justice   Fair treatment and meaningful involvement of all people regardless of race, color, sex, national origin, or income with respect to the development, implementation, and enforcement of environmental laws, regulations, and policies.

environmentally sustainable economic development Development that (1) *encourages* environmentally sustainable forms of economic growth that meet the basic needs of the current generations of humans and other species without preventing future generations of humans and other species from meeting their basic needs and (2) *discourages* environmentally harmful and unsustainable forms of economic growth. It is the economic component of an *environmentally sustainable society*. Compare *economic development, economic growth*.

environmentally sustainable society Society that satisfies the basic needs of its people without depleting or degrading its natural resources and thereby preventing current and future generations of humans and other species from meeting their basic needs.

environmental movement Efforts by citizens at the grassroots level to demand that political leaders enact laws and develop policies to curtail pollution, clean up polluted environments, and protect pristine areas and species from environmental degradation.

environmental resistance All the limiting factors that act together to limit the growth of a population. See *biotic potential, limiting factor*.

environmental revolution Cultural change involving halting population growth and altering lifestyles, political and economic systems, and the way we treat the environment so that we can help sustain the earth for ourselves and other species. This involves working with the rest of nature by learning more about how nature sustains itself. See *environmental wisdom worldview*. Compare *agricultural revolution, hunter-gatherers, industrial revolution, information and globalization revolution*.

environmental science Interdisciplinary study of how the earth works, how we are affecting the earth's life-support systems (environment), and how to deal with the how to deal with the environmental problems we face.

environmental scientist Scientist who uses scientific information to understand how the earth works, learn how humans interact with the earth, and develop solutions to environmental problems. Compare *conservation biologist, conservationist, ecologist, preservationist, restorationist*.

environmental wisdom worldview Beliefs that (1) nature exists for all the earth's species, not just for us, and we are not in charge of the rest of nature; (2) there is not always more, and it's not all for us; (3) some forms of economic growth are beneficial and some are harmful, and our goals should be to design economic and political systems that encourage earth-sustaining forms of growth and discourage or prohibit earth-degrading forms; and (4) our success depends on learning to cooperate with one another and with the rest of nature instead of trying to dominate and manage earth's life-support systems primarily for our own use. Compare *frontier environmental worldview,*

planetary management worldview, spaceship-earth worldview.

environmental worldview How people think the world works, what they think their role in the world should be, and what they believe is right and wrong environmental behavior (environmental ethics).

EPA U.S. Environmental Protection Agency; responsible for managing federal efforts to control air and water pollution, radiation and pesticide hazards, environmental research, hazardous waste, and solid-waste disposal.

epidemiology Study of the patterns of disease or other harmful effects from toxic exposure within defined groups of people to find out why some people get sick and some do not.

erosion Process or group of processes by which loose or consolidated earth materials are dissolved, loosened, or worn away and removed from one place and deposited in another. See *weathering*.

estuary Partially enclosed coastal area at the mouth of a river where its fresh water, carrying fertile silt and runoff from the land, mixes with salty seawater.

eukaryotic cell Cell containing a *nucleus*, a region of genetic material surrounded by a membrane. Membranes also enclose several of the other internal parts found in a eukaryotic cell. Compare *prokaryotic cell*.

euphotic zone Upper layer of a body of water through which sunlight can penetrate and support photosynthesis.

eutrophication Physical, chemical, and biological changes that take place after a lake, estuary, or slow-flowing stream receives inputs of plant nutrients—mostly nitrates and phosphates—from natural erosion and runoff from the surrounding land basin. See *cultural eutrophication*.

eutrophic lake Lake with a large or excessive supply of plant nutrients, mostly nitrates and phosphates. Compare *mesotrophic lake, oligotrophic lake*.

evaporation Conversion of a liquid into a gas.

even-aged management Method of forest management in which trees, sometimes of a single species in a given stand, are maintained at about the same age and size and are harvested all at once. Compare *uneven-aged management*.

evergreen plants Plants that keep some of their leaves or needles throughout the year. Examples are ferns and cone-bearing trees (conifers) such as firs, spruces, pines, redwoods, and sequoias. Compare *deciduous plants, succulent plants*.

evolution See *biological evolution*.

exhaustible resource See *nonrenewable resource*.

exotic species See *nonnative species*.

experiment Procedure a scientist uses to study some phenomenon under known conditions. Scientists conduct some experi-

ments in the laboratory and others in nature. The resulting scientific data or facts must be verified or confirmed by repeated observations and measurements, ideally by several different investigators.

exponential growth Growth in which some quantity, such as population size or economic output, increases by a fixed percentage of the whole in a given time period; when the increase in quantity over time is plotted, this type of growth yields a curve shaped like the letter *J*. Compare *linear growth*.

external benefit Beneficial social effect of producing and using an economic good that is not included in the market price of the good. Compare *external cost, full cost*.

external cost Harmful social effect of producing and using an economic good that is not included in the market price of the good. Compare *external benefit, full cost, internal cost*.

externalities Social benefits ("goods") and social costs ("bads") not included in the market price of an economic good. See *external benefit, external cost*. Compare *full cost, internal cost*.

extinction Complete disappearance of a species from the earth. This happens when a species cannot adapt and successfully reproduce under new environmental conditions or when it evolves into one or more new species. Compare *speciation*. See also *endangered species, mass depletion, mass extinction, threatened species*.

family planning Providing information, clinical services, and contraceptives to help people choose the number and spacing of children they want to have.

famine Widespread malnutrition and starvation in a particular area because of a shortage of food, usually caused by drought, war, flood, earthquake, or other catastrophic events that disrupt food production and distribution.

feedback loop Circuit of sensing, evaluating, and reacting to changes in environmental conditions as a result of information fed back into a system; it occurs when one change leads to some other change, which eventually reinforces or slows the original change. See *negative feedback loop, positive feedback loop*.

feedlot Confined outdoor or indoor space used to raise hundreds to thousands of domesticated livestock. Compare *rangeland*.

fermentation See *anaerobic respiration*.

fertilizer Substance that adds inorganic or organic plant nutrients to soil and improves its ability to grow crops, trees, or other vegetation. See *commercial inorganic fertilizer, organic fertilizer*.

financial resources Cash, investments, and monetary institutions used to support the use of natural resources and human resources to provide economic goods and services. Compare *human resources, manufactured resources, natural resources*.

first law of energy   See *first law of thermodynamics*.

first law of thermodynamics   In any physical or chemical change, no detectable amount of energy is created or destroyed, but in these processes energy can be changed from one form to another; you can't get more energy out of something than you put in; in terms of energy quantity, you can't get something for nothing (there is no free lunch). This law does not apply to nuclear changes, in which energy can be produced from small amounts of matter. See also *second law of thermodynamics*.

fishery   Concentrations of particular aquatic species suitable for commercial harvesting in a given ocean area or inland body of water.

fish farming   Form of aquaculture in which fish are cultivated in a controlled pond or other environment and harvested when they reach the desired size. See also *fish ranching*.

fish ranching   Form of aquaculture in which members of a fish species such as salmon are held in captivity for the first few years of their lives, released, and then harvested as adults when they return from the ocean to their freshwater birthplace to spawn. See also *fish farming*.

fissionable isotope   Isotope that can split apart when hit by a neutron at the right speed and thus undergo nuclear fission. Examples are uranium-235 and plutonium-239.

floodplain   Flat valley floor next to a stream channel. For legal purposes, the term often applies to any low area that has the potential for flooding, including certain coastal areas.

flows   See *throughputs*

flyway   Generally fixed route along which waterfowl migrate from one area to another at certain seasons of the year.

food chain   Series of organisms in which each eats or decomposes the preceding one. Compare *food web*.

food web   Complex network of many interconnected food chains and feeding relationships. Compare *food chain*.

forest   Biome with enough average annual precipitation (at least 76 centimeters, or 30 inches) to support growth of various tree species and smaller forms of vegetation. Compare *desert, grassland*.

fossil fuel   Products of partial or complete decomposition of plants and animals that occur as crude oil, coal, natural gas, or heavy oils as a result of exposure to heat and pressure in he earth's crust over millions of years. See *coal, crude oil, natural gas*.

fossils   Skeletons, bones, shells, body parts, leaves, seeds, or impressions of such items that provide recognizable evidence of organisms that lived long ago.

free-access resource   See *common-property resource*.

Freons   See *chlorofluorocarbons*.

freshwater life zones   Aquatic systems where water with a dissolved salt concentration of less than 1% by volume accumulates on or flows through the surfaces of terrestrial biomes. Examples are **(1)** *standing* (lentic) bodies of fresh water such as lakes, ponds, and inland wetlands and **(2)** *flowing* (lotic) systems such as streams and rivers. Compare *biome*.

frontier environmental worldview   Viewing undeveloped land as a hostile wilderness to be conquered (cleared, planted) and exploited for its resources as quickly as possible. Compare *environmental wisdom worldview, planetary management worldview, spaceship-earth worldview*.

frontier forest   See *old-growth forest*.

frontier science   Preliminary scientific data, hypotheses, and models that have not been widely tested and accepted. Compare *consensus science*.

full cost   Cost of a good when its internal costs and its estimated short- and long-term external costs are included in its market price. Compare *external cost, internal cost*.

functional diversity   Biological and chemical processes or functions such as energy flow and matter cycling needed for the survival of species and biological communities. See *biodiversity, ecological diversity, genetic diversity, species diversity*.

fundamental niche   The full potential range of the physical, chemical, and biological factors a species can use if there is no competition from other species. See *ecological niche*. Compare *realized niche*.

fungicide   Chemical that kills fungi.

game species   Type of wild animal that people hunt or fish for, for sport and recreation and sometimes for food.

gamma rays   A form of ionizing electromagnetic radiation with a high energy content emitted by some radioisotopes. They readily penetrate body tissues. See also *alpha particle, beta particle*.

GDP   See *gross domestic product*.

gene mutation   See *mutation*.

gene pool   The sum total of all genes found in the individuals of the population of a particular species.

generalist species   Species with a broad ecological niche. They can live in many different places, eat a variety of foods, and tolerate a wide range of environmental conditions. Examples are flies, cockroaches, mice, rats, and human beings. Compare *specialist species*.

genes   Coded units of information about specific traits that are passed on from parents to offspring during reproduction. They consist of segments of DNA molecules found in chromosomes.

gene splicing   See *genetic engineering*.

genetic adaptation   Changes in the genetic makeup of organisms of a species that allow the species to reproduce and gain a competitive advantage under changed environmental conditions. See *differential reproduction, evolution, mutation, natural selection*.

genetically modified organism (GMO)   Organism whose genetic makeup has been modified by genetic engineering.

genetic diversity   Variability in the genetic makeup among individuals within a single species. See *biodiversity*. Compare *ecological diversity, functional diversity, species diversity*.

genetic engineering   Insertion of an alien gene into an organism to give it a beneficial genetic trait. Compare *artificial selection, natural selection*.

geographic isolation   Separation of populations of a species for long times into different areas.

geology   Study of the earth's dynamic history. Geologists study and analyze rocks and the features and processes of the earth's interior and surface.

geothermal energy   Heat transferred from the earth's underground concentrations of dry steam (steam with no water droplets), wet steam (a mixture of steam and water droplets), or hot water trapped in fractured or porous rock.

globalization   Broad process of global social, economic, and environmental change that leads to an increasingly integrated world. See *information and globalization revolution*.

global warming   Warming of the earth's atmosphere because of increases in the concentrations of one or more greenhouse gases primarily as a result of human activities. See *greenhouse effect, greenhouse gases*.

GNP   See *gross national income*.

grassland   Biome found in regions where moderate annual average precipitation (25 to 76 centimeters, or 10 to 30 inches) is enough to support the growth of grass and small plants but not enough to support large stands of trees. Compare *desert, forest*.

greenhouse effect   A natural effect that releases heat in the atmosphere (troposphere) near the earth's surface. Water vapor, carbon dioxide, ozone, and several other gases in the lower atmosphere (troposphere) absorb some of the infrared radiation (heat) radiated by the earth's surface. This causes their molecules to vibrate and transform the absorbed energy into longer-wavelength infrared radiation (heat) in the troposphere. If the atmospheric concentrations of these greenhouse gases rise and they are not removed by other natural processes, the average temperature of the lower atmosphere will increase gradually. Compare *global warming*.

greenhouse gases   Gases in the earth's lower atmosphere (troposphere) that cause the greenhouse effect. Examples are carbon dioxide, chlorofluorocarbons, ozone, methane, water vapor, and nitrous oxide.

green manure   Freshly cut or still-growing green vegetation that is plowed into the soil to increase the organic matter and humus available to support crop growth. Compare *animal manure*.

green revolution Popular term for introduction of scientifically bred or selected varieties of grain (rice, wheat, maize) that, with high enough inputs of fertilizer and water, can greatly increase crop yields.

gross domestic product (GDP) Total market value in current dollars of all goods and services produced *within a country* during a year. Compare *gross national income, gross world product.*

gross national income (GNI) Total market value in current dollars of all goods and services produced *within* and *outside* a country during a year plus net income earned abroad by a country's citizens. Formerly called gross national product (GNP). Compare *gross domestic product, gross world product.*

gross national income in purchasing power parity (GNI PPP) Market value of a country's GNI in terms of the goods and services it would buy in the United States. This is a better way to compare the standards of living among countries.

gross primary productivity (GPP) The rate at which an ecosystem's producers capture and store a given amount of chemical energy as biomass in a given length of time. Compare *net primary productivity.*

gross world product (GWP) Market value in current dollars of all goods and services produced in the world each year. Compare *gross domestic product, gross national income.*

ground fire Fire that burns decayed leaves or peat deep below the ground surface. Compare *crown fire, surface fire.*

groundwater Water that sinks into the soil and is stored in slowly flowing and slowly renewed underground reservoirs called aquifers; underground water in the zone of saturation, below the water table. Compare *runoff, surface water.*

gully reclamation Restoring land suffering from gully erosion by seeding gullies with quick-growing plants, building small dams to collect silt and gradually fill in the channels, and building channels to divert water away from the gully.

habitat Place or type of place where an organism or population of organisms lives. Compare *ecological niche.*

habitat fragmentation Breakup of a habitat into smaller pieces, usually as a result of human activities.

half-life Time needed for one-half of the nuclei in a radioisotope to emit its radiation. Each radioisotope has a characteristic half-life, which may range from a few millionths of a second to several billion years. See *radioisotope.*

hazard Something that can cause injury, disease, economic loss, or environmental damage. See also *risk.*

hazardous chemical Chemical that can cause harm because it is flammable or explosive, can irritate or damage the skin or lungs (such as strong acidic or alkaline substances), or can cause allergic reactions of

the immune system (allergens). See also *toxic chemical.*

hazardous waste Any solid, liquid, or containerized gas that can catch fire easily, is corrosive to skin tissue or metals, is unstable and can explode or release toxic fumes, or has harmful concentrations of one or more toxic materials that can leach out. See also *toxic waste.*

heat Total kinetic energy of all the randomly moving atoms, ions, or molecules within a given substance, excluding the overall motion of the whole object. Heat always flows spontaneously from a hot sample of matter to a colder sample of matter. This is one way to state the second law of thermodynamics. Compare *temperature.*

herbicide Chemical that kills a plant or inhibits its growth.

herbivore Plant-eating organism. Examples are deer, sheep, grasshoppers, and zooplankton. Compare *carnivore, omnivore.*

heterotroph See *consumer.*

high-input agriculture See *industrialized agriculture.*

high-quality energy Energy that is concentrated and has great ability to perform useful work. Examples are high-temperature heat and the energy in electricity, coal, oil, gasoline, sunlight, and nuclei of uranium-235. Compare *low-quality energy.*

high-quality matter Matter that is concentrated and contains a high concentration of a useful resource. Compare *low-quality matter.*

high-throughput economy The situation in most advanced industrialized countries, in which ever-increasing economic growth is sustained by increasing the rate at which matter and energy resources are used, with little emphasis on pollution prevention, recycling, reuse, reduction of unnecessary waste, and other forms of resource conservation. Compare *low-throughput economy, matter-recycling economy.*

homeostasis Maintenance of favorable internal conditions in a system despite fluctuations in external conditions. See *constancy, inertia, resilience.*

host Plant or animal on which a parasite feeds.

human capital See *human resources.*

human resources Physical and mental talents of people used to produce, distribute, and sell an economic good. Compare *financial resources, manufactured resources, natural resources.*

humus Slightly soluble residue of undigested or partially decomposed organic material in topsoil. This material helps retain water and water-soluble nutrients, which can be taken up by plant roots.

hunter-gatherers People who get their food by gathering edible wild plants and other materials and by hunting wild animals and fish. Compare *agricultural revolution, environmental revolution, industrial revolution, information and globalization revolution.*

hydrocarbon Organic compound of hydrogen and carbon atoms. The simplest hydrocarbon is methane (CH_4), the major component of natural gas.

hydroelectric power plant Structure in which the energy of falling or flowing water spins a turbine generator to produce electricity.

hydrologic cycle Biogeochemical cycle that collects, purifies, and distributes the earth's fixed supply of water from the environment to living organisms and then back to the environment.

hydropower Electrical energy produced by falling or flowing water. See *hydroelectric power plant.*

hydrosphere The earth's (1) liquid water (oceans, lakes, other bodies of surface water, and underground water), (2) frozen water (polar ice caps, floating ice caps, and ice in soil, known as permafrost), and (3) small amounts of water vapor in the atmosphere. See also *hydrologic cycle.*

identified resources Deposits of a particular mineral-bearing material of which the location, quantity, and quality are known or have been estimated from direct geological evidence and measurements. Compare *undiscovered resources.*

igneous rock Rock formed when molten rock material (magma) wells up from the earth's interior, cools, and solidifies into rock masses. Compare *metamorphic rock, sedimentary rock.* See *rock cycle.*

immature community Community at an early stage of ecological succession. It usually has a low number of species and ecological niches and cannot capture and use energy and cycle critical nutrients as efficiently as more complex, mature communities. Compare *mature community.*

immigrant species See *nonnative species.*

immigration Migration of people into a country or area to take up permanent residence.

indicator species Species that serve as early warnings that a community or ecosystem is being degraded. Compare *keystone species, native species, nonnative species.*

industrialized agriculture Using large inputs of energy from fossil fuels (especially oil and natural gas), water, fertilizer, and pesticides to produce large quantities of crops and livestock for domestic and foreign sale. Compare *subsistence farming.*

industrial revolution Use of new sources of energy from fossil fuels and later from nuclear fuels, and use of new technologies, to grow food and manufacture products. Compare *agricultural revolution, environmental revolution, hunter-gatherers, information and globalization revolution.*

industrial smog Type of air pollution consisting mostly of a mixture of sulfur dioxide, suspended droplets of sulfuric acid formed from some of the sulfur dioxide, and a variety of suspended solid particles. Compare *photochemical smog.*

inertia Ability of a living system to resist being disturbed or altered. Compare *constancy, resilience*.

infant mortality rate Number of babies out of every 1,000 born each year that die before their first birthday.

infiltration Downward movement of water through soil.

information and globalization revolution Use of new technologies such as the telephone, radio, television, computers, the internet, automated databases, and remote sensing satellites to enable people to have increasingly rapid access to much more information on a global scale. Compare *agricultural revolution, environmental revolution, hunter-gatherers, industrial revolution*.

inherent value See *intrinsic value*.

inland wetland Land away from the coast, such as a swamp, marsh, or bog, that is covered all or part of the time with fresh water. Compare *coastal wetland*.

inorganic compounds All compounds not classified as organic compounds. See *organic compounds*.

inorganic fertilizer See *commercial inorganic fertilizer*.

input Matter, energy, or information entering a system. Compare *output, throughput*.

input pollution control See *pollution prevention*.

insecticide Chemical that kills insects.

instrumental value Value of an organism, species, ecosystem, or the earth's biodiversity based on its usefulness to us. Compare *intrinsic value*.

integrated pest management (IPM) Combined use of biological, chemical, and cultivation methods in proper sequence and timing to keep the size of a pest population below the size that causes economically unacceptable loss of a crop or livestock animal.

intercropping Growing two or more different crops at the same time on a plot. For example, a carbohydrate-rich grain that depletes soil nitrogen and a protein-rich legume that adds nitrogen to the soil may be intercropped. Compare *monoculture, polyculture, polyvarietal cultivation*.

intermediate goods See *manufactured resources*.

internal cost Direct cost paid by the producer and the buyer of an economic good. Compare *external benefit, external cost, full cost*.

interplanting Simultaneously growing a variety of crops on the same plot. See *agroforestry, intercropping, polyculture, polyvarietal cultivation*.

interspecific competition Members of two or more species trying to use the same limited resources in an ecosystem. See *competition, competitive exclusion principle, intraspecific competition*.

intertidal zone The area of shoreline between low and high tides.

intraspecific competition Two or more organisms of a single species trying to use the same limited resources in an ecosystem. See *competition, interspecific competition*.

intrinsic rate of increase (r) Rate at which a population could grow if it had unlimited resources. Compare *environmental resistance*.

intrinsic value Value of an organism, species, ecosystem, or the earth's biodiversity based on its existence, regardless of whether it has any usefulness to us. Compare *instrumental value*.

inversion See *temperature inversion*.

invertebrates Animals that have no backbones. Compare *vertebrates*.

ion Atom or group of atoms with one or more positive (+) or negative (−) electrical charges. Compare *atom, molecule*.

ionizing radiation Fast-moving alpha or beta particles or high-energy radiation (gamma rays) emitted by radioisotopes. They have enough energy to dislodge one or more electrons from atoms they hit, forming charged ions in tissue that can react with and damage living tissue. Compare *nonionizing radiation*.

isotopes Two or more forms of a chemical element that have the same number of protons but different mass numbers because they have different numbers of neutrons in their nuclei.

J-shaped curve Curve with a shape similar to that of the letter J; can represent prolonged exponential growth.

kerogen Solid, waxy mixture of hydrocarbons found in oil shale rock. Heating the rock to high temperatures causes the kerogen to vaporize. The vapor is condensed, purified, and then sent to a refinery to produce gasoline, heating oil, and other products. See also *oil shale, shale oil*.

keystone species Species that play roles affecting many other organisms in an ecosystem. Compare *indicator species, native species, nonnative species*.

kilocalorie (kcal) Unit of energy equal to 1,000 calories. See *calorie*.

kilowatt (kW) Unit of electrical power equal to 1,000 watts. See *watt*.

kinetic energy Energy that matter has because of its mass and speed or velocity. Compare *potential energy*.

K-selected species Species that produce a few, often fairly large offspring but invest a great deal of time and energy to ensure that most of those offspring reach reproductive age. Compare *r-selected species*.

K-strategists See *K-selected species*.

lake Large natural body of standing fresh water formed when water from precipitation, land runoff, or groundwater flow fills a depression in the earth created by glaciation, earth movement, volcanic activity, or a giant meteorite. See *eutrophic lake, mesotrophic lake, oligotrophic lake*.

landfill See *sanitary landfill*.

land-use planning Process for deciding the best present and future use of each parcel of land in an area.

late successional plant species Mostly trees that can tolerate shade and form a fairly stable complex forest community. Compare *early successional plant species, mid-successional plant species*.

latitude Distance from the equator. Compare *altitude*.

law of conservation of energy See *first law of thermodynamics*.

law of conservation of matter In any physical or chemical change, matter is neither created nor destroyed but merely changed from one form to another; in physical and chemical changes, existing atoms are rearranged into different spatial patterns (physical changes) or different combinations (chemical changes).

law of tolerance The existence, abundance, and distribution of a species in an ecosystem are determined by whether the levels of one or more physical or chemical factors fall within the range tolerated by the species. See *threshold effect*.

LD$_{50}$ See *median lethal dose*.

LDC See *developing country*.

leaching Process in which various chemicals in upper layers of soil are dissolved and carried to lower layers and, in some cases, to groundwater.

less developed country (LDC) See *developing country*.

life cycle cost Initial cost plus lifetime operating costs of an economic good. Compare *full cost*.

life expectancy Average number of years a newborn infant can be expected to live.

limiting factor Single factor that limits the growth, abundance, or distribution of the population of a species in an ecosystem. See *limiting factor principle*.

limiting factor principle Too much or too little of any abiotic factor can limit or prevent growth of a population of a species in an ecosystem, even if all other factors are at or near the optimum range of tolerance for the species.

linear growth Growth in which a quantity increases by some fixed amount during each unit of time. Compare *exponential growth*.

liquefied natural gas (LNG) Natural gas converted to liquid form by cooling to a very low temperature.

liquefied petroleum gas (LPG) Mixture of liquefied propane (C_3H_8) and butane (C_4H_{10}) gas removed from natural gas and used as a fuel.

lithosphere Outer shell of the earth, composed of the crust and the rigid, outermost part of the mantle outside the asthenosphere; material found in earth's plates. See *crust, mantle*.

loams Soils containing a mixture of clay, sand, silt, and humus. Good for growing most crops.

logistic growth Pattern in which exponential population growth occurs when the population is small, and population growth decreases steadily with time as the population approaches the carrying capacity. See *S-shaped curve*.

low-input agriculture See *sustainable agriculture*.

low-quality energy Energy that is dispersed and has little ability to do useful work. An example is low-temperature heat. Compare *high-quality energy*.

low-quality matter Matter that is dilute or dispersed or contains a low concentration of a useful resource. Compare *high-quality matter*.

low-throughput economy Economy based on working with nature by **(1)** recycling and reusing discarded matter, **(2)** preventing pollution, **(3)** conserving matter and energy resources by reducing unnecessary waste and use, **(4)** not degrading renewable resources, **(5)** building things that are easy to recycle, reuse, and repair, **(6)** not allowing population size to exceed the carrying capacity of the environment, and **(7)** preserving biodiversity. See *environmental worldview*. Compare *high-throughput economy, matter-recycling economy*.

low-waste society See *low-throughput economy*.

LPG See *liquefied petroleum gas*.

macroevolution Long-term, large-scale evolutionary changes among groups of species. Compare *microevolution*.

magma Molten rock below the earth's surface.

malnutrition Faulty nutrition, caused by a diet that does not supply an individual with enough protein, essential fats, vitamins, minerals, and other nutrients needed for good health. Compare *overnutrition, undernutrition*.

mangrove swamps Swamps found on the coastlines in warm tropical climates. They are dominated by mangrove trees, any of about 55 species of trees and shrubs that can live partly submerged in the salty environment of coastal swamps.

mantle Zone of the earth's interior between its core and its crust. Compare *core, crust*. See *lithosphere*.

manufactured capital See *manufactured resources*.

manufactured resources Manufactured items made from natural resources and used to produce and distribute economic goods and services bought by consumers. These include tools, machinery, equipment, factory buildings, and transportation and distribution facilities. Compare *financial resources, human resources, natural resources*.

manure See *animal manure, green manure*.

mass The amount of material in an object.

mass depletion Widespread, often global period during which extinction rates are higher than normal but not high enough to classify as a mass extinction. Compare *background extinction, mass extinction*.

mass extinction A catastrophic, widespread, often global event in which major groups of species are wiped out over a short time compared with normal (background) extinctions. Compare *background extinction, mass depletion*.

mass number Sum of the number of neutrons (n) and the number of protons (p) in the nucleus of an atom. It gives the approximate mass of that atom. Compare *atomic number*.

mass transit Buses, trains, trolleys, and other forms of transportation that carry large numbers of people.

material efficiency Total amount of material needed to produce each unit of goods or services. Also called *resource productivity*. Compare *energy efficiency*.

matter Anything that has mass (the amount of material in an object) and takes up space. On the earth, where gravity is present, we weigh an object to determine its mass.

matter quality Measure of how useful a matter resource is, based on its availability and concentration. See *high-quality matter, low-quality matter*.

matter-recycling economy Economy that emphasizes recycling the maximum amount of all resources that can be recycled. The goal is to allow economic growth to continue without depleting matter resources and without producing excessive pollution and environmental degradation. Compare *high-throughput economy, low-throughput economy*.

mature community Fairly stable, self-sustaining community in an advanced stage of ecological succession; usually has a diverse array of species and ecological niches; captures and uses energy and cycles critical chemicals more efficiently than simpler, immature communities. Compare *immature community*.

maximum sustainable yield See *sustainable yield*.

MDC See *developed country*.

median lethal dose (LD$_{50}$) Amount of a toxic material per unit of body weight of test animals that kills half the test population in a certain time.

megacity City with 10 million or more people.

meltdown The melting of the core of a nuclear reactor.

mesosphere Third layer of the atmosphere; found above the stratosphere. Compare *stratosphere, troposphere*.

mesotrophic lake Lake with a moderate supply of plant nutrients. Compare *eutrophic lake, oligotrophic lake*.

metamorphic rock Rock produced when a preexisting rock is subjected to high temperatures (which may cause it to melt partially), high pressures, chemically active fluids, or a combination of these agents. Compare *igneous rock, sedimentary rock*. See *rock cycle*.

metastasis Spread of malignant (cancerous) cells from a tumor to other parts of the body.

metropolitan area See *urban area*.

microclimates Local climatic conditions that differ from the general climate of a region. Various topographic features of the earth's surface such as mountains and cities typically create them.

microevolution The small genetic changes a population undergoes. Compare *macroevolution*.

microorganisms Organisms such as bacteria that are so small that they can be seen only by using a microscope.

micropower systems Systems of small-scale decentralized units that generate 1–10,000 kilowatts of electricity. Examples include microturbines, fuel cells, and household solar panels and solar roofs.

midsuccessional plant species Grasses and low shrubs that are less hardy than early successional plant species. Compare *early successional plant species, late successional plant species*.

mineral Any naturally occurring inorganic substance found in the earth's crust as a crystalline solid. See *mineral resource*.

mineral resource Concentration of naturally occurring solid, liquid, or gaseous material in or on the earth's crust in a form and amount such that extracting and converting it into useful materials or items is currently or potentially profitable. Mineral resources are classified as *metallic* (such as iron and tin ores) or *nonmetallic* (such as fossil fuels, sand, and salt).

minimum-tillage farming See *conservation-tillage farming*..

minimum viable population (MVP) Estimate of the smallest number of individuals necessary to ensure the survival of a population in a region for a specified time period, typically ranging from decades to 100 years.

mixture Combination of one or more elements and compounds.

model An approximate representation or simulation of a system being studied.

molecule Combination of two or more atoms of the same chemical element (such as O$_2$) or different chemical elements (such as H$_2$O) held together by chemical bonds. Compare *atom, ion*.

monoculture Cultivation of a single crop, usually on a large area of land. Compare *polyculture, polyvarietal cultivation*.

more developed country (MDC) See *developed country*.

mountaintop removal Type of surface mining that uses explosives, massive shovels, and even larger machinery called draglines to remove the top of a mountain to expose seams of coal underneath a mountain. Compare *area strip mining, contour strip mining*.

multiple use Use of an ecosystem such as a forest for a variety of purposes such as timber harvesting, wildlife habitat, watershed protection, and recreation. Compare *sustainable yield*.

municipal solid waste Solid materials discarded by homes and businesses in or near urban areas. See *solid waste*.

mutagen Chemical or form of radiation that causes inheritable changes (mutations) in the DNA molecules in the genes found in chromosomes. See *carcinogen, mutation, teratogen*.

mutation A random change in DNA molecules making up genes that can yield changes in anatomy, physiology, or behavior in offspring. See *mutagen*.

mutualism Type of species interaction in which both participating species generally benefit. Compare *commensalism*.

native species Species that normally live and thrive in a particular ecosystem. Compare *indicator species, keystone species, nonnative species*.

natural capital See *natural resources*.

natural gas Underground deposits of gases consisting of 50–90% by weight methane gas (CH_4) and small amounts of heavier gaseous hydrocarbon compounds such as propane (C_3H_8) and butane (C_4H_{10}).

natural greenhouse effect Heat buildup in the troposphere because of the presence of certain gases, called greenhouse gases. Without this effect, the earth would be nearly as cold as Mars, and life as we know it could not exist. Compare *global warming*.

natural ionizing radiation Ionizing radiation in the environment from natural sources.

natural law See *scientific law*.

natural radioactive decay Nuclear change in which unstable nuclei of atoms spontaneously shoot out particles (usually alpha or beta particles) or energy (gamma rays) at a fixed rate.

natural rate of extinction See *background extinction*.

natural recharge Natural replenishment of an aquifer by precipitation, which percolates downward through soil and rock. See *recharge area*.

natural resources The earth's natural materials and processes that sustain other species and us. Compare *financial resources, human resources, manufactured resources*.

natural selection Process by which a particular beneficial gene (or set of genes) is reproduced in succeeding generations more than other genes. The result of natural selection is a population that contains a greater

proportion of organisms better adapted to certain environmental conditions. See *adaptation, biological evolution, differential reproduction, mutation*.

negative feedback loop Situation in which a change in a certain direction provides information that causes a system to change less in that direction. Compare *positive feedback loop*.

nekton Strongly swimming organisms found in aquatic systems. Compare *benthos, plankton*.

net energy Total amount of useful energy available from an energy resource or energy system over its lifetime, minus the amount of energy **(1)** used (the first energy law), **(2)** automatically wasted (the second energy law), and **(3)** unnecessarily wasted in finding, processing, concentrating, and transporting it to users.

net primary productivity (NPP) Rate at which all the plants in an ecosystem produce net useful chemical energy; equal to the difference between the rate at which the plants in an ecosystem produce useful chemical energy (gross primary productivity) and the rate at which they use some of that energy through cellular respiration. Compare *gross primary productivity*.

neutral solution Water solution containing an equal number of hydrogen ions (H^+) and hydroxide ions (OH^-); water solution with a pH of 7. Compare *acid solution, basic solution*.

neutron (n) Elementary particle in the nuclei of all atoms (except hydrogen-1). It has a relative mass of 1 and no electric charge. Compare *electron, proton*.

niche See *ecological niche*.

nitrogen cycle Cyclic movement of nitrogen in different chemical forms from the environment to organisms and then back to the environment.

nitrogen fixation Conversion of atmospheric nitrogen gas into forms useful to plants by lightning, bacteria, and cyanobacteria; it is part of the nitrogen cycle.

noise pollution Any unwanted, disturbing, or harmful sound that impairs or interferes with hearing, causes stress, hampers concentration and work efficiency, or causes accidents.

nondegradable pollutant Material that is not broken down by natural processes. Examples are the toxic elements lead and mercury. Compare *biodegradable pollutant, degradable pollutant, slowly degradable pollutant*.

nonionizing radiation Forms of radiant energy such as radio waves, microwaves, infrared light, and ordinary light that do not have enough energy to cause ionization of atoms in living tissue. Compare *ionizing radiation*.

nonnative species Species that migrate into an ecosystem or are deliberately or accidentally introduced into an ecosystem by humans. Compare *native species*.

nonpersistent pollutant See *degradable pollutant*.

nonpoint source Large or dispersed land areas such as cropfields, streets, and lawns that discharge pollutants into the environment over a large area. Compare *point source*.

nonrenewable resource Resource that exists in a fixed amount (stock) in various places in the earth's crust and has the potential for renewal by geological, physical, and chemical processes taking place over hundreds of millions to billions of years. Examples are copper, aluminum, coal, and oil. We classify these resources as exhaustible because we are extracting and using them at a much faster rate than they were formed. Compare *renewable resource*.

nontransmissible disease A disease that is not caused by living organisms and does not spread from one person to another. Examples are most cancers, diabetes, cardiovascular disease, and malnutrition. Compare *transmissible disease*.

no-till farming See *conservation-tillage farming*.

nuclear change Process in which nuclei of certain isotopes spontaneously change, or are forced to change, into one or more different isotopes. The three principal types of nuclear change are natural radioactivity, nuclear fission, and nuclear fusion. Compare *chemical change, physical change*.

nuclear energy Energy released when atomic nuclei undergo a nuclear reaction such as the spontaneous emission of radioactivity, nuclear fission, or nuclear fusion.

nuclear fission Nuclear change in which the nuclei of certain isotopes with large mass numbers (such as uranium-235 and plutonium-239) are split apart into lighter nuclei when struck by a neutron. This process releases more neutrons and a large amount of energy. Compare *nuclear fusion*.

nuclear fusion Nuclear change in which two nuclei of isotopes of elements with a low mass number (such as hydrogen-2 and hydrogen-3) are forced together at extremely high temperatures until they fuse to form a heavier nucleus (such as helium-4). This process releases a large amount of energy. Compare *nuclear fission*.

nucleus Extremely tiny center of an atom, making up most of the atom's mass. It contains one or more positively charged protons and one or more neutrons with no electrical charge (except for a hydrogen-1 atom, which has one proton and no neutrons in its nucleus).

nutrient Any food or element an organism must take in to live, grow, or reproduce.

nutrient cycle See *biogeochemical cycle*.

oil See *crude oil*.

oil shale Fine-grained rock containing various amounts of kerogen, a solid, waxy mixture of hydrocarbon compounds. Heating the rock to high temperatures converts the kerogen into a vapor that can be condensed to form a slow-flowing heavy oil called shale oil. See *kerogen, shale oil*.

old-growth forest Virgin and old, second-growth forests containing trees that are often hundreds, sometimes thousands of years old. Examples include forests of Douglas fir, western hemlock, giant sequoia, and coastal redwoods in the western United States. Compare *second-growth forest, tree plantation.*

oligotrophic lake Lake with a low supply of plant nutrients. Compare *eutrophic lake, mesotrophic lake.*

omnivore Animal that can use both plants and other animals as food sources. Examples are pigs, rats, cockroaches, and people. Compare *carnivore, herbivore.*

open-pit mining Removing minerals such as gravel, sand, and metal ores by digging them out of the earth's surface and leaving an open pit. Compare *area strip mining, contour strip mining, dredging, mountaintop removal, subsurface mining.*

open sea The part of an ocean that is beyond the continental shelf. Compare *coastal zone.*

ore Part of a metal-yielding material that can be economically and legally extracted at a given time. An ore typically contains two parts: the ore mineral, which contains the desired metal, and waste mineral material (gangue).

organic compounds Compounds containing carbon atoms combined with each other and with atoms of one or more other elements such as hydrogen, oxygen, nitrogen, sulfur, phosphorus, chlorine, and fluorine. All other compounds are called *inorganic compounds.*

organic farming Producing crops and livestock naturally by using organic fertilizer (manure, legumes, compost) and natural pest control (bugs that eat harmful bugs, plants that repel bugs, and environmental controls such as crop rotation) instead of using commercial inorganic fertilizers and synthetic pesticides and herbicides. See *sustainable agriculture.*

organic fertilizer Organic material such as animal manure, green manure, and compost, applied to cropland as a source of plant nutrients. Compare *commercial inorganic fertilizer.*

organism Any form of life.

other resources Identified and undiscovered resources not classified as reserves. Compare *identified resources, reserves, undiscovered resources.*

output Matter, energy, or information leaving a system. Compare *input, throughput.*

output pollution control See *pollution cleanup.*

overburden Layer of soil and rock overlying a mineral deposit. Surface mining removes this layer.

overfishing Harvesting so many fish of a species, especially immature fish, that there is not enough breeding stock left to replenish the species, such that it is not profitable to harvest them.

overgrazing Destruction of vegetation when too many grazing animals feed too long and exceed the carrying capacity of a rangeland or pasture area.

overnutrition Diet so high in calories, saturated (animal) fats, salt, sugar, and processed foods and so low in vegetables and fruits that the consumer runs high risks of diabetes, hypertension, heart disease, and other health hazards. Compare *malnutrition, undernutrition.*

oxygen-demanding wastes Organic materials that are usually biodegraded by aerobic (oxygen-consuming) bacteria if there is enough dissolved oxygen in the water.

ozone depletion Decrease in concentration of ozone (O_3) in the stratosphere. See *ozone layer.*

ozone layer Layer of gaseous ozone (O_3) in the stratosphere that protects life on earth by filtering out most harmful ultraviolet radiation from the sun.

PANs Peroxyacyl nitrates. Group of chemicals found in photochemical smog.

parasite Consumer organism that lives on or in and feeds on a living plant or animal, known as the host, over an extended period of time. The parasite draws nourishment from and gradually weakens its host; it may or may not kill the host. See *parasitism.*

parasitism Interaction between species in which one organism, called the parasite, preys on another organism, called the host, by living on or in the host. See *host, parasite.*

parts per billion (ppb) Number of parts of a chemical found in 1 billion parts of a particular gas, liquid, or solid.

parts per million (ppm) Number of parts of a chemical found in 1 million parts of a particular gas, liquid, or solid.

parts per trillion (ppt) Number of parts of a chemical found in 1 trillion parts of a particular gas, liquid, or solid.

passive solar heating system System that captures sunlight directly within a structure and converts it into low-temperature heat for space heating or for heating water for domestic use without the use of mechanical devices. Compare *active solar heating system.*

pasture Managed grassland or enclosed meadow that usually is planted with domesticated grasses or other forage to be grazed by livestock. Compare *feedlot, rangeland.*

pathogen Organism that produces disease.

PCBs See *polychlorinated biphenyls.*

per capita GNI Annual gross national income (GNI) of a country divided by its total population at mid-year. It gives the average slice of the economic pie per person. Used to be called per capita GNP. See *gross national income.*

per capita GNI in purchasing power parity (per capita GNI PPP: The GNI PPP divided by the total population at mid-year. This is a better way to make comparisons of people's economic welfare among countries. See *per capita GNI.*

percolation Passage of a liquid through the spaces of a porous material such as soil.

perennial Plant that can live for more than 2 years. Compare *annual.*

permafrost Perennially frozen layer of the soil that forms when the water there freezes. It is found in arctic tundra.

permeability The degree to which underground rock and soil pores are interconnected and thus a measure of the degree to which water can flow freely from one pore to another. Compare *porosity.*

perpetual resource An essentially inexhaustible resource on a human time scale. Solar energy is an example. See *renewable resource.* Compare *nonrenewable resource, renewable resource.*

persistence How long a pollutant stays in the air, water, soil, or body. See also *inertia.*

persistent pollutant See *slowly degradable pollutant.*

pest Unwanted organism that directly or indirectly interferes with human activities.

pesticide Any chemical designed to kill or inhibit the growth of an organism that people consider undesirable. See *fungicide, herbicide, insecticide.*

petrochemicals Chemicals obtained by refining (distilling) crude oil. They are used as raw materials in manufacturing most industrial chemicals, fertilizers, pesticides, plastics, synthetic fibers, paints, medicines, and many other products.

petroleum See *crude oil.*

pH Numeric value that indicates the relative acidity or alkalinity of a substance on a scale of 0 to 14, with the neutral point at 7. Acid solutions have pH values lower than 7, and basic or alkaline solutions have pH values greater than 7.

phosphorus cycle Cyclic movement of phosphorus in different chemical forms from the environment to organisms and then back to the environment.

photochemical smog Complex mixture of air pollutants produced in the lower atmosphere by the reaction of hydrocarbons and nitrogen oxides under the influence of sunlight. Especially harmful components include ozone, peroxyacyl nitrates (PANs), and various aldehydes. Compare *industrial smog.*

photosynthesis Complex process that takes place in cells of green plants. Radiant energy from the sun is used to combine carbon dioxide (CO_2) and water (H_2O) to produce oxygen (O_2) and carbohydrates (such as glucose, $C_6H_{12}O_6$) and other nutrient molecules. Compare *aerobic respiration, chemosynthesis.*

photovoltaic cell (solar cell) Device which converts radiant (solar) energy directly into electrical energy.

physical change Process that alters one or more physical properties of an element or a compound without altering its chemical composition. Examples are changing the size and shape of a sample of matter (crushing ice and cutting aluminum foil) and changing a sample of matter from one physical state to another (boiling and freezing water). Compare *chemical change, nuclear change.*

phytoplankton Small, drifting plants, mostly algae and bacteria, found in aquatic ecosystems. Compare *plankton, zooplankton.*

pioneer community First integrated set of plants, animals, and decomposers found in an area undergoing primary ecological succession. See *immature community, mature community.*

pioneer species First hardy species, often microbes, mosses, and lichens that begin colonizing a site as the first stage of ecological succession. See *ecological succession, pioneer community.*

planetary management worldview Beliefs that **(1)** we are the planet's most important species; **(2)** there is always more, and it's all for us; **(3)** all economic growth is good, more economic growth is better, and the potential for economic growth is limitless; and **(4)** our success depends on how well we can understand, control, and manage the earth's life-support systems for our own benefit. See *spaceship-earth worldview.* Compare *environmental wisdom worldview.*

plankton Small plant organisms (phytoplankton) and animal organisms (zooplankton) that float in aquatic ecosystems.

plantation agriculture Growing specialized crops such as bananas, coffee, and cacao in tropical developing countries, primarily for sale to developed countries.

plasma An ionized gas consisting of electrically conductive ions and electrons. It is known as a fourth state of matter.

plates See *tectonic plates.*

plate tectonics Theory of geophysical processes that explains the movements of lithospheric plates and the processes that occur at their boundaries. See *lithosphere, tectonic plates.*

point source Single identifiable source that discharges pollutants into the environment. Examples are the smokestack of a power plant or an industrial plant, drainpipe of a meatpacking plant, chimney of a house, or exhaust pipe of an automobile. Compare *nonpoint source.*

poison A chemical that in one dose kills exactly 50% of the animals (usually rats and mice) in a test population (usually 60 to 200 animals) within a 14-day period. See *median lethal dose.*

politics Process through which individuals and groups try to influence or control government policies and actions that affect the local, state, national, and international communities.

pollutant A particular chemical or form of energy that can adversely affect the health, survival, or activities of humans or other living organisms. See *pollution.*

pollution An undesirable change in the physical, chemical, or biological characteristics of air, water, soil, or food that can adversely affect the health, survival, or activities of humans or other living organisms.

pollution cleanup Device or process that removes or reduces the level of a pollutant after it has been produced or has entered the environment. Examples are automobile emission control devices and sewage treatment plants. Compare *pollution prevention.*

pollution prevention Device or process that prevents a potential pollutant from forming or entering the environment or sharply reduces the amount entering the environment. Compare *pollution cleanup.*

polychlorinated biphenyls (PCBs) Group of 209 different toxic, oily, synthetic chlorinated hydrocarbon compounds that can be biologically amplified in food chains and webs.

polyculture Complex form of intercropping in which a large number of different plants maturing at different times are planted together. See also *intercropping.* Compare *monoculture, polyvarietal cultivation.*

polyvarietal cultivation Planting a plot of land with several varieties of the same crop. Compare *intercropping, monoculture, polyculture.*

population Group of individual organisms of the same species living in a particular area.

population change An increase or decrease in the size of a population. It is equal to (Births + Immigration) − (Deaths + Emigration).

population density Number of organisms in a particular population found in a specified area or volume.

population dispersion General pattern in which the members of a population are arranged throughout its habitat.

population distribution Variation of population density over a particular geographic area. For example, a country has a high population density in its urban areas and a much lower population density in rural areas.

population dynamics Major abiotic and biotic factors that tend to increase or decrease the population size and age and sex composition of a species.

population size Number of individuals making up a population's gene pool.

porosity Percentage of space in rock or soil occupied by voids, whether the voids are isolated or connected. Compare *permeability.*

positive feedback loop Situation in which a change in a certain direction provides information that causes a system to change

further in the same direction. Compare *negative feedback loop.*

potential energy Energy stored in an object because of its position or the position of its parts. Compare *kinetic energy.*

poverty Inability to meet basic needs for food, clothing, and shelter.

ppb See *parts per billion.*

ppm See *parts per million.*

ppt See *parts per trillion.*

precautionary principle When there is scientific uncertainty about potentially serious harm from chemicals or technologies, decision makers should act to prevent harm to humans and the environment. See *pollution prevention.*

precipitation Water in the form of rain, sleet, hail, and snow that falls from the atmosphere onto the land and bodies of water.

predation Situation in which an organism of one species (the predator) captures and feeds on parts or all of an organism of another species (the prey).

predator Organism that captures and feeds on parts or all of an organism of another species (the prey).

predator-prey relationship Interaction between two organisms of different species in which one organism, called the *predator,* captures and feeds on parts or all of another organism, called the *prey.*

preservationist Person concerned primarily with setting aside or protecting undisturbed natural areas from harmful human activities. Compare *conservation biologist, conservationist, ecologist, environmentalist, environmental scientist, restorationist.*

prey Organism that is captured and serves as a source of food for an organism of another species (the predator).

primary consumer Organism that feeds on all or part of plants (herbivore) or on other producers. Compare *detritivore, omnivore, secondary consumer.*

primary pollutant Chemical that has been added directly to the air by natural events or human activities and occurs in a harmful concentration. Compare *secondary pollutant.*

primary productivity See *gross primary productivity, net primary productivity.*

primary sewage treatment Mechanical sewage treatment in which large solids are filtered out by screens and suspended solids settle out as sludge in a sedimentation tank. Compare *advanced sewage treatment, secondary sewage treatment.*

primary succession Ecological succession in a bare area that has never been occupied by a community of organisms. See *ecological succession.* Compare *secondary succession.*

probability A mathematical statement about how likely it is that something will happen.

producer Organism that uses solar energy (green plant) or chemical energy (some bacteria) to manufacture the organic compounds it needs as nutrients from simple inorganic compounds obtained from its environment. Compare *consumer, decomposer*.

prokaryotic cell Cell that doesn't have a distinct nucleus. Other internal parts are also not enclosed by membranes. Compare *eukaryotic cell*.

proton (p) Positively charged particle in the nuclei of all atoms. Each proton has a relative mass of 1 and a single positive charge. Compare *electron, neutron*.

pure capitalism See *pure free-market economic system*.

pure command economic system System in which all economic decisions are made by the government or some other central authority. Compare *capitalist market economic system, pure free-market economic system*.

pure free-market economic system System in which all economic decisions are made in the market, where buyers and sellers of economic goods interact freely, with no government or other interference. Compare *capitalist market economic system, pure command economic system*.

pyramid of energy flow Diagram representing the flow of energy through each trophic level in a food chain or food web. With each energy transfer, only a small part (typically 10%) of the usable energy entering one trophic level is transferred to the organisms at the next trophic level. Compare *pyramid of biomass, pyramid of numbers*.

radiation Fast-moving particles (particulate radiation) or waves of energy (electromagnetic radiation). See *alpha particle, beta particle, gamma rays*.

radiation temperature inversion Temperature inversion that typically occurs at night in which a layer of warm air lies atop a layer of cooler air nearer the ground as the air near the ground cools faster than the air above it. As the sun rises and warms the earth's surface, the inversion normally disappears by noon and disperses the pollutants built up during the night. See *temperature inversion*. Compare *subsidence temperature inversion*.

radioactive decay Change of a radioisotope to a different isotope by the emission of radioactivity.

radioactive isotope See *radioisotope*.

radioactive waste Waste products of nuclear power plants, research, medicine, weapon production, or other processes involving nuclear reactions. See *radioactivity*.

radioactivity Nuclear change in which unstable nuclei of atoms spontaneously shoot out "chunks" of mass, energy, or both at a fixed rate. The three principal types of radioactivity are gamma rays and fast-moving alpha particles and beta particles.

radioisotope Isotope of an atom that spontaneously emits one or more types of radioactivity (alpha particles, beta particles, gamma rays).

rain shadow effect Low precipitation on the far side (leeward side) of a mountain when prevailing winds flow up and over a high mountain or range of high mountains. This creates semiarid and arid conditions on the leeward side of a high mountain range.

rangeland Land that supplies forage or vegetation (grasses, grasslike plants, and shrubs) for grazing and browsing animals and is not intensively managed. Compare *feedlot, pasture*.

range of tolerance Range of chemical and physical conditions that must be maintained for populations of a particular species to stay alive and grow, develop, and function normally. See *law of tolerance*.

rare species A species that has naturally small numbers of individuals (often because of limited geographic ranges or low population densities) or has been locally depleted by human activities.

realized niche Parts of the fundamental niche of a species that are actually used by that species. See *ecological niche, fundamental niche*.

recharge area Any area of land allowing water to pass through it and into an aquifer. See *aquifer, natural recharge*.

recycling Collecting and reprocessing a resource so that it can be made into new products. An example is collecting aluminum cans, melting them down, and using the aluminum to make new cans or other aluminum products. Compare *reuse*.

reforestation Renewal of trees and other types of vegetation on land where trees have been removed; can be done naturally by seeds from nearby trees or artificially by planting seeds or seedlings.

reliable runoff Surface runoff of water that generally can be counted on as a stable source of water from year to year. See *runoff*.

renewable resource Resource that can be replenished rapidly (hours to several decades) through natural processes. Examples are trees in forests, grasses in grasslands, wild animals, fresh surface water in lakes and streams, most groundwater, fresh air, and fertile soil. If such a resource is used faster than it is replenished, it can be depleted and converted into a nonrenewable resource. Compare *nonrenewable resource* and *perpetual resource*. See also *environmental degradation*.

replacement-level fertility Number of children a couple must have to replace them. The average for a country or the world usually is slightly higher than 2 children per couple (2.1 in the United States and 2.5 in some developing countries) because some children die before reaching their reproductive years. See also *total fertility rate*.

reproduction Production of offspring by one or more parents.

reproductive isolation Long-term geographic separation of members of a particular sexually reproducing species.

reproductive potential See *biotic potential*.

reserves Resources that have been identified and from which a usable mineral can be extracted profitably at present prices with current mining technology. See *identified resources, undiscovered resources*.

reserve-to-production ratio Number of years reserves of a particular nonrenewable mineral will last at current annual production rates. See *reserves*.

resilience Ability of a living system to restore itself to original condition after being exposed to an outside disturbance that is not too drastic. See *constancy, inertia*.

resource Anything obtained from the living and nonliving environment to meet human needs and wants. It can also be applied to other species.

resource partitioning Process of dividing up resources in an ecosystem so that species with similar needs (overlapping ecological niches) use the same scarce resources at different times, in different ways, or in different places. See *ecological niche, fundamental niche, realized niche*.

resource productivity See *material efficiency*.

respiration See *aerobic respiration*.

response The amount of health damage caused by exposure to a certain dose of a harmful substance or form of radiation. See *dose, dose-response curve, median lethal dose*.

restoration ecology Research and scientific study devoted to restoring, repairing, and reconstructing damaged ecosystems.

restorationist Scientist or other person devoted to the partial or complete restoration of natural areas that have been degraded by human activities. Compare *conservation biologist, conservationist, ecologist, environmental scientist, preservationist*.

reuse Using a product over and over again in the same form. An example is collecting, washing, and refilling glass beverage bottles. Compare *recycling*.

riparian zones Thin strips and patches of vegetation that surround streams. They are very important habitats and resources for wildlife.

risk The probability that something undesirable will result from deliberate or accidental exposure to a hazard. See *risk analysis, risk assessment, risk-benefit analysis, risk management*.

risk analysis Identifying hazards, evaluating the nature and severity of risks (*risk assessment*), using this and other information to determine options and make decisions about reducing or eliminating risks (*risk management*), and communicating information about risks to decision makers and the public (*risk communication*).

risk assessment Process of gathering data and making assumptions to estimate short- and long-term harmful effects on human health or the environment from exposure to hazards associated with the use of a particu-

lar product or technology. See *risk-benefit analysis.*

risk-benefit analysis Estimate of the short- and long-term risks and benefits of using a particular product or technology. See *risk assessment.*

risk communication Communicating information about risks to decision makers and the public. See *risk, risk analysis, risk-benefit analysis.*

risk management Using risk assessment and other information to determine options and make decisions about reducing or eliminating risks. See *risk, risk analysis, risk-benefit analysis, risk communication.*

rock Any material that makes up a large, natural, continuous part of earth's crust. See *mineral.*

rock cycle Largest and slowest of the earth's cycles, consisting of geologic, physical, and chemical processes that form and modify rocks and soil in the earth's crust over millions of years.

r-selected species Species that reproduce early in their life span and produce large numbers of usually small and short-lived offspring in a short period of time. Compare *K-selected species.*

r-strategists See *r-selected species.*

rule of 70 Doubling time (in years) = 70/percentage growth rate. See *doubling time, exponential growth.*

runoff Fresh water from precipitation and melting ice that flows on the earth's surface into nearby streams, lakes, wetlands, and reservoirs. See *reliable runoff, surface runoff, surface water.* Compare *groundwater.*

rural area Geographic area in the United States with a population of less than 2,500. The number of people used in this definition may vary in different countries. Compare *urban area.*

salinity Amount of various salts dissolved in a given volume of water.

salinization Accumulation of salts in soil that can eventually make the soil unable to support plant growth.

saltwater intrusion Movement of salt water into freshwater aquifers in coastal and inland areas as groundwater is withdrawn faster than it is recharged by precipitation.

sanitary landfill Waste disposal site on land in which waste is spread in thin layers, compacted, and covered with a fresh layer of clay or plastic foam each day.

scavenger Organism that feeds on dead organisms that were killed by other organisms or died naturally. Examples are vultures, flies, and crows. Compare *detritivore.*

science Attempts to discover order in nature and use that knowledge to make predictions about what should happen in nature. See *consensus science, frontier science, scientific data, scientific hypothesis, scientific law, scientific methods, scientific model, scientific theory.*

scientific data Facts obtained by making observations and measurements. Compare *scientific hypothesis, scientific law, scientific methods, scientific model, scientific theory.*

scientific hypothesis An educated guess that attempts to explain a scientific law or certain scientific observations. Compare *scientific data, scientific law, scientific methods, scientific model, scientific theory.*

scientific law Description of what scientists find happening in nature repeatedly in the same way, without known exception. See *first law of thermodynamics, law of conservation of matter, second law of thermodynamics.* Compare *scientific data, scientific hypothesis, scientific methods, scientific model, scientific theory.*

scientific methods The ways scientists gather data and formulate and test scientific hypotheses, models, theories, and laws. See *scientific data, scientific hypothesis, scientific law, scientific model, scientific theory.*

scientific model A simulation of complex processes and systems. Many are mathematical models that are run and tested using computers.

scientific theory A well-tested and widely accepted scientific hypothesis. Compare *scientific data, scientific hypothesis, scientific law, scientific methods, scientific model.*

secondary consumer Organism that feeds only on primary consumers. Compare *detritivore, omnivore, primary consumer.*

secondary pollutant Harmful chemical formed in the atmosphere when a primary air pollutant reacts with normal air components or other air pollutants. Compare *primary pollutant.*

secondary sewage treatment Second step in most waste treatment systems in which aerobic bacteria break down up to 90% of degradable, oxygen-demanding organic wastes in wastewater. This usually involves bringing sewage and bacteria together in trickling filters or in the activated sludge process. Compare *advanced sewage treatment, primary sewage treatment.*

secondary succession Ecological succession in an area in which natural vegetation has been removed or destroyed but the soil is not destroyed. See *ecological succession.* Compare *primary succession.*

second-growth forest Stands of trees resulting from secondary ecological succession. Compare *old-growth forest, tree farm.*

second law of energy See *second law of thermodynamics.*

second law of thermodynamics In any conversion of heat energy to useful work, some of the initial energy input is always degraded to a lower-quality, more dispersed, less useful energy, usually low-temperature heat that flows into the environment; you can't break even in terms of energy quality. See *first law of thermodynamics.*

sedimentary rock Rock that forms from the accumulated products of erosion and in some cases from the compacted shells, skeletons, and other remains of dead organisms. Compare *igneous rock, metamorphic rock.* See *rock cycle.*

seed-tree cutting Removal of nearly all trees on a site in one cutting, with a few seed-producing trees left uniformly distributed to regenerate the forest. Compare *clearcutting, selective cutting, shelterwood cutting, strip cutting.*

selective cutting Cutting of intermediate-aged, mature, or diseased trees in an uneven-aged forest stand, either singly or in small groups. This encourages the growth of younger trees and maintains an uneven-aged stand. Compare *clear-cutting, seed-tree cutting, shelterwood cutting, strip cutting.*

septic tank Underground tank for treating wastewater from a home in rural and suburban areas. Bacteria in the tank decompose organic wastes, and the sludge settles to the bottom of the tank. The effluent flows out of the tank into the ground through a field of drainpipes.

sexual reproduction Reproduction in organisms that produce offspring by combining sex cells or *gametes* (such as ovum and sperm) from both parents. This produces offspring that have combinations of traits from their parents. Compare *asexual reproduction.*

shale oil Slow-flowing, dark brown, heavy oil obtained when kerogen in oil shale is vaporized at high temperatures and then condensed. Shale oil can be refined to yield gasoline, heating oil, and other petroleum products. See *kerogen, oil shale.*

shelterbelt See *windbreak.*

shelterwood cutting Removal of mature, marketable trees in an area in a series of partial cuttings to allow regeneration of a new stand under the partial shade of older trees, which are later removed. Typically, this is done by making two or three cuts over a decade. Compare *clearcutting, seed-tree cutting, selective cutting, strip cutting.*

shifting cultivation Clearing a plot of ground in a forest, especially in tropical areas, and planting crops on it for a few years (typically 2–5 years) until the soil is depleted of nutrients or the plot has been invaded by a dense growth of vegetation from the surrounding forest. Then a new plot is cleared and the process is repeated. The abandoned plot cannot successfully grow crops for 10–30 years. See also *slash-and-burn cultivation.*

slash-and-burn cultivation Cutting down trees and other vegetation in a patch of forest, leaving the cut vegetation on the ground to dry, and then burning it. The ashes that are left add nutrients to the nutrient-poor soils found in most tropical forest areas. Crops are planted between tree stumps. Plots must be abandoned after a few years (typically 2–5 years) because of loss of soil fertility or invasion of vegetation from the surrounding forest. See also *shifting cultivation.*

slowly degradable pollutant Material that is slowly broken down into simpler chemicals or reduced to acceptable levels by natural physical, chemical, and biological processes. Compare *biodegradable pollutant, degradable pollutant, nondegradable pollutant.*

sludge Gooey mixture of toxic chemicals, infectious agents, and settled solids removed from wastewater at a sewage treatment plant.

smart growth Form of urban planning that recognizes that urban growth will occur but uses zoning laws and an array of other tools to prevent sprawl, direct growth to certain areas, protect ecologically sensitive and important lands and waterways, and develop urban areas that are more environmentally sustainable and more enjoyable places to live.

smelting Process in which a desired metal is separated from the other elements in an ore mineral.

smog Originally a combination of smoke and fog but now used to describe other mixtures of pollutants in the atmosphere. See *industrial smog, photochemical smog.*

soil Complex mixture of inorganic minerals (clay, silt, pebbles, and sand), decaying organic matter, water, air, and living organisms.

soil conservation Methods used to reduce soil erosion, prevent depletion of soil nutrients, and restore nutrients already lost by erosion, leaching, and excessive crop harvesting.

soil erosion Movement of soil components, especially topsoil, from one place to another, usually by wind, flowing water, or both. This natural process can be greatly accelerated by human activities that remove vegetation from soil.

soil horizons Horizontal zones that make up a particular mature soil. Each horizon has a distinct texture and composition that vary with different types of soils. See *soil profile.*

soil permeability Rate at which water and air move from upper to lower soil layers. Compare *porosity.*

soil porosity See *porosity.*

soil profile Cross-sectional view of the horizons in a soil. See *soil horizon.*

soil structure How the particles that make up a soil are organized and clumped together. See also *soil permeability, soil texture.*

soil texture Relative amounts of the different types and sizes of mineral particles in a sample of soil.

solar capital Solar energy from the sun reaching the earth. Compare *natural resources.*

solar cell See *photovoltaic cell.*

solar collector Device for collecting radiant energy from the sun and converting it into heat. See *active solar heating system, passive solar heating system.*

solar energy Direct radiant energy from the sun and a number of indirect forms of energy produced by the direct input. Principal indirect forms of solar energy include wind, falling and flowing water (hydropower), and biomass (solar energy converted into chemical energy stored in the chemical bonds of organic compounds in trees and other plants).

solid waste Any unwanted or discarded material that is not a liquid or a gas. See *municipal solid waste.*

spaceship-earth worldview View of the earth as a spaceship: a machine that we can understand, control, and change at will by using advanced technology. See *planetary management worldview.* Compare *environmental wisdom worldview.*

specialist species Species with a narrow ecological niche. They may be able to live in only one type of habitat, tolerate only a narrow range of climatic and other environmental conditions, or use only one type or a few types of food. Compare *generalist species.*

speciation Formation of two species from one species because of divergent natural selection in response to changes in environmental conditions; usually takes thousands of years. Compare *extinction.*

species Group of organisms that resemble one another in appearance, behavior, chemical makeup and processes, and genetic structure. Organisms that reproduce sexually are classified as members of the same species only if they can actually or potentially interbreed with one another and produce fertile offspring.

species diversity Number of different species and their relative abundances in a given area. See *biodiversity.* Compare *ecological diversity, genetic diversity.*

species equilibrium model See *theory of island biogeography.*

spoils Unwanted rock and other waste materials produced when a material is removed from the earth's surface or subsurface by mining, dredging, quarrying, and excavation.

S-shaped curve Leveling off of an exponential, J-shaped curve when a rapidly growing population exceeds the carrying capacity of its environment and ceases to grow.

stability Ability of a living system to withstand or recover from externally imposed changes or stresses. See *constancy, inertia, resilience.*

stewardship View that because of our superior intellect and power or because of our religious beliefs, we have an ethical responsibility to manage and care for domesticated plants and animals and the rest of nature. Compare *environmental wisdom worldview, planetary management worldview.*

stratosphere Second layer of the atmosphere, extending about 17–48 kilometers (11–30 miles) above the earth's surface. It

contains small amounts of gaseous ozone (O_3), which filters out about 95% of the incoming harmful ultraviolet (UV) radiation emitted by the sun. Compare *troposphere.*

stream Flowing body of surface water. Examples are creeks and rivers.

strip cropping Planting regular crops and close-growing plants, such as hay or nitrogen-fixing legumes, in alternating rows or bands to help reduce depletion of soil nutrients.

strip cutting A variation of clear-cutting in which a strip of trees is clear-cut along the contour of the land, with the corridor narrow enough to allow natural regeneration within a few years. After regeneration, another strip is cut above the first, and so on. Compare *clear-cutting, seed-tree cutting, selective cutting, shelterwood cutting.*

strip mining Form of surface mining in which bulldozers, power shovels, or stripping wheels remove large chunks of the earth's surface in strips. See *area strip mining, contour strip mining, surface mining.* Compare *subsurface mining.*

subatomic particles Extremely small particles—electrons, protons, and neutrons—that make up the internal structure of atoms.

subduction zone Area in which oceanic lithosphere is carried downward (subducted) under the island arc or continent at a convergent plate boundary. A trench ordinarily forms at the boundary between the two converging plates. See *convergent plate boundary.*

subsidence Slow or rapid sinking of part of the earth's crust that is not slope-related.

subsidence temperature inversion Inversion of normal air temperature layers when a large mass of warm air moves into a region at a high altitude and floats over a mass of colder air near the ground. This keeps the air over a city stagnant and prevents vertical mixing and dispersion of air pollutants. See *temperature inversion.* Compare *radiation temperature inversion.*

subsistence farming Supplementing solar energy with energy from human labor and draft animals to produce enough food to feed oneself and family members; in good years there may be enough food left over to sell or put aside for hard times. Compare *industrialized agriculture.*

subsurface mining Extraction of a metal ore or fuel resource such as coal from a deep underground deposit. Compare *surface mining.*

succession See *ecological succession, primary succession, secondary succession.*

succulent plants Plants, such as desert cacti, that survive in dry climates by having no leaves, thus reducing the loss of scarce water. They store water and use sunlight to produce the food they need in the thick, fleshy tissue of their green stems and branches. Compare *deciduous plants, evergreen plants.*

sulfur cycle Cyclic movement of sulfur in different chemical forms from the environment to organisms and then back to the environment.

superinsulated house House that is heavily insulated and extremely airtight. Typically, active or passive solar collectors are used to heat water, and an air-to-air heat exchanger is used to prevent buildup of excessive moisture and indoor air pollutants.

surface fire Forest fire that burns only undergrowth and leaf litter on the forest floor. Compare *crown fire, ground fire*. See *controlled burning*.

surface mining Removing soil, subsoil, and other strata and then extracting a mineral deposit found fairly close to the earth's surface. See area strip mining, contour strip mining. See *area strip mining, contour strip mining, dredging, mountaintop removal, open-pit mining*. Compare *subsurface mining*.

surface runoff Water flowing off the land into bodies of surface water. See *reliable runoff*.

surface water Precipitation that does not infiltrate the ground or return to the atmosphere by evaporation or transpiration. See *runoff*. Compare *groundwater*.

survivorship curve Graph showing the number of survivors in different age groups for a particular species.

sustainability Ability of a system to survive for some specified (finite) time.

sustainable agriculture Method of growing crops and raising livestock based on organic fertilizers, soil conservation, water conservation, biological pest control, and minimal use of nonrenewable fossil-fuel energy.

sustainable development See *environmentally sustainable economic development*.

sustainable living Taking no more potentially renewable resources from the natural world than can be replenished naturally and not overloading the capacity of the environment to cleanse and renew itself by natural processes.

sustainable society A society that manages its economy and population size without doing irreparable environmental harm by overloading the planet's ability to absorb environmental insults, replenish its resources, and sustain human and other forms of life over a specified period, usually hundreds to thousands of years. During this period, it satisfies the needs of its people without depleting natural resources and thereby jeopardizing the prospects of current and future generations of humans and other species.

sustainable yield (sustained yield) Highest rate at which a potentially renewable resource can be used without reducing its available supply throughout the world or in a particular area. See also *environmental degradation*.

synergistic interaction Interaction of two or more factors or processes so that the combined effect is greater than the sum of their separate effects.

synergy See *synergistic interaction*.

synfuels Synthetic gaseous and liquid fuels produced from solid coal or sources other than natural gas or crude oil.

synthetic natural gas (SNG) Gaseous fuel containing mostly methane produced from solid coal.

system A set of components that function and interact in some regular and theoretically predictable manner.

tar sand Deposit of a mixture of clay, sand, water, and varying amounts of a tar-like heavy oil known as bitumen. Bitumen can be extracted from tar sand by heating. It is then purified and upgraded to synthetic crude oil. See *bitumen*.

tectonic plates Various-sized areas of the earth's lithosphere that move slowly around with the mantle's flowing asthenosphere. Most earthquakes and volcanoes occur around the boundaries of these plates. See *lithosphere, plate tectonics*.

temperature Measure of the average speed of motion of the atoms, ions, or molecules in a substance or combination of substances at a given moment. Compare *heat*.

temperature inversion Layer of dense, cool air trapped under a layer of less dense, warm air. This prevents upward-flowing air currents from developing. In a prolonged inversion, air pollution in the trapped layer may build up to harmful levels. See *radiation temperature inversion, subsidence temperature inversion*.

teratogen Chemical, ionizing agent, or virus that causes birth defects. Compare *carcinogen, mutagen*.

terracing Planting crops on a long, steep slope that has been converted into a series of broad, nearly level terraces with short vertical drops from one to another that run along the contour of the land to retain water and reduce soil erosion.

terrestrial Pertaining to land. Compare *aquatic*.

territoriality Process in which organisms patrol or mark an area around their home, nesting, or major feeding site and defend it against members of their own species.

tertiary (higher-level) consumers Animals that feed on animal-eating animals. They feed at high trophic levels in food chains and webs. Examples are hawks, lions, bass, and sharks. Compare *detritivore, primary consumer, secondary consumer*.

tertiary sewage treatment See *advanced sewage treatment*.

theory of evolution Widely accepted scientific idea that all life forms developed from earlier life forms. Although this theory conflicts with the creation stories of many religions, it is the way biologists explain how life has changed over the past 3.6–3.8 billion years and why it is so diverse today.

theory of island biogeography The number of species found on an island is determined by a balance between two factors: the *immigration rate* (of species new to the island) from other inhabited areas and *extinction rate* (of species established on the island). The model predicts that at some point the rates of immigration and extinction will reach an equilibrium point that determines the island's average number of different species (species diversity).

thermal inversion See *temperature inversion*.

thermocline Zone of gradual temperature decrease between warm surface water and colder deep water in a lake, reservoir, or ocean.

threatened species Wild species that is still abundant in its natural range but is likely to become endangered because of a decline in numbers. Compare *endangered species*.

threshold effect The harmful or fatal effect of a small change in environmental conditions that exceeds the limit of tolerance of an organism or population of a species. See *law of tolerance*.

throughput Rate of flow of matter, energy, or information through a system. Compare *input, output*.

throwaway society See *high-throughput economy*.

time delay Time lag between the input of a stimulus into a system and the response to the stimulus.

tolerance limits Minimum and maximum limits for physical conditions (such as temperature) and concentrations of chemical substances beyond which no members of a particular species can survive. See *law of tolerance*.

total fertility rate (TFR) Estimate of the average number of children who will be born alive to a woman during her lifetime if she passes through all her childbearing years (ages 15–44) conforming to age-specific fertility rates of a given year. In simpler terms, it is an estimate of the average number of children a woman will have during her childbearing years.

totally planned economy See *pure command economic system*.

toxic chemical Chemical that is fatal to humans in low doses or fatal to more than 50% of test animals at stated concentrations. Most are neurotoxins, which attack nerve cells. See *carcinogen, hazardous chemical, mutagen, teratogen*.

toxicity Measure of how harmful a substance is.

toxicology Study of the adverse effects of chemicals on health.

toxic waste Form of hazardous waste that causes death or serious injury (such as burns, respiratory diseases, cancers, or genetic mutations). See *hazardous waste*.

traditional intensive agriculture Producing enough food for a farm family's survival and perhaps a surplus that can be sold. This type of agriculture uses higher inputs of labor, fertilizer, and water than traditional subsistence agriculture. See *traditional subsistence agriculture*. Compare *industrialized agriculture*.

traditional subsistence agriculture Production of enough crops or livestock for a farm family's survival and, in good years, a surplus to sell or put aside for hard times. Compare *industrialized agriculture, traditional intensive agriculture*.

tragedy of the commons Depletion or degradation of a potentially renewable resource to which people have free and unmanaged access. An example is the depletion of commercially desirable fish species in the open ocean beyond areas controlled by coastal countries. See *common-property resource*.

transform fault Area where the earth's lithospheric plates move in opposite but parallel directions along a fracture (fault) in the lithosphere. Compare *convergent plate boundary, divergent plate boundary*.

transmissible disease A disease that is caused by living organisms (such as bacteria, viruses, and parasitic worms) and can spread from one person to another by air, water, food, or body fluids (or in some cases by insects or other organisms). Compare *nontransmissible disease*.

transpiration Process in which water is absorbed by the root systems of plants, moves up through the plants, passes through pores (stomata) in their leaves or other parts, and evaporates into the atmosphere as water vapor.

tree farm See *tree plantation*.

tree plantation Site planted with one or only a few tree species in an even-aged stand. When the stand matures it is usually harvested by clear-cutting and then replanted. These farms normally are used to grow rapidly growing tree species for fuelwood, timber, or pulpwood. See *even-aged management*. Compare *old-growth forest, second-growth forest, uneven-aged management*.

trophic level All organisms that are the same number of energy transfers away from the original source of energy (for example, sunlight) that enters an ecosystem. For example, all producers belong to the first trophic level, and all herbivores belong to the second trophic level in a food chain or a food web.

troposphere Innermost layer of the atmosphere. It contains about 75% of the mass of earth's air and extends about 17 kilometers (11 miles) above sea level. Compare *stratosphere*.

true cost See *full cost*.

undergrazing Reduction of the net primary productivity of grassland vegetation and grass cover from absence of grazing for long periods (at least 5 years). Compare *overgrazing*.

undernutrition Consuming insufficient food to meet one's minimum daily energy needs for a long enough time to cause harmful effects. Compare *malnutrition, overnutrition*.

undiscovered resources Potential supplies of a particular mineral resource, believed to exist because of geologic knowledge and theory, although specific locations, quality, and amounts are unknown. Compare *identified resources, reserves*.

uneven-aged management Method of forest management in which trees of different species in a given stand are maintained at many ages and sizes to permit continuous natural regeneration. Compare *even-aged management*.

upwelling Movement of nutrient-rich bottom water to the ocean's surface. This can occur far from shore but usually occurs along certain steep coastal areas where the surface layer of ocean water is pushed away from shore and replaced by cold, nutrient-rich bottom water.

urban area Geographic area with a population of 2,500 or more. The number of people used in this definition may vary, with some countries setting the minimum number of people at 10,000–50,000.

urban growth Rate of growth of an urban population. Compare *degree of urbanization*.

urban heat island Buildup of heat in the atmosphere above an urban area. The large concentration of cars, buildings, factories, and other heat-producing activities produces this heat.

urbanization See *degree of urbanization*.

urban sprawl Growth of low-density development on the edges of cities and towns. See *smart growth*.

utilitarian value See *instrumental value*.

vertebrates Animals that have backbones. Compare *invertebrates*.

volcano Vent or fissure in the earth's surface through which magma, liquid lava, and gases are released into the environment.

water cycle See *hydrologic cycle*.

waterlogging Saturation of soil with irrigation water or excessive precipitation so that the water table rises close to the surface.

water pollution Any physical or chemical change in surface water or groundwater that can harm living organisms or make water unfit for certain uses.

watershed Land area that delivers water, sediment, and dissolved substances via small streams to a major stream (river).

water table Upper surface of the zone of saturation, in which all available pores in the soil and rock in the earth's crust are filled with water.

watt Unit of power, or rate at which electrical work is done. See *kilowatt*.

weather Short-term changes in the temperature, barometric pressure, humidity, precipitation, sunshine, cloud cover, wind direction and speed, and other conditions in the troposphere at a given place and time. Compare *climate*.

weathering Physical and chemical processes in which solid rock exposed at earth's surface is changed to separate solid particles and dissolved material, which can then be moved to another place as sediment. See *erosion*.

wetland Land that is covered all or part of the time with salt water or fresh water, excluding streams, lakes, and the open ocean. See *coastal wetland, inland wetland*.

wilderness Area where the earth and its community of life have not been seriously disturbed by humans and where humans are only temporary visitors.

wildlife All free, undomesticated species. Sometimes the term is used to describe only free, undomesticated animal species.

wildlife management Manipulation of populations of wild species (especially game species) and their habitats for human benefit, the welfare of other species, and the preservation of threatened and endangered wildlife species.

wildlife resources Wildlife species that have actual or potential economic value to people.

wild species Species found in the natural environment. Compare *domesticated species*.

windbreak Row of trees or hedges planted to partially block wind flow and reduce soil erosion on cultivated land.

wind farm Cluster of small to medium-sized wind turbines in a windy area to capture wind energy and convert it into electrical energy.

worldview How people think the world works and what they think their role in the world should be. See *environmental wisdom worldview, planetary management worldview, spaceship-earth worldview*.

zero population growth (ZPG) State in which the birth rate (plus immigration) equals the death rate (plus emigration) so that the population of a geographic area is no longer increasing.

zone of aeration Zone in soil that is not saturated with water and that lies above the water table. See *water table, zone of saturation*.

zone of saturation Area where all available pores in soil and rock in the earth's crust are filled by water. See *water table, zone of aeration*.

zoning Regulating how various parcels of land can be used.

zooplankton Animal plankton. Small floating herbivores that feed on plant plankton (phytoplankton). Compare *phytoplankton*.

INDEX

Meat production, 391–93
  environmental impact of, 391–92
  rangelands and pasture for, 391, 392–93
  sustainable, 392, 393
  U.S. public rangeland and, 417–18
Mechanical weathering, 175
Median lethal dose, **206**
Megacities, 239
Megalopolis, 239, 240*f*
Mercury
  cycling of, in aquatic environments, 369*f*
  health threat of, 367–68
  preventing and controlling inputs of, 370*f*
Mesotrophic lakes, **134**
Metallic mineral resources, 8, 178
Metamorphic rock, **177**, 178*f*
Metastasis, **208**
Methane hydrate as source of natural gas, 489
Methanol as fuel, 527, 528*f*
Metropolitan areas. *See* Urban (metropolitan) areas
Mexico City, environmental problems of, 240
Microclimates, topography and production of, **112***f*
Microevolution, **95**
  natural selection and, 95–96, 97*f*
  process of, 95
Microlending (microfinance), 29
Microlivestock, 390
Micronutrients, human nutrition and, 383, 385
Microorganisms (microbes), ecological role of, 65. *See also* Bacteria; Fungi; Viruses
Micropower systems, **533–34**
  centralized power system versus, 533*f*
  decentralized power system using, 534*f*
Microscale experiments, 35
Middle East, water conflicts in, 305
Midgley, Thomas, Jr., 297
Midsuccessional plant species, **153**
Migration, effects of, on population size, 232. *See also* Immigration *and* emigration
Migratory Bird Hunting and Conservation Stamp Act of 1934, 475
Military security, 36
Mimicry, 149
Mineral, **177**
Mineral resources, 8, **178**, 479
  classification of nonrenewable, 178*f*
  finding, removing, and processing nonrenewable, 179–83
  life cycle of, 183*f*
  limits on extraction and use of, 181–83
  substitutes for, 186
  supplies of, 183–86
Minimum-tillage farming, 197
Minimum viable population (MVP), **162**
Mining
  environmental impact of, 180–81, 182*f*
  of lower-grade ores, 185
  major methods of, 179*f*, 180, 181*f*
  using microbes for, 185–86
  in United States on public lands, 184–85
Minnesota Mining and Manufacturing Company (3M), 355–56
Model(s), **47**
  of complex systems, 49
Molina, Mario, 297–98
Monoculture, 119*f*, 377, 378, **380**
  tree plantations as, 419*f*

Montague, Peter, 340
  on environmentalism and pollution prevention, 354–55
Montreal Protocol, 301
Mosquito
  Asian tiger, 460
  malaria and, 170, 214*f*
Moths, camouflage coloration in peppered, 96*f*
Motor scooters as transportation, 246*f*
Motor vehicles
  alternatives to, 246*f*, 247*f*
  average fuel economy in, 512*f*
  fuel cell, 295, 513, 514*f*
  ethanol and methanol as fuel for, 528, 529*f*
  hybrid gas-electric, 295, 512, 513*f*
  reducing air pollution from, 276*f*
  reducing use of, 246
  saving energy in use of, 511–12
  in United States, 245–46
Mountains
  ecological importance of, 124
  human impact on, 124*f*
  rain shadow effect and, 112*f*
Mountaintop removal, **180**
Muir, John, 137, 439
Municipal solid waste (MSW), **348**
  burning, 363–64
  land disposal of, 364, 365*f*
  recycling, 358–62
Muskegs (acidic bogs), 124
Mutagens, 95, **208**
Mutations, 95, 208
  in pathogenic bacteria, 211–12
Mutualism, 140
  nutritional, 150–51
  protection and, 151*f*
Mycorrhizae fungi, 151*f*
Myers, Norman, 453
  on biodiversity crisis, species extinction, and future of evolution, 102–3
Nash, Roderick, 441
Nashua River, Massachusetts, 327
National Academy of Sciences, U.S., 15
National ambient air quality standards (NAAAQS), 273
National Forest System, U.S., 414, 416*f*
  economic, ecological, and recreational value of, 428
  logging companies in, 426–27
  management of, 428, 429*f*
National Marine Fisheries Service (NMFS), 466
National Park System, U.S., 414–15, 416*f*, 433–34
  improving management of, 434
  threats to, 433–34
National Priorities List (NPL), U.S., 371
National Public Lands Grazing Campaign, 391
National Resource Lands, U.S., 414, 416*f*
National security, expanding concept of, 36
National Wilderness Preservation System, U.S., 415, 440–41
National Wildlife Refuges, U.S., 414, 416*f*, 471, 486, 487*f*
Native species, **143**
Natural capital, 2, **20**
  atmosphere as, 256*f*
  biodiversity hotspots, 438*f*, 439*f*, 467*f*
  earth's, 3*f*
  endangered and threatened species as, 450–51*f*

energy resources, nonrenewable, 479*f*, 484*f*
groundwater system, 308*f*
P. Hawken on, 4–5
marine animals, 462–63, 468, 469*f*, 470*f*
in marine systems, 127*f*
medicinal plants in tropical forests, 430*f*
mineral resources, 177, 178*f*
ocean currents, looping of, 288*f*
rivers, 312*f*, 326*f*
soil resources, 188–92
songbirds, 457*f*
water budget, 307*f*
wilderness, 440*f*
Natural gas, **489**, 490*f*
  synthetic, 492
Natural greenhouse effect, 70, 84
Natural hazards, 186–88
Natural radioactive decay, 57
Natural recharge, **307**
Natural resources, 2, **20**. *See also* Resource(s)
Natural Resources Conversation Service, U.S., 173
Natural Resources Council (NRDC), 342
Natural selection, **95**
  microevolution by, 95, 96*f*
  three types of, 95–96, 97*f*
Nature Conservancy, work of, 435
Nature reserves, 435–41
  amount of land needed for, 435
  biosphere reserves, 437
  corridors connecting, 437
  in Costa Rica, 435, 436*f*
  principles for establishing and managing, 436
  top priority areas for, 437–38, 439*f*
  size of, 436–37
  wilderness as, 438–41
Neem tree, 430–31
Negative feedback loop, **50**
Nekton, **125**
Nervous system, chemical hazards to human, 208
Net energy, **481–82**
Net energy efficiency, 508, 509*f*
Net energy ratios for select types of energy systems, 482*f*
Netherlands, ecological footprint of, 8*f*
Net primary productivity (NPP), **79**, 378
  estimated average, of select ecosystems, 80*f*
Neutrons (n), **51**
Niche, **98**. *See also* Ecological niche
Nigeria, demographic factors in Brazil, U.S., and, 230*f*
Nile Basin Water Resources, 305
Nitrate fertilizers, 289
Nitrate ions, 85
Nitrification, 85
Nitrite ions, 85
Nitrogen cycle, **84***f*–85
Nitrogen fixation, 84–85
Nitrogen oxide (NO_2) as air pollutant, 259*t*
Noise pollution, **243**, 244*f*
Nondegradable pollutants, 57
  in water, 332
Nonionizing radiation, **54**
Nonmetallic mineral resources, 8, 178
Nonnative species, **143–44**
  characteristics of invader, 460*f*
  examples of, 458*f*
  as extinction threat, 457–60
  killing in ship ballast, 461
  in national parks, 434
  reducing threat of, 460

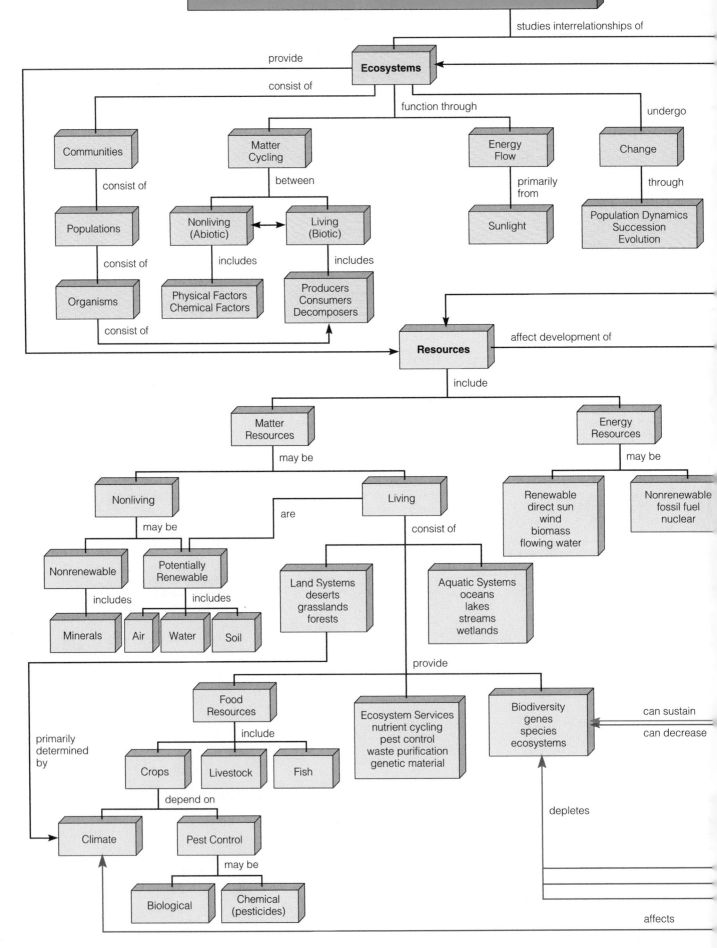

ENVIRONMENTAL SCIENCE: CONCEPTS AND CONNECTIONS

studies interrelationships of

Ecosystems

provide

consist of

function through

undergo

Communities

Matter Cycling

Energy Flow

Change

consist of

between

primarily from

through

Populations

Nonliving (Abiotic)

Living (Biotic)

Sunlight

Population Dynamics
Succession
Evolution

consist of

includes

includes

Organisms

Physical Factors
Chemical Factors

Producers
Consumers
Decomposers

consist of

Resources

affect development of

include

Matter Resources

Energy Resources

may be

may be

Nonliving

are

Living

Renewable
direct sun
wind
biomass
flowing water

Nonrenewable
fossil fuel
nuclear

may be

consist of

Nonrenewable

Potentially Renewable

Land Systems
deserts
grasslands
forests

Aquatic Systems
oceans
lakes
streams
wetlands

includes

includes

Minerals

Air

Water

Soil

provide

Food Resources

Ecosystem Services
nutrient cycling
pest control
waste purification
genetic material

Biodiversity
genes
species
ecosystems

can sustain

can decrease

primarily determined by

include

Crops

Livestock

Fish

depletes

depend on

Climate

Pest Control

may be

Biological

Chemical (pesticides)

affects

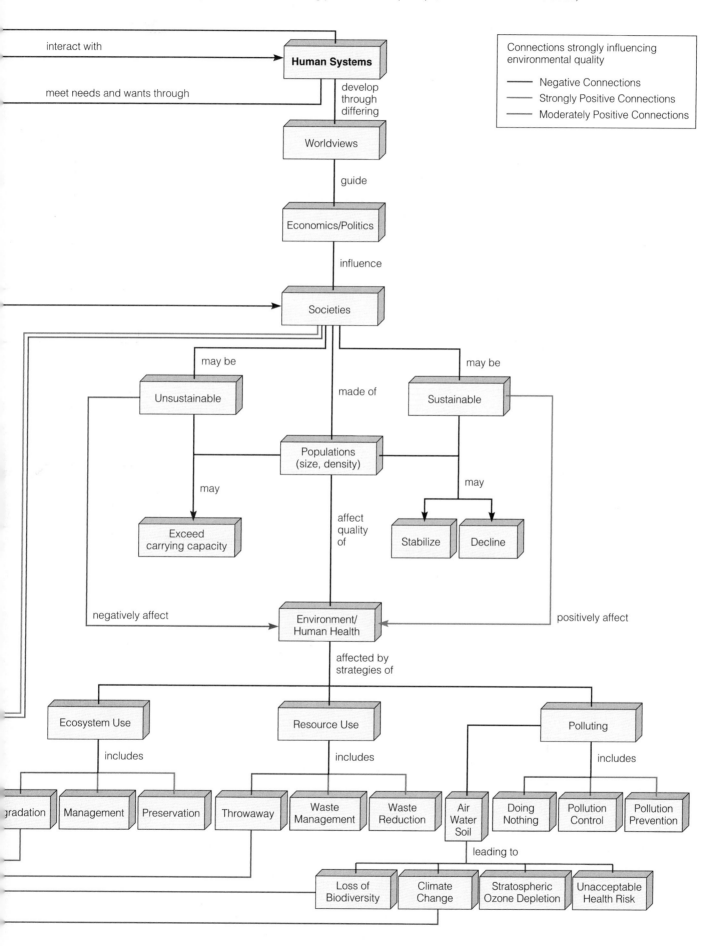